BIOLOGY

BIOLOGY

RUTH BERNSTEIN • **STEPHEN BERNSTEIN**

University of Colorado at Boulder   *University of Colorado at Boulder*

WCB   **Wm. C. Brown Publishers**

Dubuque, Iowa • Melbourne, Australia • Oxford, England

Book Team

Editor *Carol J. Mills*
Production Editors *Cathy Ford Smith and Karen L. Nickolas*
Designer *K. Wayne Harms*
Art Editor *Kathleen Huinker Timp*
Photo Editor *John C. Leland*
Permissions Coordinator *Vicki Krug*

Wm. C. Brown Publishers

President and Chief Executive Officer *Beverly Kolz*
Vice President, Publisher *Kevin Kane*
Vice President, Director of Sales and Marketing *Virginia S. Moffat*
Vice President, Director of Production *Colleen A. Yonda*
National Sales Manager *Douglas J. DiNardo*
Marketing Manager *Craig Johnson*
Advertising Manager *Janelle Keeffer*
Production Editorial Manager *Renée Menne*
Publishing Services Manager *Karen J. Slaght*
Royalty/Permissions Manager *Connie Allendorf*

A Times Mirror Company

Copyedited by Laura Beaudoin

Cover Photo: © Frans Lanting/Minden Pictures

Back Cover Excerpt: From Douglas Chatwick,
The Fate of the Elephant. Copyright © 1992 Sierra Club Books,
San Francisco, CA. Reprinted by permission.

Photo Research by Kathy Husemann

The credits section for this book begins on page 689 and
is considered an extension of the copyright page.

Printed in the United States of America by Times Mirror Higher Education Group, Inc.,
2460 Kerper Boulevard, Dubuque, IA 52001

10 9 8 7 6 5 4 3 2 1

Publisher's Note to the Instructor

Full-Color Customization

Your options are unlimited! You can now create the ideal text for your course. Take a look at the contents of *Biology*. Are there some chapters you don't address in your course? If so, select just those chapters that you do cover. Then let Wm. C. Brown Publishers professionally print and bind the *full-color* general biology text that is right for your classroom.

Of course, you can still select the standard, full-color text. It's your choice. Contact your Wm. C. Brown Publishers sales representative or call 800–228–0634 for more information.

Recycled Paper

Biology is printed on recycled paper stock. All of its ancillaries, as well as all advertising pieces, will also be printed on recycled paper. In addition, when possible, the inks used in printing are soy based.

Our goal in using these materials for *Biology* and its ancillary package is to minimize the environmental impact of our products.

Brief Contents

Contents

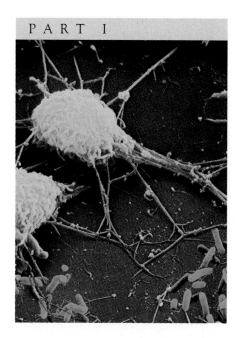

PART I

Biology of Cells 21

1 Atoms and Molecules 23

2 The Chemistry of Life 39

Contents

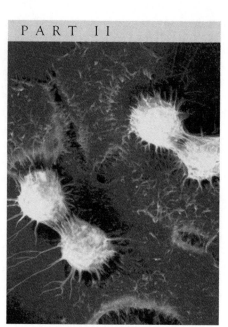

PART II

Cellular Reproduction and Inheritance 121

9 The Physical Basis of Inheritance 155

10 Patterns of Inheritance 175

11 From Genes to Proteins 191

12 Genetic Engineering 211

PART III

Biology of Plants 229

13 Plant Organization 231

Contents

PART IV

Biology of Animals 309

PART V

**Evolution and
Ecology 567**

30 The Process of Evolution 569

31 Biodiversity 591

Contents

32 Population Ecology 617

33 Ecosystems 635

34 Human Ecology 663

📼 Life Science Animations

The following illustrations in *Biology* have been correlated to the Life Science Animations videotapes:

📼 Tape 1—Chemistry, the Cell, and Energetics

Concept 1   Formation of an Ionic Bond (TA 1.5)
Concept 2   Journey into a Cell (fig. 3.5)
Concept 3   Endocytosis (fig. 4.10)
Concept 4   Cellular Secretion (fig. 4.12)
Concept 5   Glycolysis (fig. 5.4)
Concept 6   Oxidative Respiration (incl. Krebs Cycle) (fig. 5.11)
Concept 7   The Electron Transport Chain & the Production of ATP (figs. 5.9 and 5.10)
Concept 8   The Photosynthetic Electron Transport Chain & Production of ATP (figs. 6.7 and 6.9)
Concept 9   C3 Photosynthesis (Calvin Cycle) (fig. 6.10)
Concept 10   C4 Photosynthesis (TA 6.11)
Concept 11   ATP as an Energy Carrier (fig. 5.1)

📼 Tape 2—Cell Division/Heredity/Genetics/Reproduction and Development

Concept 12   Mitosis (fig. 7.5)
Concept 13   Meiosis (fig. 8.4)
Concept 14   Crossing Over (fig. 8.5)
Concept 15   DNA Replication (fig. 7.4)
Concept 16   Transcription of a Gene (fig. 11.3)
Concept 17   Protein Synthesis (fig. 11.6)
Concept 18   Regulation of Lac Operon
Concept 19   Spermatogenesis (fig. 28.12)
Concept 20   Oogenesis (fig. 28.5)
Concept 21   Human Embryonic Development (fig. 29.5)

📼 Tape 3—Animal Biology I

Concept 22   Formation of Myelin Sheath (TA 24.3)
Concept 23   Saltatory Nerve Conduction (fig. 25.1)
Concept 24   Signal Integration (fig. 25.2)
Concept 25   Reflex Arcs (TA 24.1)
Concept 26   Organ of Static Equilibrium (fig. 25.13)
Concept 27   The Organ of Corti (fig. 25.11)
Concept 28   Peptide Hormone Action (cAMP) (fig. 23.4)
Concept 29   Levels of Muscle Structure (fig. 26.10)
Concept 30   Sliding Filament Model of Muscle Contraction (fig. 26.11)
Concept 31   Regulation of Muscle Contraction (fig. 26.13)
Concept 32   The Cardiac Cycle & Production of Sounds (fig. 19.7)
Concept 33   Peristalsis (fig. 18.6)
Concept 34   Digestion of Carbohydrates (figs. 18.1 and 18.2)
Concept 35   Digestion of Proteins (fig. 18.1)

📼 Tape 4—Animal Biology II

Concept 36   Digestion of Lipids (fig. 18.1, TA 18.4)
Concept 37   Blood Circulation (fig. 19.3)
Concept 38   Production of Electrocardiogram
Concept 39   Common Congenital Defects of the Heart
Concept 40   A, B, O Blood Types (table 10.2)
Concept 41   B-Cell Immune Response (fig. 22.9)
Concept 42   Structure and Function of Antibodies (fig. 22.5)
Concept 43   Types of T-cells (figs. 22.7 and 22.8)
Concept 44   Relationship of Helper T-cells & Killer T-cells
Concept 45   Life Cycle of Malaria

📼 Tape 5—Plant Biology/Evolution/Ecology

Concept 46   Journey into a Leaf (fig. 13.10)
Concept 47   How Water Moves through a Plant (fig. 14.1)
Concept 48   How Food Moves From a Source to a Sink
Concept 49   How Leaves Change Color and Drop in Fall

Concept 50   Mitosis and Cell Division in Plants (fig. 7.6)
Concept 51   Carbon and Nitrogen Cycles (figs. 33.6 and 33.7)
Concept 52   Energy Flow through an Ecosystem (fig. 33.5)
Concept 53   Continental Drift and Plate Tectonics (fig. 31.19)

📼 Tape 6—Physiological Concepts

Concept 54   Lock and Key Model of Enzyme Action (fig. 2.11)
Concept 55   Osmosis (fig. 4.6)
Concept 56   Active Transport Across a Cell Membrane (fig. 4.8)
Concept 57   Cellular Secretion
Concept 58   Electron Transport Chain and Oxidative Phosphorylation (fig. 5.9)
Concept 59   Conduction of Nerve Impulses (fig. 24.6)
Concept 60   Temporal and Spatial Summation
Concept 61   Synaptic Transmission (fig. 24.7)
Concept 62   Visual Accommodation (TA 25.2)
Concept 63   Action of Steroid Hormone on Target Cells (fig. 23.2)
Concept 64   Action of T3 on Target Cells
Concept 65   Cyclic AMP Action
Concept 66   Life Cycle of HIV (fig. 11.13)

(Concepts 54–66 correspond to Concepts 1–13 in the *Physiological Concepts of Life Science* videotape.)

💿 Explorations CD-ROM Correlations

The following chapters in *Biology* have been correlated to the sixteen topic modules in *Explorations in Human Biology* CD-ROM by George Johnson:

1. Cystic Fibrosis (chap. 4)
2. Active Transport
3. Life Span and Lifestyle
4. Muscle Contraction (chap. 26)
5. Evolution of the Heart (chap. 19)
6. Smoking and Cancer (chap. 7)
7. Diet and Weight Loss (chap. 18)
8. Nerve Conduction (chap. 24)
9. Synaptic Transmission (chap. 24)
10. Drug Addiction (chap. 24)
11. Hormone Action (chap. 23)
12. Immune Response (chap. 22)
13. AIDS (chap. 28)
14. Constructing a Genetic Map (chap. 9)
15. Heredity in Families (chap. 10)
16. Pollution of a Freshwater Lake (chap. 33)

The following chapters in *Biology* have been correlated to the second CD-ROM by George Johnson entitled *Explorations in Cell Biology, Metabolism, and Genetics*:

1. How Proteins Function: Hemoglobin
2. Cell size (chap. 7)
3. Active Transport (chap. 5)
4. Cell-Cell Interactions (chap. 23)
5. Mitosis: Regulating the Cell Cycle (chap. 7)
6. Cell Chemistry: Thermodynamics (chap. 1)
7. Enzymes in Action: Kinetics (chap. 2)
8. Oxidative Respiration (chap. 5)
9. Photosynthesis (chap. 6)
10. Genetic Re-assortment During Meiosis (chap. 9)
11. Constructing a Genetic Map (chap. 9)
12. Heredity in Families (chap. 10)
13. Gene Segregation within Families (chap. 10)
14. DNA Fingerprinting: You Be the Judge (chap. 12)
15. Reading DNA (chap. 12)
16. Gene Regulation
17. Make a Restriction Map (chap. 12)

Chapter Outline and Objectives

Each chapter opens with a page-referenced Chapter Outline and a list of Objectives that introduce the student to important topics within the chapter and provide quick access to the material. These introductory features are followed by a prologue, which links the chapter material to everyday life.

C H A P T E R

25

The Senses

Objectives

In this chapter, you will learn
–how senses adapt an animal to its particular way of life;
–the way in which an eye detects objects;
–how cells in the nose and mouth detect chemicals;
–how an ear detects sounds and changes in body position;
–how the skin, muscles, and joints alert an animal to touch, pain, temperature, and position of body parts.

An animal's senses provide information about the environment.

Chapter-by-Chapter Customization

A separate set of chapter page numbers appears at the top of the page for the benefit of adopters who wish to customize *Biology*.

Cross References

Within the narrative, cross references to foundational material presented in earlier chapters are identified with an icon (📖 *see page 62*) followed by a page number.

A eukaryotic cell has a variety of organelles—specialized compartments that contain the biochemical pathways (📖 *see page 51*) (enzymes, substrates, and other materials) for performing a particular task. In this way, the cell's biochemical pathways are organized according to function and kept separate from one another. The breakdown of glucose (to provide energy), for example, is isolated from the synthesis of polysaccharides.

Thematic Presentation

Throughout the text, four biological themes are emphasized, and each of the five parts opens with "Exploring Themes," an overview presenting one or more of these concepts that relate to the chapters within that part.

Exploring Themes

Theme 1
The fundamental unit of life is the cell.
An animal is formed by many cells (chapter 17).

Theme 2
All cells carry out the same basic processes.
All animal cells synthesize molecules, import and export materials, acquire energy from organic molecules, reproduce, and synthesize proteins according to instructions within their genes (chapter 17).

Theme 3
The cells of a multicellular organism produce adaptations: anatomical, physiological, and behavioral traits that promote survival and reproduction.
Each cell of an animal is specialized for a particular function: cells of the digestive tract process food (chapter 18); red blood cells transport oxygen (chapter 19); lung cells import oxygen and export carbon dioxide (chapter 20); kidney cells excrete wastes (chapter 21); white blood cells kill disease microbes (chapter 22); endocrine cells secrete hormones (chapter 23); nerve cells convey information (chapters 24 and 25); muscle cells change length (chapter 26); endocrine, nerve, and muscle cells coordinate behaviors (chapter 27); and cells of the ovary and testes produce eggs and sperm, which carry genetic information to the next generation (chapters 28 and 29).

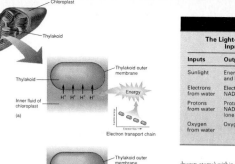

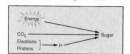

Table 6.2		
The Light-Dependent Reactions: Inputs and Outputs		
Inputs	**Outputs**	**Fate of Outputs**
Sunlight	Energy in ATP and NADPH	Energy in glucose
Electrons from water	Electrons in NADPH	Part of hydrogen atoms in glucose
Protons from water	Protons in NADPH and lone protons (H⁺)	Part of hydrogen atoms in glucose
Oxygen from water	Oxygen gas	Released from cell or used in cellular respiration

Dramatic Visuals Program Tied to Animations

Colorful, informative photographs and illustrations enhance the learning program of this text. This icon (▭) identifies figures that are depicted in the Life Science Animations video-tape series, which is made up of 53 animations. The tapes include:

Tape 1 Chemistry, the Cell, and Energetics
Tape 2 Cell Division/Heredity/Genetics/Reproduction and Development
Tape 3 Animal Biology I
Tape 4 Animal Biology II
Tape 5 Plant Biology/Evolution/Ecology

Figure 6.9 The proton gradient. (*a*) Energy released from the electron transport chain is used by transport proteins to pump protons (H⁺) across a membrane, from the outside to the inside of a thylakoid. (*b*) As the protons diffuse through a channel protein in the membrane, they release energy that is transferred to ATP.

formation of NADPH, which provides chemical energy, an electron, and a hydrogen atom for building sugar. Moreover, some of the oxygen gas that forms is used by the cell in cellular respiration. The rest diffuses out of the cell and into the surrounding environment, where it is available for use by animals and other organisms. The materials that enter and leave the light-dependent reactions are summarized in table 6.2.

The Light-Independent Reactions

The light-independent reactions do not require light; they build sugar from carbon dioxide, using the ATP and NADPH formed during the light-dependent reactions. Energy within molecules of ATP and NADPH is used to build covalent bonds within a sugar molecule. Hydrogen atoms and electrons (which join with protons to form more hy-

drogen atoms) within molecules of NADPH become incorporated into the structure of a sugar molecule:

Sugar is built by a biochemical pathway called the Calvin-Benson cycle, which is located within the inner fluid (cytosol) of a photosynthetic bacterium or within the inner fluid (stroma) of a chloroplast (see fig. 6.2).

The **Calvin-Benson cycle** is a biochemical pathway that builds a three-carbon sugar from carbon dioxide, hydrogen atoms, and chemical energy. The pathway is a cycle in which the molecule that begins the pathway—ribulose bisphosphate (RuBP)—becomes an end product that begins the same pathway again:

The cycle, shown in figure 6.10, begins when carbon dioxide joins with RuBP, which is a five-carbon molecule. (The enzyme that catalyzes this reaction, called RuBP carboxylase, is the most plentiful protein on earth.) The

113

In-Text Line Art

Simple line art placed within the narrative keeps the reader focused and typifies the drawings created by instructors as they lecture.

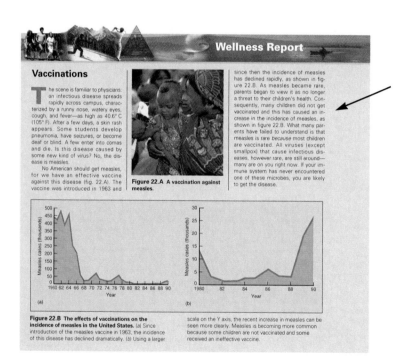

Wellness Report

Vaccinations

The scene is familiar to physicians: an infectious disease spreads rapidly across campus, characterized by a runny nose, watery eyes, cough, and fever—as high as 40.6° C (105° F). After a few days, a skin rash appears. Some students develop pneumonia, have seizures, or become deaf or blind. A few enter into comas and die. Is this disease caused by some new kind of virus? No, the disease is measles.

No American should get measles, for we have an effective vaccine against this disease (fig. 22.A). The vaccine was introduced in 1963 and

since then the incidence of measles has declined rapidly, as shown in figure 22.B. As measles became rare, parents began to view it as no longer a threat to their children's health. Consequently, many children did not get vaccinated and this has caused an increase in the incidence of measles, as shown in figure 22.B. What many parents have failed to understand is that measles is rare *because* most children are vaccinated. All viruses (except smallpox) that cause infectious diseases, however rare, are still around—many are on you right now. If your immune system has never encountered one of these microbes, you are likely to get the disease.

Figure 22.A A vaccination against measles.

Figure 22.B The effects of vaccinations on the incidence of measles in the United States. (a) Since introduction of the measles vaccine in 1963, the incidence of this disease has declined dramatically. (b) Using a larger scale on the Y axis, the recent increase in measles can be seen more clearly. Measles is becoming more common because some children are not vaccinated and some received an ineffective vaccine.

Boxed Readings

Three types of boxed readings occur within the narrative: Research Reports, Wellness Reports, and Environment Reports. Research Reports present the logic of research projects; Wellness Reports deal with the biological basis of fitness and wellness; and Environment Reports motivate student interest in biodiversity and ecology.

Student Study Aids

Each chapter ends with a Summary, page-referenced Key Terms, Study Questions, Critical Thinking Problems, and an annotated list of Suggested Readings organized for both introductory-level and advanced-level students.

New Technology CD-ROM

References to two new WCB Explorations CD-ROMS are found at the ends of many of the chapters. When chapter material relates to the topic of a CD-ROM module on *Explorations in Human Biology* or *Explorations in Cell Biology, Metabolism, and Genetics,* the student is presented with a brief topic overview and directed to the interactive CD-ROM module.

✪ EXPLORATION

Interactive Software: Diet and Weight Loss

This interactive exercise explores how diet and exercise interact to determine whether we gain or lose weight. The program provides a pathway for energy, from food to either the synthesis of ATP or the storage of fat. The student alters the kinds and amounts of foods eaten as well as the amount of energy used in exercise. The effects of these three variables are then determined by measuring changes in the amount of stored fat.

Preface

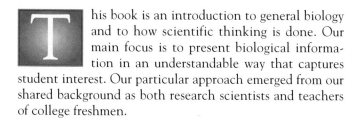

This book is an introduction to general biology and to how scientific thinking is done. Our main focus is to present biological information in an understandable way that captures student interest. Our particular approach emerged from our shared background as both research scientists and teachers of college freshmen.

Our Experience

As research scientists, we view biology as a series of exciting questions that challenge us to pursue answers. We want to share this excitement with students and point out the many interesting questions yet to be answered. We also view biological phenomena as solutions to problems that organisms have encountered and then solved by means of evolved adaptations. All forms of life, for example, have the problem of moving certain materials into and out of their cells. The solutions to this problem are found in different parts of the organism—in lungs, digestive tracts, kidneys, roots, and leaves.

As teachers, we have more than twenty-five years of experience teaching introductory biology, usually a two-semester course covering most of the standard topics of biology, from cells to ecosystems. The experience taught us how to integrate different aspects of biology in ways that help students learn. Our students were a special group, identified by the university as high-risk students because they were not well prepared for this level of work. Our task was to bring these students up to university level in terms of their biological knowledge. Consequently, our classes involved more pedagogy and more one-on-one interactions than other university courses. Through this experience, we have come to understand the types of problems encountered by these students as well as by their teachers. Over the years, we have developed ways to motivate our students to think as scientists do and to understand sophisticated biological concepts.

Unifying Concepts and Themes

We have found that it is essential to present the vast quantity of biological information in a conceptual framework. This framework allows each new topic to build on previous ones. New information can then enlighten rather than confuse the student. For example, once the idea of adaptations to particular environments is developed, the details of plant and animal biology become variations on a familiar concept.

Throughout *Biology*, our explanations are written as clearly and precisely as possible, with the intention that students truly understand the material rather than simply memorize it. We've used standard biological terms, but only those essential for communication of the ideas.

Four biological themes are emphasized:

1. The fundamental unit of life is the cell.
2. All cells carry out the same basic processes.
3. The cells of a multicellular organism produce adaptations: anatomical, physiological, and behavioral traits that promote survival and reproduction.
4. Adaptations evolve in response to interactions with other organisms and with the physical environment.

The four themes are introduced, one after another, as we build a framework of biological knowledge. Each theme, once introduced, continues to function in the chapters that follow.

The Learning System

Each chapter begins with a brief outline and a list of objectives, written in everyday language (e.g., "In this chapter you will learn how a cell uses its genes to build proteins"). This introductory material is followed by a prologue, which is designed to link the chapter material with everyday life. Within the chapter, we use three levels of headings to organize and subdivide the material and to focus the topic information. Each chapter ends with a summary, list of key terms, and two sets of questions: a set of study questions, designed to reveal whether the important concepts have been learned, and a set of critical thinking problems, designed to encourage students to use the concepts in new ways. We've also added a list of suggested readings arranged according to introductory level (for the beginning student) and advanced level (for the advanced student or instructor).

We make extensive use of in-text sketches, similar to the ones we draw on the blackboard while lecturing. They are particularly useful in explaining the actions of invisible topics, such as chemical reactions, protein receptors, and chromosomes. As biologists, we all have many such mental images to help us remember and understand. We find it very helpful to students when we communicate these learning aids.

The Boxed Essays

Throughout *Biology,* boxed essays are used to show how science is done and to relate chapter material to everyday life. We have included three kinds of boxed essays: Research Reports, Wellness Reports, and Environment Reports.

The Research Reports deal with the logic of classical and ongoing research projects. They are based on hypothesis formation and testing. These real examples of scientific thought portray the uncertain but exciting feel of research—the frustrations and joys that accompany the search for truth. In addition, logic trees, introduced in the first chapter, are used to show where each research project is positioned within the flow of scientific thought. The reports are designed to stimulate curiosity and critical thinking—to encourage the reader to consider alternative hypotheses and ways of testing them.

The Wellness Reports deal with the biological basis of fitness and wellness. Their purpose is to motivate students to learn about their own bodies through their study of biology. We link, for example, the warm-up phase of exercise with cellular respiration, antibiotics with protein synthesis, jet lag with hormone secretion, and drug addiction with neurotransmitters. Wherever possible, these reports involve the cellular basis of bodily function, a depth of knowledge students would probably not otherwise acquire.

The Environment Reports are inserted into the chapters on biodiversity and ecology. To our delight, we find that today's generation of students is sincerely concerned about the environment and seeks sophisticated knowledge about environmental problems. The Environment Reports are used to meet this need as well as to motivate interest in the fundamental basis of populations and ecosystems.

Chapter Organization

The introductory chapter sets the stage for the study of biology as a science. It describes the scientific method and shows that science, like life itself, is a dynamic process. This first chapter also introduces fundamental concepts used in later chapters: the hierarchical organization of life, the basics of Darwin's theory of evolution, the nature of adaptations, and the kingdoms of life.

The chapters then follow a conventional sequence that begins with cells and ends with ecosystems. Parts I and II develop our first two themes: the basic similarity of all cells with regard to structure, chemical processes, reproduction, and gene expression. Parts III and IV deal with our third theme: the ways in which the cells of a plant or animal are organized to form adaptations that promote survival and reproduction. Part V then develops our fourth theme: the origin of adaptations and their roles in biodiversity and ecology.

While we find this sequence works best for us, we are aware that not all biology courses are arranged in this way and, furthermore, that few courses include all the chapter topics. Hence, we have attempted to make each unit versatile in terms of sequence. Where material in previous chapters is essential background, we have used an icon that directs the reader to specific pages where this information has been presented.

The *Biology* Ancillary Package

Whether used by instructor or student, *Biology* comes equipped with unique and useful tools to enhance teaching and learning. You will find many of them correlated directly to corresponding material throughout the text.

Useful Learning Aids for Students

1. Student Study Guide. This illustrated study guide, by Mary Vick of Hampton College, focuses students on important chapter material that is often difficult to learn without further reinforcement. Each chapter in the study guide contains a variety of features designed to help students study wisely, including: chapter outline, review questions, key terms, objective self-test (includes true-false, multiple-choice, matching, and completion), and critical thinking questions (with answers provided).
2. Laboratory Manual. Written by the faculty, Department of Biology, Wm. Paterson College, and coeditors Marty Hahn, Stephen Vail, and Jane Voos, this manual contains 30 student-tested laboratory exercises.
3. *Biology* Study Cards. A set of 200 two-sided study cards containing definitions of key terms, clearly labeled illustrations, and pronunciation guides is handy for student study anytime, anyplace.

Useful Teaching Tools for Instructors

1. Instructor's Manual/Test Item File. Prepared by Linda K. Butler and Stephen W. Taber of the University of Texas, Austin, the IM/TIF features a test item file as well the following elements for each chapter: objectives, key concepts, key terms

with definitions, chapter outline, answers to study questions, answers to critical thinking problems, and a list of audio-visual resources.

2. Microtest III. A computerized test bank containing the questions found in the test item file.

3. Extended Lecture Outline Software. An expanded outline for each chapter is available on Macintosh and Windows platforms. Instructors can add or delete material and print out for lecture use or student handouts.

4. Transparency Acetates. A set of 150 full-color transparencies of images found in the text is available to adopters.

5. Transparency Masters. A set of 100 one-color, labeled line drawings of images in the text is available to adopters.

New Technology

Explorations CD-ROMs

Learn biological processes—using movement, color, sound, and interaction. Two new interactive CD-ROMs, *Explorations in Human Biology* and *Explorations in Cell Biology, Metabolism,* and *Genetics* by George Johnson, include 16 modules each that are either of high student interest, such as the unit on drug addiction, or are difficult concepts to understand, such as the unit on active transport.

When an Exploration correlates to topic material in a chapter in *Biology,* you'll find the CD-ROM icon (◐) followed by a description of the module at the end of that particular chapter. A correlation chart that lists the Exploration topics and corresponding chapters in *Biology* can be found on page (*xvii*).

Life Science Lexicon CD-ROM

This new interactive CD-ROM, by Will Marchuk of Red Deer College, offers students a glossary of over 1,200 key science terms and definitions; word construction through the use of prefixes, root words, and suffixes; descriptions of eponyms; more than 1,000 full-color images for selected terms; and over 200 questions for student quizzing.

Life Science Animation Videotapes

Tied directly to text figures, these videotapes bring movement and multi-dimension to over 60 images of often difficult-to-understand physiological processes that can only be shown in a limited, step-by-step way in the textbook. However, with the use of the animation tapes, these visual moving representations of key processes assist the instructor and enhance student learning.

Figures within the text that are correlated to animations are easily identified by the videotape icon (▭).

BioSource Videodisc

Our new BioSource videodisc features nearly 10,000 full-color illustrations and photos, many from Wm. C. Brown Publishers' general biology textbooks. Approximately twenty minutes of life science animations are also included on the videodisc.

Biology StartUp Software

This set of interactive Macintosh tutorials for students portrays difficult, biological concepts in a new way, using definitions, color illustrations and animations, and pronunciations spoken by a human voice. Quizzes are located at the ends of major sections.

Acknowledgments

We dedicate this book to our students and colleagues, who have made our study of biology such a pleasure.

We gratefully acknowledge the contributions made to the development and production of this book by many talented and dedicated persons at Wm. C. Brown Publishers. First, we wish to thank Kevin Kane for encouraging us to begin the book and for helping us to develop its contents. We thank also our developmental editors, Marge Kemp and Diane Beausoleil, who solicited the help of excellent reviewers and contributed in so many ways to the manuscript development. Of the many people who contributed to the making of the text, we want to give particular thanks to our supervising editor, Carol Mills, who competently and compassionately solved each problem as it appeared. In addition, we thank the production editors, Cathy Smith and Karen Nickolas, whose intense efforts made the book possible. We also give particular thanks to our copy editor, Laura Beaudoin, who painstakingly read and improved every sentence of the manuscript. For acquiring the photographs, we gratefully acknowledge the diligence of our photo editor, John Leland, and his assistant, Kathy Husemann, who always made the extra effort to acquire the perfect photo. We want to thank Kathleen Timp, art editor, Vicki Krug, permissions coordinator, and Wayne Harms, designer, who contributed so much to this book.

We gratefully acknowledge the assistance of the following reviewers whose constructive criticism proved invaluable to the development of this text:

Tania Beliz
College of San Mateo

James Brenneman
University of Evansville

Richard D. Brown
Brunswick Community College

W. Barkley Butler
Indiana University of Pennsylvania

Thomas Butterworth
Western Connecticut State University

John S. Campbell
Northwest College

Kirit D. Chapatwala
Selma University

R. Dean Decker
University of Richmond

Michael Dennis
Montana State University–Billings

Peggy Rae Dorris
Henderson State University

Gibril O. Fadika
Saint Paul's College

John Philip Fawley
Westminster College

Mary B. Fields
Ursinus College

Lynn R. Firestone
Ricks College

Lisa Floyd-Hanna
Prescott College

Rosemary H. Ford
Washington College

Clarence E. Fouché
Virginia Intermont College

Roseli O. Friedmann
Florida A & M University

Paul Gurn
Naugatuck Valley College

R. L. Hackler
Baptist Bible College

Christine Hagelin
Los Medanos College

Deborah L. Hanson
Indiana University-Purdue University-Columbus

Thomas S. Helget
Inver Hills Community College

Christopher J. Harendza
Montgomery County Community College

Catherine Hurlbut
Lock Haven University

M. Ian Jenness
Davis & Elkins College

Tim Knight
Ouachita Baptist University

Victor L. Landry
Duquesne University

Dewey H. Lewis
Coastal Carolina Community College

Jerri K. Lindsey
Tarrant County Junior College, Northeast Campus

Jon R. Maki
Eastern Kentucky University

Jennifer Marsh
University of Louisville

Henri Roger Maurice
Barat College

Donald Mellinger
Kutztown University of Pennsylvania

Richard Mortensen
Albion College

Rudolph Prins
Western Kentucky University

Bruce Reid
Kean College of New Jersey

Quentin Reuer
University of Alaska-Anchorage

Elizabeth Ritch
College of New Caledonia

Russell Robbins
Drury College

Steve Rothenberger
University of Nebraska at Kearney

Ramsey L. Sealy
Bossier Parish Community College

Richard H. Shippee
Vincennes University

K. Dale Smoak
Piedmont Technical College

Amy Sprinkle
University of Kentucky Community College System

Robert H. Tamarin
Boston University

Paula J. Thompson
Florida Community College at Jacksonville

David Thorndill
Essex Community College

Robin Tyser
University of Wisconsin-LaCrosse

Mary Vick
Hampton University

Arthur L. Vorhies
Ohio University-Chillecothe

Theodosia R. Wade
Oxford College of Emory University

Danny B. Wann
Carl Albert State College

Alan Wasmoen
Iowa Central Community College

Uko Zylstra
Calvin College

Biology is the study of life. Here a biologist is examining a month-old spotted owl.

Biology as a Science

Objectives

In this chapter, you will learn
–how scientists find out about nature;
–the role of biology in society;
–about the cellular basis of life;
–how the forms of life are classified;
–about the theory of evolution.

Biological knowledge pervades our culture. It forms the basis of modern medicine and agriculture and is used to make practical decisions. Consider, for example, how our knowledge of genetic material is used in court. In 1989, a man was convicted of rape. He had been identified positively by two victims and sentenced to two life terms plus 335 years in prison. Four years later, biological tests showed that his DNA (hereditary material) did not match the DNA of sperm collected from the victims right after the rapes. The chance that this man committed the rapes was 1 in 300 million. He was innocent without doubt. Real information, as provided in this case by our knowledge of DNA molecules, has become invaluable in helping juries make correct decisions.

Biological knowledge is a product of the scientific process, a way of thinking developed during the sixteenth century. What would our lives be like without the scientific process? We would still try to cure illnesses and to understand the world around us, for it is our nature to be curious and progressive. But without a process of rigorous testing, in which false conclusions are eliminated, knowledge would come only from trial-and-error learning. Our culture would be based largely on irrational fears, omens, and signs, as it was prior to the advent of science. New medical therapies, for instance, would be tried at random. If ingesting a particular herb or applying blood-sucking leeches coincided with an improvement in health, then a "cure" would be declared and the procedure added to the physician's repertoire. There would be no understanding of what produced the illness and therefore no therapy directed at the underlying cause.

Modern medicine, developed by the scientific process, is based on an understanding of how the human body functions. We know how DNA molecules direct the development of traits, and we use this knowledge to treat genetic disorders. We know how the human body fights disease microbes, and we use this knowledge to develop antibiotics and vaccines. Real understanding comes only from rigorous scientific testing.

How do scientists acquire knowledge? How does the scientific process differ from trial-and-error learning? Why is science more successful at finding truths than other forms of thought? Why do scientists sometimes change their minds? The goal of this introductory chapter is to answer these questions—to explain the nature of science and the foundations of modern biology.

The Realm of Biology

Biology is the study of life. It encompasses all aspects of life, from the chemistry of genes to the neural basis of memory to the mating behaviors of squirrels. The goal of biological research is to explain and clarify relationships within and between all forms of life. The product of this research, biological knowledge, can then be used by our society in a variety of ways.

History of Science

Humans have always studied nature. Our distant ancestors observed the habits of antelope and other game, the seasonal cycles of edible plants, and the environments in which these foods were found. Then, approximately 10,000 years ago, they began to use this knowledge to alter nature; they developed agriculture. These practical people solved problems by testing alternatives: the best seeds for planting were discovered by growing many varieties and seeing which ones gave the best yield. They went on to invent tools, like the wheel and fork, by this process of testing different ideas.

Meanwhile, scholars were struggling to understand the human body and nature. They were interested not only in describing the natural world but in understanding it. They wanted to know, for example, why children resembled their parents and why maggots came from rotting meat. These early scholars tried to answer questions by reasoning and debate. Finally, in the sixteenth century, scholars realized that the method of testing alternatives employed by farmers and toolmakers could also be used to reveal cause-and-effect relations in natural phenomena. Instead of debating their ideas, they tested them with experiments and further examinations of nature. At this point, the scientific process was complete and modern science was born.

Biology and Society

Most modern societies support the work of scientists. In return, scientists contribute an understanding of the world that is as near as possible to the truth. Scientists are different from other scholars in that they deal only with phenomena that can be observed and measured. Phenomena that involve miracles and beliefs, such as religion and philosophy, lie outside the realm of science.

Modern biologists seek to understand all aspects of the world within and around them (fig. I.1). They earn their livings mainly as teachers and researchers. Some, however, are applied biologists who use biological knowledge to solve practical problems. Major areas of applied biology include medicine, agriculture, fisheries, conservation, and genetic engineering. All of us are applied biologists to some extent. Choosing a healthy diet or maintaining modern levels of hygiene, for example, requires the use of biological information.

(a)

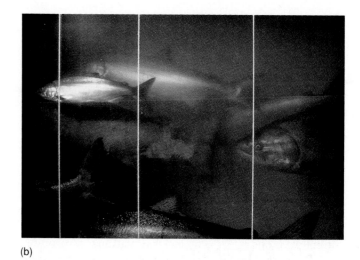

(b)

Figure I.1 Examples of what biologists do. (*a*) Ecologists studying variations in plant structure. (*b*) A behaviorist observing the migration of salmon on the Columbia River. (*c*) A geneticist shooting DNA into plant cells with a DNA particle gun.

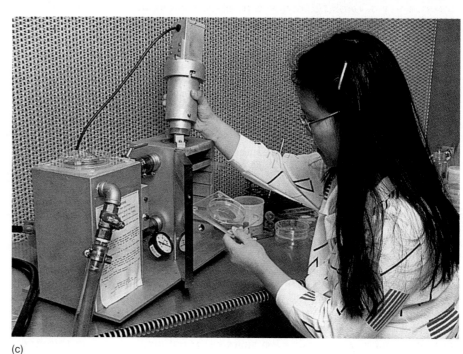

(c)

In recent years, biological knowledge has given us extraordinary power to manipulate our physical selves (even our genes) and the world around us. It has raised many ethical questions. Should we be allowed to choose the sex of our child? Should we transfer genes from one person to another? Should tissues from aborted fetuses be used to treat adult diseases? Should we develop human embryos in test tubes? Should we be able to clone these embryos? Answers to these questions involve value judgments, about which scientists have no special knowledge. Ethical questions are best answered by the society as a whole.

Say, for example, your desk light goes out. You consider the possibilities: the electricity is out in the house, you have accidentally dislodged the plug, or the lightbulb has burned out. Next you test each alternative. You check other lights in the house, then look at the plug, and finally try replacing the bulb. This is the scientific way of thinking.

The Scientific Process

Scientific thinking is a systematic way of seeking the truth, an ordinary human activity based on rationality, logic, and skepticism. You use it every day to solve practical problems.

Outline of the Scientific Process

The **scientific process** involves four activities: observing, hypothesizing, predicting, and testing. Usually many scientists work on different aspects of a problem at the same time. Often the activities are spread out over long periods

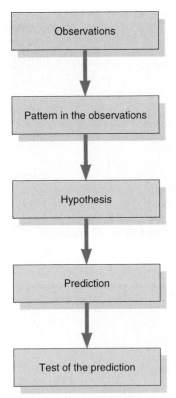

Figure I.2 The scientific process.

of time, with scientists testing hypotheses that were developed from observations made decades or centuries ago. Each scientist builds on the work of others, and the entire scientific community places great emphasis on weeding out false ideas. The scientific process is diagrammed in figure I.2 and described following.

Observations An observation is information taken from the real world; it is made directly with human senses or by way of instruments. Observations are analyzed for pattern, and if a pattern is found then it is a topic for further investigation. You might observe, for example, the characteristics of many animals and notice that warm-blooded animals with hair also suckle their young. Or you might observe medical records and notice that the incidence of lung cancer has doubled in the past fifty years. Such patterns in the observations lead to hypotheses.

Hypotheses There are two kinds of hypotheses: descriptive and explanatory. A **descriptive hypothesis** is an overall statement about the observations. Thus, for example, from observing many (but not all) animals you might conclude that "all warm-blooded animals that suckle their young have hair." A generalization summarizes and gives meaning to the individual observations. It also provides a basis for predicting things not yet observed.

An **explanatory hypothesis** is a guess about what caused the pattern in the observations; it is a tentative explanation of something. Such a hypothesis is presented as an answer to a question about cause and effect. The observation that lung cancer has doubled in the past fifty years might lead to the question, What has caused the increase in lung cancer? A hypothesis then might be, The increase in lung cancer is caused by an increase in cigarette smoking. A scientist develops as many alternative hypotheses as possible to explain the observations so as not to miss the true one. A second hypothesis about lung cancer, for example, might involve air pollution. Alternative hypotheses keep the mind open to all possibilities and help design better experiments for testing the hypotheses.

The thought process that takes a scientist from specific observations to a generalized statement (descriptive hypothesis) or explanation (explanatory hypothesis) is called **inductive logic.** There are no rules for this kind of thought; it is the most creative step in the scientific process, involving imagination, intuition, and luck. For example, August Kekulé first visualized the molecular structure of benzene, a ring of six carbon atoms, while dozing in front of his fireplace. In the flames he saw the shape of a long snake biting its own tail, which gave him the idea of a ring of atoms. A hypothesis does not have to be developed logically; the only requirement of a good hypothesis is that it be testable.

Predictions A hypothesis is tested by seeing whether its predictions hold true. A prediction is typically an "if-then" statement. For example: if all animals that are warm-blooded and suckle their young have hair, then a dolphin must have hair (it does, although not much). Or: if lung cancer is caused by cigarette smoking, then something in the smoke must convert healthy lung cells into cancer cells.

A prediction is developed by **deductive logic;** it is a logical consequence of the hypothesis. Deductive logic is not creative and does not involve guesswork. Its outcome—the prediction—must be true if the hypothesis is true.

Tests of the Predictions The predictions of a hypothesis are tested by further observations and by experiments. This fourth and final activity is crucial in the search for truth. Scientists try to design experiments that will disprove a false hypothesis. It is usually impossible to prove a true hypothesis, since some of the alternative hypotheses may have the same prediction and there is always the possibility that one of these alternative hypotheses has not yet been thought of.

Suppose, for example, you have a hypothesis that the sun revolves around the earth. A prediction of this hypothesis is that the sun will appear on one side (e.g., the eastern horizon) of the earth each morning and disappear on the other side (the western horizon) in the evening. The prediction is true but the hypothesis is not true—a second hypothesis, that the earth revolves around the sun, gives the same prediction and is the true hypothesis.

Thus, if a prediction is found to be true, the hypothesis is supported but is not proven. By contrast, if a prediction is found to be untrue, the hypothesis cannot be true. By disproving alternative hypotheses, the truth is approached by a process of elimination. A hypothesis that has survived all attempts at falsification is accepted as probably true.

Example: Hasler's Study of Homing by Salmon

Pacific salmon have remarkable abilities for navigation. A young salmon hatches out of its egg in a freshwater stream and, after a year or so of growth in an adjoining lake, swims downstream to the Pacific Ocean. The young fish then migrates thousands of kilometers to specific feeding grounds, where it spends several years growing and maturing. When reproductively mature, the salmon returns to fresh water to breed. When biologists tagged many of these fish, an astonishing fact was revealed: each fish returned to the same stream where it was born! A diagram of how Arthur Hasler, of the University of Wisconsin, used these observations to initiate a scientific investigation is shown in figure I.3.

How does a fish recognize its home stream after several years at sea? The first thing to consider is the fish's sense organs, as it is through these structures that an animal acquires information about its environment. A salmon has at least three senses: vision (eyes); smell (nose); and sound (water pressures detected by the ears and other structures). When a young salmon leaves its home stream, one of these senses must take in information about its surroundings and relay it to memory centers in the brain. When the mature salmon returns from the ocean to fresh water, this information must be retrieved and used to guide the fish to the exact same stream it left several years earlier.

Considering the obvious sense organs of fish and the kinds of information available to them, several hypotheses can be developed about how a salmon finds its home stream. Thus, a salmon may find its home stream by recognizing familiar visual landmarks, odors in the water, or pressures of water flow. In addition to these hypotheses, there may be others we have not yet thought of.

Dr. Hasler tested these hypotheses by preventing fish from using their sense organs. His experimental design is shown in figure I.4. He began with the visual landmark hypothesis, predicting that if a salmon uses visual landmarks to recognize its home stream, then a salmon that cannot see will not be able to find its way home. He tested the prediction with two groups: an experimental group of salmon that were blindfolded and a control group of salmon that were not blindfolded. When the salmon were given a choice of two streams—their home stream or a foreign stream, both groups entered their home streams with equal success. Thus, the hypothesis that salmon use visual landmarks cannot be true.

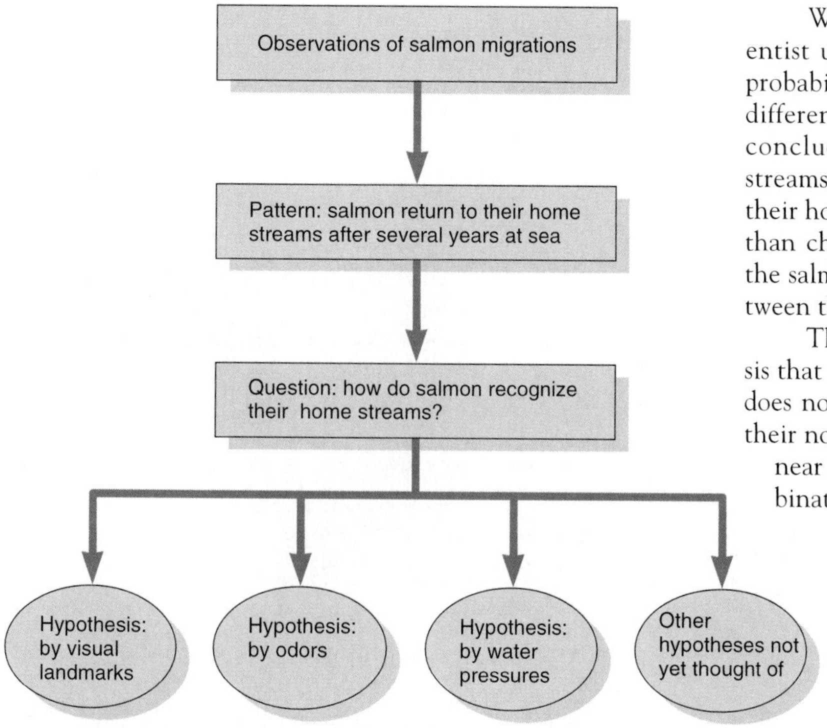

Dr. Hasler then proceeded to the second hypothesis. If a salmon uses the odors of chemicals in the water to recognize its home stream, then a salmon that cannot smell will not be able to find its way home. He used two groups of salmon: an experimental group with plugged noses and a control group without plugged noses. He obtained the following result:

	Number that entered home stream	Number that entered foreign stream
control group	65 (89%)	8 (11%)
experimental group	64 (60%)	28 (40%)

Clearly, the salmon with unplugged noses found their way home better than the salmon with plugged noses. But were the salmon with plugged noses unable to recognize their home stream? Does the 60% return to the home stream represent a random choice?

When experimental results are not obvious, the scientist uses **statistics,** a branch of mathematics based on probabilities, to decide whether the results indicate real differences. Hasler analyzed his data with statistics and concluded that the salmon with plugged noses chose streams at random—the numbers of fish that swam into their home stream and the foreign stream were no different than chance. In other words, there was no evidence that the salmon with plugged noses could tell the difference between their home stream and a foreign stream.

The result of this experiment supported the hypothesis that salmon recognize the smell of their home stream. It does not prove the hypothesis, however, because plugging their noses may have affected some other sense organ in or near their noses. Also, the salmon may be using a combination of smell and some other cue.

Continuation of the Scientific Process

The salmon studies did not end with the two experiments described here. Dr. Hasler and other biologists tested the odor hypothesis in various ways, using both laboratory and field experiments, and they examined other hypotheses.

Figure I.3 Hypotheses development in the investigation of salmon homing.

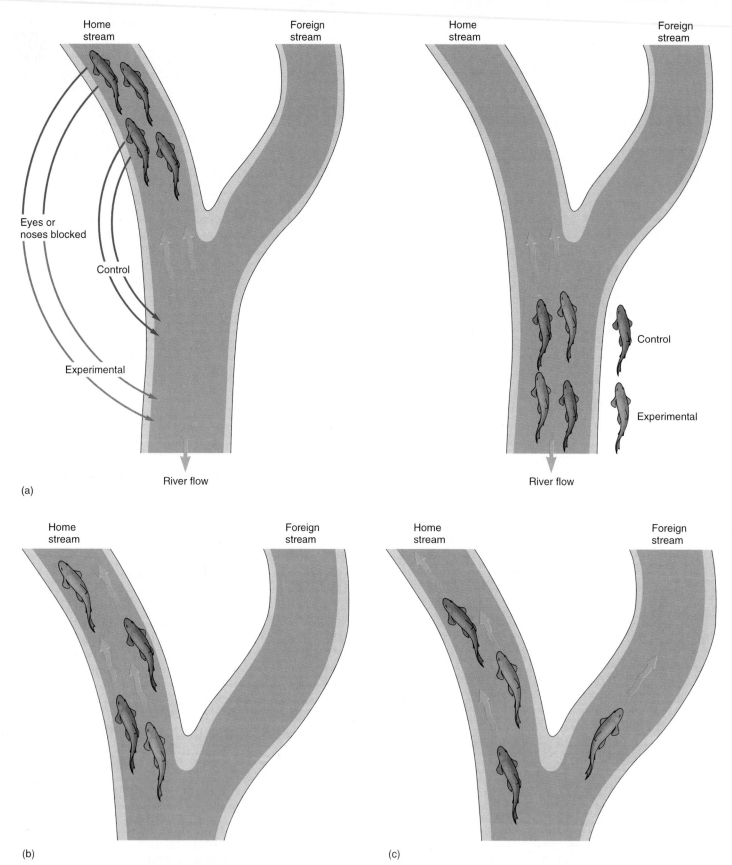

(a)

(b)

(c)

Figure I.4 Hasler's experimental design. (*a*) Fish that had moved into their home stream were transported back downstream so they had to choose a stream again. They were divided into two groups: a control group, which was not altered in any way, and an experimental group, which had one kind of sense organ blocked. (*b*) If the fish did not use that particular sense organ to recognize their home stream, then all the fish—both control and experimental—should return to their home stream. (*c*) If the fish did use that particular sense organ, then all the controls should return to their home stream but the experimental fish should choose at random—about half should choose their home stream and half the foreign stream.

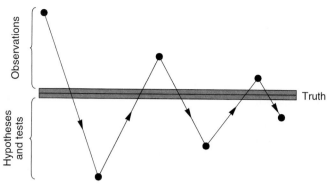

Figure I.5 The scientific method is a continuous process.
With each progression from observations to hypotheses and tests, the scientific version of reality comes closer to the truth.

Results of these experiments support the odor hypothesis: it is probably true that salmon recognize their home stream by its particular smell.

Scientific studies, like Hasler's, are never really finished: new observations are made through the invention of better instruments; additional hypotheses are developed; and better experiments are designed for testing the predictions. Each progression through the scientific process brings the researchers nearer to the truth (fig. I.5). Scientists do not fail when they change their minds.

Most hypotheses are related to other hypotheses and form part of a theory. A **theory** is a set of related hypotheses supported by many observations and experiments. Hasler's odor hypothesis, for example, is part of a theory that includes hypotheses about how salmon navigate in the open ocean and how they find the particular river that connects with their home stream. The hypotheses of a theory are in various phases of testing—some are confirmed and others are still tentative.

A generalization, hypothesis, or theory that has been tested repeatedly and not proven false is called a **law.** Even a law may turn out to be incorrect; someone may devise a test that proves the law false. If this happens, the hypotheses and theories that led to the establishment of the law must be rejected or revised.

Organization of Life

Imagine a walk through a woods. Which objects that you see are living? What criteria do you use to decide what is living and what is not? You might say that living things move, but the stream moves and the trees do not. You might say that living things grow, but a layer of dead leaves grows in the autumn and adult insects do not. Life is hard to define exactly.

Table I.1

Signs of Life

Organized in a complex manner

Maintains a particular internal environment

Acquires and uses energy to do work

Responds to stimuli

Moves at least part of its body

Grows and develops

Reproduces itself

Possesses adaptations to its local environment

Biologists define life not by a single property but rather by the possession of several properties. All forms of life share some of the same signs, but not all signs of life are found in all forms of life. Thus, for example, an object may be considered alive if it moves when touched and if it reproduces itself. Another object may be considered alive if it reproduces and grows, even though it does not move when touched. Important signs of life are listed in table I.1 and discussed in later chapters. One of the signs, organization, is described here because it has determined the arrangement of chapters in this book.

Hierarchy of Biological Organization

A conspicuous feature of life is its hierarchy of biological organization. Every aspect of life is a component of something else. Biological molecules form cells; cells form individuals; individuals form populations; populations form food webs; food webs form landscapes. A diagram of these relationships is provided in the inside front cover of this book. As an example, the position of a grasshopper in life's organization is shown in figure I.6a.

Each level of organization has properties not found in lower levels. A cell controls its internal environment but a molecule does not; an individual reproduces but an organ does not; a population evolves but an individual does not; a food web recycles its atoms but a population does not.

The Cell The basic unit of life is the cell. These small "bags" of chemicals (about ten would fit on a pencil dot) were first seen in 1663 by the English biologist Robert Hooke. Peering through a crude microscope, which could magnify only thirty times, he saw tiny compartments in cork (tree bark) (fig. I.7a). Hooke named them cells

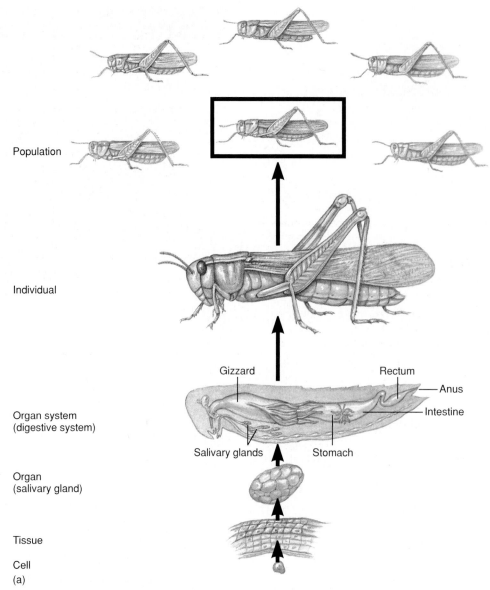

Population

Individual

Organ system
(digestive system)

Gizzard

Rectum

Anus

Intestine

Salivary glands

Stomach

Organ
(salivary gland)

Tissue

Cell

(a)

Figure I.6 A grasshopper in the hierarchy of physical organization. (*a*) An individual is formed of cells, tissues, organs, and organ systems. An individual, in turn, is a component of a population. (*b*) A population of grasshoppers is a component of a food web.

because they reminded him of the rows of small rooms, called cells, in a monastery. He went on to observe similar structures in a variety of plant materials.

A few years later a Dutch scientist, Anton van Leeuwenhoek ground lenses that could magnify almost 300 times. He discovered the same tiny structures in pond water and semen (fig. I.7*b*). Unlike the rigid rows of cells in cork, the organisms and sperm in these fluids moved about "in the most delightful manner." Neither Hooke nor van Leeuwenhoek realized they were observing the basic units of life.

As the microscope became more widely used, cells were observed in a great variety of living materials. Finally,

in the nineteenth century, all these individual sightings were brought together in three descriptive hypotheses that became the cell theory:

1. All living things are made of cells.
2. The cell is the basic unit of life—the simplest level of organization that can maintain the living state.
3. Cells arise only from preexisting cells.

Modern microscopes, which can magnify more than 100,000 times, reveal that a cell has a complex structure (fig. I.7*c*). It is much more than a compartment for holding chemical reactions. The cell theory now consists of many hypotheses, most of them firmly established.

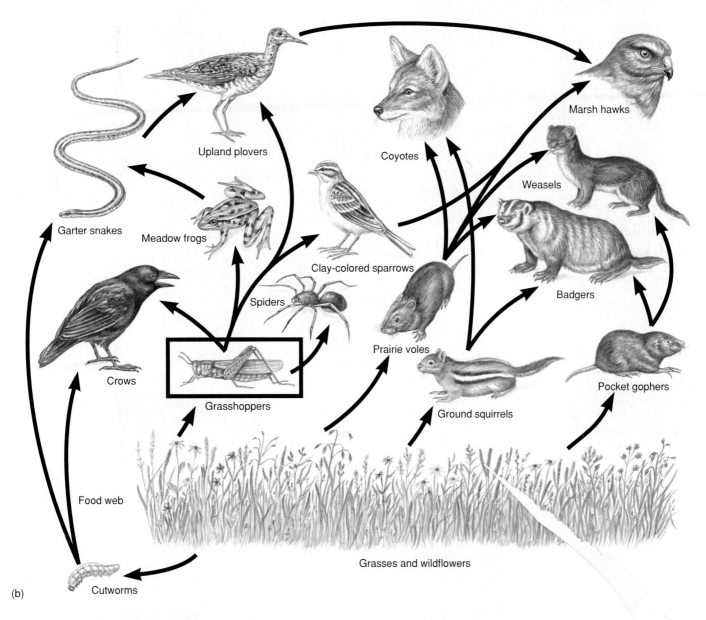

Marsh hawks

Upland plovers

Coyotes

Weasels

Garter snakes

Meadow frogs

Clay-colored sparrows

Badgers

Spiders

Crows

Prairie voles

Pocket gophers

Grasshoppers

Ground squirrels

Food web

Grasses and wildflowers

(b)   Cutworms

In the past few decades we have learned that all cells consist of similar molecules that perform the functions of life in similar ways. All forms of life are based on the same cellular foundation. With this new view of life, many colleges merged their departments of botany, zoology, and microbiology to form single departments of biology. Similarly, botanists, zoologists, and microbiologists began to refer to themselves as biologists.

The Individual and Higher Levels An individual organism is the basic unit of survival and is formed of one or more cells. Individuals that consist of only one cell are known as unicellular organisms. Individuals that consist of more than one cell are multicellular organ-

isms. The cells of most multicellular organisms are organized into tissues, organs, and organ systems, as described in later chapters.

Every individual is a member of a **species,** which is a group of similar organisms capable of mating with one another and producing healthy offspring. Each species is subdivided into several populations, with each population living in a different geographic location. The deer mice in a particular meadow are members of the same population, but deer mice on the other side of a mountain or river belong to another population.

A population is part of a food web, which is a set of relations based on who eats whom in a particular place. Deer mice, for example, eat seeds of the field

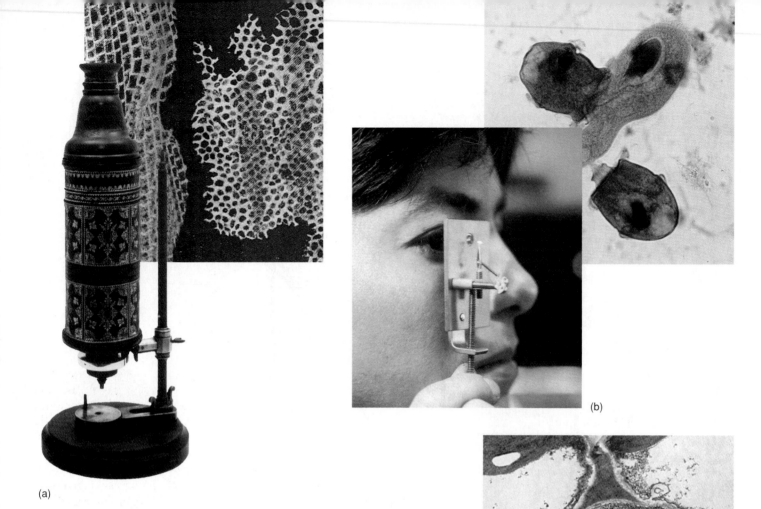

(a)

(b)

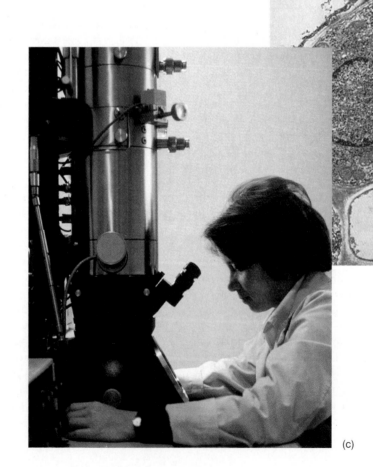

Figure I.7 Cells. (*a*) The outlines of cells in cork, magnified 30 times by the microscope of Robert Hooke (1630–1703). (*b*) Unicellular organisms (made of just one cell) in pond water, magnified 270 times by the microscope of Anton van Leeuwenhoek (1632–1723). (*c*) A plant cell, magnified 30,000 times by a modern microscope.

(c)

Figure I.8 A wolf and a coyote have similar traits.

thistle and in turn are eaten by red foxes. The three populations—deer mice, field thistles, and red foxes—are components of the food web of the field.

A food web and aspects of the nonliving environment form an ecosystem. Similar ecosystems, in turn, form a much larger unit of nature called a biome. A hillside in the prairie, for example, is an ecosystem that together with other similar units of nature form the grassland biome. Still higher, all the biomes of the earth form the biosphere—the outer layer of earth where life is found.

Hierarchy of Classification

Humans have always classified nature; placing things in groups is a characteristic activity of the human mind. Developing a hierarchy of classification is a form of inductive logic: observations of individual organisms are used to form groups (descriptive hypotheses) of organisms with similar traits. A goal of classification is to group organisms according to their degree of relatedness. Recently developed techniques for analyzing the structure of DNA molecules (hereditary material) make this task much less subjective: organisms with more similar DNA molecules are more closely related than organisms with less similar DNA molecules. Our classification system is based on the species, which is a real unit of nature. Categories higher than the species level are products of the human mind.

Species and Genus   A species is a group of individuals capable of mating with one another and producing healthy offspring. All wolves, for example, belong to the same species named *lupus*. Similar species are placed in the same **genus** (pl., genera). The wolf and coyote (fig. I.8), for example, have been placed in the same genus named *Canis*. The name of the genus is combined with the name of the species to give each kind of organism a unique identification. The scientific names of the wolf and coyote are *Canis lupus* and *Canis latrans*. Both the genus and the species

names are written in italics (or underlined), with the genus capitalized and the species not capitalized.

Both genus and species names are needed to identify an organism because the same species name may be used for different organisms (there are more species than there are descriptive words). The name *maculata*, for example, which means spotted, is used for many animals, including a cat (*Genetta maculata*), a hornet (*Vespula maculata*), and a beetle (*Diaperis maculata*). The combination of genus name and species name, however, is unique for each kind of organism.

This two-word system of naming organisms, called the binomial system, was invented in the eighteenth century by Carolus Linnaeus (fig. I.9), a Swedish botanist. (His real name was Nils Ingermarsson, but when he became a well-known botanist he took what he considered a more appropriate identification, choosing his last name from a linden tree that grew near his home.) The Linnaeus system, first published in 1737, was readily adopted around the world.

Why not just call an organism by its common name—wolf, for example, or coyote? One reason is that the common name of an organism varies from language to language. The word for rabbit, for example, is *lapin* in French, *Kaninchen* in German, and *coniglio* in Italian. Scientific names, by contrast, are always in Latin. Another reason is that even in the same language some organisms have more than one common name. *Bidens frondosa*, for example, is a plant known as both beggar-tick and stick-tight. Unlike common names, scientific names are exact and international. They mean just one kind of organism.

Higher Levels   Species are grouped into broader, more inclusive categories, as illustrated in figure I.10. Similar species are placed in the same genus and similar genera are placed in the same **family.** The genus most similar to *Canis* is *Vulpes*, which consists of the foxes. These two genera, along with other genera of doglike animals, constitute the family Canidae (the family name is capitalized, but not italicized, and usually ends in *ae*).

Figure I.9 Carolus Linnaeus (1707–78), inventor of the binomial system of classification.

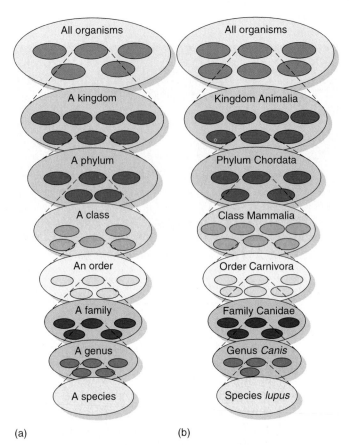

(a)   (b)

Figure I.10 The hierarchy of classification. (*a*) Each category (except kingdom) is a component of a higher category. (*b*) The specific classification of the wolf.

Similar families are placed in the same **order.** The family Canidae is similar to six other families—those with bears, raccoons, weasels, mongooses, hyaenas, and cats. All seven families constitute the order Carnivora (capitalized but not italicized).

Similar orders are placed in the same **class.** The order Carnivora is included, along with eighteen other orders (including those that contain horses, elephants, mice, and primates such as ourselves) in the class Mammalia. This class consists of all animals that are warm-blooded, have hair, and suckle their young.

Similar classes are placed in the same **phylum.** All animals with backbones—fishes, amphibians, reptiles, birds, and mammals—belong to the phylum Chordata. Similar phyla (the plural of phylum), in turn, are placed in the same **kingdom.**

Kingdoms   The most controversial category of classification is the kingdom. Biologists argue about how many kingdoms there should be and what types of organisms ought to be in each kingdom. A modern system, used in this book, recognizes six kingdoms: Archaea, Eubacteria, Protista, Fungi, Plantae, and Animalia, based on cell structure, number of cells, and mode of nutrition. Organisms are placed in these kingdoms according to the following criteria:

1. There are two basic types of cells: simple ones, called **prokaryotic cells,** and complex ones, called **eukaryotic cells** (fig. I.11). The first step in forming kingdoms is to separate organisms with these two types of cells.
2. All organisms with prokaryotic cells are unicellular, with similar cell structures. These simple cells, known as bacteria, form two kingdoms, based on fundamental differences in chemistry. Organisms in the kingdom Archaea are adapted to live in harsh environments—marshes, animal intestines, hot springs, and vents in the ocean floor. They are

chemically distinct from all other forms of life. Organisms in the kingdom Eubacteria are chemically similar to eukaryotic cells and do not inhabit such extreme environments. Some, like the bacteria that cause human diseases, feed on the molecules of other organisms; others build their own molecules from the products of photosynthesis, in which sugars are synthesized from carbon dioxide, water, and sunlight.

3. All organisms with eukaryotic cells that are unicellular during most of their lives are placed in a third kingdom, named Protista.

4. The remaining organisms—all eukaryotic and multicellular—are grouped according to mode of nutrition. Organisms that build their own molecules from the products of photosynthesis are placed in the kingdom Plantae. Organisms that feed on other organisms are placed in the kingdom Fungi if they have cell walls and feed by absorption (release digestive juices onto their food and then absorb the digested components). The kingdom Fungi includes mushrooms, yeasts, and molds. Organisms that feed on other organisms are placed in the kingdom Animalia if they do not have cell walls and feed by ingesting their food and then digesting it internally. The kingdom Animalia includes sponges, jellyfish, worms, insects, fish, and mammals.

These six kingdoms are outlined in the inside back cover of this book. They will become more familiar as you progress in your study of biology.

Evolution: Theoretical Foundation of Biology

The theory of evolution explains what we see in nature. It answers the questions: Why does each kind of organism have a particular set of traits? How did differences among species come about? Why are some species very similar to one another while others are very different? The theory of evolution is the theoretical foundation of modern biology.

The theory of evolution was developed by Charles Darwin, an Englishman who published his ideas (in 1859) in a book called *On the Origin of Species by Means of Natural Selection*. Since Darwin, the theory has been refined and strengthened by additional observations and tests.

Darwin was not the first person to think that the traits of a species change through time or even the only person to explain how such changes might take place. Many nineteenth-century biologists thought that species might change through time, and another Englishman, Alfred Russel Wallace, had the same idea as Darwin about how they change. Darwin is given credit for the

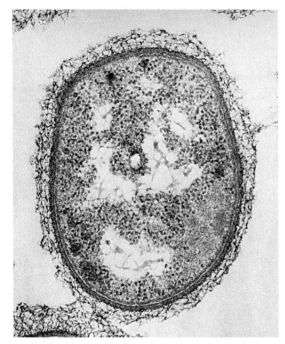

(a)

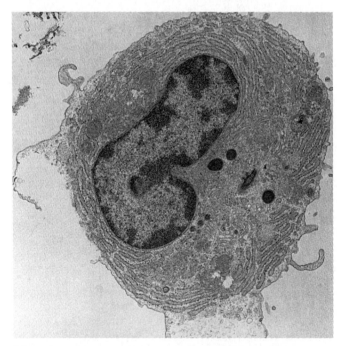

(b)

Figure I.11 The two basic kinds of cells. (*a*) A prokaryotic cell is a simple cell with few internal structures. (*b*) A eukaryotic cell is a larger, more complex cell with many internal structures.

theory because he explained it more thoroughly than anyone else and he provided a vast amount of evidence in its support.

Darwin developed his ideas and gathered most of his evidence while on a five-year journey (1831–36) aboard a

(a)

(b)

(c)

Figure I.12 Darwin and his voyage. (a) Charles Darwin (1809–82) at the age of thirty-one. (b) A replica of Darwin's ship, HMS *Beagle*. (c) The route taken by the *Beagle* on its five-year voyage, from 1831 to 1836.

British ship, HMS *Beagle* (fig. I.12). The goal of the voyage was to map the oceans of the world, particularly the coastal waters of South America. As was customary on such long trips, a naturalist was employed to record the plants and animals and to collect specimens for museums. The naturalist on the *Beagle* was Charles Darwin, a young man of twenty-two at the time.

While on the voyage, Darwin spent as much time as possible on land, often leaving the ship in one harbor and boarding again at a different harbor. He observed that cer-

14

Front paw
of bobcat

Bobcat

Front paw
of lynx

(a)   Lynx   (b)   (c)

Figure I.13 Species change from place to place in ways that make them better suited to their environment. (*a*) The bobcat and lynx are very similar species. The bobcat lives in less snowy places than the lynx. (*b*) The lynx has much larger paws, for walking on snow, than the bobcat. (*c*) Photo of a lynx walking on snow.

tain animals were restricted to one particular place and that fossils resembling these animals were found in the same place. He found in Argentina, for example, living armadillos as well as fossil armadillos that were somewhat different from the living ones. Could the fossil forms be ancestors of the living forms? He also noticed that when a species was found in many places, it had somewhat different traits in each place. This pattern was particularly evident on islands. Darwin was puzzled by these observations: if each species was created by God, as believed at the time, then it should be perfect and unchanging. He wrote, "Every character . . . has a tendency to vary by slow degrees." He noted that the slight differences in traits seemed to adjust the species to local conditions (fig. I.13).

After returning to England, Darwin thought about what he had seen and concluded that the traits of organisms are not fixed. Instead, a species changes over the generations in response to different environments. He called this idea "descent with modification" and suggested that different species originated in this way from a common ancestor. The mechanism of change—what caused a change in traits—that he proposed was natural selection.

Darwin's Theory of Evolution by Natural Selection

Darwin's theory consists of four descriptive hypotheses and two predictions. Each hypothesis summarizes a vast number of observations. Each prediction states what must be true if the hypothesis is true. The theory is outlined in table I.2, diagrammed in figure I.14, and described in more detail following.

Reproductive Potential Darwin's first hypothesis was: Every organism has the potential to leave more than one offspring during its lifetime.

This hypothesis is supported by millions of observations since Darwin. A pair of mice has many young in each litter and many litters in a lifetime. A pair of robins rears many chicks. An oak tree releases millions of seeds. On the average, each organism can produce more than one offspring during its lifetime.

Constancy of Numbers Darwin's second hypothesis was: The number of individuals within each species remains fairly constant over time.

Table I.2

Outline of Darwin's Theory of Evolution

Hypothesis 1	Every organism has the potential to leave more than one offspring during its lifetime.
Hypothesis 2	The number of individuals within each species remains fairly constant over time.
Prediction A	If hypotheses 1 and 2 are true, then not all individuals realize their reproductive potential. There is a struggle for existence (or reproduction).
Hypothesis 3	Individuals of a species vary in terms of their traits.
Hypothesis 4	Some of these variations are inherited.
Prediction B	If prediction A is true and hypotheses 3 and 4 are true, then individuals with certain traits are better suited to their environments and so leave more offspring than individuals with other traits. Inherited traits that promote successful reproduction become more common in future generations of the species. This process is called natural selection.

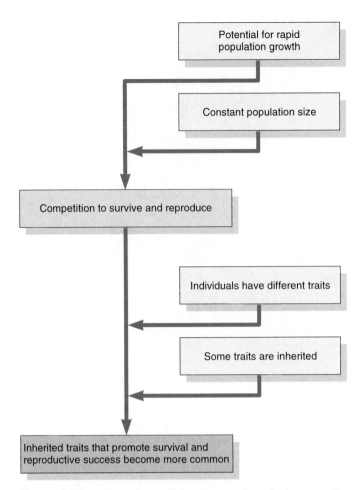

Figure I.14 A flow chart of the theory of evolution.

This hypothesis is also supported by countless observations. In general, the mice and robins in a woods do not become increasingly more abundant over the years, nor do the trees of a forest or flowers of a meadow become more crowded. Most of the time, the number of individuals remains approximately the same from year to year.

Struggle for Existence According to the first hypothesis, the number of individuals within each species should be steadily increasing, yet according to the second hypothesis, the number of individuals remains approximately the same:

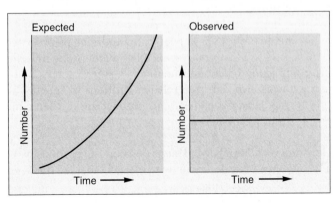

Combining these two hypotheses, Darwin made his first prediction: If every organism has the potential to leave more than one offspring and if the number of individuals remains fairly constant, then many individuals do not realize their reproductive potential—they die young or fail to reproduce as adults. He called this prediction the "struggle for existence."

Individual Differences Darwin's third hypothesis was: Individuals of a species vary in terms of their traits.

We see this most clearly in humans, but it is true of all species. When examined closely, crows exhibit individual differences as do mice, rabbits, wolves, and other animals. The oak trees in a forest and the daisies in a meadow also have traits that vary from one individual to another.

Inheritance of Traits Darwin's fourth hypothesis was: Some of these variations are inherited. This statement was based on observations that offspring resemble their parents. Darwin did not know about the mechanisms of inheritance.

We now know why offspring resemble their parents: the development of many traits is controlled by genes, which pass (within eggs and sperm) from parents to off-spring. Offspring resemble their parents because they inherit from them the genes for their traits.

Natural Selection   Darwin's second prediction was: If there is a struggle for existence (prediction from hypotheses 1 and 2) and if individuals of a species vary in terms of their traits (hypothesis 3), then certain individuals will have traits that make them more successful at surviving and reproducing than other individuals. If some of these variations are inherited (hypothesis 4), then traits that promote success will become more common in future generations, because individuals possessing them will leave more offspring. The particular traits that make an individual more successful depend on local conditions of the environment.

This second prediction is Darwin's statement of **natural selection.** Nature (the environment) selects the individuals who survive and reproduce more successfully, based on the kinds of traits that help an organism find resources and evade enemies. Successful traits that develop in response to genes become more common through time as individuals with those genes leave more offspring (fig. I.15).

Darwin's work reveals his remarkable mind. While most people have an aptitude for either inductive logic (from observations to hypotheses) or deductive logic (from hypotheses to predictions), Darwin excelled at both kinds of thinking. A keen observer, he developed hypotheses about nature and then deduced natural selection from the hypotheses.

Products of Evolution

Evolution is a change in genetic composition, and therefore in traits, of a species in response to the local environment. The change is brought about by natural selection. In viewing nature, we see two products of evolution: adaptations and species diversity.

Adaptations   An **adaptation** is an anatomical, physiological, or behavioral trait that is controlled by genes and that improves an individual's chances of surviving and of leaving offspring in a particular environment.

Examples of anatomical adaptations are shown, for a woodpecker, in figure I.16. An example of a physiological adaptation is the muscle chemistry of rabbits, which enables a rapid sprint to safety. An example of a behavioral adaptation is the cooperative behavior of wolves, which helps a pack work together as they hunt game that is too large for a single wolf to capture.

Species Diversity   When different populations of a species become reproductively isolated, due to geographic separation or some other impediment to interbreeding, each isolated population follows a unique course of evolution and may become a new species. In this way, evolution

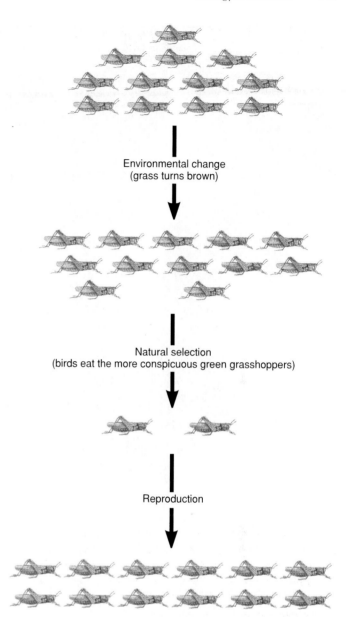

Figure I.15 The process of natural selection. In this example, a population of grasshoppers living in green grass consists mainly of green grasshoppers. A few brown ones appear from time to time but are quickly seen and eaten by birds. The environment changes when the grasses dry up and turn brown. Natural selection then occurs as birds eat more green than brown grasshoppers because they are more conspicuous on the brown grass. After the grasshoppers reproduce, the population consists of brown grasshoppers.

generates new species. More than 1.4 million species have now been described and named, yet biologists believe that most species—perhaps as many as 30 million—have not yet been described. Many unknown species may live in the dense vegetation of tropical forests; most are probably insects, yet one of the recently discovered species is a primate (fig. I.17).

The number of species in the world today is the outcome of two processes: species formation and species

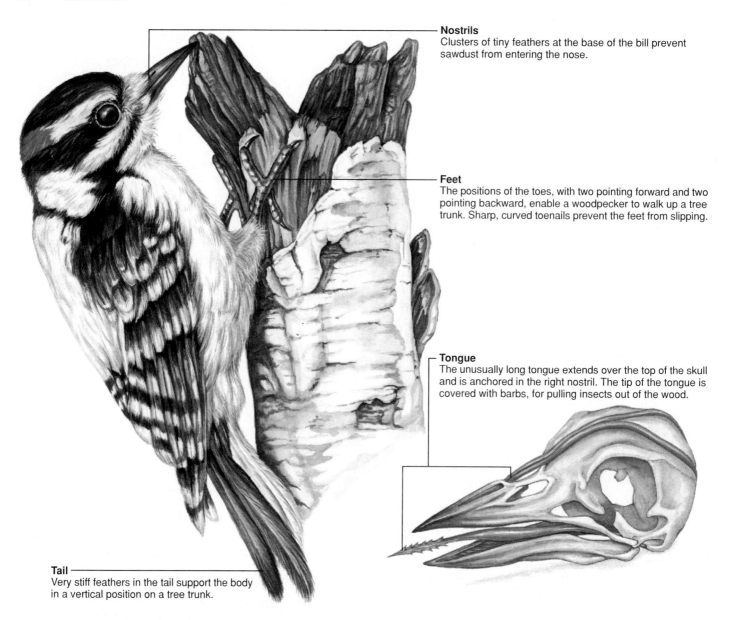

Nostrils
Clusters of tiny feathers at the base of the bill prevent sawdust from entering the nose.

Feet
The positions of the toes, with two pointing forward and two pointing backward, enable a woodpecker to walk up a tree trunk. Sharp, curved toenails prevent the feet from slipping.

Tongue
The unusually long tongue extends over the top of the skull and is anchored in the right nostril. The tip of the tongue is covered with barbs, for pulling insects out of the wood.

Tail
Very stiff feathers in the tail support the body in a vertical position on a tree trunk.

Figure I.16 Physical adaptations of a woodpecker.

Figure I.17 A new species of primate. This monkey was discovered recently in the Amazon rain forest of Brazil. It has been named *Callithrix mauesi.*

extinction. (Species do not merge, since they cannot interbreed.) New species form by accident, as described in chapter 30. Sometimes species go extinct by accident, as when all individuals are killed by a comet hitting the earth. Most often, species go extinct because they cannot evolve fast enough to cope with changes taking place in their environment. Human activities cause extinctions by changing the environment too rapidly for evolution to keep pace.

SUMMARY

1. Biology is the study of life. Some biologists study life while others use biological knowledge to solve problems. Biologists have no special insights into ethics.

2. The scientific process is a systematic way, developed in the sixteenth century, of finding out about nature. The process consists of four kinds of activities: observations, hypotheses, predictions from the hypotheses, and tests of the predictions. It is an unending process that approaches the truth. A set of related hypotheses forms a theory. A hypothesis or theory that has been tested repeatedly and not disproven becomes a law.

3. Life occurs in a hierarchy of physical organization, from biological molecules to landscapes. The basic unit of life is the cell. The basic unit of survival is the individual. Higher levels of organization include populations and food webs.

4. Biologists have organized life into a hierarchy of classification. The basic unit of classification is the species. Similar species are placed in the same genus. Each kind of organism has a genus name and a species name. Higher levels of classification include the family, order, class, and phylum. The highest category is the kingdom. A modern system of classification divides all organisms into six kingdoms.

5. Darwin's theory of evolution consists of four hypotheses and two predictions. Hypothesis 1: Every organism has the potential to leave more than one offspring during its lifetime. Hypothesis 2: The number of individuals in a species remains fairly constant over time. Prediction A: If hypotheses 1 and 2 are true, then not all individuals realize their reproductive potential. Hypothesis 3: Individuals of a species vary in terms of their traits. Hypothesis 4: Some of these variations are inherited. Prediction B: If prediction A is true and hypotheses 3 and 4 are true, then individuals with certain traits are better suited to their environments and so leave more offspring than individuals with other traits. Natural selection occurs and certain traits become more common in the species.

6. Evolution is a change in the genetic composition of a species in response to the local environment. One product of evolution is adaptation. Another product of evolution is species diversity.

KEY TERMS

Adaptation (page 17)
Class (page 12)
Deductive logic (page 4)
Descriptive hypothesis (heye-PAH-the-siss) (page 4)
Eukaryotic (YOO-kahr-ee-AH-tik) **cell** (page 12)
Evolution (page 17)
Explanatory hypothesis (page 4)
Family (page 11)
Genus (JEE-nuss) (page 11)
Inductive logic (page 4)
Kingdom (page 12)
Law (page 7)
Natural selection (page 17)
Order (page 12)
Phylum (FEYE-lum) (page 12)
Prokaryotic (PRO-kahr-ee-AH-tik) **cell** (page 12)
Scientific process (page 3)
Species (SPEE-seez) (page 9)
Statistics (stah-TIH-stiks) (page 5)
Theory (page 7)

STUDY QUESTIONS

1. Outline the scientific process. What is the function of an experiment? What is the difference between a hypothesis and a theory; a hypothesis and a law?

2. State the cell theory. What role did the microscope play in its development? When was the cell theory developed?

3. What is meant by the "hierarchy of physical organization"? Give several examples of this hierarchy.

4. Describe the binomial system of classification.

5. Outline Darwin's theory of evolution. Why is it called a theory rather than a hypothesis or law?

6. Choose an organism and describe three of its adaptations.

CRITICAL THINKING PROBLEMS

1. While resting in a meadow, you observe that the honeybees visit only blue irises, and you wish to find out why they choose these particular flowers over all the others available. Design an experiment to test the hypothesis that it is the blue color that attracts the bees and not their shape or the concentration of sugar in their nectar.

2. You observe that two species of fruit flies feed on the apples in your orchard. Design an experiment to test the hypothesis that the abundance of each species is limited by competition for food with the other species.

3. According to stars of the zodiac, someone born under the sign of Leo (July 23 to August 22) is brave. Design a way of testing the following hypotheses:

   Everyone born under the sign of Leo is brave.
   Everyone who is brave is born under the sign of Leo.

4. Hasler's experiments supported but did not prove one of his hypotheses about how salmon recognize their home stream. Why can't we consider the hypothesis proven?

5. Describe your own anatomical, physiological, and behavioral adaptations for surviving on a very hot day.

6. Darwin used the phrase "struggle for existence." Why would the phrase "struggle for reproduction" have been more accurate?

SUGGESTED READINGS

Introductory Level:

Clar, R. W. *The Survival of Charles Darwin: A Biography of a Man and an Idea*. New York: Random House, 1985. A detailed portrait of Darwin and the social environment of his time, as well as descriptions of his voyage on HMS *Beagle* and subsequent decades of observation, thought, and experimentation. Includes an analysis of Darwin's influence on modern biology.

Janovy, J., Jr. *On Becoming a Biologist*. New York: Harper & Row, 1985. An excellent introduction to biology as a science and profession, with chapters on the practice of biology, teaching and learning, and responsibilities. Includes an annotated bibliography.

Moore, J. A. *Science as a Way of Knowing: The Foundations of Modern Biology*. Cambridge, MA: Harvard University Press, 1993. A history of science from our hunter-gatherer ancestors to Darwin, as well as chapters on the history of evolutionary theory, genetics, and embryology.

Root-Bernstein, R. A. *Discovering*. Cambridge, MA: Harvard University Press, 1989. An entertaining account of the creative process of inductive logic.

Strahler, A. N. *Understanding Science: An Introduction to Concepts and Issues*. Buffalo, NY: Prometheus Books, 1992. Insights into the philosophy and sociology of science, with explanations of what scientists do, why they do it, and how they go about it.

Advanced Level:

Baker, J. J. W., and G. E. Allen. *Hypothesis, Prediction, and Implication in Biology*. Reading, MA: Addison-Wesley, 1968. Originally designed as a supplementary text for courses in general biology, this classic book explains the scientific process as it applies to biology.

Mayr, E. *The Growth of Biological Thought*. Cambridge, MA: The Belknap Press of Harvard University Press, 1982. An exploration of the history of biological thought, the philosophical background of biology as compared with mathematics and physical sciences, and the analysis of evolutionary phenomena.

Pancher, A. L. *Classification, Evolution, and the Nature of Biology*. New York: Cambridge University Press, 1992. A history of the method behind biological classification, emphasizing its role in the development of evolutionary theory.

Ro, R. N. *Bioburst: The Impact of Modern Biology on the Affairs of Man*. Baton Rouge, LA: Louisiana State University Press, 1986. A description of recent discoveries in biology and their impact on medicine, agriculture, genetics, and how we view ourselves. The author, a physician, also describes ways of coping with the ethical, moral, and social questions raised by the recent discoveries.

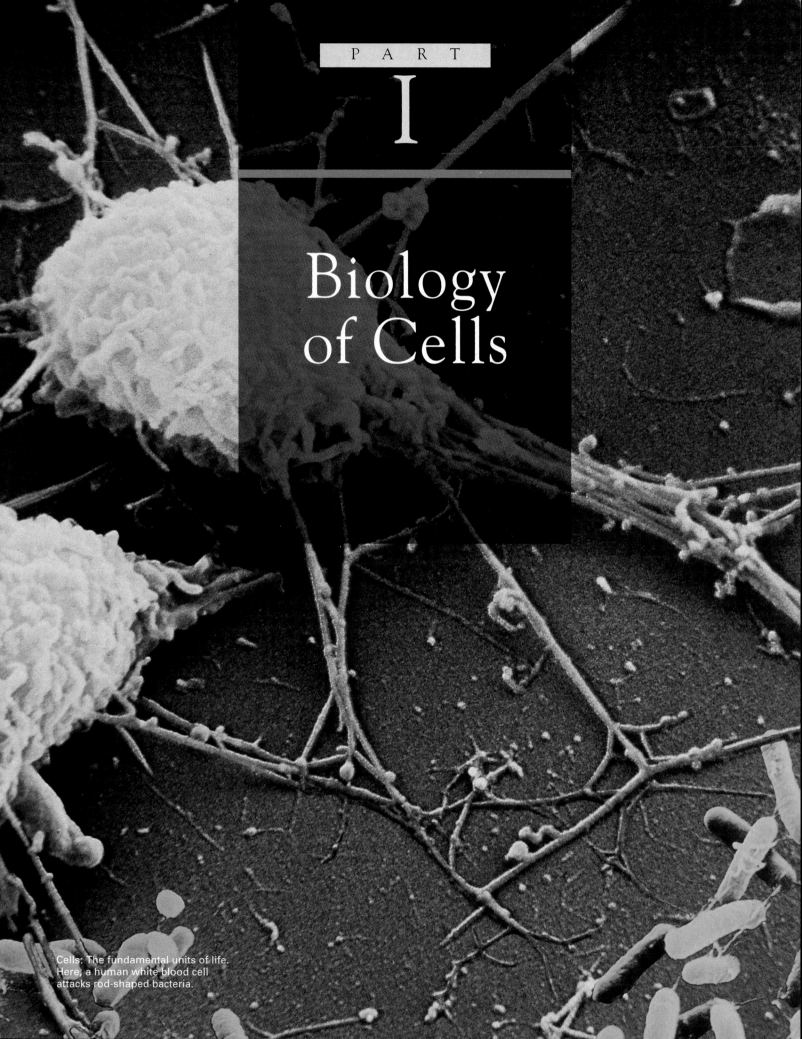

Biology of Cells

Cells: The fundamental units of life. Here, a human white blood cell attacks rod-shaped bacteria.

PART ONE

Exploring Themes

Theme 1

The fundamental unit of life is the cell.

A cell is formed of large organic molecules: polysaccharides, lipids, proteins, and nucleic acids (chapters 1 and 2). These molecules, in turn, are organized into membranes and other cell structures (chapter 3). The outer membrane controls what enters and leaves the cell (chapter 4).

Theme 2

All cells carry out the same basic processes.

Chemical reactions within a cell are governed by enzymes (chapter 2). Cells synthesize molecules (chapter 3), import and export materials (chapter 4), and acquire energy for cellular work from the chemical bonds of organic molecules (chapter 5). Some cells have chemical pathways that form organic molecules from inorganic molecules and sunlight (chapter 6).

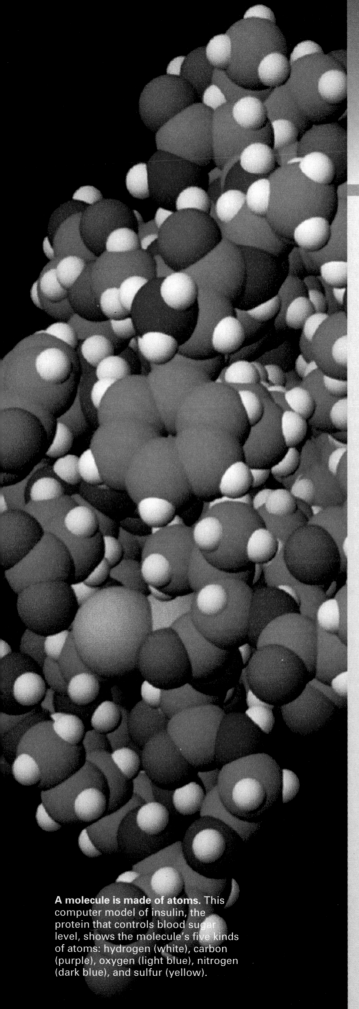

A molecule is made of atoms. This computer model of insulin, the protein that controls blood sugar level, shows the molecule's five kinds of atoms: hydrogen (white), carbon (purple), oxygen (light blue), nitrogen (dark blue), and sulfur (yellow).

Atoms and Molecules

Chapter Outline

Objectives

In this chapter, you will learn
–how the structure of an atom determines what it does;
–how atoms gain and lose electrons to become ions;
–how atoms join to form molecules;
–why some molecules dissolve in water;
–why some molecules are acids and bases;
–what happens during a chemical reaction.

ife is a chemical phenomenon. Your own body is a mixture of chemicals in about 46 liters (12 gal) of water. Your energy, movements, sensations, and thoughts are brought about by interactions among these chemicals. When the amounts of chemicals are just right, you enjoy good health and vitality. When they are not, you become ill and may even die.

Consider the life of an eighteenth-century sailor. While danger of drowning in a shipwreck was always present, an equally severe threat to a sailor's life was disruption of the body chemistry. The diet on board ship consisted chiefly of salt pork and hard crackers; it lacked many chemicals essential to health. The absence of one chemical in particular—vitamin C—was life-threatening.

An insufficient amount of vitamin C leads to scurvy, a disorder with horrible symptoms: small blood vessels burst open, teeth fall out, joints swell, muscles tear, bones break, wounds fail to heal, and old scars reopen. Scurvy can be prevented by simply adding a small amount of vitamin C to the diet. Englishmen are still called "limeys" because nineteenth-century English sailors drank lime juice, rich in vitamin C, to prevent scurvy. The chemical was first purified in 1928 and since then has been available in tablet form.

The physical components of life, such as skin and bones, are formed of chemicals. The activities of life, such as digestion and muscle contractions, are chemical reactions. Even such extraordinary activities as appreciating music and understanding mathematics are the consequences of interactions among chemicals within our bodies.

Atoms

All physical things are made of chemicals. Chemicals, in turn, are groups of **atoms** bonded together. Atoms are exceedingly small structures—so small that if a million were lined up, they would span a distance less than the width of the period at the end of this sentence.

An atom is not changed by being part of something. The atoms within your body, as well as the atoms in everything around you, originated billions of years ago in stars. Before being part of you, your atoms were part of the earth and of many living things. They became part of you when atoms in your father's sperm and your mother's egg formed a fertilized egg (your beginning), when atoms in the food your mother ate became part of your developing body, and as the food you eat becomes incorporated into your body. An atom lasts forever. After being part of you, your atoms will become parts of other living and nonliving things.

Table 1.1

The Particles of an Atom

Particle	Location	Electric Charge	Effect on Atom
Proton	Inside nucleus	Positive	Contributes weight Governs number of electrons
Neutron	Inside nucleus	No charge	Contributes weight
Electron	Outside nucleus	Negative	Governs chemical activity

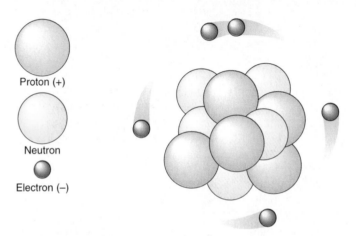

Figure 1.1 Structure of an atom. An atom consists of a small nucleus, packed with protons and neutrons and surrounded by orbiting electrons. Here, the nucleus is drawn much larger in relation to the electrons than it really is.

Physical Structure

An atom is made of many particles, but only three—protons, neutrons, and electrons (table 1.1)—play roles in the chemistry of life. These particles are the smallest units in the hierarchy of biological organization (see the inside front cover of this book).

The protons and neutrons of an atom are packed into a tiny space, called the nucleus, around which the electrons revolve (fig. 1.1). The electrons are much smaller than the nucleus, but they are so widely separated that they occupy much more space. If, for example, an atom were expanded to the size of a basketball, then the nucleus would occupy a space smaller than a pencil dot in the center and the electrons would occupy the remainder of the ball.

The Nucleus   The nucleus of an atom consists of protons, neutrons, and other particles that are poorly understood. A

proton is a particle with a positive electric charge. Two or more protons repel one another and resist being close together. (They are held together within the nucleus by other forces.) Each kind of atom is distinguished from other kinds by its number of protons. The hydrogen atom, for example, always has one proton and the carbon atom always has six.

The number of protons is the most important characteristic of an atom; it influences the number of electrons in the atom, which in turn controls the chemical activity of the atom (as will be explained later). The number of protons also affects an atom's mass, which is its weight corrected for the pull of gravity. (The pull of gravity is proportional to distance from the center of the earth. When comparing objects on the earth's surface, mass is the same as weight because the pull of gravity is the same everywhere.)

A **neutron** is a particle with no electric charge; it neither attracts nor repels other particles in the atom. Every atom except hydrogen has at least one neutron. Neutrons have no effect on an atom's chemical activity, but they do add to its mass.

Atoms with the same number of protons and different numbers of neutrons are isotopes. The carbon atom, for example, always has six protons, but it exists as three different isotopes with six, seven, or eight neutrons.

The Electrons An **electron** is a particle with a negative electric charge. Electrons repel one another but are attracted to protons. It is this attaction between opposite charges—the negative charge of an electron and the positive charge of a proton—that holds the electrons to the nucleus of an atom. The negative charge of an electron is of equal magnitude to the positive charge of a proton. An atom has the same number of electrons as it has protons and so is neutral in charge; the negative charges of its electrons cancel out the positive charges of its protons. An electron is almost weightless and has almost no effect on the mass of an atom.

As the electrons whirl around the nucleus, they remain as far as possible from one another (because of their mutual repulsion) and as close as possible to the nucleus (because of their attraction to the positively charged protons).

Kinds of Atoms There are ninety-two kinds of natural atoms (others have been made in laboratories). The different kinds of atoms can be arranged in a series according to how many protons each atom has in its nucleus. The series goes from hydrogen, with just one proton, to uranium, with ninety-two protons. Each atom in the series contains one more proton than the preceding atom. A substance consisting entirely of atoms of the same kind is called an element.

Each atom in the series has a name. Combinations of one or two letters are used as symbols for the different kinds of atoms. When one letter is used, it is always capi-

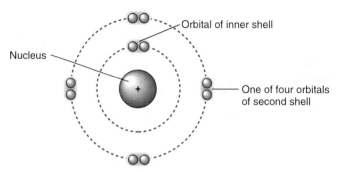

Figure 1.2 The distribution of electrons. Each electron travels within a shell and an orbit. Each shell is a particular distance from the nucleus, and each orbital is a particular position within a shell. The single orbital of the innermost shell is indicated simply by a single spot (shown here in red) occupied by one or a pair of electrons. The four orbitals of other shells are indicated in the same manner—each orbital is a particular place (shown in red) that holds one or a pair of electrons. The four orbitals are positioned as far apart as possible.

talized. The symbol for carbon, for example, is C. When two letters are used, the first is a capital letter and the second is a lowercase letter. The symbol for chlorine, for example, is Cl.

Chemical Properties

The chemical properties of an atom include its arrangement of electrons and its chemical activities. These two properties are interrelated, since the locations of an atom's electrons govern its chemical activities.

Arrangement of Electrons Each electron travels around the nucleus within a pathway defined by a shell and an orbital. A **shell** is a space that is a particular distance from the nucleus:

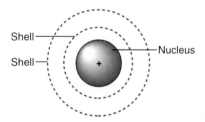

Within each shell, an electron follows a path known as an orbital. Each electron does not, however, travel along the exact same path every time it revolves around the nucleus. It is impossible to know, for any given time, the precise location of an electron within its orbital. Realistic orbitals are difficult to visualize. We use instead simplistic models in which the shells are drawn as circles around the nucleus and the orbitals within each shell are positions, equidistant from one another, on the circles (fig. 1.2).

The electrons of an atom occupy shells and orbitals in an orderly pattern based on the following four rules:

Rule 1:   Electrons occupy shells as close to the nucleus as possible because of their attractions to positive charges of the proton(s).

Rule 2:   The first shell, nearest the nucleus, has a single orbital. Larger, more distant shells have 4 orbitals each.[1]

Rule 3:   An orbital can hold at most two electrons.

Rule 4:   Whenever possible, an electron travels alone in its orbital.

The hydrogen atom (H), with one proton, and the helium atom (He), with two protons, have no shells other than a first shell (rule 1) and no orbitals other than the single one in the first shell (rules 2 and 3). The single electron of the hydrogen atom travels alone and the two electrons of the helium atom travel as a pair within the single orbital of the first shell:

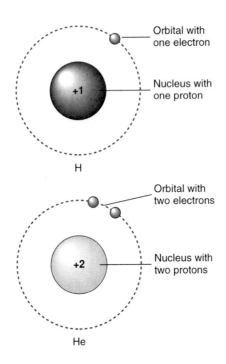

Atoms with between three and ten electrons have a first shell, occupied by two electrons (rules 1, 2, and 3), and a second shell, occupied by the remaining electrons. The second shell holds from one to eight electrons, traveling in one, two, three, or four orbitals (rules 2 and 3). Oxygen (O), with eight protons (fig. 1.3), has two electrons in its first shell and six in its second shell: four of the electrons in the second shell travel in pairs in two of the

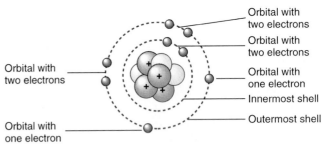

Figure 1.3 Model of an oxygen atom. The dense structure at the center of the atom is the nucleus, which in oxygen contains eight positively charged protons and eight neutrons. The nucleus is drawn larger, in relation to the electrons, than it really is. The tiny circles around the nucleus represent the eight electrons arranged in shells: the first (innermost) shell of oxygen has two electrons traveling in the same orbital and the second shell has six electrons: four traveling in pairs and two traveling alone in their orbitals.

orbitals (rule 3); each of the other two electrons travels unpaired in its own orbital (rule 4). The allocations of electrons to shells for some of the biologically important atoms are shown in figure 1.4.

Chemical Activities   The chemical activity of an atom—how it interacts with other atoms—is governed by the number of electrons in its outermost shell. Because strict rules govern the arrangement of electrons (see rules 1–4), the number of electrons in the outermost shell is set by the total number of electrons in the atom (see fig. 1.4). An atom with eight electrons, for example, has six electrons in its outermost shell because two of its eight electrons are in the first shell. An atom with twenty electrons has two electrons in its outermost shell because two of its electrons are in its first shell, eight in its second, and eight in its third shell, leaving two electrons for its fourth and outermost shell.

An atom interacts with other atoms in ways that give it a full outermost shell of electrons. (A full outermost shell contains two electrons if it is the only shell in the atom, as in helium, and eight electrons if it has more than one shell.) An atom fills its outermost shell by acquiring, getting rid of, or sharing electrons.

Two different kinds of atoms with the same number of electrons in their outermost shells interact with other atoms in the same way; they need to gain, lose, or share the same number of electrons in order to have filled outermost shells. For example, both sodium (Na), with eleven protons, and potassium (K), with nineteen protons, have one electron in their outermost shell (see fig. 1.4). The chemical activities of sodium and potassium are similar: they interact with other atoms in ways that get rid of the solitary electron in the outermost shell. After losing the electron, the next shell closer to the nucleus becomes the outermost shell and it is full.

---

1. The pattern of electron distribution is more complex in very large atoms. Because life is formed of relatively small atoms, we need not concern ourselves with more complex patterns.

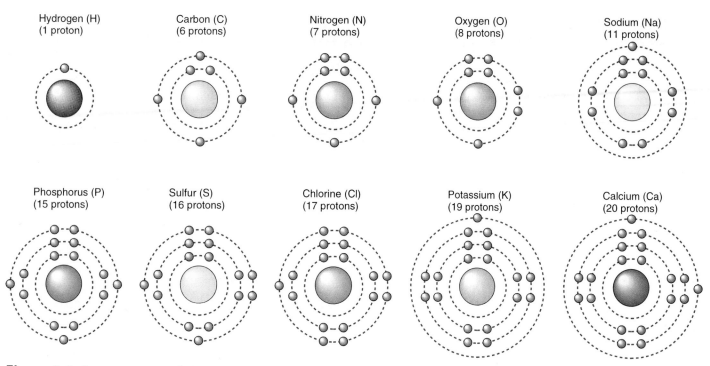

Figure 1.4 Arrangements of electrons in the atoms most abundant in living things. Circles around the nucleus represent shells, and dots in the shells represent electrons. The first shell holds at most two electrons, and other shells hold up to eight electrons each. Shells nearer the nucleus fill first.

Atoms acquire full outermost shells in two ways: (1) by forming ions, in which electrons are gained or lost and (2) by forming chemical bonds, in which electrons are transferred or shared.

Ions

Some atoms acquire full outermost shells of electrons by gaining or losing electrons. Whether a particular atom gains or loses electrons depends on which process involves fewer electrons: it is easier for an atom with an outermost shell of seven electrons to add one more electron than to lose seven, and it is easier for an atom with an outermost shell of one electron to lose that electron (making the next shell closer to the nucleus a full outermost shell) than to pick up seven more.

An **ion** is an atom with either more electrons or fewer electrons than protons; it has acquired a full outermost shell by gaining electrons or by losing electrons. When an atom loses an electron, it has more protons than electrons and becomes an ion with a positive charge. When an atom picks up an electron, it has more electrons than protons and becomes an ion with a negative charge. Whenever one atom loses an electron, to become a positively charged ion, another atom gains the electron, to become a negatively charged ion. An electron cannot leave one atom unless another atom can receive it.

The sodium atom (Na), with one electron in its outermost shell, readily loses this electron to a chlorine atom (Cl), which has seven electrons in its outermost shell:

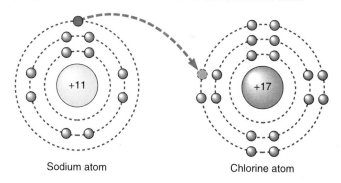

Sodium atom   Chlorine atom

The newly formed sodium ion has eleven protons and only ten electrons; it has a net charge of +1. The symbol for the sodium ion is Na^+. The newly formed chloride ion (Cl) has seventeen protons and eighteen electrons; it has a net charge of −1. The symbol for a chloride ion is Cl^-.

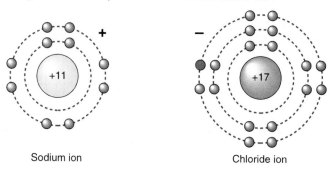

Sodium ion   Chloride ion

Table 1.2

Some Biologically Important Ions

Name of Ion	Symbol	Electric Charge	One of Its Roles in Living Things
Hydrogen	H^+	+1	Transfers energy from one molecule to another
Sodium	Na^+	+1	Sends signals along nerves
Potassium	K^+	+1	Holds water in cells
Magnesium	Mg^{++}	+2	Activates certain enzymes
Calcium	Ca^{++}	+2	Triggers muscle contractions
Chloride	Cl^-	−1	Neutralizes positive ions

When an atom loses one or more electrons and becomes a positively charged ion, the name of the ion is the name of the atom plus the word *ion*. Thus, there are hydrogen ions (H^+), sodium ions (Na^+), and calcium ions (Ca^{++}: two electrons have been lost). When an atom gains one or more electrons to become a negatively charged ion, the name of the ion is the name of the atom altered to end in *ide* and followed by the word *ion*. Thus, there are fluoride ions (Fl^-), chloride ions (Cl^-), and iodide ions (I^-).

Atoms that readily form ions in water play important roles in the chemistry of life. They are essential to such phenomena as muscle contractions, the flow of signals along nerve cells, and the opening and closing of flowers. Some ions with important biological roles are listed in table 1.2.

Three particularly interesting ions are lithium (Li^+), sodium (Na^+), and potassium (K^+). These ions are chemically similar, for they carry the same electric charge. Yet sodium and potassium ions are essential to life whereas lithium ions play no role in normal activities. What is surprising is that lithium is an effective treatment for some mental disorders. How this unusual treatment was discovered and the effect of lithium ions on brain cells are described in the accompanying Research Report.

Molecules

An atom rarely exists as an isolated structure. Instead, atoms join with other atoms to form **molecules.** In the hierarchy of biological organization (see inside front cover of this book), molecules are the next level above atoms. Atoms combine in many ways to form a great variety of molecules; there are many more kinds of molecules than there are atoms.

A molecule is described by a chemical formula, which symbolizes the kinds of atoms and how many of each kind. Water, for example, is a molecule consisting of two hydrogen atoms and one oxygen atom: its chemical

Table 1.3

Three Kinds of Chemical Bonds

Name of Bond	Force of Attraction	Strength
Ionic bond	Opposite charges of ions	Intermediate
Covalent bond	Shared electrons traveling around the nuclei of both atoms	Strongest
Hydrogen bond	Opposite charges of polar molecules	Weakest

formula is H_2O. Carbon dioxide is a molecule made of one carbon atom and two oxygen atoms: its chemical formula is CO_2. Insulin, a large molecule that controls blood sugar levels, has a chemical formula of $C_{254}H_{377}N_{65}O_{76}S_6$. A model of the insulin molecule is shown in this chapter's opening photograph.

In a chemical formula, the subscript to the right of the symbol of an atom (as in the *2* of H_2O) indicates how many atoms of that type are in the molecule. When there is no subscript after the symbol (as in the C of CO_2), it means there is only one atom of that type in the structure. (The number 1 is never written in chemical shorthand.)

The atoms of a molecule are held together by a force of attraction called a chemical bond. The formation of a chemical bond is the second way, in addition to the formation of ions, that an atom attains a full outermost shell of electrons. The biological roles of chemical bonds are (1) to increase the variety of chemical structures, by combining atoms in many ways, and (2) to store and release energy, as described later in this chapter. The three main types of chemical bonds are ionic bonds, covalent bonds, and hydrogen bonds. The differences among these bonds are summarized in table 1.3 and described following.

Research Report

Lithium and the Chemistry of Mania

More than 2 million Americans suffer from manic depression (or bipolar illness), a brain disorder characterized by periods of extreme elation (mania) alternating with periods of severe depression. During the elated phase, the person is extremely talkative, illogical, and emotional. During the depressed phase, the person is unresponsive and withdrawn.

An unusual treatment for manic depression was discovered by a young Australian physician, John Cade, in the 1940s. Dr. Cade was searching for a way to treat severe mood swings in his patients. He developed the hypothesis that people with manic depression are unable to cleanse their bodies of uric acid (a molecule similar to the urea in urine), and as this waste product accumulates it interferes with the mind. Too much uric acid, he thought, induced the manic phase of manic depression.

Cade tested his hypothesis by injecting uric acid into some guinea pigs (the experimental group) and a harmless chemical into others (the control group). Because uric acid does not dissolve well in water, he combined it with lithium ions (Li^+) to make a water-based injection. Guinea pigs in the experimental group, which Cade expected to develop mania, were injected with the solution of lithium ions and uric acid. Animals in the control group, which he expected to show no response, were injected with a solution of lithium ions and carbonate ions (CO_3^-, the harmless chemical) (fig. 1.A).

The results were surprising. Both groups of animals showed the same, striking change in behavior: their normal frantic behavior while being han-

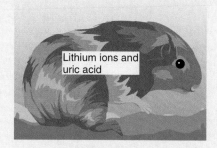 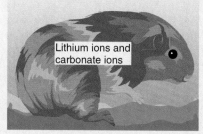

Experimental animals   Control animals

Figure 1.A The experiment that revealed the effects of lithium ions on behavior. The experiment was designed to test the hypothesis that uric acid causes manic depression. Carbonate ions were injected into the control group, as they were known to have no effect on behavior. Lithium ions were used to dissolve uric acid in the experimental group and were added also to the control group so that the experiment would test only the effects of uric acid.

dled became so submissive that when Dr. Cade turned them onto their backs, they just lay there gazing placidly at him. He concluded that uric acid does not cause mania, since the experimental group did not become hyperactive. Instead, lithium ions, which both groups received, seemed to act as a tranquilizer.

Dr. Cade wondered about the significance of his results. Could this small ion, containing just three protons, three neutrons, and two electrons, calm the emotions of mentally ill people? He then gave lithium ions to manic depressive patients who had been confined for years to mental asylums. For most treated patients, the change in behavior was miraculous; after just a few days of treatment they were able to return to their families and jobs. Today, many manic depressives are treated successfully with lithium ions.

How does lithium alter the attitudes and emotions in the human mind? Other mind-altering drugs consist of larger molecules with more

complex structures than the simple lithium ion. Moreover, lithium is rare on earth, found chiefly in rocks, and is not a normal component of living systems. The clue to its biological effects probably lies in its similarity to sodium and potassium ions—all three ions carry an electric charge of +1. Sodium and potassium ions both play crucial roles in the flow of electric signals through the brain, but neither resembles lithium in its beneficial effects on mentally disturbed people.

Researchers throughout the world are attempting to discover how lithium affects the chemistry of brain cells. One hypothesis is that lithium ions boost the level of serotonin (as does the drug called Prozac), a molecule that normally slows down hyperactive nerve cells. In manic depression, some cells are hyperactive during the manic phase and others during the depressed phase. As one researcher put it, by increasing the amount of serotonin, lithium acts "like a traffic cop who makes sure you're not going too fast or too slow."

Ionic Bonds

An **ionic bond** develops when a positively charged ion and a negatively charged ion collide and are held together by the mutual attraction of their opposite charges. A familiar example of a molecule held together by an ionic bond is sodium chloride, the salt we use to flavor our food:

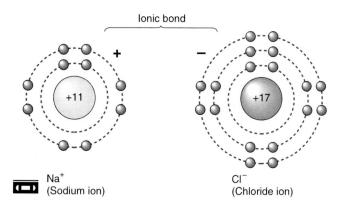

Na$^+$
(Sodium ion)

Cl$^-$
(Chloride ion)

Another example of a molecule formed by ionic bonds is calcium chloride (CaCl$_2$), formed of a calcium ion (Ca^{++}) and two chloride ions (Cl$^-$). In this case, two ionic bonds hold the atoms together:

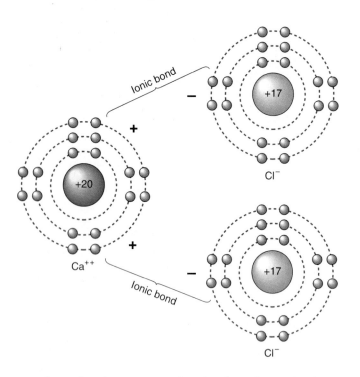

Ca^{++}

Cl$^-$

Cl$^-$

Ionic bonds are strong bonds when the molecule is dry but weak when it is in water. Table salt, for example, remains as a molecule when in the dry environment of a saltshaker but splits into its component sodium and chloride ions when placed in water. Since the chemistry of life takes place in water, ionic bonds are relatively weak bonds in living systems.

Covalent Bonds

A **covalent bond** is the mutual attraction between two atoms that share a pair of electrons. It is the strongest type of chemical bond.

A covalent bond involves sharing a pair of electrons, one from each atom. Prior to formation of the bond, each electron of the pair travels by itself in an orbital of an atom. The bond develops when the two orbitals of the two atoms merge to form a single orbital that encircles both nuclei. The two electrons then travel around the two nuclei and are shared by the two atoms (fig. 1.5). By sharing, each atom gains an electron that contributes to the filling of its outermost shell.

An atom can form covalent bonds with other atoms until its outermost shell is full of electrons. Carbon, for example, with four electrons in its outermost shell, needs four more electrons to fill the shell and so it forms four covalent bonds with other atoms (fig. 1.6).

Two atoms may share more than one pair of electrons—more than one covalent bond forms between them. A double covalent bond occurs when two atoms share two pairs of electrons. It is stronger than a single covalent bond. Ethylene (C$_2$H$_4$), a molecule in plants that hastens the ripening of fruit, has such a double bond between its two carbon atoms (fig. 1.7). A triple covalent bond occurs when two atoms are held together by three pairs of shared electrons. This type of bond, the strongest of all bonds, forms between the two nitrogen atoms of nitrogen gas (N$_2$, the most abundant molecule in air).

Hydrogen Bonds

A third type of chemical bond is the **hydrogen bond,** in which two polar molecules are held together by their opposite charges. A polar molecule is electrically neutral as a whole, because of equal numbers of electrons and protons,

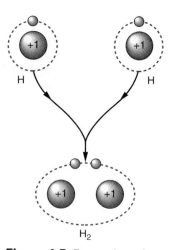

H   H

H$_2$

Figure 1.5 Formation of a covalent bond. Two hydrogen atoms are joined by a covalent bond. The two atoms come so close to each other that the orbitals of their electrons merge. In the newly formed molecule (hydrogen gas, or H$_2$), the electrons of both atoms travel around both nuclei.

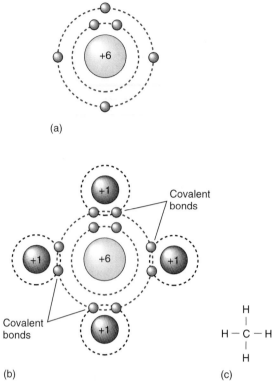

(a)

(b)

Covalent
bonds

Covalent
bonds

H
|
H — C — H
|
H

(c)

Figure 1.6 Covalent bonds formed by carbon. (*a*) Carbon, with four electrons in its outermost shell, can form four covalent bonds. (*b*) Here, carbon has developed four covalent bonds with four hydrogen atoms. Each pair of shared electrons travels around the nuclei of both the hydrogen atom and the carbon atom. The newly formed molecule, methane (CH_4), is a gas that forms in stagnant water and the digestive tracts of animals. (*c*) Molecules with covalent bonds are typically drawn like this one, of methane, with the covalent bonds represented by straight lines.

but is asymmetric with regard to the way its electrons are distributed around the nuclei: the electrons spend most of their time on one side of the molecule. This side is slightly negative and the other side, where the electrons spend less time, is slightly positive.

A polar molecule forms when two kinds of atoms, one considerably larger than the other, are joined by a covalent bond. (Most often these two atoms are hydrogen and either oxygen or nitrogen.) The larger atom attracts the electrons more strongly, so that at any one time an electron shared by the two atoms is more likely to be near the larger atom than the smaller one. The asymmetry of electron distribution gives the larger atom a slight negative charge and the smaller atom a slight positive charge.

In the water molecule, electrons shared by the oxygen and hydrogen atoms spend more time on the oxygen (larger) side of the molecule and less time on the side

(a)

H   H
 \   /
  C = C
 /   \
H   H

(b)

(c)

Figure 1.7 A double covalent bond. (*a*) This molecule, ethylene (C_2H_4), has a double covalent bond in which two pairs of electrons are shared by two carbon atoms. All four electrons involved in the bond travel around the nuclei of both carbon atoms. The double bond involves two of the four electrons in the outermost shell of the carbon atom. The remaining two electrons form two covalent bonds with two hydrogen atoms. (*b*) A more typical drawing in which the covalent bonds are represented by lines: one line for a single covalent bond and two lines for a double covalent bond. (*c*) Fruits ripened in the presence of ethylene, a gas released from the plant during the fruiting season.

31

where the hydrogen atoms (smaller) are located. Consequently, the side with the oxygen atom has a slight negative charge and the side with the hydrogen atoms has a slight positive charge:

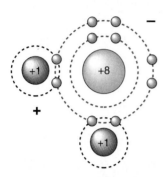

When two or more polar molecules lie near each other, they are mutually attracted by their opposite charges. This attraction, called a hydrogen bond, holds the molecules together. Hydrogen bonds between water molecules, shown in figure 1.8, give water many of its unique properties.

Hydrogen bonds are much weaker than covalent bonds. Many hydrogen bonds acting in concert, however, can provide a strong link. In living systems, hydrogen bonds hold together (1) molecules (like DNA) that are rearranged periodically; (2) different segments of very large molecules (like some proteins), thereby stabilizing their three-dimensional shapes; and (3) molecules of water.

Molecules as Ions

We saw earlier how atoms become ions by gaining or losing electrons. Similarly, some molecules have a tendency to gain or lose electrons. Such a molecule, with an unequal number of electrons and protons, is called a molecular ion.

The atoms of a molecular ion remain together and enter chemical reactions as a group. Five biologically important ions of this sort are the nitrate (NO_3^-), ammonium (NH_4^+), sulfate ($SO_4^=$), phosphate ($PO_4^{\equiv}$), and hydroxyl (OH^-) ions. When not part of living things, nitrogen atoms exist in the soil and water as components of nitrate ions and ammonium ions, and sulfur atoms exist as components of sulfate ions. Plants use these molecular ions to build proteins and other biological molecules. Phosphorus atoms exist in the soil, water, and living things as phosphate ions. They combine with other molecules to form genes, membranes, and other important structures. Hydroxyl ions are important in reducing the acidity of blood and other body fluids, as described in the next section.

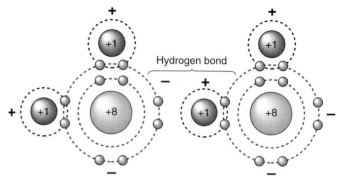

Figure 1.8 A hydrogen bond. Some molecules are polar—they have a slight positive charge on one side and a slight negative charge on the other side. A water molecule, shown here, is slightly negative on the side without the hydrogen atoms and slightly positive on the side with the hydrogen atoms. The positive side of one water molecule attracts the negative side of another water molecule. This attraction is the hydrogen bond.

Watery Solutions

Living things consist primarily of water. Your own body is at least 70% water. While you can survive for weeks if you do not eat, you will die within a few days if you do not drink. You need water because the chemistry of life can only take place in water.

Water as a Solvent

A **solvent** is a liquid that dissolves solids; the components of the solid become spread throughout the liquid. (Sugar, for example, dissolves in water.) Water is a powerful solvent—so much so that pure water rarely exists in nature. Even rainwater picks up and dissolves materials from the air prior to reaching the earth.

Molecules with electric charges dissolve in water. Such molecules are either polar molecules or molecules held together by ionic bonds. A polar molecule, as described earlier, carries weak positive and negative charges because of its asymmetry of electron distribution. A molecule formed of ionic bonds is neutral as a whole, with every positive charge matched by a negative charge, yet there remain slight charges on the ions that can attract (weakly) oppositely charged ions.

The charges on molecules that are polar or formed by ionic bonds are attracted to the charges on water molecules. This mutual attraction of charges separates and mixes the two kinds of molecules. Water molecules surround and separate the polar molecules or ions. With water molecules between them, the dispersed molecules or ions cannot come together to re-form a solid.

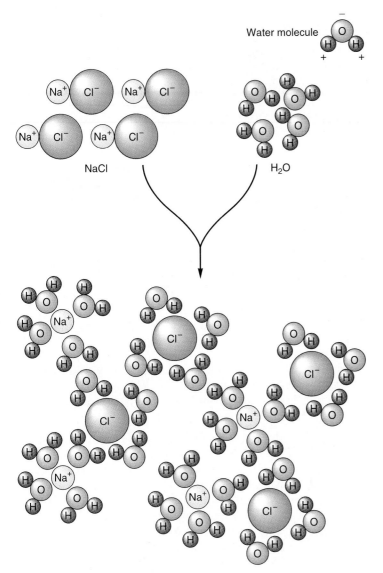

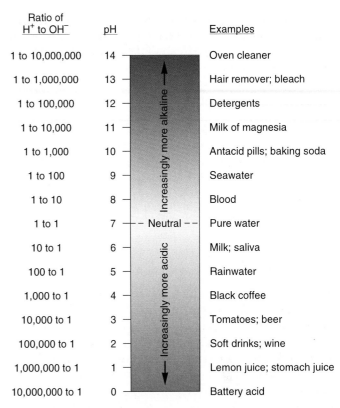

Figure 1.10 The pH scale. The pH scale ranges from 0, which has 10 million hydrogen ions for every hydroxyl ion and is extremely acidic, to 14, which has one hydrogen ion for every 10 million hydroxyl ions and is extremely alkaline. When the pH of a solution is 7, it has equal numbers of hydrogen ions and hydroxyl ions.

Figure 1.9 Water as a solvent. Solids dissolve when water molecules surround and separate charged molecules or ions. Here, table salt (NaCl) is dissolved in water. When a molecule of salt is placed in water, the electric charges on the water molecules weaken the ionic bond that holds the sodium chloride molecule together. The two ions, sodium (Na^+) and chloride (Cl^-), come apart and remain separated as the negatively charged ends of water molecules surround the positively charged sodium ion and the positively charged ends of water molecules surround the negatively charged chloride ion.

Consider what happens when table salt is mixed with water (fig. 1.9). The sodium and chloride ions in table salt are held together by ionic bonds. When salt molecules are placed in water, mutual attractions between the charges on the salt and water molecules weaken the ionic bonds so that the sodium ions detach from their chloride partners.

The negatively charged sides of water molecules then surround each sodium ion, and the positively charged sides surround each chloride ion. The two ions, sodium and chloride, become physically isolated from each other and cannot re-form the sodium chloride molecule.

Molecules that dissolve in water are dispersed throughout a solution, where they can react with other molecules. Molecules that do not dissolve, because they carry no electric charge, remain in a cluster (a solid) and have minimal contact with other molecules.

Acidity of Watery Solutions

Some solutions, like lemon juice, are acidic whereas others, like soapy water, are alkaline. Whether a watery solution is acidic or alkaline depends on its concentration of hydrogen ions (H^+) in relation to hydroxyl ions (OH^-). This ratio is expressed as the solution's **pH** (the letters stand for *potential of hydrogen*). The pH scale, shown in figure 1.10, ranges from 0 (most acidic) to 14 (most alkaline). Pure water is neutral, with a pH of 7, because it has equal amounts of hydrogen and hydoxyl ions and 7 is mid-way between 0 and 14. (Some water molecules come apart in liquid water to form H⁺ and OH⁻.) The pH scale is a negative logarithmic scale, which means that each one-point rise in pH represents a ten-fold decrease in the ratio of hydrogen ions to hydroxyl ions.

Acids and Bases   Molecules that alter the pH of a solution are either acids or bases. An **acid** is a molecule that adds hydrogen ions to water. Hydrochloric acid (HCl), for example, releases its hydrogen ion when placed in water:

$$HCl \xrightarrow{\text{In water}} H^+ + Cl^-$$

A **base** is a molecule that removes hydrogen ions from water. Some bases, like sodium hydroxide (NaOH, or lye), release hydroxyl ions that combine with the hydrogen ions:

$$NaOH \xrightarrow{\text{In water}} Na^+ + OH^-$$
$$H^+ + OH^- \longrightarrow H_2O$$

Other bases, like ammonia (NH_3), pick up hydrogen ions and bond them to their own molecular structures:

$$NH_3 + H^+ \longrightarrow NH_4^+$$
$$\text{Ammonia} \qquad\qquad \text{Ammonium}$$
$$\text{ion}$$

Buffers   A **buffer** is a substance that counteracts changes in pH; it removes hydrogen ions from acidic solutions and removes hydroxyl ions from alkaline solutions. Buffers are essential to life. In humans, for example, they prevent deviations in the blood pH that would otherwise prove fatal.

The most common buffer in living things is the combination of carbonic acid (H_2CO_3) and its bicarbonate ion (HCO_3^-). When the body fluid becomes too acidic, bicarbonate ions combine with the excess hydrogen ions to form carbonic acid. This molecule is unstable and quickly breaks down into carbon dioxide and water:

$$H^+ + HCO_3^- \longrightarrow H_2CO_3 \longrightarrow CO_2 + H_2O$$
$$\text{Hydrogen} \quad \text{Bicarbonate} \qquad \text{Carbonic} \qquad \text{Carbon} \quad \text{Water}$$
$$\text{ion} \qquad\quad \text{ion} \qquad\qquad \text{acid} \qquad\qquad \text{dioxide}$$

When the body fluid becomes too alkaline, the excess hydroxyl ions combine with carbonic acid to form bicarbonate ions and water:

$$OH^- + H_2CO_3 \longrightarrow HCO_3^- + H_2O$$
$$\text{Hydroxyl} \quad \text{Carbonic} \qquad\qquad \text{Bicarbonate} \quad \text{Water}$$
$$\text{ion} \qquad\quad \text{acid} \qquad\qquad\qquad \text{ion}$$

By shifting between carbonic acid and bicarbonate ions, the buffer eliminates excess hydrogen ions and hydroxyl ions from the solution.

Chemical Reactions

In a chemical reaction, one kind of molecule is converted into another kind. Bonds in the original molecule break and different bonds, joining different atoms, form in the new molecule. Hydrochloric acid (HCl) and sodium hydroxide (NaOH), for example, react to form sodium chloride and water:

A chemical reaction is described by a chemical equation, in which the reactants (the molecules that react) are shown on the left-hand side and the products (the molecules that form) are shown on the right-hand side. An arrow between the reactants and products means "interact to form" or "yields." An important reaction in living things, for example, is cellular respiration:

$$C_6H_{12}O_6 + 6O_2 \longrightarrow 6CO_2 + 6H_2O +$$
$$\text{Sugar} \quad\;\; \text{Oxygen} \qquad\quad \text{Carbon} \quad\; \text{Water}$$
$$\text{gas} \qquad\qquad\quad \text{dioxide}$$

Energy

In this reaction, sugar and oxygen gas interact to form carbon dioxide and water. The 6 in front of the O_2, CO_2, and H_2O means six molecules of oxygen gas enter into the reaction, and six molecules of carbon dioxide and six molecules of water are products of the reaction. (If no number is given, as in the sugar, only one molecule is involved.) The number of atoms is always the same on both sides of the equation. Thus, in this example, each side of the equation has six carbon atoms, twelve hydrogen atoms, and eighteen oxygen atoms. Atoms are not created or destroyed by ordinary chemical reactions.

Electrons and Energy

Energy is the force that makes it possible to do work—to move an object against an opposing force. **Potential energy** is stored energy that is capable of moving an object. **Kinetic energy** is the movement itself.

Two kinds of energy that are important in biology are chemical energy and heat energy. **Chemical energy** is potential energy stored in the positions of electrons within molecules and ions. **Heat energy** is kinetic energy in the form of random movements of molecules and ions. Whenever chemical energy is used to do work, some of it is converted to heat energy and is no longer available to do work.

The amount of chemical energy carried by an electron depends on how close the electron is to an atomic nucleus: the farther an electron is from a nucleus, the more chemical energy it holds. When an electron moves closer to a nucleus, some of its energy is released:

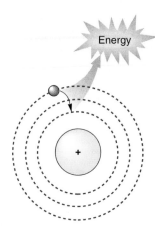

An electron moves farther from a nucleus by acquiring additional chemical energy:

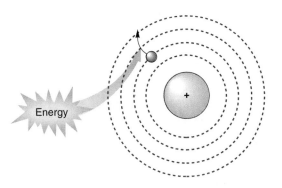

The energy that supports all biological work comes from these changes in electron position. Sometimes the change is only a slight shift toward or away from the nucleus, as in the formation of a covalent bond. Other times the change is more profound: an electron moves completely out of one atom and into another. The formation of an ion, for example, involves such an electron transfer.

Energy Changes During Chemical Reactions

Every chemical reaction involves a change in chemical energy. Some reactions move electrons closer to atomic nuclei and release energy. Other reactions move electrons farther from atomic nuclei and require an input of energy. In living systems, these energy changes occur when covalent bonds are broken and formed and when electrons are transferred from one atom to another.

The formation of a covalent bond moves electrons farther from atomic nuclei and so requires an input of energy. The breakage of a covalent bond moves electrons closer to atomic nuclei and so releases energy. Whether a chemical reaction requires or releases energy depends on the number of covalent bonds formed in relation to the number broken. A reaction that builds larger molecules (with more bonds) from smaller ones (with fewer bonds) requires an input of energy:

Smaller molecules + Energy ⟶ Larger molecule

A reaction that converts a larger molecule (with more bonds) into smaller ones (with fewer bonds) releases energy:

Larger molecule ⟶ Smaller molecules + Energy

Chemical reactions in which electrons are transferred from one atom to another also involve energy changes. Energy must be added to move an electron out of one atom and into another atom that holds it farther from its nucleus:

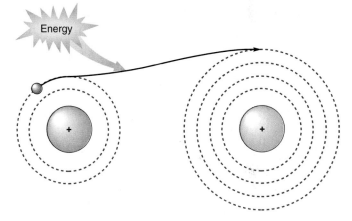

Energy is released when an electron moves out of one atom and into another one that holds it closer to its nucleus:

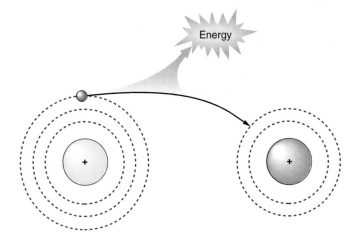

Energy-consuming and energy-releasing reactions are typically coupled in living things: a reaction that requires an input of energy occurs next to, and at the same time as, a reaction that releases energy. In this way, energy is transferred efficiently from where it is released from storage to where it is used to do work.

Rates of Reactions

For a chemical reaction to occur, the reactants must collide with one another in a way that breaks their chemical bonds. How fast a reaction proceeds depends on the rate at which the reactants collide and the orientation of the reactants when they come together.

Reactants collide more frequently when they are closer together and when they move more rapidly. In living systems, reactants are packed closer together by adding more reactants to the solution. The speed at which the reactants move about is increased by adding heat—the warmer the solution, the more rapidly they move and the more often they run into one another. Warm-blooded animals, such as ourselves, maintain high rates of chemical activity by maintaining high body temperatures.

The orientation of the reactants when they collide influences the likelihood that bonds will be broken. Certain materials, called **catalysts,** latch onto reactants and hold them in positions that make it easier for the bonds to break. The catalyst is not changed by the reaction and can be used again and again. In living systems, the catalysts are usually proteins called enzymes.

In addition to a high concentration of reactants, a warm environment, and the presence of catalysts, most chemical reactions need an input of energy before they can proceed rapidly. An energy-consuming reaction, of course, requires an input of energy, but an energy-releasing reaction also needs a small boost of energy in order to proceed rapidly. Without this boost, only some of the reactants contain sufficient energy to interact and the reaction proceeds slowly. With an energy boost, most of the reactants contain sufficient energy and the reaction proceeds rapidly. The quantity of energy needed to boost a chemical reaction is the reaction's **energy of activation** (fig. 1.11).

The interaction of oxygen gas with paper to form carbon dioxide and water, for example, is an energy-releasing reaction. (It is the same reaction as cellular respiration: paper is made of cellulose molecules, which are long chains of sugar molecules.) Without an energy boost, the reaction proceeds slowly; paper disintegrates after many centuries of exposure to air. Once primed with the heat of a match flame, however, oxygen molecules in the air collide more frequently with molecules in the paper, and the reaction

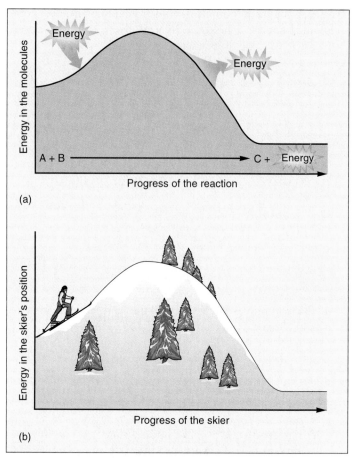

Figure 1.11 The energy of activation. Most chemical reactions require an input of energy, called the energy of activation, before they can proceed rapidly. The energy boost activates more molecules so the reaction proceeds more rapidly. (*a*) The amount of energy in the reactants (A and B) and in the product (C) as the reaction proceeds (A + B ⟶ C). (This is an energy-releasing reaction: the products have less energy than the reactants.) (*b*) The energy of activation is similar to the amount of energy needed to bring a skier to the top of a hill. Once that energy is put into the activity, the skier can move down the hill without further input of energy.

proceeds rapidly. As the paper near the match burns, heat released from the reaction itself triggers further burning in other regions of the paper. In this example, the flame provides the energy of activation.

The energy of activation is almost always in the form of heat. By speeding up the movements of reactants, heat causes more collisions and thus more breakages of bonds. A catalyst lowers the energy of activation (reduces the amount of heat needed to initiate the reaction) by making each collision more effective in breaking bonds.

SUMMARY

1. All things are made of atoms. An atom's nucleus contains protons, which are positively charged particles, and neutrons, which are particles without charge. Electrons, which are negatively charged particles, travel around the nucleus. The number of electrons in an atom equals the number of protons. The ninety-two naturally occurring atoms form a series, each with one more proton than the previous one.

2. The chemical activity of an atom depends on how many electrons it has in its outermost shell, which depends on its total number of electrons (and protons). An atom interacts with other atoms in ways that give it a full outermost shell of electrons.

3. Some atoms acquire full outermost shells by forming ions. An ion is an atom or molecule that has gained or lost electrons; it has an electric charge.

4. Some atoms acquire full outermost shells by forming chemical bonds. A chemical bond is an attractive force that holds atoms together, forming a molecule. An ionic bond is the mutual attraction of oppositely charged ions. A covalent bond is the sharing of a pair of electrons by two atoms. A hydrogen bond is the mutual attraction between two polar molecules.

5. The chemistry of life takes place in water, which is a powerful solvent because its molecules are polar. The acidity, or pH, of a watery solution depends on its ratio of hydrogen ions to hydroxyl ions. An acid raises the concentration of hydrogen ions and a base lowers it. A buffer is a substance that counteracts changes in pH.

6. A chemical reaction occurs when new molecules develop from previous molecules; it involves breaking and forming bonds.

7. Energy stored in electrons supports all biological work. An electron farther from an atomic nucleus has more energy than an electron closer to the nucleus. Electrons move closer to or farther from nuclei when bonds change. In general, energy must be added to convert smaller molecules into larger ones, and energy is released when larger molecules become smaller ones.

8. A chemical reaction occurs when the reactants collide and their chemical bonds break. A reaction can be speeded up by adding more of the reactants, heat, or a catalyst. Most reactions require an initial energy boost, called the energy of activation, to get started.

KEY TERMS

Acid (page 34)
Atom (AEH-tum) (page 24)
Base (page 34)
Buffer (page 34)
Catalyst (KAEH-tah-list) (page 36)
Chemical energy (page 34)
Covalent (koh-VAY-lent) **bond** (page 30)
Electron (ee-LEK-trahn) (page 25)
Energy of activation (page 36)
Heat energy (page 34)
Hydrogen bond (page 30)
Ion (EYE-ahn) (page 27)
Ionic (eye-AH-nik) **bond** (page 30)
Kinetic (kih-NEH-tik) **energy** (page 34)
Molecule (MAH-leh-kyool) (page 28)
Neutron (NOO-trahn) (page 25)
pH (pee-AYTCH) (page 33)
Potential energy (page 34)
Proton (PRO-tahn) (page 25)
Shell (page 25)
Solvent (SOL-vent) (page 32)

STUDY QUESTIONS

1. Draw the positions of electrons in an oxygen atom (eight protons). Draw the positions of electrons in a potassium atom (nineteen protons).

2. Explain why an ion carries an electric charge. Describe the ionic bond.

3. Draw a water molecule, showing the electrons involved in its covalent bonds.

4. Explain why water molecules form hydrogen bonds with one another. How do water molecules dissolve other molecules?

5. Describe what happens when an acid is added to pure water. Describe what happens when a base is added to pure water.

6. Explain why some chemical reactions require an input of energy and other reactions release energy.

7. Explain why a reaction is speeded up by addition of (a) more reactants, (b) heat, and (c) a catalyst.

CRITICAL THINKING PROBLEMS

1. Is it likely that we will discover more kinds of atoms? Why or why not?
2. You have a pond that has become too acidic, from acid rain, for survival of your fish. What kind of chemical could you add to the water to counteract the acidity? Explain how this chemical would alter the pH.
3. Why is water critical to life?
4. Burning paper or wood is the same chemical reaction as cellular respiration in that sugar combines with oxygen gas to produce carbon dioxide, water, and energy. Explain how the burning of forests relates to the greenhouse effect, in which the air accumulates too much carbon dioxide. Why would planting more trees help reduce the greenhouse effect?
5. From the name, what do you think "ionizing radiation" does to living things?
6. Explain why Dr. Cade added lithium ions to the solution injected into his control group of guinea pigs.

SUGGESTED READINGS

Introductory Level:

Atkins, P. W. *Molecules*. New York: Scientific American Library, 1987. A beautifully illustrated introduction to molecules that requires no prior knowledge of chemistry.

Sackheim, G. *Introduction to Chemistry for Biology Students*, 4th ed. Redwood City, CA: Benjamin-Cummings, 1991. A self-study text for students who are studying biology but lack preparation in basic chemistry.

Salzberg, H. W. *From Caveman to Chemist: Circumstances and Achievements*. Washington, D.C.: Am Chemical, 1991. A history of chemistry, focusing on how ideas and achievements of chemists have been influenced by the societies in which they lived.

Advanced Level:

Fowler, E. "The Quest for the Origin of the Elements," *Science* 226 (1984):922–35. A review of the history and present knowledge of how atoms formed.

Lemonick, J. D. "The Ultimate Quest," *Time*, April 16, 1990, 50–56. A review of how physicists study the structure of atoms.

Mauskopf, S. H. (ed.). *Chemical Sciences in the Modern World*. Philadelphia: University of Pennsylvania Press, 1993. A collection of essays that describes the relations between inorganic chemistry and modern life, including such topics as nitrogen fixation, alar, napalm, and carcinogens.

✪ EXPLORATION

Explorations in Cell Biology, Metabolism and Genetics CD-ROM

Cell Chemistry: Thermodynamics (*See* Module 6).

The Chemistry of Life

Chapter Outline

Objectives

In this chapter, you will learn
–why life is made only of certain kinds
  of atoms;
–about life's main molecules;
–how enzymes control chemical
  reactions;
–how enzymes organize the complex
  chemistry of life.

Computer image of molecules. A drug (green and yellow) is seen binding to a DNA molecule (group of spheres).

hat a difference a few atoms make! Take for instance the hormone that governs sexual development. Add a few atoms to the basic molecule and an embryo develops into a boy; remove a few and the embryo develops into a girl. A small change in a molecule makes a profound difference in our physical appearances and reproductive roles.

The distinctions between men and women are produced by two sex hormones—testosterone in males and estrogen in females. Differences between the sexes are initiated during the eighth week after conception, when a male embryo begins to produce testosterone. In the presence of this male hormone, an embryo develops testes; in the absence, an embryo develops ovaries. Male and female differences are pronounced at birth and become more so during puberty, when the testes of boys produce large quantities of testosterone and the ovaries of girls produce large quantities of estrogen. The dramatic effects of these hormones are seen most clearly during hormone therapy: a man treated with estrogen develops breasts, broader hips, and a higher-pitched voice. A woman treated with testosterone develops a beard, larger muscles, and a lower-pitched voice.

Surprisingly, testosterone ($C_{19}H_{28}O_2$) and estrogen ($C_{18}H_{24}O_2$) are almost identical molecules. Their atoms are arranged in the same configuration, but testosterone has one more carbon atom and four more hydrogen atoms than estrogen. The large differences between a man and a woman are the consequences of small differences at the molecular level.

The Atoms of Life

The atoms of living things are not exceptional; the same kinds are found in nonliving things as well. Oxygen, for example, is a component of air, rocks, and water as well as of living things. What then distinguishes **organic molecules,** which form living things, from **inorganic molecules,** which form nonliving things?

An important distinction lies in relative size: organic molecules are typically larger than inorganic molecules. Most organic molecules consist of thousands of atoms, whereas most inorganic molecules consist of less than a dozen atoms. Another important difference is that organic molecules are made of just twenty-four kinds of atoms (table 2.1), whereas inorganic molecules may be made of any of the ninety-two kinds of atoms on earth. And of those twenty-four atoms, only six are abundant in living things: 99% of all organic molecules consists of carbon, hydrogen, nitrogen, oxygen, phosphorus, and sulfur (easily remembered as CHNOPS).

Why is life made of only certain kinds of atoms? One reason is because only certain kinds of atoms are available

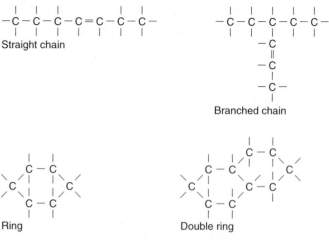

Figure 2.1 Carbon framework of organic molecules. The carbon atom, with four electrons in its outermost shell, forms four covalent bonds with other atoms. Carbon atoms join with one another to form straight chains, branched chains, and rings. These structures provide stable frameworks for large organic molecules.

where organisms live—in the earth's crust, lakes, oceans, and atmosphere. These are small atoms that, because of their light weight, have accumulated on the surface of the earth. Another reason life is built from smaller atoms is they form stronger bonds—the smaller the atom, the closer the positively charged nucleus of one atom to the negatively charged electrons of the other atom in the bond. The large sizes of organic molecules require strong bonds to hold them together.

All organic molecules contain carbon. The carbon atom is unique in its ability to develop four strong covalent bonds (see page 30) with other atoms. As carbon atoms bond with one another, they form long chains and rings that provide stable frames for large molecules (fig. 2.1).

The Macromolecules of Life

Life is extremely diverse. There are millions of different kinds of organisms, and each kind consists of millions of different kinds of individuals. The extraordinary diversity of life is generated by large molecules built of many atoms—usually hundreds or thousands. These molecules, called macromolecules, comprise more than 90% of an organism's dry weight (table 2.2), which is the weight of all its materials except water. Macromolecules are **polymers** (*poly,* many; *mer,* unit): molecules built from one or a few kinds of smaller molecules (typically about ten to twenty atoms) called **monomers** (*mono,* one). Monomers are the building blocks of polymers.

Table 2.1

The Twenty-four Atoms That Are Essential to Life

Atom	Symbol	Number of Protons	Percentage of Human Body by Weight
Hydrogen	H	1	9.5
Carbon	C	6	18.5
Nitrogen	N	7	3.3
Oxygen	O	8	65.0
Fluorine	F	9	trace
Sodium	Na	11	0.2
Magnesium	Mg	12	0.1
Silicon	Si	14	trace
Phosphorus	P	15	1.0
Sulfur	S	16	0.3
Chlorine	Cl	17	0.2
Potassium	K	19	0.4
Calcium	Ca	20	1.5
Vanadium	V	23	trace
Chromium	Cr	24	trace
Manganese	Mn	25	trace
Iron	Fe	26	trace
Cobalt	Co	27	trace
Copper	Cu	29	trace
Zinc	Zn	30	trace
Selenium	Se	34	trace
Molybdenum	Mo	42	trace
Tin	Sn	50	trace
Iodine	I	53	trace

Table 2.2

The Chemical Composition of a Human

Constituent	Percent of Total Weight	Percent of Dry Weight*
Water	70	0
Macromolecules	28.5	95
Ions and small molecules	1.5	5

*The dry weight is the weight after all the water has been removed.

All polymers are built by the same chemical reaction, called **dehydration synthesis,** in which monomers are joined by covalent bonds. Wherever a covalent bond develops between two monomers, a water molecule forms and leaves the polymer (i.e., the growing polymer is dehydrated):

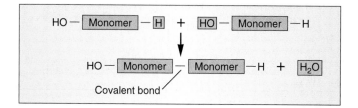

One of the monomers loses a hydroxyl group (OH) and the other loses a hydrogen atom (H). The hydroxyl group and the hydrogen atom then combine to form water (HOH, or H_2O), which is a by-product of the reaction.

The reverse reaction, known as **hydrolysis** (*hydro*, water; *lysis*, splitting), breaks a polymer down into its component monomers. A molecule of water is added for each covalent bond that is broken. The water molecule is split into its two components: the hydroxyl group attaches to one monomer and the hydrogen atom attaches to the other:

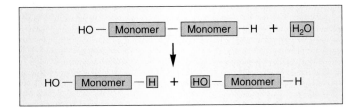

A familiar example of hydrolysis is digestion. Food consists mostly of polymers, such as proteins and starches, which must be broken down into their component monomers before they can move from the digestive tract into the bloodstream.

The immense diversity of macromolecules comes not from the variety of monomers (there are only about fifty kinds) but rather from the variety of ways in which the monomers are combined with one another. Within this huge array of macromolecules, we recognize four basic categories: polysaccharides, lipids, proteins, and nucleic acids. These macromolecules and their monomers are listed in table 2.3 and described in the next section.

Polysaccharides

A **polysaccharide** is a polymer built by dehydration synthesis from hundreds or thousands of monosaccharide monomers. These long chains of monomers are more commonly known as complex carbohydrates. A **monosaccharide** is usually a ring formed of carbon and oxygen atoms, with attached hydrogen atoms and hydroxyl groups. Monosaccharides and disaccharides (two monosaccharides bonded together) are more commonly known as sugars.

A polysaccharide forms when covalent bonds develop between hydroxyl groups on adjacent monosaccharides:

One of the hydroxyl groups combines with the hydrogen atom of the other hydroxyl group to form a water molecule, which is released at the bonding site. The remaining oxygen atom (from one of the hydroxyl groups) forms a link between the two monosaccharides. The bonds can interconnect monosaccharides to form long chains or branched structures.

By far the most common monosaccharide is **glucose,** a six-carbon sugar (fig. 2.2). As a monomer, it is a main source of energy. As a polymer, glucose forms the polysaccharides of greatest biological significance: starch, glycogen, and cellulose, shown in figure 2.3.

Starch is manufactured by plants as a way of storing energy as glucose. When excess glucose is available, it is attached to the end of a starch molecule; when glucose is needed, it is broken off the end of a starch molecule. Starch stored in the roots during the summer and fall supports growth of the stem, leaves, and flowers in the following spring. Starch stored in seeds supports growth of the plant embryos. In both cases, the stored starch is hydrolyzed into its glucose monomers, which are then used as a source of energy. We exploit the energy reserves of plants when we eat roots (carrots, radishes, and beets) and seeds (pasta, bread, and cereal). In our digestive tracts, the starch in plant material is digested and the glucose moved into our cells to supply energy for doing work.

Glycogen is synthesized by animals as a way of storing small quantities of energy in the liver and muscles. Glycogen is hydrolyzed, as needed, and its glucose monomers used to supply energy.

Cellulose is the main component of plant structures, providing rigidity to leaves, stems, and roots. It is the most abundant organic molecule on earth. Cellulose molecules are tightly linked to one another (see fig. 2.3c), which provides the strength needed to support a plant. Surprisingly, cellulose cannot be digested by most organisms. Only a few kinds of unicellular organisms (bacteria and protozoa) can digest cellulose into its glucose monomers. Some animals that eat mainly plants, such as cows and termites, get glucose from cellulose by harboring cellulose-digesting microorganisms in their digestive tracts. Humans cannot digest cellulose. Dietary cellulose, called fiber, passes unchanged through our digestive tracts. It cannot provide glucose but is useful in preventing constipation and other disorders.

Lipids

A **lipid** is a polymer built by dehydration synthesis from several kinds of smaller molecules that consist mainly of carbon and hydrogen atoms. An important characteristic of a lipid is it does not dissolve in water; a lipid is a nonpolar molecule,

Table 2.3

The Macromolecules of Life

Macromolecule	Building Blocks	Main Functions
Polysaccharides	Monosaccharides (e.g., glucose)	Store energy; support plants
Lipids		
Fats	Glycerol and fatty acids	Store energy; insulate
Phospholipids	Glycerol, fatty acids, phosphate, and R group	Form barriers (membranes)
Steroids	Carbon rings	Form hormones, vitamin D, and cholesterol (part of membranes)
Proteins	Amino acids	Form structures and enzymes
Nucleic acids	Nucleotides	Provide instructions for building proteins

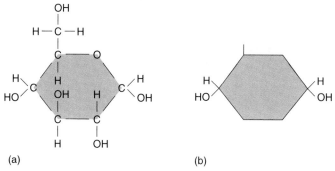

(a)   (b)

Figure 2.2 The structure of glucose. (*a*) Glucose, showing all the atoms. (*b*) Glucose, showing only the atoms of interest—those that will participate in forming bonds with other molecules.

which does not dissolve in water because it has no electric charge. The most important lipids are fats, phospholipids, and steroids.

Fats   A fat is formed, by dehydration synthesis, from a glycerol molecule and three fatty acid molecules (fig. 2.4). Fats are also known as triglycerides (*tri*, three; *glyceride*, a molecule formed of one glycerol and one fatty acid).

Glycerol contains three carbons with attached hydrogen atoms (H) and hydroxyl groups (OH). A **fatty acid** is a long chain of carbon atoms with attached hydrogen atoms. One end of the carbon chain consists of a carboxyl group (C⚌O with OH, or COOH), which makes the molecule an acid because it readily loses a hydrogen ion. A fatty acid bonds to glycerol at its carboxyl end (see fig. 2.4).

The glycerol portion of a fat is always the same. The fatty acids, by contrast, vary from one type of fat to

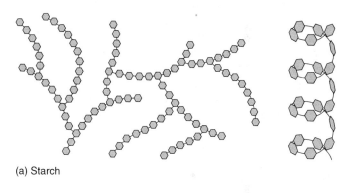

(a) Starch

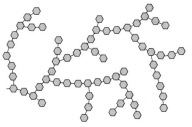

(b) Glycogen

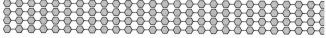

(c) Cellulose

Figure 2.3 The structures of polysaccharides. All three common polysaccharides—starch, glycogen, and cellulose—are polymers of glucose (symbol = ⬡). (*a*) Starch comes in two forms, one highly branched and the other unbranched and coiled. (*b*) Glycogen is a highly branched molecule but smaller than the branched form of starch. (*c*) Each cellulose molecule is a long, unbranched chain. Cellulose molecules bond with one another to form strong ropes.

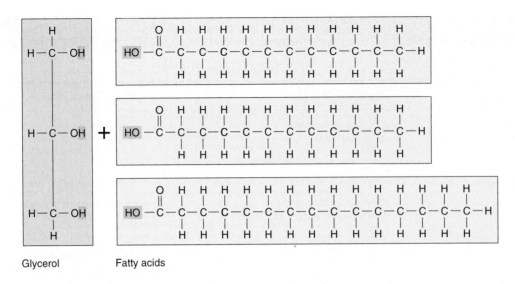

Glycerol   Fatty acids

Dehydration synthesis

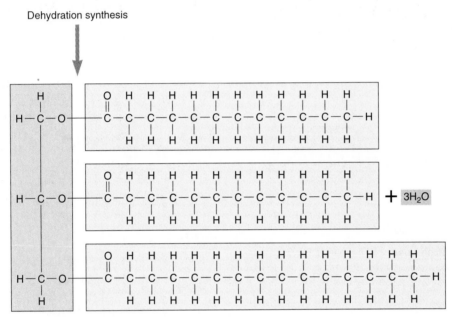

Figure 2.4 Synthesis of a fat. The building blocks of a fat are one glycerol and three fatty acid molecules. The four molecules join by dehydration synthesis to form a fat molecule and three molecules of water—one for each covalent bond that forms.

another. Fatty acids differ in the number of carbon atoms that comprise their chains (typically between twelve and twenty-four) and in the number and location of double bonds (📖 *see page 30*) between the carbon atoms. A saturated fatty acid has no double bonds between its carbon atoms; the carbon atoms are said to be saturated because there are no more places to add hydrogen atoms:

An unsaturated fatty acid has at least one double bond between its carbon atoms:

Its carbon atoms are not saturated with hydrogen atoms. If a double bond between two carbon atoms were converted to a single bond, then another hydrogen atom could be added to each carbon atom.

The more double bonds there are in the fatty acids of a fat, the freer the molecule is to move; double bonds make a fatty acid less rigid. The more movement, the lower the temperature at which the fat changes from a solid to a liquid. Fats synthesized by plants have mostly unsaturated fatty acids and are liquids (i.e., oils) at room temperature. Fats synthesized by animals have mostly saturated fatty acids and do not become liquids until heated above room temperature.

Fatty acids are concentrated sources of energy—a great deal of energy is stored within the covalent bonds between their carbon atoms. Organisms use fats chiefly as a way of storing energy. Fat is also a superb material for insulation: birds and mammals insulate their entire bodies against heat loss with layers of fat just beneath the skin.

Phospholipids   A phospholipid has almost the same molecular structure as a fat. The only difference is that a phospholipid has one of its fatty acids replaced by a phosphate ion ($PO_4^=$) and an R group (Ⓡ):

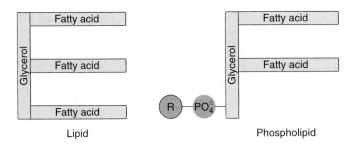

An **R group** is a small organic molecule bonded to a larger one. It is the chief way in which related molecules vary. Different kinds of phospholipids have the same basic structure, shown here, but different kinds of R groups.

The phospholipid molecule is unusual in that the end with the phosphate and R group carries an electric charge and so it dissolves in water, whereas the two fatty acids carry no charges and do not dissolve in water. Thus, one end of the molecule (the phosphate "head") is attracted to water and the other end (the fatty acid "tails") avoids water:

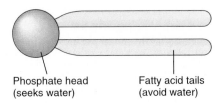

Phosphate head (seeks water)   Fatty acid tails (avoid water)

When many phospholipids are immersed in water, the phosphate heads are attracted to the water molecules and to one another, but the fatty acids are not. They form a double layer of molecules with the phosphate heads facing outward, where they are surrounded by water molecules, and the fatty acid tails packed in a water-free zone inside:

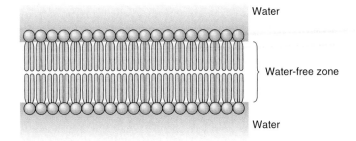

In living things, which consist mainly of water, phospholipids form such double-layered sheets of molecules. These structures, called cellular membranes, provide barriers to the flow of materials. The outer boundary of a cell, for example, consists of a double layer of phospholipid molecules. The structure and function of these membranes are described in chapter 4.

Steroids   A steroid is a lipid formed of four carbon rings. Steroids differ from one another chiefly in the kinds of R groups attached to these rings (fig. 2.5).

Steroids are categorized according to function. Some steroids are hormones, which regulate cellular activity; some are vitamins, which have a variety of functions; and others are cholesterol, which contributes to the structure of some cellular membranes. Cholesterol is also the fundamental molecule from which all other steroids are derived.

Proteins

A **protein** is formed of one or more polypeptides. A **polypeptide** is a polymer built by dehydration synthesis from many amino acids (the monomers). (Smaller molecules, built of fewer than forty-five amino acids, are called peptides.) Proteins contain carbon, hydrogen, oxygen, nitrogen, and usually sulfur atoms. They form enzymes, antibodies, hormones, basic cell structures, muscles, and many other components of life.

All **amino acids** have the same foundation:

Each amino acid has three parts: an amino group (NH_2, which usually picks up a hydrogen ion and becomes positively charged); a carboxyl group (COOH, which usually loses a hydrogen ion and becomes negatively charged); and an R group, which varies from one amino acid to another.

The R group gives each amino acid its particular identity. Only the R group varies from one amino acid to another. Twenty different amino acids, characterized by twenty different R groups, occur commonly in the proteins of living things. The R groups range in complexity from a single hydrogen atom in the simplest amino acid (glycine, common in gelatin) to elaborate rings of atoms in the most complex amino acids (fig. 2.6). Their chemical properties are varied.

Two amino acids are linked by a covalent bond (called a peptide bond) between the amino group of one and the carboxyl group of another. Many amino acids linked in this way form a long, unbranched chain that resembles the beads in a necklace:

◯ Amino acid

── Covalent bond

◯─◯─◯─◯─◯─◯─◯─◯─◯─◯

The R groups, which are not changed by the bonds between amino acids, are the chemically active parts of the protein.

Protein Diversity   An enormous variety of proteins can be built because of the many ways in which the amino acids can be arranged in a long chain. For each "bead" in the chain there are twenty possible amino acids. Thus, in a hypothetical protein made of three amino acids there are twenty choices for the first bead, twenty choices for the second bead, and twenty for the third bead. The number of possible kinds of proteins made of three amino acids is: $20 \times 20 \times 20 = 8,000$. And proteins are much larger than three amino acids—most contain hundreds of amino acids.

A great variety of proteins is essential to life, since each protein has a specific function and life is sustained by so many functions. Our own cells build about 60,000 different kinds of proteins. Moreover, the uniqueness of each organism is due mainly to its unique set of proteins.

Protein Shapes   A newly formed protein twists and folds itself into a three-dimensional shape, held by chemical bonds between the amino acids. The particular shape exposes particular R groups, which interact with other chemicals to give each protein a unique function. A protein's shape develops from several levels of organization: the primary, secondary, tertiary, and quaternary structures.

The primary structure of a protein is the specific sequence of amino acids in each of its polypeptides (fig. 2.7a). The covalent bonds between amino acids make this level of structure more stable than higher levels (covalent

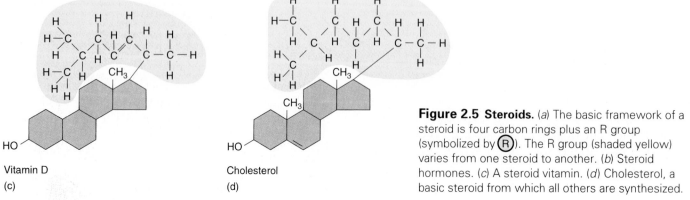

Basic framework
(a)

Estrogen   Testosterone   Progesterone   Cortisone
(b)

Vitamin D   Cholesterol
(c)   (d)

Figure 2.5 Steroids. (a) The basic framework of a steroid is four carbon rings plus an R group (symbolized by ⓡ). The R group (shaded yellow) varies from one steroid to another. (b) Steroid hormones. (c) A steroid vitamin. (d) Cholesterol, a basic steroid from which all others are synthesized.

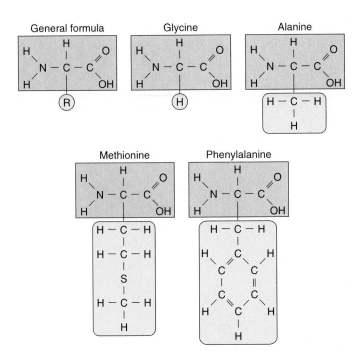

Figure 2.6 Amino acids. Only the R group (shown in yellow) varies from one amino acid to another. Shown here are the general formula for an amino acid and four kinds of amino acids, with R groups ranging from a single hydrogen atom to a more complex structure that contains a carbon ring.

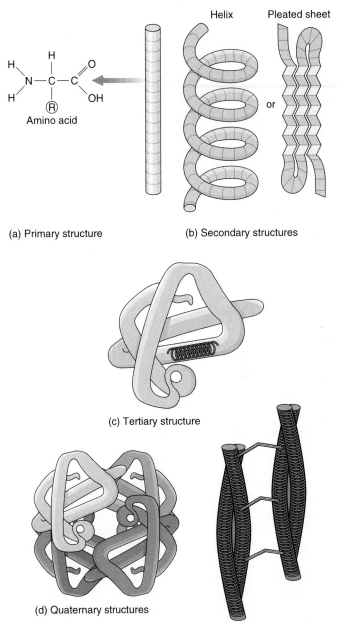

Figure 2.7 The structures of proteins. (*a*) The primary structure is the sequence of amino acids in the polypeptide chain. (*b*) The secondary structure, either a helix or a pleated sheet (silk shown here), develops as bonds form between nonadjacent amino acids. (*c*) The tertiary structure is a folding of the secondary structure (a helix is shown here) to form a spherical molecule. (*d*) The quaternary structure of a protein consists of two or more polypeptides. The final molecule is shaped like a sphere when two or more tertiary structures join and is shaped like a rope when two or more helices join. The molecule on the left is hemoglobin (a component of red blood cells) and the molecule on the right is keratin (a component of hair and fingernails).

bonds are the strongest kinds of bonds). The sequence of amino acids within a polypeptide determines all higher levels of the protein's structure.

The secondary structure of a protein develops as each polypeptide chain twists or bends to form a helix or pleated sheet (fig. 2.7*b*). Its structure is held by hydrogen bonds that form between the amino and carboxyl groups (which carry weak electric charges) in different parts of the primary structure.

A tertiary structure is found only in some proteins. It develops as the secondary structure folds back upon itself (fig. 2.7*c*) to form a spherical shape. Tertiary structures are held by bonds between the R groups of amino acids in different parts of the primary structure. Usually these are weak bonds (mostly ionic and hydrogen bonds) but some tertiary structures are held by covalent bonds between sulfur atoms (the so-called disulfide bond), which form part of the R group in cysteine (one of the twenty kinds of amino acids). Tertiary structures held by covalent bonds are less fragile than those held by weaker bonds.

The quaternary structure develops as bonds form between two or more polypeptides (fig. 2.7*d*). These bonds, of various kinds, develop between R groups on amino acids in the primary structures of different polypeptides. Only some proteins have quaternary structures. Thus, a protein may consist of a single polypeptide or it may consist of two or more polypeptides.

The shape of a protein determines its properties. Most proteins are either fibers or globes (spheres). Fibrous proteins are long coils or sheets; they have no tertiary structures but may have quaternary structures. Important fibrous proteins are keratin, a component of hair and skin; collagen, a component of tendons and bones; and elastin, a component of lungs and arteries. Globular

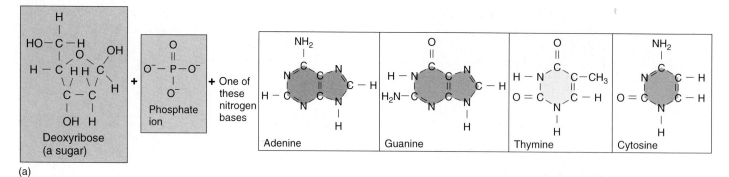

(a)

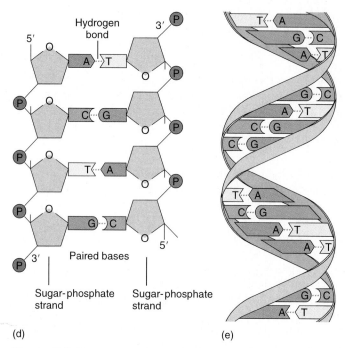

(b)

(c)

proteins are compact spheres with tertiary structures and often quaternary structures. Important globular proteins are enzymes, antibodies, and some hormones.

The tertiary and quaternary structures of proteins are fragile and tend to come apart under conditions less than optimal. This loss of shape, called denaturation, occurs as ionic and hydrogen bonds break. Salty environments (excess Na^+ and Cl^-), acidic environments (too much H^+), and alkaline environments (too little H^+) break ionic and hydrogen bonds by interfering with their electric charges. Heat causes movements within the molecule, which disrupt these relatively weak bonds. Once denatured, most proteins will not re-form their original shape. The white of an egg, for example, is a globular protein that becomes irreversibly denatured when heated.

Proteins regularly exposed to unusual environments have tertiary and quaternary structures held by covalent bonds, between the sulfur atoms of cysteine. In the highly acidic environment of the stomach, for example, dietary proteins are quickly denatured. As these proteins unfold, their primary structures are exposed to digestive enzymes, which can then digest them into their component amino acids. These enzymes are also proteins, but they are not denatured by the acid because their secondary and tertiary structures are held by covalent bonds.

Nucleic Acids

A **nucleic acid** is a polymer formed by dehydration synthesis from nucleotides (the monomers). Its atoms are carbon,

(d)

(e)

Figure 2.8 The structure of a DNA molecule. A DNA molecule is a polymer built of nucleotide monomers. (a) Each nucleotide is made from three molecules: a deoxyribose, a phosphate ion, and a nitrogenous base (adenine, guanine, thymine, or cytosine). (b) The symbol for a nucleotide. (c) The carbon atoms in deoxyribose are numbered 1' (one prime), 2', 3', 4', and 5' to help us keep track of how each molecule is oriented within the DNA molecule. (d) Nucleotides join by dehydration synthesis to form two long strands, which in turn attach to each other by forming hydrogen bonds between their bases. The joining of bases is highly specific: adenine joins only with thymine, and cytosine joins only with guanine. (e) The final shape is a double helix—a spiral formed of the double-stranded molecule. One strand is "upside down" in relation to the other strand, as seen by the positions of the 3' and 5' carbons in the deoxyribose components.

hydrogen, oxygen, nitrogen, and phosphorus. Each **nucleotide** is made from a five-carbon sugar (similar to glucose but with one less carbon), a phosphate ion ($PO_4^{\equiv}$), and a nitrogenous base (one or two rings formed of carbon and nitrogen). There are two categories of nucleic acids—deoxyribonucleic acid (DNA) and ribonucleic acid (RNA). Both carry instructions for building proteins.

Deoxyribonucleic Acid   A DNA molecule consists of two long strands of nucleotides. Each nucleotide is made of deoxyribose (the sugar from which DNA gets its name), a phosphate ion, and one of four nitrogenous bases: adenine, guanine, thymine, or cytosine (fig. 2.8a and b). A strand forms (by dehydration synthesis) when covalent bonds develop between the sugar of one nucleotide and the phosphate of another. The nitrogenous bases project outward at right angles from the strand of sugar and phosphate. The two strands of a DNA molecule are joined by hydrogen bonds between their nitrogenous bases (fig. 2.8c). While each hydrogen bond is relatively weak, the many bonds between many bases form a stable connection between the two strands of nucleotides. The entire molecule is then twisted into a helix (fig. 2.8d), called a double helix because of its two strands.

The nitrogenous bases of a DNA molecule form hydrogen bonds with one another according to specific rules known as the base-pairing rules: adenine pairs only with thymine; guanine pairs only with cytosine (fig. 2.8c and d). By contrast, no fixed rules govern the sequence of nucleotides along the length of the molecule, except that the sequence in one nucleotide strand is fixed, by the base-pairing rules, to match the sequence in the other strand. (For example, an adenine attached to one strand must be opposite a thymine in the other strand.)

A DNA molecule has three important features. First, the sugar-phosphate strands are stable structures, formed by covalent bonds, that provide a sturdy "backbone" for the molecule. Second, the bases are held firmly to the sugar-phosphate strands (by covalent bonds) but weakly to each other (by hydrogen bonds). Thus it is relatively easy for the two strands, with their attached bases, to come apart so that the molecule can make another copy of itself (as described in chapter 7). Third, the sequence of bases along each sugar-phosphate strand contains genes; they specify the order in which amino acids are linked to make proteins (as described in chapter 11). Proteins, in turn, govern the development of traits. The technique used to first describe this remarkable molecule is explained in the accompanying Research Report.

Ribonucleic Acid   An RNA molecule consists of a single strand of nucleotides. Each RNA nucleotide is made of ribose (the sugar, from which RNA gets its name), a phosphate, and one of four nitrogenous bases: adenine, guanine, cytosine, or uracil (fig. 2.9a and b). (Note that three of the four bases are the same in both DNA and RNA and the fourth base differs: in DNA it is thymine and in RNA it is uracil.)

The nucleotides of an RNA molecule are held together by covalent bonds between the sugar of one nucleotide and the phosphate of another (fig. 2.9c). As with DNA, the important part of an RNA molecule is the part that varies—the sequence of bases. The single-stranded RNA molecule is less stable than the double-stranded DNA molecule and performs more short-term tasks than the long-term storage of genetic information. These tasks involve mainly the synthesis of proteins from amino acids (as described in chapter 11).

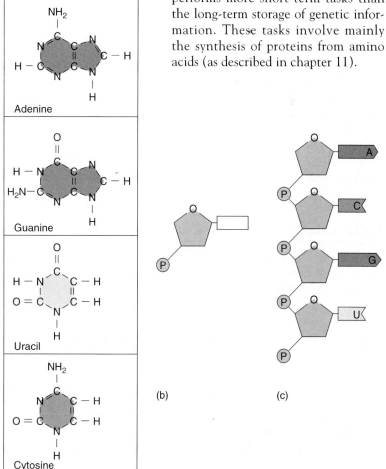

Figure 2.9 The structure of an RNA molecule. An RNA molecule is a polymer built of nucleotide monomers. (a) Each nucleotide is made from three molecules: a ribose, a phosphate, and a nitrogenous base (adenine, guanine, uracil, or cytosine). (b) The symbol for a nucleotide. (c) The nucleotides join by dehydration synthesis to form a single strand.

Research Report

Discovering the Structure of DNA

The structure of DNA was discovered in 1953 by James Watson and Francis Crick. It is considered the most significant biological breakthrough of the twentieth century, since it explained heredity in terms of molecular structure and was the first step toward genetic engineering, which has accomplished amazing feats in the past few decades.

James Watson, only twenty-four years old at the time, was an American biologist who once dreamed of becoming a bird ecologist. Francis Crick, thirty-six years old at the time, was a British physicist who knew little of genetics. This unusual research team just happened to share an office at the University of Cambridge, England. What began as a series of casual discussions led to an intense effort to figure out the structure of DNA.

As with all research, considerable work by other scientists guided Watson and Crick to the correct hypothesis. By 1953, the following observations had been made:

- Genetic material makes copies of itself.

- DNA is genetic material (at least in bacteria).
- DNA is a long molecule twisted into a helix.
- DNA contains strands of sugar and phosphate to which nitrogenous bases are attached. Of the four bases found in a DNA molecule, the amount of adenine always equals the amount of thymine, and the amount of guanine always equals the amount of cytosine.

Watson and Crick used these observations to develop alternative hypotheses about the arrangement of sugar, phosphate, and bases within the DNA molecule. They built physical models of the molecule described by each hypothesis (fig. 2.A). One model, for example, had two sugar-phosphate strands and another had three; one model had the bases projecting outward and another had them projecting inward. The model that best fit all existing observations was a double helix with paired bases on the inside (fig. 2.B). The bases join to each other in a specific manner: adenine bonds with thymine and guanine bonds with cytosine. In fact, the model fit everything that was known about the chemistry of DNA.

Figure 2.A James Watson (left) and Francis Crick in 1953, with the correct model of DNA.
Photo by A. C. Barrington Brown.

The key to the Watson-Crick model is the arrangement of bases. The sequence of bases along one strand of the molecule could hold a vast amount of information. And the paired bases between the two strands provided a mechanism for self-replication. The weak bonds (hydrogen bonds) between the bases could easily break and the molecule would come apart lengthwise. Then, if adenine pairs only with thymine and guanine

Enzymes and Biochemistry

Biochemistry is the chemistry of life. It differs from the chemistry of nonliving things in that biochemical reactions are facilitated by enzymes. An **enzyme** is a catalyst (📖 *see page 36*) that increases the rate of a chemical reaction. By making it easier for certain bonds to form and others to break, an enzyme reduces the amount of energy needed to start a reaction; it lowers the energy of activation (fig. 2.10). Most enzymes are highly specific and promote just one kind of reaction. Our own body chemistry, for example, requires more than 2,000 different kinds of enzymes. An enzyme is not altered by the reaction and can be used again and again to promote the same reaction.

Biochemical reactions require enzymes because they take place within relatively low temperatures inside organisms. One molecule of an enzyme can facilitate more than 100,000 identical reactions a second! Reactions that otherwise would take many years to complete occur within seconds when catalyzed by enzymes.

Enzyme Structure and Function

An enzyme is a globular protein[1] that catalyzes a specific reaction. Many enzymes are not active unless attached to a cofactor, which is either an ion (e.g., copper, iron, manganese and other "minerals" in our diets) or an organic molecule (often a vitamin in our diets). An enzyme brings reactants

---

1. Exceptions are certain kinds of RNA that act as enzymes, as described in chapter 11.

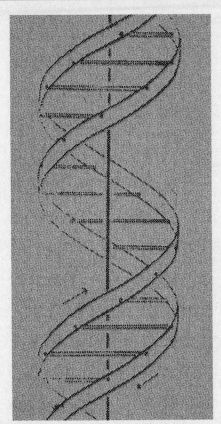

Figure 2.B Drawing of the DNA molecule as it first appeared in a scientific journal. This famous sketch was drawn from the model by Francis Crick's wife, Odile Crick.

only with cytosine, each strand could serve as a mold for building a second strand from DNA nucleotides:

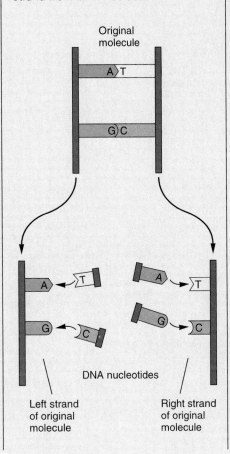

Original molecule

DNA nucleotides

Left strand of original molecule

Right strand of original molecule

Subsequent experiments (described in chapter 7) have shown that DNA duplicates itself in precisely this manner: the two strands separate and each strand serves as a template for building another strand.

Forty years later, Crick made the following remark about the discovery: "One has to reflect on, really, what a remarkable thing it is in the history at least of life on earth, because nucleic acids, certainly RNA and DNA, almost surely existed on the earth for *billions* of years. It is the basis of *all* the life forms we see, but really, it's only been in the last half-century—which is a blink, really, in time—that any form of life on earth has become *aware* of the structure of DNA."

together and orients them so that certain chemical bonds are more easily broken and formed. The reactants of an enzyme-catalyzed reaction are the enzyme's **substrates.**

An enzyme brings its substrates together by holding them within a pocket on its surface. Each enzyme catalyzes only a particular reaction because only certain substrates fit into its pocket. In some cases, the substrates and pocket fit together, like a key and a lock, before contact (fig. 2.11); in other cases, the substrates do not fit the pocket until after they have induced (by chemical reactions) a change in the pocket's shape.

The pocket of an enzyme is lined with chemically active sites, formed of R groups on the amino acids. These active sites form temporary bonds with the substrates, orienting and twisting them so as to weaken certain bonds and promote formation of others.

Enzymes are usually named according to their substrates. For example, an enzyme that adds oxygen is named oxidase, an enzyme that removes phosphate is named phosphatase, and an enzyme that digests lipid is named lipase. The ending *-ase* is simply added to the name of the molecule upon which the enzyme acts.

Biochemical Pathways

The chemistry of life is organized into biochemical pathways. Each pathway is a sequence of chemical reactions controlled by a sequence of enzymes. The product of one enzyme-facilitated reaction becomes the substrate of the next enzyme in the sequence. Many biochemical pathways contain as many as thirty reactions promoted by thirty different enzymes. A hypothetical pathway is shown in

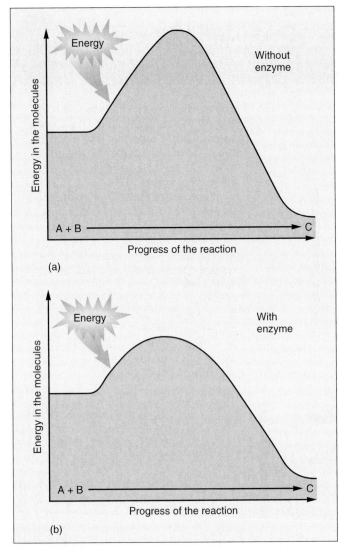

Figure 2.10 Effect of an enzyme on the energy of activation. (a) Without an enzyme, the interaction of molecules A and B to form molecule C requires a great deal of energy to make it proceed rapidly. (b) With an enzyme, the reaction proceeds rapidly with much less energy input.

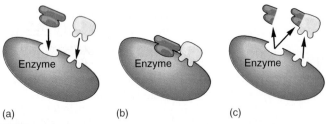

▆▆ **Figure 2.11 Catalysis by an enzyme.** An enzyme lowers the energy of activation by bringing molecules together and orienting them so that bonds are more easily broken and formed. (a) An enzyme captures its substrates (the reactants). (b) The enzyme orients the reactants, twisting them into positions that weaken existing chemical bonds and promote formation of new bonds. (c) The enzyme releases the newly formed molecules. It is unchanged by the reaction and can be used again and again.

figure 2.12. An organism's metabolism consists of all the biochemical pathways that are taking place within its body.

Each organism has a unique set of biochemical pathways governed by a unique set of enzymes. (The enzymes, in turn, are products of a unique set of genes.) The presence or absence of a particular enzyme has a noticeable effect on the organism. Some people, for example, have the enzyme that digests milk sugar and others do not. This aspect of human health is described in the accompanying Wellness Report.

Control of Biochemical Pathways

Chaos would prevail if an organism had no way of controlling its biochemical pathways. The end product of one pathway, for example, might be eliminated right away by another pathway. An organism regulates its biochemical pathways by controlling its enzyme synthesis and enzyme activity.

Control of Enzyme Synthesis   One way an organism controls its biochemical pathways is through the amount of enzymes present. When larger quantities of the end product are needed, more copies of each enzyme in the pathway are synthesized. The end product is then synthesized simultaneously by many identical pathways. Conversely, if the end product is no longer needed, synthesis of the pathway's enzymes is shut down.

One factor known to control how much of an enzyme is synthesized is the amount of its substrates available. If no substrates are present, then no enzyme is synthesized; if a lot of the substrates are present, then a lot of the enzyme is synthesized. An example of this type of control is the synthesis of glycogen (a polymer) from glucose (a monomer) in cells of the liver. Excess glucose in these cells stimulates the cell to synthesize the enzymes that convert it into glycogen for storage.

Control of Enzyme Activity   A biochemical pathway facilitated by enzymes that are already present is regulated by feedback inhibition. (A feedback system is one in which the output controls the input.) In feedback inhibition of a biochemical pathway, high concentrations of the end product inactivate the first enzyme in the pathway. Most cells, for example, convert glucose into ATP, a molecule that transports energy from glucose to places where it is needed. High concentrations of ATP in a cell inactivate the first enzyme in the biochemical pathway that converts glucose to ATP.

One way the end product inactivates the first enzyme is by changing its shape. In this type of control, the end product binds to the enzyme and, in so doing, changes the shape of the enzyme so it can no longer hold the substrate.

Why Most People Cannot Digest Milk

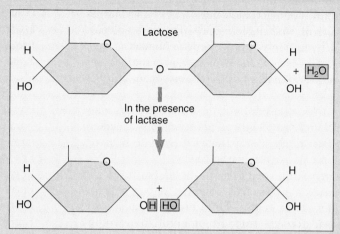

Most adults lack the enzyme that digests milk sugar. When they consume dairy foods, most of the milk sugar passes undigested into their large intestine, where it is digested and used by bacteria, which release gas in the process. Such people experience intestinal cramps, gas, and diarrhea after drinking milk or eating other dairy products, such as ice cream and cheese. They are said to be lactose-intolerant. Lactose is the scientific name for milk sugar, and lactase is the name of the lactose-digesting enzyme (see figure).

Actually, the inability to digest milk is normal for an adult mammal. Mammals, such as deer, squirrels, and mice, living in their natural environments do not drink milk as adults. Only babies drink milk. As they grow older, they gradually stop synthesizing lactase.

Humans are mammals, but about 30% of us can digest milk sugar as adults. People with this ability, which is inherited, are descended from ancestors who, for thousands of years, kept domestic cattle and depended on dairy products as adult foods. Anyone who continued to synthesize lactase into their adult years was better nourished and had more children. The children, in turn, inherited the ability to prolong lactase production and so to digest milk sugar throughout their lives.

People with the ability to digest lactose as adults are chiefly descendants of northern Europeans, some Mediterranean peoples, and the Masai of East Africa. People without the ability include descendants of Asians, Africans, and Native Americans. In addition, about half of all Hispanics and a fifth of all Caucasians are lactose-intolerant.

Dairy foods are major components of the typical American diet. If you cannot digest milk there are two things you can do: (1) restrict your intake of dairy foods to fermented products—yogurt, buttermilk, or kefir (fermented cow's milk), since much of the milk sugar in these foods has already been digested by the bacteria used in fermentation, and (2) consume milk and milk products that have been treated with lactase. You can either buy the products already treated or buy the enzyme and add it to dairy foods.

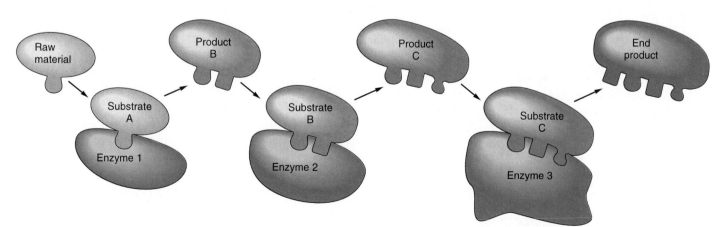

Figure 2.12 A biochemical pathway. A biochemical pathway is a sequence of reactions, each catalyzed by a specific enzyme. The product of enzyme 1 becomes the substrate of enzyme 2 and so on until the end product is synthesized.

SUMMARY

1. Living things are made of twenty-four kinds of atoms—small ones that are available on the earth's surface and that form strong bonds. All organic molecules contain carbon.

2. The four kinds of macromolecules are polysaccharides, lipids, proteins, and nucleic acids. They are polymers built by dehydration synthesis from monomers.

3. A polysaccharide is a chain of monosaccharides. Important polysaccharides are starch, glycogen, and cellulose.

4. A lipid is a fat, phospholipid, or steroid. A fat is made of glycerol and fatty acids. A phospholipid is made of glycerol, fatty acids, a phosphate ion, and an R group. A steroid is made of four carbon rings.

5. A protein is one or more chains of amino acids. Twenty kinds of amino acids combine to form an enormous variety of proteins. The primary, secondary, tertiary, and quaternary structures of a protein determine its shape and function.

6. A nucleic acid is one or two chains of nucleotides. A nucleotide consists of a sugar, a phosphate ion, and a nitrogenous base. DNA is a double strand of nucleotides, with each nucleotide formed of the sugar deoxyribose, a phosphate ion, and one of the following bases: adenine, thymine, guanine, or cytosine. The two strands are held together by weak bonds between their bases (adenine bonds with thymine; guanine bonds with cytosine). RNA is a single strand of nucleotides, with each nucleotide consisting of the sugar ribose, a phosphate ion, and one of the following bases: adenine, uracil, guanine, or cytosine.

7. Biochemical reactions are catalyzed by enzymes, which are proteins with particular shapes for holding particular substrates. An enzyme speeds up a reaction by bringing substrates together and holding them in a position that helps break and form bonds.

8. The chemistry of life is organized into biochemical pathways—long sequences of reactions, each reaction governed by a specific enzyme. An organism's metabolism consists of all its active biochemical pathways.

9. An organism controls which of its pathways are active. Some pathways are controlled by the rate at which their enzymes are synthesized, whereas others are controlled by activating or inactivating existing enzymes.

KEY TERMS

Amino (ah-MEE-no) **acid** (page 45)
Dehydration synthesis (page 41)
Enzyme (EN-zeyem) (page 50)
Fatty acid (page 43)
Glucose (GLOO-kohz) (page 42)
Hydrolysis (heye-DRAH-lih-sis) (page 42)
Inorganic (IN-or-GAEH-nik) **molecule** (page 40)
Lipid (LIH-pid) (page 42)
Monomer (MAH-noh-muhr) (page 40)
Monosaccharide (MAH-noh-SAH-kur-eyed) (page 42)
Nucleic (noo-KLAY-ik) **acid** (page 48)
Nucleotide (NOO-klee-oh-teyed) (page 48)
Organic (or-GAEH-nik) **molecule** (page 40)
Polymer (PAH-leh-muhr) (page 40)
Polypeptide (PAH-lee-pehp-teyed) (page 45)
Polysaccharide (PAH-lee-SAH-kur-eyed) (page 42)
Protein (page 45)
R group (page 45)
Substrate (page 51)

STUDY QUESTIONS

1. Explain why organic molecules are made of small atoms and why all organic molecules contain carbon.

2. Describe how a polymer is built from monomers.

3. What are three important polysaccharides? Which one is built by animals, and what is its function? Which one is important in our diet?

4. Why do phospholipid molecules line up in a row when placed in water?

5. Amino acids can combine to form an almost unlimited variety of proteins. What is it about the structure of a protein that permits such variety?

6. Describe how an enzyme speeds up a chemical reaction.

7. Draw an imaginary biochemical pathway and show two ways in which the rate at which the pathway proceeds is controlled.

CRITICAL THINKING PROBLEMS

1. Give one reason why plants, especially in cooler climates, would not be able to function well if all their fats were saturated fats.
2. Many proteins are denatured by temperatures exceeding 44° C (112° F). Explain how this characteristic of proteins relates to the danger of an unusually high fever.
3. Some infectious diseases are treated by injections of antibodies, which are proteins with specific functions determined by their quaternary structures. Explain why antibodies cannot be administered in pill form.
4. Most plant fertilizers contain ions that provide nitrogen, sulfur, and phosphorus. Why does a plant require each of these atoms for growth?
5. Explain why a DNA molecule always contains equal amounts of adenine and thymine.

SUGGESTED READINGS

Introductory Level:

Armstrong, F. B. *Biochemistry*, 2d ed. New York: Oxford University Press, 1982. A textbook written for college students who are not majoring in chemistry.

Atkins, P. W. *Molecules*. New York: W. H. Freeman and Company, 1987. An interesting description of molecules, especially the ones that affect human lives.

Dressler, D., and H. Potter. *Discovering Enzymes*. New York: W. H. Freeman and Company, 1991. An introduction to biochemistry for the nonscientist, describing first the history of chemical experimentation and then how enzymes function, including descriptions of human diseases.

Goodsell, D. S. *The Machinery of Life*. New York: Springer-Verlag, 1993. An introduction, for the nonscientist, to biochemistry and cells.

Vogel, S. "The Shape of Proteins," *Discover*, October 1988, 38–43. A description, for the nonscientist, of how biochemists study proteins, how proteins fold themselves into elaborate shapes, and the evolution of proteins.

Advanced Level:

Branden, C., and J. Tooze. *Introduction to Protein Structure*. New York: Garland Press, 1991. A description of protein structure, function, and evolution with special chapters on enzymes and research methods.

Kornberg, A. *For the Love of Enzymes: The Odyssey of a Biochemist*. Cambridge, MA: Harvard University Press, 1989. A history of the origins and development of research on DNA.

Perutz, M. *Protein Structure: New Approaches to Disease and Therapy*. New York: W. H. Freeman and Company, 1992. A discussion of how our knowledge of protein structure is helping us in medicine, with examples that include the use of antibodies in treating leukemia, regulatory proteins in treating the genes that cause cancer, and drugs that interfere with HIV enzymes.

Stryer, L. *Biochemistry*. New York: W. H. Freeman and Company, 1988. A leading textbook in biochemistry.

⊛ EXPLORATION

Interactive Software: What Enzymes Do

This interactive exercise explores the factors that govern the rate of an enzyme-catalyzed reaction. The student varies the temperature as well as the concentrations of the substrate and the enzyme, and then observes how each of these variables affects the rate at which the reaction proceeds.

Cells
An Overview

Chapter Outline

Objectives

In this chapter, you will learn
–about the variety of cells that form
  organisms;
–the structure and function of simple
  cells, as seen in bacteria;
–the structure and function of more
  complex cells, as seen in all other
  forms of life.

Cells are the functional units of life.
These round, flat structures are cells
that were taken from the skin of a
human face and dyed green.

I f you examined the parts of your body under a microscope you would find that each part— skin, bones, blood, liver, brain, or muscle—is made up of tiny compartments called cells. If you extended your observations to other living things— friends, dogs, grasshoppers, trees, and the organisms in a pinch of soil—you would find that they, too, are formed of these same tiny structures. Cells, in turn, are built of polysaccharides, lipids, proteins, and nucleic acids—the macromolecules of life.

If you could look closer at the interior of a living cell, you would find it teeming with activity, with a steady flow of materials from one part of the cell to another. Two major operations are always underway: new proteins are synthesized from amino acids and energy is released from sugars and fats. To accomplish these tasks, materials are pushed and pulled along "highways" that crisscross the cell interior. From time to time the cell boundary bulges inward as it imports raw materials or bulges outward as it exports products. Occasionally, the entire cell undergoes dramatic changes in activities and shape as it divides to form two cells. A cell is a biochemical "factory," performing tasks as dictated by its genes.

Kinds of Cells

All living things are made of cells (fig. 3.1), and almost everything that happens in an organism is the consequence of what happens inside its cells. All the properties of life—complexity of organization, internal constancy, manipulation of energy, response to stimuli, movement, growth, development, reproduction, and adaptation— come from the properties of cells.

In the past few decades, enormous progress has been made in understanding the cell and how it works. This knowledge has profoundly changed our view of life, for it has revealed that all cells, regardless of their source, are remarkably alike. The enormous diversity of life, from bacteria to humans, is generated by cells with strikingly similar structures and biochemical pathways.

We have learned so much about these basic units of life that researchers can take cells from an organism and grow them in a dish. This technique is used not only to learn more about cells but to diagnose medical conditions. Cancer, for example, is diagnosed by removing cells (a biopsy) from a person and growing them in the laboratory. Genetic disorders in an embryo, while still in the womb, are also revealed by growing samples of the embryo's cells in this way. These new techniques have generated an ethical question: To whom do the cells belong—the person

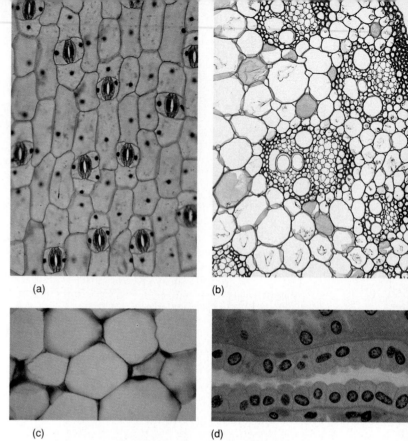

(a)

(b)

(c)

(d)

Figure 3.1 All forms of life are made of cells. (*a*) Rows of cells in a plant leaf. The outer boundary and an internal structure (the nucleus) have been stained in each cell. The oval structures are openings through which carbon dioxide enters the leaf. (*b*) Rows of cells in a plant stem. The outer boundaries have been stained. (*c*) A cluster of human fat cells, in which only the outer boundaries are stained. (*d*) Rows of cells in a human kidney. The darker structures are the cell nuclei.

they were taken from or the medical researcher? This interesting question is explored in the accompanying Research Report.

While all cells are similar, they are not identical. The cell of a unicellular organism, such as a bacterium or protozoan, are generalized cells for they must perform all the activities involved in survival and reproduction. Some unicellular organisms, like bacteria, manage this by having different activities occur at different times; others, like amoeba and paramecia, have specific compartments within the cell for each kind of activity. The cells of a multicellular organism are specialized, with each kind of cell contributing in a limited way to the survival and reproduction of the whole organism. Like a society with its specialized workers, such as farmers, grocers, and doctors, the cells of a multicellular organism form a community of specialized biochemical factories—liver cells, brain cells, muscle cells, and many more.

Research Report

Cells for Sale

How would you feel if a doctor used some of your cells to make money? Would you feel that the cells, although removed from your body, belong to you? Should you at least receive a share of the profits?

Cells are removed from patients routinely for medical tests. And sometimes they are multiplied in a laboratory and used for purposes other than to find out what was making the patient ill. The most famous cells are those of Henrietta Lacks, who, in 1951, had cells from her cervix (lower part of uterus) removed and sent to a laboratory in order to find out if she had cervical cancer. She did, and her cells—the so-called HeLa cells—were sold to a research institute. They have been grown in the laboratory since 1951 and used in many experiments. Henrietta Lacks' cells have taught us a lot about cancer. This is just one example of a research practice that has now become an industry. Cells left over after medical tests may be sold to biotechnology companies, which in turn use them to make products they can sell.

Figure 3.A A valuable cell. The scalloped projections on this white blood cell, taken from Sean's spleen, are not seen on normal cells.

One man whose cells were used to make money feels he should get a share of the profits. This man, who we will call Sean Burton, had some of his cells removed for laboratory testing in order to find out why he was so sick. The cells turned out to be abnormal, both in appearance and in what they do. They were covered with scalloped projections, as shown in figure 3.A, and manufactured unusually large amounts of the protein that triggers multiplication of white blood cells. Since white blood cells are the body's main defense against infections, the proteins from Sean's cells could be of great therapeutic value—injected into the blood, they would boost the number of white blood cells and so help a person overcome a lingering infection or even cancer.

Medical researchers are now growing Sean's cells and harvesting the protein. Sean's doctor, who removed the cells, has even patented them and named himself as the inventor. He will make a nice profit from the sale of these unique cells.

Sean Burton sued his doctor. The cells, he claimed, belong to him and not to the doctor who removed them. They are stolen property. If anyone profits, it should be the person from whom the cells were taken. The California Court of Appeals decided in Sean's favor, but the California Supreme Court later ruled that Sean did not own cells removed from his body. The U.S. Supreme Court refused to hear the case.

There are two fundamental kinds of cells in the world: the prokaryotic cells of the kingdoms Archaea and Eubacteria and the eukaryotic cells of the kingdoms Plantae, Animalia, and Fungi (see inside back cover of this book). The differences between them are summarized in table 3.1 and described following.

The Prokaryotic Cell

A **prokaryotic cell** (*pro*, before; *karyote*, nucleus) is a relatively simple cell in terms of its structure. Each cell forms an entire organism; prokaryotic cells never form multicellular organisms. The first cell on earth was most likely a simple cell of this kind. The oldest fossil cells, found in rocks nearly 3.5 billion years old (the earth itself is roughly 4.6 billion years old), resemble modern prokaryotic cells.

Prokaryotic organisms in the kingdom Archaea are quite distinct from other forms of life. They live in unusual environments, such as hot springs and ocean vents, and have unusual biochemical pathways. Some live in our intestines, where they feed on cellulose and release a foul-smelling gas. Prokaryotic organisms in the kingdom Eubacteria are found almost everywhere, in enormous abundance and diversity (fig. 3.2). These organisms, known

Table 3.1

Prokaryotic versus Eukaryotic Cells

Characteristic	Prokaryotic Cell	Eukaryotic Cell
Size (diameter)	About 0.001 mm	0.01 to 0.1 mm
DNA	Single "naked" molecule	Combined with other molecules to form chromosomes
Nuclear envelope	Absent	Present
Ribosomes	Present	Present
Flagella	Present in some	Present in some
Organelles other than ribosomes and flagella	Absent	Present

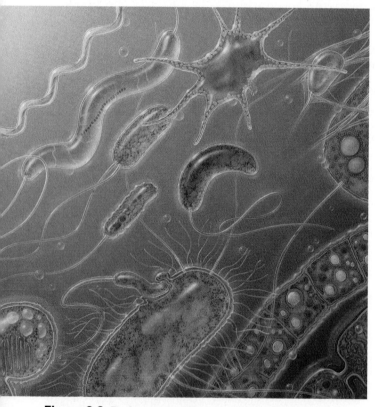

Figure 3.2 Prokaryotic cells come in many forms. All prokaryotic cells are unicellular, but some adhere to one another to form long chains of cells. Some cells are round, some are oblong, and others are shaped like spirals. Some have no flagellum and others have one or more.

collectively as bacteria, are described following. The major components of a bacterium are a cell wall, plasma membrane, DNA molecule, and cytosol.

Cell Wall and Plasma Membrane

A cell wall forms the outer boundary of a bacterium. It is a strong, porous, and flexible structure that supports and protects a bacterium. An important protective function is to prevent the cell from swelling. In most environments, a bacterium without a cell wall swells with water until it bursts. (Certain molecules of our immune system take advantage of this trait by digesting holes in the walls of disease bacteria, causing them to burst and die.)

Two kinds of cell walls are found in bacteria. Gram-positive bacteria (so-named because they retain a certain dye) have cell walls formed of a single layer (fig. 3.3a). This kind of wall is a rigid framework made of peptidoglycan, a macromolecule formed of short polypeptides and sugars that is found only in bacteria. Examples of diseases caused by gram-positive bacteria are urinary tract infections, staph infections (e.g., boils and pimples), and strep throat.

Gram-negative bacteria (so-named because they do not retain a certain dye) have cell walls formed of two layers: an inner layer like the one described for gram-positive bacteria and an outer layer formed of phospholipids, glycolipids (polysaccharides attached to lipids), and proteins (fig. 3.3b). The outer layer makes gram-negative bacteria less vulnerable to a person's immune system. It is also responsible for a life-threatening condition known as septic shock or blood poisoning, in which invading bacteria are attacked by the immune system. As the cell walls of gram-negative bacteria break apart, glycolipids are exposed and trigger a severe reaction by the immune system. In the United States, septic shock is the thirteenth leading cause of death.

Inside the cell wall is a **plasma membrane,** a sheet of lipids and proteins that encloses the contents of the cell. Every cell has a plasma membrane. This important structure, which controls what enters and leaves a cell, is described in chapter 4.

Some bacteria have one or more **flagella.** These long, whiplike structures extend from the cell wall and propel the cell through its fluid environment (fig. 3.4).

DNA

A bacterium has a single, circular molecule of DNA that carries genes for all but a few of its proteins. It is attached at one point to the inner side of the plasma membrane. The molecule is considered to be "naked" in that it is not part of a larger structure, the chromosome, found in eukaryotic cells.

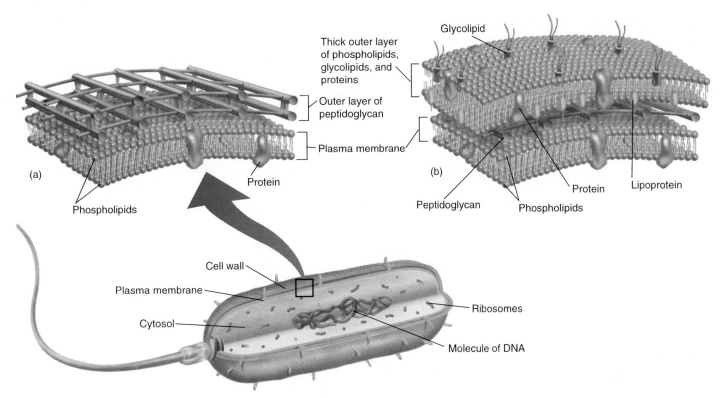

Figure 3.3 The cell walls of bacteria. (*a*) The cell wall of a gram-positive bacterium is a framework of peptidoglycan molecules. (*b*) The cell wall of a gram-negative bacterium has two layers: an inner framework of peptidoglycan molecules and an outer layer of phospholipids, glycolipids, and proteins. The glycolipids cause septic shock by triggering an extreme reaction by the immune system.

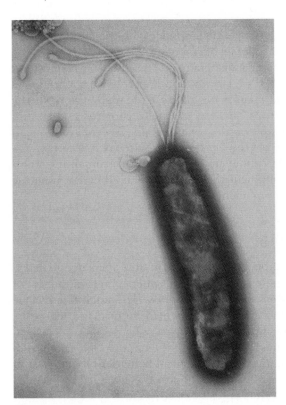

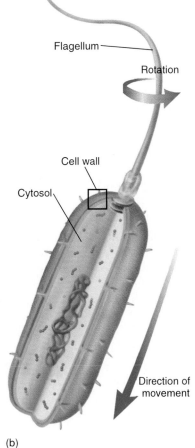

Figure 3.4 Flagella. (*a*) The bacterium suspected of causing stomach ulcers has four flagella. (*b*) The flagellum of a bacterium originates inside the cell and extends outward through the cell wall. By rotating, it pushes the bacterium forward.

In addition to the main DNA molecule, a bacterium may have tiny loops of DNA, called plasmids, that can travel from one bacterium to another. Each plasmid carries genes for two or three proteins, often unusual ones like the proteins that confer resistance to antibiotics. Such traits, coded for by genes on plasmids, can quickly spread through a population of bacteria as well as from one species to another.

Cytosol

The **cytosol** is the jellylike solution (70% water) of the cell, in which a variety of molecules and ions are suspended or dissolved. A typical cytosol contains hundreds of macromolecules, mainly proteins and RNA, thousands of small molecules, including amino acids, glucose, fatty acids, and nucleotides, and about 50,000 ions. All cells have a cytosol. The most conspicuous feature of a prokaryotic cell's cytosol is its **ribosomes**—spherical structures, formed of RNA and proteins, that serve as sites for protein synthesis.

61

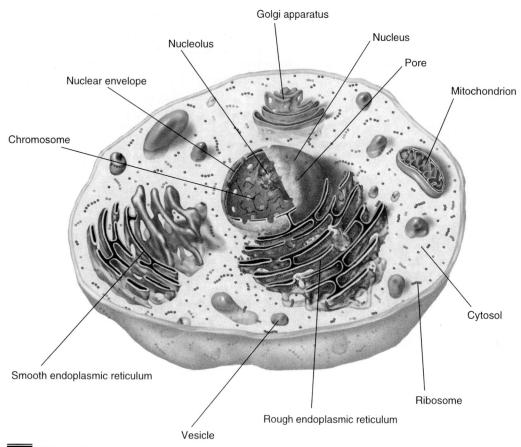

Golgi apparatus

Nucleus

Nucleolus

Pore

Nuclear envelope

Mitochondrion

Chromosome

Cytosol

Smooth endoplasmic reticulum

Ribosome

Rough endoplasmic reticulum

Vesicle

Figure 3.5 The structures of a typical eukaryotic cell.
The organelles have been cut open to show internal structures.

The Eukaryotic Cell

A **eukaryotic cell** (*eu*, true; *karyote*, nucleus) is larger (at least 10 times wider and 1,000 times larger by volume) and has many more internal structures than a prokaryotic cell (fig. 3.5). Most conspicuously, it has a membrane-enclosed region (the nucleus) that contains the DNA molecules. Eukaryotic cells appear in the fossil record about 2.1 billion years ago—more than a billion years after the first cells.

All organisms in the kingdoms Protista, Plantae, Animalia, and Fungi (see inside back cover of this book) have eukaryotic cells. The protists live as unicellular organisms whereas plants, animals, and fungi are multicellular organisms.

A eukaryotic cell has a variety of **organelles**—specialized compartments that contain the biochemical pathways (see page 51) (enzymes, substrates, and other materials) for performing a particular task. In this way, the cell's biochemical pathways are organized according to function and kept separate from one another. The breakdown of glucose (to provide energy), for example, is isolated from the synthesis of polysaccharides. A eukary-

otic cell is like a mansion with different rooms for different activities, whereas a prokaryotic cell is like a one-room apartment.

We begin this section with a brief description of membranes, since most organelles are formed of membranes. We then go on to describe the structure and function of each kind of organelle. A summary of the organelles is provided in table 3.2.

Membranes

Cellular membranes are sheets of lipids and proteins. These fundamental structures, described in detail in chapter 4, have many important functions: they provide barriers to the flow of materials; provide surfaces for embedding the enzymes of a biochemical pathway; receive signals from chemical messengers, such as hormones and neurotransmitters; and recognize disease microbes.

The plasma membrane is the cell's outer boundary; it holds the cell's contents and controls what enters and leaves. The membranes of most organelles work in the same way: they hold the organelle contents and control the inflow and outflow of materials. An enclosing membrane enables a cell or organelle to accumulate certain materials and to eliminate others, thereby maintaining an internal environment that is optimal and different from the external environment.

The Nucleus

The most prominent organelle within a eukaryotic cell is the nucleus—a relatively large, spherical region that is enclosed within a nuclear envelope. The **nuclear envelope** consists of two membranes, which lie next to each other, and is pierced with many pores (see fig. 3.5).

The nuclear envelope protects the DNA molecules by isolating them from chemical activities taking place in the rest of the cell. The pores in the envelope are guarded by proteins that permit certain macromolecules to pass into and out of the nucleus.

Table 3.2

Organelles Within Eukaryotic Cells

Organelle	Function
Nuclear envelope	Isolates DNA
Ribosome	Synthesizes proteins used in cytosol
Rough endoplasmic reticulum	Synthesizes and transports proteins used in organelles or outside cell
Smooth endoplasmic reticulum	Synthesizes polysaccharides and lipids; transports materials
Vesicle	Isolates biochemical pathways; transports materials
Golgi apparatus	Modifies proteins
Mitochondrion	Releases energy from molecules (cellular respiration)
Plastid	Builds sugars (photosynthesis) or stores macromolecules
Vacuole	Transports and stores materials; makes a plant cell rigid
Cytoskeleton	Provides shape, movement, support, and internal transport

Chromosomes   All cellular activities are controlled ultimately by genes, which are contained within DNA molecules. Genes provide instructions for building proteins from amino acids. The proteins, in turn, form enzymes, structures, and molecules like antibodies that perform special functions.

The DNA molecules of a eukaryotic cell are combined with other molecules to form chromosomes. A **chromosome** is a long molecule of DNA wrapped around special proteins called histones, which provide a sort of scaffold for the DNA. Lipids and other molecules also contribute to the structure of a chromosome. Each eukaryotic cell has more than one chromosome (a human cell, for example, has forty-six chromosomes). When a cell is not dividing, its chromosomes are elongated and spread throughout the nucleus.

Nucleoli   A **nucleolus** is a spherical structure within the nucleus (see fig. 3.5). It has no enclosing membrane to separate it from the chromosomes. The function of a nu-

cleolus is to build ribosomes (see description following) from RNA and proteins. It synthesizes the RNA, picks up protein that has moved into the nucleus via pores in the nuclear envelope, and then combines the two macromolecules to form a ribosome. Newly formed ribosomes move out of the nucleus through the same pores. A typical nucleus has two nucleoli.

The Cytoplasm

The **cytoplasm** includes all cellular materials outside the nucleus. It is packed with organelles that are suspended within the cytosol.

Ribosomes   As in bacteria, the ribosomes of a eukaryotic cell are sites of protein synthesis. They consist of RNA and protein, not enclosed within a membrane. The genetic instructions for a protein, located within a DNA molecule, are carried by an RNA molecule from the nucleus to a ribosome, where they are used to assemble amino acids into a protein. The role of ribosomes in protein synthesis is explained in chapter 11.

Some of the ribosomes are dispersed throughout the cytosol and others are embedded on membranes of the endoplasmic reticulum (see following). The dispersed ribosomes synthesize proteins that are used within the cytosol. The membrane-bound ribosomes synthesize proteins that will be transported to other organelles or outside the cell.

Endoplasmic Reticulum   The **endoplasmic reticulum** (*endoplasmic*, internal cytoplasm; *reticulum*, network) is an elaborate system of membranes that extends throughout the cytoplasm (see fig. 3.5). Each membrane is folded to form an internal canal, which may connect to other organelles or to the plasma membrane. The endoplasmic reticulum provides surfaces for synthesizing macromolecules and canals for transporting the newly synthesized macromolecules to other regions of the cell. A eukaryotic cell has two types of endoplasmic reticulum: rough and smooth.

The rough endoplasmic reticulum has ribosomes embedded within its outer surface. These ribosomes synthesize proteins that are used in biochemical pathways that take place in other organelles or are used outside the cell (e.g., digestive enzymes in the intestines and hormones in the blood). Once synthesized, a protein enters the internal canal of the endoplasmic reticulum, where it is shipped to an appropriate organelle or to the plasma membrane for export.

The smooth endoplasmic reticulum has no ribosomes and does not build proteins. Its outer surface holds the enzymes for synthesizing lipids and polysaccharides, and its internal canals transport these macromolecules from one part of the cell to another. In addition, the smooth endoplasmic reticulum may perform specialized tasks related to a particular type of cell. The smooth endoplasmic reticulum of a muscle cell, for example, stores and releases the calcium ions that trigger muscle contractions.

Vesicles   A **vesicle** is a tiny bag formed of membranes. Some vesicles transport materials from one part of the cell to another. Other vesicles hold the enzymes of specific biochemical pathways.

A special form of vesicle, the lysosome, contains digestive enzymes that break apart macromolecules. The main function of a lysosome is to destroy bacteria and to decompose organelles that are damaged. During times of starvation, lysosomes digest nonessential parts of the cell to provide monomers for energy and for building essential organelles. The most spectacular activity of lysosomes occurs during formation of an embryo, which in some structures involves an orderly self-destruction of cells. A hand, for example, begins as a rounded club. Separate fingers emerge as cells between the fingers are destroyed by their own lysosomes. These cells are programmed, by their genes, to "commit suicide" when the embryo reaches a particular age.

The Golgi Apparatus   A **Golgi apparatus** is seen as a stack of several flattened vesicles (see fig. 3.5). It receives proteins that have been synthesized in the rough endoplasmic reticulum and packaged within transport vesicles formed of pieces from the endoplasmic reticulum. Enzymes within the Golgi apparatus then modify these proteins by adding such molecules as sugars, fatty acids, or phosphates. The modified proteins are repackaged into transport vesicles for shipment to their final destinations (fig. 3.6). Most enzymes synthesized on the rough endoplasmic reticulum are modified by the Golgi apparatus before entry into their organelles. Similarly, digestive enzymes and hormones are modified by the Golgi apparatus prior to shipment out of the cell and into the digestive tract or bloodstream.

Mitochondria   A **mitochondrion** is a membrane-enclosed organelle (see fig. 3.5), called the powerhouse of the cell, that holds the biochemical pathways of cellular respiration—the pathways that release energy from glucose and other fuels:

$$C_6H_{12}O_6 + 6O_2 \longrightarrow 6CO_2 + 6H_2O + \text{Energy}$$

Glucose Oxygen gas   Carbon dioxide Water

Cellular respiration is described in detail in chapter 5.

A typical eukaryotic cell has hundreds of mitochondria. Each one consists of an outer enclosing membrane and an inner membrane with elaborate folds that extend into the inner fluid. The outer membrane controls what enters and leaves the mitochondrion. The inner membrane is packed with many enzymes involved in cellular respiration. Other enzymes of this important process are suspended within the inner fluid of the mitochondrion.

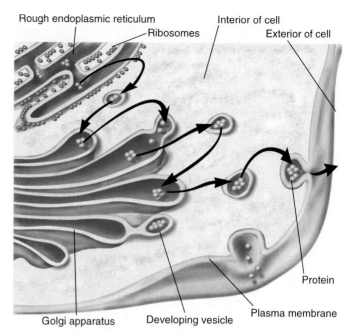

Rough endoplasmic reticulum   Interior of cell
Ribosomes   Exterior of cell
Protein
Golgi apparatus   Developing vesicle   Plasma membrane

Figure 3.6 The Golgi apparatus modifies proteins.
Proteins (yellow) synthesized on the rough endoplasmic reticulum are carried within transport vesicles (formed of a small piece of the endoplasmic reticulum) to the Golgi apparatus. There they are modified as they pass through the system of membranes and then packaged into transport vesicles and delivered to where they are needed.

Plastids   Cells of plants (fig. 3.7) and algae have membrane-enclosed structures called plastids. A **plastid** does one of two things: it either holds the biochemical pathways of photosynthesis or stores macromolecules.

The plastids that specialize in photosynthesis contain the light-absorbing pigment chlorophyll (*chloro*, green; *phyll*, leaf) and are called chloroplasts. In addition to chlorophyll, these organelles contain other light-absorbing pigments as well as the enzymes of photosynthesis. In photosynthesis, light, carbon dioxide, and water are converted into glucose (or some other sugar) and oxygen gas:

$$6CO_2 + 6H_2O + \text{Energy} \longrightarrow C_6H_{12}O_6 + 6O_2$$

Carbon dioxide Water   Glucose Oxygen gas

A chloroplast has an outer membrane that controls what enters and leaves the organelle and an inner membrane that extends inward to form parallel membranes—some spanning the entire organelle and others restricted to narrow stacks, as shown in figure 3.7. These internal membranes hold the pigments, as well as many of the enzymes, of photosynthesis. The rest of the enzymes are suspended within the internal fluid of the chloroplast. A detailed description of photosynthesis is provided in chapter 6.

Other plastids store macromolecules, mainly starch. These kinds of plastids are abundant in the roots and un-

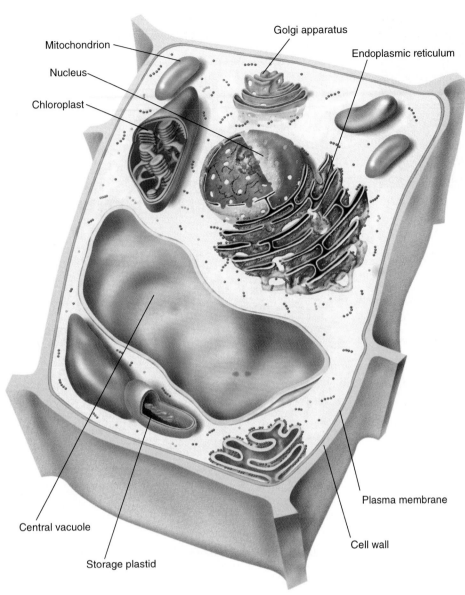

Mitochondrion

Nucleus

Chloroplast

Golgi apparatus

Endoplasmic reticulum

Plasma membrane

Cell wall

Central vacuole

Storage plastid

Figure 3.7 A plant cell. A plant cell has, in addition to the structures shown in figure 3.5, plastids (both chloroplasts and storage plastids), a central vacuole, and a cell wall.

Water accumulates in the central vacuole, expanding the cell and making it rigid, as described in chapter 4.

The Cytoskeleton

The **cytoskeleton** is a web of fibrous proteins that extends throughout the cell (fig. 3.8). Some are contractile proteins that function like muscles: they alter the shape of a cell, move the whole cell from one place to another, and push or pull organelles (e.g., vesicles and mitochondria) and materials from one part of the cell to another. Some proteins of the cytoskeleton provide a scaffold to support the cell and maintain its shape. Other proteins provide an intracellular transport network formed of tracks along which organelles and materials are transported.

The Cell Surface

The outer surface of a cell—its boundary with the rest of the world—is modified to perform a variety of functions. The cell-surface structures and their functions are summarized in table 3.3 and described following.

The Cell Wall   The cells of plants, fungi, and some protists are encased within walls. These cell walls are fairly rigid structures, made chiefly of polysaccharides, that completely surround the plasma membranes (see fig. 3.7). The structure of a eukaryotic cell wall is quite different from that of a prokaryotic cell wall.

derground stems of overwintering plants, where large quantities of energy are stored (as glucose in starch) for use in the spring.

Vacuoles   A **vacuole** is a membrane-enclosed bag that resembles a vesicle but is much larger. Vacuoles perform many functions. In unicellular organisms, they serve as food vacuoles, which digest food, and as contractile vacuoles, which pump excess water out of the cell. Among multicellular organisms, vacuoles are most prominent in plants.

A plant cell is characterized by a central vacuole, which stores materials and gives support. A typical central vacuole (see fig. 3.7) contains water, sugars, organic acids, proteins, inorganic ions, pigments, and toxic waste products that might otherwise interfere with cellular activities.

Table 3.3

Structures on Cell Surfaces

Structure	Function
Cell wall	Supports and protects cell
Cilia and flagella	Move the cell or surrounding fluid
Glycocalyx	Provides adhesion and cell recognition
Plasmodesma	Connects cytoplasms of adjacent plant cells
Gap junction	Connects cytoplasms of adjacent animal cells

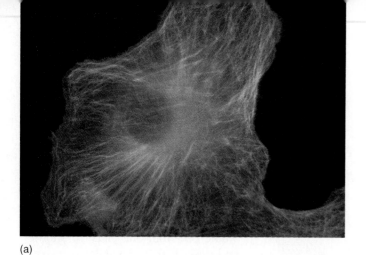

(a)

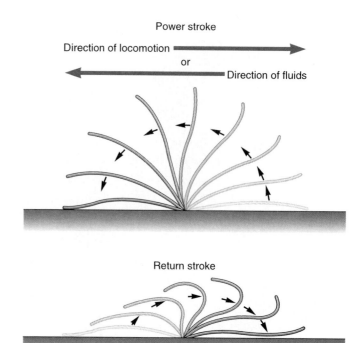

Power stroke

Direction of locomotion

or

Direction of fluids

Return stroke

(a)

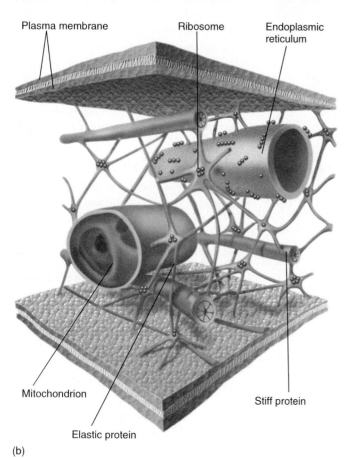

Plasma membrane   Ribosome   Endoplasmic reticulum

Mitochondrion

Stiff protein

Elastic protein

(b)

Figure 3.8 The cytoskeleton. (*a*) Photograph of a cell in which part of the cytoskeleton is stained red. (*b*) Sketch of the cytoskeleton showing the components that hold organelles in place.

The functions of a cell wall are to support the cell, to protect it from injury, and to bind together the many cells of a plant or fungus. Another important function is to prevent a cell from bursting when it accumulates water. A plant cell in fresh water, for example, accumulates water in its central vacuole and enlarges until its plasma membrane presses against the cell wall. If the cell wall were not there, the cell would continue to swell until it burst.

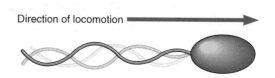

Direction of locomotion

(b)

Figure 3.9 Cilia and flagella. (*a*) A cilium generates movement by becoming stiff during its power stroke and limp during its return stroke. (*b*) The flagellum of a eukaryotic cell generates waves along its length, which propel the cell forward.

Cilia and Flagella Some eukaryotic cells have hairlike projections on their surfaces. When the projections are short and numerous they are called **cilia;** when they are long and few they are called flagella.

The function of cilia and flagella is to create movement: either to move the cell through a fluid or to move a fluid over the cell (fig. 3.9). A cilium waves to-and-fro, with a forceful stroke in one direction and a limp stroke in the other direction, like an oar on a boat. Groups of cilia move in synchrony. A flagellum generates waves that travel down its length and propel the cell in one direction.

Cilia and flagella on unicellular or tiny multicellular organisms move the entire organism in a directed fashion through its watery environment. Cilia on the cells of larger organisms sweep fluids past them. Cilia on cells that line

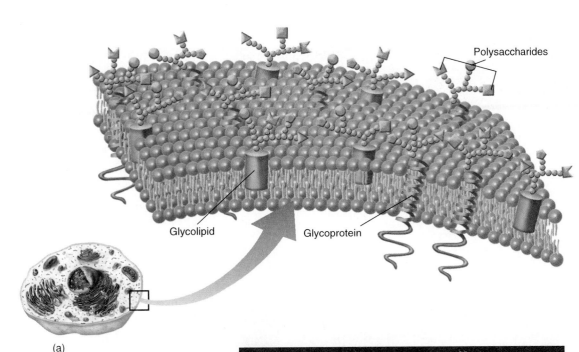

Figure 3.10 The glycocalyx.
(*a*) The glycocalyx is formed of glycoproteins (polysaccharides attached to proteins) and glycolipids (polysaccharides attached to lipids). (*b*) The glycocalyx makes a cell appear fuzzy.

Polysaccharides

Glycolipid   Glycoprotein

(a)

(b)

the respiratory tract, for example, move fluids from the lungs to the nose and mouth.

The Glycocalyx Most eukaryotic cells are covered with a sugary coat, the **glycocalyx** (*glyco*, sugar; *calyx*, coat). It is formed of glycoproteins (macromolecules made of small polysaccharides and proteins) and glycolipids (macromolecules made of small polysaccharides and lipids), as shown in figure 3.10. The polysaccharide components extend outward from the cell surface to give the cell a fuzzy appearance. The protein components extend inward and connect with the cytoskeleton.

One function of the glycocalyx is to help the many cells of an animal, which have no cell walls, bind together to form solid tissues such as those of the skin, liver, or muscle. Another function is cell recognition: each kind of cell has a unique array of polysaccharides on its surface, which gives it a unique identity. Cells recognize one another by these surface molecules. A sperm recognizes an egg of its own species in this way.

The glycocalyx, with its roles in adhesion and recognition, is of particular importance during embryonic devel-

opment, when similar cells form tissues and join with other kinds of cells to form organs. Disease bacteria also use the glycocalyx to find the particular cells they feed on, as described in the accompanying Wellness Report.

Cell-to-Cell Junctions The many cells of a plant or animal communicate with one another and exchange materials through direct connections between adjacent cells.

Plant cells are interconnected by extensions of the cytoplasm called **plasmodesmata** (sing., plasmodesma). These channels of communication extend through the cell walls of adjoining cells and allow a direct flow of materials from one cell to the next (fig. 3.11*a*).

Animal cells are interconnected by **gap junctions,** which are protein pipes that extend through the plasma membranes of adjacent cells (fig. 3.11*b*). Narrower than a plasmodesma, a gap junction permits only small materials, such as monomers and ions, to pass directly between animal cells. It opens and closes in response to changes in conditions within the cells.

How Disease Bacteria Recognize Their Host Cells

Disease microbes are finicky about the kinds of cells they infect. Some species of bacteria infect only cells of the lungs, causing pneumonia or tuberculosis; others infect only the urethra and bladder, causing urinary tract infections; still others infect only the genitals, causing gonorrhea or chlamydia. Similarly, bacteria that infect humans typically do not infect other animals. Only humans, for example, are susceptible to *Neisseria gonorrhoeae*, the bacterium that causes gonorrhea. How do bacteria recognize the particular cells they infect?

Disease bacteria recognize the glycocalyx of their host cells. This surface material is formed of polysaccharides that extend from proteins and lipids within the plasma membrane (see fig. 3.10). The polysaccharides are extremely varied; each kind of body cell has a unique array. The polysaccharides on lung cells differ from the ones on skin cells or liver cells.

In order to infect a cell, a bacterium must not only recognize its host cell but adhere to it (fig. 3.B). Otherwise it would be swept away by the body's fluids. Bacteria in the windpipe are swept in fluids upward, by cilia, toward the nose or mouth, where they are expelled through the nose or swallowed and destroyed by stomach acids. Bacteria in the urinary tract are washed out in urine. A bacterium hangs onto its host cell by forming chemical bonds between proteins on its surface and polysaccharides on the host cell; the two molecules fit together as a key fits a lock.

Now that we know how bacteria recognize and adhere to their host cells, medical researchers are trying to use this knowledge to combat bacterial infections. They are developing drugs that bind with the host cell's polysaccharides, thereby preventing bacteria

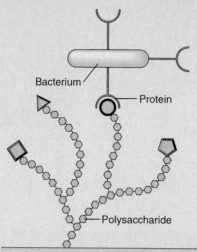

(a)

Host cell plasma membrane

Bacterium

Protein

Polysaccharide

from doing so, and drugs that bind with the bacterium's proteins (fig. 3.C).

New methods of preventing bacterial infections are particularly important because bacteria are becoming more and more resistant to antibiotics. Antibiotics are extremely important weapons in our fight against diseases. Their significance is often

Figure 3.B Bacteria recognize the glycocalyx of their host cells. (*a*) A bacterium binds with a polysaccharide on the surface of its host cell. (*b*) These rod-shaped bacteria have attached to their host cells, which line the urinary tract of a human.

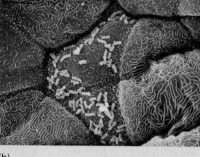

(b)

overlooked because they have been so effective. However, before widespread use of these miracle drugs began in the 1940s, millions of people died from pneumonia, infections during childbirth, tuberculosis, syphilis, and other diseases caused by bacteria. We need to find new ways of combating these diseases.

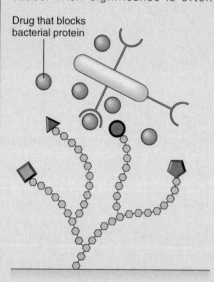

Drug that blocks bacterial protein

Host cell plasma membrane

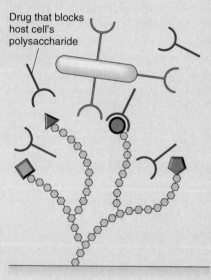

Drug that blocks host cell's polysaccharide

Host cell plasma membrane

Figure 3.C New drugs for treating bacterial infections. Bacterial infections may soon be treated with drugs that block adhesion of the bacterium to its host cell. One kind of drug binds with the bacterial protein (*left*) and another kind binds with the polysaccharide on the host cell (*right*). Both drugs prevent the bacterial protein from binding with the host cell's polysaccharide.

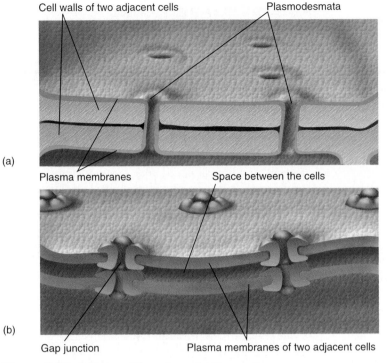

Cell walls of two adjacent cells   Plasmodesmata

(a)

Plasma membranes   Space between the cells

(b)

Gap junction   Plasma membranes of two adjacent cells

Figure 3.11 Cell-to-cell junctions. (*a*) Plant cells are connected by plasmodesmata, which are extensions of the cytoplasm that pass through the cell wall. (*b*) Animal cells are connected by gap junctions, which are protein pipes that connect the cytoplasms of adjacent cells.

SUMMARY

1. Cells are the functional units of life. In multicellular organisms they are more specialized, in structure and function, than in unicellular organisms. Two main kinds of cells are distinguished: prokaryotic and eukaryotic.

2. Prokaryotic cells are simple cells that form unicellular organisms in the kingdoms Archaea and Eubacteria. A bacterium has a cell wall, which supports and protects the cell, and a plasma membrane, which controls what enters and leaves. Some bacteria have flagella that propel them through fluids. Most of the genes are within a single circular molecule of DNA. The cytosol of a bacterium is 70% water and 30% molecules and ions. Suspended within the cytosol are ribosomes, where proteins are synthesized, and enzymes, which catalyze chemical reactions.

3. A eukaryotic cell is larger than a prokaryotic cell and contains a variety of organelles that segregate the biochemical pathways. Membranes form the outer boundary (plasma membrane) as well as most organelles.

4. The nucleus of a eukaryotic cell contains chromosomes and nucleoli. The chromosomes contain the DNA molecules. The nucleoli synthesize ribosomes.

5. The cytoplasm of a eukaryotic cell consists of a cytosol and the following organelles: ribosomes; rough endoplasmic reticulum, which synthesizes proteins (on its ribosomes) and provides transport canals; smooth endoplasmic reticulum, which synthesizes polysaccharides and lipids and also provides transport canals; vesicles, which transport materials and isolate biochemical pathways; Golgi apparatus, which modifies proteins and ships them to their final destinations; and mitochondria, which carry out cellular respiration.

6. Plant cells have, in addition to the organelles listed previously, plastids, which carry out photosynthesis or store macromolecules; and vacuoles, which isolate materials and make the cells rigid.

7. The cytoskeleton of a eukaryotic cell is a protein framework that gives the cell shape, mobility, and support.

8. The outer surface of a eukaryotic cell may contain a cell wall, cilia or flagella, a glycocalyx to provide adhesion and identity, and plasmodesmata or gap junctions for transporting materials between cells.

KEY TERMS

Chromosome (KROH-moh-sohm) (page 63)
Cilium (SILL-ee-um) (page 66)
Cytoplasm (SEYE-toh-PLAEH-zum) (page 63)
Cytoskeleton (SEYE-toh-SKEL-eh-ton) (page 65)
Cytosol (SEYE-toh-sol) (page 61)
Endoplasmic reticulum (EN-doh-PLAZ-mik reh-TIK-yoo-lum) (page 63)
Eukaryotic (YOO-kahr-ee-AH-tik) **cell** (page 62)
Flagellum (flah-JEL-um) (page 60)
Gap junction (page 67)
Glycocalyx (GLEYE-koh-KAY-liks) (page 67)
Golgi (GOHL-jee) **apparatus** (page 64)
Mitochondrion (MEYE-toh-KOHN-dree-un) (page 64)
Nuclear (NOO-klee-ahr) **envelope** (page 62)
Nucleolus (noo-KLEE-oh-lus) (page 63)
Organelle (page 62)
Plasma (PLAZ-mah) **membrane** (page 60)
Plasmodesmata (PLAEH-zmoh-dez-MAH-tuh) (page 67)
Plastid (page 64)
Prokaryotic (PROH-kahr-ee-AH-tik) **cell** (page 59)
Ribosome (REYE-boh-zohm) (page 61)
Vacuole (VAEH-kyoo-ohl) (page 65)
Vesicle (VEE-zih-kahl) (page 64)

STUDY QUESTIONS

1. What kinds of organisms have prokaryotic cells?
2. Describe how the genes of a prokaryotic cell are arranged.
3. Draw a prokaryotic cell and a eukaryotic cell, indicating the positions of the cell wall, DNA molecules, cytosol, mitochondria, ribosomes, endoplasmic reticulum, and Golgi apparatus.
4. What are two main functions of membranes? What is the advantage of having organelles in a cell?
5. Distinguish between a cilium and a flagellum.
6. What are two functions of a glycocalyx?

CRITICAL THINKING PROBLEMS

1. One hypothesis about the origin of mitochondria is that they were once bacteria that invaded a larger cell. What are some predictions of this hypothesis that you could test?
2. Explain why certain cells of the pancreas, which synthesize and release digestive enzymes (proteins), have more rough endoplasmic reticulum and more Golgi apparati than other cells.
3. When a tadpole metamorphoses into an adult frog, its tail gradually disappears. What kind of organelle do you think might be involved in this change? Why?
4. Some materials within our immune systems, which fight disease organisms, digest holes in the cell walls of bacteria. How would this type of action harm the bacteria?
5. Muscle cells have many more mitochondria than fat cells. What accounts for this difference?
6. Sperm have flagella whereas eggs do not. What accounts for this difference?

SUGGESTED READINGS

Introductory Level:

Allen, W., M. Darton, M. Page, P. Powell, and S. Shepherd (eds.). *The Cell: A Small Wonder*. New York: Torstar Books, 1985. An excellent review of the history of cell biology, cell structure and function, and the specialized cells within a human.

Goodsell, D. S. *The Machinery of Life*. New York: Springer-Verlag, 1992. An unusually well-illustrated book about the cell and how it works, with separate chapters on prokaryotic cells and eukaryotic cells. Particularly useful in providing guides to the shapes and sizes of molecules and organelles.

Thomas, L. *The Lives of a Cell*. New York: Viking Press, 1974. An entertaining collection of essays relating to the biology of cells and other topics.

Advanced Level:

Amos, L. A., and W. B. Amos. *Molecules of the Cytoskeleton*. New York: Guilford Publishing, 1991. A description of the proteins that make up the cytoskeleton and their effects on cell growth and division.

Balows, A., H. G. Truper, M. Dworkin, W. Harder, and K. Schleifer (eds.). *The Prokaryotes: A Handbook on the Biology of Bacteria: Ecophysiology, Isolation, Identification, Applications*, 2d ed. New York: Springer-Verlag, 1991. Descriptions of the molecular differences between the two kingdoms, the diversity of prokaryotes, and their importance in nature.

Holtzman, E. *Lysosomes*. New York: Plenum Publishing, 1989. A well-written description of the history of and current research on these fascinating vesicles.

Rothman, J. E. "The Compartmental Organization of the Golgi Apparatus," *Scientific American*, September 1985, 74–89. A description of how the Golgi apparatus modifies and transports proteins.

Sharon, N., and H. Lis. "Carbohydrates in Cell Recognition," *Scientific American*, January 1993, 82–89. A description of the glycocalyx, particularly its role in bacterial infections.

Cellular Membranes

Chapter Outline

Objectives

In this chapter, you will learn
–how a membrane forms a flexible
  boundary that coordinates activities
  of the cell with events occurring
  outside the cell;
–how a membrane controls what enters
  and leaves a cell;
–how lipids and small molecules pass
  through the framework of a
  membrane;
–how small molecules and ions pass
  through proteins within a membrane;
–how macromolecules are transported
  through a membrane.

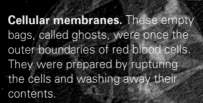

Cellular membranes. These empty
bags, called ghosts, were once the
outer boundaries of red blood cells.
They were prepared by rupturing
the cells and washing away their
contents.

E ach cell of your body is a delicate sack filled with a unique set of molecules and ions. These materials, which have been accumulated by your cells, are essential to your physical structures and biochemical processes. Natural forces work continuously to disperse these cellular contents back into the environment: there is a tendency for materials to move from areas of high concentration (you) to areas of low concentration (your environment). How, then, do you maintain your particular set of molecules and ions?

Each cell has an outer boundary with unusual properties: it holds some materials inside the cell yet at the same time allows other materials to pass into and out of the cell. It is not an impenetrable barrier, since your cells must have a regular exchange of materials with the environment. Raw materials are continually imported and wastes are exported.

Delicate sheets of molecules, called membranes, form the boundaries of cells. They hold cells together, control what kinds of materials enter and leave, and much more. A cell's outer membrane is its interface between the internal and external environment; it coordinates the activities inside a cell with changing conditions outside the cell.

Structure of the Membrane

A **cellular membrane** is a sheet of macromolecules that encloses all cells and most organelles. It is called a plasma membrane when it forms the outer boundary of a cell. All membranes, regardless of the kind of organism or cell, are built in the same way of two kinds of macromolecules: phospholipids, which provide the framework; and proteins, which provide the membrane with its specific functions. Animal membranes also contain cholesterol, which stabilizes the sheet of phospholipids.

The unique structure of a cellular membrane is often described as a fluid mosaic. It is a fluid because the phospholipids form a fluid rather than a solid. It is a mosaic because of the arrangement of diverse proteins within the sheet of phospholipids.

The Phospholipid Bilayer

The framework of a membrane is a double layer of phospholipid (see page 45) molecules. Each molecule consists of a glycerol, two fatty acids, a phosphate, and an R group (fig. 4.1a).

Phospholipids assemble themselves into a sheet of closely packed molecules when immersed in water, the fluid within which cellular materials are dissolved or suspended. All the molecules become oriented in the same way because one end of each molecule carries an electric charge and the other end does not. The phosphate "head"

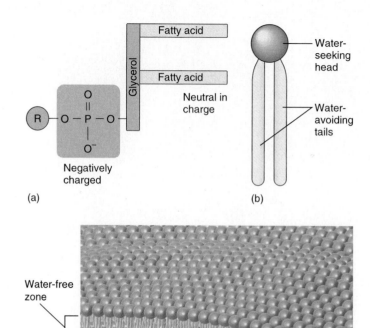

Figure 4.1 The phospholipid bilayer of a cellular membrane. (*a*) All cell membranes are formed of phospholipid molecules. The phosphate end of each molecule carries an electric charge; it dissolves in water. The fatty acid tails of each molecule are neutral in charge and do not dissolve in water. (*b*) The symbol for a phospholipid molecule. The phosphate head seeks water while the fatty acid tails avoid water. (*c*) Phospholipid molecules form a double layer of molecules, called the phospholipid bilayer, when immersed in water. The phosphate ends face outward, next to the water, and the fatty acid tails face inward, where they squeeze out water molecules to form a water-free zone.

of a fatty acid is negatively charged (because it loses a hydrogen ion in water) and readily mixes with water, whereas the fatty acid "tails" have no charge and do not mix with water. The phosphate head seeks water and the fatty acid tails avoid water (fig. 4.1b). When many phospholipids are near one another in water, they line up one after another to form two rows of molecules. All the water-seeking (hydrophilic) heads face outward, in contact with water, and all the water-avoiding tails face inward, where they squeeze out water molecules to form a water-free zone (fig. 4.1c). Adjacent molecules are held together by weak bonds. This double layer of phospholipids is the framework of all cellular membranes. The capacity of these unusual molecules to

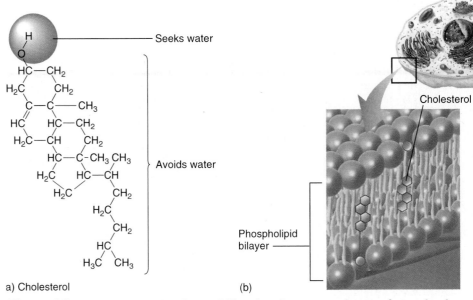

a) Cholesterol   (b)

Figure 4.2 Cholesterol molecules stabilize the plasma membrane of an animal cell. (*a*) One end of the cholesterol molecule carries an electric charge and dissolves in water. The remainder of the molecule does not dissolve in water. The carbon rings of the cholesterol make it a rigid structure. (*b*) A cholesterol molecule positions itself between phospholipids of a membrane. Their electrically charged ends form weak bonds with the surrounding water. The rest of the molecule, which avoids water, takes a position within the water-free zone occupied by the fatty acid tails of the phospholipids.

assemble themselves into membranes must have played a large role in the formation of the first cell, more than 3.5 billion years ago.

Surprisingly, the membrane is a fluid rather than a solid. (Molecules in a fluid move from place to place whereas molecules in a solid are fixed in place.) The phospholipids form an oil because their fatty acids are unsaturated (contain some double bonds). The fatty acid tails wave back and forth within the membrane, and entire molecules occasionally move sideways and exchange places with each other.

Cholesterol

The fluid structure of the phospholipid bilayer needs some kind of brace to hold its form. The plasma membranes of bacteria, plants, fungi, and many protists are bolstered by cell walls. The plasma membranes of animals are bolstered instead by cholesterol (a steroid), which form a more flexible boundary that accommodates animal movement. (It is for this reason that meat contains a lot of cholesterol.)

One end of the cholesterol molecule carries an electric charge and the other end does not (fig. 4.2*a*). Therefore, cholesterol orients itself within a membrane just like a phospholipid, with its water-avoiding end on the inside

and its water-seeking end on the outside of the phospholipid bilayer (fig. 4.2*b*).

Cholesterol is a fairly rigid molecule—its four carbon rings are fused and not free to rotate. When inserted between two phospholipids, cholesterol holds the fatty acids in place. With less movement of the fatty acids, the membrane becomes less fluid. Moreover, the fatty acids are more tightly packed when their movement is restricted by cholesterol. Small gaps, which normally appear and disappear between the moving fatty acids, develop less often and the membrane becomes less permeable to small molecules.

Proteins

Proteins are embedded within the phospholipid bilayer of a membrane. Unlike the phospholipids and cholesterol that form the framework, the proteins of a membrane are not arranged in a regular pattern: the numbers, types, and positions of proteins vary from membrane to membrane and from time to time within the same membrane. The proteins give a membrane its specific functions.

The plasma membrane of a cell has many kinds of proteins (a red blood cell, for example, has more than fifty kinds). Some, the glycoproteins, have attached sugars and others do not; some extend all the way through the membrane and others do not (fig. 4.3).

Membrane proteins perform crucial tasks: they transport certain materials into and out of a cell or organelle; identify materials (e.g., bacteria) outside the cell; transmit information about events outside the cell to appropriate centers of activity inside the cell; recognize other cells; anchor the cytoskeleton; and hold the enzymes of biochemical pathways (fig. 4.4). These vital components of membranes are described in later chapters as their functions become relevant. In this chapter, we discuss the membrane proteins that help transport materials into and out of a cell or organelle.

Three kinds of proteins are important in membrane transport: channel proteins, which form passageways through a membrane; transport proteins, which convey materials from one side of a membrane to the other; and receptor proteins, which latch onto certain materials and help carry them into or out of a cell. These proteins and modes of transport are described in the following sections.

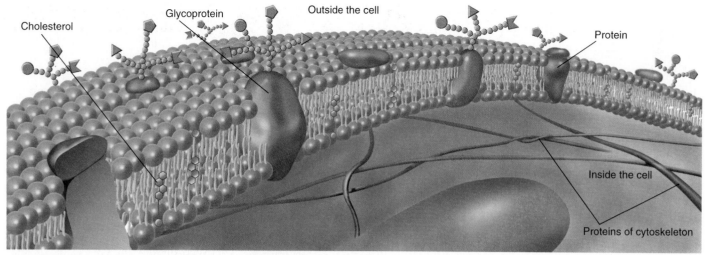

Figure 4.3 The proteins of a cellular membrane. A variety of proteins, some with attached sugars, are embedded within the phospholipid bilayer of a membrane. They give each membrane its specific properties.

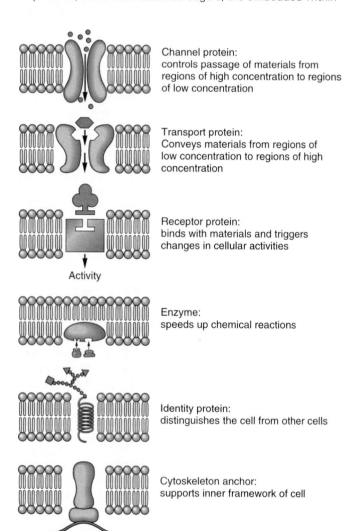

Channel protein:
controls passage of materials from regions of high concentration to regions of low concentration

Transport protein:
Conveys materials from regions of low concentration to regions of high concentration

Receptor protein:
binds with materials and triggers changes in cellular activities

Activity

Enzyme:
speeds up chemical reactions

Identity protein:
distinguishes the cell from other cells

Cytoskeleton anchor:
supports inner framework of cell

Figure 4.4 Membrane proteins and their functions.

Transport Across Membranes

A cellular membrane is a selectively permeable barrier: some materials cross freely and others cross only at certain times. Passage is controlled by the phospholipid bilayer itself and by the particular array of proteins embedded in the bilayer.

The phospholipid framework of a membrane is impermeable to most materials. Only materials that are very small or carry no electric charge can penetrate this barrier. Very small molecules, such as water and oxygen, slip through temporary gaps between adjacent phospholipids. Molecules without electric charge dissolve in lipids; they diffuse right through the fatty acid core of the membrane. Steroid (see page 45) hormones and the fat-soluble vitamins (A, D, E, and K) enter and leave cells in this way.

Larger molecules and ions cannot penetrate the phospholipid framework of a membrane. They cross only when assisted by proteins. These materials include enzymes, protein hormones, inorganic ions (e.g., Na$^+$ and K$^+$), and organic ions (e.g., amino acids and glucose, which lose hydrogen ions and become negatively charged in water).

Cells with different kinds of proteins in their plasma membranes accumulate different kinds of materials. Liver cells, for example, accumulate glucose for storing energy, and muscle cells accumulate amino acids for building muscle proteins. Moreover, each cell changes the proteins in its membranes as its activities demand the import and export of different materials.

Table 4.1

Membrane Transport

Kind of Transport	Kinds of Materials	Direction	Requirements
Simple diffusion	Steroids, fats, O_2, CO_2, H_2O	High to low concentrations	None
Facilitated diffusion	Glucose, ions	High to low concentrations	Channel protein, receptor protein
Active transport	Amino acids, ions	Low to high concentrations	Transport protein, energy
Bulk transport	Nonsteroid hormones, enzymes, mucus, microbes, neurotransmitters, wastes	Into or out of cell as needed	Energy, sometimes receptor proteins

Materials cross cellular membranes in four ways: simple diffusion, facilitated diffusion, active transport, and bulk transport. These modes of transport are summarized in table 4.1 and discussed in the following sections.

Simple Diffusion

Diffusion is the tendency of materials (molecules and ions) that form a fluid or gas to move down a concentration gradient: from regions of higher concentration to regions of lower concentration. Perfume molecules escaping from an open bottle, for example, move away from one another. The molecules continue to diffuse until they are evenly dispersed throughout the room.

Materials in watery solutions are in continuous motion, a form of kinetic energy known as heat. As they move randomly about, they collide and bounce away in different directions:

◯ Molecule or ion

◯ One particular molecule or ion

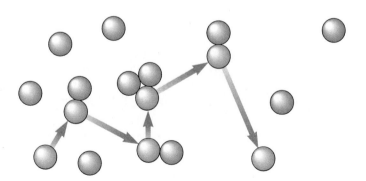

Molecules or ions that are closer together experience many more collisions, and are more likely to ricochet out of the region, than molecules or ions that are far apart. The higher frequency of collisions in regions of higher concentration moves these materials into regions of lower concentration:

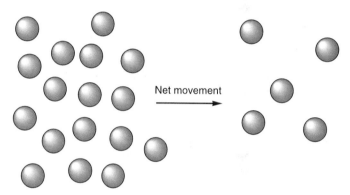

Net movement

Some molecules pass directly through the phospholipid bilayer of a membrane by simple diffusion; they move from the side of the membrane where they are more concentrated to the side where they are less concentrated. Such diffusion takes place spontaneously as long as the material can pass freely through the phospholipid bilayer. Molecules that move in this way are those without electric charges, such as lipids, and very small molecules, such as carbon dioxide and water.

Transport across a membrane by simple diffusion requires neither membrane proteins nor an expenditure of energy on the part of the cell. It is a slow process that can be speeded up by raising the temperature (more rapid movements and so more collisions) and by increasing the concentration gradient (fig. 4.5).

Table 4.2

Terms Used to Describe Osmosis

Term	Concentration of Dissolved Materials	Concentration of Water	Net Direction of Water Flow
Isotonic environment	Same inside and outside the cell	Same inside and outside the cell	None
Hypertonic environment	Higher outside the cell	Lower outside the cell	Out of cell
Hypotonic environment	Lower outside the cell	Higher outside the cell	Into cell

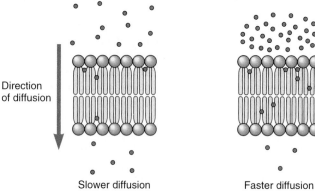

Direction of diffusion

Slower diffusion   Faster diffusion

Figure 4.5 Simple diffusion. Lipids and very small molecules diffuse through the phospholipid bilayer, moving from regions of higher concentration to regions of lower concentration. The greater the difference in concentration, on the two sides of the membrane, the more rapidly diffusion proceeds.

Osmosis   The diffusion of water through a cellular membrane has a special name, **osmosis.** Water molecules are small enough to diffuse through gaps in the phospholipid framework, whereas most molecules and ions are not. Water diffuses from the side of the membrane where it is more concentrated (i.e., fewer dissolved materials) to the side where it is less concentrated (i.e., more dissolved materials). The direction of osmosis is extremely important to a cell.

Whether water enters or leaves a cell depends on the relative concentrations of dissolved materials inside and outside the cell. Three terms describe these relative concentrations: isotonic, hypertonic, and hypotonic. (They are somewhat confusing terms because they indicate the direction that water diffuses yet refer to concentrations of dissolved materials instead of concentrations of water.) These concentrations and their effects are defined in table 4.2 and described following.

A cell immersed in a solution with the same concentration of dissolved materials as its own cytosol (inner fluid) is said to be in an **isotonic** (*iso*, same; *tonicity*, concentration of dissolved materials) environment. There is

no net flow of water into or out of the cell. Most animal cells are in isotonic environments; the blood and kidneys ensure that the fluid around the cells has the same concentration of water as the fluid inside the cells.

A cell immersed in a very salty solution is in a **hypertonic** (*hyper*, higher) environment. The concentration of dissolved materials is higher outside than inside the cell (i.e., the concentration of water is lower outside than inside). Under these conditions, water diffuses out of the cell, from a region of higher water concentration to a region of lower water concentration:

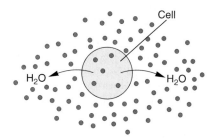

Cell

H_2O H_2O

As a cell loses water by osmosis, it shrivels and dies.

Cells are rarely in hypertonic environments. Marine organisms survive in relatively salty water without shriveling by maintaining internal fluids with the same water concentration as seawater. They have higher concentrations of ions and molecules in their cells than other organisms and are isotonic to their salty environment.

A cell immersed in a solution with a lower concentration of dissolved materials is in a **hypotonic** (*hypo*, lower) environment: the concentration of water is higher (because of so few dissolved materials) outside the cell than inside. Under these conditions, water diffuses into the cell:

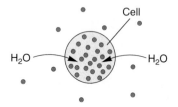

Cell

H_2O H_2O

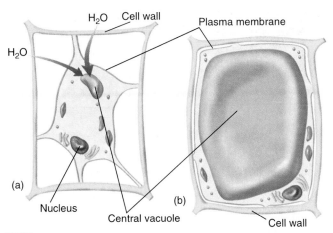

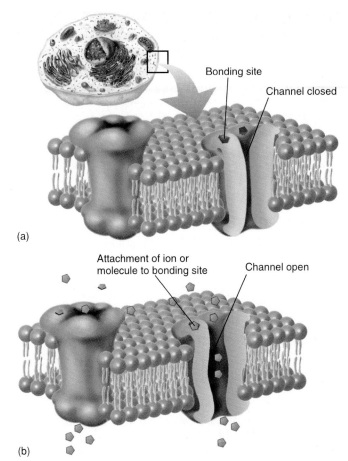

◼️◻️ Figure 4.6 Turgor pressure. The inside of a plant cell has a higher concentration of dissolved materials than fresh water. (*a*) Fresh water moves, by osmosis, into a plant cell and is stored in a central vacuole. (*b*) The cell swells with water but its expansion is limited by the surrounding wall. Instead of bursting it becomes firm, a phenomenon known as turgor pressure.

A cell in a hypotonic environment swells with water and may burst.

Organisms that inhabit soil, lakes, and streams live in modified rainwater, which is a hypotonic environment. Most organisms in these environments have cell walls that restrict swelling and so protect the cell from bursting. Other organisms, without cell walls, survive by eliminating excess water. Many protists have special organelles, called contractile vacuoles, that accumulate water and pump it out of their cells. Freshwater animals have kidneys that remove excess water (as well as waste materials) from their bodies.

Turgor Pressure   A characteristic of plant cells is **turgor pressure,** which is the outward pressure of a swollen cell against its cell wall (fig. 4.6). It develops because plant cells are usually in hypotonic environments. Water enters the cell by osmosis and is stored in a central vacuole (📖 *see page 65*), where it cannot dilute the cytosol. Water continues to enter until the expanding cell meets resistance from the rigid cell wall. At this point the outward pressure from the incoming water matches the inward pressure from the cell wall. An equilibrium is reached in which the cell still has a lower water concentration than its environment but can receive no more water. The cell becomes firm, or turgid, with water.

Turgor pressure enables a plant to hold its stem upright and to keep its leaves and flowers fully expanded. Without such controlled swelling, the plant wilts and dies.

Facilitated Diffusion

In **facilitated diffusion,** large molecules and ions diffuse through channels within membrane proteins. The process requires no input of energy on the part of the cell, since

Figure 4.7 Facilitated diffusion. Some molecules and ions diffuse through pores in channel proteins. (*a*) A channel protein is closed (has no opening) unless stimulated. (*b*) The channel protein opens for diffusion in response to attachment by a particular molecule or ion to a receptor protein. Sometimes the receptor protein and channel protein are the same molecule, as shown here, and sometimes they are separate molecules in the membrane.

materials move down a concentration gradient. Materials that cross membranes in this way include glucose, K^+, and Na^+. (Because glucose combines immediately with other molecules in the cell, its concentration there remains relatively low.) Facilitated diffusion occurs more rapidly when the temperature is higher, when the concentration gradient is steeper, and when more of its special proteins are in the membrane.

Facilitated diffusion requires two proteins: a channel protein and a receptor protein, which may be part of the same molecule or separate proteins in the membrane (fig. 4.7). The channel protein controls passage of the material by changing its shape, opening to form a channel and closing to form a barrier. The receptor protein forms a temporary bond with the material to be transported (or with a material that signals a need for diffusion). Formation of this bond alters the channel protein, causing it to form an opening through which the specific material can diffuse.

Facilitated diffusion has two advantages over simple diffusion. First, it permits the passage of materials that cannot penetrate the phospholipid bilayer. Second, it is controlled: a cell synthesizes only certain membrane proteins, which allow only certain types of material to pass. Facilitated diffusion makes the membrane selectively permeable.

Researchers have discovered recently that cystic fibrosis, an inherited disorder, is caused by a faulty channel in the plasma membrane. The role of facilitated transport in this disease is described in the accompanying Wellness Report.

Active Transport

The movement of materials through a membrane from regions of low concentration to regions of high concentration (i.e., up a concentration gradient) is called **active transport.** It is the only way a cell or organelle can accumulate high concentrations of materials, such as amino acids and potassium ions. Active transport requires a membrane protein, called the transport protein. It also requires an expenditure of energy by the cell, since materials are moved in a direction that is opposite to the direction they would diffuse spontaneously. Much of the energy expended by a cell is for active transport.

Transport proteins, embedded in the phospholipid framework of a membrane, convey materials from one side of the membrane to the other. Each transport protein is specific to a particular type of material. Exactly how these proteins transport materials is not fully understood. One hypothesis suggests that changes in chemically active sites within the protein cause transport (fig. 4.8). One side of the protein, on one side of the membrane, apparently forms a bond with its material. Formation of this bond alters the protein's shape, forming a channel through which the transported material enters. The first bond then breaks and another forms within the protein at its opposite end, on the other side of the membrane. The new bond pulls the transported material through the temporary channel in the transport protein.

Coupled Systems   Some transport proteins carry more than one material at the same time. Both materials may cross the membrane in the same direction or one may be transported into the cell while another is transported out.

The membrane protein that brings amino acids into a cell also forms a channel for facilitated diffusion of potassium ions out of a cell. The advantage of this coupled transport is that energy gained from the diffusion of one is used to actively transport the other. Potassium ions move down a concentration gradient as they diffuse out of a cell, releasing energy in the process. This same energy then drives the active transport of amino acids into the cell.

The plasma membranes of animal cells have sodium-potassium pumps, which are transport proteins that

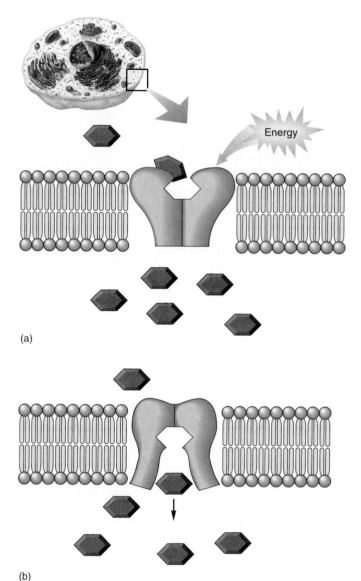

(a)

(b)

Figure 4.8 Active transport. Some molecules and ions are moved by transport proteins across the membrane, from a region of low concentration to a region of high concentration. (*a*) The material to be transported binds with its transport protein, causing the protein to form a cavity. The material enters the cavity. (*b*) The carrier protein moves the material (probably by breaking the first bond and forming a second bond on its other side) across the membrane to the other side, where the material is present in higher concentration. The transport requires an input of energy.

simultaneously bring potassium ions into the cell and remove sodium ions from the cell (fig. 4.9). They maintain a higher concentration of potassium ions inside the cell than outside and a higher concentration of sodium ions outside the cell than inside. This particular distribution of ions stores energy, both in the form of concentration gradients and electrical potentials (described following). The advantage of storing energy in this way is that it can be released

Ion Channels and Cystic Fibrosis

One in every 2,500 babies born in the United States has cystic fibrosis. Approximately one in every twenty-five white Americans carries a copy of the gene for cystic fibrosis; the disorder develops when a child inherits two copies of the gene, one from each parent. Children with cystic fibrosis suffer from a variety of symptoms: incessant coughing and gagging, respiratory infections, poor digestion, and intestinal obstruction. They rarely live beyond the age of thirty. (Chopin, the brilliant composer and pianist, died at age thirty-nine from the disorder.)

In 1989 the gene responsible for cystic fibrosis was discovered, and within a few years its protein was described. The normal version of this gene carries instructions for building a membrane protein, found in the plasma membranes of epithelial cells—the closely packed cells that line body surfaces. It forms a channel that facilitates the diffusion of chloride ions out of the cell.

The mutated version of this gene carries instructions for a protein that does not fold properly and so cannot carry out its normal functions. This faulty protein, named *cystic fibrosis transmembrane conductance regulator*, or CFTR, cannot open to form a channel for the passage of chloride ions. Consequently, chloride ions accumulate inside the cells and draw in water by osmosis. The surrounding mucus, a sticky fluid that covers most internal surfaces, dehydrates and becomes unusually stiff and sticky.

In someone with cystic fibrosis, thick mucus accumulates on internal surfaces: air passages, lungs, digestive tract, small tubes, and internal organs (fig. 4.A). Health problems develop most commonly from obstructed air passages, mucus-coated lungs, plugged tubes that carry digestive materials from the pancreas to the small intestine, and mucus-covered surfaces within the diges-tive tract. Men often become infertile as thick mucus blocks the tubes that transport sperm out of the body. Disease microbes thrive in the thick mucus of the lungs—95% of all cystic fibrosis patients die from an overwhelming lung infection.

Discovery of the gene and then its protein were major breakthroughs in biomedical research. Geneticists have now isolated the normal gene for chloride channels and are studying various methods of gene therapy—of transferring the normal gene into defective cells, where it can direct the synthesis of its normal protein. Initial results are promising. Genetic engineers have spliced the healthy human gene onto the genes of cold viruses (which have been crippled by the researchers so they can no longer cause their disease). Cold viruses were chosen because they have special adaptations for entering human lung cells, which are their normal hosts. Early experiments on volunteers, who suffer from cystic fibrosis, indicate that, as hoped, the genetically engineered viruses carry copies of the human gene into human cells. Moreover, the transplanted gene directs synthesis of the normal channel protein. Medical researchers believe they will develop an effective treatment for cystic fibrosis within the next decade.

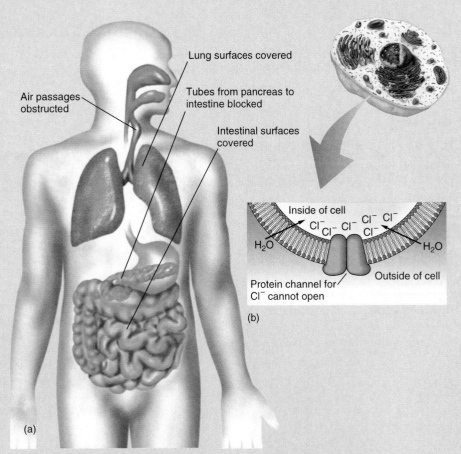

Figure 4.A Cystic fibrosis. (*a*) This genetic disorder causes the mucus on internal surfaces to be less fluid than normal. The thick mucus coats the lungs and clogs tubes in various parts of the body. (*b*) The cause of the disorder is a faulty protein in the plasma membranes of cells that line internal surfaces. The normal protein opens to allow diffusion of chloride ions out of the cell. In cystic fibrosis, the faulty protein remains closed and chloride ions accumulate inside the cell. They draw water into the cell and dehydrate the surrounding mucus.

immediately: when the membrane becomes permeable to these ions, they move down their gradients and release energy. (Plant cells, mitochondria, and chloroplasts use proton pumps, involving hydrogen ions, to store energy in this way.)

Membrane Potentials   The inside of an animal cell is more negatively charged than the outside. This difference in electric charge is maintained by transport proteins within the plasma membrane, which keep more negatively charged ions (mainly proteins and phosphate groups that have lost hydrogen ions) inside the cell than outside and more positively charged ions (mainly sodium) outside the cell than inside. If free to move, negatively charged ions would diffuse out of the cell and positively charged ions would diffuse into the cell:

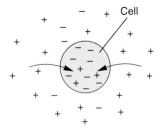

Such a flow of ions, caused by both diffusion and attractions between opposite charges, would be electric energy—a form of kinetic energy. Because cellular membranes are normally impermeable to ions, however, the flow does not take place and the energy remains potential energy.

Potential energy caused by positive charges in one place and negative charges in another is measured in volts. (A flashlight battery, for example, usually carries 1.5 volts.) The magnitude of potential energy across a plasma membrane is between 20 and 200 millivolts (1 millivolt = 0.001 volt). This electrical property of a cell is its **membrane potential.**

The membrane potential can be neutralized or reversed only if the plasma membrane changes its permeability, which it does by activating channel proteins or transport proteins. Such controlled movement of ions across plasma membranes is vital to many cellular functions. Electric signals in nerve cells, for example, are generated by allowing sodium and potassium ions to cross plasma

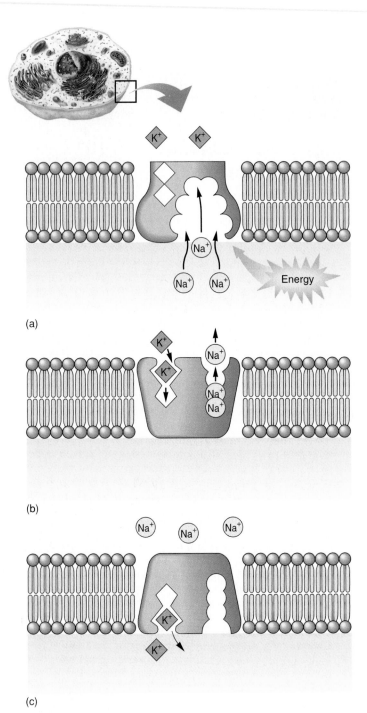

Figure 4.9 A coupled system of transport. Sodium ions and potassium ions are actively transported together. (*a*) The carrier protein binds with three sodium ions inside the cell. (*b*) Next, the protein deposits the sodium ions outside the cell while at the same time it picks up two potassium ions from outside the cell. (*c*) Last, the carrier protein deposits the potassium ions inside the cell.

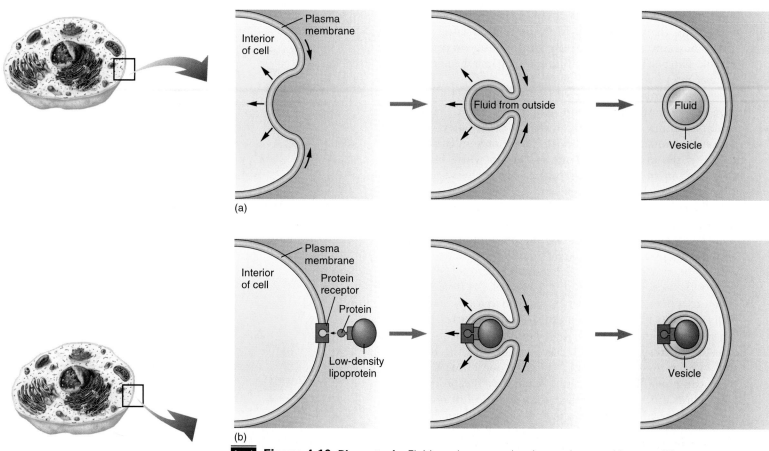

Figure 4.10 Pinocytosis. Fluids and macromolecules are imported into a cell by pinocytosis. (*a*) When only fluid is imported, the plasma membrane simply envelopes a tiny droplet of extracellular fluid, indents, and forms a fluid-filled vesicle inside the cell. (*b*) When pinocytosis involves a receptor protein, the process is known as receptor-mediated pinocytosis. Here, a protein on a low-density lipoprotein (which contains cholesterol) binds with its receptor protein in the plasma membrane of a liver cell. This binding stimulates endocytosis of the low-density lipoprotein.

membranes. Similarly, chemical energy is transferred during cellular respiration and photosynthesis by controlled movements of hydrogen ions across membranes of the mitochondria and chloroplasts.

Bulk Transport

Macromolecules and fluids cross cellular membranes by **bulk transport,** in which the transported materials are contained within vesicles and do not mix with other materials of the cytosol. There are two forms of bulk transport: **endocytosis** (*endo*, into; *cyt*, cell; *osis*, process), in which materials move into a cell, and **exocytosis** (*exo*, out of), in which materials move out of the cell. Both forms require an expenditure of energy on the part of the cell.

Endocytosis   There are two forms of endocytosis, depending on the kind of material brought into the cell. Endocytosis of a fluid is known as **pinocytosis** (*pino*, drinking); endocytosis of a solid is known as **phagocytosis** (*phago*, eating).

When only fluid is brought into the cell, endocytosis does not require a membrane protein (fig. 4.10*a*). The plasma membrane simply indents to form a pocket around some of the extracellular (*extra*, outside) fluid. The pocket then pinches off from the rest of the plasma membrane, forming a vesicle inside the cell. This vesicle contains a tiny droplet of the extracellular fluid.

When pinocytosis involves macromolecules as well as fluid, the process requires a membrane protein and is called receptor-mediated endocytosis. It begins when a macromolecule in the extracellular fluid attaches to a receptor protein in the plasma membrane. The plasma membrane then indents to form a pocket around the macromolecule and some of the surrounding fluid (fig. 4.10b). The pocket pinches off from the plasma membrane and becomes a vesicle inside the cell.

Some viruses use receptor-mediated endocytosis to invade their host cells. The virus responsible for AIDS, for example, enters human cells in this way. Its outer coat contains a protein that matches a receptor protein (named CD4) in the plasma membrane of certain white blood cells. When these two proteins combine, the virus is brought into the cell by endocytosis.

Perhaps the best-studied example of receptor-mediated endocytosis involves the transport of excess cholesterol into liver cells for destruction. The blood carries excess cholesterol in small packets, called low-density lipoproteins (LDLs), formed of cholesterol, phospholipids, and proteins. Transfer of an LDL from the blood to a liver cell begins when a specific protein in the LDL attaches to a specific protein receptor within the plasma membrane (see fig. 4.10b). The attachment stimulates endocytosis and the LDL is brought into the liver cell. The rate at which liver cells remove cholesterol from the blood depends on the number of receptors in their plasma membranes. When there are too few receptors, cholesterol accumulates in the blood and tends to clog the arteries. People with no receptors at all, due to a genetic disorder, are likely to have heart attacks while teenagers or young adults.

Phagocytosis works in a manner similar to pinocytosis of fluids. The presence of a large cluster of macromolecules stimulates the plasma membrane to engulf the materials, indent, and form a vesicle inside the cell. The vesicle typically fuses with a lysosome, where the molecules are digested and their components used by the cell. Many unicellular organisms feed this way, importing food within vesicles (called food vacuoles) that fuse with lysosomes containing digestive enzymes. Likewise, our own white blood cells eliminate disease microbes by phagocytosis (fig. 4.11).

Exocytosis   Exocytosis is the reverse of endocytosis. Transport vesicles enclosing materials to be exported are generated from internal membranes, most often from the endoplasmic reticulum or Golgi apparatus. The vesicles then travel to the plasma membrane, fuse with it, and release their cargo outside the cell (fig. 4.12).

Unicellular organisms use exocytosis to export waste materials, such as undigested portions of food. Animals use it to secrete special materials. (**Secretion** is the synthesis and controlled release of an organic molecule by a cell.) Digestive enzymes are secreted into the stomach and small intestine. Hormones are secreted into the bloodstream. Neurotransmitters are secreted into gaps between nerve cells.

The Membrane Cycle   Membranes are in a continual state of flux as fragments are added and removed during

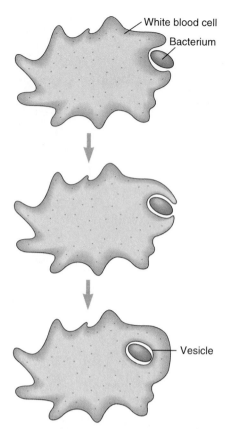

Figure 4.11 Phagocytosis by a white blood cell. A white blood cell, part of our immune system, eliminates harmful bacteria by phagocytosis. It surrounds the bacterium with extensions of its cytoplasm and then engulfs it within folds of its plasma membrane. The plasma membrane pinches off to form a vesicle, containing the bacterium, inside the white blood cell. The vesicle will fuse with a lysosome, where enzymes will digest the bacterium.

bulk transport. In endocytosis, fragments from the plasma membrane become incorporated into vesicles and then the membranes of other organelles. In exocytosis, fragments from the membranes of organelles become incorporated into the plasma membrane. Some animal cells recycle their entire plasma membrane every hour. For some reason, as yet unknown, this continual removal and addition of fragments does not interfere with a membrane's specific functions.

SUMMARY

1. A cell membrane is a sheet of macromolecules that forms a selectively permeable boundary to a cell or organelle. The framework is a phospholipid bilayer, which, in animals, is fortified with cholesterol to make it less fluid. Proteins interspersed within the phospholipid bilayer give a membrane its specific functions. Materials move through a membrane by simple diffusion, facilitated diffusion, active transport, and bulk transport.

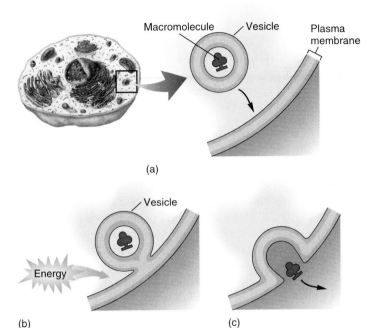

(a)

Vesicle

Energy

(b)   (c)

▶ Figure 4.12 Exocytosis. Macromolecules are exported from a cell by exocytosis. (*a*) A macromolecule to be exported is packaged within a vesicle, which travels to the inner surface of the plasma membrane. (*b*) The vesicle fuses with the plasma membrane. (*c*) The plasma membrane ruptures, releasing the macromolecule outside the cell.

2. In simple diffusion, lipids and very small molecules pass directly through the phospholipid bilayer, from regions of higher concentration to regions of lower concentration. It requires no expenditure of energy. The diffusion of water through a membrane is called osmosis. In plants, osmosis of fresh water into the cells causes turgor pressure.

3. In facilitated diffusion, ions and small molecules pass through channel proteins in the membrane. It requires no expenditure of energy. Each protein permits only certain materials to pass.

4. In active transport, transport proteins carry ions and small molecules across the membrane, from regions of lower concentration to regions of higher concentration. Each protein carries only certain materials. Active transport requires an expenditure of energy. Some transport systems are coupled, with each protein carrying more than one material at the same time. Active transport of ions generates the membrane potential.

5. In bulk transport, macromolecules and fluids cross a membrane while enclosed within a vesicle. It requires an expenditure of energy. Endocytosis is bulk transport of materials into a cell. It is called pinocytosis if the materials are fluids, receptor-mediated endocytosis if molecules in the fluids attach to receptors in the membrane, and phagocytosis if they are solids. Exocytosis is

bulk transport of materials out of a cell. Bulk transport cycles membrane fragments between the plasma membrane and the membranes of organelles.

KEY TERMS

Active transport (page 78)
Bulk transport (page 81)
Cellular membrane (page 72)
Diffusion (dih-FYOO-shun) (page 75)
Endocytosis (EN-doh-seye-TOH-sis) (page 81)
Exocytosis (EX-oh-seye-TOH-sis) (page 81)
Facilitated diffusion (page 77)
Hypertonic (HEYE-pur-TAH-nik) (page 76)
Hypotonic (HEYE-poh-TAH-nik) (page 76)
Isotonic (EYE-soh-TAH-nik) (page 76)
Membrane potential (page 80)
Osmosis (ahz-MOH-sis) (page 76)
Phagocytosis (FAY-goh-seye-TOH-sis) (page 81)
Pinocytosis (PEE-noh-seye-TOH-sis) (page 81)
Secretion (seh-KREE-shun) (page 82)
Turgor (TUR-gur) **pressure** (page 77)

STUDY QUESTIONS

1. Draw the phospholipid framework of a cell membrane. Indicate the positions of water-seeking and water-avoiding regions of the molecules. Why do phospholipids form a double membrane when placed in water? What makes a membrane a fluid? What makes it a barrier?

2. What molecules in membranes give them their specific functions? Describe the activity of one of these molecules.

3. Explain why a cell swells or shrinks when placed in rainwater or seawater.

4. Describe turgor pressure. What kinds of cells exhibit this property? What is its function?

5. Describe three ways in which membrane proteins control what enters and leaves a cell.

6. Describe what happens to portions of the plasma membrane during endocytosis and exocytosis.

CRITICAL THINKING PROBLEMS

1. Describe how the peculiar properties of phospholipids, with one end neutral and the other carrying an electric charge, probably contributed to formation of the first cells on earth.

2. Before refrigeration, people salted their meat for storage. Why does salting meat protect against the bacteria that cause food poisoning?

3. Why does a house plant wilt when it does not receive enough water?

4. Why don't the fat-soluble vitamins (A, D, E, and K) require membrane proteins to enter an animal cell?

5. Digestive enzymes are secreted into the small intestine by cells that line this part of the digestive tract. These proteins are synthesized on ribosomes in the endoplasmic reticulum, modified by the Golgi apparatus, and exported by exocytosis. Trace what happens, during this process, to membranes of the endoplasmic reticulum that form the transport vesicles for digestive enzymes.

SUGGESTED READINGS

Introductory Level:

Bretscher, M. S. "The Molecules of the Cell Membrane," *Scientific American*, October 1985, 100–108. A description of how the phospholipids, cholesterol, and proteins are arranged within a membrane and of membrane dynamics during bulk transport.

Dautry-Varsat, A., and H. F. Lodish. "How Receptors Bring Proteins and Particles into Cells," *Scientific American*, May 1984, 52–58. Discussion of the role of receptor proteins in endocytosis, using, as one example, the LDL receptors in plasma membranes.

Advanced Level:

Lauffenburger, D. A., and J. J. Linderman. *Receptors: Models for Binding, Trafficking, and Signaling.* New York: Oxford University Press, 1993. A detailed examination of how materials outside the cell bind with receptors in the plasma membrane, how receptors help transport materials into and out of a cell, and how receptors pass information into a cell.

————. *Science* 258 (1992):917–69. A series of review articles dealing with various aspects of cellular membranes.

Singer, S. J., and G. Nicolson. "The Fluid-Mosaic Model of the Structure of Cell Membranes," *Science* 175 (1972):720–31. The classical paper that first described the current model of membrane structure.

Yeagle, P. *The Membranes of Cells.* Orlando: Academic Press, 1987. A good reference for details about the structure and function of cellular membranes.

✪ EXPLORATION

Interactive Software: Cystic Fibrosis

This interactive exercise explores the relation between membrane transport and concentrations of materials inside and outside a cell. The student studies the effects of ion concentrations on osmosis by seeing what happens when the transport of chloride ions is blocked, due to a faulty channel protein, as occurs in the condition known as cystic fibrosis.

Cellular respiration provides energy for all your physical and mental activities.

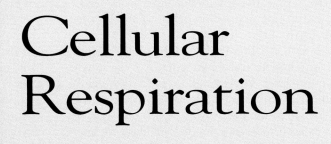

Cellular Respiration

Chapter Outline

Objectives

In this chapter, you will learn
–how a cell releases chemical energy from molecules;
–how a cell transports the energy to wherever it is needed;
–how a cell releases chemical energy when oxygen is present and when it is not;
–how a cell uses a variety of molecules as sources of energy.

Your body requires a continuous supply of energy to carry out life's activities—walking, thinking, digesting, and even sleeping. This energy comes from molecules in your food, which are broken into their smaller components within the digestive tract and transported, via the blood, to your body cells. There, chemical energy in the food molecules is released by a series of chemical reactions known as cellular respiration. Releasing chemical energy from food molecules is analogous to burning gasoline in a car engine.

Your body and a car engine burn similar fuels. Cells burn glucose, fatty acids, and amino acids, which are the digested components of carbohydrates, fats, and proteins in food. A car burns remnants of glucose, fatty acids, and amino acids, which were once components of carbohydrates, fats, and proteins in plants and algae. These remnants of molecules, which now appear as gasoline and other fossil fuels, were trapped within fossils that formed millions of years ago. Body cells and car engines combine these molecules with oxygen to form carbon dioxide, water, and energy. Only a portion of this energy can be captured and used to do work; the rest inevitably escapes as heat. Both your body and a car engine become overheated when worked too hard.

An important difference between your body and a car engine is the way in which energy is released from fuel. Your cells use enzymes to gradually release energy in a series of small steps. A car engine burns its fuels in one big step. By releasing energy gradually, your cells capture a much larger portion of the energy in fuels than a car engine. (If your body ran on gasoline, you would get about 900 miles to the gallon.) On the other hand, your body "wastes" fuel because it is kept idling. If your body were turned off, like a car engine, the cells would die and so would you. All cells require a continuous input of energy just to remain alive.

Energy and Life

Energy is the force that makes it possible to do **work**—to bring about movement against an opposing force. Work is done when an electron is moved away from an atomic nucleus, when a bicyclist climbs uphill, and when a ball is thrown into the air.

When you jump into the air, for example, your body moves against the force of gravity. This movement is accomplished by contractions of your leg muscles, a form of work that requires energy. The energy that moves your muscles is released from chemical reactions within your muscle cells—reactions that alter the positions of electrons in molecules.

Almost everything that a cell does requires energy. Cells work just to maintain themselves: they expend energy to maintain their structures and to retain their accumulation of ions and molecules. Without a continuous input of energy, a cell quickly disintegrates and its components disperse into the surrounding environment.

Chemical Energy

A cell uses chemical energy (see page 34) to do work. This form of potential energy is stored within the positions of electrons in atoms and molecules. Recall from chapter 1 that when an electron is farther from the nucleus, it holds more energy than an electron closer to the nucleus. Consequently, moving an electron farther from its nucleus is a form of work and requires energy, and moving an electron closer to its nucleus releases energy that can be used to do work. Electrons move closer to or farther from their nuclei when chemical bonds are altered and when electrons are transferred.

Changes in Chemical Bonds   Chemical reactions involve the formation and breakage of chemical bonds. When a bond forms, electrons move farther from their nuclei; formation of a bond requires an input of energy. When a bond breaks, electrons move closer to their nuclei; breakage of a bond releases energy.

A larger molecule has more chemical bonds than a smaller one. When a large molecule is synthesized from smaller ones, more bonds are formed than are broken. The synthesis requires an input of energy:

Smaller molecules + Energy ⟶ Large molecule

The energy put into the reaction becomes chemical energy within chemical bonds of the large molecule.

Conversely, whenever a large molecule is broken down into smaller ones, more bonds are broken than formed. The breakdown releases energy:

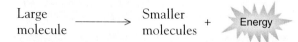

Large molecule ⟶ Smaller molecules + Energy

Some of the energy released from the reaction becomes heat and the rest can be used to do work.

Transfers of Electrons   Sometimes electrons move completely out of one atom and into another. Such electron transfers are known as **oxidation-reduction reactions.** An electron travels from one atom, where it is farther from the nucleus, to another atom, where it can lie nearer the atom's nucleus. Some of the energy carried by the electron is released during the reaction. The atom that loses the electron is said to be oxidized and the atom that acquires the electron is said to be reduced:

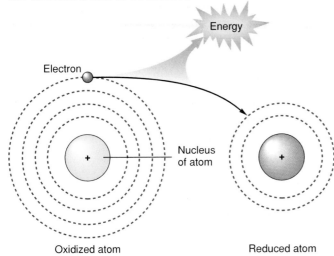

Oxidations and reductions always occur together. An electron cannot leave one atom without being accepted by another atom—it must have someplace to go to.

An oxidation-reduction reaction may involve more than one electron. Sometimes two or more electrons are transferred at the same time, and sometimes an entire hydrogen atom (both the electron and the proton) is transferred from one molecule to another.

Adenosine Triphosphate

Energy within a cell is transported from the place where it is released to the place where it is needed in a special molecule, **adenosine triphosphate** (abbreviated ATP). This organic molecule, present in all forms of life, is built from three smaller molecules: adenine (the same nitrogenous base found in DNA and RNA), ribose (the same sugar as in RNA), and three phosphate ions (fig. 5.1a). Its name reflects its structure: adenosine is the term for a molecule formed of adenine and ribose, and triphosphate refers to the three phosphates attached to adenosine.

Why is chemical energy transferred to ATP for transport to the work site? Because it is more efficient to transport energy in smaller packets. In cellular respiration (the main topic of this chapter), the chemical energy within a single molecule of glucose is transferred to thirty-eight molecules of ATP prior to transport to other parts of the cell. These smaller packets of energy can be distributed throughout the cell more efficiently than if they remained in a single molecule of glucose (just as a fleet of smaller trucks can deliver goods to a variety of destinations more efficiently than a single larger truck).

The energy-transporting portion of ATP is a bond between the terminal and middle phosphates. This bond

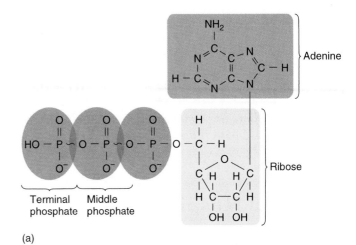

(a)

(b)

Figure 5.1 The structures of adenosine triphosphate (ATP) and adenosine diphosphate (ADP). (a) ATP is built from a molecule of adenine, a molecule of ribose, and three phosphate ions. The wavy lines between phosphates indicate bonds that hold relatively large amounts of energy. (b) Energy is released to do work when the terminal phosphate is removed. The remaining molecule then has only two phosphates and is ADP. Energy is stored when ADP is converted back into ATP.

holds a large amount of energy yet is easily broken. When the bond is broken, and the terminal phosphate removed, the remaining molecule is **adenosine diphosphate** or ADP (*di* refers to the two phosphates left on the molecule). This molecule is illustrated in figure 5.1b.

Energy carried within ATP is released to do biological work when adenosine triphosphate is broken apart by hydrolysis (see page 42) to form adenosine diphosphate and a phosphate ion:

$$ATP + H_2O \longrightarrow ADP + Phosphate\ ion + Energy$$

In the opposite reaction, energy released from glucose and other monomers is used to build ATP by dehydration synthesis (📖 *see page 41*) from ADP and a phosphate ion:

$$ADP + \text{Phosphate ion} + \text{Energy} \longrightarrow ATP + H_2O$$

A cell always contains large quantities of ATP, ADP, and phosphate ions. These three components of the energy-transfer system are used over and over again.

Cellular Respiration

Cellular respiration is the biochemical process in which chemical energy within organic molecules is transferred to the bonds of ATP. It involves both kinds of energy-releasing reactions: changes in chemical bonds and transfers of electrons. Approximately 40% of the energy released during these reactions becomes chemical energy in ATP. The remaining 60% becomes heat (whenever chemical energy is transferred, some is inevitably lost as heat).

Almost any organic molecule can be a fuel for cellular respiration. In our initial discussions, we consider only glucose. The breakdown of glucose always begins with glycolysis, in which the molecule is split in two. What happens next depends on the availability of oxygen gas. If oxygen is not present, then the two halves of a glucose molecule are converted into two-carbon and three-carbon waste products, and the entire process is called fermentation. If oxygen is present, then the two halves of a glucose molecule follow a longer pathway, in which glucose is converted to carbon dioxide and water, and the entire process is called aerobic respiration.

A flow diagram of the alternate pathways is shown in figure 5.2, and the locations of these pathways are shown in figure 5.3. The two main kinds of cellular respiration—fermentation and aerobic respiration—are described in the remainder of the chapter.

Glycolysis

When glucose is the fuel, cellular respiration always begins with **glycolysis** (*glyco*, sugar; *lysis*, splitting), in which a molecule of glucose is split into two smaller molecules. The series of biochemical reactions is catalyzed by enzymes suspended within the cytosol (the cellular fluid outside the organelles).

Covalent bonds in glucose are broken to convert glucose (a six-carbon molecule) into two molecules of pyruvic

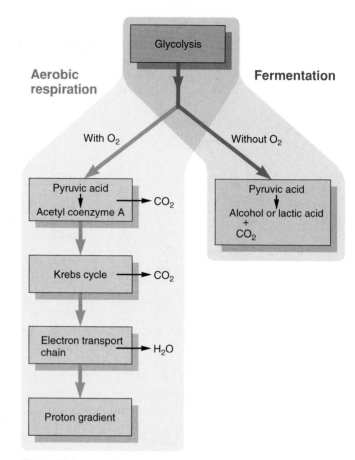

Figure 5.2 The two biochemical pathways of cellular respiration. Cellular respiration can take place with or without oxygen gas. Both pathways begin with glycolysis.

acid (a three-carbon molecule) and four hydrogen atoms. Energy is released as the chemical bonds are changed:

$$C_6H_{12}O_6 \longrightarrow 2C_3H_4O_3 + 4H + \text{Energy}$$

Glucose   Pyruvic acid

These molecular changes are diagrammed in figure 5.4. Some of the released energy is used to build four molecules of ATP from ADP and phosphate ions. The rest becomes heat. Since two molecules of ATP are used to split glucose, the net gain in energy is two molecules of ATP.

The removal of hydrogen atoms from glucose is an oxidation reaction. Since every oxidation must be accompanied by a reduction, the hydrogen atoms or their electrons are passed to an **electron carrier.** In glycolysis the electron carrier is a positively charged molecule called NAD⁺ (nicotinamide adenine dinucleotide). Each NAD⁺

Figure 5.3 Cellular respiration: locations of the biochemical pathways. (*a*) Photograph of several mitochondria sliced open to view their interiors. Each mitochondrion has an outer and an inner membrane. The inner membrane is deeply folded to increase its surface area for holding enzymes and other materials used in cellular respiration. (*b*) Sketch of a mitochondrion, showing where the different pathways of cellular respiration take place.

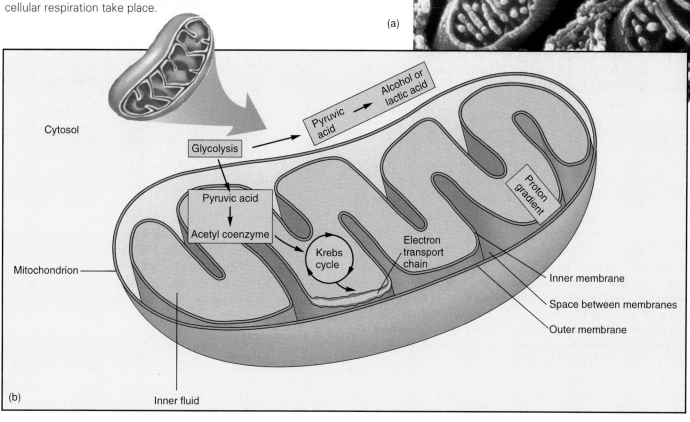

(a)

(b)

Cytosol

Pyruvic acid → Alcohol or lactic acid

Glycolysis

Pyruvic acid

Acetyl coenzyme

Krebs cycle

Electron transport chain

Proton gradient

Mitochondrion

Inner membrane

Space between membranes

Outer membrane

Inner fluid

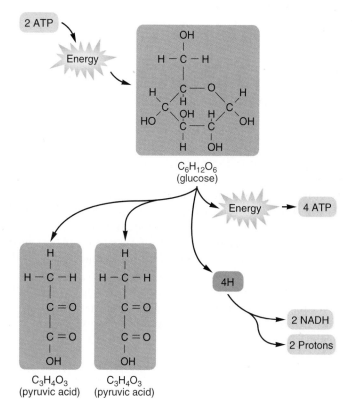

2 ATP

Energy

OH

H — C — H

C

H   O   H

C   C

HO   OH   OH   OH

C   C

H   OH

$C_6H_{12}O_6$
(glucose)

Energy → 4 ATP

4H

2 NADH

2 Protons

H
H — C — H
C = O
C = O
OH

$C_3H_4O_3$
(pyruvic acid)

H
H — C — H
C = O
C = O
OH

$C_3H_4O_3$
(pyruvic acid)

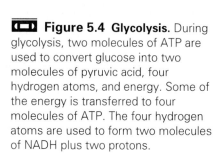 **Figure 5.4 Glycolysis.** During glycolysis, two molecules of ATP are used to convert glucose into two molecules of pyruvic acid, four hydrogen atoms, and energy. Some of the energy is transferred to four molecules of ATP. The four hydrogen atoms are used to form two molecules of NADH plus two protons.

is reduced as it picks up an electron (which neutralizes the positive charge) and a hydrogen atom. This reduction reaction forms a molecule of NADH:

$$NAD^+ + Electron + H \longrightarrow NADH$$

Of the four hydrogen atoms stripped from glucose, two are transferred intact to two NAD^+ molecules. The other two are broken apart, with their electrons joining the two NAD^+ molecules and their protons moving into the cytosol of the cell. (An unattached proton is the same as a hydrogen ion, abbreviated H^+.)

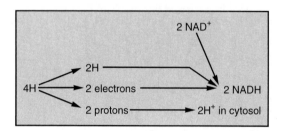

Two things are accomplished by glycolysis. First, it yields a net gain of two ATP molecules, which can then transport energy for cellular work. Second, it transfers hydrogen atoms from glucose to NADH (plus protons). In fermentation, there is no further advantage to forming NADH. In aerobic respiration, the electrons within NADH are used later to release more energy.

What happens next to the pyruvic acid and NADH molecules depends on whether or not oxygen participates in the reactions. We look first at cellular respiration in the absence of oxygen gas.

Fermentation

Fermentation is glycolysis followed by the conversion of pyruvic acid and NADH into waste products plus NAD^+. All the chemical reactions take place within the cytosol.

Fermentation usually occurs in one of two ways. In one pathway, glucose is converted to ethyl alcohol and carbon dioxide. In the other pathway, often called anaerobic respiration, glucose is converted to lactic acid. Which pathway is taken depends on the type of organism—yeasts that feed on grapes, for example, produce alcohol as a waste product whereas our own muscle cells produce lactic acid. In either case, no more energy is harvested after glycolysis, which yields two molecules of ATP for every glucose molecule. The two pathways of fermentation are summarized in table 5.1, diagrammed in figure 5.5, and described following. For each pathway, we describe the fate of a single molecule of pyruvic acid, but keep in mind that glycolysis forms two molecules of pyruvic acid from every glucose molecule.

Table 5.1

Summary of Fermentation

Phase	What Happens
1. Glycolysis	Glucose is split to form pyruvic acid, ATP, NADH, and protons
2. Formation of alcohol	Pyruvic acid, NADH, and protons are converted to ethyl alcohol, carbon dioxide, and NAD^+
or	
Formation of lactic acid	Pyruvic acid, NADH, and protons are converted to lactic acid and NAD^+

From Glucose to Alcohol   In some yeasts and bacteria, glycolysis is followed by the conversion of pyruvic acid and hydrogen atoms into ethyl alcohol and carbon dioxide (see fig. 5.5a). It involves two steps after glycolysis.

In the first step, each molecule of pyruvic acid is split to form a molecule of acetaldehyde and a molecule of carbon dioxide:

$$C_3H_4O_3 \longrightarrow C_2H_4O + CO_2$$

Pyruvic acid   Acetaldehyde Carbon dioxide

In the second step, two hydrogen atoms are added to the acetaldehyde molecule to form ethyl alcohol:

$$C_2H_4O + 2H \longrightarrow C_2H_5OH$$

Acetaldehyde   Ethyl alcohol

The two hydrogen atoms come from a molecule of NADH (contributes one entire hydrogen atom and one electron) plus an unattached proton in the cytosol. (NADH and the proton were formed from hydrogen atoms in glucose during glycolysis.) The transfer of these hydrogen atoms to acetaldehyde converts NADH to NAD^+:

$$NADH + H^+ \longrightarrow NAD^+ + 2H$$

Proton

These chemical reactions after glycolysis generate no ATP. What is accomplished is the liberation of NAD^+ so it can be used again in glycolysis. If cellular respiration ended with glycolysis, NAD^+ would remain tied up as NADH and glycolysis would come to a halt.

The overall equation for this pathway of fermentation is:

$$C_6H_{12}O_6 \longrightarrow 2C_2H_5OH + 2CO_2 + \text{Energy}$$

Glucose   Ethyl alcohol Carbon dioxide

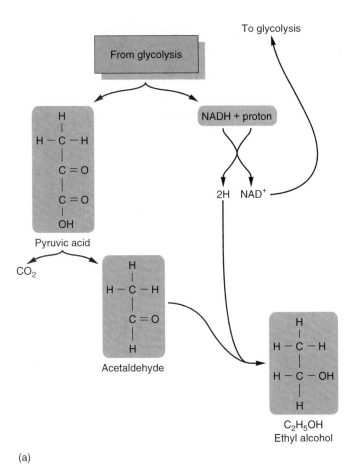

(a)

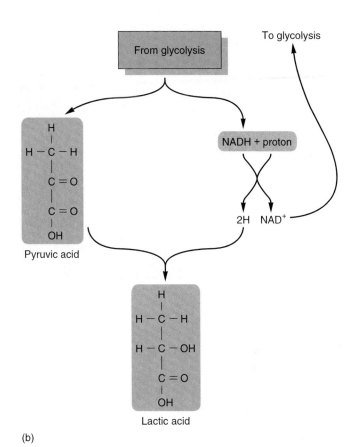

(b)

Figure 5.5 Fermentation. Cellular respiration in the absence of oxygen gas begins with glycolysis and then usually follows one of two pathways. (*a*) In one pathway, pyruvic acid is broken down into acetaldehyde and carbon dioxide. Each NADH is combined with a proton to form a molecule of NAD$^+$, which reenters glycolysis, and two hydrogen atoms, which join

acetaldehyde to form ethyl alcohol. (*b*) In another pathway, pyruvic acid is converted to lactic acid by adding two hydrogen atoms, which form a hydrogen proton combined with an electron and a hydrogen atom from NADH. After losing a hydrogen atom and an electron, the NAD$^+$ reenters glycolysis.

Human cells do not convert glucose to alcohol. We exploit this pathway, however, when we use yeasts to make wine or bread. Yeasts added to grapes convert grape sugar into ATP, ethyl alcohol, and carbon dioxide. The ATP is used by the yeasts, but the alcohol and carbon dioxide are excreted. In making wine, we retain the alcohol and allow the carbon dioxide to evaporate. Yeasts added to flour convert the glucose in starch to ATP, ethyl alcohol, and carbon dioxide. They use the ATP and excrete the alcohol and carbon dioxide. Carbon dioxide causes the bread dough to rise, while the alcohol evaporates during baking.

From Glucose to Lactic Acid   Some cells convert pyruvic acid to lactic acid when oxygen is not available. The entire sequence of chemical reactions, from the initial splitting of glucose to the formation of lactic acid, is the second pathway of fermentation (see fig. 5.5*b*).

Pyruvic acid is converted to lactic acid by adding hydrogen atoms, which were removed from glucose during glycolysis and are now present within NADH and as unattached protons (H$^+$):

$$NADH + H^+ \longrightarrow NAD^+ + 2H$$

$$\underset{\substack{\text{Pyruvic}\\\text{acid}}}{C_3H_4O_3} + 2H \longrightarrow \underset{\substack{\text{Lactic}\\\text{acid}}}{C_3H_6O_3}$$

No energy is transferred to ATP after glycolysis. What is accomplished by this final step of fermentation is the retrieval of NAD$^+$ so it can be used again in glycolysis.

The overall equation for the conversion of glucose to lactic acid is:

$$\underset{\text{Glucose}}{C_6H_{12}O_6} \longrightarrow \underset{\text{Lactic acid}}{2C_3H_6O_3} + \boxed{Energy}$$

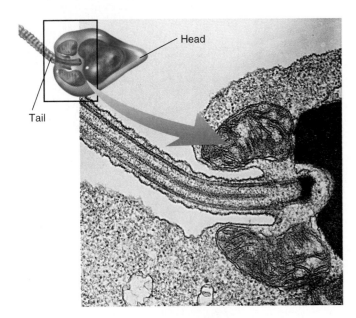

Figure 5.6 Mitochondria within a sperm. A sperm has many mitochondria to provide energy for its swim through the female reproductive tract toward an egg.

This form of cellular respiration generates ATP (from glycolysis) in our muscle cells when they are working faster than the blood can deliver oxygen. An all-out sprint, for example, is fueled almost entirely in this way. Some yeasts and bacteria also form ATP by this pathway. The ones that convert milk sugar into lactic acid are used to make cheese and yogurt.

Aerobic Respiration

In **aerobic respiration,** oxygen gas helps release energy from an organic monomer. All the bonds between the carbon atoms in the monomer are broken and all the hydrogen atoms are removed. The overall equation for aerobic respiration is:

$$C_6H_{12}O_6 + 6O_2 \longrightarrow 6CO_2 + 6H_2O + \text{Energy}$$

Glucose   Oxygen   Carbon   Water
          gas   dioxide

This more complete breakdown of glucose releases almost twenty times more energy than fermentation.

Aerobic respiration of glucose begins with glycolysis in the cytosol. The products of this reaction—pyruvic acid, NADH, and protons—enter a mitochondrion, where they participate in a series of reactions. More than 100 enzymes are involved, some embedded within the inner membrane and others suspended within the inner fluid of the mitochondrion. Cells with a great demand for energy, such as the sperm shown in figure 5.6, are packed with mitochondria.

The biochemical reactions of aerobic respiration can be grouped into five main phases: glycolysis; the conversion of pyruvic acid to acetyl coenzyme A; the Krebs cycle; the electron transport chain; and the proton gradient (see

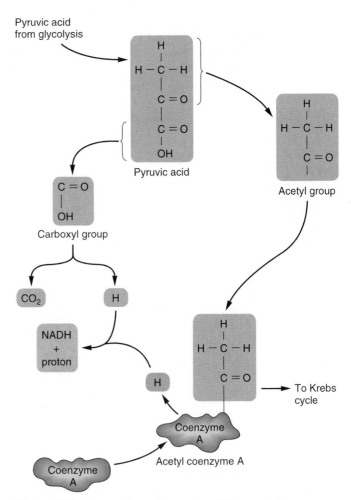

Figure 5.7 Conversion of pyruvic acid to acetyl coenzyme A. Pyruvic acid, formed during glycolysis, is broken into a carboxyl group and an acetyl group. The carboxyl group breaks down further into carbon dioxide, which diffuses out of the cell, and a hydrogen atom. The acetyl group joins with coenzyme A to form acetyl coenzyme A, releasing a hydrogen atom in the process. The two hydrogen atoms join with NAD^+ to form NADH and a proton. Acetyl coenzyme A moves on to the Krebs cycle.

fig. 5.2). Glycolysis was described earlier and the other four phases are described following.

From Pyruvic Acid to Acetyl Coenzyme A   The second phase in aerobic respiration, after glycolysis, is conversion of pyruvic acid to acetyl coenzyme A. It happens within the inner fluid of the mitochondrion (see fig. 5.3b). In the description following, we follow the fate of a single pyruvic acid molecule. Keep in mind, however, that two were formed during glycolysis.

The conversion begins with removal of a carboxyl group $\left(\begin{smallmatrix} & O \\ C & \\ & OH \end{smallmatrix}\right)$ from pyruvic acid, leaving a two-carbon molecule called an acetyl group. Another molecule, coenzyme A, then attaches to the acetyl group to form acetyl coenzyme A, removing a hydrogen atom in the process

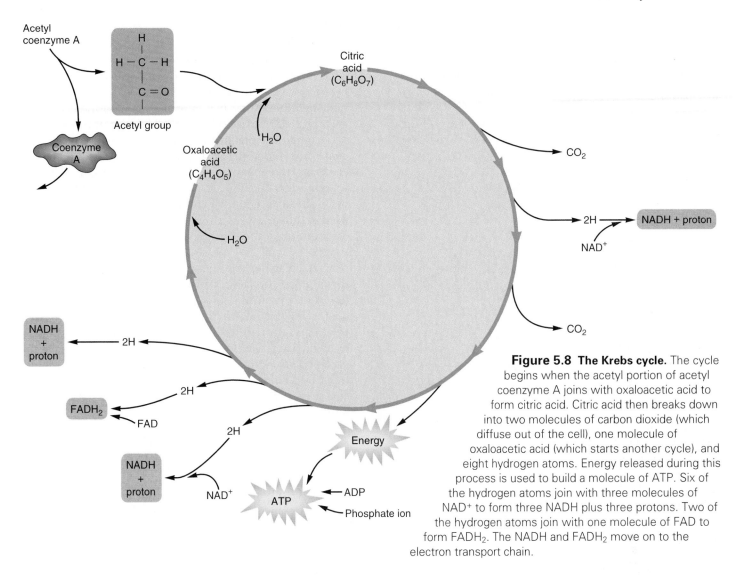

Figure 5.8 The Krebs cycle. The cycle begins when the acetyl portion of acetyl coenzyme A joins with oxaloacetic acid to form citric acid. Citric acid then breaks down into two molecules of carbon dioxide (which diffuse out of the cell), one molecule of oxaloacetic acid (which starts another cycle), and eight hydrogen atoms. Energy released during this process is used to build a molecule of ATP. Six of the hydrogen atoms join with three molecules of NAD^+ to form three NADH plus three protons. Two of the hydrogen atoms join with one molecule of FAD to form $FADH_2$. The NADH and $FADH_2$ move on to the electron transport chain.

(fig. 5.7). For each molecule of pyruvic acid, the reaction forms one molecule of acetyl coenzyme A, one molecule of carbon dioxide (from the carboxyl group), and two hydrogen atoms (one from oxidation of the carboxyl group and the other from oxidation of coenzyme A):

$$C_3H_4O_3 + \text{Coenzyme A} \longrightarrow C_2H_3O - \text{Coenzyme A} + 2H$$

Pyruvic acid   Acetyl

Carbon dioxide is a waste product that diffuses out of the cell (and, in us, is breathed out through the lungs). The two hydrogen atoms combine with NAD^+ to form NADH plus a proton (H^+):

$$NAD^+ + 2H \longrightarrow NADH + H^+$$

Three things are accomplished by the conversion of pyruvic acid to acetyl coenzyme A: (1) two hydrogen atoms are transferred to NAD^+ (plus a proton), forming NADH that can be used later in the electron transport chain; (2) pyruvic acid is converted to an acetyl group that will be used in the Krebs cycle; and (3) coenzyme A

attaches to and activates the acetyl group, preparing it for participation in the Krebs cycle.

The Krebs Cycle Energy and hydrogen atoms within the acetyl portion of acetyl coenzyme A are released during the **Krebs cycle** (also known as the citric acid cycle), shown in figure 5.8. This third phase of aerobic respiration takes place within the inner fluid of the mitochondrion (see fig. 5.3). The biochemical pathway forms a cycle that begins and ends with the same four-carbon molecule, oxaloacetic acid:

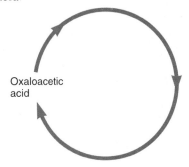

Oxaloacetic acid

The acetyl portion of acetyl coenzyme A enters the cycle by combining with oxaloacetic acid to form a six-carbon molecule called citric acid. Citric acid is broken apart as it moves through the cycle: part of it becomes another four-carbon molecule (oxaloacetic acid) for the next turn of the cycle and the rest forms carbon dioxide (a waste product) and hydrogen atoms. Some of the energy released during the process is used to build ATP from ADP and a phosphate ion:

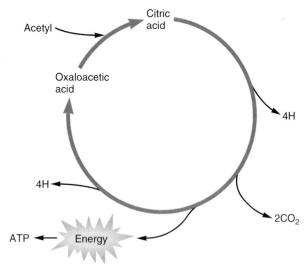

The eight hydrogen atoms removed from citric acid are picked up by two kinds of carrier molecules: NAD^+, described earlier, and a similar molecule called FAD (flavin adenine dinucleotide):

$$3 \ NAD^+ \ 6H \longrightarrow 3 \ NADH + 3H^+$$

$$FAD + 2H \longrightarrow FADH_2$$

NADH and $FADH_2$ carry hydrogen atoms and electrons to the electron transport chain. The protons, originally part of hydrogen atoms in citric acid, contribute to the proton gradient that participates in the final phase of aerobic respiration.

What has been accomplished by the Krebs cycle is the transfer of energy from an acetyl group to ATP and the transfer of hydrogen atoms and electrons to NADH and $FADH_2$. The energy transferred to ATP (one molecule per acetyl group, or two molecules per glucose) is available for use by the cell. The hydrogen atoms and electrons within NADH and $FADH_2$ will release energy in the electron transport chain.

Electron Transport Chain   The **electron transport chain** is a series of oxidation-reduction reactions. Electrons are transported along a chain of molecules (fig. 5.9), with the atoms in each molecule holding the electrons closer to the nucleus than atoms in the previous molecule. Energy is released every time an electron moves from one molecule to

the next. This fourth phase of aerobic respiration takes place on proteins embedded within the inner surface of the mitochondrion's inner membrane (see fig. 5.3).

Electrons are brought to the electron transport chain by the two carrier molecules—NADH and $FADH_2$—formed during earlier phases of aerobic respiration. A summary of their formation is provided in table 5.2.

When the electrons reach the final molecule in the chain, they combine with protons (formed earlier when electrons were removed from hydrogen atoms) and oxygen to form water (see fig. 5.9). Oxygen is said to be the terminal electron acceptor of aerobic respiration. The reason we need to inhale oxygen gas (O_2) is to provide some place for electrons to go after they have moved through the electron transport chain.

A great deal of energy is released as the electrons travel along the chain of molecules. The energy is transferred to a proton gradient (the final phase), where it is used to build ATP. Anything that blocks this flow of electrons cuts off the cell's energy supply. A lack of oxygen, for example, stops the flow of electrons because they have no place to go. When someone suffocates, their electron transport chain becomes clogged with electrons. Cyanide also stops the flow of electrons. This deadly poison is described in the accompanying Wellness Report.

What is accomplished by the electron transport chain is the release of energy from electrons that were originally part of the glucose molecule.

Proton Gradient   Energy released from the electron transport chain is used to pump protons (H^+) across the inner membrane of the mitochondrion. These protons, released during the formation of NADH, are transported from the inner fluid of the mitochondrion to the space between the inner and outer membranes (fig. 5.10a). They are carried by transport proteins (see page 73), called proton pumps, embedded within the inner membrane of the mitochondrion.

As protons are pumped across the membrane, they become more concentrated within the intermembrane space than within the inner fluid of the mitochondrion (fig. 5.10b). This difference in proton concentration is a gradient that holds potential energy.

Potential energy within the gradient becomes kinetic energy when the protons flow down the gradient. The protons cross the membrane through channel proteins (see page 73). They are moved by three forces: diffusion, mutual repulsion (because of their positive charges), and attraction to negative ions within the inner fluid. The flow of protons releases energy, which is used to build ATP from ADP and phosphate (fig. 5.10c).

Formation of ATP from a proton gradient, as described previously, is known as **chemiosmotic phosphorylation.** The term *chemi* refers to the chemical gradient of

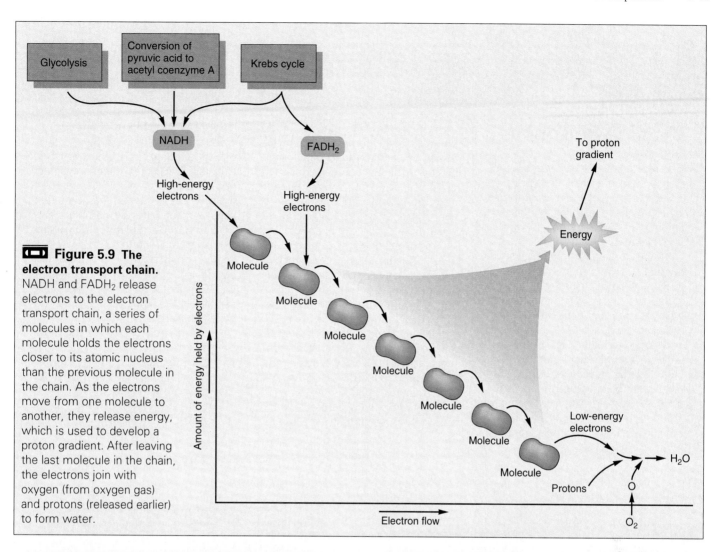

Figure 5.9 The electron transport chain. NADH and FADH$_2$ release electrons to the electron transport chain, a series of molecules in which each molecule holds the electrons closer to its atomic nucleus than the previous molecule in the chain. As the electrons move from one molecule to another, they release energy, which is used to develop a proton gradient. After leaving the last molecule in the chain, the electrons join with oxygen (from oxygen gas) and protons (released earlier) to form water.

Table 5.2

Summary of Energy Production, per Glucose Molecule,* by Aerobic Respiration

Phase	NADH		FADH$_2$		ATP	
	Formed	Used	Formed	Used	Formed	Used
Glycolysis	2	0	0	0	4	2
Pyruvic acid to acetyl coenzyme A	2	0	0	0	0	0
Krebs cycle	6	0	2	0	0	0
Electron transport chain	0	10	0	2	0	
Proton gradient	0	0	0	0	34	0

*Glycolysis of glucose produces two molecules of pyruvic acid, whereas in the text we followed just one molecule. All calculations shown here have been doubled to adjust for two molecules of pyruvic acid.

Wellness Report

Cyanide Poisoning

In recent years, individuals have illegally placed cyanide in packages of food, pop, or over-the-counter medicines. Cyanide is a deadly poison: a tiny amount can kill a human within a few minutes.

Cyanide is a C≡N group in sodium cyanide (NaCN) or potassium cyanide (KCN). (When combined with an acid, these powders form cyanide gas, which is used in gas chambers to carry out death sentences.) Cyanide kills by interfering with the electron transport chain of aerobic respiration (see fig. 5.9). It binds, irreversibly, with the last molecule in the electron transport chain—a protein called cytochrome a₃. Once cytochrome a₃ is joined to cyanide, it cannot accept electrons.

When electrons can no longer flow past cytochrome a₃, they accumulate, as cars in a traffic jam, within the chain. The electron transport chain can no longer release energy. Furthermore, no energy can be released in other phases of cellular respiration, because the electron carriers NAD⁺ and FAD become tied up as NADH and FADH₂. All phases of aerobic respiration come to a halt. Glycolysis, conversion of pyruvic acid to acetyl coenzyme A, and the Krebs cycle all depend on the presence of NAD⁺ and/or FAD⁺, and the proton gradient depends on energy from the electron transport chain. Cyanide stops energy production, which kills body cells within minutes.

Cyanide occurs naturally in the seeds of some plants. Lima beans,

chick peas, cashews, and almonds, as well as the seeds of apples, pears, peaches, lemons, limes, apricots, cherries, and plums contain cyanide. The advantage to the plant is that animals tend not to eat cyanide-laden seeds, thereby allowing the plant embryo to survive. Each seed contains only small amounts of this deadly poison, yet small children have died from eating too many of them. An adult would have to consume about sixty apricot pits or a thousand cashews to get a lethal dose.

By blocking cellular respiration, cyanide prevents a cell from releasing energy for doing work, and a cell must work continuously just to stay alive. Cyanide is a deadly poison.

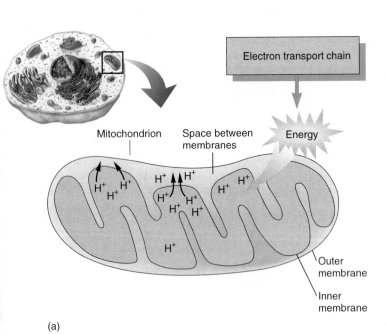

(a)

(b)

(c)

Figure 5.10 The proton gradient. (a) Energy from the electron transport chain is used by transport proteins to pump protons (H⁺) across a membrane, from the inner fluid to the space between the outer and inner membranes of the mitochondrion. (b) A concentration gradient develops, with many more protons on one side of the membrane than on the other side. The gradient is a form of potential energy. (c) A channel protein opens and the protons flow through the membrane. Energy stored in the gradient is released and used to build ATP from ADP and phosphate ions.

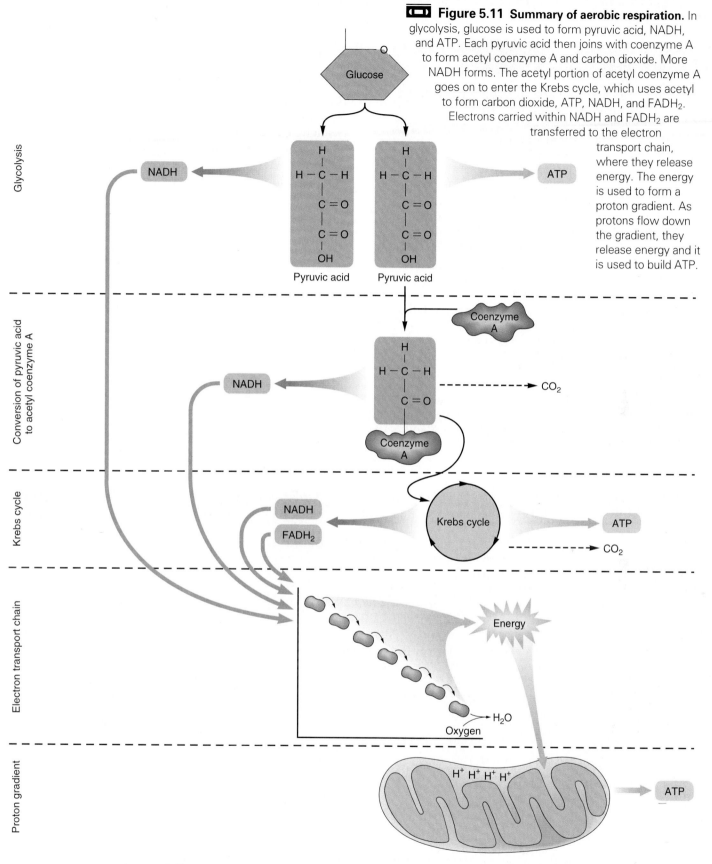

Figure 5.11 Summary of aerobic respiration. In glycolysis, glucose is used to form pyruvic acid, NADH, and ATP. Each pyruvic acid then joins with coenzyme A to form acetyl coenzyme A and carbon dioxide. More NADH forms. The acetyl portion of acetyl coenzyme A goes on to enter the Krebs cycle, which uses acetyl to form carbon dioxide, ATP, NADH, and FADH₂. Electrons carried within NADH and FADH₂ are transferred to the electron transport chain, where they release energy. The energy is used to form a proton gradient. As protons flow down the gradient, they release energy and it is used to build ATP.

protons; *osmotic* refers to diffusion of the protons across a membrane; and *phosphorylation* refers to the addition of a phosphate ion to ADP.

What is accomplished by this phase of aerobic respiration is the transfer of energy from electrons to a proton gradient and then to molecules of ATP. Most of the energy from aerobic respiration comes from this final phase.

For every molecule of glucose, thirty-four molecules of ATP are generated from energy within the proton gradient, whereas only four molecules are generated (with a net gain of two molecules) during previous phases (see table 5.2).

The five phases of aerobic respiration are reviewed in figure 5.11 and table 5.3.

Table 5.3

Summary of the Phases of Aerobic Respiration (Using Glucose as a Fuel)

Phase	What Happens
Glycolysis	Glucose is used to form pyruvic acid, ATP, NADH, and protons
Conversion of pyruvic acid to acetyl coenzyme A	Pyruvic acid is used to form acetyl coenzyme A, carbon dioxide, NADH, and protons
Krebs cycle	Acetyl is used to form carbon dioxide, ATP, NADH, $FADH_2$, and protons
Electron transport chain	Electrons from NADH and $FADH_2$ release energy during a series of oxidation-reduction reactions
Proton gradient	Energy from electron transport chain forms a gradient of protons; the protons flow down the gradient and release energy that is transferred to ATP

Fuels for Cellular Respiration

Fermentation uses only monosaccharides (glucose, fructose, or galactose) as fuels. Energy for an all-out sprint, for example, comes almost entirely from glucose. Such rapid movement can be sustained for just a short period of time (40–50 seconds) because each glucose molecule provides only two molecules of ATP and only small quantities of glucose are stored.

Aerobic respiration, by contrast, can use three kinds of fuels: monosaccharides, fatty acids, and amino acids. These fuels are compared in table 5.4 and described in the following paragraphs. The phases of aerobic respiration that the fuels enter are diagrammed in figure 5.12.

In addition to normal fuels, aerobic respiration can use the alcohol in wine, beer, whiskey, and other alcoholic beverages. The problems that arise when a person drinks too much alcohol are described in the accompanying Wellness Report.

Monosaccharides

Glucose, fructose, and galactose enter aerobic respiration at glycolysis. They are the best fuels for aerobic respiration because their chemical energy is rapidly transferred to ATP

Table 5.4

Comparisons of Fuels Used in Cellular Respiration

Fuel	Advantages	Disadvantages
Monosaccharides	Used in both fermentation and aerobic respiration Forms ATP rapidly No toxic wastes	Limited stores in animals
Fatty acids	Large amount of energy per molecule Large stores in animals Normally no toxic wastes	Used only in aerobic respiration Forms ATP slowly May lower blood pH
Amino acids	Large amounts in animals (in muscles)	Used only in aerobic respiration Breaks down the muscles Nitrogenous wastes

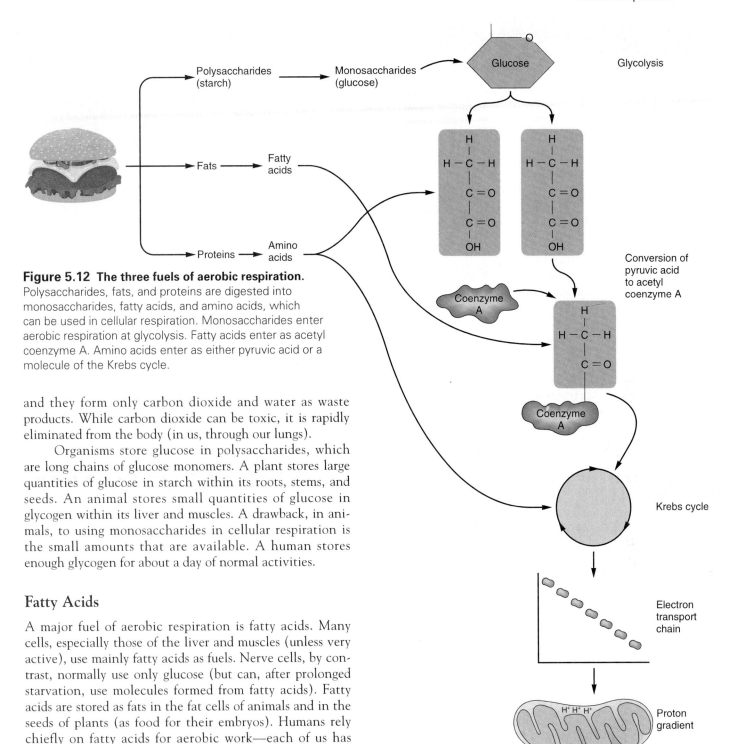

Figure 5.12 The three fuels of aerobic respiration.
Polysaccharides, fats, and proteins are digested into
monosaccharides, fatty acids, and amino acids, which
can be used in cellular respiration. Monosaccharides enter
aerobic respiration at glycolysis. Fatty acids enter as acetyl
coenzyme A. Amino acids enter as either pyruvic acid or a
molecule of the Krebs cycle.

and they form only carbon dioxide and water as waste
products. While carbon dioxide can be toxic, it is rapidly
eliminated from the body (in us, through our lungs).

Organisms store glucose in polysaccharides, which
are long chains of glucose monomers. A plant stores large
quantities of glucose in starch within its roots, stems, and
seeds. An animal stores small quantities of glucose in
glycogen within its liver and muscles. A drawback, in animals, to using monosaccharides in cellular respiration is
the small amounts that are available. A human stores
enough glycogen for about a day of normal activities.

Fatty Acids

A major fuel of aerobic respiration is fatty acids. Many
cells, especially those of the liver and muscles (unless very
active), use mainly fatty acids as fuels. Nerve cells, by contrast, normally use only glucose (but can, after prolonged
starvation, use molecules formed from fatty acids). Fatty
acids are stored as fats in the fat cells of animals and in the
seeds of plants (as food for their embryos). Humans rely
chiefly on fatty acids for aerobic work—each of us has
enough stored, as fat, to walk about 1,600 kilometers
(1,000 mi).

Alcohol and the Liver

Ethyl alcohol is a waste product of cellular respiration in yeasts. To these simple organisms, alcohol is toxic and they quickly excrete it; to us it is intoxicating. Because alcohol diffuses through the phospholipid framework of cellular membranes, it can enter every cell of the body. While alcohol affects every cell of the body, it is the effects on brain cells that we seek. Alcohol relaxes most people, makes them more sociable, and helps them have a good time.

Some people drink too much alcohol, a habit that can seriously damage every cell of the body. At first, the only signs of this problem are changes in behavior, as listed in table 5.A. Later, however, the widespread damage to cells becomes apparent: a bluish nose from broken blood vessels in the skin; loss of mental agility and physical coordination caused by damage to brain cells; various signs of malnutrition; and a potbelly. The worse problems, however, are not easily seen for they involve damage to the liver.

The liver performs many crucial functions. One function is to convert toxic molecules into harmless molecules. More than 95% of the alcohol consumed enters the liver for detoxification. The liver becomes overwhelmed with this task and can no longer perform its normal functions.

The alcohol in a drink is carried by the blood from the stomach and intestines to the liver. While in the bloodstream, some of the alcohol is exhaled as it passes through the lungs, causing the odor of alcohol on the breath; some of it passes to the muscles, where it damages muscle cells; and some of it passes to the brain, where it produces its effects. Alcohol that enters the liver is broken down into acetaldehyde and hydrogen:

$$C_2H_5OH \longrightarrow C_2H_4O + 2H$$

Alcohol   Acetaldehyde

Table 5.A

Some Questions to Identify the Alcohol-Dependent Person

1. Do you occasionally drink heavily after a disappointment, a quarrel, or when the boss gives you a bad time?

2. When you are having trouble or feel under pressure, do you always drink more heavily than usual?

3. Have you noticed that you are able to handle more liquor than you did when you were first drinking?

4. Did you ever wake up the "morning after" and discover that you could not remember part of the evening before, even though your friends tell you that you did not "pass out"?

5. When drinking with other people, do you try to have a few extra drinks when others will not know it?

6. Are there certain occasions when you feel uncomfortable if alcohol is not available?

7. Have you recently noticed that when you begin drinking you are in more of a hurry to get the first drink than you used to be?

8. Do you sometimes feel a little guilty about your drinking?

9. Are you secretly irritated when your family or friends discuss your drinking?

10. Have you recently noticed an increase in the frequency of your memory "blackouts"?

11. When you are sober, do you often regret things you have done or said while drinking?

12. Have you often failed to keep the promises you have made to yourself about controlling or cutting down on your drinking?

13. Do more people seem to be treating you unfairly without good reason?

14. Do you eat very little or irregularly when you are drinking?

15. Do you get terribly frightened after you have been drinking heavily?

National Council on Alcoholism and Drug Dependence, Inc., New York, NY.

The acetaldehyde is then converted to an acetyl group and shipped to other body cells for conversion into acetyl coenzyme A, which enters cellular respiration at the Krebs cycle.

Strangely, it is the hydrogen atoms that damage liver cells. They join with NAD^+ to form NADH and protons. Recall that NAD^+ contributes electrons to the electron transport chain. When a person drinks excessively, the electrons transferred from alcohol to NADH flood the electron transport chains of liver cells, as diagrammed in figure 5.A. The chains can no longer accept electrons from the normal pathways of cellular respiration—glycolysis, conversion of pyruvic acid to acetyl coenzyme A, and the Krebs cycle. As these normal pathways slow down, the normal fuels—glucose, fatty acids, and amino acids—accumulate and the liver cells convert them into fats.

Normally, the liver ships fats, via the blood, to special fat cells else-

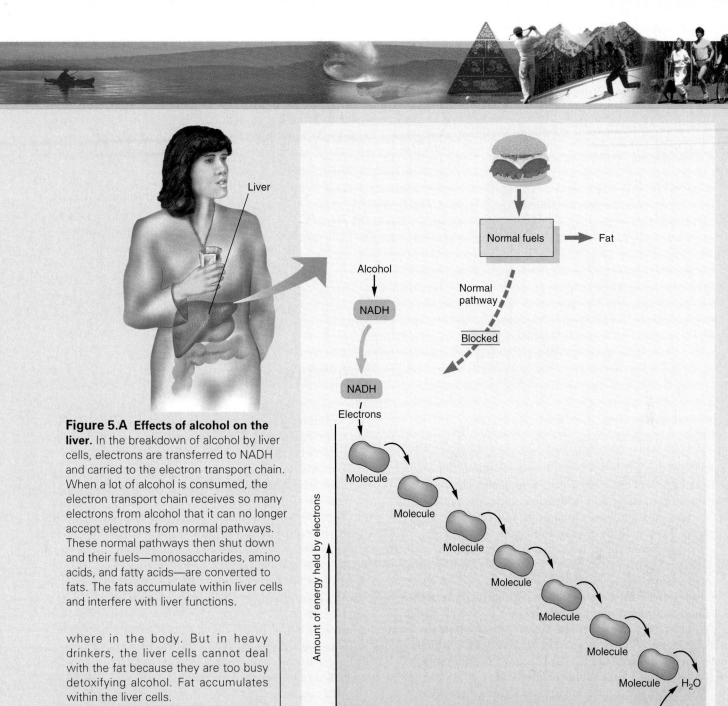

Figure 5.A Effects of alcohol on the liver. In the breakdown of alcohol by liver cells, electrons are transferred to NADH and carried to the electron transport chain. When a lot of alcohol is consumed, the electron transport chain receives so many electrons from alcohol that it can no longer accept electrons from normal pathways. These normal pathways then shut down and their fuels—monosaccharides, amino acids, and fatty acids—are converted to fats. The fats accumulate within liver cells and interfere with liver functions.

where in the body. But in heavy drinkers, the liver cells cannot deal with the fat because they are too busy detoxifying alcohol. Fat accumulates within the liver cells.

In time, the liver cells become engorged with fat and no longer perform their normal functions. In this stage, called alcoholic hepatitis, the liver is swollen and tender. The person does not feel well or look healthy. The skin and whites of their eyes develop a yellowish tinge, a clear sign of liver damage.

An alcoholic's health continues to deteriorate as the liver cells begin to die. Scar tissue replaces the dead cells and impedes the flow of blood through this crucial organ. This condition, known as cirrhosis of the liver, interferes severely with liver function.

The liver normally synthesizes blood proteins (which hold water in the blood), supplies the blood with sugar, and eliminates toxic molecules. An alcoholic is likely to develop a potbelly as fluid accumulates within the abdominal cavity. The fluid leaks out of blood vessels in the liver, because blood stagnates within scarred regions and at the same time does not contain enough proteins to hold water inside the vessels. Diabetes mellitus develops when the damaged liver fails to control how much sugar it releases into the blood. The most serious problem, the one that kills, is nitrogen poisoning from an inability of liver cells to convert ammonia into urea for excretion by the kidneys. Cirrhosis of the liver is the seventh leading cause of death in the United States.

Fatty acids are long chains of carbon atoms with attached hydrogen atoms. Transport proteins within the membranes of a mitochondrion carry fatty acids into the inner fluid of the mitochondrion. There, the fatty acids are dismantled: carbon atoms are broken off, two at a time, to form an acetyl group, which joins coenzyme A to form acetyl coenzyme A:

Acetyl group

Acetyl coenzyme A then enters the Krebs cycle.

The advantages of fatty acids are they contain a great deal of chemical energy—more than twice as many calories per gram than carbohydrates or proteins—and they are stored in large amounts, especially in animals. The drawback is a slower rate of ATP formation. Transport proteins cannot bring fatty acids into the mitochondria fast enough to form ATP as rapidly as it forms from monosaccharides.

A very active cell, which requires a very high rate of ATP formation, uses more monosaccharides and fewer fatty acids than a less active cell. During a jog, cells in the leg muscles use a mixture of about 70% fatty acids and 30% glucose. When running faster, the muscles switch to a higher percentage of glucose in order to meet their demands for more ATP. A racer must balance speed, which comes from a limited supply of glucose, with endurance, which comes from a large supply of fatty acids.

Another disadvantage of fatty acids is they may alter the blood chemistry. Liver cells provide other cells with fatty acids for cellular respiration. Before shipment through the bloodstream, the liver cells break fatty acids into acetyl groups and then transform these groups into several kinds of molecules called ketone bodies. Ketone bodies can be toxic. Under normal conditions, they occur in the blood at very low concentrations and cause no health problems. But when the body relies mainly on fatty acids as fuels, ketone bodies in the blood may reach toxic levels. Ketones are acids, which lower the blood pH and can seriously disturb bodily functions. Health problems from ketone bodies develop from a high-fat diet, starvation (in which energy comes from fat stores), or untreated diabetes mellitus (in which glucose cannot enter cells).

Amino Acids

Cells use amino acids as fuels when excess amino acids are present, as occurs from a high-protein diet, and when other fuels are not available, as happens during starvation, when muscle proteins are broken down to supply some of the fuel (most comes from fat, as long as there are still fat stores available).

An amino acid enters cellular respiration after conversion to pyruvic acid or one of the molecules within the Krebs cycle itself. Regardless of how an amino acid enters cellular respiration, its nitrogen atom is always removed in the form of ammonia (NH_3). Ammonia is toxic to cells and must be eliminated as soon as it forms. In humans, ammonia is converted into urea, a less toxic form of nitrogen, and then excreted by the kidneys in urine. A high-protein diet or starvation can harm the kidneys as they work hard to rid the body of urea.

SUMMARY

1. Cellular respiration releases energy from an organic monomer for cellular work. Energy is released when electrons move closer to a nucleus, which happens when chemical bonds are changed and electrons are transferred. Sixty percent of the released energy becomes heat and 40% is captured within ATP, which transports the energy to where it is used.

2. When glucose is the fuel, cellular respiration begins with glycolysis, which splits glucose into two molecules of pyruvic acid and two hydrogen atoms. Some of the released energy is captured within ATP. The hydrogen atoms combine with NAD^+ to form NADH and unattached protons.

3. In fermentation (without O_2), glycolysis is followed by conversion of pyruvic acid into NAD^+ (for recycling) and waste products. No ATP forms after glycolysis. In one pathway, the waste products are alcohol and carbon dioxide. In another pathway, the waste product is lactic acid.

4. In aerobic respiration, glycolysis is followed by the breakdown of pyruvic acid into carbon dioxide, water, and energy. Except for glycolysis, these pathways are located within the cell's mitochondria.

5. The first phase of aerobic respiration uses pyruvic acid to form acetyl coenzyme A, carbon dioxide, NADH, and a proton.

6. The second phase, the Krebs cycle, uses the acetyl part of acetyl coenzyme A to form carbon dioxide, ATP, NADH, $FADH_2$, and protons.

7. The third phase is the electron transport chain, in which electrons carried by NADH and $FADH_2$ release energy as they move along a chain of molecules.

8. The fourth phase is formation of a proton gradient from energy released during electron transport. As the protons flow down the gradient, they release energy that is transferred to ATP.

9. Fermentation uses only monosaccharides as a fuel, but aerobic respiration can use monosaccharides, fatty acids, and proteins. Monosaccharides produce nontoxic wastes and form ATP rapidly but in animals are not stored in large quantities. Fatty acids form ATP slowly, contain a lot of energy, and are stored in large amounts but may alter the blood pH. Amino acids form a toxic waste product, ammonia, which is converted to urea and removed from the body by the kidneys.

KEY TERMS

Adenosine diphosphate (page 87)
Adenosine triphosphate (ah-DEH-noh-seen TREYE-FOHSS-fayte) (page 87)
Aerobic (AER-oh-bik) respiration (page 92)
Chemiosmotic (KEE-me-oz-MAH-tik) phosphorylation (page 94)
Electron carrier (page 88)
Electron transport chain (page 94)
Energy (page 86)
Fermentation (page 90)
Glycolysis (gleye-KAH-lih-sis) (page 88)
Krebs cycle (page 93)
Oxidation-reduction (OX-ih-DAY-shun) reaction (page 86)
Work (page 86)

STUDY QUESTIONS

1. Briefly describe the two main ways in which chemical energy is released from glucose.
2. How is energy transported within a cell?
3. Why can an all-out sprint continue for only a short time (less than a minute)?
4. Explain the origin of alcohol in wine.
5. Where does the proton gradient form? Where do the protons come from? Where does the energy come from? How is the energy within the gradient released? What is the energy used for?
6. Why does aerobic respiration yield so many more ATP molecules per glucose molecule than fermentation?

CRITICAL THINKING PROBLEMS

1. What is the relation, in humans, between aerobic respiration and breathing (taking in oxygen gas and releasing carbon dioxide)? What is the relation between fermentation and breathing?
2. Why does intense physical exercise cause us to overheat?
3. Your body makes NAD^+ and FAD from niacin and riboflavin, which are B vitamins. These vitamins are required in the diet in tiny amounts as compared with carbohydrates, proteins, and fats. Why do you think your cells need only tiny amounts of these materials each day?
4. Suppose you are given two muscles from a human leg, to analyze under the microscope. In one of the muscles, the cells were packed with mitochondria, whereas in the other muscle there were very few mitochondria. What could you conclude about the functions of these two muscles?

SUGGESTED READINGS

Introductory Level:

Asimov, I. *Life and Energy*. New York: Avon Publishers, 1962. A thorough description of the forms of energy and how chemical energy is used in living systems. The book is written for laypeople and is remarkably easy to read.

Carter, L. C. *Guide to Cellular Energetics*. San Francisco: W. H. Freeman and Company, 1973. A workbook that takes the reader step by step through cellular respiration. It includes descriptions, illustrations, problems, and a glossary.

Gibbons, B. "Alcohol, the Legal Drug," *National Geographic* 181, no. 2, February 1992, 2–35. An essay on the history, sociology, and biology of alcohol as a beverage, including the effect of alcohol on the liver.

Advanced Level:

Alberts, B., D. Bray, J. Lewis, M. Raff, K. Roberts, and J. Watson. *Molecular Biology of the Cell*, 2d ed. New York: Garland Publishing, 1989. An excellent description of how cells acquire and use energy.

Stryer, L. *Biochemistry*, 3d ed. San Francisco: W. H. Freeman and Company, 1988. An advanced, well-written explanation of biochemical pathways, including those of cellular respiration.

⚙ EXPLORATIONS

Interactive Software: Active Transport

This interactive exercise explores the relation between ATP availability and active transport of materials across cellular membranes. The student varies the concentrations of ATP and amino acids and observes the effects on a coupled system of transport involving amino acids, sodium ions, and potassium ions.

Interactive Software: Oxidative Respiration

This interactive exercise explores how energy within organic molecules is used to drive proton pumps and how these pumps, in turn transfer energy to ATP. The student is provided with a diagram of a mitochondrial membrane, and then alters the concentrations of oxygen, organic molecules and existing ATP molecules to see how these factors govern the rate at which protons cross the membrane and generate new molecules of ATP.

Photosynthesis

Chapter Outline

Objectives

In this chapter, you will learn
–the importance of photosynthesis to
  all forms of life;
–how plants, algae, and photosynthetic
  bacteria convert light energy into
  chemical energy;
–how these organisms then use
  chemical energy to build sugars from
  carbon dioxide;
–what happens to the newly
  synthesized sugars.

Photosynthesis, the foundation of
life, gives nature its green color.

hy is nature so green? Why not blue, purple, pink, or a rainbow of colors? Why does the land become more colorful in autumn? Why are northern winters so drab?

It is the green of photosynthesis that colors the land. This biochemical process builds sugars from sunlight, carbon dioxide, and water. The energy of sunlight is captured by a green pigment called chlorophyll. Virtually all life on earth depends upon this pigment. Nearly a billion tons of chlorophyll are synthesized annually by photosynthetic organisms: plants, algae, and certain kinds of bacteria.

Nature is predominantly green when conditions favor photosynthesis. Where winters are cold, the shorter days of autumn bring a change in color: the soft green colors of summer are replaced by a glorious burst of yellows, golds, oranges, reds, and purples. Pigments of these colors were present in the leaves all summer, for many contribute to photosynthesis, but were concealed by the more abundant chlorophyll. As plants prepare for winter, chlorophyll disintegrates and the glorious colors of autumn appear. After a few weeks, these pigments also break down and the useless leaves are shed.

Winter is a drab time of the year in northern deciduous forests and grasslands. The grayish brown outlines of the trees, the brownish skeletons of fallen leaves, and the grayish yellow stems of grasses indicate the absence of photosynthesis. The machinery of this biochemical process, which pervades nature, has been dismantled.

Overview of Photosynthesis

Photosynthesis is a sequence of reactions that converts sunlight into chemical energy and then uses the chemical energy to build an organic molecule (a sugar) from inorganic molecules (carbon dioxide and water). The chemical bonds of sugar hold chemical energy that can be used by cells to do work. Most often the sugar that is built is glucose, and the overall equation for photosynthesis is:

$$6CO_2 + 6H_2O + \text{Light} \longrightarrow C_6H_{12}O_6 + 6O_2$$

Carbon dioxide   Water   Glucose   Oxygen gas

Photosynthesis is fundamental to life. Its products, the sugars, supply most forms of life with chemical energy and carbon skeletons. The chemical energy is needed to do cellular work, and the carbon skeletons are frameworks for building all organic molecules.

Photosynthesis and Aerobic Respiration

A by-product of photosynthesis is oxygen gas (O_2), which most forms of life must have to survive. Oxygen gas is used in aerobic cellular respiration, the series of reactions that transfer chemical energy from organic monomers to ATP (☐ *see page 87*) (the energy-transport molecule). When aerobic respiration burns glucose as a fuel, the overall reaction is the reverse of photosynthesis:

$$C_6H_{12}O_6 + 6O_2 \longrightarrow 6CO_2 + 6H_2O + \text{Energy}$$

An important difference between the two reactions is that the energy used in photosynthesis is light energy and the energy released in cellular respiration is chemical energy and heat. The two reactions complement each other: photosynthesis uses the products of cellular respiration (carbon dioxide and water), and cellular respiration uses the products of photosynthesis (sugar and oxygen gas). This relationship is diagrammed in figure 6.1.

Evolution of Photosynthesis

The first cells on earth, which appeared more than 3.5 billion years ago, had neither photosynthesis nor aerobic respiration. They acquired organic molecules, as sources of chemical energy and building materials, from the surrounding waters. (The first organic molecules were apparently synthesized from gases by inorganic processes.) The cells released chemical energy from these molecules by fermentation, since prior to photosynthesis there was not enough oxygen gas for aerobic respiration.

The first photosynthetic organisms appeared later, approximately 3.3 billion years ago. They completely changed the environment and therefore the course of evolution. Oxygen gas released from photosynthesis was at first toxic to cells. In time, however, most cells evolved the capacity to use this environmental pollutant in cellular respiration. Aerobic respiration, which gleans much more energy from sugars than fermentation, appeared approximately 2.5 billion years ago. Today, most organisms not only use oxygen gas but die when deprived of it.

Locations of Photosynthesis

Plants, algae, and a few kinds of bacteria perform photosynthesis. In bacteria, which are prokaryotic (simple) cells (☐ *see page 59*), the biochemical pathways of photosynthesis are located on indentations of the plasma membrane and within the inner fluid (fig. 6.2*a*). In plants and algae,

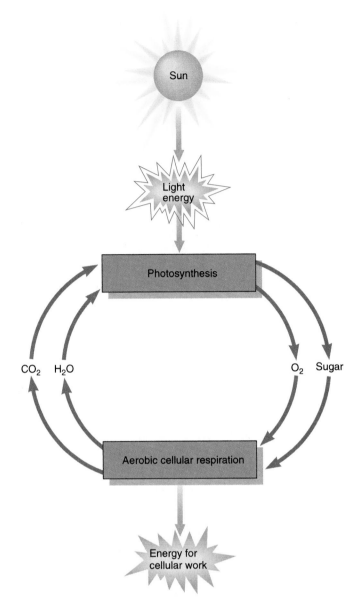

Figure 6.1 Photosynthesis and cellular respiration complement each other. The products of photosynthesis are used in aerobic cellular respiration, and the products of aerobic cellular respiration are used in photosynthesis.

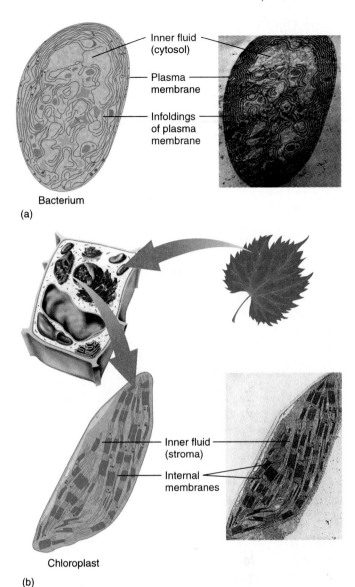

Figure 6.2 The sites of photosynthesis. Molecules that convert sunlight to chemical energy are embedded within membranes, and molecules that use this chemical energy to build sugar are suspended within fluids. (*a*) In a photosynthetic bacterium, these different phases of photosynthesis occur on infoldings of the plasma membrane and within the inner fluid (cytosol). (*b*) In the cell of a plant or alga, photosynthesis takes place within organelles called chloroplasts. Inside a chloroplast, the different phases of photosynthesis occur on membranes and within the inner fluid (stroma).

which have eukaryotic (complex) cells (📖 *see page 62*), photosynthesis occurs within specialized organelles called **chloroplasts** (fig. 6.2*b*).

Chloroplasts may once have been photosynthetic bacteria. The two structures are remarkably similar. What may have happened, more than 2 billion years ago, is that a photosynthetic bacterium invaded a larger cell and the two cells developed a mutualistic relationship: the larger cell acquired sugars from the smaller cell, and the smaller cell acquired raw materials from the larger cell.

Photosynthesis is a sequence of more than seventy biochemical reactions that occur in two phases: (1) the light-dependent reactions, in which light energy is converted to chemical energy; and (2) the light-independent reactions, in which the chemical energy is used to build sugar from carbon dioxide (fig. 6.3).

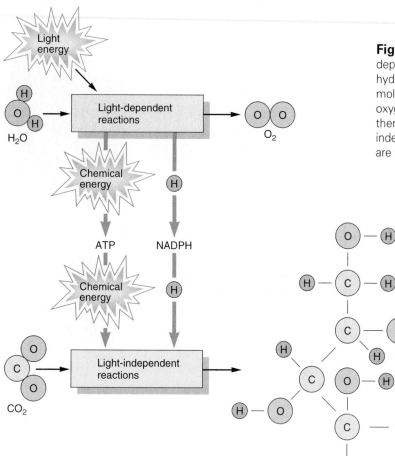

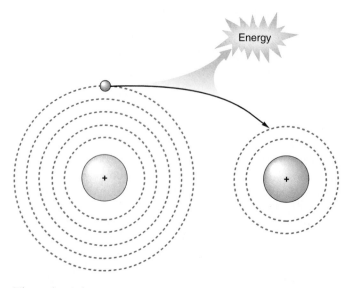

Sugar ($C_6H_{12}O_6$)

Figure 6.3 Overview of photosynthesis. The light-dependent reactions use light energy from the sun and hydrogen atoms from water to build two kinds of carrier molecules: ATP and NADPH. Oxygen gas (formed from the oxygen in water) is released in the process. ATP and NADPH then carry chemical energy and hydrogen atoms to the light-independent reactions, where ATP, NADPH, and carbon dioxide are used to build sugar.

The Light-Dependent Reactions

The light-dependent reactions require light; they convert light energy into chemical energy, which is captured within ATP and NADPH (a molecule that transports hydrogen atoms and electrons). Only chemical energy can do biological work.

Light energy becomes chemical energy when it moves electrons farther from their atomic nuclei:

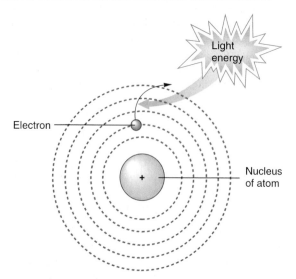

Recall that the farther an electron is from a nucleus the more chemical energy (see page 86) it holds.

The additional energy held by the displaced electron is released when the electron moves closer to the nucleus of another atom:

The released energy is captured and used to form special carrier molecules, ATP or NADPH, for later use in the light-independent reactions.

The light-dependent reactions of photosynthesis take place on membranes, where enzymes and other molecules that promote the reactions are embedded. In chloroplasts of plants and algae, these membranes are flattened sacs, called **thylakoids,** which are arranged in stacks called **grana** (fig. 6.4).

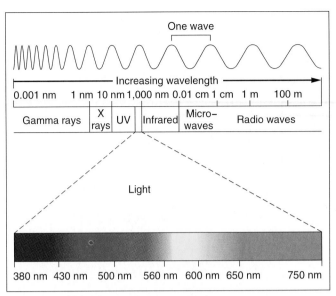

Figure 6.5 The forms of radiant energy. This arrangement of radiant energy, from shorter waves of higher energy to longer waves of lower energy, is known as the electromagnetic spectrum. The shorter waves are measured in nanometers, abbreviated nm (l nm = l billionth of a meter), and the longer waves are measured in meters. Photosynthesis makes use of intermediate wavelengths, which are called light.

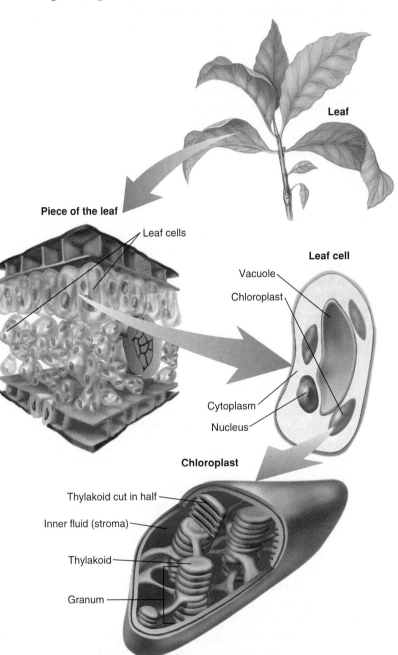

Figure 6.4 Location of photosynthesis in a plant. The biochemical pathways of photosynthesis are located in cells within the leaves. These cells contain many chloroplasts and are green in color because of the green pigment chlorophyll. The enlarged chloroplast shown here has been cut open to show its internal structure: flattened sacs, called thylakoids, that are arranged in stacks, called grana (sing., granum). The light-dependent reactions occur on the membranes of the thylakoids and the light-independent reactions occur within the fluid surrounding the grana. This inner fluid of the chloroplast is called the stroma.

The Capture of Light Energy

Energy from the sun is radiant energy, which travels in units, called photons, of different wavelengths. The shorter the wavelength, the more energy it holds. Light is radiant energy that travels in waves of intermediate lengths (fig. 6.5) and holds intermediate amounts of energy.

Light that strikes an object, such as a leaf, may be transmitted, reflected, or absorbed. Transmitted light passes through the object. Reflected light bounces off the object and is seen as a color. Absorbed light has its energy converted into the energy of motion: movements of electrons, atoms, or molecules within the object. Only absorbed light is available for photosynthesis.

Pigments   An organic molecule that absorbs light is a **pigment.** The bonds of a pigment are unusual: electrons of the carbon atoms are not held tightly by the atomic nuclei, so they oscillate from one carbon atom to another.

Light waves contain sufficient energy to lift an oscillating electron completely out of a pigment. When this happens, light energy becomes chemical energy. The displaced electron, which carries the transformed energy, is said to be an excited electron.

There are many kinds of pigments. Each kind absorbs certain wavelengths of light more than others. The main pigments in plants are chlorophylls and carotenoids, with several forms of each type. Chlorophylls appear green in color because they reflect more green wavelengths and

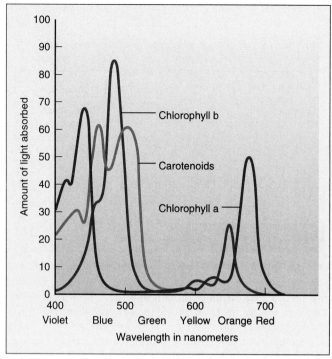

Figure 6.6 Wavelengths of light absorbed by chlorophylls and carotenoids. Plants absorb mostly violet, blue, orange, and red light. They reflect mostly green and yellow light, which gives the leaf its green or greenish yellow color. In this graph, the amount of light of each wavelength that is absorbed by a pigment is shown as the percent of total light absorbed by that pigment.

absorb more red and blue wavelengths. Carotenoids appear yellow, orange, and red because they reflect these wavelengths and absorb mostly blue and green wavelengths. The wavelengths absorbed by the two kinds of pigments, working together, are shown in figure 6.6.

Primary Electron Acceptors   An unusual feature of photosynthesis is that when an electron is lifted out of a pigment it is captured by a special molecule called a **primary electron acceptor.** This molecule holds the electron far from the pigment and also far from its own atomic nuclei. In this way, the primary electron acceptor preserves the energy carried by the excited electron.

Other kinds of pigments, such as the ones in your skin and clothing, are not associated with molecules that can catch and hold the excited electrons. When these pigments absorb light, their electrons move out of the molecule and then drop right back into it. As the electrons return to the pigment, the chemical energy they briefly held is converted to heat. Your skin and clothing become warm.

The transfer of an electron from a pigment to a primary electron acceptor is an oxidation-reduction reaction (see page 86): the pigment is oxidized (loses an electron) and the electron acceptor is reduced (gains an electron).

Formation of ATP and NADPH   From the primary electron acceptor, an excited electron travels either (1) along an electron transport chain, resulting in the formation of ATP, or (2) into an electron carrier, to form NADPH.

The **electron transport chain,** shown in figure 6.7, is a chain of molecules in which atoms in each molecule hold the transported electrons closer to their nuclei than atoms in the previous molecule. Energy is released with each transfer of an electron from one molecule to the next in the chain. The passage of an electron along the electron transport chain is a series of oxidation-reduction reactions. Some of the released energy is transferred to molecules of ATP, which later are used to build sugar in the light-independent reactions. The rest becomes heat.

Excited electrons that do not enter the electron transport chain are transported by an **electron carrier** to the light-independent reactions, where they, too, are used to build sugar. The electron carrier of photosynthesis is $NADP^+$ (nicotinamide adenine dinucleotide phosphate). This molecular ion is the same as NAD^+, an electron carrier in cellular respiration, except that it has an added phosphate group. $NADP^+$ picks up two excited electrons and a proton to form NADPH:

$$NADP^+ + 2 \text{ electrons} + 1 \text{ proton} \longrightarrow NADPH$$

(One of the electrons neutralizes the positive charge on $NADP^+$ and the other combines with the proton to form the hydrogen atom.)

The Pathway of Electrons

The molecules that participate in the light reactions work together in units called **photosystems.** Each photosystem has a certain combination of chlorophylls and other pigments that absorb specific wavelengths of light. A photosystem consists of (1) several hundred antenna pigments, (2) a reaction center, and (3) a primary electron acceptor. Hundreds of photosystems operate at the same time within a chloroplast or a photosynthetic bacterium.

The **antenna pigments** gather energy and funnel it to a special pigment called the reaction center. Antenna pigments are chlorophylls and carotenoids that absorb light but do not actually lose electrons. When electrons in these pigments absorb light, they are lifted out of the molecule

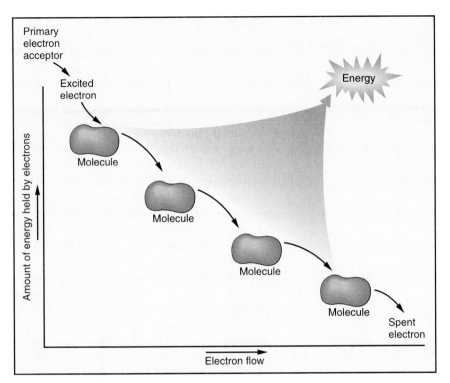

Figure 6.7 Electron transport chain. An excited electron travels from the primary electron acceptor to the first molecule in the electron transport chain. The atoms of each molecule in the chain hold the traveling electron closer to their nuclei, and energy is released every time the electron passes from one molecule to the next.

and then fall right back into it, releasing energy that is transferred to adjacent molecules and eventually to the reaction center.

The **reaction center** of a photosystem is a chlorophyll molecule that loses electrons. The energy it receives, directly from sunlight and indirectly from antenna pigments, lifts electrons out of the molecules and into the primary electron acceptor of the photosystem.

The primary electron acceptor of a photosystem captures excited electrons lost by the reaction center and passes them to either the electron transport chain or the electron carrier.

Photosystems I and II Photosynthesis involves two kinds of photosynthetic units: photosystem I and photosystem II. They absorb light differently and process electrons and energy in different ways (table 6.1).

The reaction center of photosystem I is a chlorophyll molecule called P700, which absorbs most strongly the light waves that are 700 nanometers long (see fig. 6.5).

The reaction center of photosystem II is a chlorophyll molecule called P680, which absorbs most strongly the light waves that are 680 nanometers long. Excited electrons in photosystem I are transferred to NADPH, whereas in photosystem II the electrons are transferred through an electron transport chain to the reaction center of photosystem I.

Photosystem I can function alone, but it is usually connected to photosystem II for a more efficient harvest of light energy. Plants adjust the relative amounts of each photosystem in response to different light conditions. The two systems are linked by the electron transport chain, as shown in figure 6.8. The flow of electrons through the linked photosystems is described following.

Linked Photosystems Electrons travel in pairs along a fixed path from photosystem II to photosystem I. Light absorbed by the antenna pigments and reaction center of photosystem II lifts electrons out of the reaction center.

The electrons displaced from the reaction center are replaced by two electrons from water: a water molecule is split into an oxygen atom and two hydrogen atoms. The hydrogen atoms, in turn, are broken apart to form two electrons and two protons:

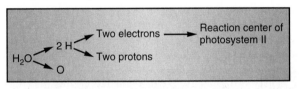

Table 6.1		
Comparisons of the Two Photosystems		
Component	**Photosystem I**	**Photosystem II**
Reaction center	Chlorophyll P700	Chlorophyll P680
Source of electrons	Photosystem II	H_2O
Where the electrons end up	NADPH	Reaction center of photosystem I
Where the energy ends up	NADPH	ATP

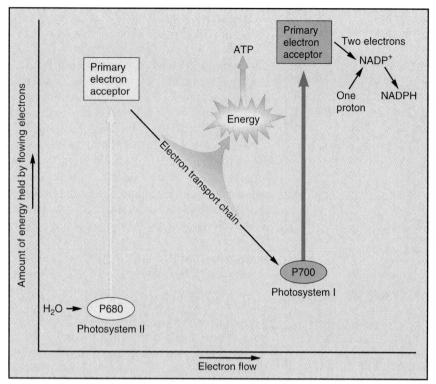

to pump protons (H^+) across the thylakoid membrane (from the outside to the inside of the thylakoid) of a chloroplast (fig. 6.9a). A proton gradient is established, with higher concentrations of protons accumulating inside the thylakoid.

When a channel protein (see page 73) in the membrane opens, protons travel through the channel to the other side. They are moved by three forces: diffusion (high to low concentrations), repulsion of positive charges within the thylakoid, and attraction to negative charges outside the thylakoid. As the protons move through the membrane, they release energy that is used to build ATP from ADP and phosphate ions (fig. 6.9b).

The formation of ATP from a proton gradient is known as **chemiosmotic phosphorylation.** The term *chemi* refers to the chemical gradient of protons, *osmotic* refers to diffusion through a membrane, and *phosphorylation* refers to the addition of a phosphate ion to ADP.

From photosystem II, the "spent" electrons from the electron transport chain enter the reaction center of photosystem I, where they replace excited electrons lifted by light energy out of this pigment. Again, the electrons travel in pairs. Excited electrons from the reaction center of photosystem I are caught by a primary electron acceptor and passed to the electron carrier, $NADP^+$. The two excited electrons plus a proton combine with $NADP^+$ to form a molecule of NADPH.

Before ending this section, let us see what happened to the components of the water molecule that started the process. Recall that the two hydrogen atoms of a water molecule were stripped of their electrons, which moved into the reaction center of photosystem II. One of the remaining protons later joined one of the electrons to form a hydrogen atom within a molecule of NADPH. The other proton will join, during the light-independent reactions, the other electron picked up by NADPH to form a second hydrogen atom. The oxygen atom of the water molecule combines with another oxygen atom (from another water molecule simultaneously stripped of its electrons) to form oxygen gas (O_2), which diffuses out of the cell:

Figure 6.8 Pathway of electrons through the two photosystems. Light absorbed by the pigments of photosystem II (shown in yellow) lifts electrons out of the reaction center (P680) and into the primary electron acceptor. These excited electrons are immediately replaced in the reaction center by electrons stripped from a water molecule. From the primary electron acceptor, the excited electrons travel along an electron transport chain, where their extra energy is released and used to build ATP. The spent electrons enter the reaction center (P700) of photosystem I (shown in blue), where they replace electrons that have been lifted, by the energy of sunlight, out of the reaction center and into the primary electron acceptor. The electrons move from the primary electron acceptor to $NADP^+$, which is converted to NADPH.

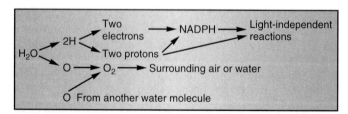

Two things have been accomplished by the light-dependent reactions: (1) the formation of ATP, which provides chemical energy for building sugar, and (2) the

Through this process, water continually feeds electrons into photosystem II:

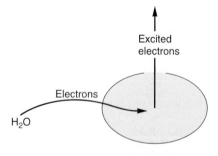

Excited electrons lifted out of the reaction center are caught by the primary electron acceptor and passed to the electron transport chain.

Energy is released as electrons are transferred from molecule to molecule along the electron transport chain. This energy is used by transport proteins (see page 73)

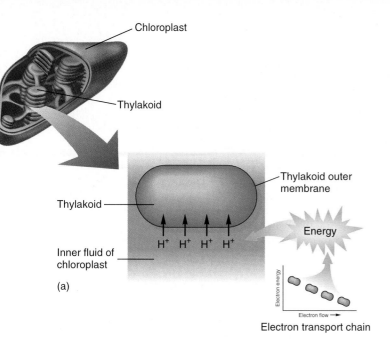

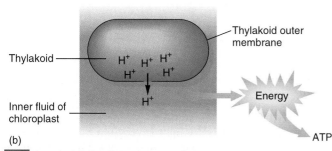

Figure 6.9 The proton gradient. (*a*) Energy released from the electron transport chain is used by transport proteins to pump protons (H⁺) across a membrane, from the outside to the inside of a thylakoid. (*b*) As the protons diffuse through a channel protein in the membrane, they release energy that is transferred to ATP.

formation of NADPH, which provides chemical energy, an electron, and a hydrogen atom for building sugar. Moreover, some of the oxygen gas that forms is used by the cell in cellular respiration. The rest diffuses out of the cell and into the surrounding environment, where it is available for use by animals and other organisms. The materials that enter and leave the light-dependent reactions are summarized in table 6.2.

The Light-Independent Reactions

The light-independent reactions do not require light; they build sugar from carbon dioxide, using the ATP and NADPH formed during the light-dependent reactions. Energy within molecules of ATP and NADPH is used to build covalent bonds within a sugar molecule. Hydrogen atoms and electrons (which join with protons to form more hy-

Table 6.2

The Light-Dependent Reactions: Inputs and Outputs

Inputs	Outputs	Fate of Outputs
Sunlight	Energy in ATP and NADPH	Energy in glucose
Electrons from water	Electrons in NADPH	Part of hydrogen atoms in glucose
Protons from water	Protons in NADPH and lone protons (H⁺)	Part of hydrogen atoms in glucose
Oxygen from water	Oxygen gas	Released from cell or used in cellular respiration

drogen atoms) within molecules of NADPH become incorporated into the structure of a sugar molecule:

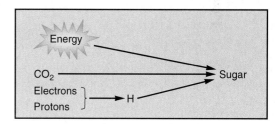

Sugar is built by a biochemical pathway called the Calvin-Benson cycle, which is located within the inner fluid (cytosol) of a photosynthetic bacterium or within the inner fluid (stroma) of a chloroplast (see fig. 6.2).

The **Calvin-Benson cycle** is a biochemical pathway that builds a three-carbon sugar from carbon dioxide, hydrogen atoms, and chemical energy. The pathway is a cycle in which the molecule that begins the pathway—ribulose bisphosphate (RuBP)—becomes an end product that begins the same pathway again:

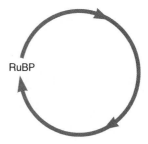

The cycle, shown in figure 6.10, begins when carbon dioxide joins with RuBP, which is a five-carbon molecule. (The enzyme that catalyzes this reaction, called RuBP carboxylase, is the most plentiful protein on earth.) The

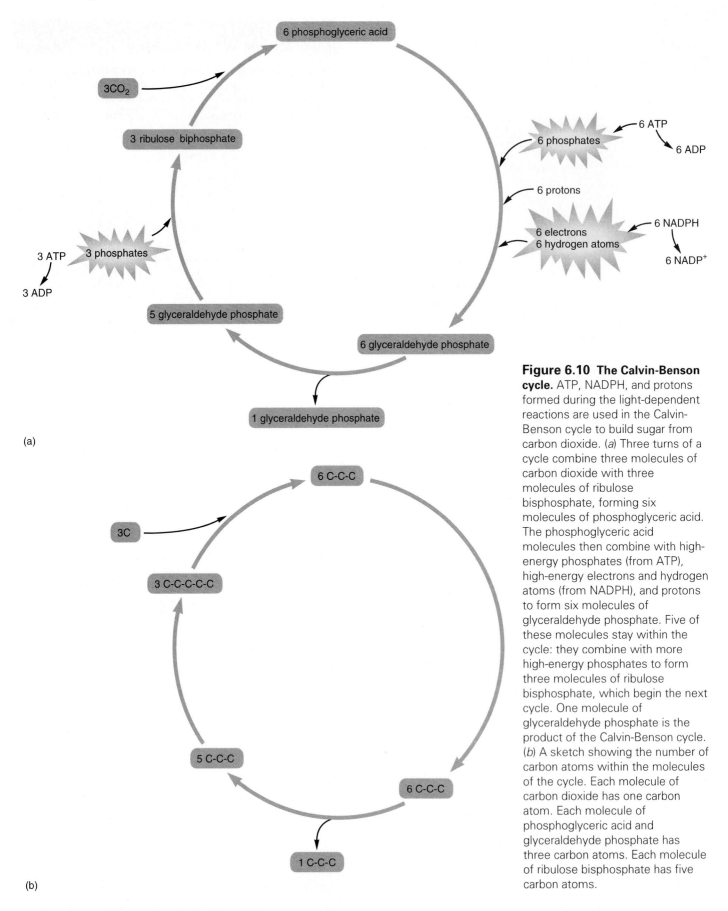

(a)

(b)

Figure 6.10 The Calvin-Benson cycle. ATP, NADPH, and protons formed during the light-dependent reactions are used in the Calvin-Benson cycle to build sugar from carbon dioxide. (*a*) Three turns of a cycle combine three molecules of carbon dioxide with three molecules of ribulose bisphosphate, forming six molecules of phosphoglyceric acid. The phosphoglyceric acid molecules then combine with high-energy phosphates (from ATP), high-energy electrons and hydrogen atoms (from NADPH), and protons to form six molecules of glyceraldehyde phosphate. Five of these molecules stay within the cycle: they combine with more high-energy phosphates to form three molecules of ribulose bisphosphate, which begin the next cycle. One molecule of glyceraldehyde phosphate is the product of the Calvin-Benson cycle. (*b*) A sketch showing the number of carbon atoms within the molecules of the cycle. Each molecule of carbon dioxide has one carbon atom. Each molecule of phosphoglyceric acid and glyceraldehyde phosphate has three carbon atoms. Each molecule of ribulose bisphosphate has five carbon atoms.

product of this union, a six-carbon molecule, immediately breaks apart to form two molecules of phosphoglyceric acid (PGA), which is a three-carbon molecule:

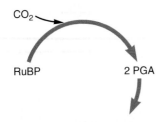

Each PGA then picks up a phosphate group from ATP (along with the energy it carries) and two hydrogen atoms (along with the energy carried within their excited electrons). These two hydrogen atoms come from NADPH (which supplies one hydrogen atom and one electron) and an unattached proton. The two PGA are converted by these reactions to two molecules of glyceraldehyde phosphate (GP):

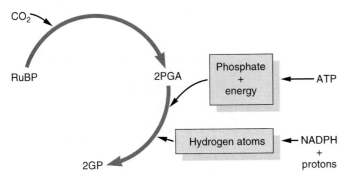

Three molecules of carbon dioxide are processed by three turns of the cycle. They are picked up by three molecules of RuBP to form six molecules of PGA. These molecules, in turn, form six molecules of glyceraldehyde phosphate.

One of the six molecules of glyceraldehyde phosphate is the product of the reaction—a three-carbon sugar that leaves the cycle. The other five molecules remain within the cycle; they are converted to three molecules of RuBP for another turn of the cycle.

The materials that enter and leave the Calvin-Benson cycle are listed in table 6.3. The unmodified Calvin-Benson cycle is known as the C_3 pathway, since the first stable molecule in the cycle (PGA) has three carbon atoms. Some plants use a modified version of the cycle, called the C_4 pathway, because in that cycle the first stable molecule has four carbon atoms. How this modified cycle was discovered is described in the accompanying Research Report.

Fates of the Products

Sugars built by photosynthesis provide most forms of life with chemical energy for cellular work and carbon skeletons for building other monomers. The monomers become building blocks for assembling polymers. Amino

Table 6.3

The Light-Independent Reactions: Inputs and Outputs of the Calvin-Benson Cycle		
Inputs	**Outputs**	**Fates of the Outputs**
RuBP	RuBP	Begins another cycle
CO₂ Energy from ATP and NADPH Electrons from NADPH Protons from NADPH and H⁺	Glyceraldehyde phosphate	Converted to glucose or another monomer

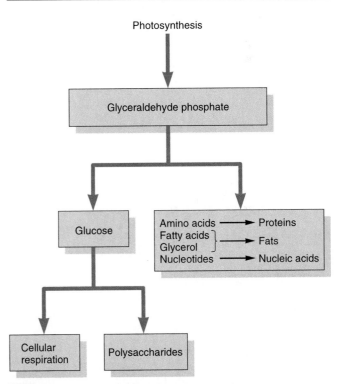

Figure 6.11 Fates of the photosynthetic products.
The immediate product of photosynthesis, glyceraldehyde phosphate, is converted into monomers. One kind of monomer, glucose, is used as a fuel in cellular respiration and as a building block in polymers called polysaccharides. Other kinds of monomers, such as amino acids, are used to build other kinds of polymers—proteins, fats, and nucleic acids (DNA and RNA).

acids are the monomers of proteins; fatty acids and glycerol are the monomers of fats; nucleotides are the monomers of DNA and RNA. The fates of sugars manufactured by photosynthesis are diagrammed in figure 6.11 and described following.

115

Research Report

Discovery of the C₄ Pathway

Biologists sometimes miss what happens in nature because they work with only a few kinds of organisms that are easy to grow in the laboratory. Early experiments on the light-independent reactions of photosynthesis are a good example of just such a mistake. These experiments were conducted on unicellular algae and spinach in the late 1940s, chiefly at the University of California, Berkeley. They revealed that algae and spinach use the same biochemical pathway to build sugars from carbon dioxide. This pathway was named the Calvin-Benson cycle after its two discoverers.

Algae and spinach are not at all closely related. One is a unicellular aquatic organism and the other is a multicellular terrestrial organism. Since these very different organisms have the same biochemical pathway, biologists hypothesized that all algae and plants have the same pathway. The hypothesis seemed so reasonable that no one questioned it at the time.

Further research on the Calvin-Benson cycle, however, revealed a peculiar phenomenon: the cycle is inhibited by oxygen gas released from the light-dependent reactions. This observation did not make sense—why would a product of the first part of photosynthesis inhibit a reaction in the second part?

Oxygen gas inhibits the Calvin-Benson cycle especially when conditions are hot and dry. The reason is that a plant partly closes tiny openings, called stomata, under these conditions to reduce the evaporation of water. A side effect is that very little carbon

dioxide can enter the leaf and very little oxygen gas can leave. The concentration of carbon dioxide declines and the concentration of oxygen gas rises.

As oxygen becomes more abundant than carbon dioxide within the chloroplasts, it latches onto the enzyme that normally binds carbon dioxide to RuBP. Oxygen rather than carbon dioxide then enters the Calvin-Benson cycle:

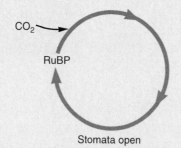

Stomata open

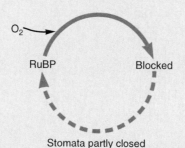

Stomata partly closed

The cycle shuts down and synthesis of sugars comes to a halt.

If all plants have the Calvin-Benson cycle and if oxygen gas inhibits the cycle, how do plants synthesize sugar while their stomata are closed? How does corn, for example, grow so rapidly during the hot, dry summers of Iowa and Illinois? (See fig. 6.A.)

Some researchers suggested that perhaps plants in hot, dry climates do not use the Calvin-Benson cycle. They questioned the hypothesis that all plants and algae use the same cycle

Figure 6.A The biochemistry of corn is adapted to hot climates. Corn uses the C₄ pathway to pump carbon dioxide into the Calvin-Benson cycle. This pathway conserves water.

and began to search for a different biochemical pathway—one not inhibited by oxygen gas. Such a pathway was found in the 1960s by M. D. Hatch in Australia and C. R. Slack in England. They discovered that sugarcane, a plant that grows in hot, dry climates, fixes carbon (i.e., joins carbon dioxide with an organic molecule) in a way that is different from carbon fixation in algae and spinach.

The Hatch-Slack pathway avoids oxygen inhibition by fixing carbon in leaf cells that do not capture light energy and so do not develop high concentrations of oxygen. Moreover, the Hatch-Slack pathway uses a different enzyme to pick up carbon dioxide and

bind it to RuBP. This enzyme binds only with carbon dioxide, even when there is hardly any in the leaf; it will not bind with oxygen gas. Carbon dioxide fixed in the nonphotosynthetic cells is pumped in massive quantities into photosynthetic cells, where it monopolizes the enzyme that binds it to RuBP. Consequently, carbon dioxide rather than oxygen joins RuBP and the Calvin-Benson cycle proceeds rapidly—even with the stomata nearly closed:

These two ways of picking up carbon dioxide are known as the C_3 and C_4 pathways. In the C_3 pathway, the first stable molecule (phosphoglyceric acid) after carbon fixation has three carbon atoms. Most plants use this pathway. In the C_4 pathway, the first stable molecule has four carbon atoms. Many plants in hot, dry climates use this pathway. (A third way of fixing carbon, known as crassulacean acid metabolism, or CAM, has been discovered in cactuses and other plants that live in deserts. It is much like the C_4 pathway, but CAM plants further reduce water loss by opening their stomata only at night.)

The C_4 pathway is particularly common in grasses, a category of plants that includes corn and sugarcane. The relation between the pathway of carbon fixation and climate is illustrated in figure 6.B. The advantage of the C_4 pathway in warm, dry climates is that it enables a plant to conserve water by partially closing its stomata while it builds sugars.

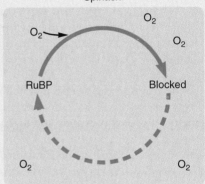

Photosynthetic cell when stomata of leaf are partly closed

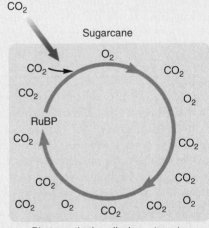

Photosynthetic cell when stomata of leaf are partly closed

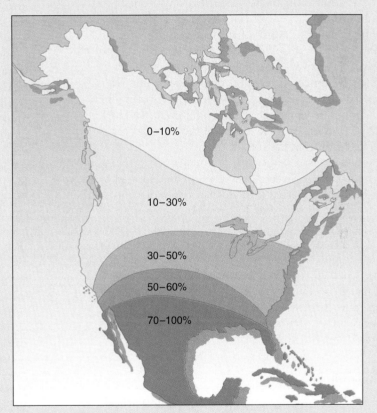

Figure 6.B The distribution of C_4 grasses in North America. In more southern regions, where the climate is warmer and water evaporates more rapidly, more than half the grass species use the water-conserving C_4 pathway. In more northern regions, where the climate is cooler and water evaporates more slowly, a smaller percentage of the grass species use the C_4 pathway.

From Peter H. Raven and George B. Johnson, *Biology, Updated Version,* 3d ed. Copyright © 1995 Wm. C. Brown Communications, Inc., Dubuque, Iowa. All Rights Reserved. Reprinted by permission.

The Monomers

The immediate product of the Calvin-Benson cycle is glyceraldehyde phosphate. This three-carbon sugar forms the foundation of monomers. It may be used directly to build amino acids, fatty acids, and other kinds of monomers or it may combine with a second glyceraldehyde phosphate to form glucose.

Glyceraldehyde Phosphate   Some of the glyceraldehyde phosphate formed by photosynthesis is converted into monomers other than glucose. Some of these monomers, such as glycerol and fatty acids, consist of the same kinds of atoms—carbon, hydrogen, and oxygen—as glyceraldehyde phosphate and so are easy to build from this sugar. Other monomers, such as amino acids and nucleotides, require the addition of other kinds of atoms, such as nitrogen and sulfur, to the carbon skeleton.

Photosynthetic organisms acquire atoms from their physical environment. As we have seen in this chapter, the carbon and oxygen atoms of sugars come from carbon dioxide in the air (or dissolved in water) and the hydrogen atoms come from water molecules. Other atoms, such as nitrogen, sulfur, phosphorus, and potassium, come from ions dissolved in the soil or in the surrounding water. The roots of plants, for example, absorb ammonium (NH_4^+), nitrate (NO_3^-), potassium (K^+), sulfate ($SO_4^=$), and phosphate ($PO_4^=$) ions from the soil.

Glucose   Most of the glyceraldehyde phosphate formed by photosynthesis is converted into glucose: two molecules of the three-carbon sugar, glyceraldehyde phosphate, join to form one molecule of the six-carbon sugar, glucose. Glucose, in turn, is used as an immediate source of energy or to build larger molecules.

About half of the glucose formed by photosynthesis becomes fuel for cellular respiration in the plant, alga, or bacterium. Energy released from this pathway supports cellular work. Glucose that is not used right away to supply energy is used to build polymers.

The Polymers

Polymers are built from monomers, which photosynthetic organisms synthesize from glyceraldehyde phosphate. Polysaccharides, such as cellulose and starch, are built from glucose; fats are built from glycerol and fatty acids; proteins are built from amino acids; and nucleic acids are built from nucleotides. These polymers are used to store energy and to form membranes, enzymes, genes, and other components of cells as the organism grows and reproduces.

When a photosynthetic organism is eaten by another organism, its polymers are digested into their component monomers. The consumer uses these monomers as sources of energy (in cellular respiration) and as building materials for constructing its own kinds of polymers. Only photosynthetic organisms (and a few chemosynthetic bacteria) can build an organic monomer from inorganic materials. Other forms of life, such as mushrooms, worms, sparrows, wolves, and humans, must make use of the monomers synthesized by plants, algae, and photosynthetic bacteria.

SUMMARY

1. In photosynthesis, light energy is converted to chemical energy and then used to build sugars. The process provides photosynthetic, as well as other, forms of life with chemical energy, carbon skeletons, and oxygen gas. Photosynthesis has two stages: the light-dependent reactions and the light-independent reactions.

2. The light-dependent reactions convert light energy to chemical energy. Light energy is captured when it lifts an electron out of a pigment. The excited electron holds the energy, as chemical energy, in its new position within a primary electron acceptor. It then passes to either an electron transport chain, where its energy is transferred to ATP, or to the electron carrier $NADP^+$.

3. In the light-dependent reactions, electrons travel from photosystem II to photosystem I. Each system consists of antenna pigments, a reaction center, and a primary electron acceptor. In photosystem I, the reaction center is P700 and excited electrons are transferred to NADPH. In photosystem II, the reaction center is P680 and energy is passed, via an electron transport chain and a proton gradient, to ATP. Electrons travel from water through photosystem II (forming ATP) and then through photosystem I (forming NADPH). After removal of electrons from water, the remaining oxygen atom is released as part of O_2 and the remaining protons later rejoin the electrons to reform hydrogen atoms.

4. In the light-independent reactions, ATP, NADPH, and protons formed during the light-dependent reactions are used to build sugars from carbon dioxide. The biochemical pathway is known as the Calvin-Benson cycle. The cycle begins when ribulose bisphosphate picks up carbon dioxide to form phosphoglyceric acid, which acquires energy and hydrogen atoms from ATP, NADPH, and protons to form glyceraldehyde phosphate, a three-carbon sugar.

5. The immediate product of photosynthesis is glyceraldehyde phosphate, which photosynthetic organisms convert into monomers. The monomers are used as fuels (in cellular respiration) and as building blocks of polymers. When a photosynthetic organism is consumed, its polymers are digested into monomers and used by the consumer as fuels and as building blocks of polymers.

KEY TERMS

Antenna pigment (page 110)
Calvin-Benson cycle (page 113)
Chemiosmotic phosphorylation (KEE-mee-oz-MAH-tik FOS-for-uh-LAY-shun) (page 112)
Chloroplast (KLOR-oh-plast) (page 107)
Electron carrier (page 110)
Electron transport chain (page 110)
Grana (GRAEH-nuh) (page 109)
Photosynthesis (FOH-toh-SIN-thuh-sis) (page 106)
Photosystem (page 110)
Pigment (PIG-muhnt) (page 109)
Primary electron acceptor (page 110)
Reaction center (page 111)
Thylakoid (THEYE-luh-koyd) (page 109)

STUDY QUESTIONS

1. Write the overall equation for photosynthesis. What are the two main uses for the sugar that is synthesized?
2. Describe how light energy is converted to chemical energy during photosynthesis.
3. Describe how ATP forms during the light-dependent reactions. Describe how NADPH forms during the light-dependent reactions.
4. Describe what happens to the water used in photosynthesis.
5. What molecule in the Calvin-Benson cycle picks up carbon dioxide? What molecules provide energy and hydrogen atoms for the cycle? Where do these molecules come from?
6. What is the immediate product of the Calvin-Benson cycle? What happens to this product?

CRITICAL THINKING PROBLEMS

1. Chloroplasts may once have been photosynthetic bacteria. What kinds of evidence would you look for to support this hypothesis?
2. Explain why an increase in plant biomass would reduce the greenhouse effect (i.e., global warming caused by an increase in carbon dioxide within the atmosphere).
3. Photosynthesis provides sugars and oxygen gas to most forms of life. What do animals, such as earthworms and people, use the sugars for? What do they use the oxygen gas for?

4. When nature is not changing, the total rate of cellular respiration in all organisms is equal to the total rate of photosynthesis in all photosynthetic organisms. Explain why this must be so.

SUGGESTED READINGS

Introductory Level:

Coleman, G., and W. J. Coleman. "How Plants Make Oxygen," *Scientific American*, February 1990, 50–58. A description of the research that led to our understanding of how electrons move from a water molecule to the reaction center of a photosystem.

Hall, D., and K. Rao. *Photosynthesis*. Baltimore: Arnold Press, 1987. An introduction to photosynthesis and how it was discovered.

Moore, P. D. "Evolution of Photosynthetic Pathways in Flowering Plants," *Nature* 295 (1982): 647–48. Comparisons of the C_3, C_4, and CAM pathways of carbon fixation—the advantages and evolutionary histories.

Youvan, D. C., and B. L. Marrs. "Molecular Mechanisms of Photosynthesis," *Scientific American*, June 1987, 42–48. A description of how the proton gradient is established during the light reactions.

Advanced Level:

Bogorad, L., and I. K. Vasil. *The Photosynthetic Apparatus: Molecular Biology and Operation*. San Diego: Academic Press, 1991. A thorough treatment of what we know about the machinery of photosynthesis in both eukaryotic and prokaryotic cells.

Taiz, L., and E. Zeiger. *Plant Physiology*. Redwood City, CA: Benjamin-Cummings, 1991. A good source of detailed information about the biochemistry of photosynthesis.

❀ EXPLORATION

Interactive Software: Photosynthesis

This interactive exercise explores the route that electrons take as they travel from chlorophyll through the two photosystems. The student is provided with a diagram of a chloroplast membrane and is able to modify the intensity and wavelengths of light to see how these variables govern the production of ATP and NADPH.

Cellular Reproduction and Inheritance

Cell division: one cell becomes two cells. These cancer cells are abnormal in that they cannot stop dividing.

Exploring Themes

Theme 1

The fundamental unit of life is the cell.

The DNA molecules of a cell contain genes—instructions for building proteins, which, in turn, determine the traits of the organism (chapters 7–12).

Theme 2

All cells carry out the same basic processes.

A cell reproduces by forming two identical cells (chapter 7) or four genetically different sperm or eggs (chapter 8). The traits of an offspring, which develops from a fertilized egg, are determined by the genes carried by the sperm and the egg (chapter 9) and by how the two sets of genes interact with each other (chapter 10). Genes determine traits by directing the synthesis of proteins (chapter 11). Genetic engineers manipulate traits by manipulating genes (chapter 12).

Mitosis and the Cell Cycle

Objectives

In this chapter, you will learn
–why a cell divides instead of growing
  larger;
–how DNA molecules are arranged in
  chromosomes;
–how DNA molecules make exact
  copies of themselves;
–how each new cell gets a complete
  copy of the DNA;
–when and where mitosis occurs in
  living things;
–how bacteria reproduce.

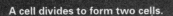

A cell divides to form two cells.

You began life as a single cell, created when your father's sperm merged with your mother's egg. Your "starter" cell, a fertilized egg, contained the unique combination of genes that governed development of your unique traits, such as the color of your eyes and the shape of your nose. The fertilized egg then began to divide: it duplicated all its genes and split into two cells, passing one complete set of genes to each cell. These two cells, in turn, duplicated the genes they received and thus you became four cells. The four cells formed eight cells, which formed sixteen cells, and so on until they formed the trillions of cells that constitute your body.

When and how a cell divides is programmed within its genes. During your own development, the cells forming each region of your body divided until they had achieved their programmed goal. Cells forming your liver, for example, divided until they had developed into a liver of a particular size. Cell division is normally under strict control.

We have much to learn about the genes that control cell division. Most of what we know comes from research on cancer, the devastating disease characterized by cells that cannot stop dividing. A cancer cell has no sense of when a structure is complete; it continues to divide and would do so forever if its vast number of descendants didn't eventually overcrowd the body and prevent normal cells from carrying out their functions.

Why does a cell divide? How does it ensure that each new cell receives a complete set of genes? How does it know when cell division is no longer appropriate? In this chapter we explore the process of cell division.

Limits to Cell Size

A cell is too small to be seen with the naked eye. Eukaryotic cells (see page 62) are typically between 0.01 and 0.1 mm in diameter; they become visible as tiny specks when clustered in a group of about ten cells. Prokaryotic cells (see page 59) are even smaller, about 0.001 mm in diameter, and can be seen as specks when present in groups of more than 100. Why are cells so small? Why do they divide when they reach a certain size rather than continue to grow?

The main reason is that when a cell grows beyond a certain size, it can no longer import and export materials rapidly enough to remain active. Cellular activity is pro-

portional to cell volume. Import and export, which occur through the plasma membrane, are proportional to surface area:

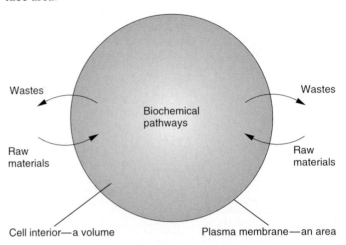

Wastes

Wastes

Biochemical pathways

Raw materials

Raw materials

Cell interior—a volume

Plasma membrane—an area

It is a law of geometry that as a three-dimensional structure becomes larger, its surface area enlarges at a slower rate than its volume. As the cell grows, then, its plasma membrane enlarges but not as rapidly as its internal volume enlarges. At some point during this growth, the plasma membrane becomes too small to keep up with the increased demands of the cell interior for new materials and for waste disposal. When this happens, the cell divides to form two smaller cells. Each smaller cell now has a more efficient ratio of plasma membrane to cell volume.

Cell Division in Eukaryotic Cells

Each cell must have its own set of genes in order to build its own proteins. Thus, a dividing cell replicates all its DNA molecules, which contain genetic instructions, and places one complete set in each new cell.

Equal segregation of the two sets of DNA molecules is complicated in eukaryotic cells, for there are many molecules of DNA located within many chromosomes. Division of these cells requires a highly coordinated "dance" of the chromosomes, called mitosis, to ensure equal division of the genetic material. When a cell is not dividing, its chromosomes are too long and thin to be visible as separate structures. When a cell prepares to divide, its chromosomes shorten and thicken so they can move more easily

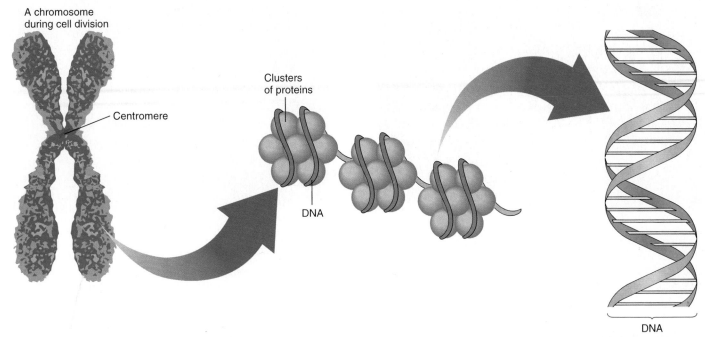

A chromosome during cell division

Centromere

Clusters of proteins

DNA

DNA

Figure 7.1 A chromosome as seen while the cell is dividing. The chromosomes duplicate themselves and then coil tightly before moving into different cells. Each chromosome is double except for one spot, the centromere, where the two parts remain held together. Each copy of the chromosome consists of a long molecule of DNA wrapped around proteins.

from one place to another without become entangled. It is only during these times that the chromosomes are visible as distinct structures.

The Karyotype

An organism's **karyotype** is its set of chromosomes as seen during cell division. At this time, each chromosome is distinct and has already been replicated but not yet separated (fig. 7.1). A karyotype describes the number of chromosomes in each cell as well as the length and shape of each chromosome.

The number of chromosomes is the same for all cells (except sperm or eggs) of an individual. A human cell, for example, contains forty-six chromosomes regardless of whether the cell is part of the skin, liver, brain, or muscle. Moreover, the number of chromosomes in each cell is the same in all the individuals of a species (except in species where chromosome number determines whether the individual is male or female). Chromosome number is similar for closely related species but does not vary in a consistent way with complexity of the species: a simple organism may have many more or many fewer chromosomes than a complex organism. The chromosome numbers of some organisms are given in table 7.1.

Table 7.1

Number of Chromosomes in the Cells of Some Organisms

Organism	Number of Chromosomes
Human	46
Chimpanzee	48
Dog	36
Cat	38
Chicken	78
Frog	26
Mosquito	6
Honeybee	32
Corn	20
Horsetail (a kind of plant)	216
Adder's tongue fern	1,262

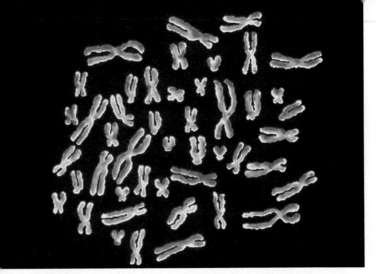

(a)

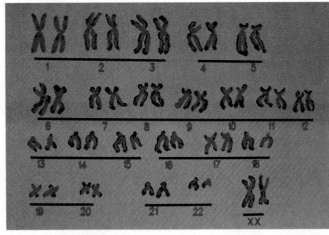

(b)

Figure 7.2 A human karyotype. (*a*) Photograph of the chromosomes taken from a human cell while it was dividing. Each chromosome is short, thick, and already duplicated (the two copies are held together by a centromere). For every chromosome of a particular length and shape there is another just like it. (*b*) The chromosomes are cut out of the photograph and arranged by pairs. This karyotype is of a female, with two similar sex chromosomes called X chromosomes. In a male the sex chromosomes are dissimilar and called the X and the Y chromosomes.

The length of a chromosome reflects the amount of DNA within the chromosome and how tightly the DNA is coiled. The shape of a chromosome depends on the position of the **centromere** (*centro*, central; *mere*, part), a constricted region that holds the duplicated chromosomes together until they move into separate cells. The centromere may be at the end, in the middle, or elsewhere within the chromosome:

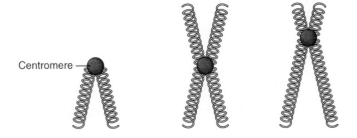

Centromere —

A karyotype is constructed by taking a sample of cells from the organism and growing them in a laboratory dish. In humans, the cells used are typically white blood cells or skin cells. The cells are treated with a chemical that stops cell division when the chromosomes are short and thick. They are then stained to highlight the chromosomes and a photograph is taken (fig. 7.2*a*). The chromosomes are cut out of the photograph and arranged according to similarities in length and shape (fig. 7.2*b*). Karyotypes are essential for identifying genetic disorders caused by chromosome breakage as well as inheritance of too many or too few chromosomes.

The chromosomes of most eukaryotic cells occur in pairs. (Each chromosome of a pair, in turn, is doubled during early phases of cell division.) The two chromosomes of each pair have the same length and shape. One of the chromosomes was inherited from the father and the other from the mother. In a human karyotype, the pairs of chromosomes are numbered from one to twenty-two: number 1 is the longest and number 22 is the shortest of the pairs. In addition, there is a pair of sex chromosomes, making a total of twenty-three pairs, or forty-six chromosomes. In a woman, the two sex chromosomes look the same and are called X chromosomes. In a man, the two sex chromosomes are quite different: one resembles the sex chromosomes in a female and is called an X chromosome; the other is much shorter and called the Y chromosome. The significance of chromosome pairs and of sex chromosomes is discussed in chapter 9.

The Cell Cycle

The **cell cycle** is a controlled sequence of events in which a small cell grows larger and then divides to form two smaller cells. The dividing cell is called the parental cell, and the newly formed cells are called the daughter cells:

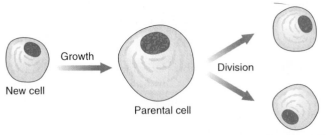

(The term daughter simply means the product of cell division; it does not imply female.) During the cycle, a cell grows to its maximum size and duplicates each chromosome. It then divides in two, with each daughter cell receiving a copy of every chromosome together with half the cytoplasm. In a rapidly dividing human cell, the entire cycle lasts about twenty-four hours.

The cell cycle has three main stages: interphase, mitosis, and cytokinesis. This cycle is diagrammed in figure 7.3, summarized in table 7.2, and described in the following sections.

Interphase   The stage of the cell cycle when the chromosomes are too long and thin to be visible is called **interphase.** A cell enters this stage as soon as it forms by cell division from a previously existing cell. The new cell is small, having received only half the cytoplasm of its parental cell. Interphase can be subdivided into three shorter periods: G_1, S, and G_2.

Early interphase, called the G_1 phase, is a time of biochemical activity during which the cell synthesizes materials (other than DNA) and grows to normal size. It synthesizes enzymes, which control biochemical pathways, and builds ribosomes, membranes, mitochondria, and other organelles. A typical cell spends most of its time in early interphase.

The S phase of interphase is a time of DNA synthesis. A cell entering this phase has committed itself to becoming two cells. Each chromosome becomes double: its DNA molecule makes an exact copy of itself, which is then wound around a protein scaffold. The two DNA molecules and their proteins remain joined by a centromere. The cell is now ready to enter mitosis.

Precise replication of DNA is an essential property of life. When Watson and Crick presented their hypothesis on the structure of DNA in 1953 (see page 48), they were particularly excited about its potential for duplication. They suggested a semiconservative model in which

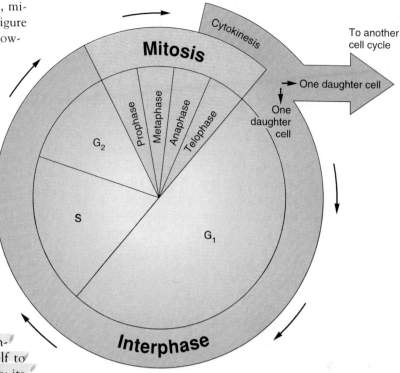

Figure 7.3 The cell cycle. The cell cycle is the period of time between when a cell forms (from another cell) and when it divides to form two daughter cells. The cycle has three phases: interphase, mitosis, and cytokinesis. Interphase, in turn, consists of a G_1 stage, when the cell synthesizes materials and grows; an S stage, when it synthesizes DNA, and a G_2 stage, when it repairs errors in the newly synthesized DNA. Mitosis consists of four stages: prophase, metaphase, anaphase, and telophase.

Table 7.2

Stages of the Cell Cycle

Stage	Main Events
Interphase	Synthesis and repair
G_1 stage	Synthesis of materials; growth of the cell
S stage	Replication of the DNA molecules
G_2 stage	Repair of errors in the newly synthesized DNA
Mitosis	Formation of two nuclei from one
Prophase	Chromosomes shorten; nuclear envelope disappears; spindle apparatus forms
Metaphase	Chromosomes line up between the two poles; spindle fibers attach to centromeres
Anaphase	Centromeres divide; sister chromatids become chromosomes and are pulled by the spindle to opposite poles
Telophase	Spindle apparatus disappears; nuclear envelope forms around each set of chromosomes
Cytokinesis	Cytoplasm splits to form two cells, each with a nucleus and about half the cellular material

the original molecule is partly conserved. According to this model, the DNA molecule splits lengthwise to expose its sequence of bases. Nucleotides then join these exposed bases, according to the base-pairing rules, to form two new molecules. This model, now considered to be correct, is shown in figure 7.4. The elegant experiment that confirmed the semiconservative model is described in the accompanying Research Report.

The S phase of interphase is followed by the G_2 phase, a time of DNA repair. Special enzymes correct errors in the newly synthesized molecules of DNA. The two copies of each chromosome, called **sister chromatids,** remain held together by the centromere, which is a short segment of the chromosome that has not yet been replicated:

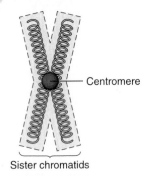

The entire structure, consisting of two chromatids held together by a centromere, is still referred to as a single chromosome.

Mitosis During **mitosis,** the sister chromatids separate and move into two nuclei. Each nucleus receives one of the sister chromatids from each chromosome:

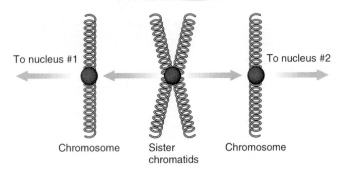

Mitosis is characterized by movements of the chromosomes from one part of the cell to another. It can be subdivided into four phases: prophase, metaphase, anaphase, and telophase (fig. 7.5). The process is, however, a continuous one with each phase flowing smoothly into the next. It takes less than an hour to complete.

In the first phase of mitosis, called prophase (*pro,* before), the stage is set for chromosome movements. Prior to prophase, the chromosomes are so long and thin that they do

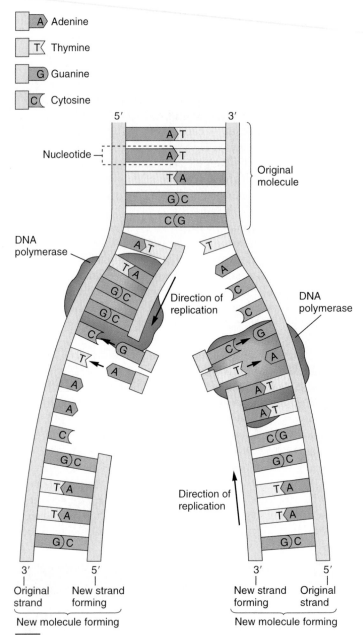

Figure 7.4 DNA replication. DNA replication begins when an enzyme pries open the double helix at a particular site. Another enzyme, DNA polymerase, then moves along the molecule, joining unattached nucleotides to attached nucleotides according to the base-pairing rules: adenine with thymine and guanine with cytosine. As the unattached nucleotides join with nucleotides in the strands, they also join with one another to form two new strands of nucleotides. Replication always proceeds from the 5′ to the 3′ end of each original strand, which means that the new strand on the left of this molecule must form in a series of short segments as the original molecule gradually opens up.

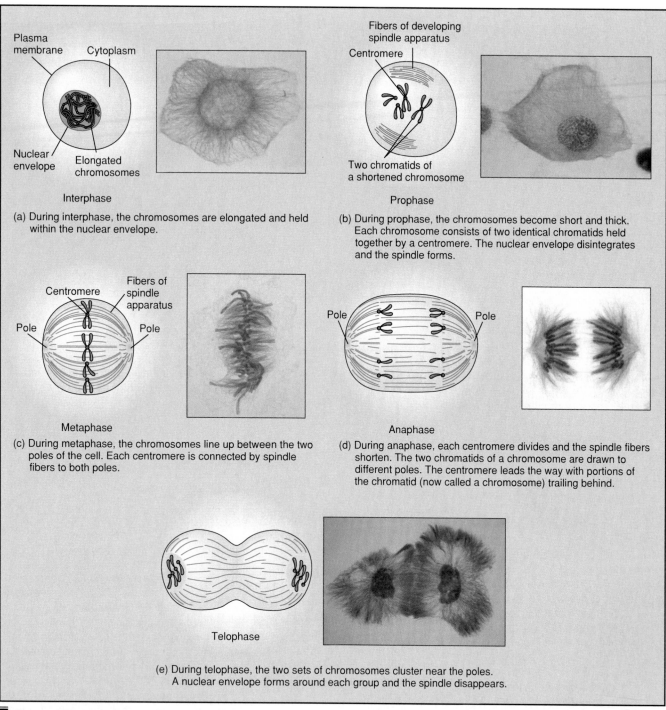

Interphase

(a) During interphase, the chromosomes are elongated and held within the nuclear envelope.

Prophase

(b) During prophase, the chromosomes become short and thick. Each chromosome consists of two identical chromatids held together by a centromere. The nuclear envelope disintegrates and the spindle forms.

Metaphase

(c) During metaphase, the chromosomes line up between the two poles of the cell. Each centromere is connected by spindle fibers to both poles.

Anaphase

(d) During anaphase, each centromere divides and the spindle fibers shorten. The two chromatids of a chromosome are drawn to different poles. The centromere leads the way with portions of the chromatid (now called a chromosome) trailing behind.

Telophase

(e) During telophase, the two sets of chromosomes cluster near the poles. A nuclear envelope forms around each group and the spindle disappears.

Figure 7.5 Interphase and mitosis. Fewer chromosomes are drawn in the sketches than appear in the photographs. The chromosomes are colored blue and the spindle apparatus is colored red.

Research Report

Discovering How DNA Replicates

James Watson and Francis Crick published their model of the DNA molecule in 1953. The final sentence of the paper, considered the greatest understatement in scientific literature, is, "It has not escaped our notice that the specific base-pairing we have proposed immediately suggests a possible copying mechanism for the genetic material."

This possible copying mechanism became the semiconservative hypothesis of DNA replication. It proposes that the molecule splits in half lengthwise to expose its sequences of bases. Unattached nucleotides in the cell then move in and bind with the exposed bases in each half of the molecule, as shown in figure 7.4. After replication, each molecule consists of an original strand and a newly synthesized strand of nucleotides.

The semiconservative hypothesis fit everything known about DNA replication. Most important, it conserves the particular sequences of bases in the molecule—the sequences that form the genes. If the sequence were disrupted, it is hard to imagine how genetic traits could be passed from cell to cell during mitosis and from parent to offspring during reproduction. The hypothesis made sense, but was it correct? DNA replication cannot, after all, be observed directly.

In 1958, Matthew Meselson and Franklin Stahl decided to tackle the problem. An outline of their logic is provided in figure 7.A. They began by

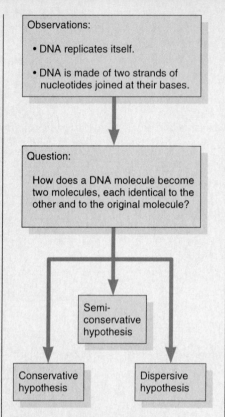

Figure 7.A The scientific logic of Meselson and Stahl.

listing all possible ways in which DNA could replicate itself. Their alternative hypotheses were as follows:

1. DNA replication is conservative, in which the original molecule remains intact and the new molecule has two newly synthesized strands of nucleotides:

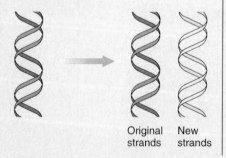

2. DNA replication is semiconservative, in which each original strand combines with a newly synthesized strand of nucleotides:

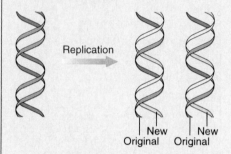

3. DNA replication is dispersive, in which the DNA molecule becomes fragmented so that original and newly synthesized strands of nucleotides are mixed along the length of both molecules after replication:

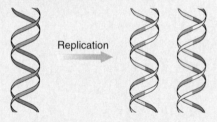

Meselson and Stahl then tested the hypotheses in an exceedingly clever fashion. They found a way of distinguishing between original strands and newly synthesized strands of nucleotides. First they grew bacteria in a medium enriched with heavy nitrogen, which has an atomic weight of 15 (from seven protons and eight neutrons) rather than ordinary nitrogen, which has an atomic weight of 14 (seven protons and seven neutrons). The bacteria synthesized nucleotides using the heavy nitrogen (nitrogen is a component of the bases).

After many generations, almost all the nitrogen in the bacterial DNA contained heavy nitrogen rather than ordinary nitrogen. The bacteria were transferred to a medium enriched with ordinary nitrogen and kept there until just one DNA replication had taken place. Then the DNA molecules were extracted from the bacteria and their relative weights determined by mixing them with a solution that formed a density gradient and spinning them in a centrifuge. After this treatment, heavier DNA settles nearer the bottom of a test tube and lighter DNA above it.

Meselson and Stahl made the following predictions:

1. If replication is conservative, then there will be two kinds of DNA molecules: heavy molecules (the original DNA) and light molecules (the newly synthesized DNA). The two kinds of molecules will form two layers in the test tube:

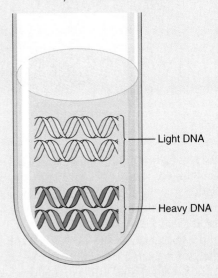

2. If replication is semiconservative, then there will be one kind of molecule and it will have an intermediate weight. Only one layer of DNA will appear in the test tube:

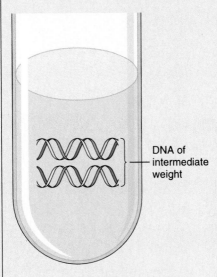

3. If replication is dispersive, then light and heavy fragments will mix and DNA molecules of intermediate weight will form. There will be just one layer of DNA in the test tube:

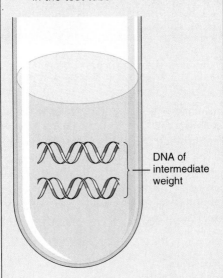

Both the semiconservative and the dispersive hypotheses predict the same outcome.

Meselson and Stahl performed the experiment and found a single layer of DNA in the test tube. The experiment eliminated the conservative hypothesis.

The two scientists then attempted to distinguish between the semiconservative and dispersive hypotheses by the following prediction: if the DNA molecules replicate one more time using nucleotides made of ordinary nitrogen, then if the semiconservative hypothesis is correct the replication will produce two kinds of DNA molecules—one kind that is light in weight and another kind that is intermediate in weight:

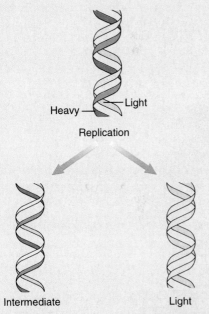

If the dispersive hypothesis is correct, the replication can produce a variety of molecules, but none will consist solely of ordinary nitrogen—there will be no band as light as the one predicted for the semiconservative hypothesis. The experiment was performed and two layers of DNA molecules were formed, one layer of light molecules and another layer of intermediate molecules. The dispersive hypothesis was disproved and the semiconservative hypothesis corroborated.

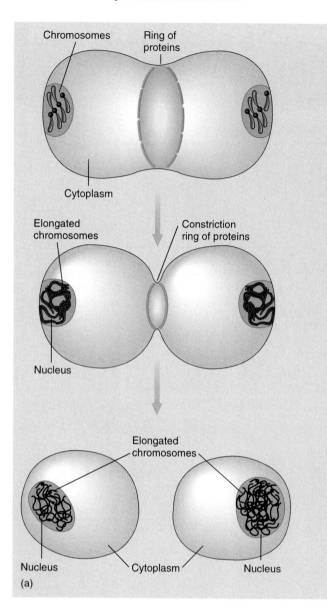

Chromosomes

Ring of proteins

Cytoplasm

Elongated chromosomes

Constriction ring of proteins

Nucleus

Elongated chromosomes

Nucleus   Cytoplasm   Nucleus

(a)

Figure 7.6 Cytokinesis. The cytoplasm divides during anaphase and telophase of mitosis. (*a*) In an animal cell, a ring of proteins constricts and pinches the cytoplasm in two. (*b*) In a plant cell, two plasma membranes are built from vesicles (hollow sacs made of membranes). A cell wall then develops between the two new membranes.

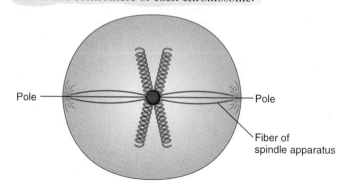

not appear as distinct structures (fig. 7.5*a*). During prophase, the chromosomes become tightly coiled, shorter, and thicker (fig. 7.5*b*). The nuclear envelope disintegrates, enabling the chromosomes to spread out through the entire cell.

Toward the end of prophase, a **mitotic spindle** develops from protein fibers that radiate outward from two centers, called spindle poles, at opposite ends of the cell. The spindle grows until it spans the entire region between the two poles:

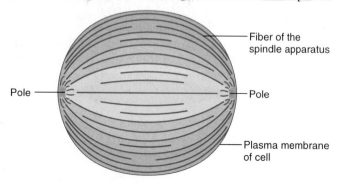

Fiber of the spindle apparatus

Pole

Pole

Plasma membrane of cell

Some of the fibers attach to chromosomes, at their centromeres, and others form a framework that maintains cell shape during mitosis. As soon as a chromosome connects with a fiber, it moves toward the spindle's equator—a region midway between the two poles.

In the second phase of mitosis, called metaphase (*meta*, middle), the chromosomes line up along the equator of the spindle (fig. 7.5*c*). Spindle fibers attach to each side of the centromere of each chromosome:

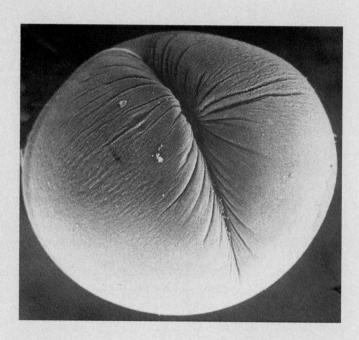

Pole

Pole

Fiber of spindle apparatus

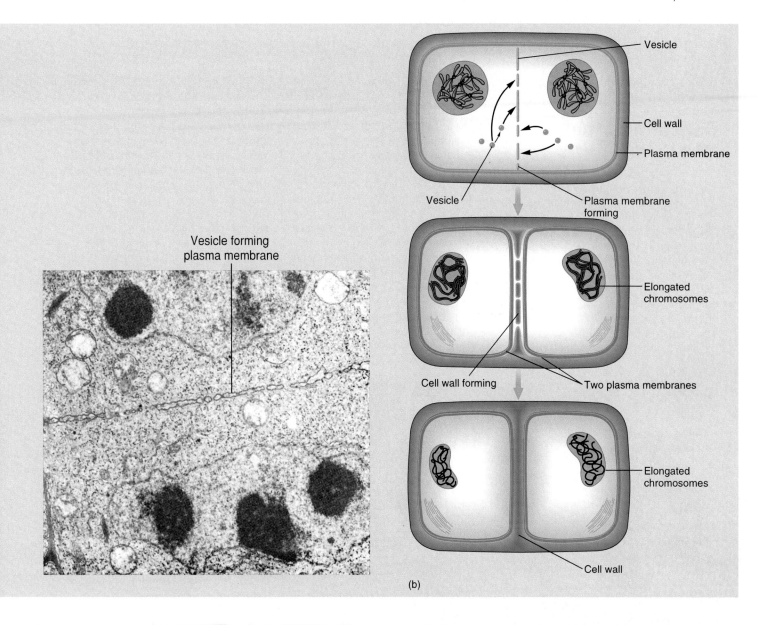

Vesicle

Cell wall

Plasma membrane

Vesicle

Plasma membrane forming

Elongated chromosomes

Cell wall forming

Two plasma membranes

Elongated chromosomes

Cell wall

(b)

Vesicle forming plasma membrane

This arrangement ensures that when the sister chromatids separate, they will be drawn to opposite poles of the spindle and not get tangled up with other chromosomes.

During the third phase of mitosis, called anaphase (*ana*, back), the sister chromatids separate as they move along the spindle to opposite poles. First, the centromeres divide in two, separating the sister chromatids of each chromosome. Each chromatid now has its own centromere and is called a chromosome. Next, each centromere moves along a spindle fiber, dissembling the fiber as it moves, toward one of the poles (fig. 7.5d). In this way, each pole receives one copy of every chromosome.

During the fourth phase of mitosis, called telophase (*telo*, end), the two sets of chromosomes become tightly packed within two nuclei, one at each pole, and the spindle disappears. Membranes then form a nuclear envelope around each set of chromosomes (fig. 7.5e).

Cytokinesis   Mitosis is accompanied by **cytokinesis** (*cyto*, cell; *kinesis*, movement), in which the cytoplasm of the cell divides. Cytokinesis begins during anaphase and continues through telophase of mitosis, while the chromosomes are moving to opposite poles.

Cytokinesis forms two cells, each with a nucleus (containing a complete set of chromosomes) and approximately half the cytoplasm. After division of the cytoplasm, the chromosomes in each daughter cell elongate once again, exposing the DNA for chemical activity during interphase. The return to interphase completes the cell cycle.

Cytokinesis in an animal cell differs from cytokinesis in a plant cell. In an animal cell, the cytoplasm splits by pinching in two (fig. 7.6a). Special proteins form a ring just beneath the plasma membrane midway between the two poles of the cell. The proteins then shorten,

Cancer and the Cell Cycle

Cancer is uncontrolled cell division. A normal cell becomes a cancer cell when it loses its ability to stop at interphase of the cell cycle. Most cancers arise from just one aberrant cell. The original cancer cell produces daughter cells with the same defect, and these daughter cells divide to form more cancer cells. At some point, a cancer cell also acquires the ability to metastasize—to burrow its way into a blood vessel or lymph vessel and invade other parts of the body (fig. 7.B).

A cancer that has metastasized can no longer be eliminated by surgery because its cells are too widely dispersed throughout the body. Cancer cells continue to multiply, in various parts of the body, until they interfere with normal body functions by monopolizing nutrients and pressing on blood vessels. When a critical function fails, the person dies. Cancer kills one out of every five adult Americans.

Cancer cells behave abnormally even under laboratory conditions. Normal cells, when grown in a dish, multi-ply until the cells cover the bottom of the dish with a single layer—no more. The physical contact of one cell with other cells signals the cell to stop dividing. This phenomenon is known as contact inhibition. Cancer cells, by contrast, continue to divide even after they contact one another. They pile up, layer after layer, within the dish.

Another abnormality seen in laboratory-grown cells is that a line of cancer cells can live forever. Normal cells are programmed to divide only a certain number of times. Certain cells in an embryo, for example, always divide just fifty times. They then remain in interphase until they die. Cancer cells, by contrast, continue to divide and divide; the line of cells does not die out. One line of cancer cells—the so-called HeLa cells (removed from a patient named Henrietta Lacks)—have been dividing continuously since 1951. The cells appear to be immortal.

Cancer begins when something goes wrong with the genes that regulate the cell cycle. Two sets of genes appear to be involved: genes that stimulate cell division and genes that suppress cell division (fig. 7.C). Whether a cell divides or remains in interphase depends on the relative activity of these two sets of genes.

Normal genes that stimulate cell division are called proto-oncogenes. They become oncogenes, and cause cancer, when too many copies of their proteins are synthesized. The excess proteins dominate the cell, ordering it to continue immediately into mitosis rather than to stop at interphase. The

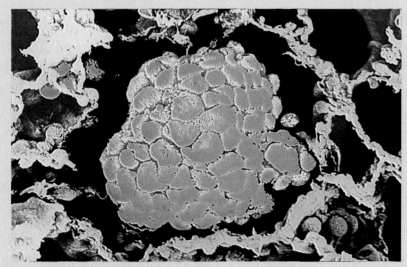

Figure 7.B Cancer cells. Cancer cells in this photograph have been tinted red. The aggregate of cells is a tumor. The individual cells migrating away from the tumor are metastasizing cancer cells, which will enter the bloodstream and spread throughout the body. A tumor can be surgically removed; the widely dispersed metastasizing cells cannot.

constricting the cell and pinching it in two. A plant cell, by contrast, has a rigid cell wall and cannot divide by pinching. Instead, the cytoplasm is divided by constructing two plasma membranes and a cell wall between the two nuclei (fig. 7.6b).

Control of the Cell Cycle

A great variety of proteins participate in the cell cycle: enzymes that promote chemical reactions, membrane re-ceptors that receive signals from outside the cell, messengers that carry signals from membrane receptors to the nucleus, and regulatory proteins that activate genes so their proteins can be synthesized. Proteins of the cell cycle are remarkably similar in all eukaryotic cells, from yeasts to humans.

Some of these proteins govern the cell cycle by stimulating DNA replication, moving a cell from the G_1 to the S phase of interphase, or by stimulating entry into mitosis. Other proteins govern the cycle by inhibiting DNA syn-

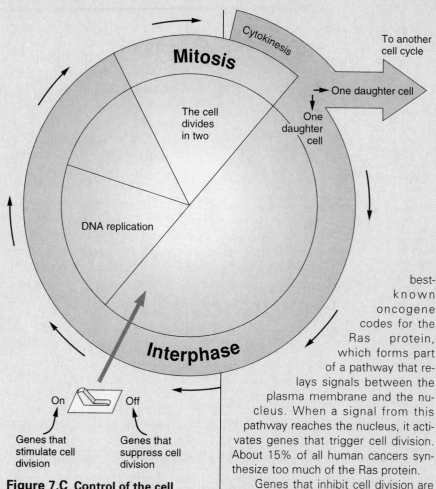

On / **Off**

Genes that stimulate cell division / Genes that suppress cell division

Figure 7.C Control of the cell cycle. Cell division is controlled by two sets of genes: one set stimulates and the other set suppresses. In cancer, defective genes may overstimulate cell division or fail to halt cell division. In either case, the control switch remains in the on position.

best-known oncogene codes for the Ras protein, which forms part of a pathway that relays signals between the plasma membrane and the nucleus. When a signal from this pathway reaches the nucleus, it activates genes that trigger cell division. About 15% of all human cancers synthesize too much of the Ras protein.

Genes that inhibit cell division are called suppressor genes. They normally switch off cell division so that the cell spends most of its time in interphase. In a cancer cell these genes fail to produce enough of their proteins to stop the cell cycle at interphase. Consequently, the cell moves directly

from one cell division to the next. The best-known suppressor gene is p53, which codes for a protein that blocks cell division. Its mutated form does not produce its protein, and so the cell does not stop at interphase. Roughly half of all human cancers contain the mutated form of p53.

A cancer cell is the outcome of a series of genetic accidents—perhaps as many as fifteen changes in the genes that control the cell cycle. Some of these changes may be inherited; others occur during the lifetime of the individual as the DNA is damaged by radiation or chemicals in the environment, such as tobacco smoke, smog, and pesticides.

In addition to damage to the genes that control the cell cycle, the development of a cancer cell may also require damage to the genes that promote DNA repair. These genes code for proteins that recognize and correct errors in the DNA molecule. If they did not function correctly, the cell would accumulate mutations rapidly.

A person's cells undergo about 10^{16} cell divisions during their lifetime. Cancer can develop anytime one of these cell cycles gets stuck in the "on" position. Most likely, these cell divisions often generate cancer cells, but they are destroyed by the immune system (white blood cells and antibodies) before they have time to proliferate into a tumor. Cancer appears to develop from both a series of altered genes and a weakened or abnormal immune system.

thesis and holding the cell in the G₁ phase of interphase. Genes (see page 49) for these proteins, called **proto-oncogenes** and **suppressor genes,** are named for their abnormal forms that cause cancer. Cancer develops when a body cell loses control of its cycle and continues to divide without cease. This condition is described in the accompanying Wellness Report.

Proto-oncogenes were first discovered in their mutated forms, with errors in their DNA that caused a cell to divide continuously. The abnormal genes were named

oncogenes (*onco,* cancer) because they cause cancer. Later, when normal forms of these genes were described, they were simply named proto-oncogenes (*proto,* original). More than 100 different proto-oncogenes have been identified.

Suppressor genes, which inhibit cell division, were given their name because they suppress the growth of tumors (aggregates of cancer cells). Cell biologists believe there are more than fifty genes of this type and have so far identified about a dozen.

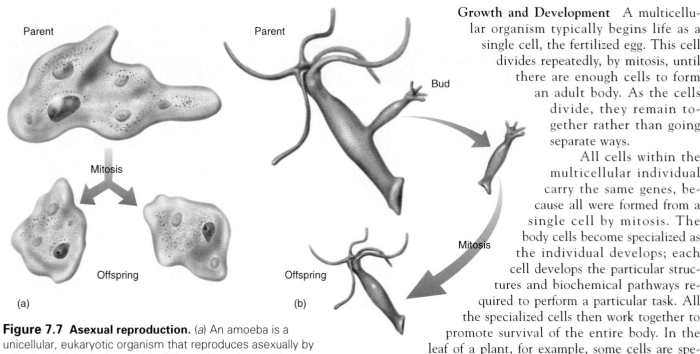

Figure 7.7 Asexual reproduction. (*a*) An amoeba is a unicellular, eukaryotic organism that reproduces asexually by mitosis. (*b*) A hydra is a multicellular, eukaryotic organism (an animal) that reproduces asexually by budding. A bud grows out of the side of the parent as body cells divide by mitosis. At some point, the bud drops off its parent and grows, through cell divisions by mitosis, into another adult.

Functions of Cell Division

Cell division by mitosis converts one cell into two cells that are genetically identical to each other and to the original cell. This kind of cell division has three functions: asexual reproduction, growth and development, and replacement of damaged cells.

Asexual Reproduction   When a single individual produces offspring, without mating, it is called **asexual reproduction** (*a*, non). The offspring receive all their genes from one parent; they are genetically identical to one another and to their parent.

Most unicellular organisms reproduce asexually (as well as sexually). An amoeba (fig. 7.7*a*), for instance, grows to a certain size and then divides by mitosis to form two amoebas. Each new cell goes its own way, growing and dividing independently of the other cell.

Some multicellular organisms reproduce asexually. What usually happens is that a fragment breaks off from the adult and then cells in the fragment divide by mitosis to build another adult. All cells in the new individual are genetically identical to one another and to the cells of their parent. A cutting from a houseplant and a bud from a hydra (fig. 7.7*b*) are examples of asexual reproduction.

Growth and Development   A multicellular organism typically begins life as a single cell, the fertilized egg. This cell divides repeatedly, by mitosis, until there are enough cells to form an adult body. As the cells divide, they remain together rather than going separate ways.

All cells within the multicellular individual carry the same genes, because all were formed from a single cell by mitosis. The body cells become specialized as the individual develops; each cell develops the particular structures and biochemical pathways required to perform a particular task. All the specialized cells then work together to promote survival of the entire body. In the leaf of a plant, for example, some cells are specialized for photosynthesis and others for protection against injury and water loss. In the intestine of an animal, some cells produce a lubricant (mucus) and others transfer the products of digestion from the digestive tract to the bloodstream. A cell becomes specialized by using only certain genes to build certain proteins—the ones that help the cell carry out its specialized task. Other genes in the cell are not used; they remain turned off and their proteins are not synthesized.

Replacement of Damaged Cells   The vitality of an animal is maintained by maintaining healthy cells. Defective cells are digested (by their own lysosomes) and replaced by new cells, formed by mitosis of other healthy cells. Cells that are particularly vulnerable to damage, such as those of the skin and lining of the digestive tract, are replaced on a regular basis.

Cell Division in Prokaryotic Cells

Prokaryotic cells form unicellular organisms in the kingdoms Eubacteria and Archaea. Cell division in these relatively simple cells is quite different from cell division in the more complex eukaryotic cells. Their DNA is contained within a single molecule, which segregates more easily into daughter cells than the many chromosomes of a eukaryotic cell. A dividing bacterium needs only to ensure that each daughter cell receives one of the replicated molecules of DNA.

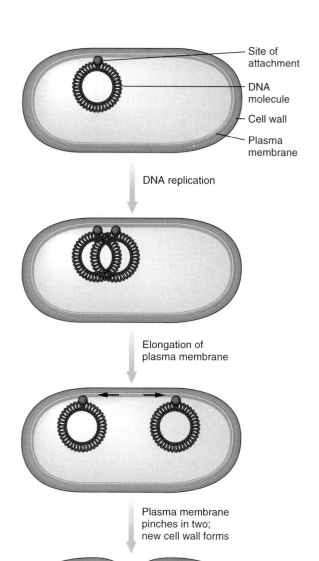

Site of attachment

DNA molecule

Cell wall

Plasma membrane

↓ DNA replication

↓ Elongation of plasma membrane

↓ Plasma membrane pinches in two; new cell wall forms

(a)

Prokaryotes divide by **binary fission** (*binary*, two; *fission*, to split) rather than by mitosis. When a bacterium reaches a certain size, it divides by binary fission (fig. 7.8). The DNA molecule replicates itself and the two molecules become attached, side by side, to the inner surface of the plasma membrane. The membrane then elongates between the two sites of attachment, separating the two molecules of DNA. The cell pinches in two between the two molecules and a new cell wall is built. Each daughter cell receives one of the DNA molecules and about half the cellular material.

SUMMARY

1. A cell grows until its interior volume becomes too large relative to its plasma membrane to continue a high level of chemical activity. At this point, the cell divides to form two smaller cells.
2. When a eukaryotic cell divides, its chromosomes become short, thick, and visible as distinct structures. The appearance of a cell's chromosomes during cell division is called its karyotype. Each species of organism has cells within a distinct karyotype.

Figure 7.8 Binary fission in a bacterium. (*a*) When a bacterium is about to divide, it replicates its DNA molecule. Both molecules become attached, side by side, to the inner surface of the plasma membrane. The plasma membrane between the two DNA molecules then elongates and pushes the molecules apart. The cell divides as proteins between the two DNA molecules constrict to pinch the cell in two. A cell wall grows between the two cells. (*b*) Photograph of two bacteria just after cell division. A new cell wall is growing between the cells and will eventually separate them completely.

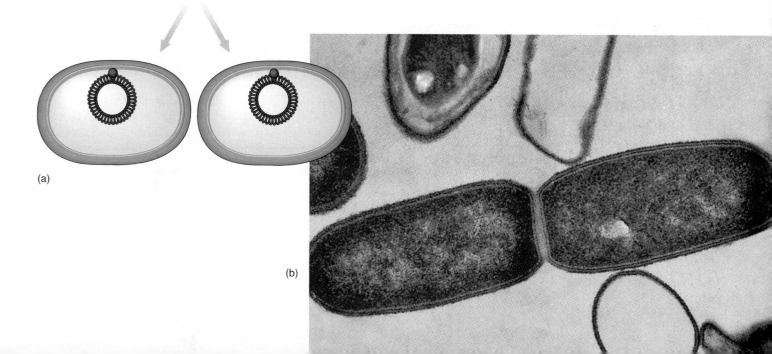

(b)

3. The cell cycle is the period between formation of a cell and its division into two cells. It includes three main phases: interphase, mitosis, and cytokinesis. Interphase has a G_1 phase, when the cell synthesizes materials and grows; an S phase, when it replicates its DNA molecules; and a G_2 phase, when it repairs errors in the replicated DNA. Mitosis segregates the two sets of DNA molecules into two nuclei. Cytokinesis splits the cell in half, with each daughter cell receiving one nucleus and half the cytoplasm. The cell cycle is controlled by two sets of genes: proto-oncogenes and suppressor genes.

4. Cell division by mitosis produces cells that are genetically identical to each other and to their parent. The process has three functions: asexual reproduction, development of multicellular organisms, and replacement of damaged cells.

5. Prokaryotic cells divide by binary fission rather than by mitosis. The DNA molecule replicates itself and each molecule attaches to the plasma membrane. The membrane then grows, separating the two DNA molecules, and the cell pinches in two.

KEY TERMS

Asexual (AY-SEK-shoo-uhl) **reproduction** (page 136)
Binary fission (page 137)
Cell cycle (page 126)
Centromere (SEN-troh-meer) (page 126)
Cytokinesis (SEYE-toh-kih-NEE-sis) (page 133)
Interphase (IN-tur-fayze) (page 127)
Karyotype (KAER-ee-oh-teyep) (page 125)
Mitosis (meye-TOH-sis) (page 128)
Mitotic spindle (page 132)
Proto-oncogenes (PROH-toh AHN-koh-jeens) (page 135)
Sister chromatids (page 128)
Suppressor genes (page 135)

STUDY QUESTIONS

1. Explain why a cell divides to form two cells rather than continue growing into a larger cell.
2. What is a karyotype? Why are all the chromosomes double in a karyotype?
3. Draw the cell cycle, showing the periods of cytokinesis, DNA replication, interphase, and mitosis.
4. Describe, using sketches, how a DNA molecule makes an exact copy of itself. Use a DNA molecule that is twelve nucleotides long.

5. Describe, using sketches, what happens to the chromosomes of a eukaryotic cell during mitosis. Use a cell with four chromosomes.
6. Compare the functions of mitosis in a human and in an amoeba (an unicellular organism).
7. Describe, using sketches, what happens to the DNA molecule of a bacterium as the cell divides.

CRITICAL THINKING PROBLEMS

1. The rate of cell division varies among the different kinds of cells in your body. Which body cells would you expect to divide rapidly? Which would you expect to seldom divide?
2. Suppose you could measure how much of each macromolecule was present within a cell. Using this measure, what would be the first indication that the cell was going to divide?
3. Suppose a cancerous cell appeared within a group of 1,000 cells (a ratio of 1 cancerous to 1,000 normal cells). The cancerous cell divided once every 24 hours, whereas the normal cells divided only when new cells were needed to replace dead ones. What would be the ratio of cancerous to normal cells after a period of 100 days?
4. One way in which cancer develops is when mutations in the suppressor genes prevent the cell from turning off cell division—the cells cannot stop in interphase. What would happen to the body if a mutation in the suppressor genes prevented the cell from turning *on* cell division?

SUGGESTED READINGS

Introductory Level:

Freedman, D. H. "Life's Off-Switch," *Discover*, July 1991, 61–67. A description of how one biologist, Arthur Kornberg, goes about his research, with particular emphasis on his study of the enzymes that control the cell cycle.

John, P. C. L. *The Cell Cycle*. Cambridge: Cambridge University Press, 1981. A review of what happens during the cell cycle.

Murray, A. W., and M. W. Kirschner. "What Controls the Cell Cycle," *Scientific American*, March 1991, 56–63. A well-illustrated article describing the cell cycle and the proteins that regulate it.

Advanced Level:

Kornberg, A., and T. A. Baker. *DNA Replication*. New York: W. H. Freeman and Company, 1992. A comprehensive review of DNA replication in eukaryotes and prokaryotes.

McIntosh, J. R., and K. L. McDonald. "The Mitotic Spindle," *Scientific American*, October 1989, 48–56. Excellent description of spindle structure and function.

Murray, A., and T. Hunt. *The Cell Cycle: An Introduction*. New York: W. H. Freeman and Company, 1993. An easy-to-read but thorough description of the cell cycle, including the enzymes, genetic controls, feedback controls, and events that convert a normal cell into a cancer cell.

Radman, R., and R. Wagner. "The High Fidelity of DNA Duplication," *Scientific American*, August 1988, 40–46. An explanation of how, through endless replications, the DNA is copied so accurately.

Varmus, H., and R. A. Weinberg. *Genes and the Biology of Cancer*. New York: Scientific American Library, 1992. A comprehensive review of the molecular biology of cancer.

❁ EXPLORATIONS

Interactive Software: Smoking and Cancer

This interactive exercise explores the relationship between smoking and lung cancer. In the exercise, tobacco smoke causes mutations in four genes that regulate cell division. The student varies the level of exposure to tobacco smoke and observes the effects on mutations and on the incidence of cancer.

Interactive Software: Cell Size

This interactive exercise explores the effects of cell size on the import of materials into a cell. The student is provided with a diagram of a cell and is able to alter its diameter and shape, observing how these variables affect the transport of materials from the surface to the center of the cell.

Meiosis and the Life Cycle

Objectives

In this chapter, you will learn
–the genetic consequences of sexual
  versus asexual reproduction;
–what happens to the chromosomes of
  a cell when it becomes a sperm or egg;
–what happens to the chromosomes of
  a sperm and egg when they become a
  fertilized egg;
–how the number of chromosomes in
  an individual's cells changes during
  the life cycle.

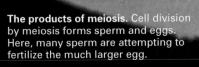

The products of meiosis. Cell division by meiosis forms sperm and eggs. Here, many sperm are attempting to fertilize the much larger egg.

Meiosis and the life cycle are features of sexual reproduction, in which offspring inherit genes from two parents rather than one. This form of reproduction apparently has great advantages over asexual reproduction, in which offspring inherit genes from just one parent, since the vast majority of organisms reproduce sexually. Humans always reproduce sexually: an egg with genes from a woman is fertilized by a sperm with genes from a man. The fertilized egg develops into a child with a combination of genes from both parents.

Sexual reproduction produces unique offspring; they are genetically different from one another and from either parent. We take for granted this particular mode of reproduction and the endless variety of individuals that it yields. Sexual reproduction, however, is not a biological rule. Some animals can reproduce asexually; their offspring are genetically identical to one another and to their single parent.

Our closest "relatives" who reproduce without sex are certain lizards, including populations of whiptail lizards that live in deserts of the Southwest (fig. 8.1). All lizards in these asexual populations are females; they form eggs that develop, in the absence of sperm, into female offspring. No males are conceived. This mode of reproduction is known as parthenogenesis (*parthenos*, virgin).

What about us? Could we reproduce without sex? Perhaps, with the help of genetic engineers. Unfertilized eggs can be artificially activated, by pricking them with a pin or applying chemicals, to develop in the absence of sperm. By such a procedure, a daughter could form with only genes from her mother. Or genes within an unfertilized egg could be removed and replaced by genes from a man's body cell, enabling a man to have a genetically identical son.

What would happen if we switched from our present mode of sexual reproduction to an asexual mode? Is sex important? Perhaps in the long run, for it generates the variety of individuals required for a species to adapt to environmental changes. But what about within our own lifetimes? What would it be like to have a society consisting only of females? What would it be like to have children who are identical to ourselves?

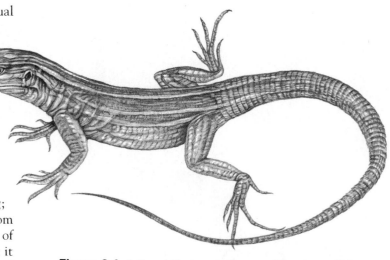

Figure 8.1 A lizard that reproduces without sex. This whiptail is a female, as are all other individuals in the population. She reproduces by parthenogenesis: her eggs develop without being fertilized by sperm. All her offspring will be females.

Sexual Reproduction

Sexual reproduction transfers various combinations of genes from two parents to their offspring; it produces offspring that are genetically different from one another and from either parent. **Asexual reproduction** occurs without sex and transfers the genes of just one parent to each offspring; it produces offspring that are genetically identical to one another and to their single parent.

Most organisms have some way of reproducing sexually—even if it does not happen on a regular basis. The simplest organisms, bacteria, occasionally acquire pieces of DNA from one another. When such cells divide, they produce new kinds of cells with genes from two different bacteria; the daughter cells are genetically different from their parents. More complex unicellular organisms, such as the paramecia shown in figure 8.2, occasionally come together and exchange genes.

Reproduction in plants, animals, and mushrooms is almost always sexual (fig. 8.3), with each offspring receiving genes from two parents. Each parent generates, by a form of cell division known as meiosis, special cells (sperm

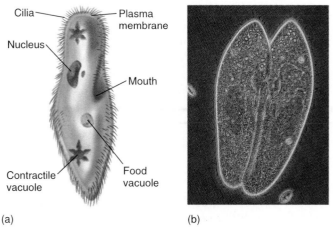

(a)   (b)

Figure 8.2 Unicellular organisms reproduce sexually. (a) A paramecium is an unicellular organism with a complex (eukaryotic) cell. (b) Here, two paramecia have come together to exchange genetic material. The offspring of this mating will receive genes from both parents.

or eggs) with half the parental number of chromosomes. Each sperm or egg has a unique set of genes. A sperm from one parent and an egg from the other parent merge, an event known as **fertilization,** to yield a cell with the same number of chromosomes as in body cells of the parent. This cell then proceeds to divide by mitosis (📖 *see page 128*) to form an offspring that is a genetic mixture of both parents.

Diploidy and Haploidy

Most multicellular organisms reproduce by means of **gametes,** which are cells specialized for fusing with each other. There are two kinds of gametes: sperm, which are small and motile, and eggs, which are large and nonmotile. A gamete is always a **haploid cell,** which means it has one set of chromosomes—one version of each chromosome. Fusion of a sperm with an egg produces a **fertilized egg,** which is a **diploid cell** with two sets of chromosomes, one from the sperm and another from the egg.

A fertilized egg (also called a **zygote**) divides by mitosis to form an aggregate of cells that is a new multicellular individual. A typical body cell, formed by mitosis from a fertilized egg, is a diploid cell.

(a)

(b)

Figure 8.3 Animals reproduce sexually. Most animal populations have two kinds of individuals: males, who produce sperm, and females, who produce eggs. (a) Most aquatic animals release their eggs and sperm into the water, where fertilization takes place. (b) The males of most land animals place their sperm inside the females, where fertilization takes place.

Roles of Fertilization and Meiosis

Fertilization combines two sets of chromosomes, one from each parent. It merges two haploid cells—the gametes—into a diploid cell:

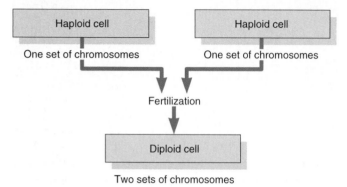

In a human, an egg carries one set of chromosomes from the mother and a sperm carries one set of chromosomes from the father. The sperm and egg unite at fertilization to produce a fertilized egg with two sets of chromosomes.

Meiosis is a form of cell division that occurs only during sexual reproduction. It converts a diploid cell into four haploid cells:

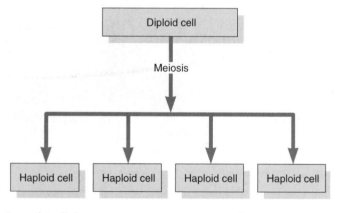

In multicellular organisms, meiosis occurs only in certain parts of the body—sexual organs within the flower of a flowering plant, for example, and the ovaries or testes of an animal.

One function of meiosis is to maintain a constant number of chromosomes in the offspring. During fertilization, the number of chromosomes is doubled. If meiosis did not correct for this doubling, then the number of chromosomes would double with each successive generation. If, for example, the 46 chromosomes in a human cell were not halved during gamete production, then both the sperm and the egg would have 46 chromosomes and the fertilized egg would have 92 chromosomes, as would all body cells of the child that developed from it. In the following generation, each gamete would have 92 chromosomes and each body cell would have 184. By only the tenth generation, each body cell would have 23,332 chromosomes. The cells would be too packed with chromosomes to function. Halving the number of chromosomes during meiosis compensates for doubling the number during fertilization.

A second function of meiosis is to shuffle the genes and chromosomes of a diploid cell, thereby increasing the variety of gametes (and offspring) that form. This function is described in chapter 9.

Characteristics of Diploid Cells

A diploid cell has two sets of chromosomes; there are two versions of each kind of chromosome. The two versions form a **homologous pair** (*homo*, same). Each of your body cells, for example, contains twenty-three homologous pairs of chromosomes. One chromosome of each pair originated in the sperm of your father and the other in the egg of your mother:

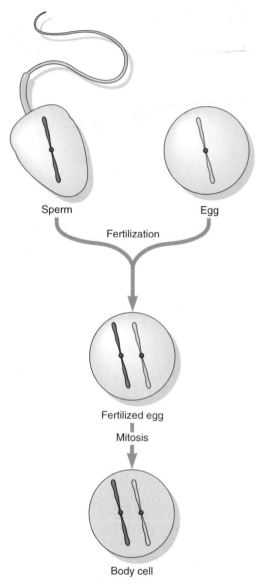

Homologous chromosomes look the same and carry copies of the same genes. The two copies of each gene are located within the same section of DNA in both of the homologous chromosomes:

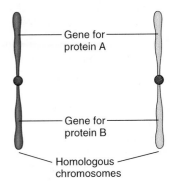

The two copies of each gene may be exactly the same or they may be slightly different versions of the gene. They have had different histories (one in the father's family line and the other in the mother's family line) and may have experienced different mutations, which are changes in DNA structure. Different versions of a gene are different instructions for building a protein, which may lead to different traits in the individual. For example, the genes that control development of the face may cause dimples in one family line but not in another line.

Advantages of Diploid Cells

A sexually reproducing organism alternates between haploid and diploid cells, since fertilization must always be compensated for by meiosis. The adult bodies of most plants and animals consist of diploid rather than haploid cells. What are the advantages of diploid cells over haploid cells?

A diploid cell has two copies of each gene. When both copies are used to synthesize a protein, a diploid cell can synthesize twice as much of the protein as a haploid cell. Where a protein is required in large quantities, as, for example, the hemoglobin in red blood cells, there is a distinct advantage in having two active forms of the gene.

The two copies of a gene may be slightly different, coding for two slightly different proteins. A diploid cell may, for example, synthesize two versions of an enzyme—one that functions optimally in cooler environments and the other in warmer environments. An organism consisting of such cells would function better over a broad range of temperatures than an organism consisting of haploid cells, with just one version of the enzyme.

For many genes, only one copy is sufficient for optimal cell function. The other copy, then, may vary without having an effect on the organism. In this way, an organism with diploid cells can preserve a variety of genes (as mutations in the "spare" copy) that can become raw material for evolutionary change. An organism formed of haploid cells, with just one copy of each gene, cannot preserve genetic variety in this way; any mutation that disrupts an essential gene is lethal.

Meiosis

Meiosis is a form of cell division that converts a diploid cell into four haploid cells. It is the physical basis of most patterns of inheritance. An understanding of meiosis is essential to an understanding of how traits are passed from one generation to the next.

Meiosis consists of two consecutive divisions: meiosis I and meiosis II. Each division has the same four phases as mitosis: prophase, metaphase, anaphase, and telophase. Meiosis I separates the homologous pairs of chromosomes, with one chromosome of each pair going to each of the two daughter cells:

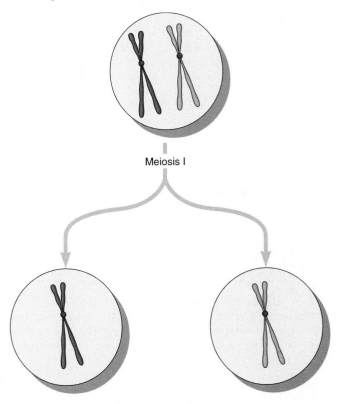

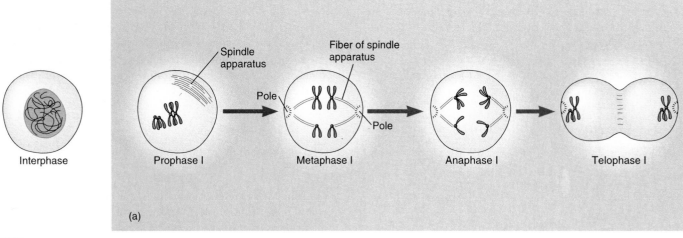

(a)

Figure 8.4 Interphase and meiosis. In interphase prior to meiosis, the chromosomes are elongated and not seen as distinct structures. At some point during interphase, the chromosomes are duplicated. Each chromosome now consists of two chromatids held together by a centromere. (*a*) In meiosis I, the homologous pairs of chromosomes separate; one version of each chromosome goes to each daughter cell. In prophase I, the chromosomes shorten, the nuclear envelope disappears, and a spindle forms. Homologous chromosomes come together. In metaphase I, the homologous chromosomes line up two by two between the two poles of the spindle. Spindle fibers connect each chromosome to one of the poles. During anaphase, the homologous chromosomes separate, with one chromosome of each pair going to one of the poles. During telophase, the two sets of chromosomes cluster near the two poles, the spindle disappears, and the cell divides in two. Each daughter cell has one version of each kind of chromosome; each chromosome consists of two chromatids. (*b*) In meiosis II, the sister chromatids that form each chromosome separate—one chromatid of each chromosome moves into each daughter cell. Both cells formed during meiosis I go through meiosis II. In prophase II, nothing much happens except that a new spindle forms. In metaphase II, the chromosomes line up one by one between the two poles of the cell. Spindle fibers attach each centromere to the two poles. In anaphase II, the centromeres divide and the spindle fibers shorten. One chromatid of each chromosome is pulled to each of the two poles. During telophase II, the two sets of chromosomes cluster near the two poles, the spindle disappears, and a nuclear envelope forms around each set of chromosomes. The cell divides in two. Meiosis I and II have formed four haploid cells, each with one copy of every kind of chromosome.

It converts a diploid cell into two haploid cells. Each haploid cell then goes through meiosis II, which, like mitosis, separates the sister chromatids:

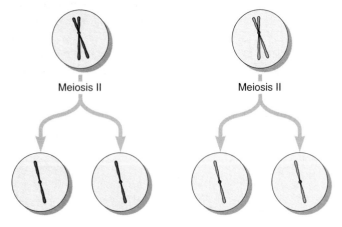

Meiosis is illustrated in greater detail in figure 8.4 and its major events are summarized in table 8.1.

Meiosis I

The DNA molecules of the cell are replicated during interphase prior to meiosis I, when the chromosomes are elongated and still enclosed within the nuclear envelope. The cell makes an exact copy of each DNA molecule, just as it does prior to mitosis. At the end of interphase, each chromosome consists of two sister chromatids held together by a centromere:

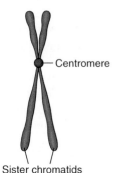

146

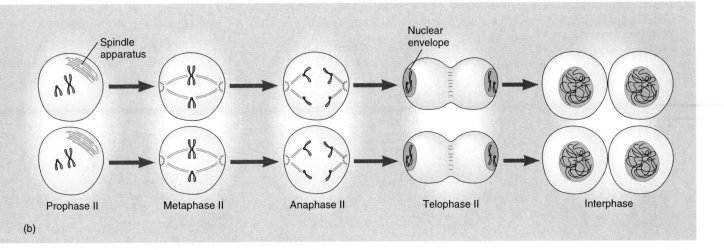

Prophase II   Metaphase II   Anaphase II   Telophase II   Interphase

(b)

Table 8.1

Important Events in Meiosis I and II

Phase	Events
Meiosis I	
Interphase	Chromosomes are duplicated (except for the centromeres); errors in the DNA are repaired
Prophase	Chromosomes shorten; homologous chromosomes pair up; crossing-over takes place
Metaphase	Homologous pairs of chromosomes line up midway between the two poles
Anaphase	The chromosomes of each homologous pair move to opposite poles
Telophase	Two nuclei form, each with one chromosome of every homologous pair, and the cell divides

Meiosis I converts one diploid cell into two genetically different haploid cells.

Phase	Events
Meiosis II	
Prophase	New spindle forms
Metaphase	Single chromosomes, each consisting of two chromatids, line up midway between the two poles
Anaphase	Centromeres divide and sister chromatids move to opposite poles
Telophase	Two nuclei form and the cell divides

Meiosis II converts two genetically different haploid cells into four genetically different haploid cells.

Early prophase of meiosis I resembles prophase of mitosis: the chromosomes become short and thick, the nuclear membrane disintegrates, and a spindle forms. Late prophase, however, differs from mitosis in that the homologous chromosomes pair up next to each other. They associate so closely that a particular gene in one chromosome is right next to its counterpart in the other chromosome:

Something unique and important takes place when the chromosomes are this close together: two chromatids, one from each chromosome, often exchange segments. The exchange of genetic material between homologous chromosomes, known as **crossing-over**, is shown in figure 8.5. It has important genetic consequences that are described in chapter 9.

▣▣ Figure 8.5 Crossing-over. Crossing-over happens during prophase of meiosis I, when homologous chromosomes come to lie next to each other. (*a*) Crossing-over is when a chromatid in one chromosome exchanges a segment with a chromatid in a homologous chromosome. (*b*) Crossing-over may occur at more than one place within a pair of homologous chromosomes. In this photograph, it is happening at five places.

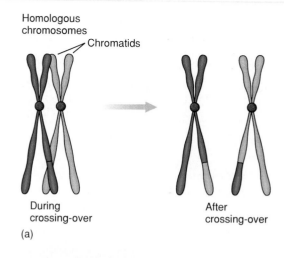

Homologous chromosomes
Chromatids

During crossing-over

After crossing-over

(a)

In metaphase of meiosis I, the chromosomes line up between the two poles of the spindle. They line up in homologous pairs—two by two rather than one by one as in mitosis:

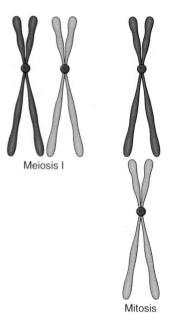

Meiosis I

Mitosis

This arrangement of chromosomes facilitates separation of the homologous chromosomes (rather than of the sister chromatids, as occurs in mitosis).

Spindle fibers then connect the centromeres of each pair to spindle poles at opposite sides of the cell:

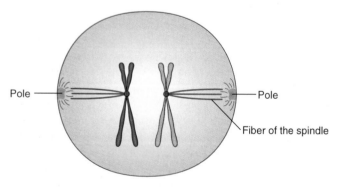

Pole

Pole

Fiber of the spindle

During anaphase, the centromeres move along the spindles toward the poles. Note that the centromeres do

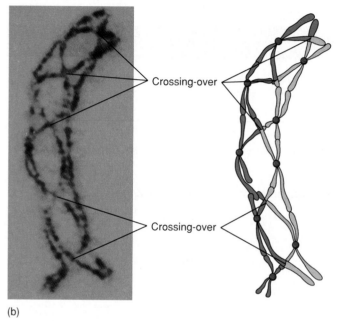

Crossing-over

Crossing-over

(b)

not divide (as they do in mitosis). The two chromosomes of each homologous pair are pulled apart as they move to opposite poles of the spindle.

As meiosis I comes to a close during telophase, the chromosomes are clustered in two distinct groups, one at each pole of the cell. Each group contains one chromosome (consisting of two chromatids) from each homologous pair. The cell then divides into two haploid cells. Both cells go directly into prophase of meiosis II, without re-forming their nuclear envelopes and without elongating their chromosomes.

Meiosis II

The two haploid cells formed during meiosis I bypass interphase and enter directly into prophase of meiosis II. Each cell contains half the original number of chromosomes, and each chromosome consists of two chromatids held together by a centromere. A new spindle forms but otherwise little happens during this prophase, since the chromosomes are already short and thick and the nuclear envelope did not reform after meiosis I.

During metaphase of meiosis II, the chromosomes of each cell line up in a column—one by one, as in mitosis, rather than two by two, as in meiosis I—and the spindle fibers connect each centromere to both poles:

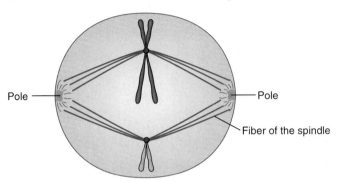

The centromere joining the two chromatids of each chromosome divides at the beginning of anaphase. Each chromatid becomes a chromosome with its own centromere. The two chromosomes, which were once sister chromatids, are pulled by their centromeres along spindle fibers to opposite poles.

The two groups of chromosomes become clustered at opposite poles of the spindle during telophase. A nuclear envelope develops around each group and the cell divides in two.

Meiosis II converts the two haploid cells formed during meiosis I into four haploid cells:

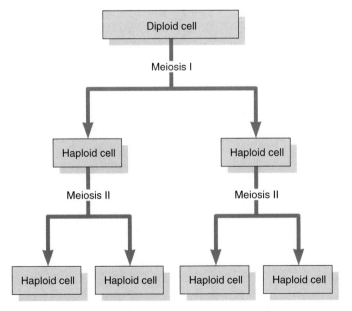

The four cells then undergo structural and chemical changes to become sperm or eggs. Meiosis in a male produces four sperm of equal size. Meiosis in a female produces one large egg, which receives all the nutrients, and three tiny cells that disintegrate.

Meiosis Versus Mitosis

There are two forms of cell division: mitosis, described in chapter 7, and meiosis, described previously in this chapter. Mitosis divides one cell into two cells that are genetically identical to each other and to the original cell. Meiosis divides one cell into four cells that are genetically different from one another and have half as many chromosomes as the original cell. To illustrate these differences, the two kinds of cell division are drawn side by side in figure 8.6. They are summarized in table 8.2.

Mitosis occurs whenever more cells of the same kind are needed. It forms new cells for asexual reproduction, for building a multicellular organism from a fertilized egg, and for replacing body cells that have been damaged. In a human, mitosis occurs in most parts of the body throughout life.

Meiosis occurs only during sexual reproduction. Its function is to produce genetically diverse offspring, which it does by producing genetically diverse haploid cells that unite, one from each parent, to form fertilized eggs. These haploid cells, the sperm and eggs, develop only in reproductively mature organisms.

In a man, meiosis takes place only in the testes. It generates sperm from the time a boy reaches puberty until he dies. In a woman, meiosis takes place within the ovaries and is very prolonged. Meiosis I actually begins in the female embryo prior to her birth, and meiosis II is not completed until after the developing egg is fertilized. A woman's ovaries release eggs, which are suspended in meiosis II, from the time of puberty until menopause (between the ages of forty-five and fifty).

The Life Cycle

At certain points in the life of every sexually reproducing individual, some of the cells change from diploid to haploid and vice versa. In an adult human, reproductive cells (diploid) in the ovaries or testes are converted into gametes (haploid) that may combine to become fertilized eggs (diploid). This alternation between diploid and haploid cells, brought about by meiosis and fertilization, is called the **life cycle** of the individual.

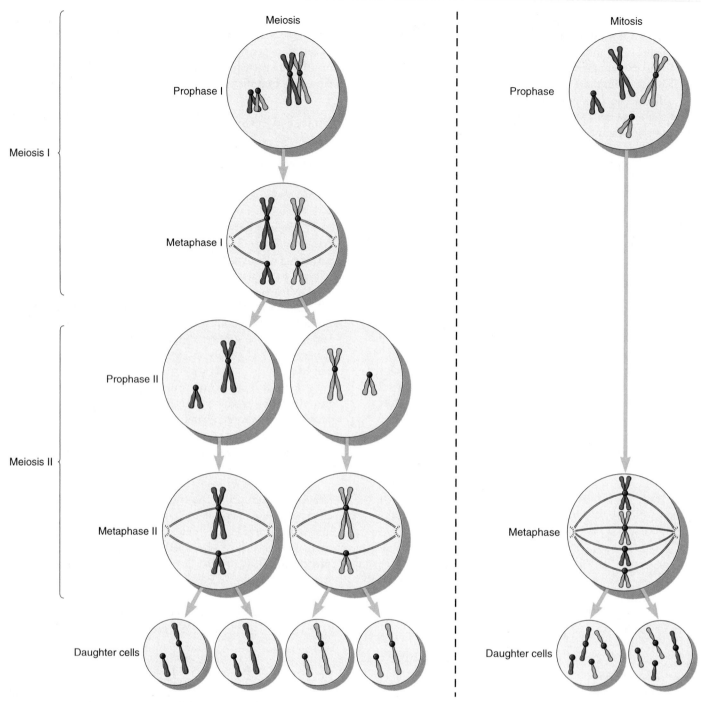

Figure 8.6 Comparison of mitosis and meiosis. The main difference between the two forms of cell division is the separation of homologous chromosomes during meiosis I, which does not happen during mitosis. Separation of the chromatids within each chromosome is the same in both kinds of cell division—mitosis is the same as meiosis II except that a cell dividing by mitosis has its original number of chromosomes whereas a cell dividing by meiosis II has only half as many chromosomes as it started with.

Table 8.2

Comparison of Mitosis and Meiosis

Mitosis	Meiosis
Forms cells that are genetically identical	Forms cells that are genetically different
One cell division	Two cell divisions
Two daughter cells	Four daughter cells
Each daughter cell has the same number of chromosomes as the original cell	Each daughter cell has half as many chromosomes as the original cell
Occurs in all cells of the body	Occurs only in the reproductive organs
Occurs throughout life	Occurs only during reproductive ages
Used in asexual reproduction, growth of a multicellular organism, and cell replacements	Used only in sexual reproduction

In us, the life span of haploid cells (sperm and eggs) is very short compared with the life span of diploid cells (body cells). In the life cycles of other species this may not be the case. The three main forms of life cycle, described following, have diploid adults, haploid adults, and both diploid and haploid forms of adults.

Life Cycles with Diploid Adults

The cells of most animals and plants are diploid during the adult phase of the life cycle, as shown in figure 8.7a. Immediately after fertilization, the diploid zygote undergoes a series of divisions by mitosis (with no change in chromosome number) to produce an adult organism composed of diploid cells. These cells then carry out the various functions necessary for adult survival. Prior to reproduction, the number of chromosomes within the reproductive cells is halved (by meiosis) to yield haploid cells. These cells exist only briefly.

In animals, haploid cells remain as single cells—the sperm and egg. Sperm develop in the testes of a male and eggs develop in the ovaries of a female. A sperm and an egg unite to form a diploid cell that divides, by mitosis, to form an adult animal.

In plants, haploid cells form multicellular structures. In flowering plants, haploid cells are produced within the flower. A haploid cell then divides by mitosis to form a small structure, made entirely of haploid cells. In the male organ of a flower, one of the haploid cells becomes a sperm; in the female organ, one of the haploid cells becomes an egg. A sperm and an egg unite to form a diploid cell that divides, by mitosis, to form an adult plant.

Life Cycles with Haploid Adults

Many seaweeds and fungi spend most of their life cycles as haploid rather than diploid cells (fig. 8.7b). The haploid adult produces gametes by mitosis (no change in chromosome number). Two gametes, from two adults, unite at fertilization to form a diploid zygote. Immediately after fertilization, the zygote undergoes meiosis to reduce the number of chromosomes to the haploid condition. Each haploid cell then divides by mitosis to produce the multicellular, haploid adult.

Diploid cells occur very briefly in the life cycle of these organisms. While diploidy appears to provide advantages when cells are chemically active, as discussed earlier, the existence of many species with haploid cells as adults indicates these advantages do not always determine the life cycle.

Life Cycles with Both Diploid and Haploid Adults

Some seaweeds and land plants have adults of both kinds: one consisting entirely of haploid cells; the other entirely of diploid cells. The life cycle of such a plant is diagrammed in figure 8.7c.

Ferns, for example, have two types of adults: a large, leafy diploid plant and a tiny, compact haploid plant. The diploid plant forms haploid cells by meiosis, and they drop to the ground and develop by mitosis into small haploid plants. These haploid ferns, which are only a centimeter or so in length, live in moist areas on the soil beneath the leaf litter. They form sperm and/or eggs by mitosis. A sperm from one plant swims to the egg of another plant, fertilizes it, and the zygote then divides again and again, by mitosis, to form another large, diploid plant.

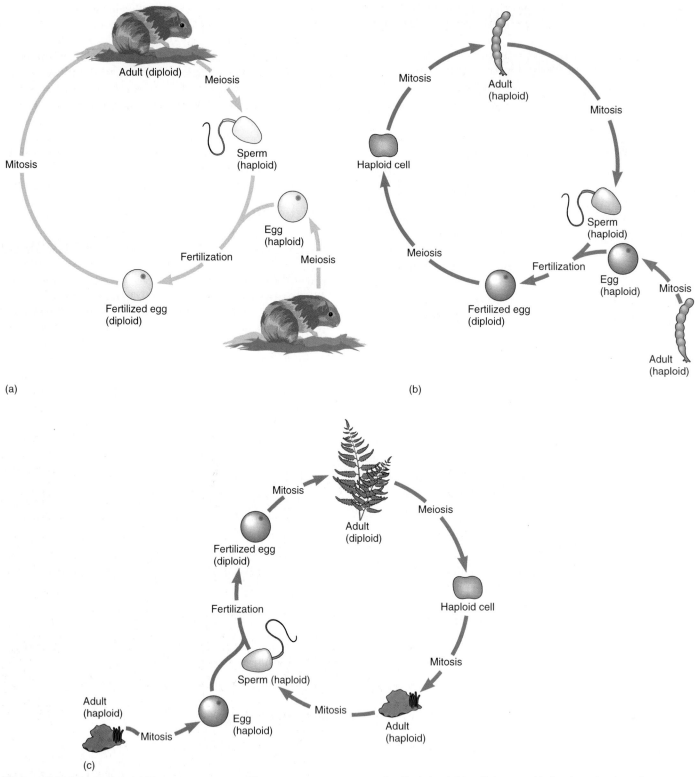

Figure 8.7 Life cycles. (*a*) The most common life cycle, as shown here for an animal, has a diploid adult that produces haploid cells by meiosis. Two haploid cells, a sperm and an egg, fuse shortly after they form to make a diploid cell (the fertilized egg), which divides by mitosis to form another diploid adult. (*b*) Many seaweeds and fungi have life cycles with a haploid adult, which produces haploid cells by mitosis. The haploid cells merge to form a diploid cell (the fertilized egg) that divides by meiosis, shortly after it appears, to form haploid cells. Each haploid cell then multiplies by mitosis to form another haploid adult. (*c*) Some plants, such as ferns and some seaweeds, alternate between diploid and haploid adults. Cells in the diploid adult divide by meiosis to form haploid cells, which then divide by mitosis to form a haploid adult. Cells in the haploid adult divide by mitosis to produce haploid cells— sperm and eggs. A sperm from one adult fuses with an egg from another adult to form a fertilized egg. This diploid cell then divides by mitosis to form another diploid adult.

SUMMARY

1. Most organisms reproduce sexually, combining the genes of two parents to form genetically diverse offspring.

2. Sexual reproduction in multicellular organisms involves fertilization and meiosis. Fertilization converts two haploid cells (eggs and sperm), each with one set of chromosomes, into a diploid cell (the fertilized egg) with two sets of chromosomes. Meiosis converts a diploid cell into four haploid cells (sperm or eggs). The adult bodies of most plants and animals are made of diploid cells, in which each cell has two versions of each chromosome (called homologous pairs). The advantage of diploidy is that each cell has two copies of each gene.

3. Meiosis consists of two consecutive cell divisions: meiosis I and meiosis II. Meiosis I separates the homologous pairs of chromosomes; it yields two haploid cells, each containing one chromosome (with two chromatids) from every homologous pair. Meiosis II separates sister chromatids in the haploid cells formed during meiosis I. It converts a haploid cell with chromosomes consisting of two chromatids into two haploid cells with chromosomes consisting of a single chromatid.

4. A life cycle is a sequence of fertilization and cell divisions (by meiosis and mitosis) that happens during a lifetime. The cells alternate between haploidy and diploidy. Most animals and plants have diploid body cells and form short-lived haploid gametes. Some seaweeds and fungi have haploid body cells; fertilization results in diploid cells that immediately undergo meiosis to form haploid cells. Some plants alternate between individuals with diploid body cells and individuals with haploid body cells.

KEY TERMS

Asexual (AY-SEK-shoo-uhl) **reproduction** (page 142)
Crossing-over (page 147)
Diploid (DIH-ployd) **cell** (page 143)
Fertilization (page 143)
Fertilized egg (page 143)
Gamete (GAEH-meet) (page 143)
Haploid (HAEH-ployd) **cell** (page 143)
Homologous (hoh-MAH-luh-gus) **pair** (page 144)
Life cycle (page 149)
Meiosis (meye-OH-sis) (page 144)
Sexual reproduction (page 142)
Zygote (ZEYE-goht) (page 143)

STUDY QUESTIONS

1. Where do the offspring's genes come from when reproduction is asexual? Where do they come from when reproduction is sexual?

2. What does it mean to say that meiosis compensates for fertilization?

3. Draw meiosis I and II, beginning with a cell that has six chromosomes.

4. When and where does meiosis occur within you? What are its products?

5. Draw the three forms of life cycle. Show where in each cycle meiosis, mitosis, and fertilization take place.

CRITICAL THINKING PROBLEMS

1. Each one of your body cells has two sets of chromosomes—there are two versions of each kind of chromosome. Explain the origin of each set. Explain what happens to the two sets when you form sperm or eggs.

2. Suppose you observe a population of lizards and suspect there are no males. You hypothesize that the females reproduce parthenogenically, in which the eggs develop without fertilization. From this hypothesis, what prediction can you make with regard to the offspring? How can this prediction be tested?

3. The cells of some organisms have only a few chromosomes, whereas others have thousands. What effect does the number of chromosomes have on genetic variability? Why?

4. Why does meiosis in a male reproductive cell produce four sperm whereas in a female it produces only one egg?

SUGGESTED READINGS

Introductory Level:

Calow, P. *Life Cycles: An Evolutionary Approach to the Physiology of Reproduction, Development and Ageing.* New York: John Wiley & Sons, 1978. Descriptions and speculations on the adaptive significance of various aspects of reproduction.

Cole, C. J. "Unisexual Lizards," *Scientific American,* January 1984, 94–100. A description of asexual reproduction in whiptail lizards.

Hartl, D. L. *Our Uncertain Heritage: Genetics and Human Diversity,* 2d ed. New York: Harper and Row, 1985. An easily understood presentation, in chapter 2, of asexual versus sexual inheritance and of meiosis.

Moore, J. A. *Science As a Way of Knowing: The Foundations of Modern Biology*. Cambridge, MA: Harvard University Press, 1993. An interesting description, in chapter 13, of how chromosomes, mitosis, and meiosis were discovered in the late nineteenth century.

Advanced Level:

Burns, G. W., and P. J. Bottino. *The Science of Genetics*, 6th ed. New York: Macmillan Publishing, 1988. A good description, in chapter 4, of the physical basis of inheritance—what happens during meiosis and how it affects movements of genes into sperm and eggs.

John, B. *Meiosis*. New York: Cambridge University Press, 1990. A survey of meiosis in plants and animals, showing that it does not always occur in textbook fashion.

Murray, A., and T. Hunt. *The Cell Cycle: An Introduction*. New York: W. H. Freeman and Company, 1993. Contains an excellent chapter on meiosis, including recombination, segregation of homologous chromosomes, and factors that control the cycle.

The Physical Basis of Inheritance

Chapter Outline

Objectives

In this chapter, you will learn
–how traits are determined by both
  genes and the environment;
–how genes arranged in chromosomes
  are shuffled during meiosis;
–how gene shuffling produces variety
  among offspring;
–the way in which sex chromosomes
  control maleness and femaleness;
–how an offspring may inherit too
  many or too few chromosomes.

Chromosomes. The physical basis of inheritance lies in the set of chromosomes passed from parent to offspring. These short, thick, duplicated chromosomes are from a dividing cell.

I t may seem strange that it took so long to discover the physical basis of inheritance. After all, humans have always had an intense interest in the unique features of family lines and in the breeding of domestic plants and animals. The main reason it took so long is that the passage of genes from one generation to the next cannot be observed directly.

In the last half of the nineteenth century, many biologists began to suspect that hereditary material was located in the chromosomes. These elongated structures become conspicuous during cell division and can be seen, through a microscope, to move from parental to daughter cells. The problem with this idea, however, was that chromosomes can be seen only during cell division. How could the physical basis of inheritance be something that appears and then disappears on a regular basis?

Careful observations of mitosis and meiosis indicated that although chromosomes disappeared from view, when they appeared again they were unchanged—each cell had a characteristic number of chromosomes and each chromosome had a characteristic length and shape. By 1900, many biologists were convinced that chromosomes represented the physical basis of inheritance.

It was not until the 1950s that genetic material was identified as DNA molecules in chromosomes. The molecular structure of DNA was then worked out and the physical nature of genetic material revealed.

Now that we know what genes are made of, we want to know more about our own genes. Where is each gene located within our twenty-three pairs of chromosomes? What is the precise sequence of bases within each gene? The Human Genome Project seeks to answer these questions. The goal of this cooperative research project, involving many laboratories, is to map and describe all the DNA of a human. Biologists involved in this vast project are preparing a library of detailed instructions for building human traits.

Of what use is all this information? Mostly it will be used to identify the physical basis of genetic disorders, which will help us diagnose and treat diseases such as cystic fibrosis, Huntington disease, sickle-cell anemia, and Tay-Sachs disease. It will also help us fight more complex disorders, such as diabetes, high blood pressure, allergies, and cancer, that are caused in part by genes.

A complete analysis of our genes may become a routine part of a medical examination. Some people question this application of science to their personal lives. If you could find out, tomorrow, the most likely cause of your death and approximately when death will come, would you want to know?

Phenotype and Genotype

In spite of recent advances in DNA technology, much of our information about genes still comes from gene products—visible traits and their patterns of inheritance. This information is confounded to some extent by environmental influences, since traits develop in response to both the genes and the environment.

An individual's **phenotype** is the physical expression of his or her genes: the anatomical, physiological, and behavioral features that can be observed. It is the consequence of both genes and environment. An individual's **genotype** is the genetic contribution to the phenotype. The term may refer to a specific set of genes that contribute to a particular trait, such as eye color, or to all the genes of the individual. (Genotype refers to portions of DNA molecules that result in traits; genome refers to all the DNA, regardless of whether or not it is expressed in traits.)

The environmental contribution to the phenotype consists of all forces other than genes that influence an individual's traits. Important environmental factors include diet, disease, physical exercise, and education. Interactions between the genotype and the environment produce the phenotype:

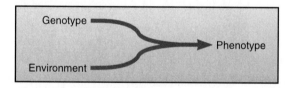

A few human traits, such as blood type, are controlled by the genotype alone. Most of our traits, however, develop in response to interactions between the genotype and the environment. The role of each factor is seen most clearly in identical twins who have been raised apart (fig. 9.1), for they have the same genotype but have grown up in different environments. Such twins are remarkably similar, not only in physical characteristics but in personality and attitude, indicating the enormous influence of genotype.

Figure 9.1 Identical twins raised apart. Jerry and Mark have the same genotype but were raised in different environments. They are remarkably alike, not only in physical traits but in personality and attitude. For instance, both became firefighters. Differences between them are attributed to their different environments.

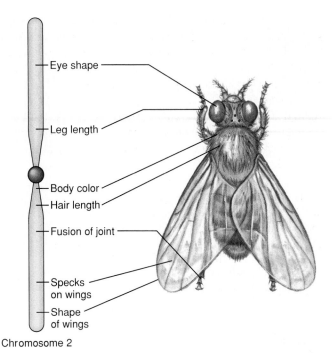

Chromosome 2

Figure 9.2 Locations of genes in a chromosome. Genes are arranged in a linear sequence along the length of the DNA molecule within the chromosome. Some of the genes within this chromosome, from a fruit fly, have been mapped.

Chromosomes and Alleles

A chromosome contains a long molecule of DNA that extends from one end of the structure to the other. Genes are short segments of the DNA molecule, arranged in a linear order within the chromosome:

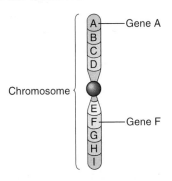

Each chromosome typically carries many thousands of genes. A gene is hard to define, and several definitions are appropriate depending on the context within which the term is used. In this chapter, we limit the definition to a structural gene, which is a segment of a DNA molecule that contains instructions for building a polypeptide (📖 *see page 45*). (A polypeptide is often the same as a protein, but sometimes more than one kind of polypeptide, coded for by more than one gene, forms a protein.)

The physical location of a gene within a chromosome is the **gene locus.** The locations of some of the genes that control the traits of fruit flies are shown in figure 9.2.

Every cell of an animal (except sperm or egg) has two sets of chromosomes—two chromosomes of every type. One set was inherited from the father, via his sperm, and the other from the mother, via her egg. The forty-six chro-

mosomes of a human body cell are twenty-three pairs of chromosomes (fig. 9.3).

The two chromosomes of a pair are called a **homologous pair** of chromosomes. They look the same and carry the kinds of genes for producing the same kinds of traits. A gene that controls whether the hairline on the forehead is a straight line or curved to form a "widow's peak," for instance, is located within a particular chromosome—one that can be identified by its length and shape. Every one of us has two copies of this chromosome and two copies of the gene, one inherited from the father and one inherited from the mother. The two copies of each gene, which form a **gene pair,** are at the same gene locus within the two homologous chromosomes:

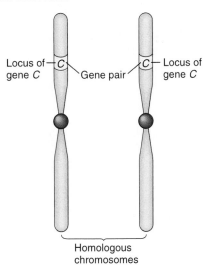

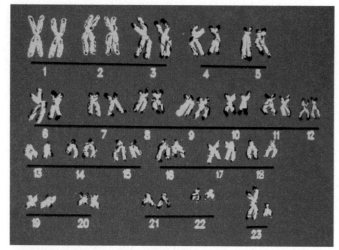

Figure 9.3 The forty-six chromosomes of a human. The chromosomes arranged in this karyotype were photographed during cell division, when they are short, thick, and duplicated (each chromosome consists of two chromatids held together by a centromere). This karyotype is of a male, since the twenty-third pair is an X chromosome and a Y chromosome. A female karyotype is shown in figure 7.2.

The two copies of a gene may be different. For example, there are two versions of the gene that controls cheek shape: one version develops dimples and the other does not. (A dimple forms when a short protein strand connects the skin to underlying muscles.) Different versions of a gene are called **alleles.** They are symbolized by different versions of a letter of the alphabet. The alleles for cheek shape for example, may be written *D* and *d*.

An individual with a gene pair consisting of two alleles is a **heterozygote** (*hetero*, different; *zygote*, fertilized egg) with regard to that locus. The genotype of a heterozygote with regard to cheek shape would be written as *Dd*. The two copies of the gene are in different chromosomes of a homologous pair:

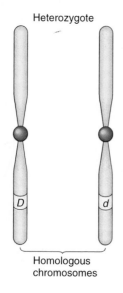

An individual with a gene pair consisting of two identical copies of an allele is a **homozygote** (*homo*, same) with regard to that locus. Their genotype is written as two identical symbols for the allele. The genotype of a homozygote with regard to cheek shape, for example, would be either *DD* or *dd*, depending on which allele was present:

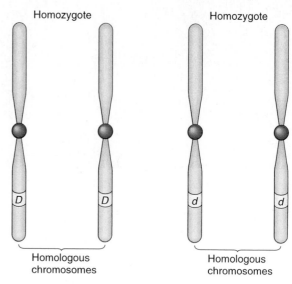

In this chapter, we consider only dominant and recessive alleles. Other ways in which the alleles of a heterozygote interact are described in chapter 10. When a dominant allele and a recessive allele are present together in a heterozygote, the **dominant allele** masks the effects of the **recessive allele.** Thus, the phenotype of the heterozygote is the same as the phenotype of the homozygote with two copies of the dominant allele. For example, the allele for dimples (*D*) is dominant over the allele for no dimples (*d*). Someone with two copies of the allele for dimples (*DD*) has dimples, as does someone with one copy of each allele (*Dd*). Only a person with two copies of the recessive allele (*dd*) has no dimples.

Meiosis and the Inheritance of Genes

How genes move from parent to offspring during reproduction depends on how chromosomes move into sperm and eggs during meiosis (see page 144). Chromosomes were first described in the late nineteenth century and recognized as the physical basis of inheritance in the early 1900s. The discoveries of chromosomes and of how they move during meiosis were major breakthroughs in our understanding of inheritance.

The science of genetics was born, however, several decades before any knowledge of chromosomes. Gregor Mendel discovered, in 1865, two fundamental laws of inheritance without knowing about either chromosomes or

Gregor Mendel

Gregor Mendel was one of our greatest biologists. He founded the science of genetics by discovering, in the 1860s, two fundamental laws of inheritance. The law of segregation describes how genes for single traits are passed from parents to offspring. The law of independent assortment describes how genes for two traits are passed simultaneously from parents to offspring.

Mendel was born, in 1822, into a farming family in a region of Europe that is now the Czech Republic. As a child he showed a remarkable talent for mathematics, but his family was too poor to send him to college. When he reached the age of twenty-one, Mendel entered a monastery in order to free himself from "the bitter struggle for existence."

The abbot (head monk) of the monastery recognized Mendel's talent and his desire to learn. He sent the young scholar to the University of Vienna to study mathematics and physics. Once more, Mendel's background thwarted his academic success—too insecure and nervous to perform well on examinations, Mendel flunked his examinations (twice!) and returned to the monastery. He devoted the next seven years of his life to the breeding of garden peas. It was through these breeding experiments that he discovered the two laws of inheritance (fig. 9.A).

Mendel was fascinated with inheritance. As a child, he had watched his father develop new strains of agricul-

Figure 9.A Gregor Mendel (1822–84) experimenting with peas in the monastery garden.

tural plants. Controlled breeding of this sort had been practiced since the dawn of agriculture, 10,000 years earlier, but prior to Mendel's work no one had explained the patterns of inheritance.

Mendel's genius was in experimental design and statistical analysis. Other scientists had performed similar breeding experiments but had worked with just a few individuals while attempting to keep track of many traits at the same time. Mendel, by contrast, concentrated on just one or two traits at a time and repeated the same experiment on thousands of pea plants.

Mendel excelled in this line of research because of his unique personality and background. His enormous patience enabled him to plan, execute,

and analyze each experiment with great care and precision. His background in mathematics helped him in two ways. First, he was accustomed to working with phenomena that he never observed directly—genes were abstract concepts to Mendel. Second, he was used to seeking patterns within numbers and so was not discouraged by the need to examine hundreds of plants in order to find a pattern of inheritance.

Mendel's carefully designed experiments revealed basic patterns in the way genes are inherited. From these patterns, he deduced the existence of physical structures of inheritance (now known as chromosomes) and the manner in which they pass from parents to offspring.

Mendel reported his spectacular discoveries to the local natural history society in 1865, and a paper describing his work was published in its journal the following year. The lack of response was disappointing. His discoveries held the key to inheritance, yet biologists did not see the significance of these elaborate experiments with peas. Even Charles Darwin, who was aware of Mendel's findings and whose theory would have benefited from them, did not appreciate the work of this brilliant but obscure monk.

Mendel gave up his unrewarding line of work, in 1868, to become abbot of the monastery. He died in 1884, his work still ignored by the scientific community. It wasn't until 1900 that biologists realized the significance of Mendel's work and their failure to recognize a genius in their midst.

meiosis. This brilliant man studied inheritance by growing plants rather than by peering at cells through a microscope. Mendel's two laws—the law of segregation and the law of independent assortment—are described here in the context of modern knowledge about chromosomes and meiosis. His life and work are described in the accompanying Research Report.

Separation of Alleles

The first pattern of inheritance discovered by Mendel was the *law of segregation*. This law describes how the two copies of each gene segregate (i.e., separate) during meiosis so that just one copy ends up in each gamete (sperm or egg). It was discovered by examining the segregation of alleles in heterozygotes, where the two copies of the gene are different and can be traced.

The physical basis of the law of segregation is the separation of homologous chromosomes during meiosis (although Mendel did not actually observe chromosomes). Two copies of a gene are located at the same locus within the two chromosomes of a homologous pair. Consider the alleles that control development of dimples in the cheeks, *D* and *d*. Prior to meiosis, the chromosomes are duplicated. The two identical copies of each chromosome, called chromatids, remain held together by a centromere:

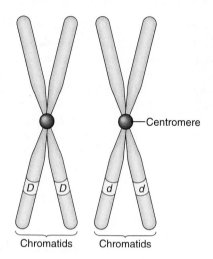

During anaphase of meiosis I, the homologous pairs of chromosomes (each consisting of two chromatids) move to opposite sides of the cell and the two alleles separate:

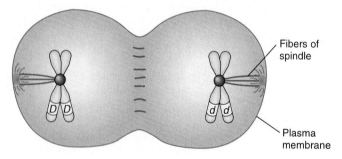

One allele (in both chromatids of a chromosome) ends up in one daughter cell, and the other allele (in both chromatids of the other chromosome) ends up in the other daughter cell. In this way, the two alleles of the heterozygote (*Dd*) become segregated—moved into different cells. After meiosis II, which separates the two chromatids within each chromosome, the daughter cells will develop into gametes—sperm in a male, eggs in a female.

Combining Alleles at Fertilization

When a sperm fertilizes an egg, two copies of each gene—one in the sperm and the other in the egg—come together to form a gene pair within the fertilized egg:

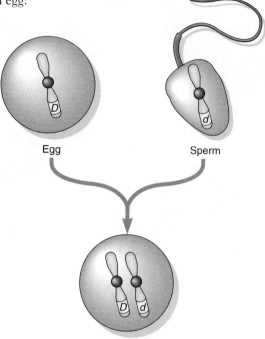

Sperm and eggs unite at random during a particular mating: which sperm joins with which egg is determined by chance. If the genotypes of the sperm and eggs are known, then the genotypes of the offspring can be predicted. We do this by using a tool called the Punnett square.

In a **Punnett square,** the kinds of sperm (with regard to the particular allele they carry) are listed in a row along the top and the kinds of eggs in a column along the left side. Take, for example, matings between a male and a female who both have the *Dd* genotype:

Kinds of sperm

		D	*d*
Kinds of eggs	*D*		
	d		

These listings represent not only the kinds of sperm and eggs but also their relative frequencies: how often one kind appears in relation to another kind. In the previous example, a male of genotype *Dd* forms two kinds of sperm: half the sperm carry the *D* allele and half carry the *d* allele. The same is true for the eggs of the *Dd* female.

By combining sperm and eggs, as happens during fertilization, we obtain the kinds of genotypes that can form from the mating and how frequently each kind will appear in relation to the others. We do this by combining each kind of sperm with each kind of egg, and writing the combined genotype in the square:

Kinds of sperm

	D	d
D	DD	Dd
d	Dd	dd

(left axis label: Kinds of eggs)

The genotype within one square has the same chance of forming as the genotype within every other square. Since one out of four squares in the Punnett square shown here has the *DD* genotype, this means that one-fourth of the offspring is likely to have that genotype. Similarly, we predict that half the offspring will have the *Dd* genotype and one-fourth will have the *dd* genotype.

We can take this analysis one step further and predict the phenotypes of the offspring. In our example, the dominant allele will mask the effects of the recessive allele in the phenotype, so a *DD* individual will have the same phenotype as a *Dd* individual. Consequently, we predict that three-fourths of the offspring will have dimples and one-fourth will not. There will be a 3:1 ratio of phenotypes.

It was this 3:1 ratio of phenotypes that led Mendel to his law of segregation. He attributed this ratio to a physical condition in which there were two copies of a gene, one dominant over the other, and the copies separate during gamete formation—one copy moving into one gamete and the other copy moving into another gamete.

The predicted ratio of phenotypes is not expected with each mating. The ratio reflects an average of many matings and a large number of offspring. If there are just four offspring, we cannot expect random encounters of sperm with eggs to produce exactly three offspring with one phenotype and one offspring with the other phenotype. Instead, the ratio of 3:1 indicates the *most likely* outcome. For each offspring born, there is a 75% chance it will have the dominant phenotype and a 25% chance it will have the recessive phenotype.

The outcome of one mating does not influence the outcome of the next mating. Thus, for example, if there is a 75% chance that an offspring will have dimples, and the first offspring does not have dimples, then the chance that the second offspring will have dimples is still 75%. Each successive mating between a particular man and woman has the same probability of producing a particular phenotype.

Independent Assortment of Alleles

The second pattern of inheritance discovered by Mendel is the *law of independent assortment*. According to this law, the inheritance of alleles for one trait in no way influences the inheritance of alleles for another trait. Whether a child inherits dimples, for example, does not influence whether he or she inherits freckles. Gene pairs for different traits assort (separate, or segregate) independently. The law of independent assortment applies to traits controlled by genes on different pairs of homologous chromosomes.

The physical basis for Mendel's second law is the movements of two pairs of homologous chromosomes, in which one pair holds the alleles for one trait and the other pair holds the alleles for the other trait.

Independent assortment of homologous chromosomes during meiosis I explains the independent inheritance of two (or more) traits. In independent assortment, the way in which a gene pair within one homologous pair of chromosomes separates and moves into gametes is independent of the way in which a gene pair within another homologous pair of chromosomes separates and moves into gametes.

Consider two gene pairs, one that controls whether or not dimples develop in the cheeks and another that controls whether or not freckles develop in the skin. Assume the person undergoing meiosis (i.e., forming sperm or eggs) is a heterozygote with regard to both gene loci (i.e., *Dd* and *Ff*) and that the two gene pairs are in different homologous chromosomes:

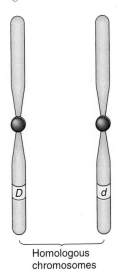

Homologous chromosomes

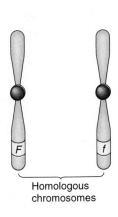

Homologous chromosomes

During meiosis I, the homologous pairs of chromosomes line up in the center of the cell and then move to opposite poles. The way the pair of chromosomes carrying the *D* and *d* alleles line up and separate in no way influences how the pair of chromosomes carrying the *F* and *f* alleles line up and separate. Two patterns are possible:

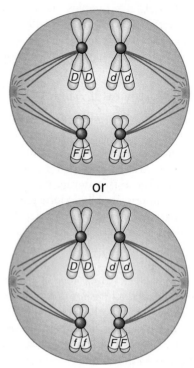

At the end of meiosis, four types of gametes are possible:

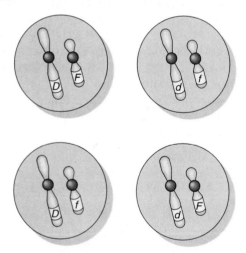

All four types have the same chance of forming. When a large number of gametes form, the four types appear in approximately equal numbers. (If the chromosomes did not assort independently, then they would always line up in the same way and only two types of gametes would form.)

The consequence of independent assortment—an increased variety of sperm and eggs—is seen only when the gene loci consist of alleles, as in the previous example.

When the loci consist of identical copies of genes, independent assortment takes place but has no consequence, since two identical copies cannot be distinguished.

The greater variety of gametes generated by independent assortment gives rise to a greater variety of offspring. Consider a mating between two people, both heterozygous for dimples and freckles. The allele for dimples (*D*) is dominant over the allele for no dimples (*d*), and the allele for freckles (*F*) is dominant over the allele for no freckles (*f*). The man will produce four kinds of sperm and the woman will produce four kinds of eggs with regard to these alleles. As shown in the Punnett square in figure 9.4, the mating could produce four kinds of phenotypes: dimples and freckles, dimples but no freckles, no dimples but freckles, no dimples and no freckles.

If this couple were able to produce a large number of offspring, then the four phenotypes would appear in a 9:3:3:1 ratio (i.e., for every offspring with no dimples and no freckles, there would be three offspring with dimples but no freckles, three offspring with freckles but no dimples, and nine offspring with both dimples and freckles).

It was this 9:3:3:1 ratio of phenotypes in Mendel's experiments that led him to the law of independent assortment. This law, an extension of the law of segregation, assumes that gene pairs for different traits separate independently of each other during gamete formation. It applies only to genes located within different pairs of homologous chromosomes.

Linked Genes

Genes that lie near each other within the same chromosome are generally inherited together; they move together during meiosis. The law of independent assortment does not apply to such genes.

Consider, for example, the genes for chin shape (fig. 9.5) and cheek shape. Assume that both gene pairs are near each other within the same homologous pair of chromosomes. Assume also that the dominant *C* allele (which produces a cleft chin) and *D* allele (which produces dimples) are in the same chromosome and that the recessive *c* allele (which produces a smooth chin) and *d* allele (which produces no dimples) are in the other chromosome of the homologous pair:

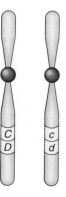

Kinds of sperm

Kinds of eggs	D;F	D;f	d;F	d;f
D;F	DD;FF Dimples; freckles	DD;Ff Dimples; freckles	Dd;FF Dimples; freckles	Dd;Ff Dimples; freckles
D;f	DD;Ff Dimples; freckles	DD;ff Dimples; no freckles	Dd;Ff Dimples; freckles	Dd;ff Dimples; no freckles
d;F	Dd;FF Dimples; freckles	Dd;Ff Dimples; freckles	dd;FF No dimples; freckles	dd;Ff No dimples; freckles
d;f	Dd;Ff Dimples; freckles	Dd;ff Dimples; no freckles	dd;Ff No dimples; freckles	dd;ff No dimples; no freckles

Figure 9.4 A Punnett square. The kinds of sperm, with regard to the genes under study, are listed in a row along the top and the kinds of eggs, with regard to the same genes, are listed in a column along the side. Combining each kind of sperm with each kind of egg gives all possible genotypes of the offspring in the proportions they are expected to appear. Phenotypes, indicated beneath the genotypes in the boxes, are deduced from the genotypes.

Figure 9.5 A cleft chin. A groove in the chin develops in response to a dominant allele.

This genotype is written as *CD/cd* to indicate that the *C* allele and the *D* allele are in one chromosome and the *c* allele and *d* allele are in the other chromosome of the homologous pair.

The alleles within each chromosome travel together: *C* moves with *D* and *c* moves with *d*. The two gene pairs do not assort independently. Only two kinds of gametes form: a gamete with the chromosome carrying *C* and *D* and a gamete with the chromosome carrying *c* and *d*. Thus, someone who inherited the *C* allele would also inherit the *D* allele—a cleft chin would always go with dimples. Likewise, a smooth chin would always go with no dimples, since someone with two copies of the *c* allele would also have two copies of the *d* allele.

Thus, the law of independent assortment does not apply to the genes that lie near each other within the same pair of homologous chromosomes.

Crossing-Over

Genes that are linked within a chromosome do not always remain linked. They sometimes move from one chromosome of a homologous pair to the other. Such an exchange between similar segments of homologous chromosomes is called **crossing-over.**

Segments of chromosomes are exchanged during late prophase of meiosis I (*see page 146*), when each chromosome consists of two identical chromatids held together by a centromere. At this time, homologous chromosomes line up so precisely that the same gene loci are right next to each other. Segments of two chromatids, one from each chromosome, sometimes break off from their original chromatid and join with the other one:

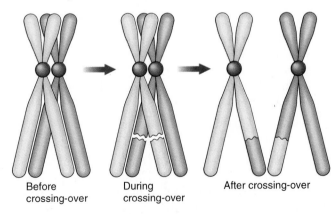

Before crossing-over   During crossing-over   After crossing-over

Matching segments of the chromatids, containing matching pairs of genes, are exchanged at this time. One gene of each pair is traded for the other. When the gene pairs consist of alleles, new combinations of traits become linked in inheritance.

Consider the gene pairs that govern dimples and cleft chin. If they are both in the same pair of homologous chromosomes, and do not experience crossing-over, then they are inherited together: a child with dimples also inherits a cleft chin; a child without dimples also inherits a smooth chin. If, however, crossing-over takes place between the two genes, then the genes become unlinked:

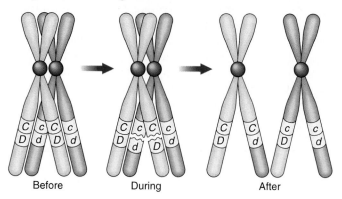

Before   During   After

Crossing-over increases the variety of gametes: some of the gametes receive copies of the original chromosomes and some receive new kinds of chromosomes from the crossing-over. Let us assume, here, that crossing-over occurred during formation of sperm but not of eggs. Then four kinds of sperm (*CD*, *cD*, *Cd*, *cd*) will join two kinds of eggs (*CD* and *cd*) to produce the following kinds of offspring:

Genotype	Phenotype
CD/CD	cleft chin and dimples
cD/CD	cleft chin and dimples ⎫ new genotypes but
Cd/CD	cleft chin and dimples ⎭ not new phenotypes
cd/CD	cleft chin and dimples
CD/cd	cleft chin and dimples
cD/cd	smooth chin and dimples ⎫ new genotypes and
Cd/cd	cleft chin and no dimples ⎭ new phenotypes
cd/cd	smooth chin and no dimples

In this example, crossing-over generated two new combinations of traits: dimples with a smooth chin and no dimples with a cleft chin.

On the average, crossing-over happens at several sites within a homologous pair during each meiosis. It is a major source of new combinations of traits.

Genetic Variability

Genetic variability refers to the different kinds of genotypes within a population (fig. 9.6). Recall that a genotype is the set of genes that an individual inherits. As we shall see in later chapters, genetic variability is the raw material of evolution.

Figure 9.6 Genetic variability among dogs. The visible differences among these dogs are due chiefly to their different genotypes.

Table 9.1	
Origins of Genetic Variability	
Genetic Event	**How the Event Generates Variability**
Mutation	Forms new alleles
Independent assortment of chromosomes	Forms new combinations of chromosomes
Crossing-over	Forms new combinations of alleles within a chromosome
Fertilization	Combines the alleles of different individuals

Different kinds of genotypes are generated by different kinds of genes and by new combinations of the genes. These important genetic phenomena, known as mutations and recombinations, are summarized in table 9.1 and described following.

Mutations

Mutations are the ultimate source of genetic variability, for they give rise to new forms of a gene. A **mutation** is a change in the structure of a DNA molecule. It develops when a mistake is made during DNA replication or when something in the environment, such as radiation, tobacco smoke, or pesticides, alters the structure of a DNA molecule.

A mutation produces a new allele—a new version of the gene (a mutation in gene *A*, for example, could produce allele *a*). The new allele is then copied, by DNA replication prior to meiosis, and passed via sperm or eggs to subsequent generations.

A population of organisms may have many alleles of a particular gene. Most human genes come in more than one form. The gene for transferrin (an iron-transporting protein in the blood), for example, has at least eighteen alleles in human populations.

Recombination

Recombination occurs when the alleles of different gene pairs combine with one another in different ways. Different combinations of alleles develop from independent assortment of chromosomes, crossing-over, and the fusion of different kinds of sperm and eggs during fertilization.

Independent assortment of chromosomes during meiosis yields new combinations of chromosomes, as shown in figure 9.7*a*. Crossing-over, shown in figure 9.7*b*, gives new combinations of alleles within homologous chromosomes. And fertilization combines alleles from different individuals (fig. 9.7*c*).

Amount of Genetic Variability

Sexual reproduction, which always involves meiosis and fertilization, generates an enormous amount of genetic variability from recombinations. The variety of genotypes that arise from independent assortment alone is vast.

If we consider only one trait per pair of homologous chromosomes, and assume that each trait is controlled by just two alleles, then the number of different kinds of sperm or eggs generated by independent assortment is 2^n, where *n* is the number of homologous chromosomes under

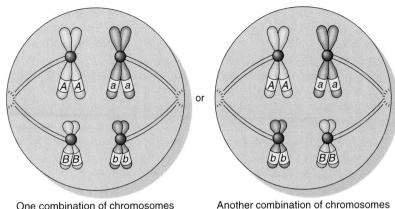

One combination of chromosomes   or   Another combination of chromosomes

(a)

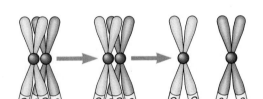

Original combination of alleles   Crossing-over   New combination of alleles

(b)

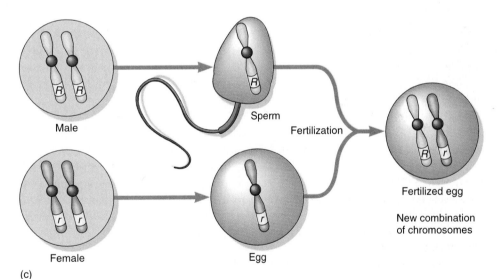

Male

Female

Sperm

Egg

Fertilization

Fertilized egg

New combination of chromosomes

(c)

Figure 9.7 Genetic variability from recombination.
Recombinations are different arrangements of alleles.
(*a*) Independent assortment of chromosomes during meiosis I generates different combinations of chromosomes in the sperm and eggs. (*b*) Crossing-over generates new combinations of alleles within a chromosome. (*c*) Fertilization brings alleles from different parents together within the fertilized egg.

165

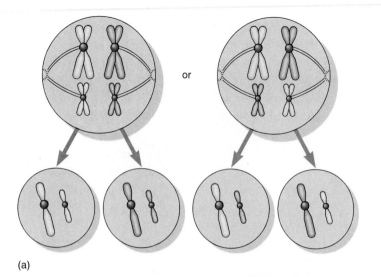

(a)

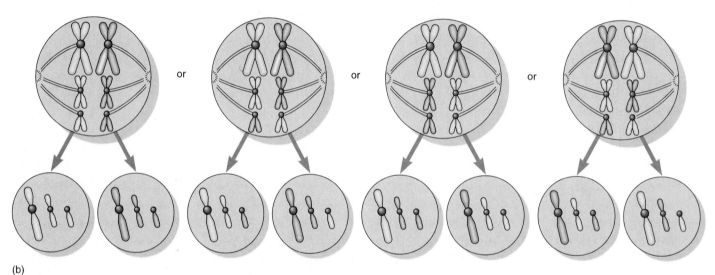

(b)

Figure 9.8 Genetic variability from independent assortment. Independent assortment during meiosis I generates new combinations of chromosomes in sperm and eggs. (*a*) Two pairs of homologous chromosomes can line up and separate in two ways, forming four (2^2) kinds of sperm or eggs. (*b*) Three pairs of homologous chromosomes can line up and separate in four different ways, forming eight (2^3) kinds of sperm or eggs.

consideration (fig. 9.8). Thus, 1 pair of homologous chromosomes yields 2 (2^1) types of sperm or eggs, 2 pairs yield 4 (2^2), 3 yield 8 (2^3), and so on. You have 23 pairs of homologous chromosomes and therefore could form 8,388,608 (2^{23}) different kinds of sperm or eggs. The actual variety is much greater because each pair of chromosomes carries alleles for more than one trait.

Sexual reproduction generates an enormous amount of genetic variability. Without sex, an organism has only mutations as a source of variability. With sex, an organism adds to mutations all the variability that occurs during meiosis (independent assortment and crossing-over) and fertilization. Genetic variability, seen as phenotypic vari-

ability, is especially apparent to us in human populations because we pay more attention to differences within our own species (as do the animals of other species). There are many more possible kinds of phenotypes than there are people on earth.

Chromosomal Inheritance

Some aspects of inheritance depend on entire chromosomes rather than on specific genes. Whether an individual is a male or a female depends on which chromosomes are inherited from the father. In addition, certain genetic disorders develop when an individual inherits too many or too few chromosomes.

Sex Determination

The particular set of chromosomes within a fertilized egg determines the sex of the individual. One pair of chromosomes, called the **sex chromosomes,** determines whether the reproductive organs develop into testes or ovaries. These organs then synthesize the sex hormones, which control the development of masculine or feminine traits.

The Sex Chromosomes   Each body cell of a human contains twenty-two pairs of homologous chromosomes, called **autosomes,** plus a pair of sex chromosomes (see fig. 9.3). In a male, the sex chromosomes are of two types—a long X chromosome and a short Y chromosome. In a female, the sex chromosomes are both X chromosomes:

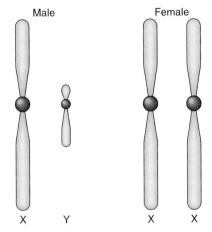

This system of sex determination is the same for all mammals. Other animals have different systems. In birds, the male has two sex chromosomes of the same kind (like two X chromosomes) and the female has two different sex chromosomes (like the X and Y). In many insects, the male has just one sex chromosome (like an X but no Y) and the female has two copies of the same sex chromosomes (like two Xs).

Most genes in the human X chromosome direct a broad range of traits and have nothing to do with femininity. (This makes sense, since the X chromosome appears in both males and females.) Examples of traits governed by genes in the X chromosome are aspects of color perception, muscle function, and blood clotting.

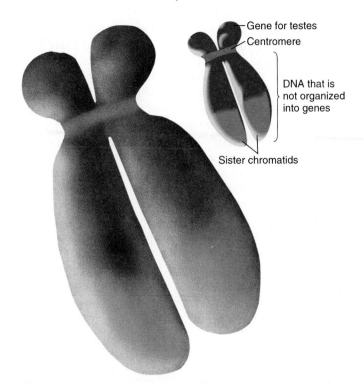

Figure 9.9   The Y chromosome. This computer-enhanced photograph shows a Y chromosome during cell division, after it has duplicated itself but before the two chromatids have separated. One tip of the chromosome, shown here in purple, contains the gene that causes testes to develop. Most of the chromosome contains DNA that is not organized into genes.

In humans, the Y chromosome contains only a few genes (fig. 9.9). The most important one, discovered in 1991, controls the formation of testes. All other aspects of male sexual development follow from this initial event. Male and female embryos develop in the same way until about seven weeks after fertilization. At that point, if the gene for testes formation is present, then the embryo forms testes and develops male traits. If the gene is absent, then the embryo forms ovaries and develops female traits. Genes within the Y chromosome tend to remain there, since crossing-over almost never occurs between the Y chromosome and the X chromosome.

Separation of Sex Chromosomes   Meiosis in a man occurs within cells of his testes. It starts with a cell that has both an X and a Y chromosome in addition to the twenty-two pairs of autosomes. During meiosis, the two sex chromosomes separate and one goes to each daughter cell (fig. 9.10a). These cells, each with twenty-two autosomes plus either an X or a Y, then develop into sperm. Half the sperm will carry an X chromosome and half will carry a Y chromosome.

Meiosis in a woman occurs in cells of her ovaries. It starts with a cell that contains twenty-two pairs of autosomes plus two X chromosomes. After meiosis, each resulting cell contains twenty-two autosomes plus one of the X chromosomes (fig. 9.10b). Four cells form, but only one develops into a mature egg. The other three cells (called polar bodies), which are much smaller and receive no yolk, degenerate. Regardless of which cell develops into the egg, it will carry an X chromosome.

Combining Sex Chromosomes at Fertilization   An egg, which always carries an X chromosome, may unite at fertilization with a sperm carrying either a Y or an X chromosome. If a Y-carrying sperm fertilizes the egg, then a male (XY) develops. If an X-carrying sperm fertilizes the egg, then a female (XX) develops:

The sex of the child is determined by the sperm rather than by the egg. A male produces equal numbers of X-carrying sperm and Y-carrying sperm because of the way the X and Y chromosomes separate during meiosis I. Thus, with every fertilization there should be an equal chance of producing a son or a daughter. However, a sperm carrying a Y chromosome is slightly smaller and moves slightly faster than a sperm carrying an X chromosome. Sperm carrying the Y chromosome therefore tend to reach the egg first, so that more males than females are conceived. At the time of fertilization, the ratio of sexes is about 160 males to every 100 females.

More males than females die during embryonic development, so at the time of birth there are only 106 males for every 100 females. The ratio of males to females continues to decline throughout life. By the ages of fifteen to nineteen, the sex ratio is equal; by the age of eighty-five, females outnumber males by a ratio of about two to one.

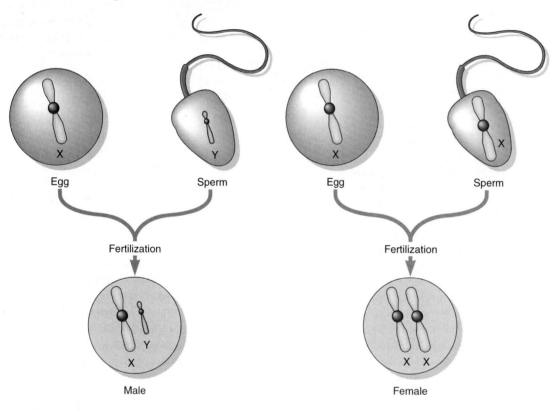

Figure 9.10 Separation of the sex chromosomes during meiosis.
While each cell holds twenty-two pairs of autosomes and one pair of sex chromosomes, only the sex chromosomes are shown here. (*a*) In a man, meiosis I separates the X chromosome from the Y chromosome. Meiosis II then separates the chromatids of each chromosome. Half the sperm carry the X chromosome and half carry the Y chromosome. (*b*) In a woman, meiosis I separates the two X chromosomes and meiosis II separates the chromatids of each chromosome. Every daughter cell carries an X chromosome. Only one of the cells develops into a mature egg. The other three cells, which are much smaller, degenerate.

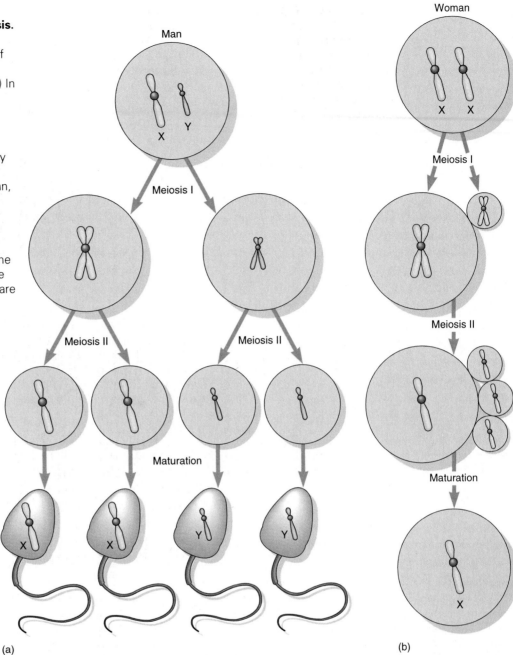

(a)

(b)

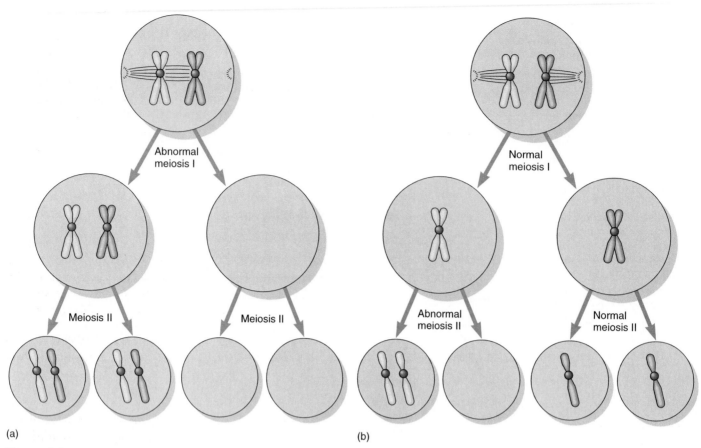

(a)

(b)

Figure 9.11 Nondisjunction of chromosomes.
Nondisjunction is when chromosomes fail to separate during meiosis. (*a*) When nondisjunction happens during meiosis I, both homologous chromosomes go to the same cell. (*b*) When nondisjunction happens during meiosis II, both chromatids of a chromosome go to the same cell. When one of these abnormal cells becomes a sperm or egg and unites with a normal sperm or egg, the offspring will have too many or too few chromosomes.

Too Many or Too Few Chromosomes

Human development is sensitive to the chemical balance provided by twenty-two pairs of autosomes and a single pair of sex chromosomes. An offspring who inherits more than or fewer than forty-six chromosomes (which occurs once in every 200 live births) is afflicted with many serious abnormalities. Such a condition, in which a single disorder causes many symptoms, is known as a syndrome. Many traits are affected by an abnormal number of chromosomes because many genes are present in inappropriate amounts.

Too many or too few chromosomes is usually the consequence of **nondisjunction** of chromosomes during meiosis, in which the chromosomes of a homologous pair fail to separate during meiosis I or the two chromatids of a chromosome fail to separate during meiosis II (fig. 9.11). After meiosis, one or two of the cells have an extra chromosome and the others lack that same chromosome. When one of these abnormal cells develops into a sperm or egg and participates in fertilization, the fertilized egg contains too many or too few chromosomes.

Sex Chromosomes An unusual number of sex chromosomes causes many abnormalities in humans, affecting both mental and sexual traits. The abnormality can develop from nondisjunction in either parent if their sex chromosomes fail to separate during meiosis I or meiosis II. When this happens in a male, four types of abnormal sperm are possible: XY, XX, YY, and sperm with no sex chromosome at all. When nondisjunction happens in a female, two types of abnormal eggs are possible: XX and eggs with no sex chromosome at all. When one of these abnormal sperm or eggs combines with a normal sperm or egg, the fertilized egg receives either three sex chromosomes or just one sex chromosome, as shown in figure 9.12.

Kinds of abnormal sperm

	XY	XX	YY	No sex chromosome
X	XXY	XXX	XYY	X

Normal egg

(a)

Kinds of normal sperm

	X	Y
XX	XXX	XXY
No sex chromosome	X	Y

Kinds of abnormal eggs

(b)

Figure 9.12 Punnett square with gametes that have unusual numbers of sex chromosomes. (a) Nondisjunction of sex chromosomes in a man can produce four kinds of abnormal sperm. When these abnormal sperm fertilize normal X-carrying eggs, four kinds of abnormal genotypes are formed. (b) Nondisjunction of sex chromosomes in a woman can produce two kinds of abnormal eggs. When these eggs are fertilized by normal sperm, four kinds of abnormal genotypes are formed.

The disorders caused by an abnormal number of sex chromosomes are described in table 9.2. Notice that whether the affected person develops into a male or female depends on the presence or absence of the Y chromosome rather than on the number of X chromosomes. Also, as might be expected, someone with only one sex chromosome can survive if it is an X but not if it is a Y: the X chromosome carries genes that are essential to survival whereas the Y chromosome does not.

Autosomes Too little or too much of the genetic material carried in autosomes also causes syndromes with pronounced abnormalities. About half of all miscarriages are of embryos with too many or too few autosomes.

A fetus with a missing autosome, regardless of which one, dies before it is born. A fetus with an extra autosome may develop and survive, especially if it is one of the smaller autosomes, but will not be normal. The condition of an extra autosome is known as a trisomy (*tri*, three; *some*, body).

The most common trisomy is Down syndrome, also known as trisomy 21 (fig. 9.13). In this condition, a child inherits three copies of chromosome 21, giving a total chromosome count of forty-seven rather than forty-six. Chromosome 21 is a small chromosome with approximately 1,500 genes. The abnormal meiosis that produces this syndrome occurs more often during formation of eggs than of sperm and increases with age of the mother, as described in the accompanying Wellness Report.

Table 9.2

Too Many or Too Few Sex Chromosomes

Sex Chromosomes	Frequency in Live Births	Name of the Syndrome	Sex	Symptoms
X	1 in 5,000	Turner	Female	Immature breasts, ovaries, and uterus; sterile; short; may have webbed neck; may be mentally retarded
Y	0			Lethal to the embryo
XXX	1 in 700	Triple-X	Female	May appear physically normal; often sterile; sometimes mentally retarded or emotionally disturbed
XXY	1 in 500	Klinefelter	Male	Undersized testes; development of breasts; sterile; short-lived; usually with normal intelligence
XYY	1 in 1,000	XYY	Male	May be entirely normal; may have intellect that is below normal but not retarded; usually tall

(a)

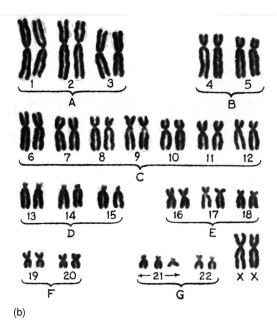

(b)

Figure 9.13 Down syndrome. (*a*) A child with Down syndrome can be recognized by features of the face. (*b*) The karyotype of someone with Down syndrome shows three copies of chromosome 21. The genes in that chromosome produce too much of their products. (*c*) The incidence of Down syndrome increases with age of the mother.

SUMMARY

1. An individual's phenotype (observable traits) develops in response to interactions between the genotype (set of genes) and the environment.

2. Genes are arranged in a linear sequence within a chromosome. Chromosomes occur in homologous pairs; there are two copies of each gene, positioned at the same locus within a pair of homologous chromosomes. When the copies are different, they are alleles and the individual is a heterozygote. When the copies are identical, the individual is a homozygote.

3. The physical basis of inheritance lies within chromosomes and their movements during meiosis I. Separation of alleles (Mendel's law of segregation) results from separation of homologous chromosomes. Independent assortment of alleles (Mendel's law of independent assortment) results from independent alignment and separation of nonhomologous chromosomes.

4. Genes that are linked within the same chromosome do not assort independently. Such genes become unlinked when a crossing-over occurs between them.

5. Genetic variability arises from mutations and events associated with sexual reproduction: independent assortment, crossing-over, and fertilization. The amount of genetic variability from sexual reproduction is enormous.

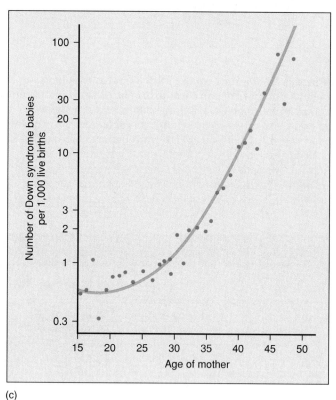

(c)

6. The sex of the offspring is determined by the kinds of sex chromosomes it inherits. A male mammal has an X and a Y chromosome; half its sperm carry the X chromosome and half carry the Y chromosome. A female mammal has two X chromosomes; all its eggs carry the X chromosome. At fertilization, an X-carrying egg is fertilized by either a Y-carrying sperm (forming a male offspring) or an X-carrying sperm (forming a female offspring).

Wellness Report

Down Syndrome and Age of the Mother

A person with Down syndrome has three copies of chromosome 21 rather than two copies in every body cell. The extra chromosome affects practically every part of the body. Signs of Down syndrome include a wide, rounded face; short, broad hands; short arms and legs; and a thick tongue. More serious medical problems arise from heart defects, mental retardation, defective lenses in the eyes, and an impaired ability to fight infections.

The chance of a woman having a child with Down syndrome increases dramatically with age. While the average incidence of Down syndrome in the United States is about 1 in 700, a woman in her twenties has only 1 chance in 1,500 of giving birth to such a child. A woman past forty has 1 chance in 80 and beyond forty-five years of age the chance becomes 1 in 44.

In the early 1900s, when families were larger, it was noted that Down syndrome appeared more often in the youngest of many children. Physicians at that time attributed the condition to an "exhausted womb." We now know the syndrome is caused by nondisjunction of chromosome 21 during meiosis and is correlated with age of the mother rather than the number of children she has had. Something causes nondisjunction to occur more frequently during meiosis in older women.

Nondisjunction of chromosomes can happen during formation of either sperm or eggs. Yet, it seems to happen mainly during egg formation, since incidence of Down syndrome (and similar chromosome disorders) increases with age of the mother but not with age of the father. Experts estimate that 95% of Down syndrome cases originate during the formation of eggs. Why should errors of meiosis occur more often in a woman than a man?

Meiosis in a man takes place rapidly. Once the DNA has been replicated, the cell proceeds rapidly through meiosis I and II. It all happens within a day. Meiosis in a woman, by contrast, is prolonged. Cells that will become eggs begin meiosis when the female is still a young embryo. (This means that the egg that will form a child begins to form while the mother is still in the uterus of *her* mother—the grandmother of the child!) Meiosis I proceeds until prophase, when homologous chromosomes come to lie near each other. Meiosis stops at this point and the chromosomes hold their positions until sometime between puberty (about age twelve) and menopause (about age forty-five to fifty), when the immature egg is released from the ovary and fertilized by a sperm. Meiosis then resumes as a sperm enters the egg.

An egg that matures in a forty-year-old woman has remained in prophase of meiosis I for more than forty years. The longer a pair of chromosomes remain next to each other, in suspended animation, the greater the chance that something will go wrong to prevent them from separating when meiosis resumes. Perhaps the chromosomes become sticky with age and cling to each other, or perhaps they become damaged by radiation or toxic chemicals during the many years they remain next to each other.

A recent hypothesis explaining the relationship between incidence of Down syndrome and maternal age is that an older body is less capable of recognizing and rejecting a defective embryo than a younger body. Thus, defective embryos may arise with equal frequency, regardless of age, but younger women miscarry these embryos more often than older women.

---

7. When homologous chromosomes or sister chromatids fail to separate during meiosis, some gametes receive too many and others receive too few chromosomes. An offspring with too many or too few chromosomes is almost always abnormal.

KEY TERMS

Allele (ah-LEEL) (page 158)
Autosome (AW-tuh-sohm) (page 167)
Crossing-over (page 163)
Dominant allele (page 158)

Gene locus (page 157)
Gene pair (page 157)
Genotype (JEE-noh-teyep) (page 156)
Heterozygote (HEH-tur-oh-ZEYE-goht) (page 158)
Homologous (hoh-MAH-luh-gus) **pair** (page 157)
Homozygote (HOH-moh-ZEYE-goht) (page 158)
Mutation (myoo-TAY-shun) (page 164)
Nondisjunction (NON-dis-JUNK-shun) (page 170)
Phenotype (FEE-noh-teyep) (page 156)
Punnett (PUH-net) **square** (page 160)
Recessive allele (page 158)
Recombination (page 165)
Sex chromosome (page 167)

STUDY QUESTIONS

1. Explain what is meant by the statement, "A human has twenty-two pairs of homologous chromosomes and a pair of sex chromosomes."
2. Describe the law of segregation.
3. Describe the law of independent assortment.
4. Describe two reasons why it is exceedingly unlikely that no two eggs of a woman (or two sperm of a man) have the same set of genes.
5. Explain why the number of boy babies is about equal to the number of girl babies.
6. In Down syndrome, each cell in the body has three copies of chromosome 21. Explain how body cells get such an abnormal number of chromosomes.

CRITICAL THINKING PROBLEMS

1. List ten environmental factors that contributed to the development of your current phenotype.
2. One hundred years ago, domestic pigs from Virginia were released onto an uninhabited island in the Caribbean. Today, the island pigs are smaller and accept a broader range of foods than their mainland relatives. What kinds of experiments could you perform to determine whether these differences in size and behavior are due to different genotypes?
3. What are some potential benefits and problems for society that could result from the Human Genome Project?
4. John Brown wants to have a son. He divorces his wife, who has given birth to four daughters and no sons, and marries another woman to improve his chances of having a son. What is wrong with his logic?

SUGGESTED READINGS

Introductory Level:

Kevles, D. J., and L. Hood (eds.). *The Code of Codes: Scientific and Social Issues in the Human Genome Project.* Cambridge, MA: Harvard University Press, 1992. A collection of essays by geneticists, social scientists, and lawyers about problems that may arise from complete knowledge of our genes.

Orel, V. *Mendel.* Oxford: Oxford University Press, 1984. A description of Mendel's personal and scientific life.

Pierce, B. A. *The Family Genetic Sourcebook.* New York: John Wiley & Sons, 1990. Written for the layperson, this book explains basic concepts of inheritance and describes, in alphabetical order, many genetic traits and all known disorders.

Advanced Level:

Lewontin, R. *Human Diversity.* San Francisco: W. H. Freeman and Company, 1982. A description of genetic diversity among humans and how it is formed and perpetuated.

Patterson, D. "The Causes of Down Syndrome," *Scientific American,* August 1987, 52–60. A review of the genes within chromosome 21 and how they contribute to the syndrome.

Singer, M., and P. Berg. *Genes and Genomes.* Mill Valley, CA: University Science Books, 1991. A sourcebook for detailed information about how genes are arranged on chromosomes.

Wachtel, S. S. (ed.). *Molecular Genetics of Sex Determination.* San Diego, CA: Academic Press, 1993. A history of the development of our understanding of sex determination, with particular emphasis on the X and Y chromosomes of mammals and the testes-forming gene.

❂ EXPLORATIONS

Interactive Software: Constructing a Genetic Map

This interactive exercise explores the effects of the locations of genes within a chromosome on the frequency of crossing-over during meiosis. The program provides a map of a chromosome, showing the relative positions of three genes within the chromosome, as well as information about the traits of offspring from matings between individuals carrying this chromosome. The student uses this information to estimate the frequency of crossing-over between the genes, and then alters the positions of the genes to see the effects of position on crossing-over frequency.

Interactive Software: Constructing a Genetic Map

This interactive exercise explores the actual locations of genes within the chromosomes of various organisms, including humans, mice, fruit flies, yeast, and protozoa. The student is provided with current maps of genes within the chromosomes of each kind of organism and can see how the locations of particular genes have changed over the course of evolution.

Interactive Software: Genetic Re-assortment during Meiosis

This interactive exercise explores the way in which the chromosomes move into eggs and sperm during meiosis. The student is provided with a set of chromosomes, each carrying alleles for observable traits, and then varies the positions of these alleles to see how they affect the combinations of traits observed in the offspring.

Patterns of Inheritance

Objectives

In this chapter, you will learn
–about the contribution of each gene
  of a gene pair;
–about inheritance when one gene pair
  determines one trait;
–about inheritance of genes located in
  the X chromosome;
–about inheritance when two or more
  gene pairs determine a trait;
–about inheritance when one gene pair
  masks the effects of other gene pairs;
–about inheritance when one gene pair
  determines many traits;
–how a gene's expression may depend
  on whether it is inherited from the
  mother or from the father.

Inheritance. Patterns of inheritance
are seen in family resemblances.

Abeautiful fashion model appeared on a televised talk show. Tall and slim with dark, lustrous hair and perfect skin, her lovely face was familiar to most viewers from the covers of magazines. The host asked her, "To what do you attribute your enormous success as a model?" She answered, "Good genes."

Her answer, which surprised the host, was a good one. While it is true she works hard to enhance her beauty—she exercises, eats only healthful foods, and takes good care of her skin and hair, she nevertheless would not have succeeded as a model unless endowed with a particular set of genes for physical traits.

We have known for centuries that physical traits, such as facial features and body build, are inherited. Now it seems that personality traits may be inherited as well. Studies of identical twins separated since birth indicate that the child's genes have a stronger influence on how the personality develops than the environment in which the child is raised. Traits that appear to depend more on genes than on environment include leadership, obedience to authority, risk-seeking, alienation, vulnerability to stress, fearfulness, shyness, and even the capacity for ecstasy from artistic experiences. These findings contradict the widespread belief that family life is the main influence on the kind of personality a child develops.

What do these findings mean to you? Do they mean your behavior is rigidly fixed? Do they mean the environment you create for your family will have no influence on the personalities of your children? The answer to both questions is no. Environment, and the experiences it provides, always influence personality. What these findings do mean, however, is that if you are nervous and apprehensive most of the time, you may be able to train yourself to be more relaxed and more trusting but are unlikely to ever enjoy living on the edge of danger. If your child is timid, you may be able to teach him or her to be more outgoing but probably cannot turn the child into a leader. While our personalities have a genetic foundation, they can be modified by our environment.

Alleles and Phenotypes

Most genes come in pairs—two copies of the gene at the same gene locus (location) within the two chromosomes of a **homologous pair** (*see page 144*). When the two copies are different, they are called **alleles** (*see page 158*).

Someone with two different alleles at a gene locus is called a **heterozygote.** Someone with two copies of the same allele at a gene locus is called a **homozygote:**

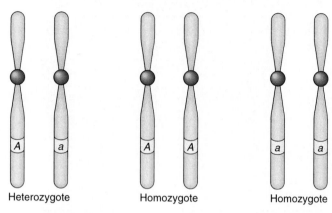

Heterozygote   Homozygote   Homozygote

Each gene pair contributes to the phenotype (observable traits) of the individual. Genes carry instructions for building polypeptides, which form proteins. Proteins, in turn, determine traits—either directly as physical structures (e.g., keratin in hair) or indirectly as enzymes that control biochemical reactions. Alleles are different versions of the instructions for building a polypeptide; they influence the same trait. Whether someone does or does not have dimples, for example, is governed by two alleles.

Gene pairs vary in the way each allele contributes to the phenotype. Two alleles may be dominant and recessive, incompletely dominant, or codominant. These kinds of allelic interactions are summarized in table 10.1 and described following.

Dominant-Recessive Alleles

One allele is dominant and the other recessive if, when present together, the effect of one allele masks the effect of the other. The allele with an effect on the trait is **dominant** over the **recessive** allele, which has no effect on the trait.

The dominant allele is abbreviated by a capital letter and the recessive allele by a lowercase letter. The symbol *Aa*, for example, indicates a heterozygote with a dominant (*A*) and a recessive (*a*) allele forming the gene pair at that particular locus.

With dominant-recessive alleles, one copy of a gene—the dominant allele—is as effective in producing the trait as two copies of the gene. The heterozygote (e.g., *Aa*), with one copy of each allele, has the same phenotype as the homozygote with two copies of the dominant allele

Table 10.1

Alleles and Their Phenotypes

Kind of Allelic Interaction	Genotype*	Phenotype
Dominant-recessive	AA	Dominant
	Aa	Dominant
	aa	Recessive
Incompletely dominant	AA	Dominant
	Aa	Intermediate
	aa	Recessive
Codominant	A_1A_1	Trait 1
	A_1A_2	Traits 1 and 2
	A_2A_2	Trait 2

*The alleles at a particular gene locus are symbolized by variations of the same letter of the alphabet. The genotype (the two copies of the gene at the locus) is then written as the two symbols, which are the same in a homozygote and different in a heterozygote.

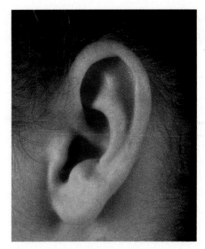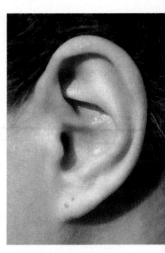

Figure 10.1 Shape of the ear lobe. In some people, the bases of the ear lobes are unattached (*left*) while in others they are attached (*right*).

(AA). Effects of the recessive allele are seen only in a homozygote with two copies of the allele (e.g., aa), when there is no dominant allele present.

Many human traits are controlled by dominant-recessive alleles. Shape of the ear lobe is an example of such a trait (fig. 10.1). The dominant allele (E) produces an unattached lobe; someone who is homozygous dominant (EE) or heterozygous (Ee) has unattached ear lobes. Someone who is homozygous recessive (ee) has attached ear lobes.

Incompletely Dominant Alleles

The alleles of a gene pair show **incomplete dominance** when the phenotype of the heterozygote (e.g., Aa) is intermediate between the phenotypes of the two homozygotes (e.g., AA and aa). In this kind of gene action, one copy of a gene is not as effective as two copies in producing the phenotype.

Hair shape in light-skinned people, for example, appears to be controlled by incomplete dominance. (Genetic control of hair shape in dark-skinned people is different.) The hair may be curly, straight, or wavy. Someone with curly hair carries two copies of the allele that makes hair curly. Someone with straight hair carries two copies of the allele that makes hair straight. Someone with wavy hair carries one copy of each allele.

Genes that control color are often incompletely dominant, because color depends on how much of each pigment is present (just like mixing paints). The dominant allele causes synthesis of the pigment (i.e., a color), and the recessive allele causes no synthesis of the pigment (i.e., white). The heterozygous individual has a unique phenotype, with less pigment than the homozygous dominant and more pigment than the homozygous recessive individual. The colors of human skin, hair, and eyes are products of incompletely dominant alleles but involve more than one gene pair. They are described later in the chapter.

Codominant Alleles

Codominant alleles are both seen, as separate entities, within the phenotype of a heterozygote. They differ from incompletely dominant alleles in that two different gene products are synthesized and both contribute equally to the phenotype.

Blood type is an example of a trait controlled by dominant alleles. The A and B polysaccharides on the surface of human blood cells {i.e., within the glycocalyx (see page 67) of the cell} are synthesized in response to two alleles, one that results in an A polysaccharide (symbolized as I^A) and one that results in a B polysaccharide (symbolized as I^B). The I^A and the I^B alleles are

Table 10.2

The ABO Blood Groups 📼

Genotype	Polysaccharide on Red Blood Cells	Phenotype: Blood Type	Kinds of Antibodies in Blood*	Can Receive Blood from	Can Donate Blood to
$I^A I^A$	A	A	Anti-B	A and O	A and AB
$I^A i$	A	A	Anti-B	A and O	A and AB
$I^B I^B$	B	B	Anti-A	B and O	B and AB
$I^B i$	B	B	Anti-A	B and O	B and AB
$I^A I^B$	A and B	AB	Neither	Anyone (universal recipient)	AB
ii	Neither A nor B	O	Anti-A and anti-B	O	Anyone (universal donor)

*An individual makes antibodies against all foreign molecules (not present naturally in the body) and against all red blood cell polysaccharides except those on its own cells.

codominant; both gene products appear in the phenotype of a heterozygote:

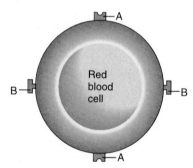

A third allele at this locus results in production of neither the A nor the B polysaccharide. This allele, symbolized as *i*, is recessive to the other two alleles.

Knowledge of the kinds of molecules on red blood cells is crucial to the success of blood transfusions, because antibodies in the bloodstream attack blood cells covered with molecules that are different from the body's own molecules. For example, someone of blood type A (genotype $I^A I^A$ or $I^A i$) has red blood cells with only A polysaccharides and makes antibodies that attack B polysaccharides:

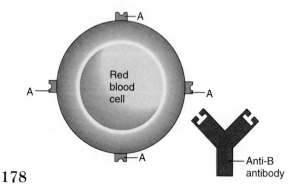

If given a transfusion from someone of blood type B (genotype $I^B I^B$ or $I^B i$) or of blood type AB (genotype $I^A I^B$), the anti-B antibodies will attach to the introduced blood cells, forming a clump that blocks the flow of blood:

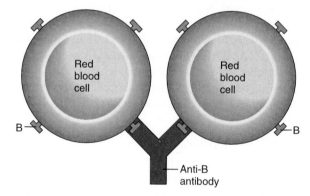

A person of blood type A can, however, receive blood from someone of type A (genotype $I^A I^A$ or $I^A i$), because the recipient's blood has no anti-A antibodies, and from someone of type O (genotype *ii*), because type O cells have neither A nor B polysaccharides for antibodies to attack. A summary of the ABO blood types and the kinds of blood attacked by their antibodies is provided in table 10.2.

Inheritance Patterns

Much of what we know about human inheritance comes from family **pedigrees,** which are diagrams of the phenotypes of family members, over several generations, with regard to a particular trait (fig. 10.2). A pedigree can be extremely valuable if it alerts living family members to a genetic disorder they may have inherited. Many genetic disorders, when detected early, can be successfully treated.

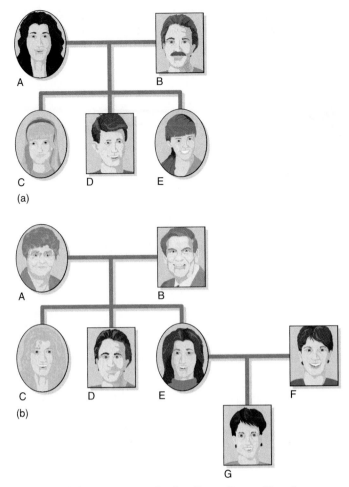

(a)

(b)

Figure 10.2 How to read a family pedigree. The diagrams illustrate the symbols used in a pedigree. (*a*) This pedigree shows two parents and their three children. *A* is the mother, *B* is the father, and *C*, *D*, and *E* are the children listed in order of birth from left to right (i.e., *C* is the oldest). (*b*) The same pedigree several years later. The youngest child, *E*, has mated with *F* and they have a son, *G*. The child *G* is the grandson of *A* and *B* and the nephew of *C* and *D*.

Family pedigrees provide information about the genetic bases of traits. Certain patterns of inheritance indicate the trait is governed by a single gene pair, whereas others indicate the involvement of many gene pairs. Some patterns indicate traits that are controlled by genes within the sex chromosomes rather than within the autosomes (the twenty-two pairs of chromosomes that are not sex chromosomes).

The patterns of inheritance associated with different relationships between genes and traits are summarized in table 10.3 and described in the following sections.

Mendelian Inheritance

The simplest pattern of inheritance, described by Gregor Mendel in the 1860s, is known as **Mendelian inheritance.** In it, a single gene pair governs the development of a single trait.

Table 10.3

Patterns of Inheritance

Pattern	Number of Gene Pairs	Number of Traits
Mendelian	One	One
X-linked recessive	One (one gene in a male)	One
Polygenic	Two or more	One
Epistatic	One gene pair that affects other gene pairs	Variable
Pleiotropic	One	Many

Traits That Characterize Families   About 4,000 human traits are known to be inherited in a Mendelian fashion. Most features of the face, for example, are products of single gene pairs in which one allele is dominant over the other. For each such facial trait, there is a single gene locus with many alleles in the human population. Each family has only a few of these alleles, and they are responsible for family resemblances—form of the nose, shape of the lips, position of the eyes, curve of the eyebrows, prominence of the cheekbones, and so on. Some of these traits are shown in figure 10.3.

A trait controlled by a single pair of dominant-recessive alleles can be recognized by examining family pedigrees. A person with the dominant trait, with at least one copy of the dominant allele, will always have a parent with the trait (fig. 10.4*a*). A person with the recessive trait, with two copies of the recessive allele, may or may not have a parent with the trait (fig. 10.4*b*).

Recessive Alleles   Each family line has its own unique collection of recessive alleles, which originated as mutations in ancestors and have been passed on from generation to generation. Each recessive allele is common within one family but may be rare in the general population. While some of these alleles are beneficial, many are harmful when they are not masked by a dominant allele.

A recessive allele may be carried, hidden by a dominant allele, in the family line for centuries before showing up in a homozygote. A child with two copies of a rare allele is likely to be born to parents who are closely related, since two people within the same family are likely to have inherited the same rare allele. Such children often have serious genetic disorders, brought about by their homozygous recessive genotypes. It is largely for this reason that most societies discourage matings between close relatives. This subject is discussed in the accompanying Wellness Report.

(a)

(b)

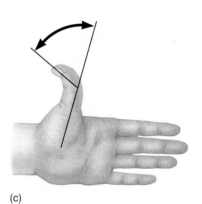

(c)

Figure 10.3 Some traits controlled by dominant-recessive alleles. (*a*) A widow's peak is dominant over a straight hairline. (*b*) Freckles are dominant over no freckles. (*c*) A hitchhiker's thumb, which can be bent far backward, is recessive to a thumb that cannot be bent in this way.

Figure 10.4 Dominant alleles and recessive alleles exhibit different patterns of inheritance. (*a*) The development of six fingers on each hand and six toes on each foot is controlled by an allele that is dominant over the allele that produces five fingers and toes. Its pattern of inheritance, characteristic of a dominant allele, is that someone with six fingers and toes always has a parent with six fingers and toes. They may or may not have children with the same trait. (*b*) The development of red hair is controlled by an allele that is recessive to the development of hair of different colors. Its pattern of inheritance is that it often skips generations: both parents of someone with red hair may have hair that is not red. One may, actually, have to go several generations back to find a red-haired ancestor.

☐ Male with the trait

☐ Male without the trait

○ Female with the trait

○ Female without the trait

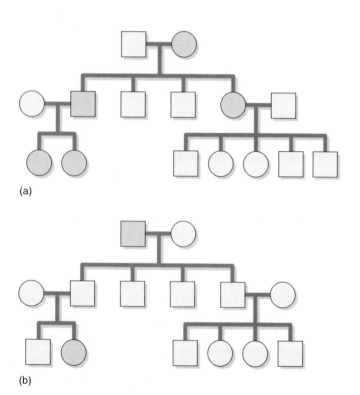

(a)

(b)

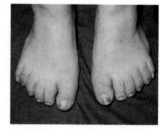

Genetic Consequences of Incest

Most cultures forbid matings between close relatives because children born of such parents often inherit serious genetic disorders. Two members of the same family are likely to carry the same harmful, recessive allele. Although such an allele is typically rare in the human population, it is common within the family line because it originated as a mutation in an ancestor and has been passed on from generation to generation.

When a man and woman are from different family lines, it is highly unlikely that both of them carry the same rare allele. Consequently, their children are not likely to inherit two copies of the same rare allele:

D = Normal allele
d = Rare, recessive allele

Father — Mother
Dd — DD

Child: Dd or DD

However, when members of the same family line mate with each other, their child is likely to inherit two copies of the same recessive allele, one from each parent, and exhibit the recessive trait:

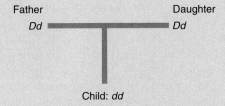

Father — Daughter
Dd — Dd

Child: dd

The chance of having a child with a serious defect is 4% when the parents are unrelated and 8% when they are first cousins; the chance rises to

(a) (b)

Figure 10.A Toulouse-Lautrec. (*a*) Henri Toulouse-Lautrec (1864–1901) was just 1.4 meters (4.5 ft) tall and always sat in a child's chair. His skeleton was deformed and his bones were extremely weak and brittle. Although he always walked with a cane, he fell often and broke his legs many times. (*b*) His best-known poster, designed in 1892.

(b) Toulouse-Lautrec, Henri de. *Divan Japonais*. 1893. Lithograph, printed in color, 31⅞″ × 24½″. The Museum of Modern Art, New York. Abby Aldrich Rockefeller Fund. Photograph © 1995, The Museum of Modern Art, New York.

20% when they are either brother and sister or parent and child. It is the recessive gene pair (e.g., *dd*) that causes abnormal children from incestuous relations.

We all carry harmful genes that are hidden in the heterozygous state (e.g., *Dd*). On the average, each person carries four to six recessive alleles that, in the homozygous state, cause severe developmental problems. The most common abnormalities are deafness, blindness, deformities of the skeleton, and various kinds of mental deficiencies.

Matings between relatives may bring together two copies of a rare and favorable recessive allele. The parents of Toulouse-Lautrec, for example, were first cousins; he may have acquired both his deformed skeleton and his outstanding artistic talent from homozygous, recessive alleles (fig. 10.A).

All humans are related to some degree. Population geneticists estimate that no human is more distantly related to any other human by more than fiftieth cousins! And within local communities, most of us are much more closely related than that. This surprising nearness of relation was brought about by our descent from the same population that, for a long period of our history, was very small and immobile. With so few mates from which to choose, most marriages were necessarily between relatives. You and your neighbors, whoever they might be, could trace your family lines back to a common ancestor who lived within the past 200,000 years. The so-called family of man really does exist.

Table 10.4

Genetic Disorders of Some Ethnic Groups

Ethnic Origin	The Chance Is	That You Are a Carrier
African	1 in 10	Sickle-cell anemia (distorted red blood cells that clog vessels)
European	1 in 25	Cystic fibrosis (thick mucus that clogs lungs)
European	1 in 80	Phenylketonuria (brain cells damaged from a buildup of phenylalanine)
Jewish with ancestors from eastern Europe	1 in 30	Tay-Sachs disease (brain cells damaged from a buildup of fats)
Italian or Greek	1 in 10	Thalassemia (not enough O₂ reaches cells because not enough hemoglobin)
White South African	1 in 330	Porphyria (not enough O₂ reaches cells because of abnormal hemoglobin)

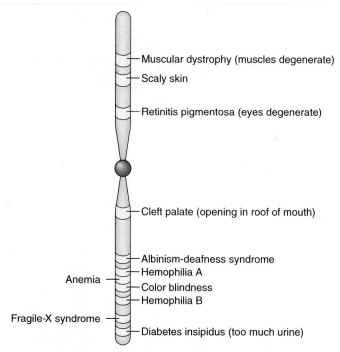

Figure 10.5 X-linked recessive alleles. A variety of recessive alleles in the X chromosome are responsible for disorders that appear most often in males. (Some of these disorders can be caused by alleles in other chromosomes as well.)

Just as each family line carries unique recessive alleles, so also does each ethnic group, since an ethnic group is a set of interrelated family lines. Every one of us is a carrier—is heterozygous for a recessive allele—of several serious disorders. The particular disorders we carry depend on our ethnic group. The chances of carrying an allele for particular disorders are shown, for several ethnic groups, in table 10.4.

Inheritance of X-Linked Traits

Whether a child is a male or a female depends on the sex chromosomes the child inherits. A boy has an X and a Y chromosome and a girl has two X chromosomes. Only a few genes, none essential to survival, are located within the Y chromosome. By contrast, many important genes are located within the X chromosome. Genes in the X chromosome are called X-linked genes, and the traits they govern are called **X-linked traits.** A male, with just one X chromosome, has just one copy of each X-linked gene. A female, with two X chromosomes, has two copies of each X-linked gene.

Recessive Alleles   A recessive allele carried within an X chromosome is expressed in a female only when present in duplicate—when the female is homozygous for that allele. Her genotype, if homozygous for the a allele, is symbolized X^aX^a. If she is heterozygous (genotype X^AX^a), the recessive allele is hidden. In a male, however, a recessive allele carried within the X chromosome is always expressed, since he has only one copy of the gene. His genotype, if it consists of the a allele, is symbolized X^aY. Only one copy of a recessive allele is needed for its trait to appear in a male, whereas two copies are needed for its trait to appear in a female. Consequently, genetic disorders caused by X-linked, recessive alleles are much more common in males than in females. A map of the human X chromosome, showing the locations of recessive alleles that cause disorders, is provided in figure 10.5.

Hemophilia is one of the disorders caused by an X-linked, recessive allele. This disease is characterized by slow formation of blood clots, so slow that a great deal of blood flows out of the blood vessels after a cut or bruise. Many genes in many chromosomes contribute to normal blood clotting, but two gene loci in the X chromosome are responsible for the two most common clotting disorders—hemophilia A and hemophilia B. A recessive allele that interferes with normal clotting can appear at either one of these two loci. When such an allele is in the X chromosome of a male, it causes hemophilia; a male has no normal allele (in a second X chromosome) to cover for a defective one. Hemophilia occurs primarily in males.

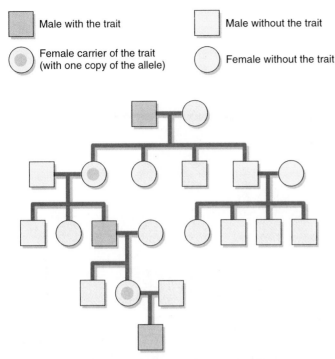

Male with the trait

Male without the trait

Female carrier of the trait
(with one copy of the allele)

Female without the trait

Figure 10.6 An X-linked pattern of inheritance. A trait caused by an X-linked, recessive allele appears much more often in males than in females. A man with an X-linked trait, such as color blindness, inherits the allele for the trait from his mother, who inherits it from her father. Thus, a color-blind man is likely to have a grandfather, on his mother's side, who is also color-blind. And he is likely to have a grandson, borne of his daughter, who is color-blind.

Inheritance Pattern A man inherits his X chromosome and all his X-linked traits from his mother. A woman inherits one X chromosome from her mother and one from her father. This pattern of inheritance can be seen in a Punnett square (see page 160):

Kinds of sperm

	X	Y
X	XX Daughter	XY Son
X	XX Daughter	XY Son

Kinds of eggs

A defective X chromosome in a man is always passed to each daughter but never to a son. A defective X chromosome in a woman has a 50% chance of being passed to each son and daughter. The inheritance pattern of X-linked traits governed by recessive alleles is distinctive: the trait first appears in a man and then reappears in a grandson—a son of the man's daughter (fig. 10.6). The

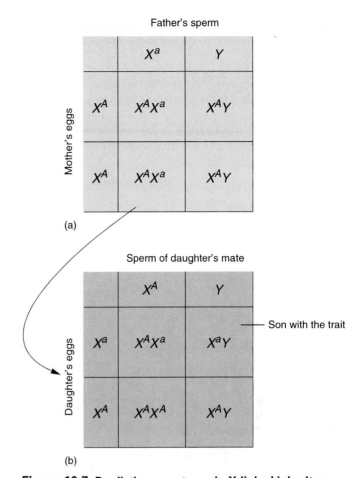

Father's sperm

	X^a	Y
X^A	$X^A X^a$	$X^A Y$
X^A	$X^A X^a$	$X^A Y$

Mother's eggs

(a)

Sperm of daughter's mate

	X^A	Y
X^a	$X^A X^a$	$X^a Y$ — Son with the trait
X^A	$X^A X^A$	$X^A Y$

Daughter's eggs

(b)

Figure 10.7 Predicting genotypes in X-linked inheritance. (a) A man with a trait caused by a recessive allele in an X chromosome (X^a in this example) will pass the allele to all his daughters but none of his sons. The daughters will be carriers: they will have the recessive allele but not exhibit the trait (unless they also receive a copy of the allele from their mother). (b) A woman who is a carrier of the allele has a 50:50 chance of passing the allele to each daughter and son. A son who inherits the allele will exhibit the trait whereas a daughter will be a carrier.

daughter receives the recessive allele (X^a) from her father and may pass it to her son, as diagrammed in the Punnett squares in figure 10.7.

This pattern of X-linked inheritance appears in the family pedigree of Queen Victoria (fig. 10.8). Queen Victoria, who reigned over England from 1837 to 1901, carried a recessive allele for hemophilia within one of her X chromosomes. She did not have the disorder herself, because she had a dominant allele for normal blood clotting in her other X chromosome. However, Queen Victoria passed copies of her defective X chromosome to one of her sons, who developed hemophilia, and two of her daughters, who passed the chromosome to their children. Since people of royal blood tend to marry others of royal blood, the defective gene passed through many royal families in Europe.

183

Figure 10.8 Queen Victoria and the hemophilia allele. Queen Victoria (1819–1901) carried the recessive allele for hemophilia. The family pedigree with regard to hemophilia. The defective X chromosome passed from Queen Victoria to two daughters, Alice and Beatrice, and one son, Leopold. They, in turn, passed copies of the chromosome to their children. Most of the men who inherited the allele for hemophilia died prematurely from uncontrolled bleeding from wounds. (This pedigree is not complete—some of the Queen's children are not shown in order to simplify the illustration.)

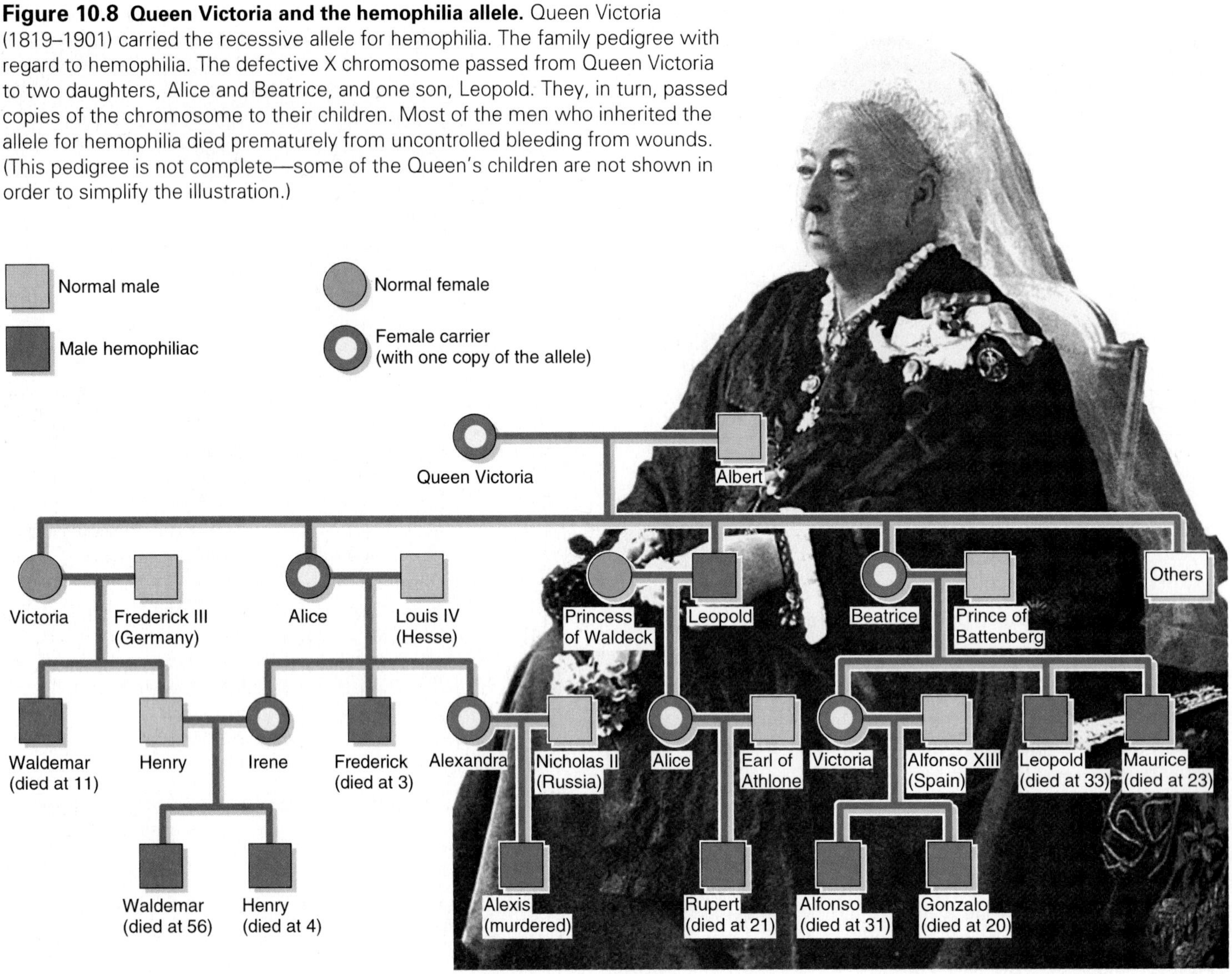

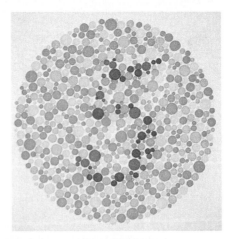

A less serious disorder produced by a recessive allele in the X chromosome is color blindness. It is inherited in the same manner as the allele for hemophilia. A simple test for this disorder is provided in figure 10.9. Recently, physicians have identified yet another X-linked disorder that they call fragile-X syndrome. Research on this serious and surprisingly common disorder is described in the accompanying Research Report.

Polygenic Inheritance

Most traits vary over a broad range of forms. Thus, we are not either short or tall but, as a population, are of all possible heights within a certain range. The same is true of skin color, hair color, eye color, body build, heart rate, and longevity. Traits that exhibit this type of variation are influenced by many gene pairs, each contributing its product to the development of the trait. The passage of such traits from one generation to the next is called **polygenic inheritance** (*poly*, many; *genic*, genes).

Figure 10.9 Are you color-blind? If you can see colors but cannot detect the number 5 within this circle, you cannot distinguish red from green. This type of color blindness, caused by a recessive allele within the X chromosome, occurs in approximately 7% of American men and almost never in women.

Research Report

Tracing Fragile-X Syndrome

Fragile-X syndrome is an inherited disorder that has only recently been recognized. In 1987, TIME magazine published the following report.[1]

Denver housewife Jeannie Lancaster suspected that something was wrong with her son a few months after his birth. "He was always fussy and would never cuddle," she remembers. "When I would go to pick him up, he would arch away." By the time he was one, he had acquired the alarming habit of banging his head against the wall, and when he began to talk, he repeated the same sound or word over and over again. Says Lancaster, "I kept asking myself, 'What am I doing wrong?'" Professional counselors offered conflicting views. A pediatrician thought the boy might be autistic. School officials said he was emotionally disturbed. Further testing indicated he was mildly retarded.

It was only after Lancaster had given birth to a second son who showed some of the same mysterious symptoms that a family doctor began to suspect the cause. He had just returned from a medical seminar on a newly recognized disorder called Fragile-X syndrome, which results from a weakness in the structure of the X sex chromosome (one of the 46 chromosomes—complex molecules containing long segments of DNA—in the nucleus of the human cell). The behavior of the Lancaster children seemed to fit the patterns described at the meeting and tests at Denver's Children's Hospital soon confirmed that they both had the defective chromosome.

Since this early report, we have learned much more about Fragile-X syndrome. It is characterized by an unusual X chromosome: one end is shaped like an exclamation point, with the point attached by only a thin thread:

This chromosomal defect impedes normal development of the brain. Children with Fragile-X syndrome are hyperactive and have tantrums when their activities are disrupted or the environment is noisy. They are slow to learn to talk and continue to have speech and language difficulties throughout their lives. The syndrome is now recognized as a leading cause of mental retardation. Girls are less often affected than boys by the chromosomal defect because they have two X chromosomes and a normal one usually compensates for a defective one.

In 1991, the gene that causes Fragile-X syndrome was identified. Surprisingly, the abnormal allele does not weaken the X chromosome, as originally believed, but rather adds sequences of bases to its length. The normal allele consists of, at most, fifty copies, one after the other, of the three-base sequence cytosine-guanine-guanine (CGG). The abnormal allele consists of many more copies of this sequence. People with Fragile-X symptoms have between 230 and several thousand copies of the CGG sequence at this locus. People with intermediate amounts, between 50 and 230 copies, do not exhibit obvious symptoms of the syndrome.

The normal version of this gene acts to "turn on" certain other genes, which in turn code for proteins required by nerve cells. The abnormal version of this gene cannot "turn on" these genes and so their proteins are not synthesized. The symptoms of Fragile-X are attributed to a lack of these proteins in nerve cells of the brain.

Inheritance of the fragile-X allele follows a peculiar pattern: symptoms intensify from one generation to the next, especially when inherited from the mother. Researchers have learned that while the fragile-X allele is in a woman's egg, it adds CGG sequences to its length; the defective allele becomes longer and longer, with a higher probability that it will produce symptoms in the child. While the allele is in a man's sperm, by contrast, the allele does not lengthen and may even shorten. We do not yet understand this strange phenomenon, known as genomic imprinting, in which an allele is altered by the environment of either an egg or a sperm but not both.

Women with family pedigrees that show more than one mentally retarded male may carry the fragile-X chromosome and should be tested to find out. A woman with the defective chromosome has a 50% chance of passing it to each of her children. The earlier the Fragile-X syndrome is diagnosed and treated, the greater the chance of overcoming it. New medications are very effective in improving concentration and diminishing the aggressive behavior seen in fragile-X children.

[1]"Tracing Fragile X Syndrome," *Time Magazine,* March 16, 1987, p. 78.

A polygenic trait exhibits a greater variety of phenotypes than a Mendelian trait because it is controlled by more loci within each individual—the greater the number of genes involved, the more ways they can combine to produce a trait.

Consider, for example, the inheritance of skin color in humans. Skin color is determined by three pigments: a brownish pigment called melanin, a yellowish pigment called carotenoid, and the hemoglobin of red blood cells within vessels of the skin. Here we will look only at inheritance of melanin.

The quantity of melanin synthesized by skin cells is governed by at least four pairs of genes (the exact number is not yet known) on four pairs of homologous chromosomes. The same gene, present in two forms (alleles), occurs at all four gene loci. One allele (M) causes skin cells to synthesize large quantities of melanin, and the other allele (m) causes skin cells to synthesize very little.

In very dark-skinned people, all four gene pairs (eight genes) consist of the allele that instructs the cell to synthesize large amounts of melanin:

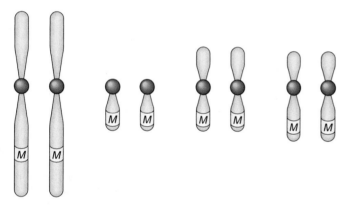

Their genotype is MM, MM, MM, MM. In very light-skinned people, all four gene pairs consist of the allele that instructs the cell to synthesize small amounts of melanin:

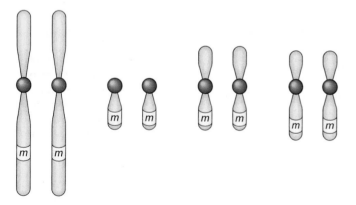

Their genotype is mm, mm, mm, mm.

People with intermediate-colored skin have some of each type of allele at the four loci. Someone with the genotype Mm, Mm, Mm, Mm, for example, has skin that is exactly intermediate between very dark and very light, as does someone with the genotype MM, Mm, mm, Mm or any other combination of four copies of the M allele and four copies of the m allele.

A child inherits four genes for skin color from each parent. When one parent has very dark skin and the other has very light skin, their child will inherit four genes for dark skin from one parent and four genes for light skin from the other parent. The child will have intermediate-colored skin. When both parents have intermediate-colored skin, with four genes of each kind, the outcome is much less predictable. The child could inherit any one of a variety of gene combinations, and the skin could be any shade from very dark to very light (fig. 10.10).

Epistasis

Epistasis occurs when the effects of one gene pair mask the effects of other gene pairs at other loci.

An albino, with no brownish pigment in the skin, hair, or eyes (fig. 10.11a) develops as a result of epistasis. The involved gene codes for an enzyme that controls synthesis of melanin. In an albino, both copies of the gene are defective and no melanin is synthesized, regardless of the kinds of alleles present at the gene loci governing the *quantity* of melanin (fig. 10.11b). A child may inherit genes for large amounts of melanin (dark skin, hair, and eyes), but if the child also inherits two copies of the albino allele (i.e., lacks the ability to synthesize melanin) he or she will have white skin, pale yellow hair, and pink eyes. (The eyes are pink because the red blood can be seen through the colorless irises.)

Red hair is another example of a trait caused by epistasis. As with albinism, the locus that gives rise to red hair is separate from the loci that control how much melanin can be synthesized by hair cells. Red hair appears only in people who inherit two copies of the allele for red hair (one from each parent) in addition to alleles at other gene loci that produce small quantities of melanin.

Pleiotropic Inheritance

Pleiotropy occurs when a single gene pair controls more than one phenotypic trait. The gene pair governs synthesis of a single kind of protein, but that protein is fundamental to the development of many traits. When such a key protein is absent or abnormal, a broad range of symptoms appear together. The medical term for a group of related symptoms is syndrome.

Marfan syndrome is an example of pleiotropic inheritance. In this disorder, a dominant allele contains instructions for an abnormal version of a protein. The normal version of this protein (most likely fibrillin) binds together and supports cells of the body. When this protein is abnormal, there is abnormal development of many structures, including bones, joints, eyes, blood vessels, and heart valves. Because the disorder is caused by a dominant

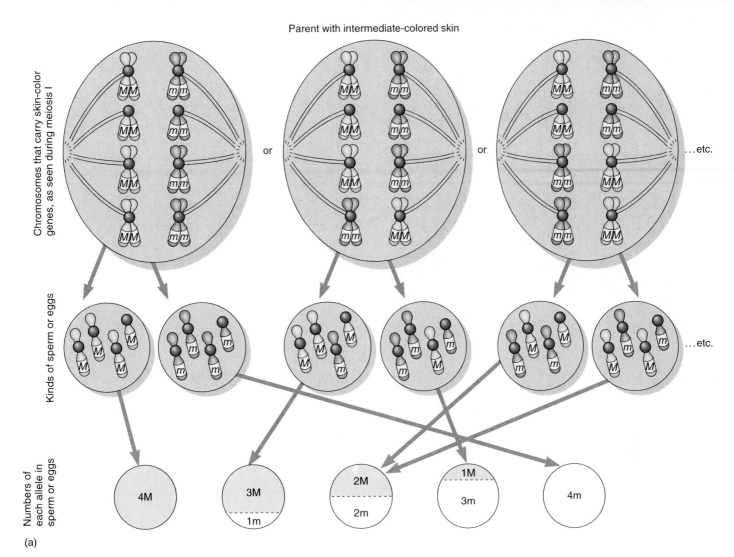

Parent with intermediate-colored skin

Chromosomes that carry skin-color genes, as seen during meiosis I

or ... or ...etc.

Kinds of sperm or eggs

...etc.

Numbers of each allele in sperm or eggs

4M | 3M / 1m | 2M / 2m | 1M / 3m | 4m

(a)

Kinds of sperm from man with intermediate-colored skin

	4M	3M 1m	2M 2m	1M 3m	4m
4M	8M Very dark	7M 1m Dark	6M 2m	5M 3m	4M 4m Intermediate
3M 1m	7M 1m Dark	6M 2m	5M 3m	4M 4m Intermediate	3M 5m
2M 2m	6M 2m	5M 3m	4M 4m Intermediate	3M 5m	2M 6m
1M 3m	5M 3m	4M 4m Intermediate	3M 5m	2M 6m	1M 7m Light
4m	4M 4m Intermediate	3M 5m	2M 6m	1M 7m Light	8m Very light

Kinds of eggs from woman with intermediate-colored skin

(b)

Figure 10.10 Polygenic inheritance: matings between people with intermediate-colored skin. (*a*) Someone with skin of an intermediate color has several alleles for large quantities of melanin (*M*) and several alleles for very little melanin (*m*). They form many kinds of sperm or eggs with regard to these alleles. The parent in this example has the genotype *Mm, Mm, Mm, Mm*. During meiosis I, the pairs of homologous chromosomes can line up and separate in many different ways—five types of sperm or eggs can be generated with regard to the alleles for skin color. (*b*) A Punnett square showing the kinds of genotypes that can form when both parents have intermediate-colored skin and the genotype shown in (*a*). The man forms five kinds of sperm and the woman forms five kinds of eggs. The sperm and eggs can combine to form children with many shades of skin, ranging from very light to very dark.

(a)

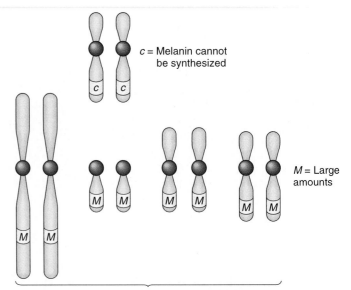

c = Melanin cannot be synthesized

M = Large amounts

Loci governing quantity of melanin in skin

(b)

Figure 10.11 An albino child. (*a*) Albinos appear in all human populations. The parents of this child have very dark skin. In Africa, an albino child is called a moon child because he or she is extremely sensitive to the sun and goes outdoors mainly at night. (*b*) The ability to synthesize melanin is determined by alleles on a pair of homologous chromosomes separate from the pairs that control the amount of melanin that is synthesized. The albino child with black parents has two copies of the allele that prevent melanin synthesis and eight copies (in four pairs of homologous chromosomes) of the allele that synthesizes large quantities of melanin. In spite of his genes for dark skin, the child has no pigment in his skin.

allele, someone with Marfan syndrome is likely to have parents and children with the same syndrome (see the pedigree in fig. 10.4*a*).

People with Marfan syndrome typically are tall and lanky, with disproportionately long arms and fingers, and have poor eyesight. The disorder is most often diagnosed, however, by a coroner: symptoms are not noticed until a blood vessel bursts. Flo Hyman, a member of the 1984 Olympic volleyball team, had the syndrome and died of a burst artery during a game. Abraham Lincoln, too, may have had Marfan syndrome.

Genomic Imprinting

A recently discovered pattern of inheritance is **genomic imprinting,** in which the expression of a gene (appearance in the phenotype) depends on whether it is inherited from the mother or the father. For example, myotonic dystrophy (a muscle disorder) and Fragile-X syndrome (described in this chapter's Research Report) are more likely to be inherited from the mother than the father. Certain forms of

Huntington disease, by contrast, develop primarily when the defective allele is inherited from the father.

We do not know for certain how genomic imprinting develops. Apparently, something within a sperm or egg turns off certain genes. Preliminary research suggests that a methyl group (–CH$_3$) attaches to a portion of the DNA and prevents it from being used to synthesize its protein. The gene inactivation is not permanent. For example, a man with the allele for myotonic dystrophy may pass the defective gene to his children. They will not develop the disorder because the gene was inactivated while in the man's sperm. The man's daughters, however, may have children with myotonic dystrophy because the gene was reactivated in their eggs.

Even more perplexing is the significance of genomic imprinting. Does it have some advantage? One hypothesis is that the normal function of gene inactivation is to neutralize foreign DNA—genes inserted into the cell by viruses. Genomic imprinting, then, occurs when a cell mistakes a piece of its own DNA for a piece of viral DNA. Why sperm misidentify certain genes and eggs misidentify others is not clear.

SUMMARY

1. Two alleles of a gene pair interact in various ways to produce a phenotype. With dominant-recessive alleles, one allele masks the effects of the other allele; a heterozygote has the same phenotype as a homozygote with two copies of the dominant allele. With incompletely dominant alleles, the heterozygote has a phenotype that is intermediate between the phenotypes of the two homozygotes. With codominant alleles, the heterozygote synthesizes two different proteins that contribute equally to the phenotype.

2. Patterns of inheritance vary, depending on how many gene pairs contribute to the trait, where these genes lie within the set of chromosomes, and how the gene products affect the trait.

3. In Mendelian inheritance, one gene pair controls one trait.

4. In X-linked inheritance, one gene pair (in the two X chromosomes of a female) or one gene (in the single X chromosome of a male) controls a trait. Traits controlled by recessive alleles in the X chromosome appear much more often in males than in females.

5. In polygenic inheritance, more than one gene pair controls a trait. The more gene pairs involved, the greater variety of phenotypes with regard to that trait.

6. In epistasis, the effects of one gene pair mask the effects of other gene pairs at other loci.

7. In pleiotropic inheritance, a single gene pair controls many traits.

8. In genomic imprinting, an allele is inactivated while in one kind of gamete—either a sperm or an egg—so that its expression depends on whether it is inherited from the mother or father.

KEY TERMS

Allele (ah-LEEL) (page 176)
Codominant allele (page 177)
Dominant allele (page 176)
Epistasis (EH-peh-STAY-sis) (page 186)
Genomic (jeh-NOH-mik) **imprinting** (page 188)
Heterozygote (HEH-tur-oh-ZEYE-gut) (page 176)
Homologous (hoh-MAH-luh-gus) **pair** (page 176)
Homozygote (HOH-moh-ZEYE-gut) (page 176)
Incomplete dominance (page 177)
Mendelian (men-DEE-lee-ahn) **inheritance** (page 179)
Pedigree (PEH-deh-gree) (page 178)
Pleiotropy (PLEYE-oh-troh-pee) (page 186)
Polygenic (PAH-lee-JEE-nik) **inheritance** (page 184)
Recessive allele (page 176)
X-linked trait (page 182)

STUDY QUESTIONS

1. Distinguish among dominant-recessive alleles, incompletely dominant alleles, and codominant alleles. Give an example of each.

2. Explain, using family pedigrees, how inheritance of a trait controlled by a dominant allele can be distinguished from inheritance of a trait controlled by a recessive allele.

3. Explain how a trait controlled by a recessive allele in an autosome can be distinguished from a trait controlled by a recessive allele in an X chromosome.

4. Explain how it is possible for two people with intermediate-colored skin to have a child that is very dark-skinned.

5. A child with red hair is born to a family with no red-haired members except for a great-grandmother on one side and a great-great-grandfather on the other side. Explain the inheritance pattern exhibited by this child.

6. Explain why people within an ethnic group tend to carry the same harmful recessive alleles.

7. Explain how someone who inherits genes for dark skin can be an albino.

CRITICAL THINKING PROBLEMS

1. A child inherits a trait that has not been seen in the family of either parent for many generations. What type of gene do you think is responsible for this trait? Why did it disappear for so many generations?

2. Why does each ethnic group have a unique set of genetic disorders?

3. A man is color-blind and there is no history of this disorder in his wife's family. What is the chance that his son will be color-blind?

4. Suppose that you discover a new genetic disorder that occurs in both humans and laboratory mice, and you would like to determine whether it is caused by an X-linked recessive allele. How would you investigate it in humans? How would you investigate it in mice?

SUGGESTED READINGS

Introductory Level:

Diamond, J. "Curse and Blessing of the Ghetto," *Discover*, March 1991, 60–65. An analysis of the origins of Tay-Sachs disease.

Neubauer, P. B., and A. Neubauer. *Nature's Thumbprint, The New Genetics of Personality.* Reading, MA: Addison-Wesley, 1990. An examination of the roles that heredity and experience play in the development of personalities.

Pierce, B. A. *The Family Genetic Sourcebook*. New York: John Wiley & Sons, 1990. A guidebook for parents that lists the various human traits and disorders, in alphabetical order, and explains what is known about their inheritance.

Shoumatoff, A. "The Mountain of Names," *The New Yorker*, May 13, 1985, 51–101. A description of genealogies and the relatedness of humans.

Advanced Level:

McKusick, V. A. *Mendelian Inheritance in Man*, 9th ed. Baltimore: Johns Hopkins University Press, 1990. A comprehensive review of genetic traits in humans and which traits are believed to be inherited in a Mendelian fashion.

Neel, J. V. *Physician to the Gene Pool*. New York: John Wiley & Sons, 1994. A fascinating account of medical genetics—part autobiography and part science—by a physician who has studied Japanese populations (the effects of radiation from atomic bombs), the genetic consequences of consanguineous marriages, the unique genes of Amerindians, and many other aspects of human genes.

Sapienza, C. "Parental Imprinting of Genes," *Scientific American*, October 1990, 52–60. A description of the observations and experiments regarding genomic imprinting as well as possible effects of imprinting development and resistance to disease.

✪ EXPLORATION

Interactive Software: Heredity in Families

This interactive exercise explores how the character and location of a gene influence its pattern of inheritance. The program provides the pedigree of a family that carries, within an X chromosome, the allele for hemophilia. The student alters the character of the allele, from recessive to dominant, and the position of the gene, from an X chromosome to an autosome, to see the effects on the pattern of inheritance.

11

From Genes to Proteins

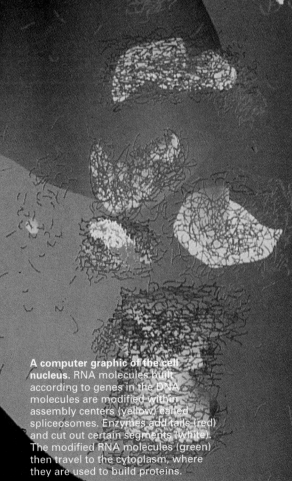

A computer graphic of the cell nucleus. RNA molecules built according to genes in the DNA molecules are modified within assembly centers (yellow) called spliceosomes. Enzymes add tails (red) and cut out certain segments (white). The modified RNA molecules (green) then travel to the cytoplasm, where they are used to build proteins.

Chapter Outline

Objectives

In this chapter, you will learn
–how DNA molecules contain
  instructions for building proteins;
–how a cell uses the instructions to
  assemble proteins from amino acids;
–that genes consist of more than
  instructions for proteins;
–that the first genes may have been
  RNA molecules;
–how mutations in the DNA produce
  faulty proteins;
–the way in which a virus uses a cell to
  build its own proteins.

An organism's proteins, which determine its traits, are built according to genetic instructions within DNA molecules. The proteins are not, however, synthesized by the DNA molecule itself. Instead, instructions for building proteins are copied into RNA molecules, which are then modified and moved to other places in the cell, where proteins are assembled from amino acids.

Genetic information within DNA molecules is like factual information within a reference book that cannot be checked out of the library. When the factual information is needed, it may be photocopied from the reference book and taken home. There the photocopy is read and then thrown away. Similarly, when information within DNA is needed for building a protein, it is copied into an RNA molecule. The RNA copy then moves to other places in the cell, where it is used to build a protein and then broken apart.

Something strange happens to the RNA copy of a gene on its way to the site of protein synthesis. Before leaving the nucleus, large sections of the RNA molecule are cut out and a new beginning and end tacked on. These changes are controlled by a cluster of enzymes, called a spliceosome (*splice*, to cut; *some*, body). Why does a gene contain parts that are discarded? How do enzymes identify these parts?

There is also something strange about the DNA molecules themselves; much of the DNA in a cell appears to have no function. Less than 15% of our DNA is involved in protein synthesis. The rest has no known function. Much of it consists of repetitive sequences: simple "words," repeated thousands of times, that vary enormously from one person to another. Do these extremely simple messages carry information? How can DNA that varies so much from person to person have a function?

Many geneticists believe that most of our DNA is "junk," which a cell has no way of eliminating from its chromosomes. Other geneticists, however, are convinced that the so-called junk DNA contains treasures. They believe that these parts of our genome contain information that coordinates the activities of all our genes. If so, learning more about this "junk DNA" will help us understand cancer and other genetic disorders.

The Genetic Language

An overall plan for every organism is encoded within its DNA molecules. The structure of DNA contains instructions for building proteins, which in turn determine the organism's traits. The instructions are written in a remarkably simple language. A DNA molecule and a protein are both linear sequences of their monomers—nucleotides in

the case of DNA(see page 48) and amino acids in the case of proteins. Thus, a sequence of nucleotides within a DNA molecule specifies a sequence of amino acids within a protein(see page 45).

Genetic Codes for Amino Acids

The genetic language was deciphered in the 1960s, shortly after the DNA molecule was described. The language is almost too simple to believe. Its letters are the four bases of DNA nucleotides: adenine, thymine, guanine, and cytosine. Its words consist of just three letters—a sequence of three bases within the DNA molecule.

The bases are symbolized by their first letters: A for adenine, T for thymine, G for guanine, and C for cytosine. A genetic word that is spelled TAC, for instance, contains the bases thymine, adenine, and cytosine.

Each sequence of three bases along one strand of a DNA molecule is the code for an amino acid. The sequence TAC, for example, specifies tyrosine (an amino acid) and the sequence GGG specifies glycine (another amino acid). There is no overlap and there are no punctuation marks within the instructions: the sequence TACGGG instructs a cell to link tyrosine with glycine:

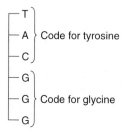

The genetic language has sixty-four words, since each word consists of three letters and any one of the four bases can form each letter: $4 \times 4 \times 4 = 64$. Of these sixty-four words, sixty-one specify amino acids. There are, however, only twenty amino acids in proteins. Some of the amino acids are represented by more than one triplet: the third base of the sequence sometimes varies. Alanine, for example, is spelled in the following ways: GCA, GCC, GCG, and GCT. One of the amino acids, methionine (specified by the triplet ATG), serves not only as a building block of proteins but also as a start signal that indicates the beginning of a protein. Three of the sixty-four words do not specify amino acids; they are stop signals that indicate the end of a protein.

Surprisingly, almost all forms of life use the same genetic language. The triplet TAC, for instance, stands for tyrosine in such diverse organisms as bacteria, dandelions, oak trees, earthworms, and humans. The universality of the code strongly suggests that all modern organisms are descended from the same ancestral organism.

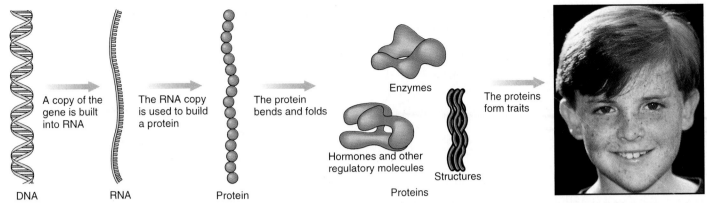

DNA   RNA   Protein

Enzymes

Hormones and other regulatory molecules

Structures

Proteins

A copy of the gene is built into RNA

The RNA copy is used to build a protein

The protein bends and folds

The proteins form traits

Figure 11.1 From genes to traits. A gene is a linear sequence of bases within a DNA molecule. When its protein is needed, the gene is copied into a linear sequence of bases within a molecule of RNA, which then determines a linear sequence of amino acids in the protein. An organism's proteins determine its traits—aspects of its physical appearance, biochemistry, and behavior.

Genetic Instructions for Proteins

A gene can be defined as one or more segments of DNA that contain instructions for building a polypeptide (📖 see page 45). A polypeptide, in turn, either is a protein or ultimately becomes part of one (some proteins consist of more than one polypeptide). A long sequence of bases along one side of the DNA molecule specifies a long sequence of amino acids within a polypeptide. A segment of DNA with the bases TACGGGTTT, for example, tells the cell to bond tyrosine (TAC) to glycine (GGG) and glycine to phenylalanine (TTT). A much longer sequence, with hundreds of bases, determines the arrangement of hundreds of amino acids within a polypeptide of average size.

An organism's genes determine the kinds of proteins it can synthesize. Proteins, in turn, control an organism's traits (fig. 11.1). Most proteins are enzymes that control the chemical activities of cells. Other proteins form structures like hair, fingernails, and muscles. Still others serve as chemical messengers, such as hormones, that regulate cellular activities.

Gene Expression

Gene expression occurs when a gene is used to build a protein. The gene is "expressed" as a protein in the cell. A protein is not built directly on a DNA molecule. Instead, a copy of the gene is built into an RNA molecule (📖 see page 49), which then carries the information to the place in the cell that specializes in building proteins from amino acids.

Gene expression has three phases: transcription, in which the gene is copied into an RNA molecule; RNA processing, in which the RNA molecule is modified; and translation, in which the RNA molecule is used to arrange amino acids within the polypeptide:

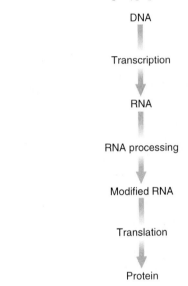

DNA

↓

Transcription

↓

RNA

↓

RNA processing

↓

Modified RNA

↓

Translation

↓

Protein

Transcription takes place within the nucleus of a eukaryotic cell, and translation takes place on ribosomes in the cytoplasm (fig. 11.2). These three phases of gene expression are summarized in table 11.1 and described following.

Transcription: Building an RNA Copy of the Gene

During **transcription,** a copy of the gene is built into a molecule of RNA for transport to the place where proteins are synthesized. The process is called transcription because information is simply copied, or transcribed, from a sequence of bases in DNA to a sequence of bases in RNA. The language is the same in both molecules. Each triplet of bases built into the RNA molecule is called a **codon** and

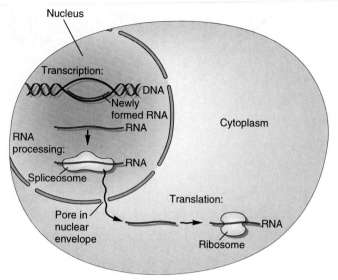

Figure 11.2 Sites of protein synthesis. Transcription, which forms an RNA copy of the gene, occurs on a DNA molecule within the nucleus. RNA processing, in which the RNA copy is modified, occurs on a spliceosome within the nucleus. Translation, in which the RNA copy is used to build a polypeptide, occurs on a ribosome in the cytoplasm.

Table 11.1

The Three Phases of Gene Expression

Name of Phase	What Happens	Where It Happens
Transcription	A copy of the gene is built into an mRNA molecule	On a DNA molecule
RNA processing	In a eukaryotic cell, the mRNA is modified	In the nucleus
Translation	The mRNA is used to assemble amino acids into a protein	On a ribosome

specifies a particular amino acid or a stop signal, as shown in table 11.2. Materials used in transcription are listed in table 11.3.

The RNA molecule that carries genetic instructions to the site of protein synthesis is called **messenger RNA** (abbreviated **mRNA**). The initial copy of the gene, built directly from DNA, is the precursor messenger RNA.

The precursor mRNA is synthesized from one strand, called the template strand, of the double-stranded DNA molecule:

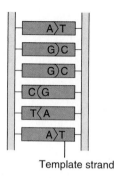

Template strand

An RNA molecule differs from a DNA molecule in having just one strand of nucleotides, using uracil in place of thymine, and using ribose instead of deoxyribose. As an mRNA is built, the sequence of nucleotides within the DNA molecule serves as a template for building an RNA molecule.

The sequence of bases in the precursor mRNA is assembled according to the base-pairing rules, in which adenine pairs with thymine (or uracil, in the RNA) and guanine pairs with cytosine. Thus, for example, wherever there is a guanine in the template strand of DNA there is a cytosine in the RNA. The sequence of bases in the precursor mRNA is complementary to the sequence of bases in the DNA template. Transcription is illustrated in figure 11.3 and described following.

Construction of a precursor mRNA begins when an enzyme called RNA polymerase binds with the first part of a gene within a DNA molecule. The tightly coiled DNA molecule loosens and unwinds, separating the two strands next to the enzyme. In this way, the initial portion of the template strand is exposed for copying (fig. 11.3*a*).

RNA polymerase then promotes the formation of hydrogen bonds (weak attractions) between bases in the template strand of DNA and bases in the unattached RNA nucleotides, which are always present in the area (fig. 11.3*b*). The temporary joining of RNA nucleotides with the template strand follows the rules of base pairing: guanine bonds with cytosine, and adenine bonds with either thymine (in DNA) or uracil (in RNA).

As hydrogen bonds form between the bases, covalent bonds (strong attractions) form between the RNA nucleotides. A continuous strand of RNA nucleotides develops (fig. 11.3*c*). As the RNA molecule grows, the hydrogen bonds holding it to the template strand of DNA break. The two strands of the DNA molecule close up behind the moving site of transcription.

Table 11.2

The Genetic Code in mRNA: Each Sequence of Three Bases Is a Codon That Specifies a Particular Amino Acid or a Stop Signal

First Base	Second Base				Third Base
	U	C	A	G	
U	UUU phenylalanine	UCU serine	UAU tyrosine	UGU cysteine	U
	UUC phenylalanine	UCC serine	UAC tyrosine	UGC cysteine	C
	UUA leucine	UCA serine	UAA stop	UGA stop	A
	UUG leucine	UCG serine	UAG stop	UGG tryptophan	G
C	CUU leucine	CCU proline	CAU histidine	CGU arginine	U
	CUC leucine	CCC proline	CAC histidine	CGC arginine	C
	CUA leucine	CCA proline	CAA glutamine	CGA arginine	A
	CUG leucine	CCG proline	CAG glutamine	CGG arginine	G
A	AUU isoleucine	ACU threonine	AAU asparagine	AGU serine	U
	AUC isoleucine	ACC threonine	AAC asparagine	AGC serine	C
	AUA isoleucine	ACA threonine	AAA lysine	AGA arginine	A
	AUG methionine/start	ACG threonine	AAG lysine	AGG arginine	G
G	GUU valine	GCU alanine	GAU asparagine	GGU glycine	U
	GUC valine	GCC alanine	GAC asparagine	GGC glycine	C
	GUA valine	GCA alanine	GAA glutamic acid	GGA glycine	A
	GUG valine	GCG alanine	GAG glutamic acid	GGG glycine	G

Note: To use this table, find the letter of the first base in the codon (*left*) and then go across this row until you are in the column headed by the letter of the second base. Then find the third base, marked at the far right of the table.

DNA   mRNA   tRNA

DNA   AAG ——→ UUC ——→ AAG = phenylalanine

A — U
G — C

Table 11.3

Some of the Materials Used in Transcription

Material	What the Material Does
Template strand of DNA	Provides recipe for building a precursor mRNA
RNA nucleotides	Building blocks of RNA
RNA polymerase	Enzyme that promotes construction of the precursor mRNA
ATP	Provides energy for forming bonds between RNA nucleotides

When the end of the gene is reached, RNA polymerase stops working and the newly formed RNA molecule breaks free of the DNA molecule. The precursor mRNA has formed. The DNA molecule returns to its double-stranded, coiled state, exactly the same as it was before transcription. The gene is preserved within the molecule of DNA.

Processing the Precursor mRNA

In a eukaryotic cell, the precursor mRNA is modified before it is used to build a polypeptide. **RNA processing,** which converts the precursor mRNA into a mature mRNA, occurs in the nucleus of a eukaryotic cell. It involves two kinds of changes: molecules are added to both ends of the RNA molecule and segments within the molecule are removed (fig. 11.4).

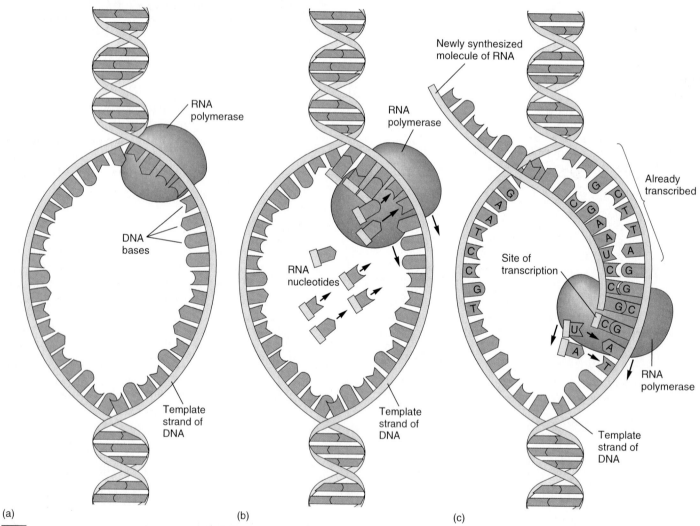

(a)   (b)   (c)

Figure 11.3 Transcription. During transcription, a copy of the gene is built into a molecule of RNA. (a) An enzyme, RNA polymerase, opens up the DNA molecule near the beginning of the gene, exposing its sequence of bases. (b) The enzyme then attaches RNA nucleotides to the template strand of the DNA molecule according to the base-pairing rules: A with T (in DNA) or U (in RNA); C with G. (c) RNA polymerase moves along the DNA molecule, forming bases in RNA nucleotides with bases in the template strand of DNA. As the bases bond with each other, the enzyme also joins adjacent RNA nucleotides to form a long chain of nucleotides that becomes an RNA molecule.

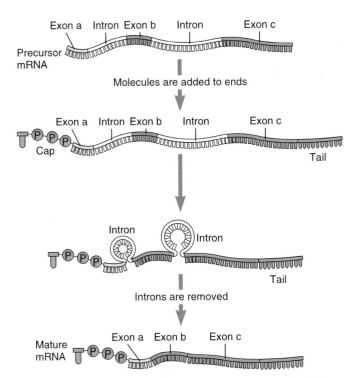

Figure 11.4 RNA processing. A precursor mRNA, built from the DNA molecule, is processed into a mature mRNA. First, a cap and a tail are added to the ends of the molecule. The cap consists of three phosphate groups and a nucleotide carrying guanine. The tail consists of many nucleotides carrying adenine. Second, portions of the molecule, called introns, are removed. The remaining portions, called exons, are spliced together to form a continuous molecule, which is used to build a protein.

Molecules are added to both ends of the precursor mRNA. A cap, formed of a modified nucleotide (consisting of one ribose, three phosphates, and one guanine with an attached $-CH_3$ group) is added to the first part of the molecule. It protects the mRNA from breakage and helps initiate protein synthesis on the ribosome. A tail, formed of 100 to 200 adenine nucleotides (consisting of one ribose, one phosphate, and one adenine), is added to the other end of the molecule. The tail protects the molecule from breakage and helps export it from the nucleus.

RNA processing also eliminates certain base sequences from the RNA copy of the gene. These changes are promoted by a group of enzymes that bind with the precursor mRNA to form a structure known as the spliceosome. The discarded segments are called **introns** (from *intervening* sequences) and the retained segments are called **exons** (from *expressed* sequences). After removal of the introns, the exons are spliced together to form a continuous molecule of mRNA.

Generally the amount of DNA in introns far exceeds the amount in exons. The function of these discarded pieces of DNA remains unclear. One hypothesis is that introns are evolutionary leftovers—instructions for building

proteins that are no longer used in modern forms of life. The cell apparently has no way of eliminating segments of DNA that it no longer needs. Another hypothesis, called the exon shuffling hypothesis, is that the function of an intron is to split a gene into fragments that can be recombined in a variety of ways; it provides a way of building several kinds of proteins from a single length of DNA.

Translation: Building the Protein

In the second phase of gene expression, called **translation,** amino acids are linked according to instructions within the mRNA. The process is called translation because the information is translated from the language of nucleic acids to the language of proteins.

Translation materials include the mRNA molecule, amino acids, enzymes, ATP, a ribosome, and transfer RNA molecules. These structures and their functions are listed in table 11.4. Before describing what happens during translation, we need to look more closely at two of its structures: the ribosome and the transfer RNA.

The Ribosome Translation occurs on the surface of a ribosome. In a prokaryotic cell, the ribosome is located right next to the DNA molecule. In a eukaryotic cell, the ribosome is in the cytoplasm and the mature mRNA must leave the nucleus, through a pore in the nuclear envelope, to reach a ribosome (see fig. 11.2).

A **ribosome** is a dense ball of RNA and proteins. It consists of two separate structures that remain apart when the ribosome is not in use but come together upon arrival of an mRNA. The mRNA molecule fits into a groove

Table 11.4

Some of the Materials Used in Translation

Material	What the Material Does
Ribosome	Holds and orients materials; promotes the formation of bonds between amino acids
Amino acids	Building blocks of proteins
mRNA	Provides instructions for linking amino acids into a protein
tRNA	Brings amino acids to the ribosome and lines them up according to instructions in the mRNA
Enzymes	Bind materials to the ribosome; bind amino acids to the tRNA; initiate translation
ATP	Provides energy for the reactions

between the two parts. A ribosome's functions are to orient the molecules used in protein synthesis and to promote the formation of bonds between amino acids. It has three sites for attachment of RNA molecules: the groove between its two parts for mRNA and two sites for transfer RNA, one known as the peptide site and the other as the amino site:

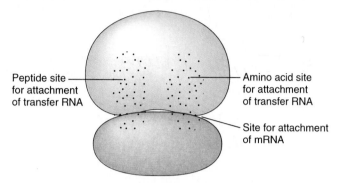

Transfer RNA   Transfer RNA (abbreviated **tRNA**) is an RNA molecule that is looped back upon itself (fig. 11.5). Its function is to transport amino acids to the ribosome and arrange them in the order dictated by the sequence of codons within the mRNA molecule.

Each tRNA has two important binding sites. One end of the molecule consists of three unpaired bases that can bind with the three unpaired bases of a codon in the mRNA. The three unpaired bases of tRNA are called an **anticodon,** because they are complementary to a codon. The anticodon joins with the codon according to the base-pairing rules: adenine with uracil and guanine with cytosine:

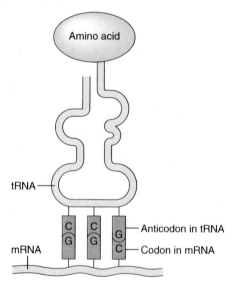

At the opposite end of the tRNA molecule is a binding site for the amino acid—the one called for by the codon in the mRNA (see table 11.2). The tRNA with the anticodon CCG, as shown in the previous illustration, carries glycine

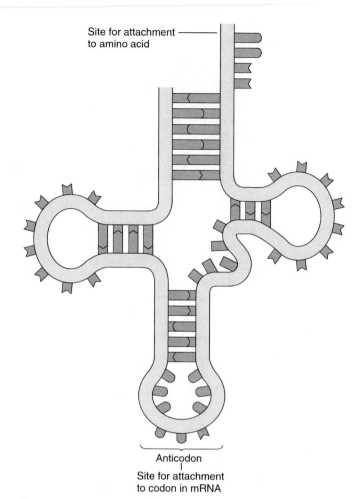

Figure 11.5 Transfer RNA. A transfer RNA molecule is a long chain of nucleotides that is folded back upon itself and held in that shape by bonds between some of the bases. The molecule has two important attachment sites: an anticodon, for attachment to a codon in the mRNA, and a site for picking up an amino acid.

since the matching codon, GGC, in the mRNA specifies this amino acid. There are at least twenty kinds of tRNA molecules, one for each of the twenty kinds of amino acids. Special enzymes join amino acids to the tRNA molecules: they recognize each anticodon and attach the appropriate amino acid to the other end of the molecule.

Assembling the Polypeptide   The polypeptide is assembled, one amino acid at a time, as the ribosome moves along the mRNA, one codon at a time. As each codon contacts the ribosome, it binds temporarily with its anticodon on a tRNA. The amino acid carried by each tRNA then bonds with the amino acid that arrived just before it. A polypeptide grows, as one amino acid after another becomes bonded to the chain of amino acids. This orderly process is illustrated in figure 11.6.

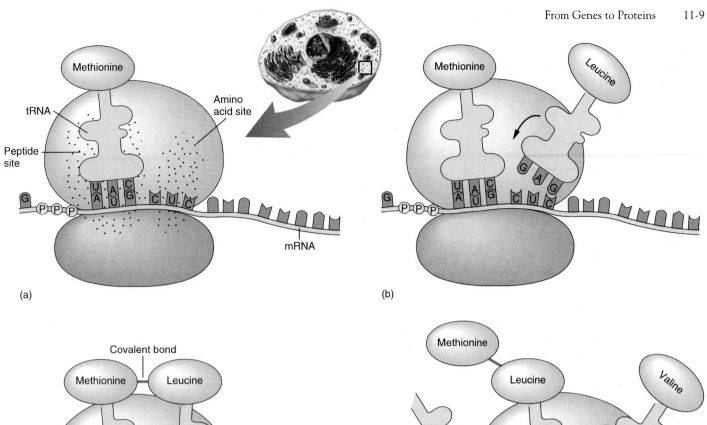

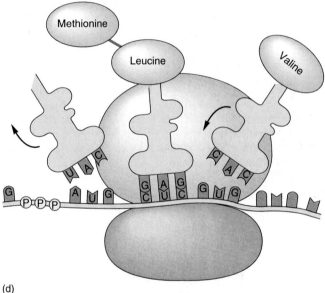

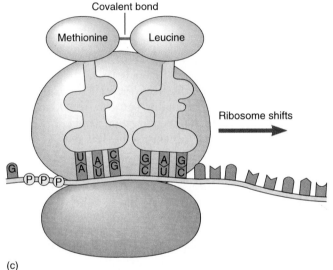

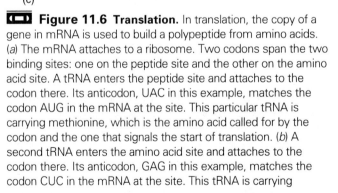

Figure 11.6 Translation. In translation, the copy of a gene in mRNA is used to build a polypeptide from amino acids. (*a*) The mRNA attaches to a ribosome. Two codons span the two binding sites: one on the peptide site and the other on the amino acid site. A tRNA enters the peptide site and attaches to the codon there. Its anticodon, UAC in this example, matches the codon AUG in the mRNA at the site. This particular tRNA is carrying methionine, which is the amino acid called for by the codon and the one that signals the start of translation. (*b*) A second tRNA enters the amino acid site and attaches to the codon there. Its anticodon, GAG in this example, matches the codon CUC in the mRNA at the site. This tRNA is carrying leucine, the amino acid called for by the codon. (*c*) A covalent bond forms between methionine, brought by the first tRNA, and leucine, brought by the second tRNA. (*d*) The ribosome then shifts one codon down the mRNA, pushing the first tRNA off the ribosome and moving the second tRNA and the next codon in mRNA up to the peptide site. A third tRNA now enters the vacant amino acid site. Its anticodon, CAC, matches the codon, GUG, at the amino acid site. GUG codes for valine, the amino acid carried by the third tRNA. The anticodon bonds with the codon and a bond forms between leucine and valine. The ribosome continues to move, codon by codon, along the mRNA molecule. The amino acids called for by the codons are joined to form a polypeptide.

Ribosomes typically occur in clusters so that each mRNA molecule is associated with several ribosomes at the same time (fig. 11.7). In this way, many copies of the polypeptide are built from a single mRNA molecule. At some point, synthesis of the polypeptide stops and the mRNA molecule breaks apart. Its nucleotides return to the nucleus for recycling. The newly formed polypeptide folds into its characteristic shape. This molecule may be the functional protein, or the polypeptide may join with other polypeptides to form the functional protein.

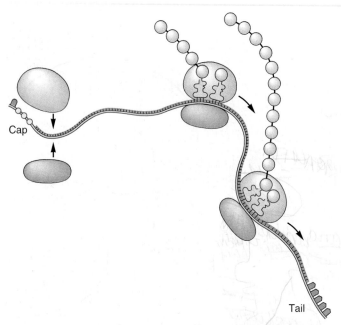

Figure 11.7 Several ribosomes may translate the same mRNA at the same time. The mRNA in this sketch is being translated by three ribosomes at the same time. The ribosome on the right has almost finished translating the mRNA: it has joined many amino acids (green). The ribosome in the middle has just gotten started and has joined five amino acids. The ribosome on the left, at the cap where the gene begins, is preparing to attach to the mRNA. Its two parts will join and hold the mRNA between them.

The machinery of translation is the same in all eukaryotic cells (see page 59), but it is slightly different from the machinery in prokaryotic cells (see page 62). Bacterial ribosomes are smaller, with different kinds of proteins and RNAs, than the ribosomes of eukaryotic cells. We exploit these differences when we use antibiotics to fight bacterial diseases, as described in the accompanying Wellness Report.

Control of Gene Expression

A cell controls which of its genes are active at any one time, synthesizing only the proteins it needs for its current activities. No more than 20% of the genes in an animal cell are expressed at any one time.

We still do not know exactly how genes are turned on in a eukaryotic cell. We know that transcription of a particular gene is initiated by a special kind of protein, called a transcription factor, that binds with a segment of DNA that controls a gene's activity. This segment of DNA is called the promoter of the gene. But how does the environment of the cell activate the transcription factor?

Response to Conditions Inside the Cell   There is continuous interaction between transcription factors and the activities of the biochemical pathways. Some genes are turned on in response to an excess of certain raw materials for a pathway. A liver cell, for example, synthesizes the enzymes that convert glucose into glycogen whenever excess glucose is present. Other genes are turned on when the products of biochemical pathways are in short supply. A brain cell, for example, synthesizes small polypeptides, called neurotransmitters, after it has secreted large quantities of them.

Response to Conditions Outside the Cell   In a multicellular organism, a cell alters its activities in response to the activities of other cells. It receives information from other cells by way of **chemical messengers**—molecules secreted by one cell that alter the activities of another cell.

Some chemical messengers, such as steroid hormones, pass directly through a cell's membranes. They travel to the cell's nucleus and assume direct control over the transcription factors of certain genes. The male hormone, testosterone, operates in this way, turning on genes, such as the ones that synthesize muscle proteins.

Most chemical messengers do not enter a cell; they effect changes indirectly, from the cell surface. In 1993, biologists discovered a major pathway that conveys signals from the cell's surface to its nucleus. This chain of molecular reactions, known as the Ras pathway, begins when a chemical messenger attaches to a receptor protein (see page 73) embedded within the plasma membrane of the cell. This attachment activates the Ras protein, which lies just inside the plasma membrane and is the first molecule in a pathway leading to the cell nucleus (fig. 11.8).

Surprisingly, the Ras pathway is identical in both humans and yeasts (and probably all other organisms with eukaryotic cells). Evolutionists attribute such identity of complex biochemistry to presence of the same pathway in a common ancestor. In the case of humans (kingdom Animalia) and yeasts (kingdom Fungi), the common ancestor existed more than a billion years ago. The Ras pathway must be crucial to the function of multicellular organisms, for it has been preserved over billions of generations of diverse lines of animals, fungi, and probably plants.

The Gene

An organism's genes are part of its **genome,** which is all the DNA molecules within its cells. Surprisingly, only

How Antibiotics Fight Diseases

Antibiotics are organic molecules that kill bacteria. In nature, they are manufactured by certain kinds of molds and bacteria as chemical weapons against each other. A mold and a bacterium compete for the same kinds of food. Each acquires more food for itself by poisoning its competitor. We take advantage of this natural warfare when we use antibiotics to treat human diseases caused by bacteria. Antibiotics such as penicillin, streptomycin, and tetracycline are administered to combat bacterial diseases like pneumonia, gonorrhea, tuberculosis, and syphilis.

Antibiotics synthesized by molds have been used for centuries. Boils on the skin, for instance, were cured by coating the area with moldy soybean curds. The fact that moldy materials kill bacteria was discovered accidently in 1928. Sir Alexander Fleming, an English biologist, was experimenting with bacteria in his laboratory and forgot to cover some of the dishes before he left on vacation. When he returned, he found mold growing in the dishes of bacteria and noticed that wherever there was mold there were no bacteria left (fig. 11.A). Fleming concluded that something in the mold destroyed the bacteria.

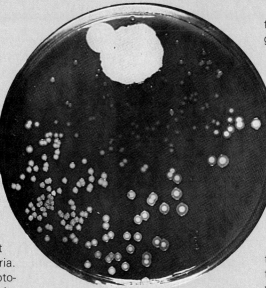

Figure 11.A   Fleming's bacteria. Colonies of bacteria are seen as white spots in the bottom half of the laboratory dish. The large white object in the top is a mold, which has secreted penicillin that has killed nearby colonies of bacteria.

More than a decade later, the bacteria-killing chemical was isolated. It was named penicillin, after *Penicillium notatum*, the species of mold that manufactures it. This particular antibiotic kills bacteria by preventing formation of a cell wall. The molecular structures of many antibiotics have now been described and are mass-produced in chemical laboratories.

Many antibiotics kill bacteria by interfering with the translation phase of gene expression. Ribosomes and other components of the protein-synthesizing machinery are not exactly the same in the prokaryotic cells of bacteria as in the eukaryotic cells of animals (and molds). Thus, any material that interferes with protein synthesis in a bacterium but not in a eukaryotic cell is a potential medicine for treating bacterial infections in an animal such as ourselves.

Many antibiotics kill bacteria by attaching to ribosomes and blocking translation of mRNA into its polypeptide. For example, tetracycline prevents tRNA molecules from binding with the ribosome; chloramphenicol prevents one amino acid from joining with another; and erythromycin prevents movement of the tRNA from the amino site to the peptide site on the ribosome. A bacterium that cannot translate its mRNA into proteins cannot survive.

Antibiotics are effective against all kinds of bacteria but not against other kinds of organisms that cause human diseases. Viruses have no ribosomes for antibiotics to attack. Protozoa, yeasts, and parasitic worms have eukaryotic cells, like our own, with ribosomes that are not affected by antibiotics. Only infections caused by bacteria can be cured with antibiotics.

about 3% of all the DNA in a human cell codes for proteins. The number of genes within this DNA is probably around 100,000. Of the remaining 97%, some control gene expression and some code for RNA molecules other than mRNA. The rest, which comprises most of the genome, remains a mystery.

We learn more about genes every day. This growing knowledge continues to change our definition of a gene. It also has changed our views about the origin of life and the very first gene. In addition, we are discovering that the human genome is more complex than we could have imagined, and that mutations occur in a great variety of ways. These topics are discussed following.

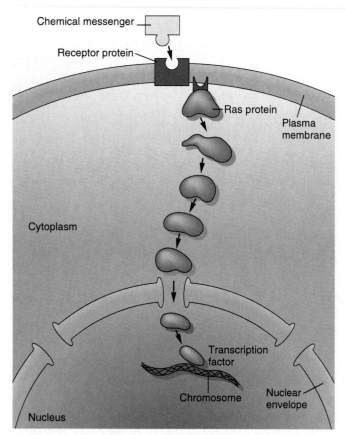

Figure 11.8 The Ras pathway. Events that occur outside a cell are communicated to the genes by way of the Ras pathway. Attachment of a chemical messenger to its receptor in the plasma membrane activates the Ras protein, which lies just inside the membrane. The Ras protein activates an adjacent protein, which then passes the signal to the next protein and the next along a series of proteins leading to the chromosome in the nucleus. There, the signal activates a transcription factor that turns on a certain gene.

Definition of a Gene

Early definitions of a gene were limited to the role of DNA in directing the synthesis of a protein. Shortly after DNA was described, a gene was defined as a segment of DNA that codes for an enzyme. When we learned that some proteins are formed of more than one polypeptide, a gene was redefined as a segment of DNA that codes for a polypeptide. More recently, we learned that a gene is not a continuous stretch of DNA but is broken up by introns. The definition of a gene then became the following: one or more segments of DNA (i.e., the exons) that together code for a polypeptide.

The modern view of DNA is that its primary function is to produce RNA molecules. Some of these RNA molecules are messenger RNA, which in turn direct the synthesis of polypeptides. Other RNA molecules produced by DNA molecules include transfer RNA and ribosomal RNA, which assist in the synthesis of polypeptides. A simple, modern definition of a **gene** is: one or more segments of DNA that together produce a functional RNA molecule. In this definition, a gene includes portions of DNA that control its expression.

The First Genes: DNA or RNA?

Since the genes of modern cells are formed of DNA, it was natural to assume that the first cells on earth, more than 3.5 billion years ago, also had genes formed of DNA. A problem with this view, however, was in understanding how the first DNA molecule synthesized the first protein. In modern cells, the synthesis of proteins requires specific enzymes. Since enzymes are proteins, it was hard to see how the first protein could be synthesized from DNA.

This view of the first genes changed, and the problem of enzymes was resolved, in 1982, with the discovery that some RNA molecules can act as enzymes. If genes were formed of RNA, they could act both as genetic instructions for building proteins and as enzymes for facilitating the biochemical reactions of protein synthesis. An RNA molecule can perform all three tasks fundamental to the origin of life: replicate itself; provide instructions for building proteins; and promote biochemical processes, including self-replication and protein synthesis.

Life probably began with RNA molecules. Later, the task of providing genetic instructions must have been transferred to DNA molecules, which are more stable (because they are double-stranded rather than single-stranded), and the task of promoting biochemical reactions was transferred to enzymes made of proteins. The function of RNA became restricted to assisting DNA in protein synthesis.

Mutations

A **mutation** is a spontaneous or induced change in the sequence of bases within a gene. A spontaneous mutation occurs as a mistake during DNA replication prior to cell division. Such copy errors are usually repaired before the cell divides, but some persist as permanent changes in the DNA. An induced mutation occurs in response to some environmental factor, such as radioactive materials, certain chemicals, and ultraviolet light. Some mutations develop when pieces of DNA, from other molecules in the cell or from viruses, are inserted within the gene.

A mutation in the DNA of a cell that forms an egg or sperm may be passed from one generation to the next. Most mutations are random; they are not related to the particular gene or the function of its protein.

A mutation alters a trait when it alters the activity of a protein. Many mutations, called silent mutations, involve amino acids that have no affect on a protein's activity. The function of an enzyme, for example, depends

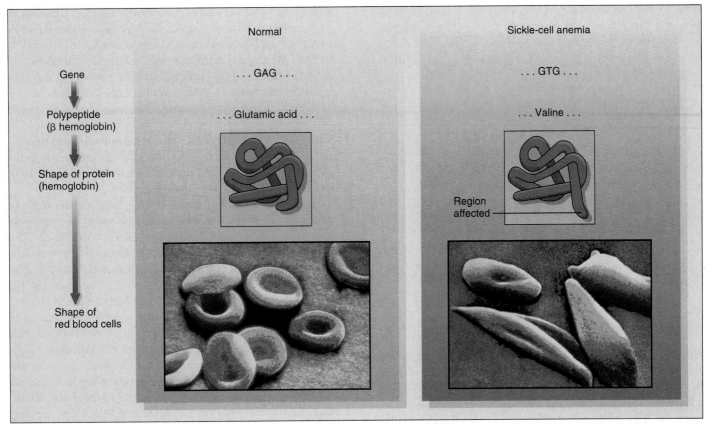

Normal

... GAG ...

... Glutamic acid ...

Gene

Polypeptide
(β hemoglobin)

Shape of protein
(hemoglobin)

Shape of
red blood cells

Sickle-cell anemia

... GTG ...

... Valine ...

Region
affected

Figure 11.9 Sickle-cell anemia. A person with sickle-cell anemia has the triplet GTG instead of GAG in the gene for hemoglobin. This base substitution causes a valine to replace a glutamic acid in hemoglobin. The abnormal hemoglobin has a slightly different shape that stiffens and distorts the red blood cells, causing them to clog small blood vessels.

largely on its three-dimensional shape, which, in turn, is controlled by chemical bonds between certain of its amino acids. Some amino acids in an enzyme are not involved in these bonds and so may have no effect on the enzyme's shape.

Point Mutations   Most mutations are small changes, called **point mutations,** in which a single base in a gene is altered. There are three kinds of point mutations: substitutions, additions, and deletions.

The most frequent kind of mutation is the substitution of one base for another within the structure of DNA. Sickle-cell anemia is the consequence of a mutation of this sort. Its symptoms develop because of abnormal hemoglobin in the red blood cells. Hemoglobin is a large protein, formed of two kinds of polypeptides encoded within two genes. Its function is to transport oxygen from the lungs to the body cells. In sickle-cell anemia, just one of the amino acids in one of the polypeptides is wrong. An incorrect amino acid is inserted because one base in the DNA has been replaced by another: a thymine has been substituted for adenine, changing a triplet from GAG to GTG. GAG spells glutamic acid and GTG spells valine. The substitution of valine for glutamic acid causes incorrect folding of

the hemoglobin, which changes a red blood cell from a flexible disc to a rigid sickle (fig. 11.9). The sickle-shaped cells tend to get stuck in small blood vessels and block normal blood flow.

A base addition shifts all the three-base groupings of a gene, causing a larger change in the protein than a base substitution. Take, for example, the instructions that begin with GCCTACAAT, which specifies alanine (GCC), tyrosine (TAC), and asparagine (AAT):

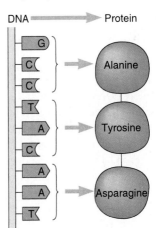

DNA → Protein

G
C
C → Alanine

T
A
C → Tyrosine

A
A
T → Asparagine

If an extra base is added to the sequence, say a thymine after the first guanine, then the entire grouping of triplets is shifted. The new sequence of bases, GTCCTACAAT, now specifies valine (GTC), leucine (CTA), and glutamine (CAA):

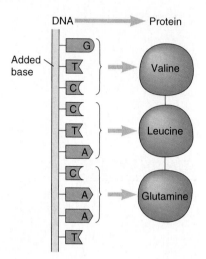

All triplets after the addition are altered and so, also, are all the amino acids within the protein. A deletion causes the same problem. By removing a base, the three-base grouping is shifted and the sequence of amino acids is changed. The addition or deletion of a base dramatically changes the kind of protein that is synthesized.

Dynamic Mutations   Some mutations are small regions of a DNA molecule that become repeated. These changes in the DNA are called **dynamic mutations,** because they continue to enlarge the gene. Short sequences of bases (usually just two or three bases) are added repeatedly to the gene. The mutation gets longer in successive generations, as more and more repeats of the same simple sequence are added.

Dynamic mutations, discovered in 1991, are known to cause several genetic disorders and are suspected in many others. Repetitive sequences were first detected in children with Fragile-X syndrome, described in chapter 10, in which the base sequence CGG continues to be added to a region of the X chromosome.

DNA Inserts   Sometimes whole segments of DNA break loose from a chromosome and insert themselves elsewhere in the genome. Such DNA inserts may break up a gene or interfere with the mechanism that turns it on or off.

A gene that moves from one chromosome to another within a cell is called a **jumping gene** (also known as a transposable element). The gene may seldom move, but when it does it can have a large effect on another gene. Researchers are puzzled by these mobile pieces of DNA. They are common in some organisms—in fruit flies, jumping genes account for more than half the "spontaneous"

mutations. We do not yet know how common these forms of mutations occur in human cells or where these mobile pieces of DNA come from. They resemble the genes of viruses. Were jumping genes once the genes of viruses that invaded human cells and became incorporated into the cell's chromosomes? Or did jumping genes originate within cells and later give rise to viruses?

Chromosomal Mutations   A **chromosomal mutation** involves changes in large segments of a chromosome. Some commonly observed chromosomal changes include translocations, nondisjunctions, and inversions.

A translocation occurs when a large segment of one chromosome breaks off and attaches to a nonhomologous chromosome. A nondisjunction occurs during cell division, when one of the daughter cells receives too many and the other receives too few chromosomes (as described in chapter 9). An inversion occurs when a segment of the DNA molecule becomes reversed within the chromosome.

Chromosomal abnormalities are not uncommon in humans. Approximately 6 of every 1,000 newborns have a major chromosomal defect that is visible through a light microscope. The actual rate at which these chromosomal mutations occur is much higher: approximately half of all miscarriages are of embryos with chromosomal abnormalities. These embryos are so defective that they die very early in development.

Viruses

A **virus** is a cluster of organic molecules that uses the cell of an organism to reproduce itself. Is it a living organism or merely a cluster of molecules? It has its own unique set of genes, which perhaps qualify it as an organism. Yet without a host cell, a virus cannot use its genes to build proteins, for it lacks the ribosomes, tRNA, and other materials essential to protein synthesis.

There are thousands of different kinds of viruses, each specialized for parasitizing a different kind of cell. Viruses that invade human cells cause such diseases as herpes, AIDS, colds, mononucleosis, measles, mumps, and the flu.

The Structure of a Virus

Every virus contains nucleic acids and proteins. Nucleic acids (either DNA or RNA) form the core of a virus and proteins form the coat. Some viruses also have an envelope, formed of lipids and other molecules, around the protein coat (fig. 11.10). A virus is tiny; it is barely the size of a ribosome. It would take approximately 10,000 cold viruses, lined up side by side, to span the width of a pencil dot.

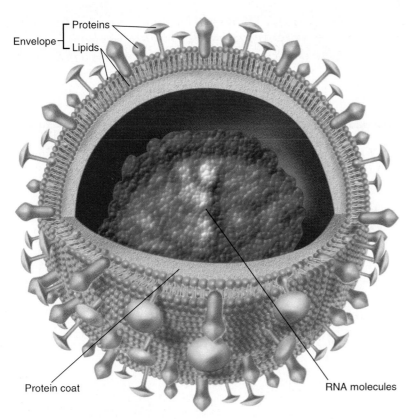

Figure 11.10 **The structure of a flu virus.** A virus is built of nucleic acids, either DNA or RNA, and a protein coat. This particular virus (cut open), which causes the flu, also has an envelope formed of lipids and proteins. Its genes are in its core, arranged in eight coils of RNA.

The nucleic acids of a virus contain instructions for building two kinds of viral proteins: coat proteins and enzymes. The coat proteins attach the virus to the outer surface of its host cell, enabling the virus to enter and leave the cell. The enzymes promote certain biochemical reactions, including the synthesis of viral DNA or RNA. Viral genes do not contain instructions for building all of the envelope. Viruses with this outermost structure construct part of it from portions of the plasma membrane of their host cell.

How a Virus Reproduces

A virus forms many copies of itself by placing its genes inside a cell and using the cell's materials to make viral proteins. The host cell's ribosomes, amino acids, enzymes, ATP, and tRNA are monopolized by the virus.

Invasion of a Cell   A virus invades a cell by first attaching itself to receptor molecules on the surface of the host cell. It then either inserts its nucleic acid into the cell, leaving its protein coat outside (fig. 11.11*a*), or completely

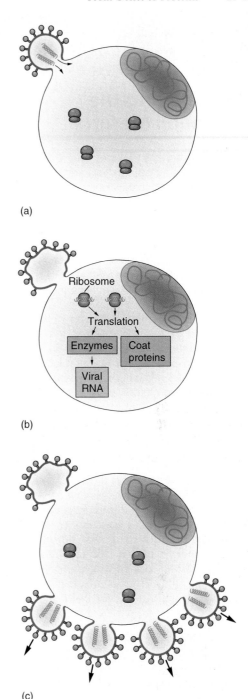

Figure 11.11 **How an RNA virus reproduces.** (*a*) The virus inserts its genes into a host cell. (*b*) The viral genes—RNA in this example—move to a ribosome, where they are translated into two categories of proteins: coat proteins and enzymes that promote synthesis of more viral RNA as well as other reactions. (*c*) The newly formed RNA molecules wrap themselves in the coat proteins to form new viruses, which leave the cell to invade other cells.

Identifying the Genetic Material

It seems strange that only a few decades ago we did not know that genes are located within DNA molecules. The search for genetic material began in earnest just after World War II, when an international group of scientists, including biologists, chemists, and physicists, decided to collaborate rather than compete. What kind of molecule, they asked, contains the genetic material?

There were four obvious hypotheses: the genetic material was in proteins, carbohydrates, nucleic acids, or lipids. An early line of research led to viruses, and here the scientists found an important clue. A virus must contain genetic information, they reasoned, since it reproduces (inside a cell) to form copies of itself. Yet many viruses consist only of proteins and nucleic acids. Thus, viral genes must be located within either proteins or nucleic acids. The hypotheses that genes were located in carbohydrates and lipids were ruled out.

Most scientists at that time favored the hypothesis that genetic information was located within proteins. After all, they argued, enzymes are proteins and it is enzymes that control what a cell does. Besides, proteins are almost unlimited in their variety. Nucleic acids, by contrast, seem to play little role in cell function and their molecular structures seem too simple to hold complex information.

For several years scientists tried to devise experiments that would differentiate between the protein hypothesis and the nucleic acid hypothesis regarding gene location. A main stumbling block was distinguishing between the effects of proteins and the effects of nucleic acids inside a cell. A breakthrough occurred in 1952, when Alfred Hershey and Martha Chase of the Cold Spring Harbor Laboratory in New York conducted an inspired and lucky experiment. They had seen photographs of viruses infecting bacteria

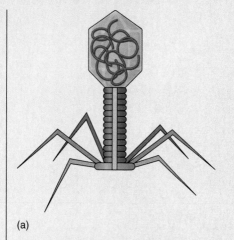

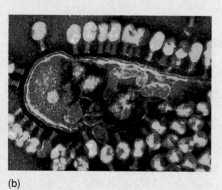

(a)

(b)

Figure 11.B A virus that infects bacteria. (*a*) The core of this virus consists of DNA; its coat, including the "appendages," is made of proteins. (*b*) Photograph of many viruses attaching to the surface of a bacterium.

(fig. 11.B), and they noticed that not all parts of a virus get into a bacterium. Some parts are left outside, stuck to the outer surface of the bacterium. The part of the virus that gets inside, they reasoned, must carry genetic information because inside a bacterium is where new viruses are formed. Which part of the virus, protein or nucleic acid, enters a bacterium? Hershey and Chase devised an experiment to find out whether viral protein or viral DNA entered a bacterium. This experiment, diagrammed in figure 11.C and described following, has been called the experiment that revolutionized biology.

Hershey and Chase labeled viruses with radioactive atoms so that either their proteins or the viral DNA gave off energy that could be detected by laboratory instruments. They developed these viruses by growing them in two separate dishes, one containing bacteria that had been grown in radioactive sulfur and the other containing bacteria that had been grown in radioactive phosphorus. The newly formed viruses incorporated the radioactive materials into their structures. In one dish, radioactive sulfur became part of the viral proteins, since viral proteins contain sulfur but not phosphorus. In the other dish, radioactive phosphorus became part of

the viral DNA, since DNA contains phosphorus but not sulfur. Thus, one group of viruses had radioactive proteins and the other group had radioactive DNA. The experiment was ready to go.

The two groups of viruses were mixed with fresh bacteria not grown in radioactive materials. After allowing enough time for the viruses to insert molecules into the bacteria, the mixtures were shaken in a blender to pull off viral materials that adhered to the surface of the bacteria. The mixture was then spun in a centrifuge to separate the heavier bacteria from the lighter viral materials that did not enter the bacteria.

The results of the experiment were conclusive. The bacteria infected by viruses that had been grown in radioactive phosphorus were radioactive, and the bacteria infected by viruses that had been grown in radioactive sulfur were not. Hershey and Chase concluded that DNA entered the bacteria but proteins did not. Moreover, the infected bacteria released newly formed viruses, confirming that the viral molecules that entered the bacteria did indeed carry instructions for building more viruses. Information about how to build a virus must be in the DNA and not in the proteins. The genetic material was identified as DNA.

(a)

Figure 11.C The Hershey-Chase experiment. (*a*) Martha Chase and Alfred Hershey in 1953, one year after conducting their classic experiment. (*b*) Step I: One group of viruses (*left*) was grown in radioactive sulfur, which became incorporated into their proteins. A second group (*right*) was grown in radioactive phosphorus, which became incorporated into their DNA. Step 2: The viruses infected bacteria. Step 3: The bacteria were shaken and spun to separate the bacteria from the smaller molecules that did not enter the bacteria. Step 4: The bacteria were examined for radioactivity. The bacteria on the left, infected with viruses made of radioactive proteins, were not radioactive; viral proteins did not enter these bacteria. The bacteria on the right, infected with viruses made of radioactive DNA, were radioactive; viral nucleic acids did enter these bacteria. Thus, the molecule that entered the bacteria was DNA.

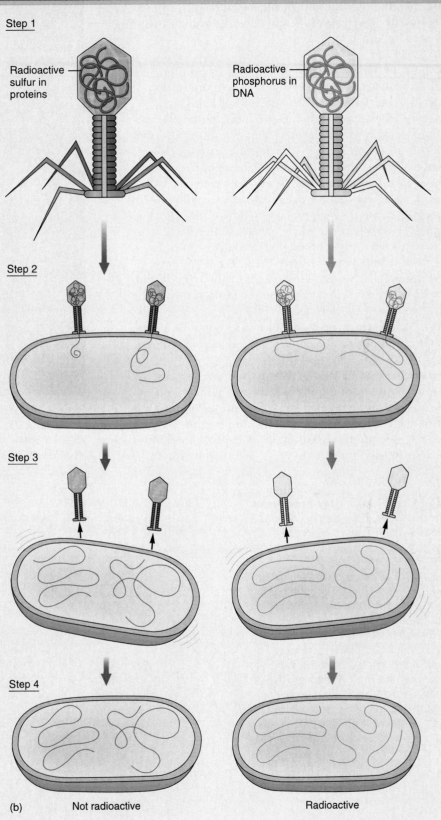

Step 1

Radioactive sulfur in proteins

Radioactive phosphorus in DNA

Step 2

Step 3

Step 4

(b)   Not radioactive   Radioactive

enters the cell and sheds its coat within. Viruses that place only their nucleic acids into the cell, leaving their coat outside, were used as research tools to find out whether genetic material was located in nucleic acids or proteins. This elegant experiment is described in the accompanying Research Report.

Synthesis of Viral Proteins   Once inside a cell, the viral nucleic acids travel to the place where transcription or translation occurs. When the viral nucleic acids are DNA molecules, they move to the site of transcription (the nucleus of a eukaryotic cell), where their information is transcribed into messenger RNA. When the nucleic acids are RNA molecules, they may bypass transcription and move directly to a ribosome for translation into proteins.

A virus uses its host cell's materials to build protein coats and enzymes, which promote synthesis of viral genes (fig. 11.11b). When the nucleic acids and proteins have been synthesized, they assemble themselves into viruses. The newly formed viruses leave the cell and go on to invade other cells (fig. 11.11c). Viruses with lipid-protein envelopes acquire portions of these structures from the cell's membrane as they leave. The host cell is usually killed, either directly from a rupture in its outer membrane or indirectly from insufficient synthesis of its own proteins.

Retroviruses   Some viruses, such as the ones that cause acquired immune deficiency syndrome (AIDS) and cat leukemia, are retroviruses (fig. 11.12). Instead of using the cell to make its proteins right away, a **retrovirus** incorporates its genes into the cell's own DNA molecules. (Some of the DNA in our own cells may have originated in this way.)

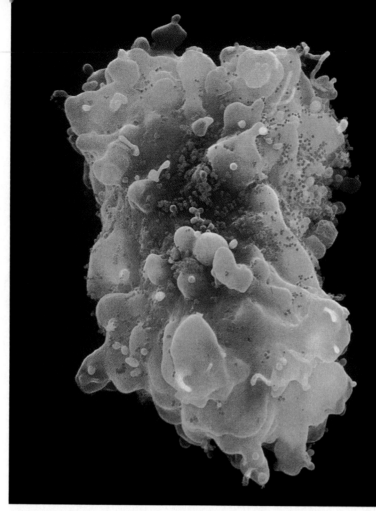

Figure 11.12 The AIDS virus with its host cell. The virus that causes acquired immune deficiency syndrome (AIDS) infects human white blood cells. Here, many copies of the virus (blue) are shown invading a white blood cell (pink).

Figure 11.13 How a retrovirus reproduces. (a) A retrovirus inserts its RNA molecules and its enzymes, reverse transcriptase and integrase, into a host cell. (b) Reverse transcriptase promotes formation of a DNA copy of the viral RNA. (c) The viral DNA moves into the nucleus, where integrase splices it into one of the chromosomes. (d) The viral DNA is transcribed and translated into viral proteins of four types: coat proteins, reverse transcriptase, integrase, and enzymes for processing viral RNA. More viral RNA is synthesized. (e) The RNA, reverse transcriptase, integrase, and coat proteins assemble themselves into retroviruses, which leave the cell to invade other cells.

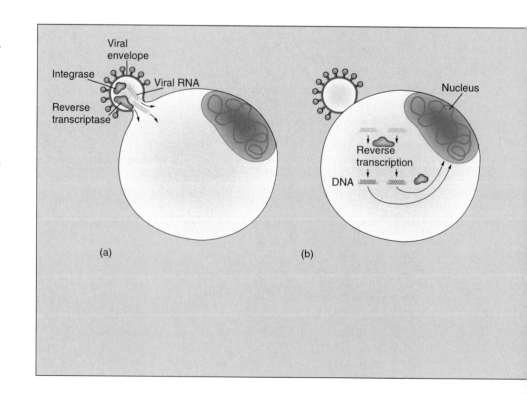

The genes of a retrovirus are made of RNA. The retrovirus inserts its RNA, along with two enzymes, into its host cell (fig. 11.13a). One of the enzymes, called reverse transcriptase, promotes the synthesis of DNA copies of the viral RNA (fig. 11.13b). The synthesis of DNA from RNA, the opposite of transcription (which goes from DNA to RNA), is called **reverse transcription.** (Because human cells do not normally undergo reverse transcription, most drugs developed to fight AIDS are designed to attack this process.) The other enzyme, called integrase, then splices the newly synthesized viral DNA into one of the cell's own molecules of DNA (fig. 11.13c). The viral DNA becomes integrated into the chromosomes.

When the cell divides, the viral DNA is replicated, along with the cell's own DNA, just prior to mitosis (see page 128). One copy of the viral genes goes to each daughter cell. An infected cell produces more infected cells. Under certain conditions that are not well understood, the viral DNA is transcribed into mRNA and then translated into viral proteins: a coat, reverse transcriptase, integrase, and the enzymes that promote the synthesis of viral RNA (fig. 11.13d). The viral RNA is then synthesized and wrapped, along with reverse transcriptase and integrase, in the proteins of the coat. The newly formed viruses leave the cell and invade other cells (fig. 11.13e).

SUMMARY

1. The genetic language, built into DNA molecules, contains instructions for assembling amino acids into polypeptides. Each sequence of three bases within one strand of a DNA molecule specifies a particular amino acid. A much longer sequence of bases specifies the sequence of amino acids within a polypeptide.

2. Gene expression occurs when a gene is used to build its polypeptide. It occurs in three phases: transcription, RNA processing, and translation.

3. During transcription, a copy of the gene is transcribed into a precursor messenger RNA. RNA processing then forms a mature mRNA by adding molecules to both ends and removing certain internal segments.

4. During translation, the sequence of codons (base triplets) in the mRNA is used to form a sequence of amino acids in a polypeptide. It happens on a ribosome, where transfer RNA molecules carrying amino acids join with appropriate codons within the mRNA molecule. As the tRNA molecules join with the mRNA, the amino acids they carry join with each other to form a polypeptide.

5. A gene is expressed when the cell requires its polypeptide. A protein, called the transcription factor, turns the gene on in response to conditions inside and outside the cell.

6. A modern definition of a gene is: one or more segments of DNA that together produce a functional RNA molecule. The first gene was probably formed of RNA, which can self-replicate, direct synthesis of proteins, and act as an enzyme.

7. A mutation is a change in the sequence of bases within a gene. A point mutation is a change of just one base. A dynamic mutation is a short sequence of bases that continues to repeat itself within a gene. A DNA insert is a piece of DNA that inserts itself into a

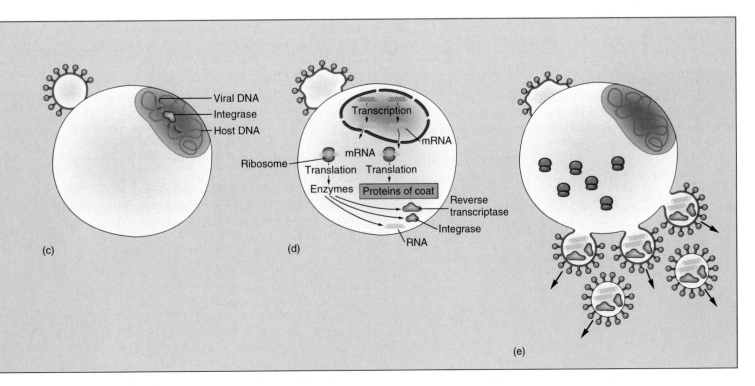

(c)

(d)

Viral DNA
Integrase
Host DNA

Transcription
mRNA
mRNA
Ribosome
Translation   Translation
Enzymes
Proteins of coat
Reverse transcriptase
Integrase
RNA

(e)

209

gene. A chromosomal mutation is a change in a large segment of a chromosome—a translocation of a segment, an extra or missing chromosome, or inversion of a segment.

8. A virus consists of nucleic acids (either DNA or RNA) enclosed within a protein coat. It reproduces by invading a cell and taking over the cell's machinery for protein synthesis. A retrovirus is unusual in that it makes DNA copies of its RNA genes and then integrates the viral DNA into the DNA of its host cell.

KEY TERMS

Anticodon (page 198)
Chemical messenger (page 200)
Chromosomal mutation (page 204)
Codon (KOH-don) (page 193)
Dynamic mutation (page 204)
Exon (EX-ahn) (page 197)
Gene (page 202)
Gene expression (page 193)
Genome (jeh-NOHM) (page 200)
Intron (IN-trahn) (page 197)
Jumping gene (page 204)
Messenger RNA (mRNA) (page 194)
Mutation (myoo-TAY-shun) (page 202)
Point mutation (page 203)
Retrovirus (page 208)
Reverse transcription (page 209)
Ribosome (REYE-boh-sohm) (page 197)
RNA processing (page 196)
Transcription (page 193)
Transfer RNA (tRNA) (page 198)
Translation (page 197)
Virus (page 204)

STUDY QUESTIONS

1. Describe the genetic code. Explain why the code has at most sixty-four words. How many words are needed to spell all the amino acids?
2. Explain how a linear sequence of nucleotides contains information for assembling a linear sequence of amino acids.
3. What is a mutation? Why does a base substitution have less of an effect on a protein than a base addition?
4. Describe how genes control traits.
5. Briefly describe the three phases of gene expression. Where does each phase occur in a eukaryotic cell?
6. Describe reproduction by a virus with RNA genes.
7. Describe reproduction by a retrovirus.

CRITICAL THINKING PROBLEMS

1. Explain why penicillin or tetracycline is not prescribed for viral infections.
2. Do you consider a virus to be a living organism or just a cluster of molecules? Why?
3. In the search for the molecular basis of genetic material, how were carbohydrates and lipids ruled out?
4. Explain why a good approach to developing a cure for AIDS involves attacking reverse transcriptase.
5. Jumping genes may have originated as viruses. If so, why are retroviruses the most likely kind of virus to form a jumping gene?

SUGGESTED READINGS

Introductory Level:

Morse, S. S. (ed.). *Emerging Viruses*. New York: Oxford University Press, 1993. A collection of essays, by a variety of professionals (historians, physicians, evolutionists, virologists, epidemiologists, etc.), on the subject of new forms of viruses and their expected impact on human populations.

Radetsky, P. *The Invisible Invaders: The Story of the Emerging Age of Viruses*. Boston: Little, Brown, and Company, 1991. An entertaining description of viruses—what they do, the history of our knowledge about them, and diseases they cause.

Varmus, H. "Reverse Transcription," *Scientific American*, September 1987, 56–64. A description of how reverse transcription was discovered and, in detail, how it works.

Advanced Level:

Gesteland, R. F., and J. F. Atkins (eds.). *The RNA World: The Nature of Modern RNA Suggests a Prebiotic RNA World*. Cold Spring Harbor, NY: Cold Spring Harbor Laboratory Press, 1993. A description of the structure and function of modern RNA, and a discussion of its likely role in the origin of life.

Lewin, B. *Genes III*, 3d ed. New York: John Wiley & Sons, 1987. A textbook on molecular genetics. Chapters 2 through 9 are advanced descriptions of protein synthesis.

Morse, S. S. (ed.). *The Evolutionary Biology of Viruses*. New York: Raven, 1993. A synthesis of evolutionary biology and virology, with particular emphasis on the rapidly evolving viruses, such as HIV, that cause human diseases.

Rhodes, D., and A. Klug. "Zinc Fingers," *Scientific American*, February 1993, 56–65. A good description of what is known about how gene expression is controlled in eukaryotic cells.

Watson, J. D., N. H. Hopkins, J. W. Roberts, J. A. Steitz, and A. M. Weiner. *Molecular Biology of the Gene*, 4th ed. Menlo Park, CA: Benjamin-Cummings, 1987. Chapters 13 through 16 are clear descriptions of how genes instruct the synthesis of proteins.

Genetic Engineering

Chapter Outline

Objectives

In this chapter, you will learn
–how human proteins are mass-
 produced within bacteria;
–that billions of copies of a gene are
 made within a few hours;
–that a gene is found within a set of
 DNA molecules;
–how the sequence of bases within a
 gene is determined;
–how a normal gene is transplanted
 into a human cell to replace an
 abnormal one;
–how a gene is transplanted from one
 species to another.

A plant that glows like a firefly.
This tobacco plant glows in the
dark because it has been given a
firefly gene. Genetic engineering
can profoundly change an
organism's traits.

Woman mates with petunia! Man mates with mouse! These announcements sound like lurid headlines in the tabloids sold near checkout counters in supermarkets. They are not. Rather, they are exaggerated descriptions of the feats of genetic engineers. A woman's gene has been placed inside a petunia. While she did not "mate" with it, the flower accepted her gene and synthesized its protein—a hormone normally synthesized only in pregnant women. Likewise, a man's gene transplanted into a mouse directed the production of a human protein that alleviates heart attacks. The protein was secreted in the mouse's milk and easily harvested for medical use.

Genetic engineers use our considerable knowledge about what a gene is and how it governs the synthesis of proteins to manipulate genes. They have given us enormous power over ourselves and nature. We can transfer genes from one kind of organism to another, combining the traits of different kinds of organisms. We can create entirely new forms of life.

With such power comes responsibility. Will we use this power to improve or to degrade human life? Will we cure genetic disorders? Will we modify our natural traits—change the human species? Will organisms carrying cancer genes or deadly viruses escape from our laboratories and invade our cells? Will unscrupulous scientists sell lethal organisms to terrorists? Will careless tampering with genes upset the balance of nature?

How we use genetic engineering is up to our society as a whole. As an educated person, you ought to know how genes can be manipulated and the consequences of such procedures. You should be aware of what scientists are doing so that you can evaluate their discoveries and make intelligent decisions about how scientific knowledge is used.

Mass-Producing Proteins and Genes

Genetic engineering is the use of laboratory techniques to manipulate genes. Some of these techniques are listed in table 12.1 and described in this chapter. The biological revolution that generated this area of applied biology began around 1970, when scientists had accumulated three pieces of information about genes. First, the language of genes is the same in all cells: a particular sequence of three bases within DNA is the code for the same amino acid regardless of the kind of cell it comes from. Second, ribosomes, mRNA, tRNA, and enzymes synthesize proteins (see page 193) in a similar way in all kinds of cells. Third, foreign pieces of DNA (as from a virus) are transcribed and translated into proteins in the same way as the cell's own DNA.

Putting this information together, biologists concluded they could manipulate the kinds of proteins a cell synthesizes. They believed they could transfer a gene from one kind of cell to another, and the host cell would synthesize the transplanted gene's protein. The host cell would not recognize the transferred gene as foreign.

Human control of protein synthesis became a reality when enzymes that modify DNA and RNA were identified and isolated. These enzymes, invaluable tools of the genetic engineer, are listed in table 12.2.

Recombinant DNA Technology

Recombinant DNA technology is used to mass-produce proteins and genes inside bacteria (or sometimes yeasts). The basic procedure used in this form of genetic engineering is diagrammed in figure 12.1 and described following.

Table 12.1

Techniques of Genetic Engineering

Name	What It Does	What It Is Used for
Recombinant DNA technology	Transfers genes to bacteria	Mass production of proteins and genes
Polymerase chain reaction	Makes copies of DNA in a test tube	Mass production of genes
DNA probe	Attaches to a gene	Marks the position of a gene
Restriction fragment-length polymorphism	Fragments DNA and measures lengths	Locates a gene by correlating a trait with a fragment length
Base sequencing	Identifies the sequence of bases in DNA	Describes recipes for proteins
Gene therapy	Transfers genes from one human to another; not passed to offspring	Replaces defective genes
Formation of transgenic organism	Transfers genes from one species to another; passed to offspring	Forms new traits that are inherited

Table 12.2

Enzymes Used in Genetic Engineering

Type of Enzyme	What the Enzyme Does
DNA polymerase	Forms bonds between DNA nucleotides
RNA polymerase	Forms bonds between RNA nucleotides
Deoxyribonuclease	Breaks bonds between DNA nucleotides
Ribonuclease	Breaks bonds between RNA nucleotides
DNA ligase	Joins DNA fragments
Restriction enzyme	Cuts DNA at specific base sequences
Reverse transcriptase	Makes a DNA copy of RNA

The procedure begins by isolating the gene that codes for the protein to be mass-produced. The segment of DNA containing the gene is removed from its natural location, usually a human cell, and attached to a piece of DNA that invades a bacterium. The attached gene is the **donor gene** and the DNA molecule that receives it is the **vector:**

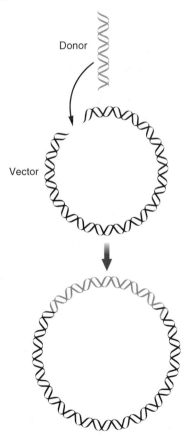

Together, the donor gene and vector are called recombinant DNA; they are a combination of two kinds of DNA.

Bacteria are used in recombinant DNA technology because they are easy to grow in the laboratory, they reproduce rapidly, and their mechanisms for turning genes on and off are well understood. Moreover, bacteria readily accept pieces of DNA from one another—almost any piece of DNA placed inside a bacterium will be used to build proteins.

Isolating the Gene   Before a gene can be transferred from one cell to another, it has to be snipped out of its DNA molecule. The "scissors" of the genetic engineer are **restriction enzymes,** which occur naturally in bacteria, where they shred the DNA molecules of viruses that infect these cells. Hundreds of restriction enzymes have been discovered; each kind cuts DNA molecules at a specific base sequence. These molecular scissors enable us to break and rejoin DNA molecules with ease.

A typical restriction enzyme consists of two identical parts arranged in mirror images, like a left and a right hand (fig. 12.2a). The two parts recognize and cut mirror-image sequences of bases. If one part of the enzyme cuts one strand between the G and the A of the base sequence GAATTC, then the other part cuts the other strand between the G and the A of the mirror-image sequence, CTTAAG:

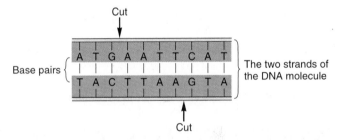

The two cuts are made near each other, usually within a span of less than ten bases.

After the enzyme has cut both strands of the DNA molecule, the bonds joining the bases in the tiny segment between the cuts break and the two strands come apart:

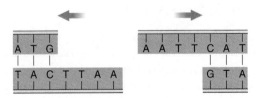

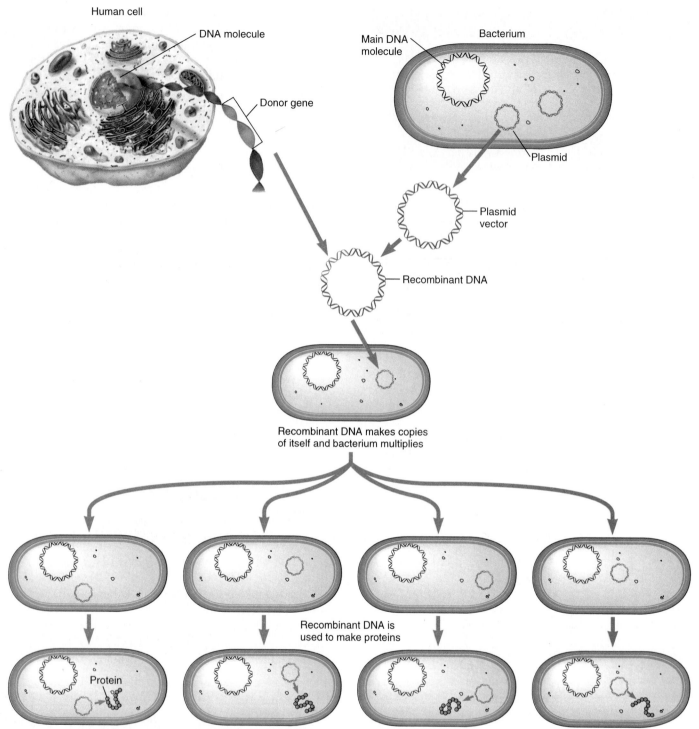

Figure 12.1 Recombinant DNA technology. A donor gene is cut from its DNA molecule and spliced into a plasmid from a bacterium. Together, they form a molecule of recombinant DNA, which invades a bacterium and replicates itself. The bacterium reproduces to form many copies of itself and the recombinant DNA. The bacterium also uses the recombinant DNA, including the donor gene, to synthesize proteins. The donor gene's protein can be harvested in massive quantities.

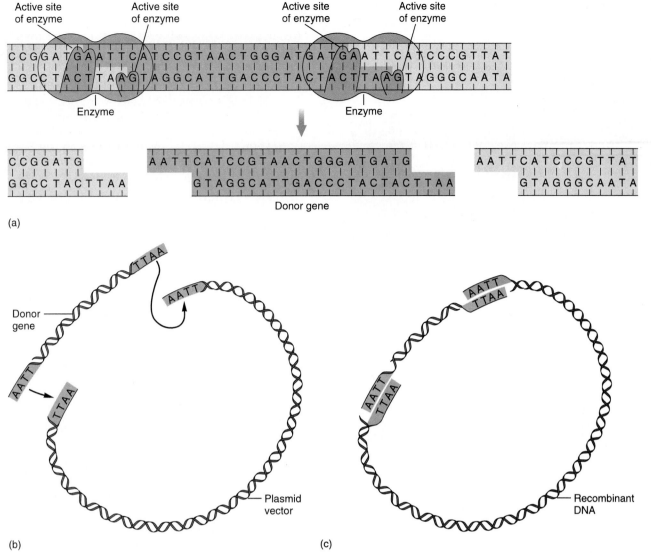

Active site of enzyme   Active site of enzyme   Active site of enzyme   Active site of enzyme

C C G G A T G A A T T C A T C C G T A A C T G G G A T G A T G A A T T C A T C C C G T T A T
G G C C T A C T T A A G T A G G C A T T G A C C C T A C T A C T T A A G T A G G G C A A T A

Enzyme   Enzyme

C C G G A T G
G G C C T A C T T A A

A A T T C A T C C G T A A C T G G G A T G A T G
G T A G G C A T T G A C C C T A C T A C T T A A

Donor gene

A A T T C A T C C C G T T A T
G T A G G G C A A T A

(a)

Donor gene → ...TTAA... ...AATT...

...AATT... →

...TTAA...

Plasmid vector

(b)

AATT/TTAA

AATT/TTAA

Recombinant DNA

(c)

Figure 12.2 Restriction enzymes. Each restriction enzyme cuts DNA at a particular sequence of bases. (*a*) Two restriction enzymes of the same kind cutting a DNA molecule. In the upper strand, they are cutting between a G and an A of the sequence GAATTC. In the lower strand they are cutting between a G and an A of the sequence CTTAAG (the mirror image of the upper strand). The cut strands separate and their base pairs come apart, leaving a short sequence of unpaired bases on the ends. The segment cut out of this molecule has the unpaired bases AATT on its upper strand and TTAA on its lower strand. (*b*) Two molecules of DNA cut with the same restriction enzyme have the same sequences of unpaired bases on their ends. They join when their unpaired bases form bonds according to the base-pairing rules: A with T and G with C. Here a donor gene, cut from a longer molecule of DNA, joins a ring of DNA called a plasmid. (*c*) The two molecules, cut with the same restriction enzyme, become a single molecule of recombinant DNA.

A restriction enzyme cuts the DNA molecule wherever its particular sequence of bases appear; it breaks a molecule of DNA into many fragments:

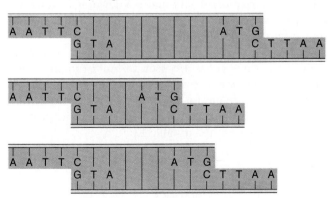

A short stretch of unpaired bases remains at the end of each fragment—AATT on one fragment and TTAA on the other fragment in the previous example. These ends are "sticky"—their unpaired bases readily bond with complementary bases (T with A; G with C) (📖 *see page 49*). An enzyme, DNA ligase, then forms covalent bonds that join the strands of the two fragments.

Two DNA molecules cut by the same restriction enzyme have the same unpaired bases at the ends of their fragments (fig. 12.2*b*). As a consequence, the unpaired bases on a fragment from one of the molecules will join with complementary unpaired bases on a fragment from the other molecule (fig. 12.2*c*), even when the two molecules are from different species of organisms.

215

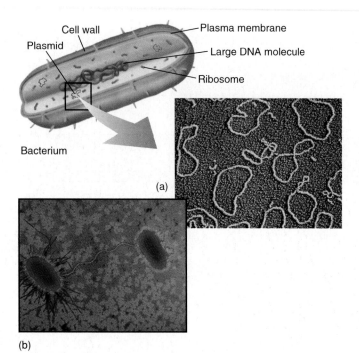

(a)

(b)

Figure 12.3 Plasmids. Bacteria contain small loops of DNA, called plasmids, in addition to their single large molecule of DNA. (a) Highly magnified photograph of plasmids. (b) In one of these bacteria, a plasmid has changed from a loop to a thread and is moving into the other bacterium. Both bacteria are of the species *Escherichia coli,* which normally lives within human intestines.

Preparing a Vector Once a gene has been isolated, it is spliced into a special kind of DNA—the vector—for transport into a bacterium. The most commonly used vector in recombinant DNA technology is a bacterial plasmid.

A **plasmid** is a tiny loop of DNA, carrying only a few genes (fig. 12.3a), that occurs naturally within bacteria. Like a virus, a plasmid uses the cell's ribosomes, RNA, and enzymes to synthesize its proteins and replicate itself. Unlike a virus, its genes code for proteins that are useful, although not essential, to the survival and reproduction of the bacterium.

A plasmid travels from one bacterium to another through a process called conjugation, the bacterial equivalent of sex (fig. 12.3b). Each bacterium readily accepts a foreign plasmid and synthesizes its proteins. Resistance of a bacterium to antibiotics, for example, is often controlled by genes within plasmids. This resistance is passed, via plasmids, from one species of bacterium to another, as described in the accompanying Wellness Report.

Genetic engineers attach donor genes to plasmids for transport into bacteria. A plasmid is removed from a bacterium and cut by the same restriction enzyme that cut the donor gene out of its DNA molecule. The two pieces of DNA have complementary bases on their cut ends. When

Table 12.3

Some Proteins Manufactured by Recombinant DNA Technology for Use in Medical Treatments

Protein	What It Does in the Body
E5 antibody	Prevents blood poisoning
Epidermal growth factor	Hastens regrowth of skin after burns and infections
Erythropoietin	Stimulates production of red blood cells
Factor VIII	Promotes blood clotting (treats hemophilia)
Growth hormone	Accelerates body growth (treats abnormally short children)
Insulin	Controls blood sugar (treats diabetes)
Interferon	Fights viral infections
Interleukin	Regulates the immune system
Relaxin	Relaxes birth canal during labor
Somatostatin	Slows body growth (treats abnormally tall children)
Thymosin	Stimulates production of white blood cells
Tissue plasminogen activator	Dissolves blood clots during heart attacks

mixed together, the donor gene joins the plasmid to form a recombinant DNA molecule (see fig. 12.1).

Mass-Producing the Protein The recombinant DNA, consisting of donor gene and plasmid vector, readily invades a bacterium. There the recombinant DNA replicates itself and its proteins are synthesized. At the same time, the bacterium divides repeatedly to form many bacteria, with each bacterium holding copies of the recombinant DNA (including the donor gene). The formation of identical bacteria from one bacterium is known as cloning, and the formation of many copies of the donor gene from one copy is known as **gene cloning.**

When the recombinant DNA molecules are not replicating, their genes are used to build proteins. Huge amounts of the donor gene's protein are synthesized. This remarkable technique for synthesizing proteins is used to mass-produce a variety of medically important proteins, some of which are listed in table 12.3.

Wellness Report

The Spread of Antibiotic Resistance Among Bacteria

Genetic engineers make use of plasmids, which are small loops of DNA that travel from one bacterium to another—even bacteria of different species. A potential danger to humans is that genes spliced into bacterial plasmids will escape from laboratory bacteria and enter bacteria that live in humans. There, such genes could damage human cells and cause serious disorders. Genetic engineers are acutely aware of this potential threat and employ laboratory techniques that prevent the escape of genetically engineered DNA.

A more serious threat from plasmids comes from the spread of antibiotic resistance, especially from the bacteria that infect domestic animals to the bacteria that infect humans. Domestic animals grow better when they are given antibiotics. No one knows for sure why this occurs, but it might be because the animals are raised in

Figure 12.A Livestock are kept in crowded pens. Cattle held in crowded feedlots gain weight more rapidly when treated with antibiotics. This method of meat production leads to antibiotic resistance in bacteria, including the kinds that infect humans.

such crowded conditions that without the medicine they would become diseased. Antibiotics, such as penicillin and tetracycline, are routinely given to cattle, poultry, and pigs (fig. 12.A).

Raising livestock with antibiotics promotes the multiplication of bacteria that are not killed by antibiotics. Some bacteria have genes that code for enzymes that inactivate one or more forms of antibiotic. When animals are fed antibiotics, the bacteria in their in-

testines that do not have the gene for resistance die, and the ones that do have the gene flourish in the absence of their competitors. In time, most of the intestinal bacteria in the animals are of the resistant form. New forms of antibiotics must then be used.

The main problem with these antibiotic-resistant bacteria is that many of the genes conferring resistance are located within plasmids; they can move from the bacteria in livestock and processed meat into the bacteria that infect humans.

Antibiotics are our main weapons against bacteria that cause human diseases. As these bacteria acquire plasmids that convey resistance to antibiotics, it becomes more and more difficult to treat the infections they cause. The bacteria responsible for gonorrhea, tuberculosis, and meningitis, for example, are now resistant to many kinds of antibiotics. Our only hope in treating these diseases is to continue to discover new antibiotics as our old ones become ineffective. This may not always be possible. Many people feel it is unwise to jeopardize human health by feeding antibiotics to domestic animals just to make them grow faster.

The Polymerase Chain Reaction

Many copies of a gene are needed to study it thoroughly. One way to get more copies to work with is through recombinant DNA technology, as described previously. A faster way is to make copies in a test tube.

The test-tube method, called the **polymerase chain reaction,** isolates a gene and then induces continuous replication of the gene. The name polymerase comes from the enzyme DNA polymerase, which forms DNA molecules from nucleotides. Once started, the replication becomes a chain reaction in which a new replication develops from each preceding one.

The Procedure   The polymerase chain reaction, diagrammed in figure 12.4, proceeds as follows. A molecule of

DNA containing the gene is placed in a test tube along with a large supply of DNA nucleotides and a heat-resistant form of DNA polymerase (harvested from bacteria that live in hot springs). The solution is heated almost to boiling, which separates the double-stranded DNA molecule into two single-stranded molecules.

Two short segments of single-stranded DNA, called primers, are added to the solution. Their sequences of base pairs are complementary to the sequences at the beginning and end of the gene. They bind with the bases at each end of the gene, bracketing the gene so that only its base sequences are replicated.

As the solution cools, the DNA nucleotides in the solution attach to the exposed bases in the single strands of the bracketed DNA: nucleotides with adenine pair with thymine in the strand, nucleotides with guanine pair with

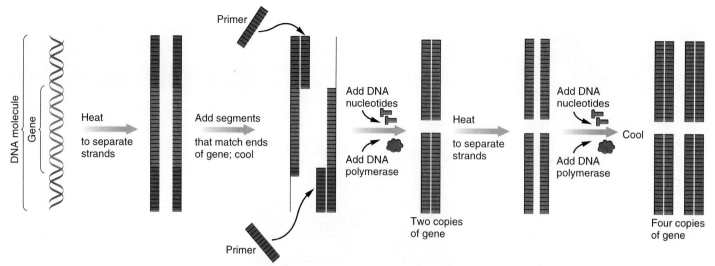

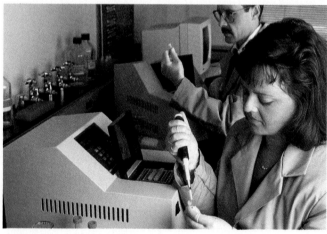

Figure 12.4 The polymerase chain reaction. DNA containing the gene is heated to make it single-stranded. Then, short pieces of single-stranded DNA (called primers), containing sequences of bases that match both ends of the gene, are added to the solution. The solution is cooled and the primers bind with their matching sequences in the DNA molecule to bracket the gene. A supply of DNA nucleotides and DNA polymerase are added and the gene replicates. The solution is then heated to separate the strands and cooled to promote another replication. Repeated heating and cooling produces as many copies of the gene as desired.

Figure 12.5 Gene-copying machine. This machine automatically carries out the DNA polymerase chain reaction. It is loaded with test tubes containing the gene and other materials. After several hours, it will have made billions of copies of the gene.

cytosine, following the base-pairing rules. DNA polymerase then connects adjacent nucleotides. Each single-stranded molecule of DNA becomes a double-stranded molecule of DNA.

The reaction proceeds rapidly. One copy of the gene becomes two copies within a few minutes. The solution is then heated and cooled again so that each newly formed copy of the gene becomes two more copies. Laboratory machines, designed to heat and cool the solutions automatically (fig. 12.5), make billions of copies of the gene in a single afternoon.

Uses of the Procedure   The polymerase chain reaction has many uses. Because it can spot and amplify a gene, the reaction is a test for the presence of small pieces of DNA, such as a particular mutation or presence of the AIDS virus. Mainly, however, the reaction is used to mass-produce genes for further study—to describe genes that cause disorders; to identify people; and even to identify organisms that have been dead for thousands or millions of years. (DNA molecules are remarkably stable, especially when dehydrated.)

A single copy of a gene, preserved within a hair, a speck of dried blood, semen, or a bone, can be multiplied into billions of copies within a few hours. The genes of Abraham Lincoln (from a lock of hair, a blood stain on a shirt, and fragments of the skull) are being analyzed to find out whether he suffered from a genetic disorder called Marfan syndrome. Even more dramatic, genes from 2,400-year-old Egyptian mummies and a 7,000-year-old man have been mass-produced for study. It may be possible to examine the genes of our Neanderthal ancestors, who lived more than 40,000 years ago, and see how these people differed from us.

To date, the oldest DNA multiplied by the polymerase chain reaction is 136 million years old. This DNA is from a weevil preserved in amber, the dehydrated and hardened resin from trees. We are still a long way from using fossilized DNA to resurrect complete organisms, such as the dinosaurs described in Michael Crichton's novel *Jurassic Park*. Nevertheless, bits and pieces of fossilized DNA can tell us a great deal about evolutionary relationships.

Describing Genes

We now have laboratory techniques for mapping the exact locations of genes within all the DNA molecules of a cell and for describing the complete sequence of bases that comprise a gene. While initial research focuses on the genes that cause major disorders, it is conceivable that in time we will accumulate a complete set of instructions for building a human. The techniques used by genetic engineers to map and describe genes are described in the following sections.

Mapping Genes

The most difficult part of genetic engineering is locating a gene. A typical cell contains enough DNA to hold more than a million genes. (If all the DNA molecules in a human cell were uncoiled and placed end to end, their total length would be 2 meters!) The genetic engineer seldom knows exactly where among all this DNA a particular gene is located. Two methods of locating genes are described here: DNA probes and restriction fragment-length polymorphisms.

DNA Probes   A **DNA probe** is a radioactively labeled (or colored) copy of one strand of a small segment of a gene. When added to a set of DNA molecules, the probe will bind with its gene and reveal (by its radioactivity) the gene's location. The technique is illustrated in figure 12.6 and described following.

A segment of the gene is heated to separate its two strands. DNA nucleotides built with radioactive atoms are added to the solution of single-stranded DNA, along with the enzyme DNA polymerase. As the solution cools, the labeled nucleotides join with unpaired bases on the single-stranded DNA molecule and DNA polymerase bonds the nucleotides to one another. The single-stranded molecules become double-stranded molecules. The newly synthesized strand made of radioactive nucleotides is the DNA probe. Its bases are complementary to the bases along one strand of the gene.

Many copies of the probe are made and added to a set of DNA molecules suspected of containing the gene. All this DNA is then heated to make it single-stranded. When cooled, the probe will combine with one of the strands of the gene to form a double-stranded DNA molecule: the sequence of bases in the probe match the sequence of bases in one strand of the gene. The location of the gene is revealed by the radioactivity on the DNA molecules.

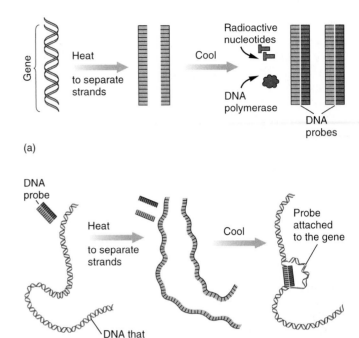

(a)

(b)

Figure 12.6 A DNA probe. (*a*) The gene is heated to make it single-stranded and then cooled in the presence of radioactive nucleotides and DNA polymerase. Bases in the nucleotides pair with bases in the single strands of DNA, and their sugar-phosphate portions are joined by DNA polymerase. Two copies of the gene form: one strand of each copy is radioactive (the probe) and the other is not. (*b*) The DNA probe, still part of a double-stranded molecule, is added to a set of DNA suspected of containing the gene. All the DNA molecules are heated to make them single-stranded. When cooled, the DNA probe combines with a strand of the gene where its bases match the sequence of bases in the gene. Radioactivity from the probe, detected by laboratory machines, reveals the location of the gene.

Restriction Fragment-Length Polymorphisms   Sometimes only the effects of a gene are known—the gene has not been isolated nor has the protein for which it codes been identified. This is the situation with most genes that cause human disorders. Such a gene is located within a set of DNA molecules by a technique known as restriction fragment-length polymorphisms, diagrammed in figure 12.7 and described following.

As described earlier, restriction enzymes cut DNA at particular sequences of bases. When a molecule of DNA is cut with one of these enzymes, the length of each fragment is the distance between these two particular sequences. In the following sketch, the length of the fragment is the distance between the sequences ATGAATT in the top strand:

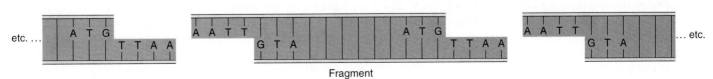

Figure 12.7 Restriction fragment-length polymorphisms. The same segment of DNA from two people—Kelly and Jason—is shown. When a particular repeat sequence in the DNA is cut with a restriction enzyme, Kelly's DNA forms three short fragments and Jason's forms a short fragment and a long one. Such fragments of different length cut from the same segment of DNA in different people are known as fragment-length polymorphisms.

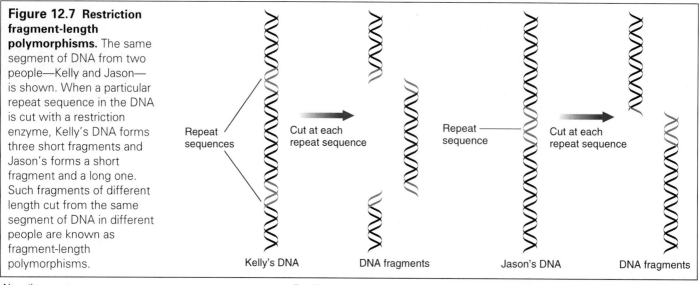

Repeat sequences

Cut at each repeat sequence

Kelly's DNA   DNA fragments

Repeat sequence

Cut at each repeat sequence

Jason's DNA   DNA fragments

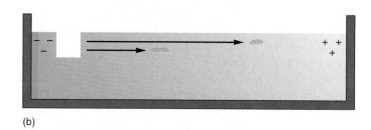

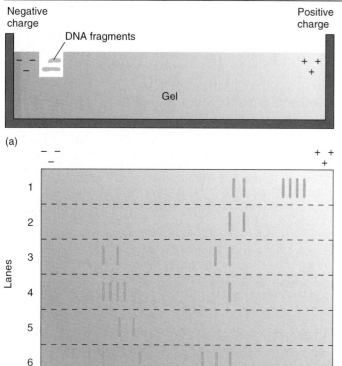

(a)

(b)

Negative charge

Positive charge

DNA fragments

Gel

(c)

Lanes

1
2
3
4
5
6

Direction of movement

Figure 12.8 Gel electrophoresis. (a) Fragments of DNA are placed in a gel that is negatively charged at one end and positively charged at the other end. The fragments, which are negatively charged, are placed at the negatively charged end and migrate toward the positively charged end. (b) Shorter fragments are lighter and move more rapidly. After a period of time, the fragments become sorted according to length, with shorter fragments nearer the positive charge. (c) This gel has been divided into six lanes to compare DNA fragments from six sources. The fragments in each lane have been sorted according to size. Lane 1, for example, has six relatively short fragments and lane 5 has two relatively long fragments.

If everyone's DNA were identical and cut by the same restriction enzyme, we would have matching sets of fragments. The lengths of our fragments differ, however, chiefly because we have different amounts of "nonfunctional" DNA—segments of DNA that do not code for proteins. These base sequences tend to be repeated. The number of times each sequence is repeated varies from person to person and gives us different lengths of DNA fragments. Variations in the length of a particular fragment are called **restriction fragment-length polymorphisms.** Restriction refers to the restriction enzyme that fragments the DNA. Polymorphism means many different forms among different individuals.

Restriction fragment-length polymorphisms can be used to locate the gene responsible for a genetic disorder. They have already been used in locating the aberrant genes that cause Huntington disease and cystic fibrosis. In the procedure, DNA molecules from members of families with histories of the genetic disorder are fragmented with restriction enzymes. The fragments are arranged according to length, using a laboratory technique known as **gel electrophoresis.**

Gel electrophoresis arranges molecules according to size (fig. 12.8). Molecules with an electrical charge move through a gel toward their opposite charge. Smaller molecules move faster than larger ones so that

Table 12.4

Correlation Between Length of a Fragment and Occurrence of a Genetic Disorder in an Imaginary Family

Family Member	Fragment Size*	Has the Disorder?
Sister Mary	Long	Yes
Brother Joe	Short	No
Brother Chris	Short	No
Father Don	Long	Yes
Mother Alice	Short	No
Grandmother Anna	Short	No
Grandfather Joe	Short	No
Grandmother Susan	Long	Yes
Grandfather James	Short	No

Conclusion: The long fragment and the gene that causes the genetic disorder are inherited together. Presence of the long fragment in an embryo or an infant can be used to predict development of the disorder.

*The long fragment and the short fragment were taken from the same segment of chromosome. The long one contains many repeat sequences whereas the short one contains few repeat sequences.

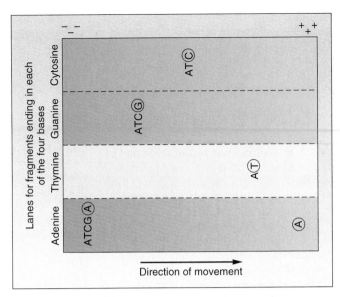

Figure 12.9 DNA sequencing. Many copies of the gene are fragmented to yield a series of fragments, each one base longer than another. The terminal base of each fragment is known. The fragments are sorted by gel electrophoresis. All fragments with the same terminal base are placed in the same lane. The fragments become sorted according to length. Knowing the lengths and the terminal bases of all the fragments are all that is needed to describe the sequence of bases in the gene. The hypothetical gene sequenced here is unnaturally short to make the technique easier to understand. The smallest fragment consists of a single nucleotide containing the base adenine. Thus, adenine is the first base in the gene. The next larger fragment has two nucleotides and ends with thymine. Thymine is the second base in the gene. The next base is cytosine, and then guanine and adenine. The base sequence of this gene is ATCGA.

after awhile the molecules become spread out in the gel according to size. DNA fragments have a negative charge and move through the gel toward a positive charge. With DNA, differences in size are due only to differences in length, since all DNA molecules have the same width.

The relative lengths of DNA fragments, determined by electrophoresis, from people with the disorder are compared to fragments from people without the disorder. A hypothetical example is shown in table 12.4. If everyone with the disorder also has the same fragment, then the defective gene is probably located near or within that fragment.

Further analysis of the fragment associated with the disorder can reveal the precise location of the gene. Both normal and defective forms of the gene can then be isolated and studied—their sequences of bases described and their contributions to cellular activity determined. This information may lead to effective ways of treating the disorder.

A dramatic use of restriction fragment-length polymorphisms is **DNA fingerprinting.** This technique for identifying people and their relatedness is described in the accompanying Research Report.

Describing the Base Sequences of Genes

A gene that has been isolated from the rest of its DNA molecule can be described in terms of its sequence of bases. In this process, called **DNA sequencing,** many copies of the gene are fragmented in a such a way that fragments of all possible lengths are formed, and the particular base at the end of each fragment is known. With these two pieces of information, the sequence of bases in the DNA can be reconstructed. This technique is diagrammed in figure 12.9 and described following.

Only one strand of the DNA is sequenced since the order of bases in the other strand can be inferred by using the base-pairing rules. To begin, one strand of the DNA is used as a template to build a complementary strand, and among the bases that go into the new strand are a few phonys—chemically altered nucleotides with bases that bind with their matching bases but halt synthesis whenever they are inserted. Synthesis proceeds in four containers, one container for phony nucleotides with each of the four kinds of bases: adenine, thymine, guanine, or cytosine.

Research Report

DNA Fingerprinting

John and Susan Gladstone had a young child and a bad marriage. They decided to divorce, but both of them wanted to keep their three-year-old daughter, Katie. Susan shocked her husband when she claimed the child was not his. Katie's biological father, she said, was the man next door. Susan planned to move in with him and take Katie with her.

John couldn't believe he wasn't Katie's father but arranged a paternity test to make sure. Samples of blood were taken from all three Gladstones as well as from the man next door, and DNA fingerprints were made of the DNA molecules in the white blood cells. A DNA fingerprint would identify with greater accuracy than any other test the biological father of Katie.

DNA fingerprinting was developed in 1985 by a British geneticist, Alec Jeffreys, at the University of Leicester. He coined the term *DNA fingerprinting* and was the first to use techniques of genetic engineering in paternity and murder cases.

Jeffreys observed that human DNA contains particular sequences of bases that are repeated many times. The repetitive sequences, which appear within genes (as introns) and between genes, do not code for proteins. The number of times the sequences are repeated varies from one person to the next. Thus, when the same DNA molecule in different people is cut with the same restriction enzyme (which cuts the DNA at the same base sequences), the lengths of the fragments vary: someone with a longer fragment has more repetitive sequences than someone with a shorter fragment. Moreover, the number of repetitive sequences, and so the fragment length, is inherited. A

Figure 12.B DNA fingerprints. Each column of bars is a DNA fingerprint. Each bar in a column represents a fragment of DNA. The vertical arrangement of bars represents the DNA fragments sorted according to length. The four fingerprints shown are from a mother (Susan), a child (Katie), and two men (the man next door and John) who claim to be Katie's father. Some of Katie's fragments are the same as her mother's and some are the same as fragments in the man next door. Her fingerprint is quite different from the fingerprint of John Gladstone. Since half the fragment lengths of a child are inherited from the mother and half from the father, it is clear from these prints that Katie's biological father is the man next door.

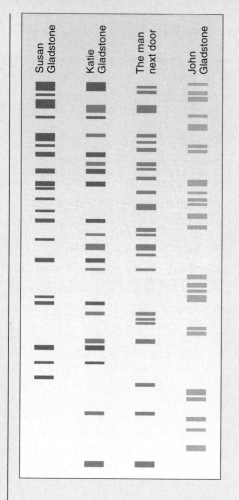

child inherits fragment lengths from both parents; the array of fragments should contain some like the mother's array and others like the father's array.

DNA molecules taken from white blood cells of the Gladstones and the man next door were fragmented by restriction enzymes and the fragments lined up according to length by gel electrophoresis. The fragments were then labeled with DNA probes (built to match the repetitive sequences) so they could be seen and photographed. The array of fragments for each person appears as a pattern of bands, similar to the bar codes used in supermarkets.

The fingerprints of Susan, Katie, and John Gladstone, as well as the man next door, are shown in figure 12.B. They reveal that the man next door is Katie's real father. Katie's fingerprint is a mixture of her mother's fragments and fragments in the fingerprint of the man next door. Note that Katie has bands found in the finger-

print of the man next door but not found in either her mother's fingerprint or John's fingerprint. John had helped raise Katie, but the man next door gave her his genes. The court ruled that Katie should live with her biological mother and father.

A DNA fingerprint is remarkably precise. Virtually everyone, except an identical twin, has a unique banding pattern. And all that is needed to identify a person is a tiny sample of blood, semen, skin, or hair. A criminal cannot afford to leave a single cell at the scene of the crime.

To see how this works, let's follow what happens in the container with adenine. Synthesis proceeds until one of the phony nucleotides is incorporated. This stops the synthesis. One DNA fragment has been built and its terminal base is adenine:

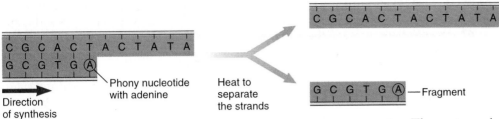

Direction of synthesis

Phony nucleotide with adenine

Heat to separate the strands

Fragment

If this process is repeated often enough, synthesis will be stopped at every place in the DNA molecule where adenine is called for. When these fragments are sorted according to length, they reveal the position of each adenine in the gene:

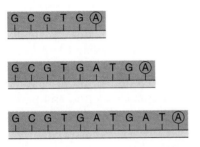

The process is repeated in the other three containers, with phony nucleotides containing thymine, guanine, and cytosine.

Gel electrophoresis arranges all the fragments within each container according to length. Knowing the fragment lengths and the terminal base in each fragment allows us to reconstruct the sequence of bases in the DNA strand. Thus, in the container containing phony adenines, the smallest fragment terminates with the first adenine in the DNA, the next smallest terminates in the second adenine, and so on. Combining information from all four containers, a complete sequence of bases in a gene can be determined.

When a gene has been sequenced, the structure of its protein can be deduced from the genetic code (📖 *see page 192*). This method is particularly valuable in identifying proteins that contribute to a disorder caused by many genes. A protein that contributes to Alzheimer's disease, for example, was discovered in this way. The method is used also to pinpoint a mutation within a gene and identify how this change affects the protein. Lastly, DNA sequences will help us understand the vast regions of DNA that appear to have no function.

The Human Genome Project

An enormous project is underway to describe our complete set of human genes—to locate all our genes and decipher their base sequences. This project was initiated in 1988 and is called the Human Genome Project (genome refers to all the DNA in a cell).

The Human Genome Project is ambitious. A human cell contains approximately 100,000 genes within DNA molecules formed of about 3 billion base pairs. The main techniques used in this project are the polymerase chain reaction, to make many copies of the DNA for many analyses; restriction fragment-length polymorphisms, to locate genes; and DNA sequencing, to describe the sequences of bases. When finished, the project will give us the complete set of instructions for making a human.

This library of genetic information will help us find new ways to diagnose and treat the more than 3,000 genetic disorders known to be caused by a single gene, as well as to fight more complex disorders, such as diabetes, heart disease, and allergies, that are caused in part by genes. Analysis of our genes may become a routine part of medical examinations.

Transplanting Genes

Genes can be transplanted from one organism to another, and, under appropriate conditions, the recipient cell synthesizes the proteins coded for by the transplanted genes. Foreign genes are not recognized; they are transcribed and translated as if they were the host cell's own genes. One reason for transplanting a gene is to correct a genetic defect. This medical treatment is called gene therapy. Another reason is to alter organisms in ways that suit human needs. The product of this kind of transplant is called a transgenic organism.

Gene Therapy

A high priority of genetic engineering is to cure genetic disorders. Every year, thousands of babies are born with defective genes that cannot produce an essential protein. The lack of just one protein can have a devastating effect on the development of a child.

The goal of **gene therapy** is to correct genetic disorders. It consists chiefly of replacing a defective gene with a

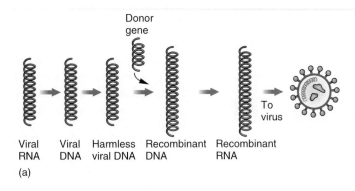

Donor
gene

To
virus

| Viral | Viral | Harmless | Recombinant | Recombinant |
| RNA | DNA | viral DNA | DNA | RNA |

(a)

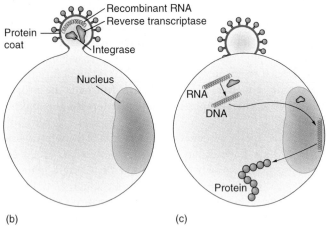

Recombinant RNA
Reverse transcriptase

Protein
coat

Integrase

Nucleus

RNA

DNA

Protein

(b)   (c)

Figure 12.10 Gene therapy. A gene is carried into a human cell by a virus, most often a retrovirus as shown here. (*a*) A DNA copy of the viral RNA is made and genes that might harm human cells are cut out. The human gene is spliced onto the viral DNA and an RNA copy is made of the recombinant DNA. (*b*) The virus inserts the recombinant RNA (donor gene plus viral genes), reverse transcriptase (converts viral RNA to DNA), and integrase (integrates viral DNA into host chromosomes) into the host cell. (*c*) A DNA copy of the recombinant RNA is synthesized and incorporated into a chromosome and its protein is synthesized. The protein corrects the genetic disorder.

normal gene. This procedure alters only the body cells—not the sperm or egg—and so the change is not inherited. The techniques of gene therapy are diagrammed in figure 12.10 and described following.

Inserting a gene into a human cell is not as easy as inserting a gene into a bacterium, for eukaryotic cells do not usually accept plasmids. The best vectors for gene therapy are viruses, since these parasites have already mastered the techniques of invading human cells. Retroviruses (see page 208) are especially good vectors because their RNA genes are transcribed into DNA and integrated into the chromosomes of the host cell. Thus, a human gene attached to the genes of a retrovirus (one that infects mice is

commonly used) can be carried into a human cell and incorporated into the cell's chromosomes. It will then be replicated, along with all the other DNA, every time the cell divides.

A danger in using viruses as vectors is that they are capable of harming cells and could spread from cell to cell and from person to person. The first step in gene therapy, therefore, is to cripple the virus by removing some of its genes. All genes with harmful effects on a cell or that enable the virus to spread are cut out.

Meanwhile, the donor gene is isolated and many copies of it are synthesized in a test tube. Each copy of the gene is spliced into a DNA molecule of the retrovirus. An RNA copy of the recombinant DNA is inserted into a virus. The retrovirus vectors are mixed with the defective human cells so that infections take place. Inside a host cell, the donor gene is copied into DNA and incorporated into a chromosome. There the donor gene directs synthesis of its protein and replicates when the cell divides.

A real stumbling block in developing gene therapy is regulating gene expression. Eukaryotic cells turn their genes on and off in complex and largely unknown ways. Most often, a gene can be transplanted into a human cell but cannot be turned on and off at appropriate times. A great effort is being made to find and understand the regions of DNA (promoters) and regulatory proteins (transcription factors) that control gene expression. The genes for curing several diseases, including hemophilia and sickle-cell anemia, are available for transfer, but the problem of getting them to produce their proteins at the right time and place has not yet been surmounted.

Gene therapy is simpler when the transplanted genes govern proteins built by white blood cells. These cells, which protect the body from infectious diseases and cancer, develop from immature cells within the bone marrow. The immature blood cells, which are not bound together, can be removed from the bone and placed in a test tube for the gene transplant (fig. 12.11). They can then be returned to the body, where they multiply and develop into white blood cells that synthesize the protein encoded by the transplanted gene. Inherited disorders of the immune system are being treated, on an experimental basis, in this way.

New methods are being developed for transplanting genes into cells without removing them from the body. A cold virus has been modified to carry corrective genes into lung cells. Disorders of the lungs, such as cystic fibrosis, may be treated by spraying such modified cold viruses into the respiratory passages. The cold sore virus, which infects nerve cells, is being modified to carry genes into brain cells as a way of treating such brain disorders as Alzheimer's and Parkinson's diseases.

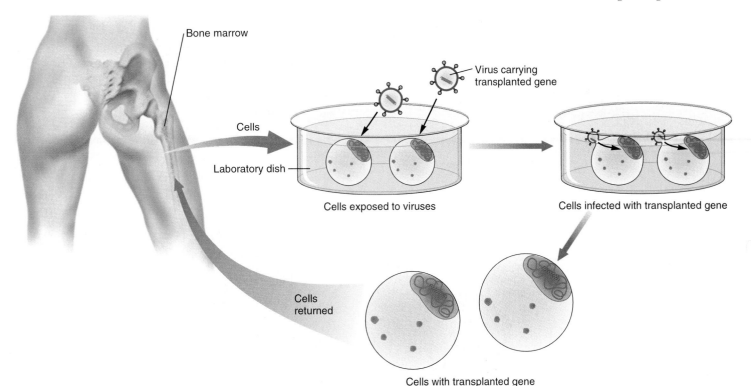

Figure 12.11 Curing blood disorders by gene therapy.

Figure 12.11 Curing blood disorders by gene therapy.
Cells in the bone marrow, which produce white blood cells, are removed and infected with retroviruses carrying copies of the normal human gene. The cells, now carrying the transplanted gene, are returned to the bone marrow where they multiply and form blood cells. Each blood cell descended from one of these bone marrow cells contains a copy of the transplanted gene and synthesizes the normal protein that cures the genetic disorder.

Transgenic Organisms

We now know how to transfer genes from one species to another—from a bacterium to a potato, from a cow to a pig, from a human to a mouse. An organism receiving genes from another species is a **transgenic organism.**

Transplanting genes into bacteria, which consist of just one cell, is done by recombinant DNA technology. Transplanting genes into every cell of a multicellular organism requires insertion of the gene (by syringe or a gun that shoots pellets coated with DNA) into a fertilized egg (fig. 12.12). All body cells, which develop from the fertilized egg, then carry the transplanted gene and it is passed on, via the sperm or egg of the adult, to the next generation.

More than seventy transgenic plants have been developed. Genes from other species have made our food plants more nutritious, more resistant to bruising and spoiling, and less vulnerable to insects, viruses, and the effects of herbicides used on weeds. One species of bacteria has been particularly useful in providing genes for proteins that kill leaf-eating insects. This bacterium, *Bacillus thuringiensis*, which normally infects the digestive tracts of insects, has genes for proteins that kill a variety of insects. Genes from this bacterium have been transplanted into tomatoes, where they protect against caterpillars (the larvae of butterflies and moths); potatoes, where they protect against potato beetles (fig. 12.13); and algae, where researchers hope the genes will kill mosquito larvae that eat the algae.

Transgenic animals have been developed as livestock and as laboratory animals for experiments. Domestic pigs have been given a gene, from cattle, for a growth hormone. The transgenic pigs grow faster to become larger and leaner than normal pigs. Unfortunately, they also have a variety of health problems, including lower fertility, arthritis, and ulcers. Researchers have made transgenic mice that are extremely prone to breast cancer. The mice are used in cancer research. Human genes have been transplanted to other mammals—goats, sheep, cows, and pigs—where their protein is synthesized and secreted in the animal's milk. Proteins already synthesized in this way include human hemoglobin (for transfusions), human antibodies, and various proteins for treating emphysema, heart attacks, and hemophilia.

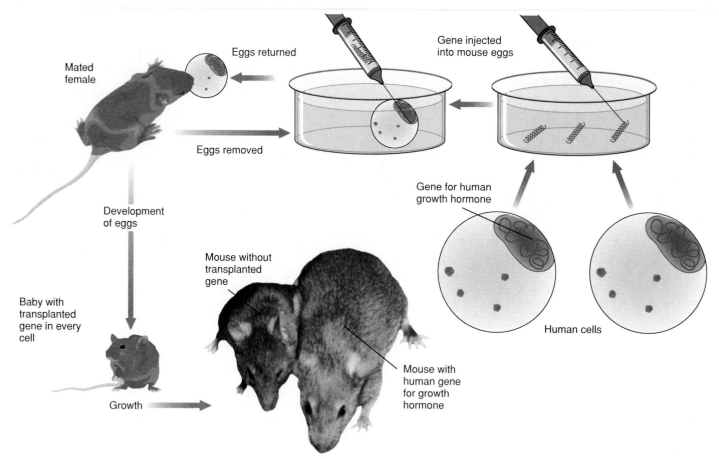

Figure 12.12 Creating a transgenic animal. In this example, a human gene that produces human growth hormone is transplanted into mice. A female mouse is mated and the fertilized eggs transferred to a dish. Copies of the human gene are injected into the mouse eggs, which are then returned to the mouse's uterus. Each fertilized egg develops into an offspring with a copy of the transplanted gene in every cell. An offspring with the gene for human growth hormone synthesizes the hormone and grows to be twice as large as a mouse without the transplanted gene. Moreover, its sperm or eggs also carry copies of the transplanted gene, so it is passed from generation to generation within the family line.

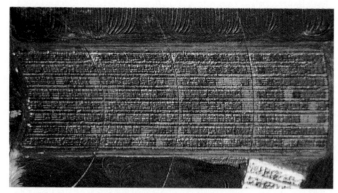

Figure 12.13 Effects of a transgenic plant. This aerial view (using infrared light) is of a potato field used in an experiment to test the effects of a bacterial gene. Green areas once held ordinary potato plants, which were consumed by potato beetles. Red areas are potato plants that were exposed to the same density of potato beetles but provided with a gene, from a bacterium, that kills potato beetles. (The white patch consists of wheat plants used in an unrelated experiment.)

SUMMARY

1. Genetic engineers use laboratory techniques to manipulate genes.
2. In recombinant DNA technology, a gene is transferred into a bacterium for mass production of its protein or gene. The donor gene is cut from its DNA molecule by a restriction enzyme and spliced into a bacterial plasmid. The plasmid invades a bacterium, where the donor gene is replicated and its protein synthesized.
3. The polyermase chain reaction makes many copies of a gene in a test tube.
4. Genetic engineers locate specific genes within the DNA of a cell. A DNA probe is a radioactively labeled, single-stranded copy of a gene that binds with a gene and reveals its position. Restriction fragment-length polymorphisms locate a gene by the appearance of certain fragments in family members with the trait but not in members without the trait.

5. DNA sequencing reveals the sequence of bases that forms a gene. The DNA is broken into many fragments, each one base longer than another and with the terminal base identified. Arranging the fragments according to length then gives the sequence of bases in the DNA.

6. The Human Genome Project is designed to map and sequence all human genes.

7. A gene can be transplanted from one cell to another, where its protein is synthesized. In gene therapy, a normal gene is carried by a virus from one human to another to replace a defective gene. The gene is not passed to offspring. In transgenic organisms, a gene is transplanted from one species to another, where it produces new traits. The gene occurs in the sperm or egg and is passed to offspring.

KEY TERMS

DNA fingerprinting (page 221)
DNA probe (page 219)
DNA sequencing (page 221)
Donor gene (page 213)
Gel electrophoresis (ee-LEK-troh-for-EE-sus) (page 220)
Gene cloning (page 216)
Gene therapy (page 223)
Genetic engineering (page 212)
Plasmid (PLAZ-mid) (page 216)
Polymerase (poh-LIH-muhr-ayz) **chain reaction** (page 217)
Recombinant (ree-KAHM-bih-nant) **DNA technology** (page 212)
Restriction enzyme (page 213)
Restriction fragment-length polymorphism (PAH-lee-MOHR-fiz-um) (page 220)
Transgenic (trans-JEE-nik) **organism** (page 225)
Vector (page 213)

STUDY QUESTIONS

1. Describe a restriction enzyme and explain how it is be used to join two pieces of DNA (e.g., a donor gene and a vector).
2. Describe how proteins are mass-produced within bacteria.
3. Describe the polymerase chain reaction.
4. What is a restriction fragment-length polymorphism? How is it used to locate a gene? How is it used in DNA fingerprinting?

5. Explain how a description of the base sequences within a gene can be used to predict the arrangement of amino acids in the gene's protein.
6. Distinguish between gene therapy and the formation of transgenic organisms.

CRITICAL THINKING PROBLEMS

1. Suppose that a new genetic disorder has been observed, but its genes have not been isolated or its proteins identified. What technique would you use to locate the abnormal gene within all the DNA molecules of a human cell? Why?
2. Suppose that a genetic disorder has been observed, but the only thing known about the gene is that part of its DNA sequence is ATTGGGCGTA. What technique would you use to locate the abnormal gene within all the DNA molecules of a human cell? Why?
3. An enormous amount of government money is being spent on the Human Genome Project. What benefits can we expect from this information?
4. Suppose you were responsible for deciding how gene therapy is used. Would you permit gene therapy for curing diseases? For improving physical appearance? For improving mental capacity? Would you permit the transfer of genes into fertilized eggs so that they will be passed from generation to generation? For each answer, explain why you would or would not give your approval.

SUGGESTED READINGS

Introductory Level:

Bains, W. *Genetic Engineering for Almost Everybody.* London: Penguin Books, 1987. A presentation of the techniques and applications of genetic engineering that is easy to read yet does not gloss over the details.

Baskin, Y. "DNA Unlimited," *Discover,* July 1990, 77–79. A history of how the polymerase chain reaction was discovered and how it is being used.

Franklin-Barbajosa, C. "The New Science of Identity," *National Geographic,* May 1992, 112–24. A description of DNA fingerprinting and discussion of its many uses.

Gasser, C. S., and R. T. Gasser. "Transgenic Crops," *Scientific American,* June 1992, 62–69. A description of genetically engineered plants and a discussion of the environmental soundness and commercial viability of such procedures.

Montgomery, G. "The Ultimate Medicine," *Discover,* March, 1990, 60–68. A history of the first gene transfer using viruses.

Advanced Level:

Grosveld, F., and G. Kollias (eds.). *Transgenic Animals*. San Diego, CA: Academic Press, 1992. A summary of the research, particularly techniques, involving the transfer of genes into animals.

Micklos, D. A., and G. A. Freyer. *DNA Science*. Cold Spring Harbor, NY: Cold Spring Harbor Laboratory Press, Carolina Biological Supply Company, 1990. Chapters 1 through 8 are a review of DNA and descriptions of the techniques used in genetic engineering. (The remaining chapters are laboratory exercises.)

Mulligan, R. C. "The Basic Science of Gene Therapy," *Science*, May 14, 1993, 926–32. A review of the techniques used in gene therapy, with particular emphasis on the difficulties in developing this technique for medical use.

Mullis, K. B., F. Ferré, and R. A. Gibbs (eds.). *The Polymerase Chain Reaction*. Cambridge, MA: Birkhäuser, 1994. A description of the technique and discussion of its application and impact in various fields of biology.

Verma, I. M. "Gene Therapy," *Scientific American*, November 1990, 68–84. A description of the difficulties and prospects of gene therapy as well as the kinds of disorders that are easier to treat in this way.

EXPLORATIONS

Interactive Software: Genetic Engineering

This interactive exercise explores the technique used to isolate a gene for cloning. The student is provided with a variety of real gene sequences and learns how to choose an appropriate restriction enzyme for removing the gene from its chromosome. The size of the gene and the number of times it appears on a chromosome can be varied to see how these variables affect the isolation of a gene by means of a restriction enzyme.

Interactive Software: Constructing a Restriction Site Map

This interactive exercise explores how a chromosome is mapped by means of restriction fragment-length polymorphisms. The student is provided with information about the relative sizes of DNA fragments, acquired from gel electrophoresis, and uses this information to construct physical maps of genes within a chromosome.

Explorations in Cell Biology, Metabolism and Genetics **CD-ROM**

DNA Fingerprinting: You be the Judge (*See* Module 14)

Biology
of Plants

All the cells of a plant work together to promote survival and reproduction. Trees along the Carp River in the Porcupine Mountains of Michigan prepare for winter.

PART THREE

Exploring Themes

Theme 1

The fundamental unit of life is the cell.
A plant is formed of many cells (chapter 13).

Theme 2

All cells carry out the same basic processes.
All plant cells synthesize molecules, import and export materials, acquire energy from organic molecules, reproduce, and synthesize proteins according to instructions within their genes (chapter 13).

Theme 3

The cells of a multicellular organism produce adaptations: anatomical, physiological, and behavioral traits that promote survival and reproduction.
Each cell of a plant is specialized for a particular function: cells of the leaves build sugars (chapter 13); cells of the xylem and phloem transport fluids (chapter 14); cells of the flower form embryos (chapter 15); cells of the meristem generate new cells for plant growth (chapter 15); and secretory cells form hormones that integrate plant activities and coordinate them with environmental events (chapter 16).

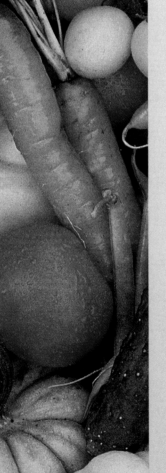

Plant organization is familiar.

Plant Organization

Objectives

In this chapter, you will learn
–how a plant differs from other
  organisms;
–that a plant consists of a variety of
  specialized cells;
–the way in which specialized cells
  form tissues;
–how tissues form leaves, stems, and
  roots;
–how the activities of a plant's cells are
  coordinated.

What does the word *plant* bring to mind? A potted geranium? A wild rose? A maple tree? Or does it bring to mind the produce section of your grocery store? After all, most of us eat some form of plant at every meal.

Vegetables are the leaves, stems, and roots of plants. Leaf blades in our diet include lettuce and spinach. Leaf stalks, which connect leaves to stems, include celery and rhubarb. We eat immature leaves (buds) in the form of brussels sprouts and sugar-storing leaves (bulbs) in the form of onions. A stem in our diet is asparagus (its leaves are tiny, flat, and pressed against the stem). Other stems are white potatoes and kohlrabi, which contain large quantities of starch. We also eat a variety of roots, such as carrots, radishes, beets, and sweet potatoes.

We even eat the reproductive organs of plants. Sexual organs (flowers) in our diet include cauliflowers, broccoli, artichokes, and figs. We eat embryos, along with their starchy food supply, in the form of seeds—corn, wheat, rice, rye, beans, peas, and peanuts. And we eat plant ovaries when we eat fruits. Oranges, melons, bananas, and berries are plant ovaries, as are squash, eggplant, cucumbers, and tomatoes.

Not all parts of a plant are edible: some parts are poisonous and others are indigestible. Of the millions of plants around us, the few that we eat were discovered by our hunter-gatherer ancestors. These people hunted wild animals and gathered wild plants for food. They learned by trial and error what was edible and nourishing and what was poisonous or indigestible. Their knowledge about wild plants has greatly influenced the kinds of plants we consume today.

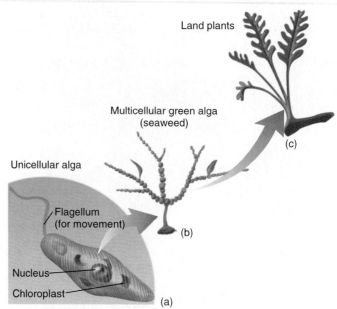

Figure 13.1 Plants evolved from unicellular algae. (*a*) A unicellular alga has a eukaryotic cell, with a variety of organelles including chloroplasts that carry out photosynthesis. The entire cell is enclosed within a cell wall. (*b*) Approximately 1.4 billion years ago, some unicellular algae evolved into multicellular green algae, or seaweeds. The body of one of these primitive plants consisted of a string of cells, with one cell attached to some underwater surface, like a rock. (*c*) Approximately 430 million years ago, some seaweeds colonized land and developed adaptations for life in this very different environment.

Origin of Plants

A plant is a multicellular organism with eukaryotic cells. All plants belong to the kingdom Plantae. They differ from multicellular organisms in other kingdoms (Animalia and Fungi) in their capacity for photosynthesis (📖 *see page 106*)—for building sugars from carbon dioxide, water, and sunlight.

From Unicellular Algae to Seaweeds

Plants evolved from unicellular, eukaryotic algae that lived in the oceans. The evolutionary sequence is diagrammed in figure 13.1. Like plants, these algae had cell walls and chloroplasts, which hold the biochemical pathways of photosynthesis. They were members of the kingdom Protista (see inside back cover of this book).

Unicellular algae were well adapted to life within the calm waters of the open ocean. Selection for multicellularity arose when they began to colonize the more turbulent

waters of coastal shores. In such waters, an organism survives and reproduces better if it can attach itself to a rock. In this way the organism becomes fixed in place and the turbulent water flows over its surface, providing it with nutrients and removing wastes. Algae rooted themselves in place by becoming multicellular, with one cell specialized for attachment and the others for photosynthesis.

These first plants were green algae, or seaweeds. Each plant consisted of just a few photosynthetic cells that remained together after cell division rather than separating to lead independent lives as unicellular algae. Plants appear in the fossil record approximately 1.4 billion years ago—almost a billion years after the first unicellular algae (see inside back cover).

Colonization of Land

Seaweeds colonized land approximately 430 million years ago. At that time the continents were flat, with gradually descending shores, and the climate was warm and humid. Small decreases in sea level periodically uncovered large regions of coastal land, exposing the plants to warm, moist air.

Most of these early land plants simply dried out and died when exposed to air. Every generation, however, a few individuals survived because they had traits that helped them retain water. These individuals, in turn, left offspring with the same traits.

In time, so many adaptations to life in the air accumulated that these new forms of plants could live entirely out of water. Their cells became specialized for different activities: roots for absorbing water and ions from the soil; leaves for absorbing light and carbon dioxide from the air; and a stem for positioning the leaves in sunlight and transporting materials between the roots and the leaves.

Cells and Tissues

A plant cell has the organelles typical of other eukaryotic cells: a nucleus, mitochondria, endoplasmic reticulum, Golgi apparatus, vesicles, and ribosomes. In addition, a typical plant cell has chloroplasts, a water vacuole, and a cell wall (fig. 13.2).

A chloroplast is a highly structured, membrane-enclosed sac that contains the pigments and enzymes of photosynthesis. A water vacuole is a water-filled sac that distends the cell and gives it shape. The vacuole may occupy most of the cell interior. The cytoplasm of a cell and its vacuole contain a high concentration of potassium ions (K^+), which continuously draw in water by osmosis (see page 76). The cell swells until expansion is resisted by the cell wall:

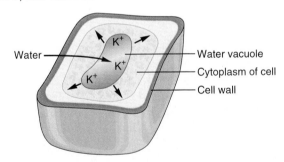

The two forces—inflow of water and resistance by the wall—generate a pressure, known as turgor pressure, that maintains the cell shape. Without the force of inflowing water, the cell collapses and the plant wilts.

In addition to its role in generating turgor pressure, the cell wall supports the cell it encloses, and all the cell walls together support the plant as a whole. Each mature plant cell has a primary wall, formed of polysaccharides—chiefly cellulose but often reinforced with pectin, which adds strength and flexibility to the wall and binds together the walls of adjacent cells.

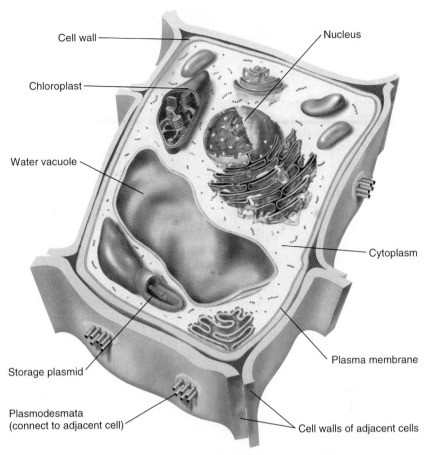

Figure 13.2 A typical plant cell. A plant cell is characterized by its chloroplasts, water vacuole, and cell wall. Some plant cells, like the one shown here, have just a primary wall and others have both a primary and a secondary wall. Adjacent cells are typically glued together by materials in their cell walls and interconnected by plasmodesmata.

Some cells have a secondary wall inside the primary wall:

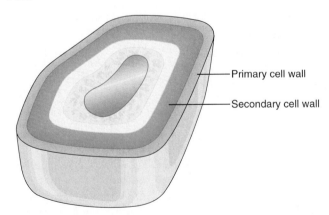

The secondary wall is stronger than the primary wall. Its cellulose foundation is reinforced with a more rigid polysaccharide called lignin. Cells that form rigid structures, such as tree trunks, have secondary walls with lignin.

Table 13.1

Major Types of Plant Cells

Cell Type	Structure	Functions
Parenchyma	Storage sac / Thin primary wall	Photosynthesis; storage; secretion
Collenchyma	Water vacuole / Thick primary wall	Support of leaves, flowers, and growing parts
Sclerenchyma	Fiber: Thick primary wall / Thick secondary wall. Sclereid: Thick primary wall / Thick secondary wall	Walls of dead cells support nongrowing parts
Tracheid cell	Pit / Thick primary wall / Thick secondary wall	Walls of dead cells form channel for transport of water
Vessel member	Hole / Thick primary wall / Thick secondary wall	Walls of dead cells form channel for transport of water
Sieve tube member	Hole / Thick primary wall / Thick secondary wall / Cytoplasm	Living cells form channel for transport of sugar and other cell products

An advantage of multicellularity is that each cell can be specialized for carrying out one or a few functions. A plant has a variety of cell types, each unique in structure and function.

Basic Cell Types

As a plant cell develops, it becomes specialized in structure and function. All the cells of a plant have the same genes; a cell becomes specialized as some of its genes are activated and others inactivated. Each kind of cell expresses only certain of its genes—synthesizes certain kinds of enzymes, membrane proteins, and other proteins. By activating only certain genes within each cell, a plant generates a variety of specialized cells. The main kinds of plant cells are described in table 13.1 and in the following paragraphs.

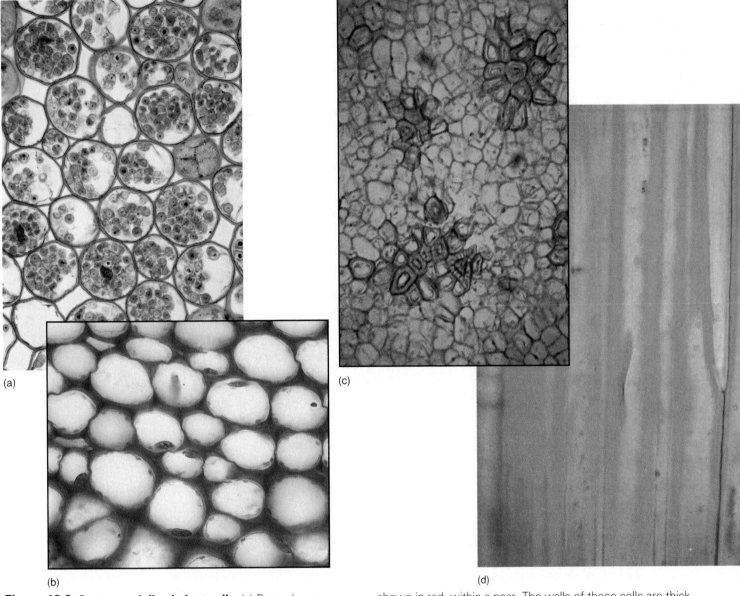

(a)

(b)

(c)

(d)

Figure 13.3 Some specialized plant cells. (a) Parenchyma cells from a root. The red objects inside the cells are grains of starch. (b) Collenchyma cells, which provide support, have thick cell walls. (c) Sclereid cells (a form of sclerenchyma cell), shown in red, within a pear. The walls of these cells are thick and hard. (d) Fiber cells (a form of sclerenchyma cell) are elongated and very strong.

Parenchyma Cells   A plant is formed chiefly of **parenchyma cells.** A typical parenchyma cell is roughly spherical in shape with a thin, flexible primary wall and no secondary wall (fig. 13.3a). It is the least specialized of plant cells and retains its ability to multiply and to develop into other kinds of cells. Formation of new cells is not, however, the major function of a parenchyma cell; it does so only to replace an injured part.

Parenchyma cells in the leaves have chloroplasts and specialize in photosynthesis. Parenchyma cells in the stem and root have sacs for storage of water, starches, and oils. They also secrete (synthesize and release) hormones, nectar, resins, and special molecules that poison or in other ways deter herbivores (animals that eat plants). We have learned to use many of these antiherbivore molecules as medicines, stimulants, and hallucinogens. The use of natural drugs is described in the accompanying Wellness Report.

Collenchyma Cells   Collenchyma cells provide flexible support to the leaves, flowers, and growing parts of a plant. A collenchyma cell has a thick primary wall, reinforced

Medicinal Effects of Plants

As our hunter-gatherer ancestors searched for edible foods, they discovered plants that produced unusual effects on their minds and bodies. In time, they learned to identify and use these plants to treat specific illnesses, to stimulate their minds, and to produce hallucinations. Each generation, the knowledge was refined and then passed on to the next generation. Hunter-gatherer societies that still exist today often use several hundred plants to treat illnesses and to alter the mind.

We now know that the active ingredients in medicinal plants evolved as chemical defenses against herbivores, especially insects. As many as 100,000 such molecules exist. These unusual molecules, evolved to alter the bodies of insects, have profound effects on human bodies as well.

Consider aspirin, a miracle drug used for the treatment of pain, fever, inflammation, and headaches. Salicin, a molecule similar to aspirin, occurs naturally in the leaves of myrtle (a Mediterranean bush), the leaves and bark of willow trees, the oil of wintergreen, and several other plants. We have evidence, in the form of written prescriptions found in Egypt, that myr-

tle leaves were used to treat uterine pains as many as 3,500 years ago. Hippocrates, of ancient Greece, prescribed myrtle leaves to reduce fever. Later, in A.D. 30, the Romans used willow leaves to treat inflammation. In the Middle Ages, a derivative of salicin was used to treat headaches. We still use salicin, in the form of aspirin, to treat pains, fever, inflammation, and headaches.

Consider another example—digitalis, a drug that strengthens heart contractions and synchronizes the beat. In 1785, a British physician reported that ingestion of dried leaves from the foxglove plant eased dropsy, a disorder in which fluid accumulates in the legs because the heart fails to pump blood adequately. The doctor explained, "I was told that this use of foxglove had long been kept a secret by an old woman in Shropshire, who had sometimes made cures after the more regular practitioners had failed." The active ingredient in foxglove is digitalis.

Today, drug companies and researchers seek medicinal plants in all corners of the world. With more than 450,000 plant species to examine, the search must necessarily be focused on plants with high probability of yielding useful drugs. How do we identify such plants? By consulting indigenous healers—local people whose ancestors have lived in the area for genera-

Figure 13.A A healer collecting medicinal plants in Belize, Central America. The leaves and flowers of this plant (*Cornutia pyramidata*) are used to treat skin rashes.

tions and accumulated knowledge about the effects of local plants on human health. The knowledge such healers possess is quite remarkable (fig. 13.A).

Samoan healers treat yellow fever (a viral disease) with brews made by steeping the wood of a certain tree in water. Researchers have discovered that the active ingredient in this brew is effective against the virus that causes AIDS. In Thailand, healers use

---

with pectin and uneven in thickness (fig. 13.3b). It has no secondary wall. Collenchyma cells form long strands within leaves, helping them to lie flat, and thin sheets just beneath the surface of growing stems, helping them to remain erect yet bend with the wind. The "strings" of a celery stalk, for example, are made of these cells.

Sclerenchyma Cells   **Sclerenchyma cells** give rigid support to nongrowing parts. They have both primary and secondary cell walls and are stronger and more rigid than the walls of other cells.

A sclerenchyma cell does not begin to function until it dies. Its cell walls, which remain after death, give sup-

Table 13.A

Drugs Developed from Plants Used by Healers

Drug	Medical Use	Plant Source
Aspirin	Reduces pain, headache, fever, and inflammation	*Filipendula ulmaria* (meadowsweet)
Codeine	Eases pain; suppresses coughing	*Papaver somniferum* (opium poppy)
Ipecac	Induces vomiting	*Psychotria ipecacuanha* (perennial herb)
Pilocarpine	Reduces pressure in the eye	*Pilocarpus jaborandi* (South American shrub)
Pseudoephedrine	Reduces nasal congestion	*Ephedra sinica* (an evergreen shrub)
Quinine	Combats malaria	*Cinchona pubescens* (a tropical tree)
Reserpine	Lowers blood pressure	*Rauwolfia serpentina* (Indian snakeroot)
Scopolamine	Eases motion sickness	*Datura stramonium* (Jimson weed)
Theophylline	Opens bronchial passages	*Camellia sinensis* (tea plant)
Vinblastine	Combats Hodgkin's disease	*Catharanthus roseus* (rosy periwinkle)

the roots of a plant related to ginger to ease stomach pains and other gastrointestinal disorders. A chemical in this plant is now known to kill parasitic worms, such as tapeworm and hookworm. In India, healers treat snakebites and mental illness with the roots of a climbing shrub called Indian snakeroot. The active ingredient in these roots is reserpine, now prescribed throughout the world for high blood pressure. In Peru, the Jivaro Indians use sap from a particular tree to make wounds heal more rapidly. The active ingredient in this sap is now being tested for its healing properties.

Materials isolated from plants now account for about 25% of all prescriptions issued in North America. Some of these drugs are listed in table 13.A. How do we thank the local healers, in various isolated regions of the world, for this invaluable knowledge? Many drug companies and research groups are now arranging to provide these peoples with a share of the royalties earned from the drugs. Many of these cultures, however, value their land and its forests more than they value money. Hence, in many places, such as Costa Rica and Samoa, the royalties from drugs are used to preserve local forests.

port and shape to the plant. There are two kinds of sclerenchyma cells: sclereid cells and fiber cells. Sclereid cells come in a variety of shapes. They may be scattered throughout a structure, as in a pear (making this fruit gritty) (fig. 13.3c), or arranged in sheets, as in a peach pit or walnut shell. Fiber cells are long, slender (fig. 13.3d), and often arranged in bundles to form very strong materials. They are most abundant in the wood and bark of trees. Plant materials with high densities of fiber cells are used to make baskets, ropes, and linen.

Tube-Forming Cells A plant transports materials through rigid tubes, of which there are three kinds: tracheids, vessels, and sieve tubes. Each tube is a stack of specialized cells.

A **tracheid** is a tube formed of tracheid cells. These long, narrow cells have small pits in the sides of their end walls, where they contact neighboring cells. The pits are thinner regions, but not actual holes, in the cell walls. A tracheid cell dies before it becomes functional, leaving only its walls to form part of a tube:

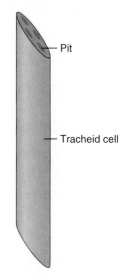

Fluids flow through the tube, passing through the pits from one dead tracheid cell to the next.

A **vessel** is a tube made of vessel members, which are shorter and broader than tracheid cells. Vessel members, like tracheid cells, are dead when functional. Holes in their end walls allow fluids to flow freely from one dead cell to the next:

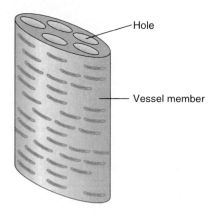

A **sieve tube** is made of sieve tube members. These cells form tubes by losing their nuclei and most organelles rather than by dying. The remaining cytoplasm lies next to the cell walls, leaving room for fluids to pass through the center. Fluids flow from one cell to the next through holes in the end walls:

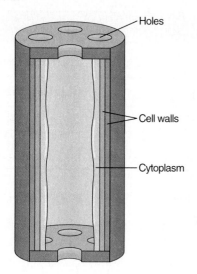

Plant Tissues

The specialized cells of a plant are organized into tissues. A **tissue** is a group of cells that lie next to one another and work together to carry out a particular task. The four basic plant tissues—vascular, dermal, ground, and meristem—are described in table 13.2 and in the following paragraphs.

Vascular Tissues Vascular tissues form tubes that transport fluids and support the plant in an upright position. These tissues are built of tube-forming cells as well as parenchyma cells (for storage of materials) and sclerenchyma cells (mostly fiber cells, for added strength). There are two kinds of vascular tissue: xylem and phloem.

The **xylem** consists of tubes that extend from the roots to the leaves. It transports water and inorganic ions from the soil to all parts of the plant. In most flowering plants, the xylem is formed of two kinds of tubes: tracheids and vessels (fig. 13.4). In ferns and conifers, the xylem is formed of tracheids only. In addition to tracheids and vessels, the xylem contains parenchyma cells and fiber cells (i.e., sclerenchyma). The parenchyma cells store starch and ions and the fiber cells provide support.

Table 13.2

Major Types of Plant Tissues

Tissue Type	Kinds of Cells	Function
Vascular		
Xylem	Tracheid cells; vessel members; parenchyma cells; sclerenchyma cells	Transport water and inorganic ions
Phloem	Sieve tube members; companion cells; parenchyma cells; sclerenchyma cells	Transport sugars and other cell products
Dermal	Parenchyma cells	Protect
Ground	Parenchyma cells	Support, store, secrete
Meristem	Immature cells	Form new parts

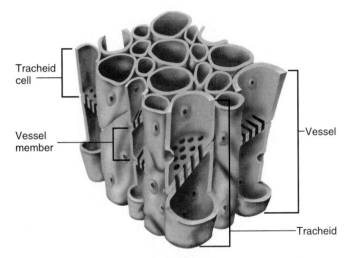

Figure 13.4 Xylem. Xylem in a flowering plant is a cluster of two kinds of tubes: tracheids and vessels. Tracheids are formed of tracheid cells and vessels are formed of vessel members. The cells are dead; only their cell walls remain to form tubes.

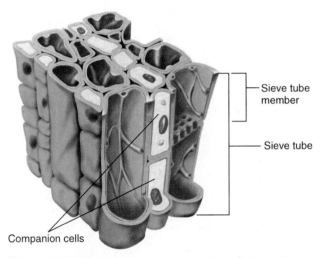

Figure 13.5 Phloem. Phloem is a cluster of sieve tubes, formed of sieve tube members. These living cells, which have lost their nuclei and other structures, receive life-sustaining materials from companion cells.

Wood consists of xylem. The wood of pines and other conifers (called softwood) is softer than the wood of flowering plants (called hardwood) because it typically lacks both vessels and fiber cells.

The **phloem** consists of tubes that transport organic molecules, especially sugars, from one part of the plant to another. It is formed of sieve tubes. Special parenchyma cells, called **companion cells,** lie next to the sieve tube members and synthesize materials for them (fig. 13.5). These two kinds of cells are interconnected by extensions of their cytoplasm called plasmodesmata. Like the xylem, the phloem also contains parenchyma cells for storing materials and fiber cells for support.

The xylem and phloem lie next to each other, forming a **vascular bundle.** In a leaf, the fiber cells of the xylem

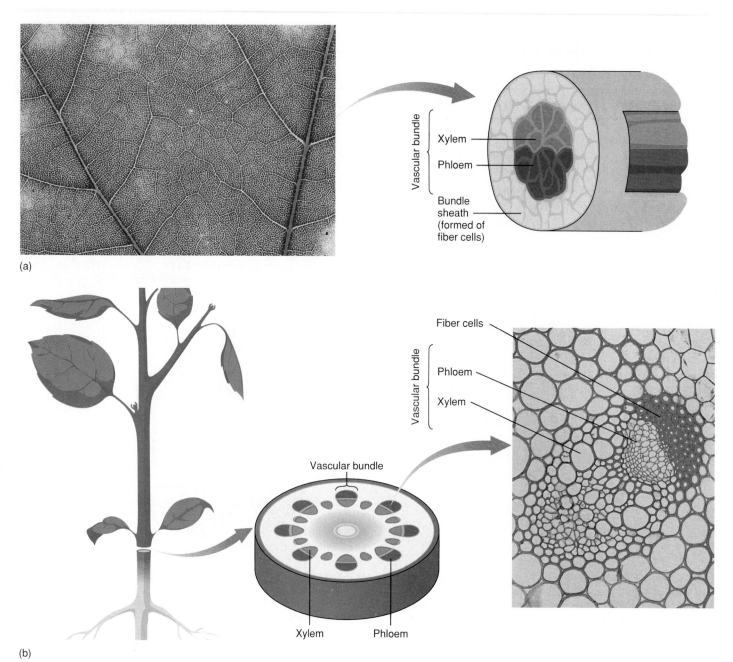

Figure 13.6 Vascular bundles. A vascular bundle consists of xylem and phloem. (*a*) The veins of a leaf are vascular bundles enclosed within bundle sheaths (formed of fiber cells).

(*b*) The vascular bundle of a stem is supported by fiber cells that lie along one side.

and phloem are arranged as a **bundle sheath** (fig. 13.6*a*), which surrounds the bundle. (The bundle and sheath are easily seen as the so-called veins in a leaf.) Elsewhere, the fiber cells lie only along the phloem side of the bundle (fig. 13.6*b*).

Dermal Tissue Dermal tissue is a plant's outer covering in contact with the environment. Its function is to protect the plant from water loss, injury, and invasions by bacteria,

fungi, and insects. This tissue consists mainly of tightly packed parenchyma cells. There are two kinds of dermal tissue: epidermis and periderm.

Epidermis is a single layer of flattened cells (fig. 13.7). These cells lack chloroplasts, enabling light to penetrate and reach the photosynthetic cells beneath them. In a young plant, epidermis covers all parts; in an adult plant, it covers the leaves and roots but is replaced by a thicker tissue in the stem.

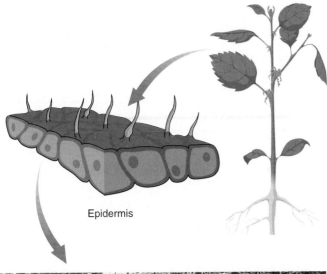

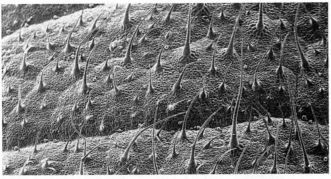

Figure 13.7 Epidermis. The surface of a leaf consists of closely packed cells called the epidermis. This leaf of a stinging nettle has some of its epidermal cells modified to form hairs that release irritating chemicals when touched.

Epidermal cells in the leaves and stem secrete a waxy coating, called the **cuticle,** that restricts water loss. The ability to secrete this coating must have been an early adaptation to life on land. Groups of epidermal cells may form elongated structures, called hairs, that reflect light (to prevent overheating) and restrict water loss. In some plants these hairs are sticky or toxic, which discourages chewing by insects. Epidermal cells of the root do not form a cuticle, for it would interfere with water uptake, but typically form elongated extensions, called root hairs, that increase the surface area for absorption of water.

Periderm replaces epidermis in the stems of adult trees and shrubs. It forms the outer bark. Periderm is several layers thick and continuously replaces itself as the stem widens with age. The characteristics of periderm are described in chapter 15.

Ground Tissue The bulk of a plant consists of **ground tissue,** formed chiefly of loosely packed parenchyma cells that fill the spaces between dermal tissue and vascular tissue. It provides support, stores starch, and secretes materials such as hormones. In the stem and root, ground tissue

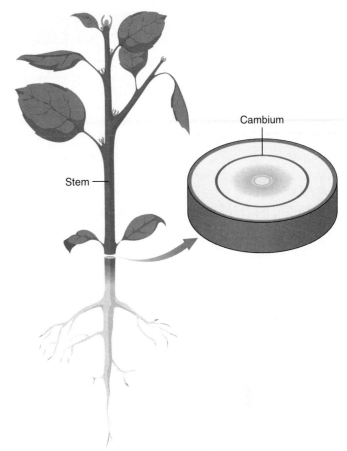

Figure 13.8 Meristem. A plant has two types of meristem: apical and cambium. Apical meristem (shaded red) occurs at the tips of the stem and its branches as well as at the tips of the root and its branches. As the stem grows taller, some apical meristem moves upward and some remains in place, between the stem and leaf, where it can form new branches, leaves, or flowers. Cambium (shaded blue) occurs as cylinders of meristem within the stem and root; its function is to widen these structures.

that lies to the inside of vascular tissue is called **pith** and ground tissue that lies to the outside is called **cortex:**

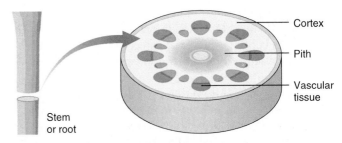

Meristem   Meristem is immature tissue located in special regions of a plant. It provides new cells for plant growth. When a meristem cell divides, one of the newly formed cells becomes part of a new structure and the other remains in the meristem, where it can continue to form more cells. A plant has two kinds of meristem: apical and cambium (fig. 13.8).

241

Table 13.3

Plant Structures: Dicotyledons Versus Monocotyledons

Part	Dicotyledon	Monocotyledon
Number of embryonic leaves	Two	One
Adult leaf	Two kinds of mesophyll; branching vascular bundles	One kind of mesophyll; parallel vascular bundles
Leaf attachment	By petiole	Typically by leaf base
Stem	Vascular bundles form a ring	Vascular bundles scattered
Root	Usually tap root; xylem forms center	Always a fibrous root; pith forms center

Apical meristem is located at or near the tip (apex) of a stem, branch, or root. It adds cells that lengthen the structure, allowing a stem or branch to grow toward light and a root to grow toward moisture. Apical meristem also adds new leaves, branches, and flowers (or cones) to the shoot. Cambium occurs as cylinders of immature cells that add new xylem, phloem, and periderm to both the shoot and root, as well as new branches to the root. These remarkable tissues are discussed in chapter 15.

The Plant Body

The major categories of plants—seaweeds, bryophytes (such as mosses), ferns, conifers, and flowering plants—have different body forms. Rather than compare all these plant structures, we focus here on the form of a typical flowering plant.

There are two major categories of flowering plants: **dicotyledons** (*di*, two; *cotyledons*, embryonic leaves) and **monocotyledons** (*mono*, one), commonly referred to simply as dicots and monocots. Examples of dicots are oak trees, raspberry bushes, violets, and dandelions. Examples of monocots are palm trees, onions, lilies, and grasses. The differences between these two categories are listed in table 13.3 and discussed here as we examine each part of the plant body.

The body of a flowering plant has a shoot and a root (fig. 13.9). The shoot consists of aboveground parts: the leaves, the stem and its branches, as well as flowers and fruits during the reproductive season. The shoot acquires resources from the sun and the air. The root consists of underground parts that anchor the plant and acquire resources from the soil.

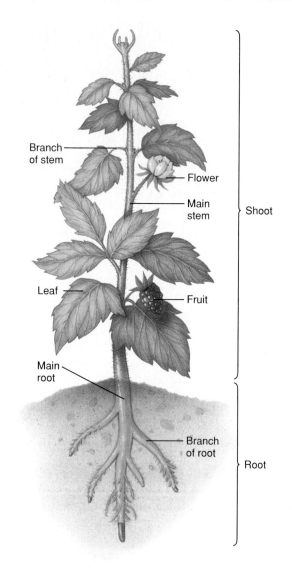

Figure 13.9 Body of a flowering plant. The plant shoot consists of leaves, a main stem and its branches, as well as flowers (which become fruits) in season. The root consists of a main root and its branches.

Research Report

Plants That Eat Animals

Our conventional view of nature is that plants are eaten by animals. But there are some plants that eat animals. Why? To obtain nitrogen for building such molecules as proteins, nucleic acids, and ATP. Carnivorous (meat-eating) plants live mainly in bogs, where soils hold very little nitrogen. An abundant source of nitrogen in their environment is animal protein, and they have evolved ways of capturing and digesting insects. Perhaps the most spectacular of these plants is the Venus's-flytrap (*Dionaea muscipula*), a native of North and South Carolina.

The leaves of a Venus's-flytrap are baited traps (fig. 13.B). The surfaces are coated with nectar, a sugary solution secreted by cells of the epidermis, that lures insects. At rest, each leaf consists of two parts connected by a hinge (the main vascular bundle). The borders of the leaf are studded with long spines. When a hungry fly, ant, or spider lands on the leaf, the two parts fold rapidly together, the spines interlock, and the insect is trapped inside:

(a)

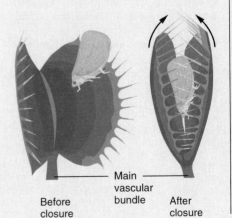

Before closure — Main vascular bundle — After closure

(b)

(c)

Figure 13.B Venus's-flytrap. (*a*) The leaves of this remarkable plant are modified for capturing insects. (*b*) When an insect lands on the leaf, its two halves suddenly fold over the insect. (*c*) Spines along the edges of the leaf interlock to form a cage with the victim inside.

The leaf closes whenever an insect touches two sensitive hairs on the leaf surface. The hairs respond by sending electric signals to other leaf cells, stimulating the leaf to fold rapidly and form a cage around the victim. There, enzymes (secreted by the epidermis) digest the insect's proteins into amino acids, which are absorbed into the leaf cells.

The cage forms within a few seconds. How can a leaf bend inward this fast? Two hypotheses have been offered (fig. 13.C). One hypothesis, first proposed by Charles Darwin in 1875, is that the upper epidermis (which forms the inside of the cage) suddenly loses water and shrinks. The other hypothesis, first proposed by W. H. Brown in 1916, is that the lower epidermis (which forms the outside of the cage) suddenly expands:

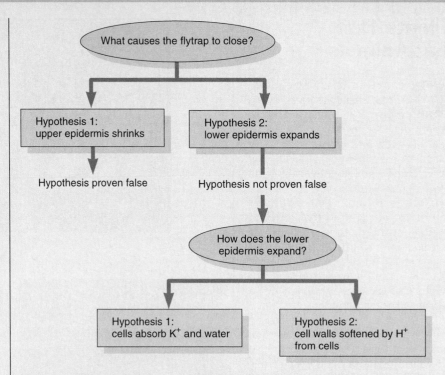

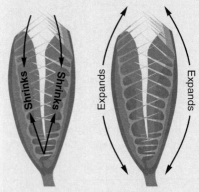

Figure 13.C Hypotheses regarding how the Venus's-flytrap works.

The hypotheses were tested by measuring whether the upper epidermis became smaller or the lower epidermis became larger when the trap closed. A clever way of doing this was to cover the epidermis with uniformly spaced dots and then to measure whether the dots moved closer together or farther apart when the trap closed. The results were clear-cut. Dots on the upper epidermis did not change during trap closure, which meant that this part of the leaf did not shrink. Dots on the lower epidermis became farther apart during trap closure, which meant that this part of the leaf expanded. Thus, Darwin's hypothesis was proven false and Brown's hypothesis was supported.

The next question is what causes expansion of the lower epidermis? Again, there are two main hypotheses (see fig. 13.C). One hypothesis, called the potassium-ion hypothesis, is that the epidermal cells absorb potassium ions (K$^+$), which draw in water by osmosis. The cells enlarge as they become more turgid (swollen with water):

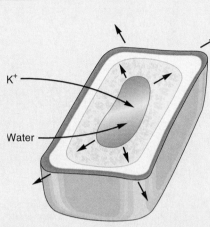

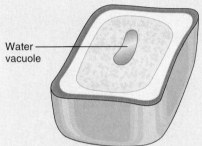

Before (shrunken)

During

After (larger and more turgid)

The other hypothesis, called the hydrogen-ion hypothesis, is that the epidermal cells secrete hydrogen ions (H$^+$) into their cell walls. The ions soften the walls, weakening their resistance to pressure from water inside the cells. Consequently, the cells expand and become less turgid:

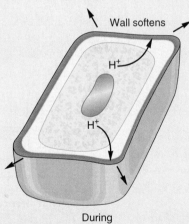

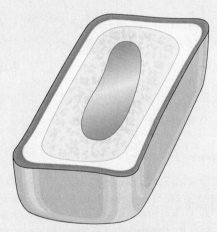

Before

During

After (larger but not more turgid)

These two hypotheses have different predictions. If the potassium-ion hypothesis is correct, then the epidermal cells should become more turgid during trap closure. If the hydrogen-ion hypothesis is correct, then the epidermal cells should become less turgid during trap closure. While the results are not conclusive, most experiments indicate that the cells become less turgid. More research is needed, however, to completely rule out the potassium-ion hypothesis.

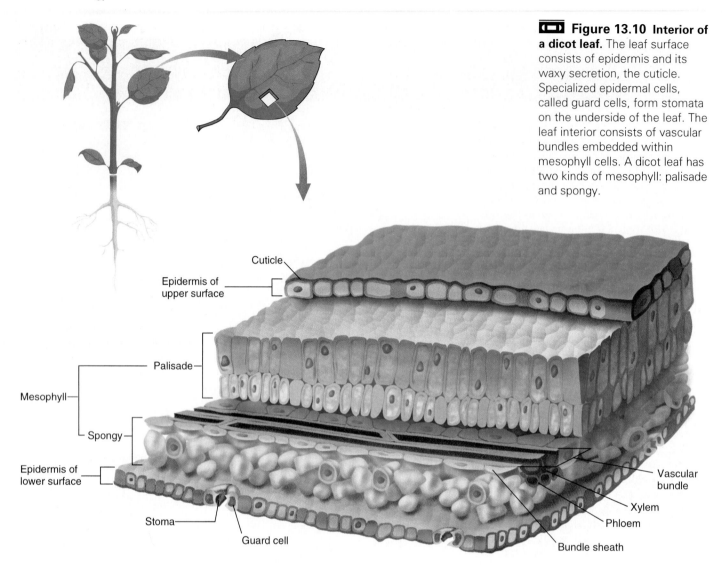

▭ **Figure 13.10 Interior of a dicot leaf.** The leaf surface consists of epidermis and its waxy secretion, the cuticle. Specialized epidermal cells, called guard cells, form stomata on the underside of the leaf. The leaf interior consists of vascular bundles embedded within mesophyll cells. A dicot leaf has two kinds of mesophyll: palisade and spongy.

Cuticle

Epidermis of
upper surface

Palisade

Mesophyll

Spongy

Epidermis of
lower surface

Stoma

Guard cell

Vascular
bundle

Xylem

Phloem

Bundle sheath

Leaves

A leaf builds sugars from sunlight, carbon dioxide, and water. To carry out this function, it absorbs light from the sun and CO_2 from the air. Water and inorganic ions are supplied by the xylem. Oxygen gas, a by-product of photosynthesis, is released into the surrounding air.

Most leaves are flat, green structures. Their flatness provides a large surface for absorbing sunlight; the green color is from chlorophyll, the main light-absorbing pigment. Each plant has leaves of a particular size, shape, number, and arrangement. A lily of the valley, for example, has two relatively large leaves positioned near the ground whereas an oak tree has approximately 700,000 smaller leaves positioned high above the ground.

Some leaves perform functions other than photosynthesis. The leaves of a cactus form spines that protect the plant from being eaten by animals (photosynthesis occurs in the stem); some of the leaves of a garden pea are tendrils that help the plant climb; and the leaves of an onion bulb are used for storing starch. The most spectacular leaves are modified for capturing insects, as described in the accompanying Research Report.

Internal Structure   The interior of a leaf consists of vascular bundles embedded within a special kind of ground tissue called **mesophyll** (*meso*, middle; *phyll*, leaf). Mesophyll cells specialize in photosynthesis and are packed with chloroplasts.

A dicot leaf has two kinds of mesophyll: palisade and spongy (fig. 13.10). Palisade mesophyll fills the upper half of the leaf. Its cells are barrel-shaped and closely packed in one or two rows just beneath the upper epidermis. Spongy mesophyll is found in the lower half of the leaf. Its more spherical cells are loosely arranged, with air spaces between the cells that allow diffusion of carbon dioxide, oxygen gas, and water vapor. A monocot leaf, by contrast, has just one kind of mesophyll, with cells that are scattered throughout the leaf.

Vascular bundles extend throughout the leaf so that each mesophyll cell is near a branch of the xylem, from which it obtains water and ions, and a branch of the phloem, to which it exports sugars. In a dicot leaf, the vascular bundles are arranged as a net, whereas in a monocot leaf they are arranged parallel to one another:

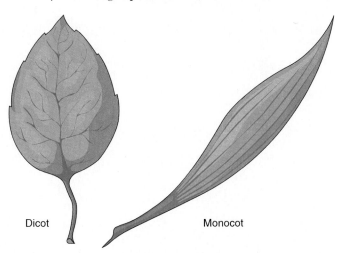

Dicot   Monocot

External Structure   The outer surface of a leaf is epidermis, which forms a barrier to diffusion of water vapor, carbon dioxide, and oxygen gas. These gases enter and exit a leaf mainly through tiny holes, called **stomata** (= mouths), in the epidermis (see fig. 13.10).

Carbon dioxide diffuses from the air through the stomata and into the air spaces between the spongy mesophyll cells. From there it diffuses into the cells. Oxygen gas, a by-product of photosynthesis, diffuses out of the cells into the air spaces and then out of the leaf through the stomata.

Gases diffuse into and out of a cell only when its surface is wet. A thin film of water coats the mesophyll cells, enabling the inward diffusion of carbon dioxide and outward diffusion of oxygen gas. The film is maintained by a high concentration of water vapor (almost 100%) within the air spaces.

A leaf that is open to the diffusion of carbon dioxide and oxygen gas is also open to the diffusion of water vapor out of the leaf. Some loss of water through the stomata is useful, for it cools the leaf and draws water up from the soil (described in chapter 14). Most plants, however, do not have access to large quantities of water and must guard against the loss of water through their leaves.

A plant closes its stomata at times to restrict water loss. Special epidermal cells, called **guard cells,** control the sizes of these openings. Guard cells open the stomata in response to light (indicating suitable conditions for photosynthesis) and to low concentrations of carbon dioxide within the leaf (indicating a need for more of this gas). They close the stomata in response to low concentrations

of water vapor (indicating a need to conserve water) and to high concentrations of carbon dioxide inside the leaf (indicating enough of this gas).

Each stoma (singular of stomata) is formed of two guard cells. As these cells swell and shrink, they open and close a pore between them. How this happens is remarkably simple. A guard cell has a thick wall on the side next to the stoma and thinner walls on the other sides:

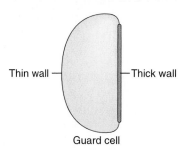

Thin wall   Thick wall

Guard cell

To open a stoma, the guard cells absorb potassium ions (K^+), which then pull in water by osmosis (both the ions and the water come from nearby cells of the epidermis). As the guard cells swell, their thin walls balloon out and their thick walls bend. An opening forms between the two swollen guard cells:

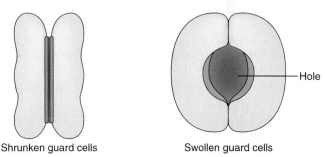

Shrunken guard cells   Swollen guard cells

Hole

To close a stoma, the guard cells pump out potassium ions, and water follows by osmosis. The cells shrink and their thick walls straighten, eliminating the hole between them. The delicate guard cells are particularly vulnerable to damage by air pollution, as shown in figure 13.11.

Plants adapted to dry environments reduce water loss in several ways. They have thick cuticles with stomata only on the shady underside of the leaf. Some have their stomata inside pits or covered with hairs that hold water vapor near the opening and slow the rate at which water diffuses out of the leaf. The most striking adaptations for conserving water involve leaf size and shape, as shown in figure 13.12. In addition, some plants have special forms of photosynthesis, called C_4 and CAM, that help conserve water.

Attachment of the Leaf   A typical dicot leaf has two parts: the blade, which absorbs light, exchanges gases, and performs photosynthesis; and the petiole (or stalk), which attaches the blade to the stem. A leaf may be simple, with just one blade, or compound, with more than

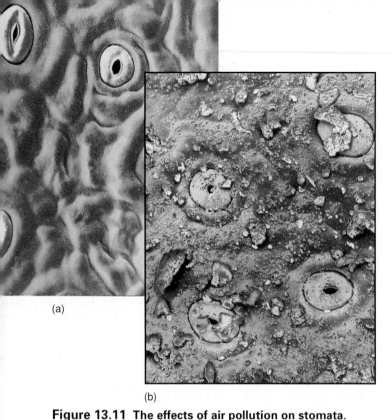

(a)

(a)

(b)

Figure 13.11 The effects of air pollution on stomata.
These photographs are close-ups of leaves from a holly tree.
The yellow structures are guard cells that form openings called
stomata. (*a*) A healthy leaf from a tree growing in a country
churchyard. (*b*) An unhealthy leaf from a tree growing near a
busy road just 9.5 kilometers (6 mi) from the churchyard. Its
stomata are clogged with pollutants from car exhausts.

one blade. A typical monocot leaf, such as a grass, has no
petiole. Instead, the base of the leaf is wrapped around
the stem:

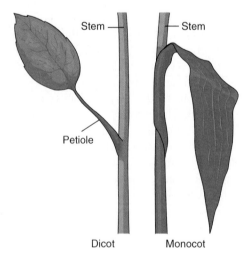

The leaves of each kind of plant are arranged in a
characteristic pattern. They may be attached in an alternate, opposite, or whorled pattern (fig. 13.13). The pattern
is an important characteristic in the classification of flowering plants.

(b)   (c)

**Figure 13.12 The size and shape of a plant's leaves are
adaptations to climate.** (*a*) A plant adapted to a wet climate
can afford to have large leaves with smooth edges. This plant
(*Cecropia peltata*) absorbs large amounts of light and heat but
has plenty of water for evaporative cooling (losing water
through its leaves). (*b*) The leaf of a tulip tree (*Liriodendron
tulipifera*) is medium-sized with deep indentations. It absorbs a
moderate amount of light and heat and can acquire a moderate
amount of water for cooling. Its smaller, irregular-shaped
leaves lose more heat by convection (air currents) than a
larger, smoother leaf. (*c*) The tiny but numerous leaves of a
mesquite tree (*Prosopis juliflora*) are an adaptation to dry
climates. Each leaf absorbs only a small amount of light and
heat and uses only a small amount of water for cooling. It
eliminates excess heat mainly by convection.

Stems

The main function of the stem and its branches is to hold
the leaves in sunlight and the flowers and seeds in positions favorable for fertilization and dispersal. A second

(a)   (b)   (c)

Figure 13.13 Attachments of leaves to the stem.
Flowering plants exhibit three patterns of leaf attachment: alternate, opposite, or whorled. (*a*) An alternate pattern, with leaf attachment staggered along the stem. (*b*) An opposite pattern, with two leaves attached at the same site but on opposite sides of the stem. (*c*) A whorled pattern, with several leaves attached at the same site on different sides of the stem.

function is to transport materials between the leaves and roots. In addition, the stems of some plants are modified for storage, as in a potato; asexual reproduction, as in the runners of a strawberry plant; photosynthesis, as in a cactus; and climbing, as in a strangler fig tree (fig. 13.14).

Internal Structure   The stem consists of vascular bundles embedded within ground tissue. The bundles transport fluids and support the plant in an upright position. The ground tissue serves mainly as a place for storing starches. Differences between the stems of dicots and monocots are shown in figure 13.15 and described following.

The vascular bundles of a dicot stem form a ring that divides the ground tissue into pith and cortex. A cut through a stem shows three concentric structures: pith, vascular bundle, and cortex. Within each vascular bundle, xylem lies to the inside of the phloem. The stems of some plants have a cylinder of collenchyma cells (part of the cortex) just beneath the epidermis. In a monocot, the vascular bundles do not form a ring but are scattered throughout the ground tissue; there is no separation into pith and cortex.

Figure 13.14 A stem modified for climbing. A strangler fig uses the trunk of a tree to support its upward growth into light. This fig has apparently strangled its host tree, preventing outward growth of the trunk. The host tree died and decomposed, leaving only the spiral-shaped stem of the strangler fig.

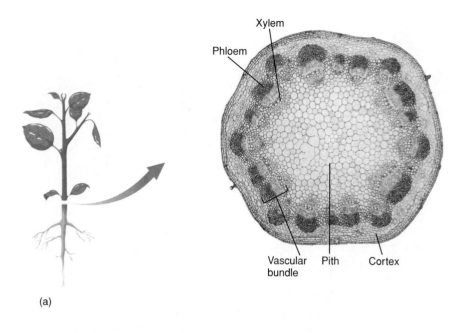

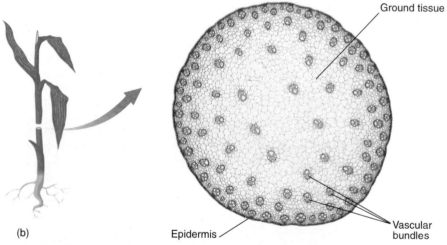

(a)

(b)

Figure 13.15 The interior of a stem. (a) A dicot stem has a central pith surrounded by a ring of vascular bundles, which is surrounded by cortex. (b) The vascular bundles of a monocot stem are scattered; they do not separate the ground tissue into pith and cortex.

External Structure The stem is covered with dermal tissue—epidermis in the stem of an herbaceous plant and periderm in a mature bush or tree. Sometimes this tissue is modified to form hairs. The height and branching pattern of the stem determines a plant's shape, which is an evolved adaptation for efficiently intercepting the sun's rays. Some of these adaptations are shown in figure 13.16.

Roots

The main functions of a root are to anchor the plant and to absorb water and nutrients. Many plants also use portions of their roots for storage of starch or water.

Internal Structure The inside of a root has special tissues for absorbing water and inorganic ions. Xylem and phloem in the stem extend without interruption into the root.

The interior of a dicot root consists of vascular tissue, pericycle, and cortex (fig. 13.17a). Xylem fills the center (there is no pith) and typically has several "arms" that extend outward for absorbing fluids. Patches of phloem lie outside the xylem. Next comes the **pericycle,** which consists of parenchyma cells that may give rise to lateral branches (*lateral,* on the side), as shown in figure 13.18. Outside the pericycle is the cortex, beginning with a special layer called the endodermis, followed by the rest of the cortex. The **endodermis** is a single layer of cells that controls what enters the xylem. The rest of the cortex is used chiefly for storage of starch.

The interior of a monocot root consists of pith, vascular tissue, pericycle, and cortex (fig. 13.17b). In the center is pith, a ground tissue that stores starch. The pith is surrounded by vascular tissue, with alternating clusters of xylem and phloem. Next is the pericycle, followed by endodermis, and then the rest of the cortex.

External Structure The outside of a root is covered with epidermis. Cells of this covering typically form long, delicate extensions called root hairs:

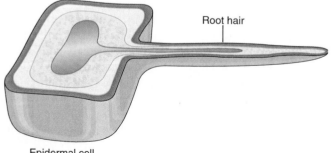

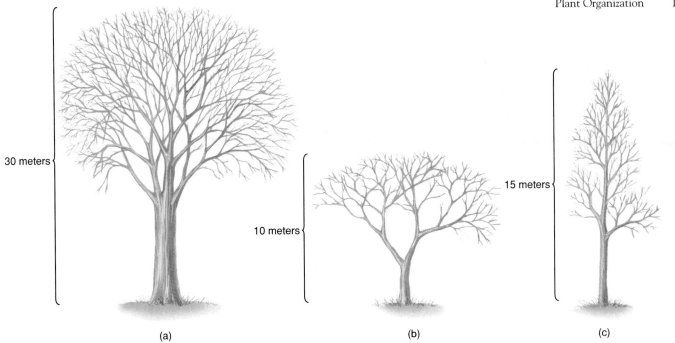

(a)   (b)   (c)

Figure 13.16 The size and shape of a stem are evolved adaptations to availability of sunlight. (*a*) An elm grows in a dense forest; its branches grow straight upward to position the leaves above other plants. (*b*) A dogwood grows beneath other trees of a forest; its branches grow outward to reach patches of sunlight that penetrate the crowns of taller trees. (*c*) A balsam poplar grows where disturbances have eliminated other kinds of trees; its short branches grow outward all along the trunk, positioning leaves to receive light from all directions.

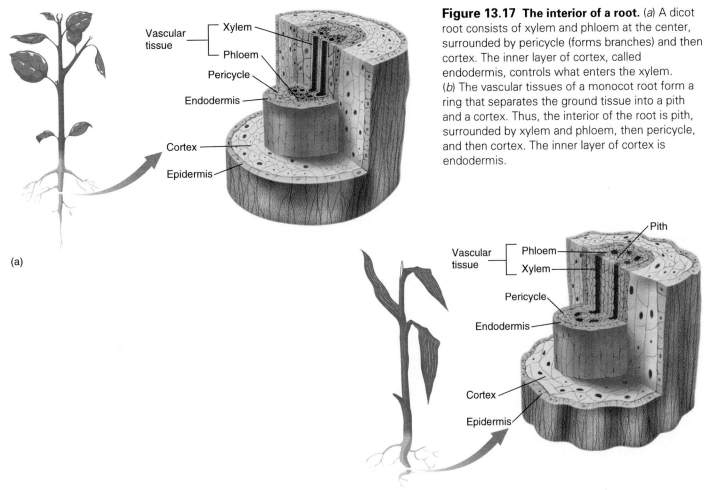

Figure 13.17 The interior of a root. (*a*) A dicot root consists of xylem and phloem at the center, surrounded by pericycle (forms branches) and then cortex. The inner layer of cortex, called endodermis, controls what enters the xylem. (*b*) The vascular tissues of a monocot root form a ring that separates the ground tissue into a pith and a cortex. Thus, the interior of the root is pith, surrounded by xylem and phloem, then pericycle, and then cortex. The inner layer of cortex is endodermis.

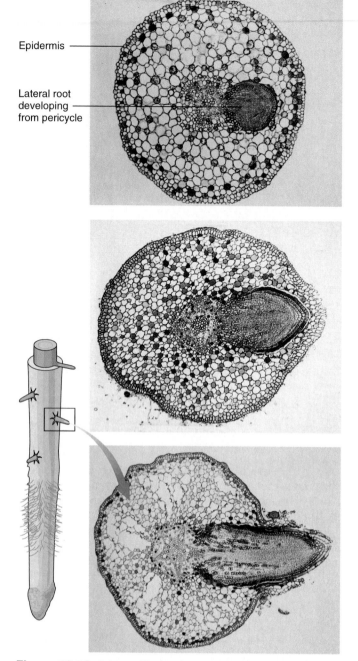

Epidermis

Lateral root
developing
from pericycle

Figure 13.18 A lateral branch developing in a dicot root.
Branches of a root develop from the pericycle. As cells in this
tissue multiply, they form a branch that grows outward
through the other tissues.

The hairs greatly increase the surface area for absorption of
water and ions. A single rye grass plant, for example, has
more than 14 billion root hairs with a combined surface area
approximately the size of a tennis court! Epidermis at the tip
of a root is modified to form a covering, the root cap, that
protects the delicate tissues as they grow through the soil.

Root shape is an adaptation for acquiring water from
different regions of the soil. There are two basic shapes:
taproots and fibrous roots (fig. 13.19). A taproot has a single main root that extends straight down, with smaller
branches that extend out the sides. A dandelion is an example of a plant with this kind of root; it taps accumula-
tions of water deep within the soil. A fibrous root has

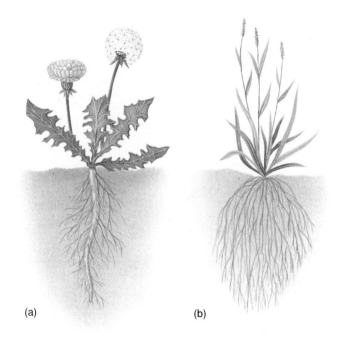

(a)   (b)

Figure 13.19 Root shapes. (*a*) Most dicots have a taproot: a
large branch that extends straight down and smaller branches
that extend sideways. (*b*) All monocots have fibrous roots,
with many branches of equal size.

many slender branches of equal size and extends outward
to absorb water from shallow regions of the soil. Crabgrass
is an example of a plant with a fibrous root. Most dicots
have taproots, whereas all monocots have fibrous roots.

Coordination of Cellular Activities

We have seen how a variety of specialized cells, each performing a particular function, contribute to the survival of
a plant. The activities of all the plant's cells are coordinated, so that what one cell does is in harmony with what
other cells are doing. This coordination is governed by
hormones. (Plants do not have nervous systems.)

A hormone is a molecule that is secreted (synthesized and released) by one cell and travels through body
fluids to other cells, where it influences their activities.
Plant hormones coordinate internal activities as well as responses to the external environment.

The cellular activities of a plant are adjusted to the
seasons: hormones initiate dormancy at the onset of unfavorable seasons; rapid growth at the onset of favorable seasons; and flower development at the onset of the
reproductive season. A plant responds to environmental
changes mainly by growth and movement. Hormones govern growth of the stem toward sunlight and of the root toward moisture. Hormones control the movements that raise
and lower leaves and that open and close flowers. Plant
hormones and their functions are described in chapter 16.

SUMMARY

1. Plants are multicellular, eukaryotic organisms. The first plants, seaweeds, evolved from unicellular, eukaryotic organisms that lived in the oceans. When they colonized land, plants evolved many specialized structures for restricting water loss.

2. A typical plant cell is characterized by chloroplasts, a water vacuole, and a cell wall (a primary wall and, in some cells, a secondary wall as well). A plant has four basic kinds of cells: parenchyma, collenchyma, sclerenchyma, and tube-forming cells.

3. A tissue is a group of specialized cells that carries out a particular task. A plant has four kinds of tissues: vascular (xylem and phloem), dermal (epidermis and periderm), ground, and meristem.

4. A plant body consists of leaves, stem, root, and reproductive organs in season. There are two kinds of flowering plants, dicotyledons and monocotyledons.

5. Most leaves are specialized for photosynthesis. The interior of a leaf consists of mesophyll cells, which carry out photosynthesis, and vascular bundles, which transport materials to and from the leaf. The exterior of a leaf is formed of epidermis studded with stomata, which are regulated by guard cells.

6. A plant stem positions the leaves and reproductive structures and transports fluids. The interior consists of vascular bundles embedded within ground tissue. The exterior of a stem is formed of epidermis or periderm.

7. A root anchors the plant and absorbs water and nutrients. The interior consists of pith (in monocots only), vascular tissue, pericycle, and cortex. The exterior is formed of epidermis that is adapted for absorbing water and ions.

8. The activities of plant cells are coordinated by hormones—chemical messengers that travel through the body fluids.

KEY TERMS

Bundle sheath (page 240)
Collenchyma (koh-LEN-kih-mah) **cell** (page 235)
Companion cell (page 239)
Cortex (page 241)
Cuticle (KYOO-tih-kul) (page 241)
Dicotyledon (DEYE-kah-tih-LEE-duhn) (page 242)
Endodermis (EN-doh-DUHR-miss) (page 250)
Epidermis (EH-pih-DUHR-miss) (page 240)
Ground tissue (page 241)
Guard cell (page 247)
Meristem (MEHR-ih-stem) (page 241)

Mesophyll (MEE-zoh-fil) (page 246)
Monocotyledon (MAH-noh-kah-tih-LEE-duhn) (page 242)
Parenchyma (pahr-EN-kih-mah) **cell** (page 235)
Pericycle (PEHR-ih-seye-kul) (page 250)
Periderm (PEHR-ih-durm) (page 241)
Phloem (FLOH-uhm) (page 239)
Pith (page 241)
Sclerenchyma (skler-EN-kih-mah) **cell** (page 236)
Sieve tube (page 238)
Stomata (stoh-MAH-tah) (page 247)
Tissue (page 238)
Tracheid (TRAY-kee-id) (page 238)
Vascular bundle (page 239)
Vessel (page 238)
Xylem (XEYE-luhm) (page 238)

STUDY QUESTIONS

1. A plant builds sugars from carbon dioxide, water, and sunlight. What adaptations does a plant have for acquiring these materials from the environment?

2. Which plant structures are made of dead cells?

3. What is the function of stomata? How do they open and close? Why is it important that a plant be able to close its stomata?

4. Draw a dicot leaf, showing the epidermis, stomata, mesophyll, and vascular bundles.

5. Draw a dicot stem, showing the epidermis, vascular bundles, and ground tissue. How does this stem differ from a monocot stem?

6. Draw a dicot root, showing the epidermis, ground tissue, pericycle, and vascular bundles. How does this root differ from a monocot root?

CRITICAL THINKING PROBLEMS

1. Describe the plant structures you have eaten in the past week, including plant foods that are leaf blades, leaf stalks, buds, starch- or sugar-storing leaves, stems, roots, and reproductive organs.

2. Suppose you were shown a strange new organism. What criteria would you use to decide whether it was a plant?

3. Explain why most leaves are green and flat.

4. If you discovered a new plant, how could you determine whether it was a dicotyledon or a monocotyledon?

5. If you were given leaves from a variety of plants, how could you determine which of the leaves were from plants adapted to very dry conditions?

SUGGESTED READINGS

Introductory Level:

Bell, A. D. *Plant Form: An Illustrated Guide to Flowering Plant Morphology*. New York: Oxford University Press, 1991. An introduction to the basic organization of flowering plants, including terminology, methods of investigation, and descriptions of each part and how different parts are organized.

Prance, G. T. *Leaves: The Formation, Characteristics, and Uses of Hundreds of Leaves Found in All Parts of the World*. New York: Crown Publishers, 1985. A beautifully illustrated book, with over 300 photographs, that describes the variations in leaf shape and color.

Rupp, R. *Blue Corn and Square Tomatoes*. Pownal, VT: Storey Communications, 1987. A fascinating account of food plants: their origins, how they have been used in the past, and aspects of their biology.

Advanced Level:

Horn, H. S. *The Adaptative Geometry of Trees*. Princeton, NJ: Princeton University Press, 1971. An analysis of the factors that determine tree shape, including such topics as leaf distribution, crown shape, and successional changes in tree shape.

Kaplan, D. R., and W. Hagemann. "The Relationship of Cell and Organism in Vascular Plants," *BioScience* 41, no. 10 (1991): 693–703. An argument that an individual cell is not a distinct entity in a plant as it is in an animal.

Tomlinson, P. B. "Tree Architecture," *American Scientist* 71 (1983): 141–49. A technical discussion of developmental patterns that produce tree form.

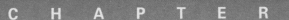

Transport and Nutrition

Objectives

In this chapter, you will learn
–what forces move water upward
  through a plant;
–what forces move sugar and other cell
  products throughout the plant;
–about the movement of soil water into
  the root;
–what inorganic ions and molecules are
  required by a plant.

Coast redwoods. These enormous trees, each more than 90 meters (300 ft) tall, transport water from their roots to their leaves and sugars from their leaves to their roots.

The coast redwood, shown in the opening photograph, is a wonder of nature—the world's tallest living thing. This spectacular tree is a form of sequoia (*Sequoia sempervirens*) that grows along the Pacific coast, from northern California to southern Oregon. A typical coast redwood is taller than a football field is long, which means that if the tree were felled at one goal line its top would extend beyond the other goal line! How do all the cells of such a lofty plant acquire water from the soil? How is water lifted, without a pump, from the roots to the uppermost leaves?

Water moves through a plant within narrow tubes. The tubes extend from the roots to the leaves, which in a coast redwood may be over 90 meters. The transport mechanism, described in this chapter, requires that each tube remain full of water. Thus, whenever water evaporates out of a leaf, the same amount is pulled in from the soil. The quantity of water that flows through a large tree is truly awesome. A coast redwood, with its enormous height and girth, lifts at least 2,000 liters (approximately 4,000 lb) of water every day!

Transport of Fluids

A plant transports fluids through two kinds of tubes. Water is transported from the soil to the leaves through the xylem (tracheids and vessels). Xylem (see page 238) contains inorganic ions, required by plant cells, that were dissolved in the soil water. Sugars and other cell products are transported from one place to another through the phloem (sieve tubes) (see page 239).

In spite of a long history of research on fluid transport in plants, we still do not know for sure just how fluids are moved through the xylem and phloem. But we have learned a great deal, and some of what we know is described in the following sections.

Ascent of Water

Almost all the water that enters a plant through its roots evaporates out through stomata (tiny holes) (see page 247) in its leaves. Evaporative water loss from plant tissues, chiefly through the stomata, is called **transpiration** and may involve enormous quantities of water. Each day, approximately 7 liters (1.5 gal) of water pass through a ragweed, 150 liters (33 gal) through an apple tree, and as many as 500 liters (110 gal) through a date palm. Why does a plant lose so much water?

A plant cannot avoid some water loss through transpiration, for when the stomata are open for uptake of carbon dioxide they are also open for evaporation of water. Beyond this unavoidable loss of water, transpiration serves two functions. First, evaporation of water cools the leaves, which tend to absorb too much heat while they absorb light. Second, the more soil water that moves through a plant, the more inorganic ions it can extract from the water. A plant needs nitrate, sulfate, phosphate, and other ions to build molecules and carry out various cellular activities.

What pulls water upward through the xylem? The explanation supported by most research, known as the **transpiration-cohesion theory** (or the tension-cohesion theory), attributes the upward movement of water to the interactions among three forces: transpirational pull, cohesion, and adhesion.

Transpirational Pull   According to the transpiration-cohesion theory, water is pulled upward through a xylem tube when a water molecule evaporating out of the top of the tube pulls up the next water molecule below it:

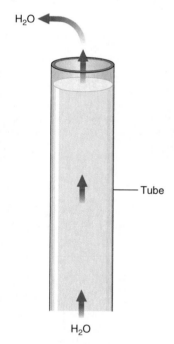

This pull is similar to what happens when you sip a drink through a straw. As you suck on the straw, you pull air upward, lowering the air pressure at the surface of the column of fluid. The liquid then flows from higher pressure, at the bottom of the straw, to lower pressure, above the liquid in the straw. In a similar fashion, removal of water from the top of a xylem tube lowers the air pressure there and creates an upward pull on the water.

Cohesion and Adhesion A water molecule is a polar molecule: one side is slightly negative and the other side is slightly positive in electric charge. Opposite sides of two water molecules, with opposite charges, attract each other. The attraction, known as the hydrogen bond (📖 *see page 30*), holds water molecules together. This tendency of molecules to cling to each other is called **cohesion.**

A similar attraction, **adhesion,** develops between water molecules and the cellulose molecules that form the wall of a xylem tube. Opposite electric charges on water molecules and on cellulose molecules attract each other, forming hydrogen bonds that hold water against the walls of the tubes:

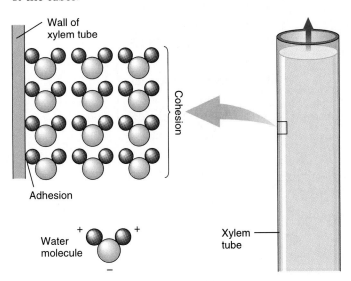

Adhesion is not a strong force; it can hold only a narrow column of water, which is why tracheids and vessels of the xylem are no wider than a single cell.

Pathway of Water Molecules Transport of water through the xylem involves a long chain of tiny movements, in which molecule after molecule displace one another (fig. 14.1). The chain of displacements begins in the leaf.

A leaf is comprised primarily of water. Water vapor saturates the air spaces between the cells. Liquid water coats the surfaces of the cells, permeates the cell interiors, and fills the xylem. The first displacement in the chain

occurs as water vapor diffuses from an air space out through a stoma and into the relatively dry air outside the leaf:

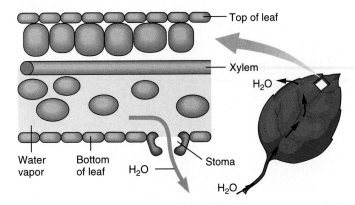

Each molecule of water vapor that leaves an air space is replaced by another molecule that evaporates from the thin film of water covering a cell. The molecule that evaporates, in turn, is replaced by a water molecule from inside the cell. The cell interior then has a lower concentration of water than its neighboring cells within the leaf. Consequently, it receives water, by osmosis (📖 *see page 76*), from a neighboring cell, which then receives water from another cell, and so on:

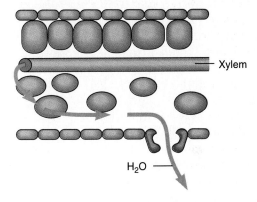

Eventually the chain of displacement reaches a xylem tube. There, a molecule of water is pulled out of the xylem to replace a molecule displaced from an adjacent cell. As the molecule of water leaves the xylem, a molecule below it is pulled up by cohesion. The next molecule down in the column also moves up, because of cohesion, as does the next and the next all the way down to the bottom of the water column in the xylem tube. As the molecules are pulled upward, they are held in the higher position by

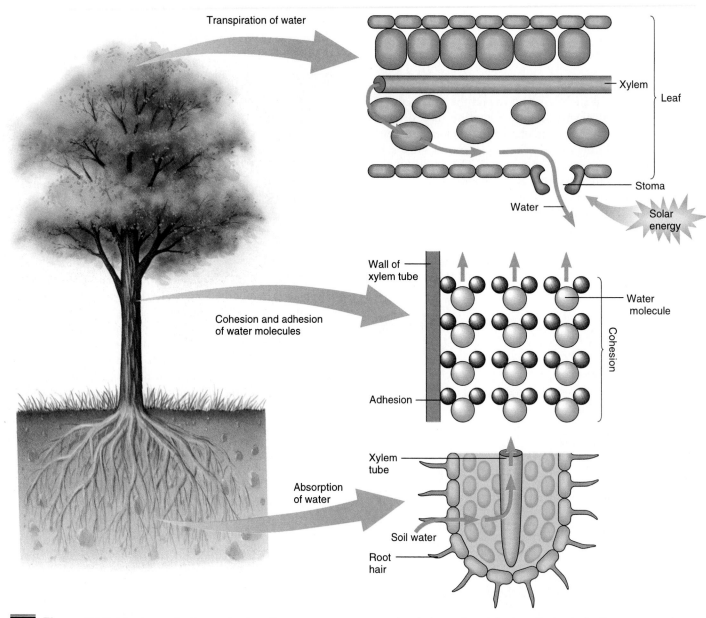

Figure 14.1 The transpiration-cohesion theory.
Water moves through a plant by a series of tiny displacements that begin with transpiration. As a water molecule evaporates from the leaf, another molecule evaporates from the film around a cell to replace it. This molecule, from the film, is replaced by a molecule inside the cell, which is replaced by a molecule from another cell and so on until the uppermost molecule in a xylem tube is pulled out. As the uppermost molecule leaves, another is pulled up by cohesion. All along the xylem tube, as one molecule is pulled up, it pulls up yet another by cohesion. The molecules are held in their new positions by adhesion to the wall of the tube. When the series of displacements reaches the root, a new molecule is pulled into the xylem tube, a process known as absorption.

adhesion. At the bottom of the xylem tube in the root, a molecule of water is drawn in from the soil to replace the last molecule pulled upward.

The transpiration-cohesion system of water transport works only when the column of water is continuous. Any break in the column stops the flow of water: the water molecules remain in position, held to the walls by adhesion, but they now form two groups that are too far apart to cohere. Air bubbles that develop in a xylem tube (from the conversion of water to ice, the chewing by insects, and

other injuries) break the column and prevent the flow. The disruption of just one xylem tube, however, has no serious consequences; a plant has many such tubes and builds new ones every year.

Remarkably, the only energy used in lifting water through a plant comes from the sun. By evaporating water from a leaf, solar energy initiates a sequence of displacements that lifts water from the roots to the leaves. Thus, a plant transports water—sometimes vast quantities to enormous heights—without expending chemical energy.

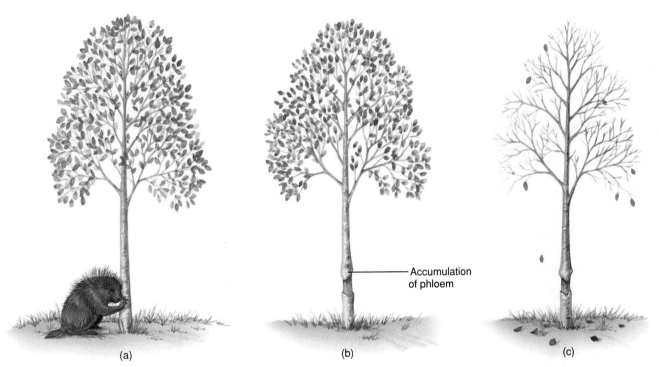

Figure 14.2 A girdled tree. (*a*) A tree is girdled when the bark, which includes the phloem, is chewed around the stem. (*b*) Some of the sugary solutions traveling downward in the phloem accumulate to form an enlarged area above the girdle. (*c*) The plant eventually dies as its roots no longer receive materials from the stem and leaves.

Transport of Cell Products

Materials exported by cells for use elsewhere in the plant are transported through the phloem. The most abundant of these materials are sugars, which are photosynthesized in some cells (mainly the leaf) and used in other cells to provide energy and to build various kinds of molecules, including starch for energy storage. Other materials commonly exported by plant cells include hormones and ions.

The phloem consists of bundles of sieve tubes formed of living cells, described in chapter 13. The activities of these cells are supported by companion cells, which are connected to the sieve tubes by plasmodesmata (see *page 67*).

The Pathway from Source to Sink   Sugars and other cell products travel in a watery solution from sources to sinks. A source is a place where sugars are added to the phloem. The main sources are leaves, which export newly synthesized sugars to other parts of the plant. A sink is a place where sugars are removed from the phloem. The main sinks are actively growing parts, such as emerging leaves and developing fruits, that require an input of sugar for energy.

Sugar is stored, usually as starch, in a storage place within the plant. This part of the plant may be either a source or a sink depending on the season. In summer, a plant stores sugars as starch in the ground tissue of its stem and roots, which in some plants are enlarged to form tubers or bulbs. During this season, the leaves are sources and the storage places are sinks. In early spring, the flow is reversed. The storage places become sources, releasing sugars from starch, and the emerging leaves become sinks until they become mature and photosynthesize all their own sugars.

A plant will die if phloem transport through its stem is blocked. This phenomenon, known as girdling, sometimes happens in trees (fig. 14.2), where phloem forms the inner layer of bark. Porcupines, rabbits, and other mammals seeking food in winter may gnaw on the bark all around the trunk at their particular feeding height. A wound of this sort disrupts the phloem but not the xylem (which lies farther inward). A swelling develops just above the girdled area as the downward flow of fluid is blocked. The tree dies several years later, after the root has exhausted its supply of starch.

We are not sure exactly how fluids are moved through the phloem. Most research supports the **pressure-flow theory.**

The Pressure-Flow Theory   According to the pressure-flow theory, fluids in the phloem are moved by gradients of water pressure that develop in response to osmosis. The source develops a high pressure and the sink develops a

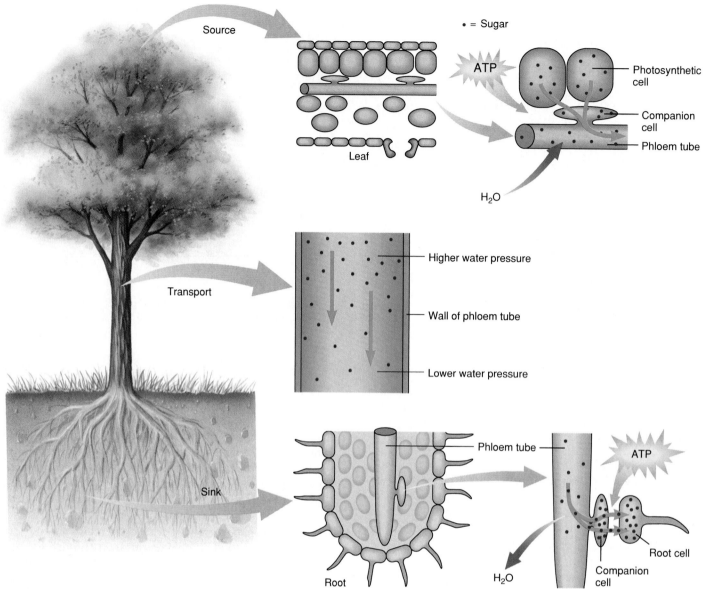

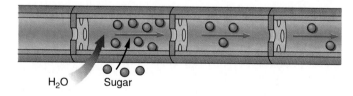

Figure 14.3 The pressure-flow theory. In the source, sugars are actively transported into companion cells. They then flow via plasmodesmata into the phloem tube. The higher concentration of sugars inside the tube pulls in water by osmosis, creating a higher water pressure. The sugary solution then flows away from this part of the phloem tube. In the sink, sugars flow into companion cells and then are pumped, by active transport, into other cells (root cells in this example). The concentration of sugars in the phloem tube decreases and water is drawn out by osmosis, creating a lower water pressure. Fluids then flow into this part of the phloem tube.

low pressure. Fluids then move from source to sink along a gradient of water pressure (fig. 14.3).

A leaf, for example, becomes a source when sugars move out of the mesophyll cells, where they are synthesized, and into regions around the sieve tubes. There they are pumped into companion cells and then flow, via plasmodesmata, into the sieve tubes.

As the concentration of sugar in a sieve tube rises, water is drawn in by osmosis (from the xylem). The inflow of water raises the water pressure, and the fluid is pushed to regions of the tube where water pressure is lower:

A sink develops wherever sugar leaves a sieve tube. In a root, for example, sugar travels from the sieve tube via plasmodesmata to a companion cell. The companion cell pumps the sugar into other cells of the root. As sugar leaves the sieve tube, water follows by osmosis (and enters the xylem). The water pressure in this part of the sieve tube then drops, and fluids flow into it from other parts:

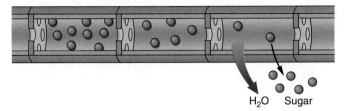

The transport of fluids within a sieve tube, from source to sink, does not itself require an input of energy; fluid moves passively from regions of higher to regions of lower pressure. Energy is required, however, to create a source and a sink—to pump in sugar at one end of the sieve tube and to pump out sugar at the other end. The "pump" is active transport (*see page 78*) by the companion cells, which lie next to the sieve tubes of the phloem.

This explanation of transport within the phloem fits most observations and experiments. Phloem transport is particularly difficult to study because most experimental procedures disrupt the structure of the sieve tubes and interfere with the flow. Some clever ways of overcoming this difficulty are described in the accompanying Research Report.

Uptake of Water and Soil Nutrients

A plant builds organic molecules from carbon dioxide, water, and inorganic ions. All the inorganic materials that a plant needs to achieve normal growth and reproduction are called nutrients. Except for carbon dioxide, which comes from the air, these materials come from the soil. Inorganic ions that a plant acquires from the soil are known as **soil nutrients.** They enter a plant through its root, along with the soil water.

From Epidermis to Xylem

Water and soil nutrients enter a root through the epidermis, mainly through hairlike extensions, called root hairs, of the epidermal cells. From the epidermis they travel through other tissues of the root to the xylem.

Soil Water   Soil water, which contains nutrients, enters the epidermis of a root and then travels toward the xylem along two pathways: through cell interiors and along cell walls (fig. 14.4).

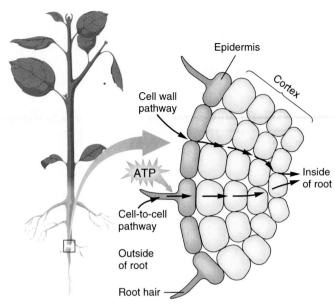

Figure 14.4 Absorption of water and nutrients. Soil water moves laterally through a root toward the xylem along two pathways. In one pathway, the inorganic ions are actively transported into epidermal cells and water follows by osmosis. The ions and water then travel, via plasmodesmata, from cell to cell toward the xylem. In the other pathway, the ions and water travel along cell walls.

Some of the soil water is absorbed into the epidermal cells. It is drawn in by osmosis, as the cell interior has a higher concentration of dissolved materials (lower concentration of water) than the soil water. From the epidermal cells, the water then moves inward toward the xylem, passing freely from cell to cell via plasmodesmata.

Most of the soil water that enters a root does not pass through the interiors of cells but travels instead along the cell walls. The walls of adjacent root cells touch so that water moves freely along cellulose "highways" toward the xylem. Water seeps along the cell walls just as it seeps through a paper towel, which, like the cell walls, is made of cellulose.

Selective Uptake of Certain Nutrients   A plant requires certain inorganic ions in higher concentrations than found in soil water. These ions are selectively absorbed by cells of the epidermis and endodermis.

Cells of the epidermis accumulate certain ions from the soil water. The ions are brought into the cells by active transport, in which special proteins within the plasma membranes pick up ions and carry them into the cell interior. This form of transport, from regions of lower concentration (soil water) to regions of higher concentration (the cell interior), requires an input of energy (as ATP). From the epidermal cells, the ions move inward by diffusion (down a concentration gradient) traveling from one cell interior to the next via plasmodesmata.

261

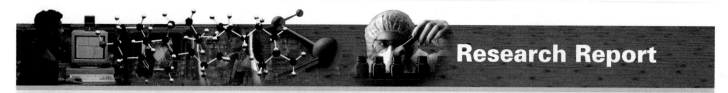

Research Report

Finding Ways to Study the Phloem

Fluid transport in phloem remained a mystery until 1953, when a clever research tool was discovered. Prior to that, research on the phloem was virtually impossible because every experimental procedure interfered with the flow of fluids within the sieve tubes. Whenever an instrument was inserted to sample the fluid, the sieve tubes immediately plugged the hole with a gummy material. This ability of a plant to repair wounds is adaptive, for it prevents sugars from leaking out of sieve tubes that have been chewed by insects and it seals off areas infected by fungi. But to researchers, the ability to repair wounds presented a stumbling block to their investigations.

In 1953, plant biologists discovered a new research tool: aphids. These tiny insects can tap the phloem without triggering the wounding response (fig. 14.A). An aphid feeds by piercing the plant with its modified mouthparts, shaped like a needle and called a stylet, until it strikes the sugary solution inside a sieve tube. The fluid then flows from a region of higher pressure (inside the sieve tube) through the stylet to a region of lower pressure (the aphid's digestive tract).

Researchers exploit aphids in collecting fluid from the sieve tubes. As an aphid feeds, its body is cut away, leaving only the stylet embedded within the phloem. The stylet serves as a miniature syringe through which fluid flows into a test tube. Phloem fluid collected in this way has been an-

(a)

(b)

Figure 14.A Aphids. (*a*) Clusters of aphids on a rosebud. Each aphid is only a few millimeters in length. (*b*) An aphid using its stylet to pierce the phloem of a plant.

alyzed and found to contain approximately 30% dry matter and 70% water. Of the dry matter, 90% is sugar and the rest is hormones, amino acids, and other organic molecules, as well as some inorganic ions.

Another way researchers study the phloem is to make the sugars radioactive, so their transport within the phloem can be traced. In this procedure, the plant leaves are covered with plastic bags filled with carbon dioxide in which the carbon atoms are radioactive. (A radioactive carbon atom has fourteen neutrons rather than twelve, and these extra neutrons break down and release energy that can be detected by photography, using X-ray

film.) The plant absorbs the radioactive carbon dioxide and uses it to build sugars. When radioactive sugars enter the phloem, their location and rates of movement can be traced by exposing the plant to X-ray film.

Prior to these radioactive experiments, researchers believed that sugars moved through the phloem by diffusion. The radioactive experiments, however, revealed that sugars flow as rapidly as 2 meters an hour. This rate is much too fast for diffusion; other forces must be involved. The researchers then turned their attention to a search for these forces, which culminated in the pressure-flow theory described in the text.

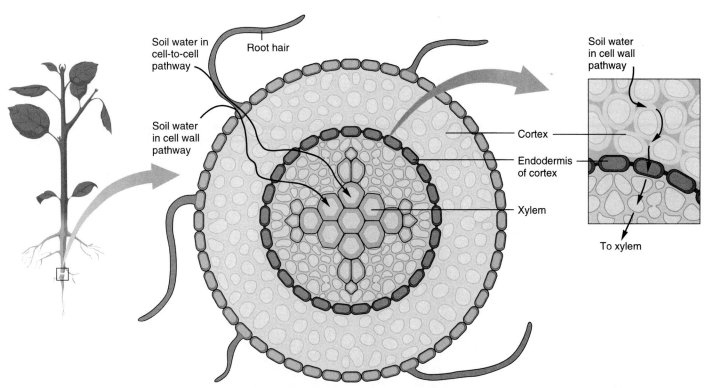

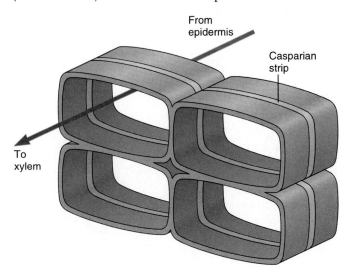

Figure 14.5 Control of ions traveling the cell wall pathway. The endodermis controls how much of each ion flows from the cell wall pathway into the xylem. The walls of endodermal cells are coated with a waterproof barrier, called the Casparian strip, that forces soil water to pass through the endodermal cells rather than continue along the cell walls. Cells of the endodermis, in turn, actively transport certain ions from the cortex to the center of the root.

Cells of the endodermis (the innermost layer of the cortex) govern the composition of water traveling along the cell wall pathway. These cells do not allow water and ions to seep passively through their walls. Instead, four of the six walls that enclose each cell are coated with a wax (called suberin) and therefore are impermeable to water:

This partial barrier to the flow of soil water is known as the **Casparian strip.** It channels soil water through the cell, forcing it to enter through one waxless side and leave by way of the other. Plasma membranes of the cells control what enters and leaves; proteins in the membranes actively transport certain inorganic ions and block the movements of others. In this way, cells of the endodermis control the kinds of ions that reach the xylem (fig. 14.5).

Entry into the Xylem

Water and minerals that pass through the endodermis continue on through the pericycle cells and into the xylem (fig. 14.6). Water enters the xylem by osmosis, because fluid inside contains a higher concentration of ions than fluid outside the xylem. This difference in ion concentration is maintained by cells next to the xylem, which actively transport ions into the xylem. The soil water then moves upward within the xylem, distributing water and inorganic ions to all the plant cells.

Figure 14.6 Uptake of ions by the xylem. Ions that pass through the endodermis travel through the pericycle to cells next to the xylem tubes. These cells actively transport ions into the tubes. Water follows by osmosis.

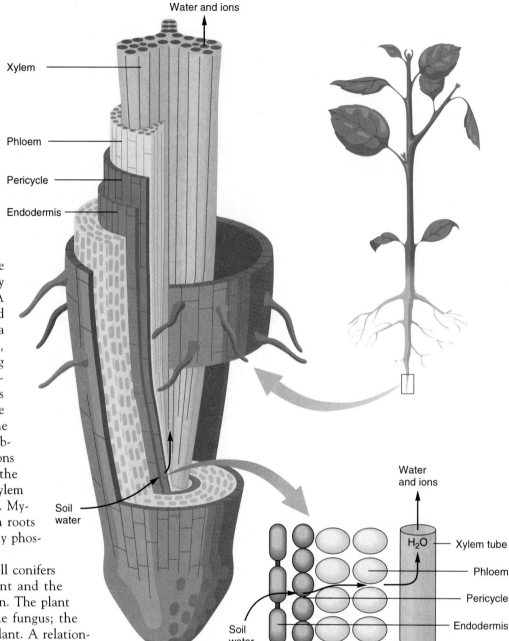

Water and ions

Xylem

Phloem

Pericycle

Endodermis

Soil water

Water and ions

H_2O

Xylem tube

Phloem

Pericycle

Endodermis

Soil water

Mycorrhizae

Many plants increase their surface area for absorbing soil water by forming mycorrhizae (fig. 14.7). A **mycorrhiza** is a structure formed of a certain kind of fungus and a plant root (*myco*, fungus; *rhiza*, root). The fungus is a mesh of long threads that absorb water and inorganic ions. Part of the fungus grows into the plant root and the rest extends outward through the soil, forming a large surface for absorption. Some of the water and ions that enter the fungus pass into the plant root, where they enter the xylem and are transported to plant cells. Mycorrhizae are more efficient than roots at absorbing scarce ions, especially phosphate, zinc, and copper.

Most flowering plants and all conifers form mycorrhizae. Both the plant and the fungus benefit from the association. The plant acquires water and ions from the fungus; the fungus acquires sugars from the plant. A relationship of this kind, in which both species benefit, is called **mutualism.**

Plant Nutrition

The growth of a seed into an adult plant seems a miracle. Where does all that material come from? Most of it comes from water in the soil and carbon dioxide in the air. The bulk of a plant is water—cells consist mainly of water, as do fluids that fill the xylem and phloem. (The consequences of these fluids, with regard to lightning, are

Figure 14.7 Mycorrhiza. A mycorrhiza is a threadlike structure formed of roots and fungi. It greatly expands the surface area for absorption of water and ions.

Wellness Report

Oaks Can Be Dangerous to Your Health

The worst place to be during a thunderstorm is near an oak tree. Oaks are hit by lightning more often than any other tree, and they are also more likely than any other tree to transfer the electricity to surrounding objects.

Lightning travels readily through water. All trees are mostly water on the inside and, during a heavy rainstorm, most trees become soaked on the outside. When lightning strikes an upper branch it generally travels downward along the outside of the trunk and into the ground (fig. 14.B). An oak tree, however, has rough bark that does not become soaked. The outer bark is arranged in thick plates with indented regions between the plates that remain dry. When lightning hits an oak tree, it zips along the rain-soaked branches until it hits a region of dry bark on the trunk. There its path is blocked and it seeks a watery detour.

One watery detour always available to the lightning is the fluid within the phloem, which forms the inner layer of bark. Lightning carries an enormous amount of energy. When it enters the phloem, it instantly heats the fluid to approximately 28,000° C (50,000° F). The fluid vaporizes and expands so rapidly that the tubes explode violently. The tree blows up.

Another watery detour available to lightning is a person standing under the tree. A human, like a tree, is mostly water. A bolt of lightning that hits a tree will enter anyone in contact with the tree. It can also enter someone quite far from the trunk of the tree: lightning that enters the phloem travels to the roots. Most trees have elaborate roots that grow horizontally beneath the soil surface. They may extend more than 20 meters (66 ft) from the base of the tree. If someone is standing on the soil above such a root during a thunderstorm, lightning may continue its watery journey into the person's feet and upward through the body.

Stay away from trees, especially oak trees, during a thunderstorm. While you may get soaked, huddled in the rain far from the tree canopy, you are less likely to be hit by lightning.

Figure 14.B A tree is particularly vulnerable to lightning.

described in the accompanying Wellness Report.) When water is removed from a plant, what remains is mainly cellulose—the large polysaccharide, formed of glucose, from which cell walls are built. Glucose, in turn, is synthesized from carbon dioxide and water.

Besides water and carbon dioxide, a plant is made of inorganic ions from the soil water—the soil nutrients. Examples of soil nutrients are nitrate (NO_3^-), sulfate ($SO_4^=$), phosphate ($HPO_4^=$), and potassium (K^+).

Soil

Soil is a complex mixture of rock particles, air, water, **humus** (decaying vegetation), plant roots, bacteria, fungi, protozoa, and soil animals (fig. 14.8). Its composition varies from place to place, depending largely on the climate, type of rock from which it formed, and size of the rock particles. Soil composition plays a large role in determining the kinds of plants that can grow in an area.

Inorganic ions first enter the soil when rain and melting snow dissolved minerals out of the rock particles.

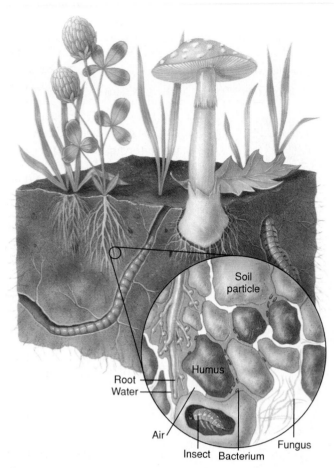

Figure 14.8 Soil. A soil is a complex mixture of plant roots, soil particles, humus, air, water, decomposers (mainly fungi and bacteria), and soil animals.

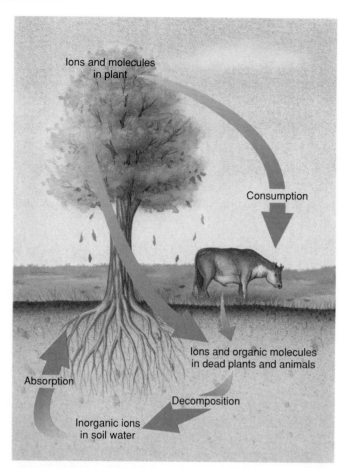

Figure 14.9 Inorganic ions cycle between the soil and the organisms. Inorganic ions in the soil are absorbed by plant roots. Some, such as the potassium ion, remain as ions in plants, where they play important roles in this form. Others become components of organic molecules in plant cells. When an animal eats a plant, the plant's ions and molecules become part of the animal. When these and other organisms die, their molecules are converted into inorganic ions in the soil, where they can be absorbed again by plants.

As the minerals dissolve, they become inorganic ions in soil water. An exception to this entry route is nitrogen, which first enters the soil when bacteria convert nitrogen gas (N_2) from the air into ammonium (NH_4^+) and nitrate (NO_3^-).

Once ions become part of the soil, they are recycled over and over again (fig. 14.9). Inorganic ions are absorbed into plants by their roots and then pass from plants to herbivores, as components of their food, or return to the soil as discarded leaves or other dead parts. Ultimately, all these ions return to the soil when the organisms die and are decomposed by bacteria and fungi. Decomposition releases inorganic ions into the soil, where they can be absorbed again by plants.

Some inorganic ions dissolve in the soil water and others cling to oppositely charged ions on the surfaces of soil particles and humus. Ions held in this way are more available to plants, for they remain in the soil rather than move downward in water until they are beyond

reach of the plant roots. Clay holds mainly ions with positive charges, whereas humus holds mainly ions with negative charges:

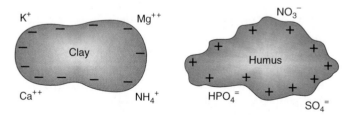

The availability of ions to plants also depends on the pH (see page 33) of the soil. When the soil water becomes more acidic or more alkaline, some of its ions form molecules that are neutral in charge. These molecules,

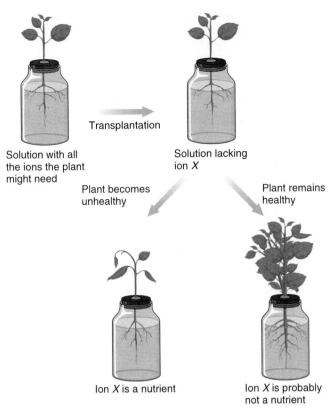

Transplantation

Solution with all
the ions the plant
might need

Solution lacking
ion *X*

Plant becomes
unhealthy

Plant remains
healthy

Ion *X* is a nutrient

Ion *X* is probably
not a nutrient

Figure 14.10 How nutrient requirements of a plant are determined. Whether or not a particular ion is a nutrient is determined by growing the plant without the ion. The test begins by growing the plant in a solution that contains all the ions that the plant might need. Then the plant is transplanted to a solution that lacks one of these ions (ion *X* shown here). If the plant becomes unhealthy, then the missing ion must be a nutrient. If the plant remains healthy, then the missing ion is probably not a nutrient.

A macronutrient occurs in relatively large amounts within a plant (more than 0.5% of the dry weight). Carbon, hydrogen, and oxygen are macronutrients, as are the following six soil nutrients: nitrogen, potassium, calcium, magnesium, phosphorus, and sulfur. Except for potassium, each macronutrient is a component of one or more organic molecules synthesized by the plant. All proteins require nitrogen and some require sulfur. Nucleic acids (DNA and RNA) and ATP are built in part from nitrogen and phosphorus. Cell membranes are built of phospholipids, which contain phosphorus. Chlorophyll contains both nitrogen and magnesium. Calcium forms part of cell walls and, as an ion, helps control what enters and leaves a cell. Potassium, used only in its ionic form (K^+), is the most abundant ion inside cells and the one that controls turgor pressure. The transfer of potassium ions into and out of guard cells, for example, controls the opening and closing of stomata.

A micronutrient occurs in tiny amounts within a plant (less than 0.5% of the dry weight). The seven micronutrients required by all plants are chlorine, iron, boron, manganese, zinc, copper, and molybdenum. Many of these ions act as coenzymes, which activate enzymes by attaching to them. Additional micronutrients required by at least some plants are cobalt, nickel, silicon, and sodium.

Nutrient Deficiencies

A plant is not healthy if it does not receive an optimal amount of each nutrient. Common signs of nutrient deficiency are listed in table 14.1.

Some nutrients are highly mobile within the plant and are transferred from older stems and leaves into developing buds and leaves as needed by these rapidly growing structures. For these nutrients, which include magnesium and phosphorus, signs of deficiency appear first in the older tissues. Other nutrients, like calcium and sulfur, are immobile and remain in older tissues so that signs of deficiency appear first in the newly formed tissues.

Nutritional Adaptations   Plants in their natural environments are adapted to the soil found there. They are genetically programmed to cope with the particular availability of each nutrient. Soils poor in calcium, for example, support plants that are very efficient at absorbing this nutrient. Wild plants in their natural soils seldom develop nutritional deficiencies.

Domestic plants grown in fields, gardens, and pots often develop nutritional deficiencies, for it is difficult to simulate their natural soils. The main way we adjust soils to meet the needs of domestic plants is by using fertilizers.

Fertilizers   Under natural conditions, soil nutrients are recycled on a regular basis. In gardens and fields, however, they are removed in the harvest rather than recycled. Most

which do not dissolve in water, are not available to a plant. Moreover, acidic soils are deficient in positively charged ions other than hydrogen (H^+), because the hydrogen ions attach to negative charges on clay and humus, thereby preventing attachment of other ions.

Soil Nutrients

Sixteen nutrients are essential for plant growth and reproduction. Three of them—carbon, hydrogen, and oxygen—come from carbon dioxide in the air and water in the soil. The remaining thirteen are soil nutrients, absorbed as inorganic ions in soil water. The particular soil nutrients required by a plant are determined by growing the plants in solutions that lack each kind of ion (fig. 14.10). If the plant becomes unhealthy in one of the solutions, then the particular ion that was absent in that solution is considered a soil nutrient. There are two categories of nutrients: macronutrients and micronutrients (table 14.1).

Table 14.1

Soil Nutrients

Nutrient	Form Absorbed by Plants	Some Functions	Signs of Deficiency
Macronutrients			
Nitrogen	NO_3^-; NH_4^+	Part of proteins, nucleic acids, and ATP	Loss of green color in lower leaves (yellow, red, or purple leaves)
Potassium	K^+	Control of osmosis; activates certain enzymes	Loss of green color in lower leaves; brown spots along edges of leaves
Calcium	Ca^{++}	Membrane function; synthesis of cell walls	Little growth of meristems
Magnesium	Mg^{++}	Part of chlorophyll; activates certain enzymes	Loss of green color between vascular bundles; patches of red or purple on older leaves
Phosphorus	$H_2PO_4^-$; $HPO_4^=$	Part of nucleic acids, phospholipids, and ATP	Stunted plant; dark green older leaves
Sulfur	$SO_4^=$	Part of certain proteins	Loss of green color in young leaves (yellow leaves)
Micronutrients			
Chlorine	Cl^-	Control of osmosis; photosynthesis	Wilted leaves; stunted roots; fewer fruit
Iron	Fe^{+++}; Fe^{++}	Part of certain enzymes; synthesis of chlorophyll	Loss of green color between vascular bundles in young leaves
Boron	$H_2BO_3^-$	Nutrient transport; synthesis of nucleic acids	Thick, dark leaves; little growth of meristems
Manganese	Mn^{++}	Activates certain enzymes; photosynthesis	Small, speckled spots on leaves
Zinc	Zn^{++}	Synthesis of hormones and proteins	Very small leaves; loss of green between vascular bundles; fewer flowers
Copper	Cu^+; Cu^{++}	Activates certain enzymes	Withered tips of leaves; dark leaves
Molybdenum	$MO_4^=$	Part of certain enzymes	Loss of green color; mottled, wilted leaves

of the nutrients in crops do not return to the soil. Farmers and gardeners replace these nutrients with fertilizers. Some fertilizers are inorganic and others are organic.

Inorganic fertilizers are manufactured in laboratories or mined from rocks. They typically consist only of nitrogen, phosphorus, and potassium. The proportions of these nutrients are indicated by a three-number code, which lists nitrogen first, then phosphorus, and finally potassium. A fertilizer marked 20-10-5, for example, contains more nitrogen and less phosphorus than a fertilizer marked 10-20-5.

The advantage of inorganic fertilizers is control over the proportions of nutrients added to the soil. If you are growing spinach, for example, you want a fertilizer high in nitrogen because this nutrient stimulates leaf growth. If you are growing tomatoes, by contrast, you want a fertilizer low in nitrogen to shift growth from the leaves to the fruits.

Organic fertilizers come from organisms. As organic wastes, such as feces, and dead organisms decompose, their inorganic ions are released into the soil water. The most common organic fertilizer is manure, the nitrogen-rich feces and urine from cows, sheep, and other domestic animals (table 14.2). The earliest farmers, thousands of years ago, learned to improve their crops by spreading manure on their fields. Native Americans used a novel form of fertilizer: they placed fish (rich in phosphorus and nitrogen) next to corn seeds as they planted. Other organic fertilizers include partially decayed vegetation (as peat or compost), seaweed, and wood ashes.

One advantage of an organic fertilizer is that its nutrients are released slowly, as organic molecules are broken down by fungi and bacteria. Such nutrients are absorbed immediately by plant roots. An inorganic fertilizer, by contrast, may enter the soil as ions too numerous for plants to absorb. A large proportion of these nutrients

Table 14.2

Pounds of Nutrients in 1,000 Pounds of Animal Manure

Animal	Nitrogen	Potassium	Phosphoric Acid*
Horse	154	150	81
Cow	137	140	92
Sheep	175	133	88
Pig	331	130	158
Chicken	293	72	119

*Phosphoric acid is H_3PO_4; a plant takes in phosphorus as $H_2PO_4^-$ or $HPO_4^=$.

travel downward through the soil to the water table. From there, they enter and pollute lakes.

Another advantage of organic fertilizers is they add humus, which helps a soil hold water and nutrients. Manure, for example, is mostly undigested plant materials that become humus in the soil.

A plant itself does not distinguish between organic and inorganic sources of nutrients. It absorbs the nutrients it needs regardless of the source. Plants grown with organic fertilizers are not healthier to eat than plants grown with inorganic fertilizers. The value of organic gardening lies in the choice of pesticides. Plants grown organically are not exposed to chemical pesticides, which may be toxic to humans. Organic gardeners eliminate weeds by pulling, hoeing, and mulching; and they control herbivorous insects by handpicking, introducing parasitic wasps, and supplying insectivorous birds, such as swallows, with bird houses.

SUMMARY

1. A plant transports water and inorganic ions through its xylem. According to the transpiration-cohesion theory, water is pulled upward by an interaction of three forces: transpiration, which lowers water pressure in the leaves; adhesion, which holds together water molecules in the xylem; and cohesion, which holds water molecules to the sides of the xylem tubes.
2. Sugars and other cell products are transported within the phloem. According to the pressure-flow theory, water is drawn into the phloem at a source (where sugars are added), creating a region of higher pressure; and water is drawn out of the phloem at a sink (where sugars are removed), creating a region of lower pressure. Fluid then moves in the phloem from regions of high to regions of low pressure.
3. A plant requires certain nutrients—inorganic molecules and ions—for its cellular activities. Inorganic ions enter a root as soil water, which travels from the epidermis toward the xylem from cell to cell or along cell walls. Uptake of inorganic ions is controlled by cells of the epidermis (in the cell-to-cell pathway) and cells of the endodermis (in the cell wall pathway). Inorganic ions that reach the xylem are actively transported into the tubes, and water follows by osmosis.
4. Many plants enlarge their surface areas for absorption of soil water by forming mycorrhizae. These structures are mutualistic relationships between the plant and a fungus.
5. Soil contains nutrients that recycle from soil to organisms to soil. Native plants are rarely deficient in nutrients, because they are adapted to the particular conditions of their soil. Domestic plants, in unnatural soils, often exhibit signs of nutrient deficiency. Fertilizers replace nutrients that have been removed from the soil.

KEY TERMS

Adhesion (page 257)
Casparian (kah-SPAH-ree-ahn) **strip** (page 263)
Cohesion (page 257)
Humus (HYOO-muss) (page 265)
Mutualism (page 264)
Mycorrhiza (meye-koh-REYE-zah) (page 264)
Pressure-flow theory (page 259)
Soil nutrients (page 261)
Transpiration (page 256)
Transpiration-cohesion theory (page 256)

STUDY QUESTIONS

1. What is transpiration? Why does a plant transpire so much?
2. Describe the transpiration-cohesion theory of water transport in plants. Include in your answer a description of transpiration, cohesion, and adhesion, as well as where energy is used and where it comes from.
3. Describe the pressure-flow theory of sugar transport in plants. Include in your answer a description of a source and a sink, as well as where energy is used and where it comes from.
4. Trace the two pathways taken by water and nutrients as they travel from the soil to the xylem of a root. What is the advantage of having soil water pass through at least one cell before reaching the xylem?
5. Explain the relations among soil particles, soil nutrients, and fertilizers.

CRITICAL THINKING PROBLEMS

1. Why can it be said the movement of water through a plant begins at the leaf?
2. Cut flowers last longer when their stems are placed under water before cutting, and then kept underwater without exposure to air. Explain why this is so.
3. Describe, using a specific plant, how different seasons affect the sources and sinks.
4. What would happen to life on earth if soil nutrients were not recycled? List several activities that interfere with the recycling.
5. Suppose that you want to transplant wild blueberries into your garden. What steps would you take to ensure that this plant receives enough nutrients to produce blueberries?

SUGGESTED READINGS

Introductory Level:

Raven, P. H., R. F. Evert, and S. E. Eichhorn. *Biology of Plants*, 5th ed. New York: Worth Publishers, 1991. An elementary text that describes the experimental evidence for both xylem and phloem transport.

Rupp, R. *Red Oaks and Black Birches*. Pownal, VT: Storey Communications, 1990. A delightful description of everything you might want to know about trees—their history, distributions, structures, and uses by humans.

Stewart, D. "Green Giants," *Discover*, April 1990, 61–64. A description of sequoias: the history of their discovery, their amazing dimensions, and how they transport water.

Zimmerman, M. H. "Piping Water to the Treetops," *Natural History*, July 1982, 6–13. A description of how trees transport water to their leaves while at the same time guarding against dehydration and injury.

Advanced Level:

Cronshaw, J., W. J. Lucas, and R. T. Giaquinta (eds.). *Plant Physiology: Vol. 1: Phloem Transport.* New York: Alan R. Liss, 1986. A detailed review of what we know about phloem transport.

Hartmann, H. T., A. M. Kofranke, V. E. Rubatzky, and W. J. Flocker. *Plant Science: Growth, Development, and Utilization of Cultivated Plants*. Englewood Cliffs, NJ: Prentice-Hall, 1988. A textbook that explains how knowledge of plant structure and growth can be applied to the cultivation of crops.

Zimmerman, M. H. *Xylem Structure and the Ascent of Sap*. New York: Springer-Verlag, 1983. A monograph that relates the structure of xylem cells to their role in transporting water.

Flowers contain a plant's sex organs. The orange spikes in these Calla lilies are female sex organs. The male organs are deeper inside the flower.

Reproduction in Flowering Plants

Objectives

In this chapter, you will learn
–about reproductive characteristics and
  life spans of plants;
–how the egg and sperm form in a
  flower;
–about the transport of pollen (sperm)
  by animals;
–how plant embryos develop;
–about the growth of mature roots and
  stems.

ur ancestors, many centuries ago, believed that flowers were nature's symbols of purity. Flowers were carried by brides, worn as corsages, and placed on coffins. When an English physician first suggested, in 1676, that flowers might be sexual organs, London society was so repulsed by the idea that the doctor was nearly driven out of town. It still surprises many of us to learn that flowers are sex organs.

The function of a flower is not to please us but rather to attract another kind of animal—one that will enter the flower of a plant, pick up pollen (which contains sperm), and place it near the egg of another plant. In return for helping the plant reproduce, the animal receives food. Some animals eat a portion of the pollen itself, but most animals feed on the sugary nectar offered as a reward by the flower. A flower advertises its food in ways that attract certain kinds of animals.

Most flowers are brightly colored because most animals that visit flowers have color vision. Bees are attracted to blue, butterflies to pink, and birds to red. Flowers that attract nocturnal animals, like moths and bats, are white because it shows up best in the dark.

Flowers also lure animals with perfumes. Insects have superb senses of smell. Bees prefer the sweet fragrance of such flowers as violets and lavender, moths prefer the heavy fragrance of gardenias and evening primroses, and flies are attracted to the smell of rotting meat, a fragrance imitated by some flowers. Bats have a keen sense of smell and are drawn to the heavy odors of large pale flowers, such as the saguaro cactus and organ-pipe cactus. Birds, on the other hand, can hardly smell at all; bird-pollinated plants, such as the Indian paintbrush and scarlet gilia, do not waste energy on perfumes.

It is only a coincidence that we appreciate the beauty of flowers. We appreciate many of the same colors and odors as the animals that plants have evolved to attract. We also enjoy the taste of nectar after it has been processed by bees into honey.

Our attraction to flowers began a very long time ago. Evidence of flowers used in a burial ritual has been found in a grave that is between 40,000 and 60,000 years old. An ancient hunter, his skull crushed, lies in the grave on a bed of bachelor buttons and hollyhocks.

Reproductive Patterns

Plants reproduce in a variety of ways. A desert poppy, for example, grows quickly from a seed after a heavy rain and then makes a "suicidal" effort to flower and develop its own seeds within just a few weeks. The tiny plant then dies. An oak tree has a more conservative life-style. It does not begin to reproduce until at least ten years old, devoting its initial years to growth instead of offspring, but then produces acorns every year for several hundred years. Each way of life is an adaptation—an inherited trait that maximizes reproductive success in a particular environment.

Inheritance: One Parent Versus Two

Many plants have the capacity to reproduce asexually, in which offspring develop from a single parental cell. All the offspring are genetically identical to the parent. The advantages of asexual reproduction are twofold: (1) a successful individual produces identical copies of itself, which will also be successful as long as the environment remains the same, and (2) reproduction can be accomplished by just one individual, who may live far from others of its kind.

All plants have the potential to reproduce sexually, in which each offspring inherits a combination of genes from two parents. Sexual reproduction generates a variety of offspring, which is an advantage when the environment changes from place to place or through time. An individual that leaves a variety of offspring has a greater chance that some of them will be well-matched to the new environment than an individual that produces just one kind of offspring.

Asexual Reproduction   Asexual reproduction in plants occurs either by vegetative reproduction or apomixis.

In **vegetative reproduction,** a severed part of the parent grows into a new individual: a leaf grows a new stem and root; a stem grows a new root; or part of a root grows a new stem. This kind of reproduction enables a plant to spread quickly throughout a favorable environment. The structures that grow into new plants, as well as examples of plants that use them, are listed in table 15.1. A spectacular example of vegetative reproduction is the quaking aspen, which, in the Rocky Mountains, has reproduced chiefly by suckers for thousands of years (fig. 15.1).

In **apomixis,** an embryo develops from a single cell in the parent rather than from a fertilized egg. This cell is located in the flower. The seeds of a dandelion, for example, usually contain embryos that develop from unfertilized cells. Other plants that reproduce by apomixis include lemon trees, orange trees, and some grasses. Like vegetative reproduction, the offspring are genetically identical to their parent. Unlike vegetative reproduction, the young are dispersed, as embryos in seeds, away from the parent.

Sexual Reproduction   Sexual reproduction involves an alternation of generations, in which a generation of individuals with haploid (see page 143) cells (one set of chromosomes) alternates with a generation of individuals

Table 15.1

Modes of Vegetative Reproduction

Part of Plant	Name of Structure	Examples
Leaf	Plantlet	African violet
Branch of a stem		
Aboveground	Stolon (or runner)	Strawberry; lily of the valley
Underground	Rhizome	Blueberry; cattail; bamboo; many grasses
Storage region of underground stem		
With leaves	Bulb	Daffodil; tulip
Without leaves	Corm	Crocus; gladiolus
	Tuber	White potato*
Root	Sucker	Aspen; blackberry; many grasses

*Potatoes are typically grown from pieces of last year's crop rather than from seeds. Each "eye" can form a new stem and root.

Figure 15.1 Aspen clones. These trees reproduce vegetatively, by suckers, to form clones of genetically identical individuals. Different clones are easily distinguished in autumn when their leaves turn unique shades at different times.

with diploid (*see page 143*) cells (two sets of chromosomes, one from each parent). The two generations are linked by meiosis (*see page 145*) and fertilization:

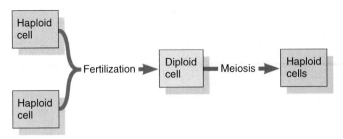

Genetic differences among offspring are generated by both meiosis and fertilization.

In plants, the individual with haploid cells is called a **gametophyte** (*gameto*, gamete; *phyte*, plant). A male gametophyte produces sperm and a female gametophyte produces eggs. A sperm and an egg unite to form a diploid cell (the fertilized egg) that develops into a **sporophyte,** which is an individual with diploid cells:

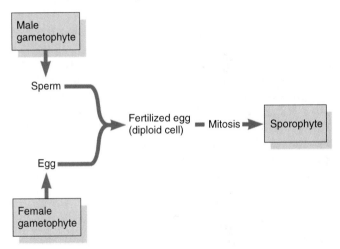

The sporophyte is the main body of a flowering plant. It produces gametophytes, which live as tiny structures within its flowers.

Life History

The life history of an organism is its pattern of maturation, reproduction, and aging. Important events include the age at first reproduction, number of offspring in each "litter," number of litters in a lifetime, and maximum life span. Each species has a unique life history. Among plants, life histories fall into three patterns: annual, biennial, and perennial.

Annuals   An **annual** plant lives for less than a year. It develops from a seed, grows, flowers, forms its own seeds, and then dies. The entire life may be over within a few weeks.

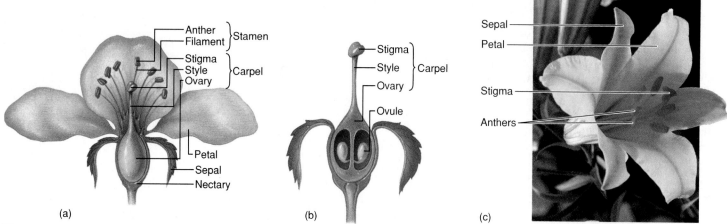

Figure 15.2 The structure of a flower. (*a*) A flower has whorls of sepals, petals, stamens, and carpels. Most flowers have a nectary, with glands that secrete nectar. (*b*) Fertilization and early development of the embryo take place inside an ovule, located within the ovary. (*c*) Photograph of a lily, showing the arrangement of its parts.

Not surprisingly, most annuals are quite small. They live in places that are favorable for only short periods of time. A desert with a short rainy season or a forest that has just burned provides a brief period of favorable conditions for an annual. We choose to grow annuals in our fields and gardens because they produce leaves, seeds, and fruits within a few months. Lettuce and peas are examples of annuals.

Biennials   A **biennial** plant lives for two years. The first year it grows from a seed but does not reproduce. Excess sugars are built and stored as starch in roots or underground stems. The following year, the starches are used to develop flowers and seeds. As soon as the seeds form, the plant dies.

Not surprisingly, we grow biennials for their large stores of starch. Carrots, radishes, and turnips are biennials. Many plants in environments with short summers, as in alpine tundra, are biennials because it takes two seasons to grow and reproduce.

Perennials   A **perennial** plant lives for many years. Trees and shrubs are perennials, as are many herbs (e.g., daisies) and grasses. They require more than two years to mature, after which they reproduce every year for many years. The oldest known perennial, a bristlecone pine, is almost 5,000 years old.

In trees and shrubs, the aboveground structures (except perhaps the leaves) remain during all seasons of the year. In other perennials, such as grasses and pansies, the aboveground structures die at the end of each growing season but the roots and underground stems survive and provide nutrients for rapid growth at the beginning of the next growing season.

A few species live for many years but reproduce only once. After growing for many years, they have a single reproductive season and then die. A century plant, for example, lives ten to seventy-five years before it blooms and dies (see fig. 16.5); a bamboo may live more than a hundred years before it reproduces and dies.

The Flowers

Flowering plants diverged from early gymnosperms (the ancestors of modern conifers) approximately 130 million years ago. The flowers represented an innovative way of reproducing, in which animals, rather than wind or water, transported sperm to an egg. Discovery of the oldest fossil flower is described in the accompanying Research Report. The innovation was so successful that today approximately 96% of all plant species are flowering plants.

Flowers are the reproductive organs of a flowering plant. They produce and enclose the male and female gametophytes, which form sperm and eggs. Flowers are also the sites of fertilization and early development of the embryo.

Flower Structure

A flower has four kinds of structures: sepals, petals, stamens, and carpels (fig. 15.2). Each structure is divided into several parts and arranged in a whorl within the flower:

Research Report

The Problem of the Missing Flowers

Until recently, most biologists accepted the hypothesis that the first flowering plants were trees similar to magnolia trees, with large cuplike flowers (fig. 15.A). Like magnolia flowers, the first flowers were believed to have both male and female organs in each flower. The hypothesis is based on the following observations: (1) flowering plants apparently evolved from gymnosperms (plants, like conifers, with cones rather than flowers), which are almost always trees; (2) magnolias resemble an extinct gymnosperm that lived just prior to the appearance of flowering plants; (3) the oldest fossil flower pollen, dating back 130 million years, closely matches the pollen of magnolia trees; and (4) the oldest fossil flowers, dating back 100 million years, look like magnolia flowers.

While all the evidence was consistent with the hypothesis, a mystery remained: why were there no fossils of leaves, stems, or flowers from the early flowering plants—why the gap between 130 million years ago, the age of the oldest pollen, and 100 million years ago, the age of the oldest flower?

This mystery may now be solved. In 1990, two scientists from Yale University—David Taylor, a biologist, and Leo Hickey, a geologist—described a fossil of a flowering plant that lived 110 million years ago. The type of plant represented by the fossil is not new—many such fossils had been found, but they were poorly preserved and classified as ferns. What is new is the quality of the fossil—so perfect that details of floral parts can be seen for the first time.

Surprisingly, this oldest fossil of a flowering plant is not at all like a magnolia tree. It is small—less than 50 centimeters (20 in) tall—and apparently lived in wet places. Each flower is tiny—just 1 millimeter wide—with either male organs or female organs. The flowers have no petals but are formed of two small leaves adjacent to the male or female organs (supporting another hypothesis—that petals evolved from leaves).

An exciting consequence of this discovery is that we may now find many more fossils of the early flowering plants. Scientists have been looking for fossils of trees with large flowers; they may have overlooked fossils of small plants with tiny flowers. The discovery also reopens the question of what kind of animal pollinated the first flowers. If the earliest flowers were large and cup-shaped, like the magnolia, then the pollinators were probably beetles, which mate on flowers of this type while they feed on the pollen. If the earliest flowers were as tiny as the 110-million-year-old fossil, then the earliest pollinators were smaller than beetles, perhaps tiny moths.

Figure 15.A A magnolia tree and its flower.

The outermost whorl of a flower consists of **sepals,** which are usually green and leaf-shaped. They enclose and protect the inner parts of a flower while it is still a bud.

The next whorl of flower parts consists of **petals.** They are usually colorful and serve to attract insects and other animals that carry pollen from one flower to another.

The third whorl of structures consists of **stamens,** which are the male reproductive organs. Each stamen has an anther and filament. The anther produces pollen grains, which are male gametophytes, and the filament holds the anther in a position favorable for dispersal of the pollen.

The innermost whorl of floral structures consists of **carpels,** which are the female reproductive organs. A carpel consists of a stigma, style, and ovary. The stigma forms the upper surface and is usually sticky. Its function is to receive pollen grains. The style connects the stigma with the ovary at the bottom; a pollen grain grows through the style on its way from the stigma to the ovary. The ovary holds the ovules, which form the female gametophytes and, if fertilized, become seeds.

In some flowers, the female organ consists of a single carpel and in others it consists of more than one carpel, often fused at the bottom. (The entire organ is sometimes referred to as a pistil regardless of its number of carpels.) The three carpels of a cucumber flower, for example, are fused at the bottom but remain separate at the top:

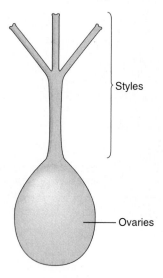

Some plants cluster their flowers into a larger structure called an inflorescence. The blossom of a daisy, sunflower, or dandelion, for example, is actually many tiny flowers bunched together.

Perfect Versus Imperfect Flowers   A flower with both stamens and carpels is a perfect flower, as it has reproductive organs of both sexes. A flower with only stamens or only pistils is an imperfect flower; it has either male or female reproductive organs. In some species, both kinds of imperfect flowers—male and female—occur on each individual; such species are said to be **monoecious** (*mon,* one;

oecious, house). Examples of monoecious plants are the squash, cucumber, pumpkin, hazel, hickory, and walnut.

A plant with both male and female reproductive organs (i.e., a plant with perfect flowers or a monoecious plant with imperfect flowers) usually has some way of preventing its eggs from being fertilized by its own sperm. Self-fertilization is typically avoided, since it would counter the advantages of sexual reproduction.

How does a plant prevent self-fertilization? In some species, male and female organs on the same plant do not mature at the same time. In other species, the positions of the organs impede the flow of pollen from an anther to a stigma of the same flower. The style, for example, may extend far above the anther so that pollen would have to fall upward to reach the stigma. A third way of preventing self-fertilization, found in more than half of all flowering plants, is to have a protein on the stigma that recognizes and rejects pollen from the same plant. The pollen of a petunia, for example, will not grow on the stigma of its own flowers.

In some plant species, each individual plant has only one kind of imperfect flower: some individuals are males and others are females. Such species, called **dioecious** (*di,* two; *oecious,* house) species, comprise about 5% of all flowering plants. They cannot self-fertilize. Examples of dioecious plants are spinach, asparagus, holly, cottonwood, and willow.

Variations in Color, Shape, Scent, and Nectar   Flowers originated as adaptations for attracting animals, which help a plant reproduce by carrying pollen from one flower to another. The transfer of pollen from the male part of one flower to the female part of another is called **pollination.** The animal that does this is a **pollinator.**

A flowering plant reproduces more successfully when its pollinator does not visit the flowers of other species, for pollen left in these flowers is wasted—it does not lead to offspring. A pollinator, on the other hand, seeks only food and would readily visit all kinds of flowers if it could. Flowering plants keep their pollinators to themselves by having flowers that attract only a particular kind of animal. For this reason, the flowers of different species vary greatly in size, shape, scent, and color, as well as the time of day when they are open and secrete nectar.

A flower that attracts hummingbirds has its petals fused to form a long tube that fits the hummingbird's bill and tongue. There is no landing platform, for the bird hovers while it feeds. The flower is red, the color seen most acutely by birds, and has no scent, for birds do not have a keen sense of smell. A hummingbird flower secretes high-calorie nectar during the day when hummingbirds feed.

A flower that attracts moths typically has a flat surface for the moths to stand on while feeding. Its nectar is located within a short tube, which is white because that is the color seen best at night when moths are active. The flower gives off a heavy odor. Moth-pollinated flowers open and secrete their nectar at night.

Figure 15.3 Most flowers are pollinated by animals.
(*a*) Hummingbird-pollinated flowers are long, red tubes without landing platforms. (*b*) Bee-pollinated flowers are often blue, with landing platforms. They are often difficult to enter; bees (unlike most pollinators) can use their front legs to push petals away from the opening. (*c*) A butterfly must have a platform and is attracted to pink flowers. (*d*) A bat, which feeds at night, is attracted to large, white flowers. The flower and stalk must be sturdy to support this relatively heavy, clumsy animal.

Other flowers are adapted to entice bees, butterflies, beetles, moths, bats, and other animals. Some of them are shown in figure 15.3. Surprisingly, some flowers have patterns that are visible only in ultraviolet (fig. 15.4), a portion of the light spectrum that we cannot see. Bees, however, are able to see ultraviolet and use these patterns to guide their way into the flower. Characteristics of flowers and their pollinators are listed in table 15.2.

Some flowering plants have evolved adaptations for using wind, in place of animals, to carry their pollen. These plants have tiny flowers without colorful petals, scents, or nectar. Most grasses are wind-pollinated, as are ragweed, birch, beech, maple, and oak. It is pollen from these plants that most often causes allergies, as described in the accompanying Wellness Report.

Formation of Gametophytes

A gametophyte is a multicellular, haploid structure that forms gametes—sperm or eggs. In flowering plants, the gametophytes are located within the flowers. Many male gametophytes form in each anther; a single female gametophyte forms in each ovule, which is inside an ovary.

Pollen Grains A **pollen grain** is the male gametophyte. Initially it consists of two cells, a tube cell and a generative cell that resides within the cytoplasm of the tube cell. The cells are enclosed within a hard coat:

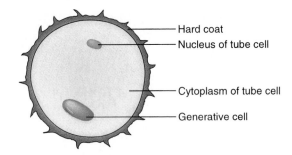

Both cells of a pollen grain play roles during fertilization, as described following. The hard coat protects the cells

(a)

(b)

Figure 15.4 A flower with ultraviolet patterns. (*a*) A marsh marigold as seen by our naked eyes. (*b*) The same flower in ultraviolet light, as seen by a bee.

Table 15.2

Kinds of Flowers and the Pollinators They Attract

Pollinator	Pollinator Characteristics	Shape of Flower	Color of Flower	Scent of Flower	Opening Time of Flower
Moth	Feeds at night; needs a flat place to stand on	Flat; easy to enter	White	Strong; sweet	Night
Butterfly	Feeds during the day; needs a flat place to stand on	Flat; easy to enter	Variable; often pink	Sweet	Day
Fly	Lays eggs on dead meat, which is mottled and has a fetid smell	Flat; easy to enter	Brown or gray; mottled	Fetid	Day
Beetle	Needs flat place to stand on	Cup-shaped; easy to enter	Dull	Fruity	Day and night
Bee	Needs a landing platform; sees blue best; does not see red; uses front legs as hands	Landing platform of fused petals; hard to enter; short tube	Variable but not red	Sweet	Day
Hummingbird	Hovers while feeding; sees red best; poor sense of smell	Long tube; no landing platform	Red; orange	None	Day
Bat	Feeds at night; heavy and clumsy	Large; sturdy	White	Musty (like a sexually receptive bat)	Night

from drying out as they travel to an egg. The surface of a pollen grain is sculpted in a way that is unique to each plant species. The coat is so tough that pollen grains from ancient soils and rocks still retain their distinct sculpture and are used to identify plants that lived in the region millions of years ago.

Ovules An **ovule** has the potential to develop into a seed. Some ovaries have many ovules and others have just one. A young ovule develops an embryo sac, which is a fluid-filled region surrounded by a layer of cells. The cells, in turn, are enclosed within layers of cells called integuments:

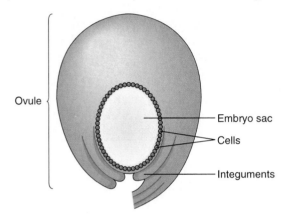

One of the cells surrounding the embryo sac migrates into the sac and undergoes a series of cell divisions (by both meiosis and mitosis) to form a mature gametophyte. The embryo sac of the gametophyte has four haploid cells. One cell has two nuclei, called polar nuclei; it will eventually contribute to formation of a tissue (the endosperm) that provides the embryo with food. Of the remaining three cells, clustered at one end of the gametophyte, one will become the egg:

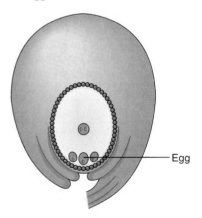

This mature gametophyte is ready to be fertilized.

Union of the Gametes

In sexual reproduction, a sperm and an egg unite to form a fertilized egg that develops into an embryo. In flowering plants it requires two steps: pollination and fertilization.

Pollination is the transfer of pollen, which contains the sperm, from an anther to a stigma. Fertilization is the penetration of a cell by a sperm nucleus followed by fusion of the sperm and egg nuclei.

Pollination Flowers are adaptations that encourage pollination by animals. An animal visits a flower to feed on pollen or nectar. As it moves into or out of the flower, it brushes against an anther and some of the pollen sticks to its body. When the animal enters another flower, some of the pollen brushes off onto the stigma.

A pollen grain on the stigma of a flower forms a pollen tube, which grows downward through the style and into the ovary. Recall that a pollen grain contains two cells. The nucleus of the tube cell directs growth of the pollen tube. The nucleus of the generative cell divides to form two sperm nuclei:

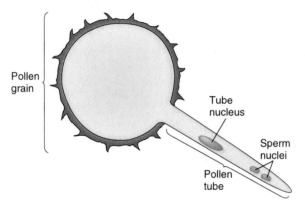

Fertilization The tip of the pollen tube grows into the ovule and delivers its two sperm nuclei. Both nuclei participate in fertilization.

Flowering plants are unique in having **double fertilization.** One fertilization involves union of a sperm nucleus and an egg; it forms a diploid cell (with two sets of chromosomes). The fertilized egg then divides by mitosis to become the embryo. The other fertilization involves union of a sperm nucleus and the cell with two polar nuclei; it forms a triploid cell (with three sets of chromosomes) that divides by mitosis to become the **endosperm,** a tissue that provides food for the growing embryo. In this remarkable way, a flowering plant does not waste food on eggs that never get fertilized. A conifer, by contrast, does not have double fertilization to initiate the development of food stores. It provides food for every egg prior to fertilization, wasting food on eggs that never get fertilized.

Development of the Embryo

An embryo is a multicellular structure capable of developing into an adult. In a flowering plant, the embryo develops inside an ovule (the female gametophyte and integuments)

Hay Fever

The most common allergy is hay fever. This miserable condition, with symptoms of a stuffed-up nose, runny eyes, and sneezing, is an inappropriate response by the immune system to an otherwise harmless material—pollen. (The medical term for hay fever is allergic rhinitis; hay fever is rarely caused by hay and rarely produces a fever.) Many kinds of pollen cause hay fever when breathed into our respiratory passages.

Most of the pollen we inhale is from wind-pollinated plants. All conifers (e.g., pines and spruces) spread their pollen by wind. Many flowering plants also are wind-pollinated, including the maple, oak, birch, elm, beech, ragweed, Russian thistle, pigweed, dock, and most grasses. These wind-pollinated flowering plants are recognized by their tiny flowers, which often lack petals.

Hay fever develops mainly in response to wind-pollinated plants because such plants release vast quantities of pollen, in order to compensate for the pollen's haphazard pathways with regard to their targets—stigmas of other flowers. According to one study, an insect-pollinated plant releases 6,000 pollen grains for every ovule that becomes fertilized (about 90% is eaten by animals), whereas a wind-pollinated plant releases a million grains for each fertilized ovule.

(a)   (b)

Figure 15.B Ragweed: the main cause of hay fever. (*a*) Clusters of ragweed flowers. (*b*) Highly magnified pollen grains from ragweed.

Different plants cause hay fever at different times of the year. In the spring, airborne pollen comes mostly from trees, such as oak, maple, elm, and beech. In late spring and early summer, most pollen is from grasses, such as timothy grass, redtop, orchard grass, fescues, and agricultural grains. In late summer, most airborne pollen is from ragweed (fig. 15.B). In fact, ragweed causes hay fever more often than any other plant. Why is this weed such a nuisance?

A hundred years ago, ragweed was a relatively uncommon plant. Now it grows almost everywhere. The prevalence of ragweed, together with its vast quantities of pollen, are what makes it such a culprit. Ragweed is an annual plant that produces immense numbers of long-lived seeds. Most soils are crammed with ragweed seeds. Whenever the existing vegetation is removed, ragweed seeds germinate and young ragweeds replace whatever used to grow there. Ragweed thrives wherever the natural vegetation has been disturbed—near farms, along roadsides, and next to houses, which are also places where humans spend their time. There is no way to escape it, for when we move to an area without ragweed, we create disturbed habitats that are soon colonized by ragweed. And when ragweed reproduces, its pollen is practically everywhere.

as it matures into a seed. The ovule is inside the ovary of a flower. Development involves three processes: cell division, cell growth, and cell differentiation.

In cell division, one cell forms two identical cells by mitosis. It begins with the fertilized egg, which divides to form two cells, which divide to form four cells, and so on until the multicellular embryo is formed. Between cell divisions, a plant cell grows rapidly as it absorbs water and expands 10 to 100 times its previous size. Cell differentiation is the formation of specialized cells—photosynthetic cells,

storage cells, support cells, and so on. During this process, certain genes in each cell are activated while others remain inactive. In a photosynthetic cell, for example, genes for building the enzymes of photosynthesis are activated whereas genes for building root hairs remain inactive.

The Seed

The triploid cell, formed when a sperm nucleus fertilized the cell with two polar nuclei, divides by mitosis to form

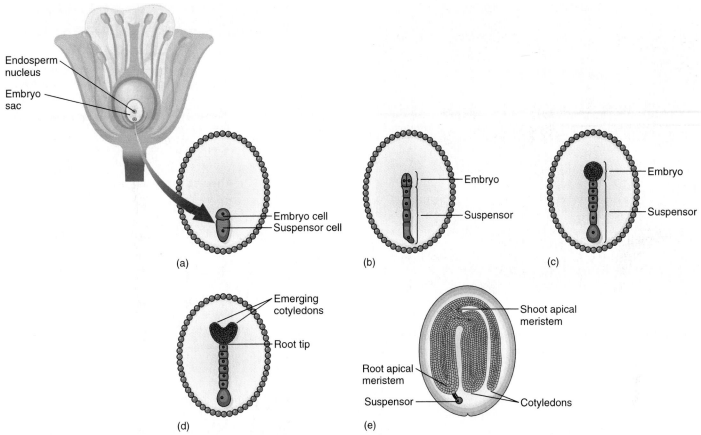

(a)   (b)   (c)

(d)   (e)

Figure 15.5 Early development of a plant embryo. (*a*) The fertilized egg divides to form a suspensor cell and an embryo cell. (*b*) The suspensor cell divides repeatedly to form a suspensor. (*c*) The embryo and suspensor enlarge by cell division and cell growth. (*d*) The embryo develops one or two cotyledons. (*e*) Once the embryo develops a shoot apical meristem and a root apical meristem, it forms a seed coat and enters a period of dormancy.

the endosperm. The cells of this tissue absorb nutrients from the xylem and phloem in the ovary and stockpile them for later use. The stored nutrients are mostly starches but also sugars, amino acids, proteins, and oils. The endosperm transfers the nutrients, as needed, to the growing embryo. Not surprisingly, we also consume these nutrients. White bread, for example, is made of starch from wheat endosperm, cornflakes are made of starch from corn endosperm, sweet corn contains sugar from corn endosperm, and coconut milk is a rich blend of oils, starches, and sugars from coconut endosperm.

When the fertilized egg is completely surrounded with endosperm, it begins to develop into an embryo (fig. 15.5). First it divides into two cells: a suspensor cell and an embryo cell. The suspensor cell then divides repeatedly to form a suspensor, which is a column of cells that connects the developing embryo to the inner wall of the ovule. The suspensor anchors the embryo and transfers nutrients into it from the endosperm. The embryonic cell divides repeatedly to form three embryonic tissues: cotyledon, root apical meristem, and shoot apical meristem.

A cotyledon is an embryonic leaf that absorbs nutrients from the endosperm and transfers them to the rest of the embryo. (It replaces the suspensor cell in this function.) A dicotyledon has two cotyledons, as shown in figure 15.5, whereas a monocotyledon has just one.

The root and shoot apical meristems add new cells to the tips of the root and stem. They continue to function throughout the life of a plant.

A seed coat develops from the integuments of the ovule. The seed, consisting of embryo, endosperm, and seed coat, then loses water and the coat hardens into a waterproof covering. At this point the embryo becomes isolated, inside the seed, from its parent. It enters a resting state called dormancy.

The Fruit

While the ovule is developing into a seed, the ovary around it is developing into a fruit (fig. 15.6). A fruit is a structure that develops from an ovary and contains seeds; watermelon, squash, peas, and peaches are examples of fruits.

Maturing ovary

Figure 15.6 The ovary of a flower develops into a fruit.
As the ovary of this zucchini squash grows into a mature
fruit, the rest of the flower is pushed outward and eventually
falls off.

The primary function of a fruit is dispersal—movement
of the seed (and embryo) away from the parent. The embryos are transported to sites where they will not compete
with their parents for sunlight, water, and soil nutrients.

Fruit size and shape vary greatly depending on their
mode of dispersal. Fruits of the maple and milkweed are
aerodynamic and drift with the wind. The pods of peas and
beans split open and eject their seeds. Coconuts float on
water. Sticktights travel on the fur of a mammal, and the
water plantain travels on the feet of a bird. (Charles Darwin, who developed the theory of evolution, once grew a
whole garden from seeds he collected from the feet of
ducks and geese.) Cherries and raspberries are eaten by animals: the fruits are digested but the seeds pass through the
digestive tract and are deposited in feces some distance
from the parent plant. Acorns are buried by squirrels and
jays, for later consumption, and the forgotten ones grow
into oaks. These and other categories of fruits are described in table 15.3. A few are illustrated in figure 15.7.

The Seedling

An embryo remains dormant inside its seed coat until environmental conditions favor development into a seedling.
Dormancy is an adaptation that enables a plant to develop
only when conditions are ideal. The seeds of most plants
can remain dormant for a couple of years, and those of
some annuals for hundreds of years. The soil is a reservoir
of dormant seeds waiting for their particular favorable conditions to germinate.

Table 15.3

Some Categories of Fruits

Category	Description	Examples
Simple fruit	Develops from one ovary in one flower	
Fleshy	Adapted for consumption by birds and mammals	Tomato; peach
Dry	Adapted for dispersal by wind, water, or on the outside of an animal	Sticktight; dandelion
Aggregate fruit	Develops from many ovaries in one flower	Raspberry; blackberry
Multiple fruit	Develops from many ovaries of many flowers	Pineapple; fig
Accessory fruit	Develops from other tissues in addition to ovary	Strawberry; apple; pear

Germination is activation of a dormant embryo. It is
precisely timed to occur when environmental conditions
are favorable, for a young plant is fragile and readily damaged by extremes of temperature and moisture. The seeds
of each species are genetically programmed to germinate in
response to specific environment cues, which are correlated with onset of a favorable season for growth of its
young. Northern plants require a few weeks of cold, which
indicate winter, followed by warmth. Desert annuals require a heavy rain. Other plants respond to a particular
combination of day length and soil temperature.

Germination begins with absorption of water, which
swells the seed and ruptures the seed coat. Cells of the embryo then begin to divide and enlarge. The first structure
to emerge from the seed is a root, and when this happens
the embryo contacts its environment and becomes a
seedling. A stem soon grows upward through the soil and
leaves appear.

Seedling growth differs in dicotyledons and monocotyledons (fig. 15.8). A dicot has two cotyledons, which
may appear aboveground (as in a bean) or remain below-
ground (as in a pea). They degenerate as soon as the stem
forms its own leaves. A monocot has just one cotyledon,
which typically remains underground. The emerging leaves
of a monocot are enclosed within a sheath, called the
coleoptile, that protects them and guides their way
through the soil toward sunlight.

As the stem and root lengthen, strands of elongated
cells develop into xylem and phloem. The seedling then
has all the structures it needs to acquire resources from the
environment. It enters a period of growth.

(a)

(b)

(c)

(d)

Figure 15.7 The function of a fruit is to disperse the seeds. (a) The parachutes of a dandelion help it float on the wind. (b) The fruit of a lupine pops open and the seeds are ejected. (c) Many fruits are carried on the fur of mammals. (d) A strawberry is eaten by a meadow vole and its seeds deposited in feces some distance from the parent plant.

Growth

Plant growth is restricted to the meristem tissues, of which a plant has several kinds (table 15.4). Every plant has a shoot apical meristem, a root apical meristem, and pericycle. A woody plant also has cambium. The shoot and root apical meristem (see fig. 13.8) are responsible for primary growth—lengthening of the stem and root as well as forming new branches on the stem, leaves, and flowers. The pericycle forms branches of the root. The cambium is responsible for secondary growth—widening of the stem and root.

Primary Growth

The embryonic processes of development are division, growth, and differentiation. In an adult plant, these processes can occur within its meristem tissues; a plant retains, throughout life, its potential to form new roots, stems, leaves, and flowers. The particular shape of the plant develops in response to genetic programming combined with local conditions of light and moisture.

The Shoot Apical Meristem The shoot apical meristem is located at the top of the main stem as well as the tips of the branches. It forms new leaves, branches, and flowers.

Shoot apical meristem appears as a dome-shaped tissue surrounded by immature leaves (fig. 15.9). When conditions favor growth, the meristem cells divide and absorb water from the xylem. The stem elongates rapidly. The cells then differentiate and become functioning parts of the plant. When conditions do not favor growth, the meristem forms a terminal bud—inactive meristem covered with immature leaves that protect it from injury and dehydration (fig. 15.10).

As the tip of the stem elongates, small pieces of meristem remain in the axils, which are regions between the leaves and the stem. These pieces of meristem, called lateral buds, may develop into flowers during the reproductive season or into branches that replace injured parts or take advantage of new sources of light. Each point on the stem where leaves and lateral buds occur is a node. The area between two successive nodes is an internode (see fig. 15.10).

The Root Apical Meristem The root apical meristem is located at the tip of each root. It is covered and protected by the root cap. Within this narrow region, all three phases of development are visible (fig. 15.11). Just above the root cap is a zone of cell division. Next is a zone of cell growth and then a zone of cell differentiation. As the root grows, new cells form beneath the root cap and become the new zone of division. The previous zone of division becomes the zone of growth; the previous zone of growth becomes the zone of differentiation; and the previous zone of differentiation becomes mature tissues of the root.

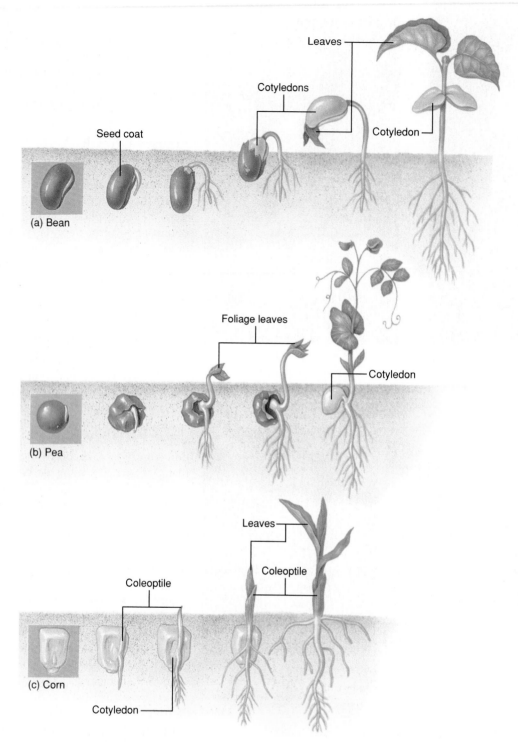

Figure 15.8 Development of dicot and monocot seedlings. (a) In some dicots, like the bean, the cotyledon appears aboveground for a short period of time. (b) In other dicots, like the pea, the cotyledons remain underground. (c) In a monocot, the cotyledon typically remains underground and a coleoptile protects the emerging stem and leaves.

Lateral roots develop from the pericycle rather than from the root apical meristem. The pericycle is a ring of tissue deep within the root, between the endodermis and the vascular tissues (see fig. 13.17). A new branch of the root grows outward through the mature tissues of the root (see fig. 13.18).

Secondary Growth

Trees and shrubs broaden their stems and roots each year in order to support their ever-increasing load of branches. Thickening of the stem and root, called secondary growth, is the product of the vascular cambium and the cork cambium (fig. 15.12).

Table 15.4

Meristem Tissues

Type of Meristem	Location	Structures It Forms
Shoot apical meristem		
Terminal	Tips of main stem and its branches	Stem and leaf (flower in annual or biennial)
Lateral	Between stem and petiole	Branch or flower
Root apical meristem	Tip of main root and its branches	Root
Pericycle	Around vascular tissue of root	Branches of root
Vascular cambium	Between xylem and phloem	Xylem and phloem
Cork cambium	In bark, between phloem and cork	Cork

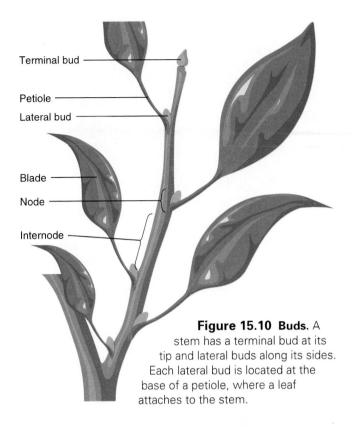

Figure 15.10 Buds. A stem has a terminal bud at its tip and lateral buds along its sides. Each lateral bud is located at the base of a petiole, where a leaf attaches to the stem.

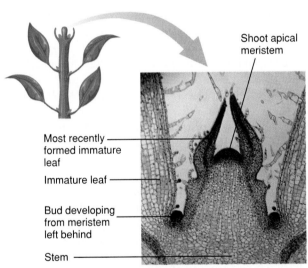

Figure 15.9 Shoot apical meristem. As the meristem grows upward, it leaves behind small pieces of meristem (a bud) that may form a branch, a leaf, or a flower.

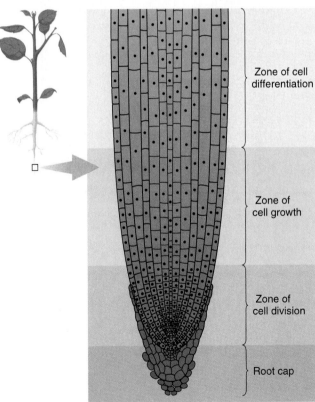

Figure 15.11 Primary growth of the root. A growing root shows three zones of development. In the first zone, just beneath the root cap, cells of the root apical meristem divide to form new cells. In the second zone, just above the meristem, the newly formed cells grow. In the third zone, the full-sized cells differentiate to form specialized cells.

285

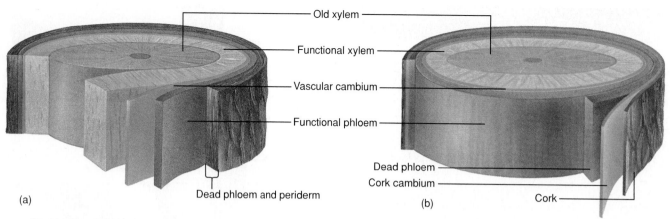

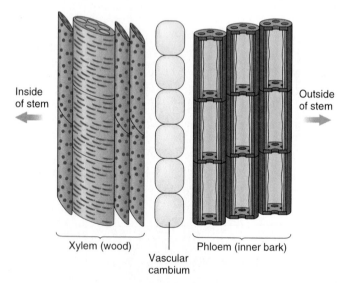

Figure 15.12 The tissues of a woody stem. (*a*) Starting in the core and moving outward, the tissues of a stem are: old xylem, functional xylem, vascular cambium (forms new xylem and phloem), functional phloem, and dead phloem and periderm. Together, the phloem and periderm form the bark. (*b*) The periderm consists of cork cambium and cork cells. The cork cambium forms new cork cells, most of which die and become part of the bark.

Vascular Cambium   The **vascular cambium** is a band of meristem between the xylem and the phloem. New cells that form on the inner side of this cambium become new xylem, called wood. New cells that form on the outer side become new phloem, or inner bark:

Most of the new cells develop into xylem and phloem for vertical transport of fluids within the plant, but some form horizontal chains of cells (called rays) that transport fluids to cells in the center and periphery of the stem.

Each time the vascular cambium forms new xylem, the cambium and all tissues external to it are pushed outward. As older phloem cells are pushed outward, they become crushed against the outer bark and die. Dead phloem becomes part of the bark (see fig. 15.12).

Xylem accumulates from year to year as new cells form and old cells remain deep inside the stem or root. The diameters of new xylem cells reflect the abundance of water at the time they were formed. When rainfall is heavy, the vascular cambium forms xylem that transports more water; the cells are broader with thinner walls. When rainfall is light, the cambium forms xylem that transports less water; the cells are more narrow, with thicker walls. These anatomical differences in xylem cells form annual rings—bands of broader xylem (rainy season) separated by bands of more narrow xylem (dry season), as shown in figure 15.13. Since each ring reflects the seasons of a year, the number of rings reflects the age of a tree.

Cork Cambium   The periderm covers the phloem of a woody stem, forming the outermost layer of bark. Periderm consists of cork cambium and cork cells (see fig. 15.12). The **cork cambium** forms new cork cells. Some of the cork cells remain alive, forming cells that store starch and other materials. Most of the cork cells do not become functional until they die, at which point they form part of the bark.

Bark consists of dead cork cells and dead phloem cells. It protects the tree from temperature extremes, dehydration, fire, and invasions by insects and disease organisms. Dead cork cells are filled with air and their walls are impregnated with suberin, the same waxy material that waterproofs cells of the endodermis. Lenticels—tiny regions of loosely packed cells—are scattered throughout the cork tissue to allow passage of oxygen and carbon dioxide.

Cork cambium produces new cork cells to keep pace with the growing tree. As the stem widens, the outer layer of cork becomes too small and splits. New cork is formed continually, by the cork cambium, to replace the cork that splits. The cork cambium of some trees occurs as a continuous cylinder, whereas in others it occurs in patches of various size and thickness. Each pattern of cork cambium splits the bark in a distinctive way, as shown in figure 15.14.

The cork cambium also repairs wounds in the bark. Wound scars are particularly distinct in smooth-barked trees, like beech, aspen, and birch, where inscriptions carved with knives remain visible for decades.

286

(a)

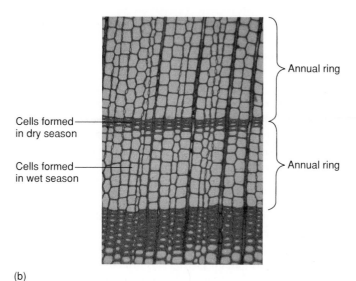

Annual ring

Cells formed in dry season

Cells formed in wet season

Annual ring

(b)

Figure 15.13 Annual rings in wood. (*a*) The annual rings of xylem in a thirty-nine-year-old larch tree. Each ring represents a year in the life of the tree. The darker rings (called heartwood) represent old xylem that is no longer functional. Lighter rings around them (called sapwood) represent functional xylem. (*b*) Magnified view of an annual ring in a pine tree. Xylem formed during the wet season has broader cells than xylem formed in the dry season.

Figure 15.14 Bark texture reflects the arrangement of cork cambium. (*a*) The cork cambium of a smooth-barked tree, like the beech shown here, forms a continuous sheath around the stem and adds cork uniformly. (*b*) The cork cambium of a Shagbark hickory occurs in blocks, rather than a continuous sheath, and forms large blocks of cork that peel off in sheets. (*c*) The cork cambium of many trees, like the persimmon shown here, is arranged in small blocks. It adds new cells unevenly, causing the cork to crack and form deep ridges.

(a)

(b)

(c)

SUMMARY

1. Many plants can reproduce asexually by vegetative reproduction or apomixis. All plants can reproduce sexually, in which there is alternation of diploid individuals (sporophytes) with haploid individuals (gametophytes). In a flowering plant, the diploid individual (sporophyte) is the main body and the haploid individuals (male and female gametophytes) are parts of a flower. A flowering plant may be an annual, biennial, or perennial.

2. Flowers are the reproductive organs of a flowering plant. Parts of a flower include sepals, petals, stamens, and carpels. A perfect flower has both stamens, which produce male gametophytes, and carpels, which produce female gametophytes. An imperfect flower has either stamens or carpels. A plant with imperfect flowers is either monoecious or dioecious.

3. The flowers of each species have a specific color, shape, scent, and nectar that attract a specific kind of pollinator. Pollinator specificity ensures that the pollen fertilizes eggs of the same species.

4. A pollen grain is a male gametophyte; at maturity, it consists of a hard coat, a tube cell, and two sperm nuclei. An ovule is a female gametophyte; at maturity it contains four cells, of which one has two polar nuclei and one is the egg, enclosed within a hard coat.

5. A flowering plant has double fertilization. One of the sperm nuclei fuses with the egg to form a fertilized egg. The other sperm nucleus fuses with the cell that has two polar nuclei to form an endosperm cell.

6. An ovule develops into a seed: the fertilized egg develops into an embryo; the endosperm cell develops into endosperm; and the integuments develop into a seed coat. The developing embryo has cotyledon(s), root apical meristem, and shoot apical meristem. The ovary, surrounding the ovule, develops into a fruit, an adaptation for seed dispersal. At this point, the seed becomes dormant until conditions are suitable for growth into a seedling.

7. When conditions become favorable, the seed germinates and becomes a seedling with a root, stem, and cotyledon(s). Primary growth involves three kinds of meristem: the shoot apical meristem extends the stem upward and forms branches, leaves, and flowers; the root apical meristem lengthens the root; and the pericycle forms lateral branches on the root.

8. Secondary growth, which broadens the stem and root, occurs only in trees and shrubs. The vascular cambium forms new xylem and phloem. Old xylem accumulates as wood and old phloem accumulates as bark. The cork cambium develops cork cells. The bark consists of phloem (both living and dead) and periderm (living cork cells, cork cambium, and dead cork cells).

KEY TERMS

Annual (page 273)
Apomixis (AEH-poh-MIX-us) (page 272)
Biennial (BEYE-eh-nee-ul) (page 274)
Carpel (CAHR-pul) (page 276)
Cork cambium (page 286)
Dioecious (deye-EE-shus) (page 276)
Double fertilization (page 279)
Endosperm (EN-doh-spurm) (page 279)
Gametophyte (gah-MEE-toh-feyet) (page 273)
Germination (page 282)
Monoecious (moh-NEE-shus) (page 276)
Ovule (OH-vyool) (page 279)
Perennial (PUR-eh-nee-ul) (page 274)
Petal (page 276)
Pollen grain (page 278)
Pollination (PAH-leh-NAY-shun) (page 276)
Pollinator (PAH-leh-NAY-tur) (page 276)
Sepal (SEE-puhl) (page 276)
Sporophyte (SPOR-oh-feyet) (page 273)
Stamen (STAY-men) (page 276)
Vascular cambium (CAM-bee-um) (page 286)
Vegetative reproduction (page 272)

STUDY QUESTIONS

1. Draw a perfect flower and label its parts. How do perfect flowers avoid self-fertilization? How does a perfect flower differ from an imperfect flower?

2. Compare a hummingbird-pollinated flower with a moth-pollinated flower. Compare a bat-pollinated flower with a bee-pollinated flower. Explain why these differences occur.

3. Why do the fruits of plants vary so greatly in size and shape?

4. Describe double fertilization. What is the advantage of double fertilization over single fertilization (fusion of a sperm nucleus and an egg)?

5. Describe the functions of shoot apical meristem, root apical meristem, pericycle, vascular cambium, and cork cambium.

CRITICAL THINKING PROBLEMS

1. Explain why the flowers of many trees and grasses are so inconspicuous.

2. What is the advantage of having a flower that attracts only a particular kind of animal?

3. Compare the advantages of animal pollination and wind pollination. Under what conditions would you expect a flowering plant to switch from animal pollination to wind pollination?

4. What role does endosperm play in your diet?

5. What can be determined about earlier climates by examining the annual rings of trees?

SUGGESTED READINGS

Introductory Level:

Barth, F. C. *Insects and Flowers: The Biology of a Partnership*. Princeton, NJ: Princeton University Press, 1991 (translated from a German edition first printed in 1982). An engaging description of pollination, pollinators, floral adaptations, as well as the senses and behavior of insect pollinators.

Bernhardt, P. *Wily Violets and Underground Orchids: Revelations of a Botanist*. New York: Wm. Morrow & Company, 1989. A delightful collection of essays, dealing mainly with unusual ways in which plants achieve pollination.

Holm, E. *The Biology of Flowers*. New York: Penguin Books, 1979. A well-illustrated summary of both structural and reproductive aspects of flowers.

Meeuse, B., and S. Morris. *The Sex Life of Flowers*. New York: Facts on File, 1984. A beautifully illustrated history of pollination biology, with chapters on floral structures, mechanisms that ensure cross-pollination, coevolution of plants and pollinators, and floral deceptions.

Advanced Level:

Johri, B. M. *Embryology of Angiosperms*. New York: Springer-Verlag, 1984. A detailed description of sexual reproduction in flowering plants.

Niklas, K. J. "Wind Pollination—A Study in Controlled Chaos," *American Scientist* 73 (1985): 462–70. A description of how wind-pollinated plants gain some control over pollen capture.

Richards, A. J. *Plant Breeding Systems*. Winchester, MA: Allen and Unwin, 1986. A description of angiosperm reproduction, especially floral structures, apomixis, and such related topics as the genetic structure of a population and the foraging methods of pollinators.

van der Pijl, L. *Principles of Dispersal in Higher Plants*, 2d ed. Berlin: Springer-Verlag, 1982. A compilation of all aspects of seed and fruit dispersal.

Hormonal Control

Objectives

In this chapter, you will learn
–how hormones control the activities
  of plant cells;
–how plants receive information from
  their environment;
–the way in which a tree prepares for
  winter;
–why a plant flowers at the appropriate
  season;
–how a plant moves part of its body;
–how internal rhythms control the
  activities of plant cells.

Plants follow seasonal schedules.
A Christmas rose (*Helleborus niger*)
reproduces on schedule in spite of an
unseasonal snowstorm.

ne of the pleasures of a seasonal environment is watching the landscape change. We welcome the warmth of springtime, when the soft green of new leaves replaces the gray of winter. Each plant awakens at a prescribed time—a sugar maple in late April, a beech in early May, a catalpa in mid-May, and a rhododendron in late May or early June.

After a burst of spring growth, the green carpet becomes sprinkled with bright colors, as plants begin their schedules of reproduction. Spring flowers include the larkspur, trillium, pasque flower, and lady's slipper. Summer brings the flowers of gladioli, clover, delphinium, and iris. In autumn we treasure the late-blooming aster, goldenrod, dahlia, and chrysanthemum.

Autumn brings the warm hues of mature fruits to the landscape. Currants, grapes, blackberries, raspberries, plums, and cherries advertise themselves to animals, which eat the fleshy parts and disperse their seeds. The leaves of trees and shrubs turn from green to magnificent reds, oranges, and yellows. This rich display of color marks the end of the growing season and the beginning of winter dormancy.

A plant anticipates each season and prepares for it well in advance. How does a plant detect the time of year? How does it then coordinate its cells to bring about the appropriate response? This chapter deals with answers to these questions.

Plant Hormones

A plant has a variety of cells: mesophyll cells, which photosynthesize sugars; leaf epidermal cells, which secrete a waxy cuticle; root epidermal cells, which absorb water and nutrients; meristem cells, which form new parts; and many other kinds of specialized cells. All the cells work together to promote survival and reproduction of the plant as a whole. In autumn, for example, a lilac prepares for winter whereas in springtime it develops flowers.

A plant's responses to the environment are governed by chemical messengers called **hormones.** A plant hormone is a molecule that is secreted by one kind of cell and that influences the activities of other cells. Hormones travel through a plant by way of the xylem, phloem, and other pathways. They are small organic molecules, typically formed of one-, two-, or three-carbon rings with attached side chains.

Each hormone influences certain kinds of cells, called its **target cells,** by binding with proteins in their plasma membranes (see page 60). This binding, like a key in a lock, signals the cell to alter its activities. Typically, a hormone signals the cell to alter (1) its rate of a normal activity or (2) the permeability of its plasma membrane to a particular material. For example, the hormone cytokinin signals a cell to synthesize more of the enzymes

Table 16.1

Plant Hormones

Hormone	Main Functions
Auxin	Elongates stem; suppresses lateral buds; effects tropisms
Gibberellin	Lengthens internodes; develops fruit; initiates flowering
Cytokinin	Stimulates cell division; stimulates development of lateral buds
Abscisic acid	Initiates dormancy; coordinates responses to stress
Ethylene	Ripens fruit; induces leaf abscission
Salicylic acid	Triggers immune response

involved in cell division, and the hormone auxin signals a cell to open its channels for the passage of hydrogen ions.

The relations between plant hormones and their target cells are complex. Each hormone controls several kinds of target cells and influences each kind in a different way. The same hormone, for example, may decrease cell division in one cell and open channels for potassium ions in another. Moreover, each plant response, such as spring growth, is the product of several hormones.

Six classes of plant hormones are known. Each class is a group of similar molecules with similar effects. Three of them—auxins, gibberellins, and cytokinins—promote growth. The others promote dormancy, fruit ripening, immune responses, and other coordinated responses to the environment. The characteristics of these hormones are summarized in table 16.1 and described in the following paragraphs.

Hormones That Promote Growth

A plant responds to its environment chiefly by growth. The main stem and branches grow toward light; its roots grow toward moisture. In spring, seeds grow into seedlings, bulbs grow into flowering plants, and trees grow new leaves. Plant growth is controlled by three hormones—auxin, gibberellin, and cytokinin.

Auxin   Auxin was the first plant hormone to be identified (in 1934). It is secreted by cells of shoot apical meristem (see page 283), located at the tips of the stem and branches. Auxin governs many plant responses, of which three are described here: suppression of the lateral buds (see page 250), elongation of the stem, and tropisms.

Auxin suppresses development of the lateral buds into new branches, leaves, or flowers. This phenomenon is

(a)

Figure 16.1 Apical dominance. (*a*) Auxin, produced in the apical meristem, inhibits development of lateral buds. The plant grows mainly upward. (*b*) When the apical

(b)

meristem is cut off, the source of auxin is eliminated and the lateral buds develop into branches.

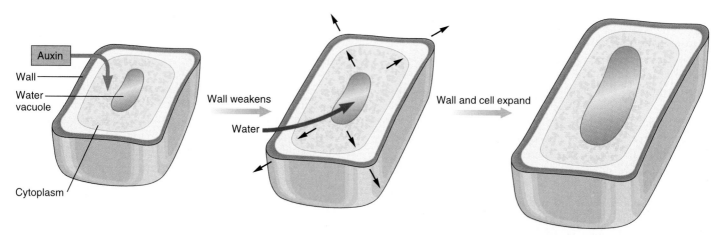

Figure 16.2 Effect of auxin on cell size. Auxin stimulates a plant cell to release hydrogen ions and certain enzymes into its cell wall. The ions and enzymes soften the cell wall. The

outward pressure of water inside the cell pushes against the softened cell wall, causing it to expand. The cell takes in more water and elongates.

known as **apical dominance,** since meristem in the stem apex (tip) dominates meristem in the lateral buds. It can be demonstrated by pinching off the top of a houseplant; without its shoot apical meristem, the plant grows lateral branches (fig. 16.1).

Auxin elongates the stem by stimulating enlargement of newly divided cells in the apical meristem. When auxin

binds to its receptors on these target cells, channels in the membrane open so that hydrogen ions can move out of the cells and into the cell walls. Certain enzymes are also transported to the cell walls. There, the ions and enzymes soften the wall structure. The weakened wall then expands from water pressure inside the cell. As the wall expands, the cell absorbs more water and elongates (fig. 16.2).

- Elongated cells on shaded side

- Stem

Figure 16.3 How auxin bends a plant toward light. The stem of this *Impatiens* plant bends toward light. Auxin accumulates on the shaded side of the stem, where it causes the cells to elongate. As the shaded side grows longer, the plant bends toward light.

Auxin governs tropisms in a similar way as it governs stem elongation. A tropism is movement of the plant, or part of the plant, toward or away from some external stimulus. A plant bends toward light, for example, because auxin stimulates cell elongation on the shaded side of the stem (fig. 16.3). The shaded side grows and the lighted side does not, causing the stem to bend toward light. Experiments that revealed the role of auxin in phototropism are described in the accompanying Research Report.

Auxins are synthesized in laboratories and used for commercial purposes. They are sprayed on unfertilized flowers to produce seedless fruits, and they are sprayed on cuttings to stimulate the growth of new roots. Auxins are also used as herbicides: when applied in large doses, they kill dicots by stimulating uncontrolled growth. Auxin herbicides can selectively remove dicots from fields of monocots; they can eliminate dandelions from grassy lawns and thistles from cornfields. Unfortunately, they may also be harmful to your health, as described in the accompanying Wellness Report.

Gibberellin A second growth hormone, gibberellin, was discovered in Japan during the 1920s, when rice fields became infected with "foolish seedling" disease. The young plants grew very tall and then fell over and died. Analysis of the plants revealed they were infected with a fungus, *Gibberella fujikuroi*, that secreted a chemical that stimulated rice seedlings to grow too fast. Later, the same chemical was found to occur naturally, in smaller concentrations, in all plants. It was recognized as a plant hormone and named gibberellin, after the fungus that led to its discovery.

Gibberellin is secreted primarily by the shoot and root apical meristems but also by leaves, buds, fruits, seeds, and embryos. Its main effect is elongation of the internodes, which are the regions between lateral buds on a stem (see fig. 15.10). The more gibberellin secreted by a plant, the longer the internodes (fig. 16.4).

A plant with very little gibberellin develops very short internodes; its leaves are stacked one on top of the other (see fig. 16.4). This form, called a rosette, is typical of many plants, including lettuce and cabbage. When it is time to reproduce, these short plants secrete large amounts of gibberellin, which cause the flower stalk to grow very rapidly. This phenomenon, known as bolting, positions the flowers high above the ground for better dispersal of pollen and seeds (fig. 16.5).

Other responses that involve gibberellin include leaf growth, flower production in some plants, and seed germination in grasses. Gibberellin also plays a role in breaking winter dormancy (described later).

Cytokinin Cytokinin, discovered in the 1950s, is named after its role in stimulating cell division (cytokinesis is the phase of mitosis in which the cytoplasm divides in two). It is secreted chiefly by root cells.

Cytokinin promotes cell division in the meristem tissues, apparently by stimulating DNA replication and synthesis of enzymes used in mitosis. Florists spray synthetic versions of this hormone on cut flowers, which have no roots and therefore little cytokinin, to make them last longer. The hormone is also involved in dormancy and growth of the lateral buds (described later).

Other Hormones

A plant responds to its environment in ways other than growth. Sometimes an appropriate response to the environment is to stop growing and enter into a dormant state. Other times an appropriate response is to ripen fruit or to defend against infections. Hormones that coordinate these responses are described following.

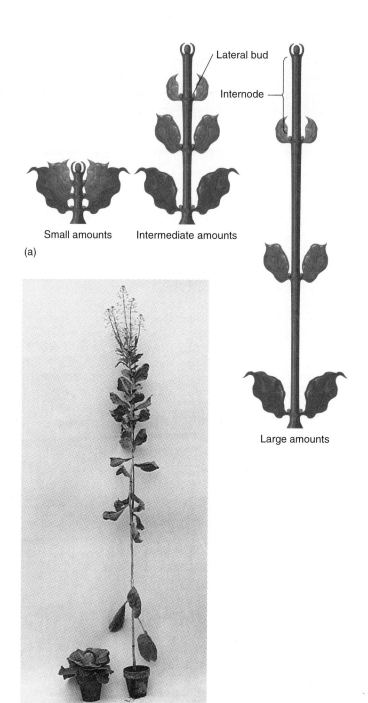

Lateral bud

Internode

Small amounts   Intermediate amounts

(a)

Large amounts

(b)

Figure 16.4 Effects of gibberellin on plant growth. (*a*) The amount of gibberellin determines the lengths of the internodes in a growing plant. (*b*) A cabbage secretes very little gibberellin and develops into a rosette (*left*) with extremely short internodes. Application of gibberellin to a cabbage plant (*right*) lengthens the internodes and profoundly changes the shape of the plant.

Figure 16.5 Flower stalk of a century plant. A century plant (genus *Agave*) secretes hardly any gibberellin during shoot growth and develops into a rosette. When its flowers begin to form, however, it secretes large amounts of gibberellin; the flower stalk bolts upward, placing the flowers in a position more favorable for pollination and seed dispersal.

Abscisic Acid   Abscisic acid suppresses growth by blocking protein synthesis. It is secreted mainly by cells in the leaves, fruits, and root caps. When first discovered, in 1963, its main effect was believed to be **abscission,** in which a leaf or fruit breaks away from the stem. We now know that abscission is not a major effect of abscisic acid, yet its misleading name persists.

By suppressing growth, abscisic acid helps to induce and maintain dormancy. It promotes formation of winter buds and a variety of physiological changes that prepare a plant for winter.

Research Report

Discovery of the First Plant Hormone

One of the first biologists to study plant hormones was Charles Darwin. He and his son Francis wondered what caused grass seedlings to bend toward light. Grasses are monocots, with leaves that grow upward through a sheath called the coleoptile. The Darwins observed that when a coleoptile was exposed to light from one side, it bent toward the light:

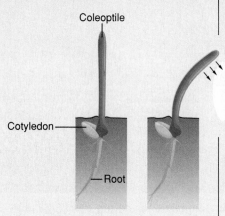

They wondered where in the seedling the light-detecting mechanism was located (fig. 16.A). In 1880, the Darwins performed an experiment to determine whether the mechanism was located in the tip or the base of the coleoptile. They placed seedlings in three groups. In the control group, seedlings were simply grown in light. In one experimental group, the coleoptile tips were covered with an opaque material through which light could not pass. In a second experimental group, the coleoptile bases were covered with the same opaque material:

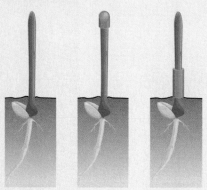

Control group   First experimental group   Second experimenta group

All the seedlings were grown in light coming from one side. The Darwins predicted that if the mechanism for light detection was located in the coleoptile tip, then seedlings in the first experimental group would not bend toward light but the others would. If the mechanism was in the coleoptile base, then seedlings in the second experimental group would not bend toward light and the others would.

The result was that all seedlings bent toward light except those in the first experimental group:

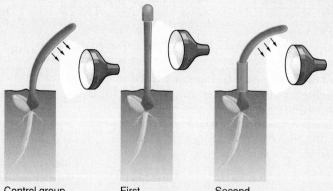

Control group   First experimental group   Second experimental group

The Darwins concluded that a mechanism for detecting the position of light was located in the coleoptile tip (a region of the plant that does not require light for photosynthesis). This mechanism, in turn, caused the coleoptile to bend toward the light.

In 1926, Fritz Went (a Dutch biologist) hypothesized that the light-detecting mechanism was a chemical that moved through the seedling and caused bending (see fig. 16.A). He performed experiments to test whether the coleoptile tip synthesizes a chemical that then moved into the base and caused bending. In a classic experiment, Went cut off the tip of a coleoptile and placed it for an hour on a block of gelatin (to absorb any chemical the tip might secrete):

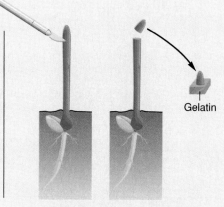

Gelatin

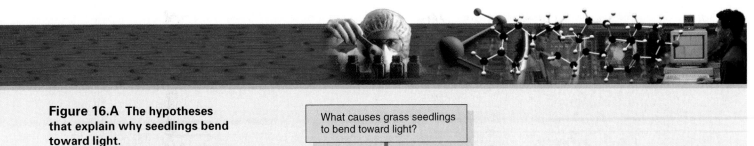

Figure 16.A The hypotheses that explain why seedlings bend toward light.

What causes grass seedlings to bend toward light?

Where is the controlling factor located in the seedling?

What is the controlling factor?

Hypothesis 1: coleoptile tip

Hypothesis 2: coleoptile base

Hypothesis 1: a chemical that moves through the plant

Other hypotheses

Supported by Darwin experiment

Disproved by Darwin experiment

Supported by Went experiment

He then placed the block of gelatin in various positions on decapitated coleoptiles grown in uniform light (a shaded container). Went predicted that if a chemical had moved from the coleoptile tip onto the gelatin, then it should now move onto the decapitated coleoptile and cause bending (even though light was not coming from any particular direction). He found that when the gelatin block was placed on top of the decapitated coleoptile, the seedling grew straight up, but when it was placed on one side, the coleoptile grew away from that side:

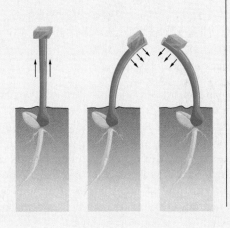

Went concluded that a chemical released onto the gelatin from the tip of the coleoptile caused one side of the coleoptile to grow more rapidly than the other side, and this uneven growth caused the seedling to bend. He named this chemical auxin, after the Greek word *auxein,* which means to increase.

We now know that cells in the coleoptile tip secrete a hormone (auxin) that moves down the coleoptile and accumulates on the side opposite the light source. In response to the hormone, cells on the shaded side elongate, making this side longer than the lighted side. The coleoptile bends toward light:

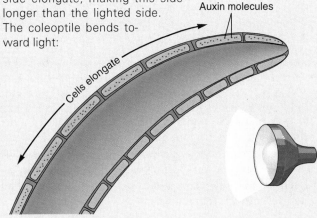

Cells elongate

Auxin molecules

Many questions remain unanswered. How does the seedling detect the direction of light? How does it turn on auxin secretion? Why does auxin accumulate only on the shaded side? Experiments have revealed that bending occurs more readily in response to blue light than to other portions of the light spectrum. Thus, the light is probably detected by a pigment that absorbs blue light. The pigment has not yet been identified. As in much of biological research, the answer to one question has generated even more questions.

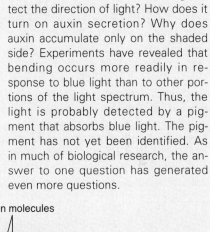

Wellness Report

Herbicides, Agent Orange, and Dioxin

In 1948 a chemical called 2,4,5-T was registered for use as an herbicide. This chemical provided an easy and relatively inexpensive way to get rid of dicots, such as dandelions and thistles, without harming monocots, such as lawn grasses and grains. In the 1960s and 1970s millions of pounds of 2,4,5-T were sprayed on roadsides, rangelands, fields, and forests.

About this time the United States was fighting the war in Vietnam and used this miracle chemical to strip the jungles of leaves (fig. 16.B) to make the enemy more visible and prevent surprise attacks. Most widely used was an herbicide called Agent Orange, which is a mixture of 2,4,5-T and a similar herbicide, 2,4-D. Over 10 million gallons of Agent Orange were sprayed on Vietnam.

Both of the herbicides in Agent Orange are synthetic versions of auxin. This plant hormone, which stimulates growth under natural conditions, kills dicots when applied in high concentrations. It disrupts normal cellular activities and causes the plants to grow themselves to death.

Whenever 2,4,5-T is synthesized commercially, a contaminant called dioxin inevitably forms. Experimental evidence indicates that dioxin is ex-

Figure 16.B The effects of Agent Orange on a jungle in Vietnam.

tremely harmful to laboratory animals, causing miscarriages, birth defects, cancer, and disorders of the heart, thyroid, brain, immune system, liver, and reproductive organs. Researchers claim it is the most powerful carcinogen (cancer promoter) they have ever tested.

Indirect evidence indicates that dioxin is also harmful to humans. In high doses, it apparently causes cancer. In low doses, it has subtle effects on the immune system and fetal development. We are all exposed to dioxin in low doses, mainly from beef and milk products—cows

feed on plants grown in soils that were contaminated with dioxin when 2,4,5-T was used as an herbicide. Medical researchers have found dioxin in the fat tissue of almost everyone tested, as well as in the milk of nursing mothers.

Herbicides made of 2,4,5-T are now banned. They are not, however, the only source of dioxin. This dangerous pollutant forms when plastic is burned and when paper is bleached. We need to be alert to new sources of dioxin and its effects on our own health as well as the health of other organisms.

Abscisic acid also coordinates a plant's responses to stress. For example, it closes a plant's stomata (□ *see page 247*) in response to drought. It does this by binding with receptors in the plasma membranes of guard cells, an action that opens channels for potassium ions. As potassium ions leave the guard cells, water follows by osmosis. The cells shrink and the opening between them closes.

Ethylene   Ethylene is the only hormone that is a gas. Its main function, first described in the 1960s, is to ripen fruit (fig. 16.6). Ethylene is secreted by most plant cells, includ-

ing those of the fruit. It diffuses not only through the plant interior but from fruit to fruit through the air.

As a fruit ripens, it secretes more and more ethylene, which accelerates ripening in nearby fruits. This phenomenon is the basis of the saying, "One bad apple spoils the barrel." A bad apple is a very ripe apple, and the ethylene it releases hastens ripening of other apples nearby. Ethylene is used commercially to ripen fruits, such as tomatoes, that are picked when they are green and hard.

Ethylene induces several changes in fruit. It converts starches to sugars and softens cell walls. It also stimulates

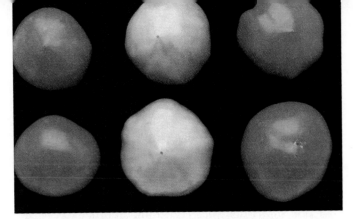

Figure 16.6 Effects of ethylene on fruit ripening. In this experiment, tomatoes in the control group (*left*) ripened normally. In one experimental group (*middle*), the tomatoes were treated with a chemical that blocked synthesis of ethylene. These fruits did not ripen. In the other experimental group (*right*), the tomatoes were treated with the same chemical as the middle group, but then sprayed with ethylene. They ripened just as well as the control group.

synthesis of molecules that give each fruit its particular taste and aroma—at least 118 such molecules are synthesized in a tomato and more than 200 in a banana. Ethylene changes fleshy fruits from green to red, orange, yellow, or purple—a color that attracts the animals that disperse its seeds.

In addition to its role in ripening fruits, ethylene prepares a tree or shrub for winter dormancy by promoting leaf abscission. Another function is to strengthen the stems of plants in windy places. The more an aspen tree is shaken, for example, the more ethylene it secretes and the stronger its stem becomes.

Salicylic Acid   The medicinal effects of salicylic acid have been known for more than a century. The leaves of certain plants, especially willow, myrtle, and wintergreen, have been used to treat a variety of illnesses (see the Wellness Report in chapter 13). Salicylic acid is similar to the active ingredient in aspirin. The function of this molecule in plants, however, has only recently been revealed.

Salicylic acid is a hormone that coordinates a plant's immune response to invasion by a disease organism. The immune response provides long-term protection against viruses, bacteria, and fungi that cause plant diseases. When one of these disease organisms invades a plant, cells near the invasion site secrete salicylic acid, which travels through the phloem to other parts of the plant. It signals plant cells to synthesize certain proteins, such as antibiotics, that provide defenses against disease organisms.

Interactions Among Hormones

Most plant responses are stimulated by a relative concentration of two or more hormones rather than by the presence of a single hormone. The pattern of shoot growth is an example of such a response.

The height of a stem and its degree of branching depend on the relative concentrations of auxin and

cytokinin in the lateral buds. Auxin, secreted by the apical meristem, flows downward and suppresses growth of the lateral buds. Cytokinin, secreted by the root, flows upward and stimulates growth of the lateral buds. Usually there is more auxin near the top of the plant, and so development is suppressed more in lateral buds near the top than near the bottom. As a consequence, trees are narrower near the top, where lateral buds have only recently been released from apical dominance, than near the bottom, where they have been far from the apex for a longer time.

Other plant responses brought about by relative concentrations of two or more hormones are dormancy and flowering, described later in this chapter.

Synchrony with the Environment

How does a plant synchronize its activities with the environment? How does it "know" when to reproduce and when to enter dormancy? As we have seen, internal coordination of responses is governed by hormones. The secretion of hormones, in turn, occurs in response to environmental changes:

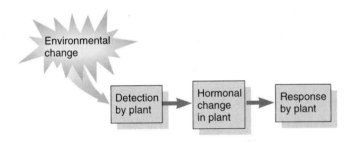

A plant grows, flowers, enters dormancy, and comes out of dormancy at the right time of the year. In most cases, these seasonal adjustments are initiated well before the season actually occurs. A plant could not, for example, afford to wait until the weather was very cold to protect its tissues from freezing. Preparations must be made in advance.

Sensing the Environment

We know that plants receive information from the environment; they respond to the time of day, the season of the year, touch, which way is up and which is down, and so on. In most cases, we do not fully understand how a plant receives this information or how the information controls hormone production.

Plants are known to respond to changes in light, and our working hypothesis is that this important source of information is received by plant **pigments,** which are organic molecules that absorb light. The best-known sensing pigment is phytochrome (*phyto*, plant; *chrome*, color).

Phytochrome   **Phytochrome** is a protein, located within the membranes of leaf cells, that changes form when it absorbs certain wavelengths of light. This change, in turn, governs the secretion of hormones that control responses to light.

Phytochrome occurs in two molecular forms: R-phytochrome and FR-phytochrome. R-phytochrome absorbs red light, which is abundant in sunlight. The absorption of red light converts R-phytochrome to FR-phytochrome:

$$R \; + \; \text{Red} \longrightarrow FR$$

FR-phytochrome absorbs far-red light, which is abundant in shade. The absorption of far-red light converts FR-phytochrome to R-phytochrome:

$$FR \; + \; \text{Far red} \longrightarrow R$$

FR-phytochrome also changes, gradually, to R-phytochrome in the absence of light:

$$FR \; + \; \text{No light} \longrightarrow R$$

Thus, the relative amounts of these two phytochromes in a leaf cell depend on the conditions of light to which the leaf has been exposed. After some time in bright sunlight, the cells of a leaf have mostly FR-phytochrome. After some time in the shade or darkness, the cells of a leaf have mostly R-phytochrome.

The relative amounts of the two phytochromes, in turn, govern hormone secretions. Gibberellin secretion, for example, is stimulated by an increase in the amount of R-phytochrome. A plant growing in the shade accumulates more R-phytochrome and secretes more gibberellin. It grows long internodes and develops into a tall plant with sparse branches and leaves. In contrast, a plant growing in bright sunlight accumulates more FR-phytochrome and secretes less gibberellin. It grows short internodes and develops into a short plant with dense branches and leaves.

In addition to stem growth, phytochrome coordinates plant activity with environmental events by influencing leaf growth, seed germination, daily rhythms of leaf movement, chloroplast formation, and other processes.

Photoperiod   A plant anticipates the seasons; it receives information that signals the approach of spring and winter, as well as the approach of its particular season of reproduction. The best indicator of the time of year is **photoperiod,** which is the length of day in relation to the length of night. Plants measure photoperiod and use the information to regulate the hormones that induce adjustments to seasons.

We do not know for sure how plants detect photoperiod. Most likely they use a pigment that absorbs blue light, since plant responses appear most sensitive to these wavelengths. Such a pigment has not yet been found.

A plant seems to measure the duration of night rather than day. When exposed to a single flash of red light during the night, a plant responds as if it were experiencing the short nights of early summer (fig. 16.7). The artificial break in the dark period signals the plant, incorrectly, that one short night has ended and another has begun. Flower growers exploit this incorrect signal to force plants to blossom outside their normal season. Bouquets of chrysanthemums, for example, are available all year round.

Seasonal Periods of Dormancy

Plants become dormant to survive prolonged periods of cold and drought. **Dormancy** is a genetically programmed reduction in cellular activity; it occurs before onset of an unfavorable season rather than in direct response to it. Dormancy prepares a plant for a predictable period of adverse conditions.

Plants vary in the kinds of structures that persist during unfavorable seasons. In an annual plant, it is a seed; in a biennial, it may be a bulb; and in a perennial it may be a root (e.g., grass) or an entire leafless plant (shrub or tree). In the discussion that follows, we consider what happens in a deciduous tree (one that drops its leaves) as it prepares for a cold winter.

Leaf Death   In more northern climates, a deciduous tree conserves water and nutrients by shedding its leaves when it is too cold to photosynthesize. The leaves begin to die in early August (fig. 16.8), in response to decreasing day length. Death begins with a decline in the enzymes of photosynthesis followed by a shutdown in production of chlorophyll, the green pigment that captures light for photosynthesis. Without chlorophyll, the leaves display the colors of other pigments—reds, oranges, and yellows, which were always present but masked by the green of chlorophyll.

Ions, amino acids, sugars, and other nutrients are transported from the dying leaves to the woody parts of the tree. There they are stored until springtime. An abscission zone of thin-walled cells then develops where each leaf is attached to the stem (fig. 16.9). It shuts off the supply of water and nutrients to the leaf. Eventually, the abscission zone dries out and the leaf falls off. A leaf scar then forms and seals off the xylem and phloem.

Leaf death is controlled by the relative concentrations of abscisic acid and growth hormones in the leaf. A fundamental change in hormone production begins at the time of the summer solstice (around June 22), when nights are shortest. The tree begins to secrete less of the growth hormones (auxin, gibberellin, and cytokinin) and more of the dormancy hormone, abscisic acid. By autumn, the large amount of abscisic acid in relation to the growth hormones stimulates transfer of nutrients out of the leaves followed by leaf shed.

Dark

Light

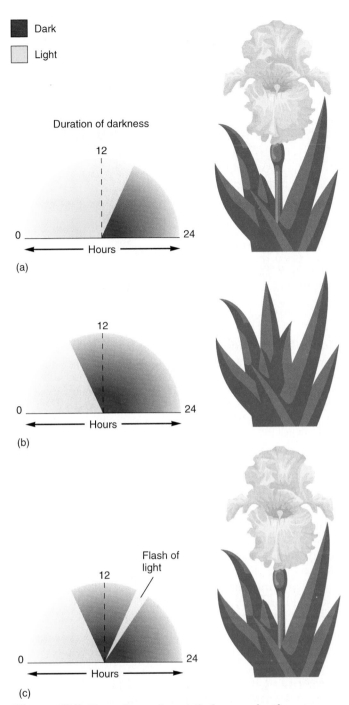

Duration of darkness

(a)

(b)

Flash of light

(c)

Figure 16.7 How plants detect their reproductive season. A plant detects the time of year by the length of night. (*a*) This plant is a long-day plant; it flowers only when nights become short (spring). (*b*) It does not flower when nights are long (autumn). (*c*) If the nights are long, but interrupted by a short flash of light, the plant will flower. The plant interprets the flash of light as a "daytime" between two short nights.

Physiological Changes   In early autumn, abscisic acid is transferred, along with the nutrients, out of the dying leaves and into woody portions of the tree. There it promotes a variety of cellular changes (called "hardening off") that prepare the tree for cold temperatures.

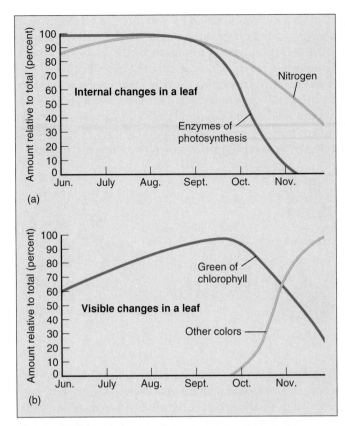

Figure 16.8 Leaf changes in preparation for winter dormancy. (*a*) In August, the enzymes of photosynthesis and the nutrients in a leaf begin to decline. (*b*) In early October, chlorophyll production begins to decline and fall colors emerge.

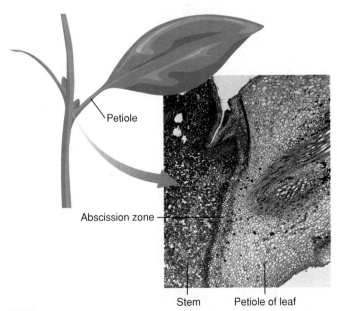

Figure 16.9 Leaf abscission. A leaf is shed when the layer of thin-walled cells, called the abscission zone, dries out in response to hormones. The layer breaks and the leaf falls off.

One change signaled by abscisic acid is the addition of sugars to the cytoplasm. Starches, which do not dissolve in water, are converted to sugars, which dissolve in water. The sugars prevent ice from forming in cells (just as salt prevents ice from forming on roads). The formation of ice inside a cell ruptures cellular membranes. Water is converted to ice when hydrogen bonds stabilize between water molecules. Dissolved sugars interfere with this process and so lower the temperature at which the water will freeze:

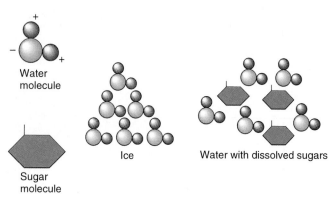

Bud Formation   Abscisic acid inhibits growth of the meristem tissues. As autumn approaches and relative concentrations of abscisic acid rise, the tree stops growing. Its shoot apical meristem and lateral buds are converted to winter buds, with miniature leaves, branches, or flowers packed in a cottony insulation and covered with waterproof scales (fig. 16.10a).

Emergence from Dormancy   A tree emerges from winter dormancy in the springtime. Each species breaks dormancy in response to a warm, moist soil after a certain day length has been reached. The plant then begins to produce more of the growth hormones and less abscisic acid. This particular balance of hormones stimulates plant growth.

The buds respond to an increase in growth hormones by swelling and bursting out of their protective scales (fig. 16.10b). The miniature leaves, branches, or flowers then begin to grow.

The nutrient reserves, removed from the dying leaves and stored since the previous autumn, are transferred to the rapidly growing structures. They are essential to spring growth, for young leaves cannot synthesize all the materials they need.

Seasonal Periods of Reproduction

All plants of a population reproduce at the same time of year, a time that varies from species to species. This rigid schedule not only ensures that a plant reproduces during favorable weather but also that it flowers when other plants of its species flower, thereby ensuring that its eggs will be fertilized by sperm from another plant of its kind. What environmental signal triggers simultaneous flowering?

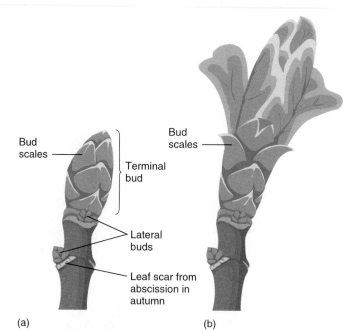

Figure 16.10 Winter buds. A winter bud protects immature leaves, branches, or flowers, as well as the delicate meristem, during cold and drought. (*a*) A winter bud is insulated with cottony material and covered with waterproof scales. (*b*) In the springtime, the bud swells with water and the scales burst open. The miniature leaves, branches, or flowers then begin to develop.

Control of Flowering   Most plants use a specific photoperiod as the environmental signal that initiates flowering. When the length of day in relation to night is just right, a plant's leaves secrete hormones that move into the shoot apical meristem, where they suppress elongation of the stem. The buds then develop into flowers. (Flowers develop from apical buds in some plants and from lateral buds in other plants.)

Flowering appears to be controlled by gibberellin and a second hormone, not yet identified but tentatively called florigen. Gibberellin triggers rapid growth of the flower stalk. Florigen apparently shifts the activities of meristem from elongation of the stem to development of flowers.

Categories of Plants   Each species of plant flowers in response to a particular photoperiod. In general, plants can be categorized as long-day plants, short-day plants, and day-neutral plants (fig. 16.11). (While plants are categorized according to day length, they are actually responding to length of the night, as discussed earlier.)

Long-day plants flower when day length exceeds some critical value. They blossom as the days become longer in spring and early summer. Examples of such plants include the iris, hollyhock, clover, and spinach. Each species has a specific critical value: spinach, for example, flowers as soon as day length exceeds fourteen hours.

Short-day plants flower when the day length drops below some critical value. These plants bloom as the days

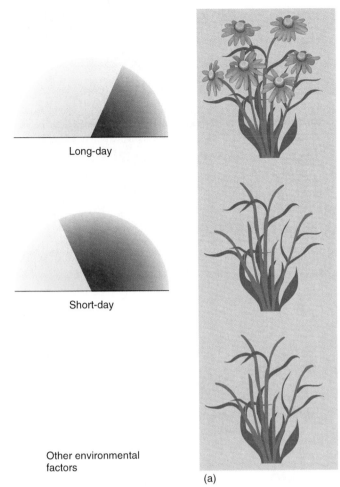

Long-day

Short-day

Other environmental factors

(a)

(b)

(c)

Figure 16.11 Three categories of flowering schedules.
(a) A long-day plant flowers when increasing day length exceeds some critical value. It blossoms in springtime.
(b) A short-day plant flowers when decreasing day length becomes shorter than some critical value. It blossoms in mid-summer or fall. (c) A day-neutral plant flowers in response to some factor other than day length.

become shorter, from mid-summer through autumn. Examples of short-day plants include ragweed, goldenrod, and most asters.

Day-neutral plants flower in response to signals other than photoperiod. These plants, which include the dandelion, snapdragon, carnation, tomato, and cucumber, apparently originated near the equator, where day length is nearly the same all year and cannot be used as an environmental signal. Some of these day-neutral plants flower in response to seasonal changes in rainfall. Others simply flower whenever they reach a certain size.

Plant Movements

A plant alters the positions of its leaves, flowers, and roots in response to environmental conditions. Some movements, called **tropisms** (*trop,* to turn toward), are toward or away from an external stimulus. Other movements, called **nastic responses,** occur in response to a stimulus but without regard to position of the stimulus. How does a plant move without muscles?

Tropisms

Movement of a plant structure toward or away from something in its environment is caused, typically, by the growth of some cells and not of others. This type of growth is controlled by auxin.

Phototropism is growth of a plant toward light. Auxin travels from the shoot apical meristem to the shaded part of the stem, where it elongates the cells. The shaded part of the stem grows more than the lighted side, and the stem bends toward the light. In this way, a plant positions its leaves in patches of sunlight. We do not know exactly how a plant detects the direction of light, but some evidence implicates flavoprotein, a pigment that is sensitive to blue light.

Gravitropism is growth with regard to the center of the earth, the source of gravitational pull. This response is particularly important to young seedlings, in which the shoots must grow upward and the roots grow downward. The shoots are said to have negative gravitropism and the roots to have positive gravitropism (fig. 16.12). Gravity seems to be detected by starch granules in the cells. The granules, which are heavier than the cellular fluids, sink to the bottom of the cell—they are pulled toward the center of the earth (fig. 16.13). In some way, perhaps by pulling on the cytoskeleton, the sinking granules alter cellular activities that culminate in a distribution of hormones that promotes growth of a root downward or a stem upward (see fig. 16.12).

Thigmotropism is growth in response to contact with a solid object. The tendrils of a pea, for example, grow around objects they touch. Heliotropism is solar tracking, in which the flowers or leaves of a plant change position in response to the sun's position. Hydrotropism is growth toward water, a characteristic of plant roots. Chemotropism is growth along a gradient of chemicals. An example of chemotropism is growth of a pollen tube toward a higher concentration of molecules in the ovary of a flower.

Nastic Responses

Nastic responses are plant movements in response to a stimulus but not oriented toward or away from the stimulus. A flower, for example, opens and closes in response to time of day (as measured by light) but the petal movements are the same each time it opens or closes. Nastic responses are caused, typically, by changes in turgor pressure. They may happen quite rapidly.

Turgor pressure is the pressure of water against the cell wall. A plant cell alters its turgor pressure by gaining or losing potassium ions. When potassium ions enter a cell, water follows by osmosis. The cell swells and becomes rigid enough to support a leaf or petal. When potassium ions leave a cell, water follows by osmosis. The cell shrinks and support for the structure is lost:

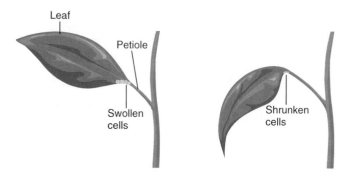

Flowers open and close by changes in turgor pressure. The petals are held up, away from the center of the flower, by swollen cells in their bases. This exposes the sexual organs of the flower to their pollinators

(a)

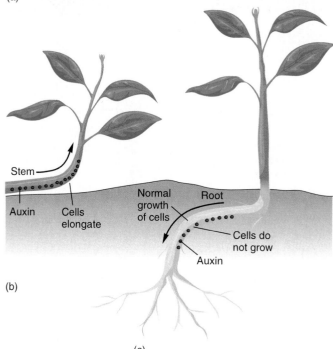

(b)

(c)

Figure 16.12 Gravitropism. (*a*) Negative gravitropism caused the stem of this pine, located on the side of a cliff, to grow away from the center of the earth. (*b*) When a stem lies horizontal to the center of the earth, auxin accumulates on its lower side and causes the cells there to elongate. The stem bends upward (negative gravitropism). (*c*) When a root lies horizontal to the center of the earth, auxin accumulates on its lower side, where it inhibits cell growth. The root bends downward (positive gravitropism).

(see page 276). The petals collapse down, over the center of the flower, when these same cells lose water and shrink. By closing at certain times, the flower's pollen and nectar become unavailable to animals other than its own pollinators.

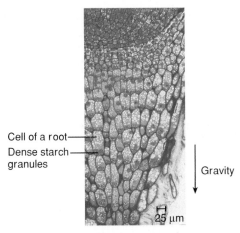

Cell of a root—
Dense starch—
granules

Gravity

25 μm

Figure 16.13 How a plant detects gravity. Gravity is apparently detected by the positions of starch granules. These granules, which are heavier than cell fluids, settle on the lower sides of the cells. This settling in some way induces a particular distribution of auxin that effects gravitropism.

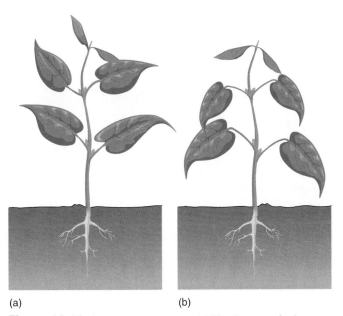

(a)   (b)

Figure 16.14 Sleep movements. (a) The leaves of a bean plant are raised at the beginning of the day, as cells in the base of each leaf take in water. (b) The leaves are lowered at the end of the day, as cells in the base of each leaf lose water.

Figure 16.15 Rapid leaf movements. This plant (*Mimosa pudica*) rapidly folds its leaves when touched. (a) The undisturbed plant, with its leaves positioned to absorb sunlight. (b) The same plant a second or two after being touched. The leaves have folded and drooped.

Some plants exhibit sleep movements, in which the leaves change positions in the morning and night. The leaves of green beans, for example, are raised at dawn and lowered at dusk (fig. 16.14); the leaves of pea plants are unfurled at dawn and folded up again at dusk. These changes in leaf position are brought about by changes in turgor pressure within clusters of cells at the leaf bases.

What is the function of sleep movements? Several hypotheses are being tested. One hypothesis is that a verti-cal or furled leaf cannot be stimulated, inappropriately, by moonlight during the night. A second hypothesis is that a vertical leaf loses less water from its stomata during the night. Yet another hypothesis is that a furled leaf retains more heat during the night.

Very rapid movement occurs in the *Mimosa pudica*, which folds its leaves when touched (fig. 16.15). A cluster of cells in the base of the petiole hold the leaf open. When the leaf is touched, electrical signals travel to these cells and stimulate them to rapidly export potassium ions. The cells shrink and the leaves fold suddenly. The function of this response may be to protect the leaf from being eaten by insects.

Control by Circadian Rhythms

We have seen how hormones coordinate plant activities with seasons, time of day, and other events occurring in the external environment. A second form of control involves **circadian rhythms** (*circa*, around; *dies*, day). These daily cycles of activity are regulated by internal "clocks" rather than by changes in the external environment. This mysterious phenomenon appears to be characteristic of all cells (including our own, as described in chapter 23).

A circadian rhythm originates within the cell itself. Its twenty-four-hour periodicity, however, is maintained by outside influences; the "clock" is apparently reset at sunrise or sunset, as detected in plant cells by phytochrome. When a plant is kept in continuous darkness, the rhythm is maintained but it drifts away from a precise twenty-four-hour cycle.

Plants display daily rhythms of cell division, biochemical processes, and other activities. A conspicuous circadian rhythm is the sleep movements described previously. As long as the plant is exposed to an environment of day and night, its leaves move to their day and night positions at the same time each day. The plant anticipates sunrise and sunset; it is not the light itself that causes the response. But when temperature, light, and other environmental factors are held constant, the cycle lengthens: the leaves are raised and lowered 1.4 hours later each day. In a week or so, the leaves are in the day position at night and the night position by day.

SUMMARY

1. The activities of plant cells are coordinated by six hormones. Auxin elongates the stem, suppresses lateral buds, and brings about tropisms. Gibberellin lengthens the internodes and flower stalks. Cytokinin stimulates cell division and development of lateral buds. Abscisic acid promotes dormancy and coordinates responses to stress. Ethylene ripens fruit and induces abscission. Salicylic acid triggers the immune response. Most plant responses to the environment are caused by relative concentrations of two or more hormones.

2. A plant responds to environmental events by sensing the event, synthesizing hormones in response to this information, and altering cellular activities in response to the hormones. Plants use pigments, such as phytochrome, to sense light. A plant anticipates seasons by measuring the photoperiod.

3. Plants enter into dormancy during unfavorable seasons. Dormancy is initiated by higher levels of abscisic acid in relation to the growth hormones. Abscisic acid triggers many changes that protect a plant during dormancy. Emergence from dormancy is initiated by higher levels of growth hormones in relation to abscisic acid.

4. Flowering occurs at a particular season and is governed by gibberellin and florigen (a hormone not yet isolated). Plants can be grouped, according to flowering season, into long-day plants, short-day plants, and day-neutral plants.

5. A tropism is movement of a plant toward or away from an environmental stimulus, such as light, gravity, sun, touch, water, and chemicals. The movement is caused by the elongation of certain cells. A nastic response is movement without regard to position of the stimulus. Examples include changes in the opening of flowers and changes in leaf position. The movement is caused by changes in turgor pressure within certain cells.

6. Some cellular activities are controlled by circadian rhythms—daily cycles of activity generated by "clocks" within the cell.

KEY TERMS

Abscission (aeb-SIH-shun) (page 295)
Apical dominance (page 293)
Circadian (sur-KAY-dee-un) **rhythm** (page 306)
Dormancy (DOHR-mahn-see) (page 300)
Hormone (HOHR-mohn) (page 292)
Nastic (NAEH-stik) **response** (page 303)
Photoperiod (FOH-toh-peer-yud) (page 300)
Phytochrome (FEYE-toh-krohm) (page 300)
Pigment (page 299)
Target cell (page 292)
Tropism (TROH-pism) (page 303)

STUDY QUESTIONS

1. A plant grows in response to auxin, gibberellin, and cytokinin. Describe the unique way in which each hormone contributes to plant growth.

2. How does interaction between auxin and cytokinin determine the pattern of shoot growth?

3. What is the function of phytochrome? How does it perform this function?

4. How does a plant detect the time of year?

5. How do changes in hormones trigger a tree's preparation for winter dormancy? What is the advantage of shedding the leaves? How do changes in hormones trigger a tree's emergence from dormancy?

6. How does a plant detect the time of year when it should reproduce? Give two advantages of reproducing at a particular time of the year.

7. What is a tropism? Describe how gravitropism is brought about.

8. What is a nastic response? Describe how such a movement is caused by changes in turgor pressure.

CRITICAL THINKING PROBLEMS

1. In the experiments performed by Fritz Went, why did the seedlings grow straight up when the gelatin block was placed on top of the decapitated coleoptiles?

2. What would happen to a lettuce plant if you gave it large doses of gibberellin during the early stages of its growth? Draw two lettuce plants, one without added gibberellin and one with added gibberellin.

3. What will happen if you place unripe tomatoes in a closed container with very ripe tomatoes? Why?

4. Suppose plants were grown by astronauts far from the gravity of a planet. Make a drawing to show how you think these plants would look. Explain why.

5. Suppose that you live in a climate with long winter nights and want your short-night plants to blossom (indoors) for a party in January. How could you force this flowering?

SUGGESTED READINGS

Introductory Level:

Evans, M. L., R. Moore, and K. H. Hasenstein. "How Roots Respond to Gravity," *Scientific American,* December 1986, 112–19. A description of research dealing with starch granules and how they might trigger hormonal changes that bend the roots.

Wilkins, M. *Plant Watching: How Plants Remember, Tell Time, Form Relationships, and More.* New York: Facts on File, 1988. An excellent, well-illustrated book on plant physiology written for the nonscientist.

Advanced Level:

Chadwick, C. M., and D. R. Garrod. *Hormones, Receptors and Cellular Interactions in Plants.* New York: Cambridge University Press, 1986. A thorough description of plant hormones and how they alter cellular activities.

Lumsden, P. J. "Circadian Rhythms and Phytochrome," *Review of Plant Physiology and Molecular Biology* 42 (1991): 351–71. A brief review of these two poorly understood topics.

Salisbury, F. B., and C. W. Ross. *Plant Physiology,* 4th ed. Belmont, CA: Wadsworth Publishing, 1992. An excellent source for information on plant hormones, responses, and biological clocks.

Sisler, E. C., and S. F. Yang. "Ethylene, the Gaseous Plant Hormone," *BioScience* 33 (1984): 238. A review of this unusual hormone: its functions and how it influences its target cells.

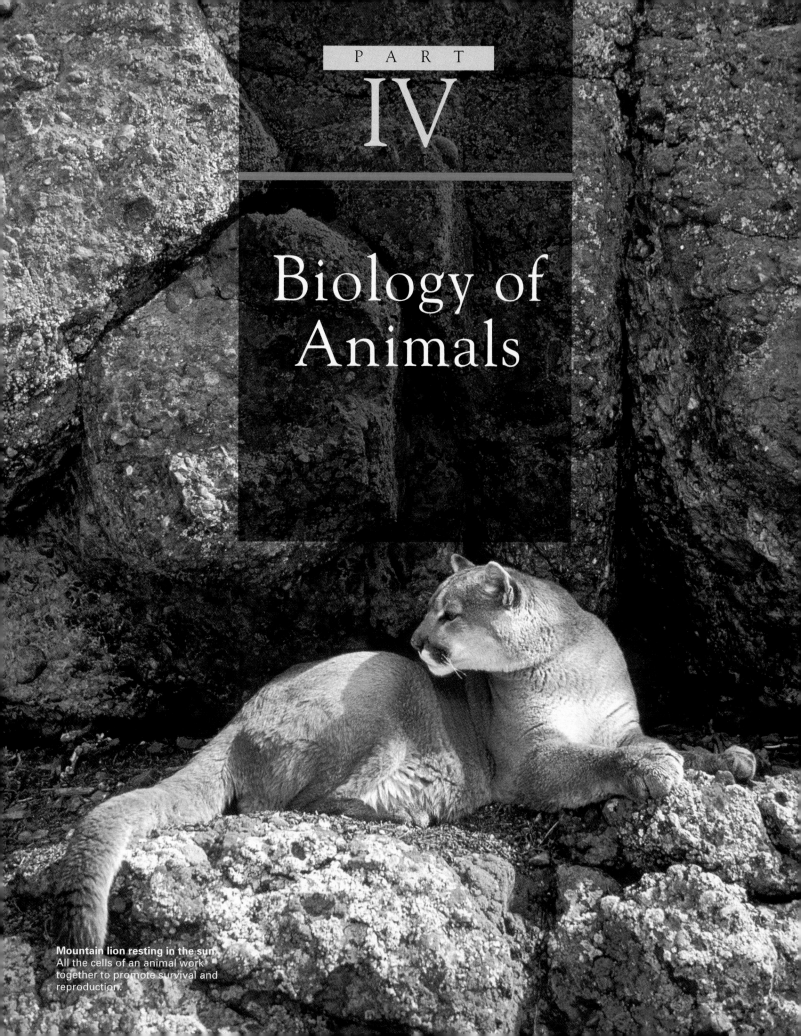

Biology of Animals

Mountain lion resting in the sun.
All the cells of an animal work
together to promote survival and
reproduction.

PART FOUR

Exploring Themes

Theme 1

The fundamental unit of life is the cell.
An animal is formed by many cells (chapter 17).

Theme 2

All cells carry out the same basic processes.
All animal cells synthesize molecules, import and export materials, acquire energy from organic molecules, reproduce, and synthesize proteins according to instructions within their genes (chapter 17).

Theme 3

The cells of a multicellular organism produce adaptations: anatomical, physiological, and behavioral traits that promote survival and reproduction.
Each cell of an animal is specialized for a particular function: cells of the digestive tract process food (chapter 18); red blood cells transport oxygen (chapter 19); lung cells import oxygen and export carbon dioxide (chapter 20); kidney cells excrete wastes (chapter 21); white blood cells kill disease microbes (chapter 22); endocrine cells secrete hormones (chapter 23); nerve cells convey information (chapters 24 and 25); muscle cells change length (chapter 26); endocrine, nerve, and muscle cells coordinate behaviors (chapter 27); and cells of the ovary and testes produce eggs and sperm, which carry genetic information to the next generation (chapters 28 and 29).

17

Animal Organization

Chapter Outline

Objectives

In this chapter, you will learn
–the ways that animals differ from
    other organisms;
–how the first animal probably
    evolved;
–how animal cells are specialized in
    structure and function;
–how specialized cells form tissues;
–that tissues form organs and organ
    systems;
–that cellular activities are coordinated
    by hormones and a nervous system.

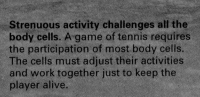

**Strenuous activity challenges all the
body cells.** A game of tennis requires
the participation of most body cells.
The cells must adjust their activities
and work together just to keep the
player alive.

Michael played in a tennis tournament on a hot summer day. His natural athletic ability, combined with years of lessons, gave him a cannonball serve as well as a powerful backhand and forehand. He knew when to charge the net and smash the ball, and he had shrewd tactics that frustrated his opponents.

While playing in the tournament, Michael was completely focused on coordinating his movements with the flight of the ball. He was not aware, however, of the more impressive coordination of events going on within his body—the urgent messages relayed from cell to cell, and the adjustments in cellular activity in response to these messages. Michael's body cells had their own programs for winning a more serious game—keeping him alive.

Strenuous physical activity, especially on a hot day, requires adjustments in almost every cell of the body. Sugar and fat reserves are mobilized to fuel the muscles; water is released onto the skin to prevent overheating; the diameters of blood vessels are adjusted to shunt blood to the more active cells; and brain cells send messages to the chest to initiate deeper breathing and to the heart to accelerate its pumping so that oxygen is delivered more rapidly to the cells. If any one of these adjustments were not perfectly executed, the person would not survive the activity. The primary function of most of our bodily activities is to keep us alive.

Origin of Animals

An animal is a multicellular, eukaryotic (📖 *see page 13*) organism that obtains energy and nutrients by eating other organisms. All animals belong to the kingdom Animalia (see inside back cover of this book). A sponge, coral, earthworm, clam, insect, fish, and human are examples of animals.

Animals differ from other kingdoms of multicellular, eukaryotic organisms—Plantae and Fungi—in the following ways: (1) an animal does not carry out photosynthesis, whereas a plant does; (2) an animal's cells have no cell walls, whereas the cells of both plants and fungi do; (3) an animal digests food inside its body, whereas a fungus digests food outside its body.

Animals arose sometime between 700 million and 1 billion years ago (see inside back cover for the geologic record of life). The oldest animal fossils, around 700 million years old, are already quite advanced and diverse.

Selection for Multicellularity

Animals evolved from a unicellular, eukaryotic organism. These forms of life had existed for perhaps a billion years before giving rise to the first animal. What advantage did this first animal have over its unicellular relatives? In general, multicellularity results in two adaptive traits: a larger body and a division of labor among the many cells.

Advantages of a Larger Body The first animal would have been the largest organism that ate other organisms in its community, which consisted of seaweeds and a great variety of unicellular organisms. It probably fed by filtering small organisms and debris from the surrounding water. Its larger body would have formed a larger filter than the body of a unicellular organism. At the same time, the first animal would have been too large for unicellular organisms to eat. This new kind of organism must have been immediately successful—consuming more food and surviving longer than its unicellular relatives.

Advantages of a Division of Labor A division of labor among the cells of a multicellular body promotes efficiency of function. Each cell can become specialized in performing only a few tasks and so perform them better than a cell that performs all tasks. As with plants, the initial advantage of a division of labor in animals may have been specialized cells that attached the organism to the bottom of the ocean. Immobilization of the body in turbulent waters, as found along the coasts where animals probably evolved, would have enhanced the flow of water over the animal's surface. The increased flow, in turn, would have brought them in contact with more particles of food.

The First Animals

The first animals left no traces of themselves. We can only speculate about their origins, body plans, and ways of life. The oldest fossils are of relatively complex animals of many kinds. They had soft bodies, leaving only imprints of their bodies, tracks, and burrows on the ocean floor.

Animals probably evolved from some kind of protozoan (kingdom Protista), since they, like animals, have eukaryotic cells and consume other organisms. (The amoeba and paramecium are examples of modern protozoa.) The main difference is that a protozoan consists of one cell and an animal consists of many cells. Perhaps this protozoan lived in a colony, which would have facilitated the conversion from a unicellular to a multicellular way of life. The first animal may have resembled a sponge, which is the simplest animal alive today and just one step beyond a colony of cells (fig. 17.1).

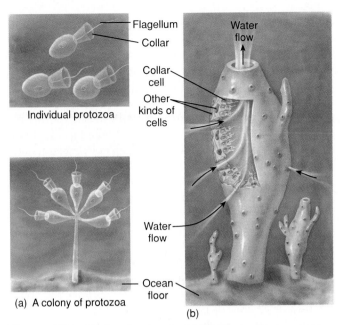

Flagellum
Collar
Individual protozoa

Water flow
Collar cell
Other kinds of cells
Water flow
Ocean floor

(a) A colony of protozoa

(b)

Figure 17.1 Origin of animals. (*a*) The first animal probably evolved from a colony of unicellular protozoa, such as a colony shown here, which exists today. Each protozoan in the colony has a flagellum that sweeps food onto a "collar." The plasma membrane of the collar then engulfs the food particle for digestion. (*b*) The first animal may have resembled a sponge, which is the simplest animal alive today. The inner surface of a sponge consists of collar cells that resemble the colonial protozoa shown in (*a*). The flagellum on each of these inner cells sweeps food onto its collar for digestion. Unlike protozoa in a colony, collar cells in a sponge share their food with other cells in the animal.

Animal Characteristics

An animal consists of many cells—your own body, for example, is formed of approximately 100 trillion cells. The cells of an animal are specialized for carrying out particular functions, such as secretion, oxygen transport, contraction, or carrying information. These specialized cells, in turn, are organized into specialized tissues and organs, which perform life-promoting tasks, such as procurement of food, digestion of food, maintenance of body temperature, and elimination of wastes. All these structures contribute to the three key characteristics of animals: heterotrophy, locomotion, and homeostasis.

Heterotrophy

Animals are **heterotrophs,** which means they acquire energy and nutrients by eating other organisms. After an animal has eaten another organism, the organic molecules within that organism are broken down into monomers such as glucose, amino acids, and fatty acids. The break-

down of large food molecules into smaller ones is called **digestion.** The animal then uses the monomers (see *page 40*) as sources of energy (in cellular respiration) and as materials for building its own kinds of polymers (polysaccharides, lipids, proteins, and nucleic acids) (see *page 40*).

An animal has specialized structures for processing food. In vertebrate animals, the mouth captures food, the stomach stores large food items, the small intestine digests food and absorbs its components, and the large intestine eliminates materials that cannot be digested. The liver and pancreas contribute substances that help process food.

Locomotion

A striking characteristic of most animals is **locomotion,** in which the entire body is moved, in a controlled fashion, from one place to another. Crawling, swimming, walking, and flying are modes of locomotion. Animals travel from place to place in search of food, shelter, and mates, as well as to evade enemies.

Locomotion is linked with heterotrophy. An animal on land needs to move about in search of food. A horse, for example, moves almost continually while grazing. An animal in water may remain in one place and feed on organisms that pass by, but it first had to travel to locate that feeding site. A barnacle, for example, remains fixed in place as an adult but travels as a juvenile in search of a suitable place to settle down.

Many aspects of an animal's body are associated with locomotion. Muscles produce the movement and a skeleton supports the body as it moves. Nerves control movements of the muscles. Structures for seeing, smelling, hearing, and tasting guide an animal in a particular direction. A brain coordinates information from the senses and controls how the body responds to this information.

Homeostasis

The activities of a cell take place within a narrow range of conditions. The temperature and concentration of ions must be just right for a cell to remain active. The environment of an animal's cells is controlled by maintaining an optimal fluid around the cells. **Homeostasis** (*homeo,* steady; *stasis,* standing) is the term for the condition in which an organism's internal environment is maintained in an optimal state.

An animal cell has two fundamental tasks: to survive on its own, as an individual unit of life, and to contribute in some specialized way to the survival of other cells within the animal. A red blood cell, for example, must generate its own ATP (usable form of energy) for doing work and transport materials into and out of its interior. In addition, it must supply other body cells with oxygen. All the cells of an animal contribute in some way toward maintaining an optimal cellular environment—toward homeostasis.

Homeostasis is most highly developed in animals. Many forces work to disturb an animal's internal environment—air that is too cold, too hot, or too dry; water that is too salty or not salty enough; the disruptive changes that accompany a sudden sprint, a heavy meal, or a disease. Many mechanisms within the animal work to counter these disturbances—to return the body to its optimal condition.

Homeostasis operates by **negative feedback:** a disturbance of the internal environment in one direction triggers a corrective response in the opposite direction. A familiar, nonliving example of negative feedback is a furnace and its thermostat. The thermostat is set at a desired room temperature. When the room temperature drops below the set temperature, the thermostat turns on the furnace. When the room temperature rises above the set temperature, the thermostat turns off the furnace. In this way, the room temperature oscillates around the desired temperature:

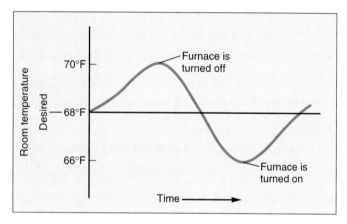

An animal has internal mechanisms that operate by negative feedback, like the furnace and thermostat, to maintain a constant internal environment. Two aspects of the internal environment—temperature and chemical composition—are described briefly here and more thoroughly in later chapters.

Body Temperature   Temperature controls the rate of cellular activity. The advantage of a warmer temperature is that chemical reactions proceed more rapidly: the organism can be more active in acquiring food and evading enemies. The disadvantage of a warmer temperature is that at some point—for most animals around 43° C (110° F)—the cells become too warm for their enzymes to function, and they die.

Animals have adaptations (📖 *see page 17*) that give them some control over their body temperature, so that their cells can maintain a high level of activity without being in danger of overheating. Anatomical

adaptations, which include body size, body shape, hair, and feathers, control whether heat tends to move into or out of the body. Physiological adaptations, such as shivering and sweating, control how much heat is generated and how much is lost. Behavioral adaptations, such as moving into sunshine or shade, control exposure to heat.

Most animals regulate their body temperatures chiefly by anatomical and behavioral adaptations. Such animals, commonly referred to as cold-blooded, can modify their body temperatures but generally cannot maintain an optimal temperature. A lizard, for example, warms its body on a cold day by basking in the sun. Its broad, flat body (anatomy) warms rapidly when positioned at right angles to the sun's rays, and regions of its brain react to cold by commanding the body to seek sunshine (behavior). By moving back and forth between sunshine and shade, a lizard may achieve an internal temperature that is near optimal. It cannot, however, maintain this optimal body temperature at all times—the sun does not shine at night nor does it provide enough warmth on overcast days. Moreover, the lizard engages in other activities and cannot bask all the time.

Mammals and birds have better control over their body temperatures than other animals. These so-called warm-blooded animals have physiological adaptations in addition to the anatomical and behavioral adaptations seen in other animals. They accomplish temperature control by generating heat in their muscles whenever their body temperature drops below optimal and by panting or sweating whenever their body temperature rises above optimal:

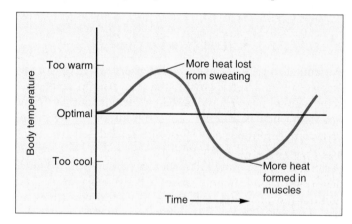

These physiological adaptations are described more fully in chapter 21.

The degree to which an animal can control its internal temperature strongly influences the degree to which it can remain active under a broad range of environmental temperatures. Different levels of control can be seen by looking at the animals in a northern meadow in both summer and winter. In the warmth of summer, all the animals are fully active. In the cold of winter, only the foxes, deer, rabbits, sparrows, chickadees, hawks, and other kinds of birds and mammals remain active. The frogs, snakes, spiders, ants, and worms are dormant—they cannot maintain internal temperatures that are warm enough to move about.

Body Chemistry   Animals regulate the chemical environment of their cells. A major way in which this is accomplished is by physiological adaptations that control the chemical composition of fluid around the cells. This critically important fluid is known as the **extracellular fluid** (*extra*, outside). Almost every cell of the body participates in keeping this fluid optimal for cellular activities.

The extracellular fluid provides the cells with oxygen, glucose, amino acids, and other materials required for survival. It also disposes of wastes formed during cellular activities. The concentrations of ions in the extracellular fluid are carefully regulated, for they govern the cell's pH (see page 33) and electric potential (see page 80). Ions also control how much water is in the cell by controlling the direction of osmosis:

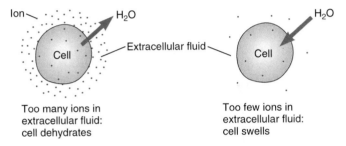

Too many ions in extracellular fluid: cell dehydrates

Too few ions in extracellular fluid: cell swells

Most animals have two kinds of extracellular fluids: interstitial fluid and plasma. **Interstitial fluid** is the liquid that surrounds the cells. **Plasma** is the noncellular part of the blood—everything except the red blood cells, white blood cells, and platelets (cell fragments involved in clotting). The chemistry of the cells is controlled by the chemistry of the interstitial fluid; materials move back and forth between the interior of a cell and the fluid surrounding the cell. The chemistry of the interstitial fluid, in turn, is controlled by the chemistry of the blood plasma; materials move back and forth between the fluid surrounding the cells and the fluid within the blood vessels (fig. 17.2).

Almost every cell of the body contributes in some way toward maintaining the blood in an optimal condition. The kidneys and liver are particularly important in this aspect of homeostasis.

As with body temperature, the degree to which an animal can control its internal chemistry determines the degree to which it can live in different environments. A Pacific salmon, for example, has physiological adaptations that give it superb control over its internal concentrations of ions. As a consequence, this salmon can live in both the saltwater of oceans and the fresh water of inland lakes and streams. A rainbow trout, by contrast, only has adaptations for controlling its internal concentration of ions when

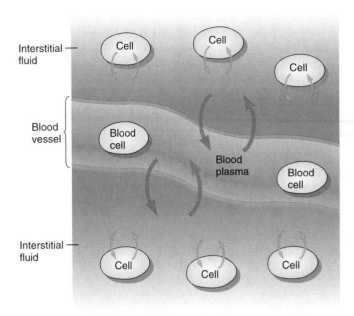

Figure 17.2 Chemical homeostasis involves two extracellular fluids: the interstitial fluid and the blood plasma. An optimal environment for cellular activities is maintained by the chemistry of the interstitial fluid: materials are exchanged between the cell interior and its surroundings. The interstitial fluid, in turn, is maintained by the chemistry of the blood plasma; materials are exchanged between the interstitial fluid and the plasma. Many organs work together to maintain the chemistry of the blood.

surrounded by fresh water. If placed in the ocean, water would flow out of its body (by osmosis) and it would die of dehydration. Therefore, rainbow trout are restricted to life in freshwater lakes and streams.

Disturbances of Homeostasis   While a human has excellent mechanisms for achieving homeostasis, these mechanisms do not always function perfectly. Homeostasis declines with age (beginning at about thirty years of age) and is weakened by disease, prolonged exposure to adverse weather, strenuous physical activity, and emotional stress.

The most common disturbances of homeostasis occur during strenuous exercise, especially on a hot day. Control of body temperature fails when the body can no longer lose heat by sweating, usually because so much water has already been lost that not enough remains to form sweat. Brain cells are most vulnerable to abnormal temperatures: a dangerously overheated person exhibits convulsions (involuntary muscle contractions) and loss of consciousness. Body chemistry becomes disturbed when too many ions are lost in sweat. The first signs of chemical imbalance from strenuous exercise are muscle cramps.

Our ability to maintain homeostasis reflects our overall health. Homeostasis is affected not only by unusual conditions, such as strenuous exercise or diseases, but also by the way we live our lives—what we eat, how much we exercise, and whether or not we get enough sleep each night. A physician evaluates all aspects of homeostasis during a physical examination. An important part of the evaluation is a blood test, for the composition of this fluid reflects how well all the body cells are contributing to homeostasis. Just what the blood test tells a physician about homeostasis is described in the accompanying Wellness Report.

Organizations of Cells

A complex animal, like yourself, has approximately 350 kinds of cells that work together to promote survival and reproduction. A simpler animal, like a sea anemone, has fewer than twenty kinds. Each cell performs a specialized task and has specialized structures and biochemical pathways for accomplishing that task. The cells of an animal are organized according to their tasks into tissues, organs, and organ systems. These organizations of cells are shown in the hierarchy of biological organization diagrammed inside the front cover of this book and discussed below.

Animal Tissues

A **tissue** is a group of cells located near one another that contribute in a similar way to the same task. The cells of a tissue are held together by a meshwork of organic molecules (proteins and polysaccharides) and by adhesions between their plasma membranes. An animal has many kinds of tissues. They are grouped, according to general appearance and function, into four basic categories: epithelial, connective, muscle, and nervous tissue (table 17.1).

Epithelial Tissue **Epithelial tissue** is a sheet of tightly packed cells that covers a surface (fig. 17.3). It covers the outer surface of an animal, the outer surfaces of internal organs, and the inner surfaces of tubes and cavities—the stomach, intestines, lungs, blood vessels, urinary bladder, and tubes associated with reproduction and urination. Epithelial tissue is further classified according to cell shape and number of layers (table 17.2).

Everything that enters and leaves the body (its internal cells and extracellular fluids) passes through epithelial tissue. This barrier between internal tissues and external environment regulates what enters and leaves the body, protects the internal tissues, and secretes materials. These three functions are described following.

The cells of epithelial tissue regulate what passes through them by taking in certain materials on one side of their structure and releasing them on the other side:

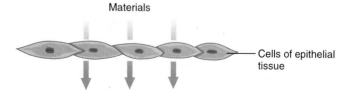

Materials

Cells of epithelial tissue

Table 17.1		
The Four Categories of Animal Tissues		
Type of Tissue	**Traits of the Cells**	**Main Locations**
Epithelial	Tightly packed	Outer layer of body and organs; inner layer of digestive tract, lungs, and blood vessels
Connective	Dispersed within a fluid	Bone, ligament, fat, tendons, blood
Muscle	Change length	Skeletal muscles, heart, wall of digestive tract
Nervous	Generate electrical currents; secrete chemical messengers	Brain, spinal cord, inner ear, nose, retina of eye

Epithelial tissue in the kidneys, for example, regulates what passes from the blood to the urine, and epithelial tissue in the intestines regulates what passes from the digestive tract to the blood.

The closely packed cells of epithelial tissue protect internal tissues by forming a barrier against water loss, mechanical damage, and invasions by viruses and microorganisms.

Certain cells within epithelial tissue are specialized for **secretion**, which is the synthesis and release of special materials for use outside the cell. They secrete hormones, mucus (a slimy fluid that lubricates surfaces), sweat, oils, and digestive enzymes. Some epithelial tissues fold inward during development and become glands (fig. 17.4), which are organs that specialize in secretion.

Connective Tissue The cells of **connective tissue** are loosely packed rather than right next to one another. Blood, tendons, and bones are examples of this kind of tissue (fig. 17.5). Connective tissue is classified according to the nature of its extracellular material, which may be a fluid, a gel, or a solid (table 17.3).

The main functions of connective tissues are to support, connect, and anchor other tissues. Bone supports the body; tendons connect muscles to bones; and connective tissue in the skin anchors epithelial tissue to underlying muscle tissue. Some connective tissues have specialized functions: fat stores energy, insulates against heat loss, and cushions body organs; blood transports materials and protects against infectious diseases.

Muscle Tissue **Muscle tissue** produces movement. Its bundles of rod-shaped cells are specialized for shortening and elongating. About half of your body mass consists of muscle tissue (fig. 17.6).

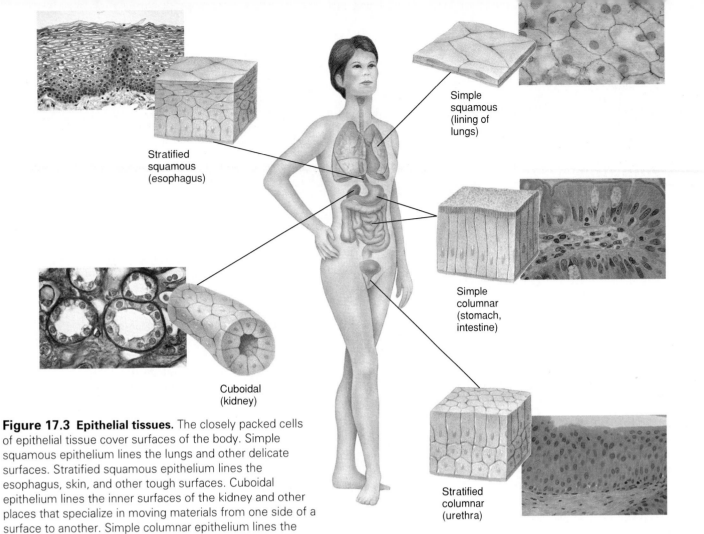

Stratified
squamous
(esophagus)

Simple
squamous
(lining of
lungs)

Cuboidal
(kidney)

Simple
columnar
(stomach,
intestine)

Stratified
columnar
(urethra)

Figure 17.3 Epithelial tissues. The closely packed cells of epithelial tissue cover surfaces of the body. Simple squamous epithelium lines the lungs and other delicate surfaces. Stratified squamous epithelium lines the esophagus, skin, and other tough surfaces. Cuboidal epithelium lines the inner surfaces of the kidney and other places that specialize in moving materials from one side of a surface to another. Simple columnar epithelium lines the stomach, intestine, and other places where secretion is important. Stratified columnar epithelium lines the urethra and other tubes that specialize in transporting fluids.

Table 17.2

Classification of Epithelial Tissue

Tissue Name	Tissue Structure	Main Locations
Simple* squamous		Lungs; inner lining of vessels
Simple cuboidal		Inner lining of kidneys
Simple columnar		Inner lining of stomach, intestine, nose
Stratified† squamous		Skin, mouth, throat
Stratified columnar		Bronchi, urethra

*Simple = one layer of cells.

†Stratified = more than one layer of cells.

Animals have three kinds of muscle tissue: skeletal, cardiac, and smooth (table 17.4). Skeletal muscle moves the skeleton, enabling locomotion. It also moves the skin, as in facial expressions, and the tongue. Cardiac muscle forms the heart, which pumps blood through the body. Smooth muscle encircles tubes and cavities, such as blood vessels, stomach, intestines, and urinary bladder. When smooth muscle shortens, it narrows the channel through the tube or cavity:

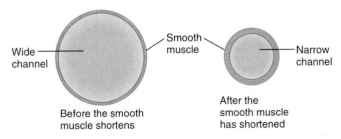

Wide channel

Smooth muscle

Narrow channel

Before the smooth muscle shortens

After the smooth muscle has shortened

Nervous Tissue Nervous tissue is a network of nerve cells that carries signals from one part of the body to another (fig. 17.7). Nerve cells are called **neurons.** A neuron specializes in generating electric currents, which carry messages from one end of the neuron to another, and

317

Evaluating Homeostasis: The Blood Test

Aroutine part of a physical examination is a laboratory analysis of the blood. The composition of blood is a measure of overall health because almost all the body cells work to keep this fluid optimal. The composition of the blood controls the composition of the interstitial fluid, which, in turn, influences the chemistry of the cells. A blood test reveals deviations from homeostasis and indicates where in the body the problem lies. It includes analyses of the cells and the plasma.

A complete blood count is an estimate of how many cells are in the blood. The total number of blood cells, both red and white, indicates whether or not your blood-forming tissues (in the bone marrow) are performing adequately.

The number of red blood cells reflects the capacity of the blood to transport oxygen. An insufficient number of these cells, a condition known as anemia, can have many causes, including internal bleeding or a diet that is deficient in protein or iron, which are needed to build red blood cells. Also measured is the sedimentation rate—the time it takes for the red blood cells to settle to the bottom of a tube. Blood cells that are clumped settle more rapidly than normal, unclumped cells. An abnormally high rate of sedimentation may indicate an abnormal biochemical pathway (e.g., rheumatoid arthritis), a certain kind of infection (e.g., tuberculosis), or cancer.

The number of white blood cells reflects the condition of the immune system, which defends the body against infections. A high white blood cell count reflects a normal response to an infection. A very high count, however, may indicate leukemia (cancer of the bone marrow, which makes white blood cells). A low white blood cell count may reflect a low-protein diet, emotional stress, or strenuous physical exercise.

There are many kinds of white blood cells. The differential count measures the proportion of one kind of white blood cell in relation to the others. An abnormal proportion or abnormal appearance is caused by certain diseases and drugs. Infectious mononucleosis ("mono"), for example, is diagnosed by unusual B cells (the cells that make antibodies)—large, irregularly shaped B cells that resemble monocytes, another kind of white blood cell.

The analysis of plasma, the noncellular component of the blood, is extensive and of great importance in finding out what is wrong and why. Recall that this fluid supplies materials to and receives materials from the interstitial fluid. Thus, the composition of the plasma reflects the composition of the interstitial fluid, which forms the immediate environment of the body cells. If an unhealthy cell secretes an unusual product or unusual amounts of a normal product, it will show up in the blood plasma. Some of the materials measured by a blood test are listed in table 17.A.

Blood tests, as well as blood pressure, pulse, and other aspects of the physical examination, are important not only for revealing the cause of a disorder but for finding out what is normal for you. Each of us has a unique set of normal values for the various components of the physical exam. By having such an exam every year or so, your doctor can establish what is normal for you. Then when you do get sick, any deviation from these normal values tells the physician exactly how your body has changed from when it was healthy.

Table 17.A

Some Materials Measured by a Blood Test

Material	Likely Cause of Abnormal Values
Blood urea nitrogen (waste product from protein breakdown)	Kidney disorder
Uric acid (waste product from the breakdown of proteins and nucleic acids)	Kidney disorder; high-protein diet
Creatinine (waste product from muscular work)	Kidney disorder
Albumin (normal protein in the plasma)	Liver or kidney disorder; insufficient protein in diet
Cholesterol (component of membranes and steroids)	Disorder of the liver or pancreas; high-fat diet
Bilirubin (waste product from the breakdown of red blood cells)	Disorder of the liver or spleen
Glucose (blood sugar)	Disorder of the pancreas
Electrolytes (ions, including sodium, potassium, phosphorus, calcium, chloride, and bicarbonate)	Disorder of the kidney, pancreas, or parathyroid glands

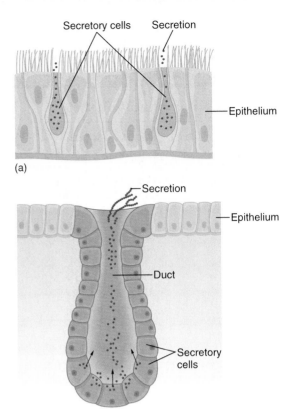

(a)

(b)

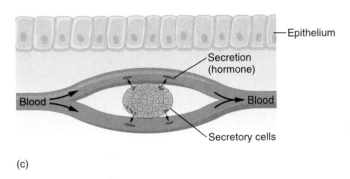

(c)

Figure 17.4 Some epithelial tissue is specialized for secretion.
(*a*) Secretory cells within epithelial tissue release their materials directly onto the tissue surface. Secretory cells within the epithelium of the digestive tract, for example, secrete digestive enzymes. (*b*) An exocrine gland is formed of epithelial tissue that has folded inward during development. A duct (small tube) remains where the tissue folded. Secretions move through this duct onto a body surface or into a cavity. Examples of exocrine glands are the salivary glands and sweat glands. (*c*) Many endocrine glands are formed of epithelial tissue that has folded inward and lost its connection to a surface or cavity. The cells secrete their materials into the bloodstream. Examples of endocrine glands are the thyroid and the adrenal cortex.

Figure 17.5 Connective tissue. The cells of connective tissue are dispersed within extracellular material. The extracellular material is a fluid in blood (the plasma), a solid in bone, and a gel in other kinds of connective tissue.

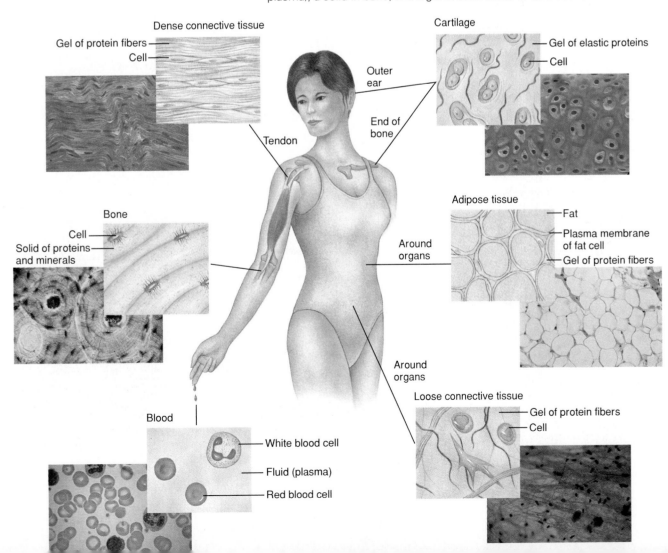

Table 17.3

Classification of Connective Tissue

Name of Tissue	Description of Extracellular Material	Main Locations
Loose connective tissue	Gel of loosely packed protein fibers	Fat deposits; smooth muscles
Dense connective tissue	Gel of tightly packed protein fibers	Tendons; dermis of skin; ligaments; lungs
Cartilage	Gel of tightly packed elastic proteins	Ends of bones; tip of nose; outer ear
Bone	Solid packed with minerals and elastic proteins	Skeleton
Blood	Fluid consisting mainly of water, ions, and proteins	Channels of blood vessels; chambers of heart

Table 17.4

Classification of Muscle Tissue

Name of Tissue	Description	Main Location
Skeletal	Can be contracted voluntarily	Attached to skeleton and facial skin; forms tongue
Cardiac	Cannot be contracted voluntarily	Walls of heart
Smooth	Cannot be contracted voluntarily	Walls of blood vessels and digestive tract

Figure 17.6 Muscle tissue. The rod-shaped cells of muscle tissue are specialized for changing their lengths. Smooth muscle appears within the walls of tubes and cavities. Cardiac muscle forms the walls of the heart. Skeletal muscle moves the skeleton as well as the facial skin (in facial expressions) and the tongue.

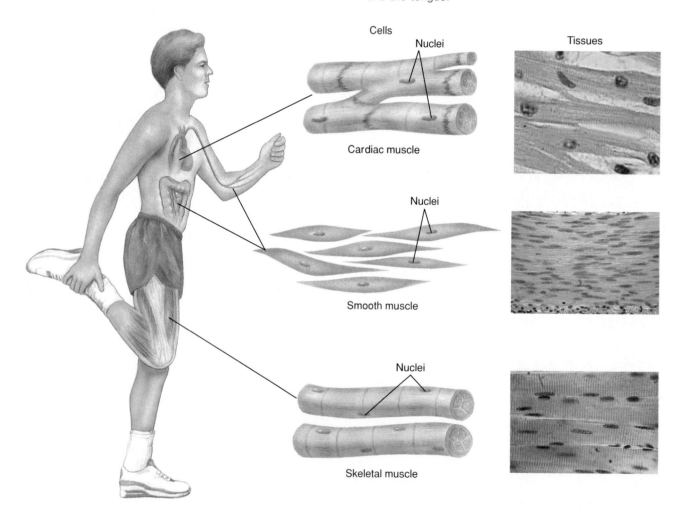

Cells

Nuclei

Tissues

Cardiac muscle

Nuclei

Smooth muscle

Nuclei

Skeletal muscle

in secreting chemical messengers, which send messages to other cells.

Nervous tissue consists of three kinds of neurons (table 17.5). Sensory neurons receive information from the environment (either outside the body or inside the body but outside the nervous system) and pass it to other neurons in the network. Interneurons receive messages from other neurons and pass them to other neurons in the network. Motor neurons receive messages from other neurons and pass commands to muscles or glands, which bring about reactions to the information:

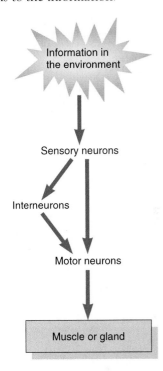

By receiving information and controlling the activities of muscles and glands, nervous tissue governs how the body responds to its environment.

Organs and Organ Systems

The next higher levels of animal organization after tissues are the organ and the organ system. A diagram on the inside cover of this book shows these levels of organization within the hierarchy of biological organization.

Organs   An **organ** consists of at least two (and usually all four) kinds of tissues that contribute in different ways to some particular task. Examples of organs are the small intestine, which processes food, and the kidney, which forms urine.

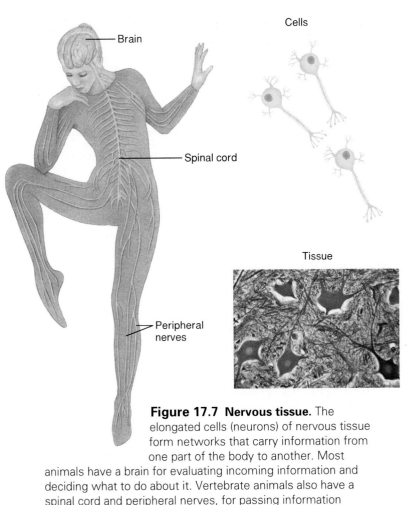

Cells

Tissue

Figure 17.7 Nervous tissue. The elongated cells (neurons) of nervous tissue form networks that carry information from one part of the body to another. Most animals have a brain for evaluating incoming information and deciding what to do about it. Vertebrate animals also have a spinal cord and peripheral nerves, for passing information between the brain and the rest of the body.

Table 17.5

Classification of Cells Within Nervous Tissue

Name of Cell	Function
Sensory neuron	Carries information from the environment to an interneuron (or sometimes a motor neuron)
Interneuron	Carries information from a sensory neuron to another interneuron or a motor neuron
Motor neuron	Carries information from an interneuron (or sometimes a sensory neuron) to a muscle or gland

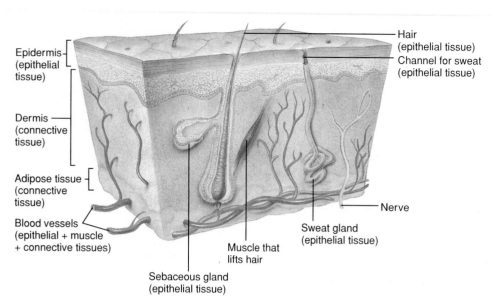

Figure 17.8 The skin: our largest organ. The skin consists of epithelial, connective, muscle, and nervous tissues.

Consider the skin (fig. 17.8), the largest organ in the human body. This outer covering of the body is approximately 2 millimeters deep—about the thickness of a heavy blanket. Its main task is to shield internal tissues from the external environment. The skin contains all four kinds of tissues: epithelial, connective, muscle, and nervous.

The epidermis, made of epithelial tissue, forms the outermost portion of the skin. The surface layer consists of dead, flattened cells that are packed with keratin, a strong protein fiber that protects the skin against water loss and injury. These cells are sloughed off on a regular basis (forming a major component of house dust) and replaced by new cells formed below. Beneath the dead cells are living cells of several kinds: squamous cells, which specialize in synthesizing keratin; melanocytes, which specialize in synthesizing melanin (the brown pigment that protects against solar radiation); and basal cells, which specialize in forming new cells. Some cells of the epidermis form sweat glands, sebaceous (oil-secreting) glands, and body hairs. Interspersed among the epidermal cells are white blood cells, which defend against disease microorganisms.

Skin cancer develops when epidermal cells divide continually, without control. A major cause of skin cancer is overexposure to sunlight for many years. Of the three forms of skin cancer, squamous cell carcinoma and basal cell carcinoma are common, especially among older people with fair skin. These cancers, which appear as scaly patches or lumps on the skin, are rarely fatal. The third form, melanoma (cancer of the melanocytes), is less com-

mon but often fatal—it appears as an irregularly shaped growth that changes color; it usually spreads to other parts of the body before being detected.

Beneath the epidermis is the dermis, a connective tissue that anchors the epidermis and holds a variety of structures: blood vessels, nerve cells, and tiny muscles that make your hairs stand on end when the skin is cold or when you are frightened. The sweat glands, sebaceous glands, and body hairs extend downward into the dermis.

Beneath the dermis is yet another form of connective tissue—fat, which insulates the body against heat loss, cushions it against injury, and stores energy. This layer of fat also holds blood vessels and nerve cells.

Organ Systems An **organ system** is a group of organs, located in different regions of the body, that contribute in different ways to a particular bodily function. The digestive system, for example, includes the following organs: mouth, esophagus, stomach, small intestine, large intestine, pancreas, and liver.

Nine organ systems work to keep an animal alive: the digestive, circulatory, respiratory, urinary, immune, endocrine, nervous, skeletal, and muscular systems (fig. 17.9). These organ systems are the topics of the following nine chapters. Yet another system of organs, the reproductive system (also shown in fig. 17.9), functions in mating and in developing offspring. This system, which differs in males and females, is described in chapters 28 and 29. All the organ systems, together with the skin, form the body of an animal.

Traditionally, medical science has isolated each organ system for study and medical care. Most physicians specialize in just one of the organ systems, as do the different wings of a hospital. The more we learn about human health, however, the more we realize that organ systems interact with one another in surprising ways. This topic is described in the accompanying Research Report.

Can Emotions Make Us Sick?

Lisa had not seen a doctor in years. She almost never got sick—not even a sniffle or a headache. But then her luck changed. First she caught a bad cold in September, and then she had a bout of flu in October followed by a nasty cough that lingered until after the New Year. She was really upset about these illnesses because they happened at a time when she needed to feel good in order to deal with her parents' divorce.

Was Lisa's series of illnesses related in some way to the divorce of her parents? Can our emotions make us more vulnerable to diseases? Can the nervous system influence the immune system?

Until recently, most scientists and physicians would have said definitely not. In traditional medicine, the immune system is treated as a separate entity because until recently there was no reason to believe that it was influenced by any other system. Now we are not so sure.

Recent studies indicate a connection between the nervous system and the immune system. They demonstrate that stress, depression, and bereavement can increase susceptibility to infectious diseases and cancer. In one study, volunteers were interviewed and then infected with the cold virus (via nose drops). People who had recently experienced emotional stress, such as a death in the family, a divorce, or job layoff, developed worse colds than the others. Another study, which monitored the emotions and white blood cells of medical students, revealed that students are much more susceptible to illnesses during final exams.

In yet another study, the husbands of women with advanced breast cancer gave blood samples to measure their white blood cells during their wives' illnesses and after their deaths. While the immune systems of the men remained normal during their wives' illnesses, they became significantly weaker after their wives died and remained weaker for at least a year. Psychologists concluded it was

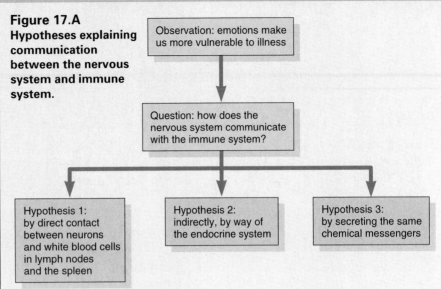

Figure 17.A Hypotheses explaining communication between the nervous system and immune system.

Observation: emotions make us more vulnerable to illness

Question: how does the nervous system communicate with the immune system?

Hypothesis 1: by direct contact between neurons and white blood cells in lymph nodes and the spleen

Hypothesis 2: indirectly, by way of the endocrine system

Hypothesis 3: by secreting the same chemical messengers

the feeling of helplessness after the death, rather than stress of the illness, that depressed the immune systems of the husbands.

Other evidence comes from the unusual phenomenon of hair turning gray within a few days after an emotional shock. Apparently, the strong emotion evokes an inappropriate response by the immune system, in which antibodies attack pigment cells in the hair follicles. The hair loses its pigment and turns gray. The fact that antibodies change their behavior in response to emotions indicates the immune system receives information from the nervous system.

We now have considerable evidence that the mind influences the immune system. We even have evidence of the reverse: the immune system appears to influence the mind. How do these two organ systems communicate with each other? What messages are sent back and forth? Researchers are beginning to explore the physical basis of the mind-body relationship.

Several hypotheses about the physical basis of communication between the nervous system and immune system are now being tested (fig. 17.A). One hypothesis is a two-way communication within certain organs: neurons in the lymph nodes and spleen secrete neurotransmitters that bind with protein receptors within the plasma membranes of white blood cells and alter their activi-

ties. In a test of this hypothesis, researchers stimulated these neurons (in laboratory animals) and found that the immune system responded by changing its activities. A second hypothesis is that the two systems communicate by way of the endocrine system: the immune system communicates with the endocrine system, which in turn communicates with the nervous system. The stress hormone epinephrine (also known as adrenalin) has been shown to depress the immune system. A third hypothesis is that neurons and white blood cells speak the same language: they secrete and respond to some of the same chemical messengers. Disease symptoms that originate in the mind (sleepiness, headache, and muscle aches) can be induced by chemical messengers secreted by white blood cells, and neurotransmitters secreted by brain cells are known to bind with membrane proteins in white blood cells.

This exciting new field of research will undoubtedly change the way physicians treat us when we are sick. If scientists can devise methods for controlling the communications between the nervous and immune systems, then new therapies can be developed. As an adjunct to traditional treatments, some doctors are already treating infectious diseases and cancer with guided imagery (visualizing wellness), relaxation techniques, psychotherapy, and humor.

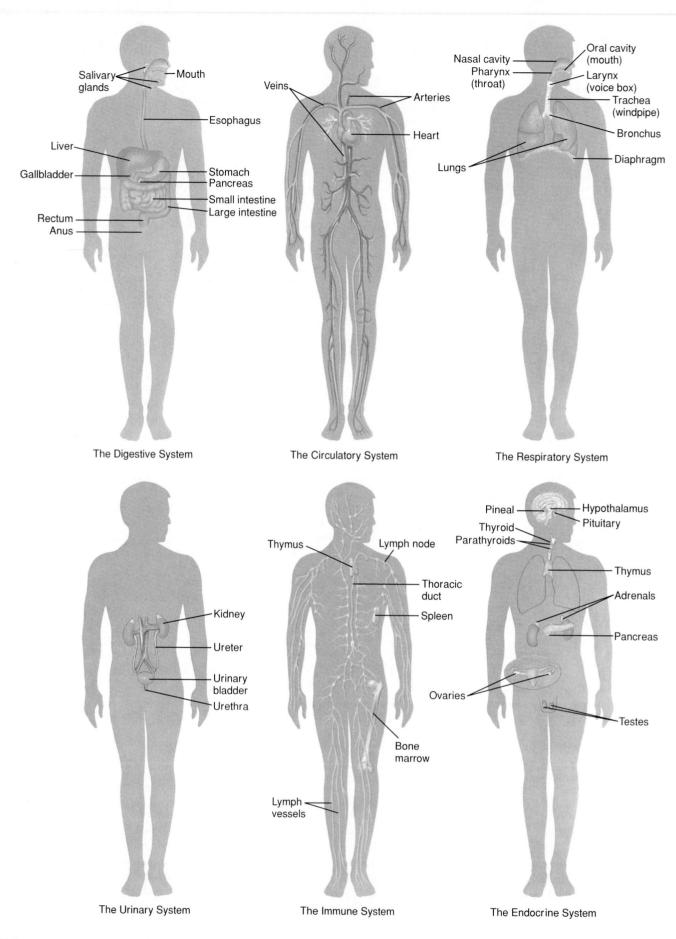

The Digestive System

Salivary glands
Mouth
Esophagus
Liver
Gallbladder
Stomach
Pancreas
Small intestine
Large intestine
Rectum
Anus

The Circulatory System

Veins
Arteries
Heart

The Respiratory System

Nasal cavity
Oral cavity (mouth)
Pharynx (throat)
Larynx (voice box)
Trachea (windpipe)
Bronchus
Diaphragm
Lungs

The Urinary System

Kidney
Ureter
Urinary bladder
Urethra

The Immune System

Thymus
Lymph node
Thoracic duct
Spleen
Bone marrow
Lymph vessels

The Endocrine System

Pineal
Hypothalamus
Pituitary
Thyroid
Parathyroids
Thymus
Adrenals
Pancreas
Ovaries
Testes

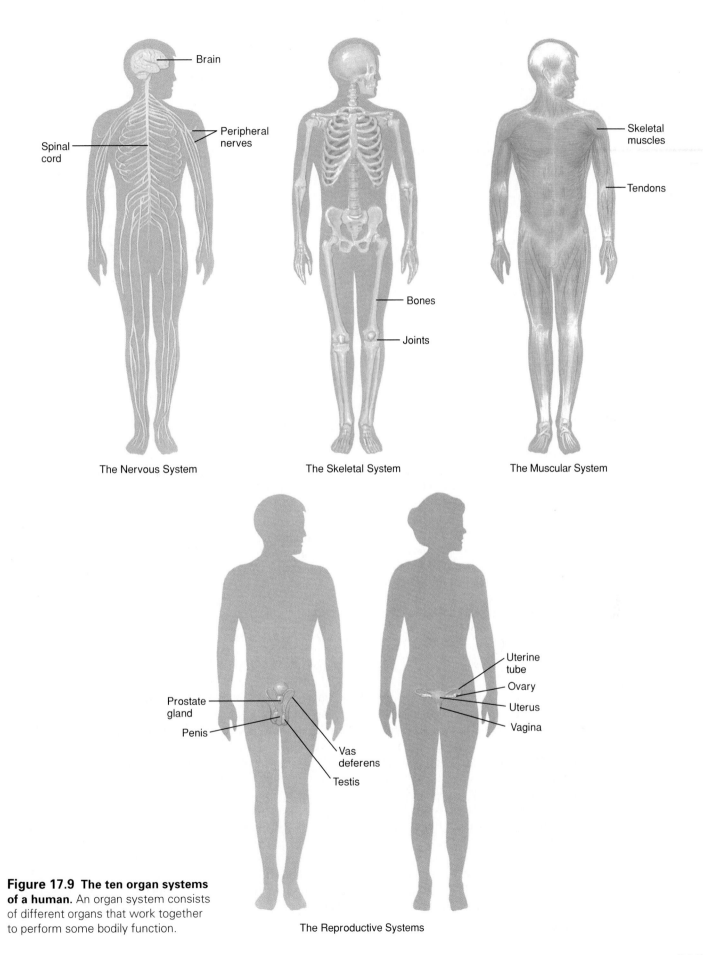

The Nervous System

The Skeletal System

The Muscular System

Figure 17.9 The ten organ systems of a human. An organ system consists of different organs that work together to perform some bodily function.

The Reproductive Systems

325

Coordination of Cellular Activities

An animal coordinates the activities of its cells so that they work together in bringing about appropriate responses to the environment. When fleeing an enemy, for example, muscle cells in the legs, heart, and rib cage contract more rapidly, liver cells convert more glycogen to glucose, and cells of the adrenal glands secrete more epinephrine (also known as adrenalin). At the same time, cells within the digestive and urinary tracts become less active. Such coordination requires internal mechanisms that (1) receive information about the external environment (e.g., food and enemies) and internal environment (deviations from optimal); (2) evaluate the information and decide upon appropriate responses; and (3) command certain body cells to carry out the responses.

Cells communicate with one another by means of chemical messengers—molecules that are secreted by one cell and influence the activities of another cell. Most chemical messengers alter the activities of their target cells by binding with receptor proteins (□ *see page 73*) within the plasma membranes. Binding of a chemical messenger with its receptor protein either alters the permeability of the membrane to certain materials or signals the cell to increase or decrease one of its normal activities.

While all cells are capable of secreting chemical messengers, certain cells are specialized in just this activity. These cells are organized into endocrine glands, which secrete chemical messengers called hormones, and a nervous system, which secretes chemical messengers called neurotransmitters. Hormonal and neural systems of control are previewed here so that an overall understanding of bodily function can develop as we examine each organ system in the following chapters. More detailed descriptions of these systems are provided in chapters 23 and 24.

Hormonal Control

An animal **hormone** is a chemical messenger that travels in the bloodstream. Hormones are secreted by specialized cells within **endocrine glands,** which are organized into an endocrine system (see fig. 17.9). Examples of hormones are growth hormone, epinephrine (or adrenalin), estrogen, testosterone, and insulin. Examples of endocrine glands are the pituitary gland, adrenal glands, ovaries, testes, and pancreas. (Exocrine glands, by contrast, secrete fluids, such as sweat, oil, saliva, and mucus, directly into body cavities or onto body surfaces.)

The endocrine system is essential to an animal's responses to its environment. Information about events occurring within the external and internal environments is detected either directly by the glands themselves or indirectly by sensory neurons that pass the information to the appropriate gland:

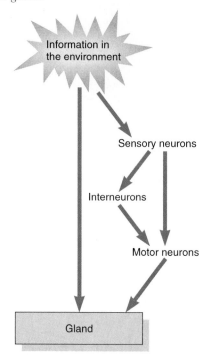

When an endocrine gland receives a certain signal, it releases its hormone into the bloodstream. There the hormone travels throughout the entire body. Only certain body cells, called the target cells of the hormone, are equipped to receive the signal. These cells then bring about appropriate behaviors or homeostatic adjustments in response to the original information.

Many aspects of homeostasis are coordinated by the endocrine system. The pancreas, for example, is a gland that secretes two hormones: insulin and glucagon. Together, they maintain an optimal amount of glucose in the blood (the so-called blood-sugar level) regardless of how much glucose enters the blood from the digestive tract or how much is expended by cellular activities (fig. 17.10). The blood-sugar level is monitored at all times by cells within the pancreas. If too much glucose is detected, the pancreatic cells secrete insulin, which signals body cells to take up and use (as a fuel) more glucose and signals the liver cells to store more glucose as glycogen (a polymer made of hundreds of glucose monomers). As glucose moves out of the blood, the blood-sugar level returns to normal. If too little glucose is detected, the pancreatic cells secrete glucagon, which signals the liver cells to convert glycogen into glucose and to release the glucose into the bloodstream. As glucose enters the blood, the blood-sugar level returns to the optimal level. Diabetes mellitus develops when the pancreas does not secrete enough insulin or body cells do not receive insulin's signal. This relatively common disorder is described in the Wellness Report in chapter 23.

Neural Control

Neural control operates more rapidly than hormonal control. Neurons carry messages as electric currents, which travel more rapidly than hormones travel through the blood. Neural control is essential to rapid locomotion—

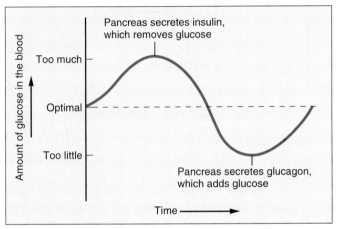

Figure 17.10 Hormonal control of blood sugar by the pancreas. When the amount of glucose in the blood rises above optimal, the pancreas secretes more insulin. This hormone removes glucose from the blood by stimulating all body cells to take in more glucose and the liver cells to convert glucose to glycogen for storage. When the amount of glucose falls below optimal, the pancreas secretes more glucagon. This hormone adds glucose to the blood by stimulating the liver cells to convert glycogen to glucose and then to add the glucose to the bloodstream.

every animal except a sponge has a nervous system. It also plays important roles in homeostasis.

A vertebrate animal (a fish, amphibian, reptile, bird, or mammal) has two main neural centers for homeostasis: the **hypothalamus** and the **medulla oblongata** in the brain, as shown in figure 17.11. We will refer to these centers often in the next few chapters and then describe them in detail in chapter 24.

Neural control involves networks of neurons. Messages travel from neuron to neuron within a network by means of chemical messengers called neurotransmitters. The distances between neurons are very short. Information about the environment (external or internal) is picked up by sensory neurons and passed (via neurotransmitters) to interneurons. In most animals, the interneurons are clustered together to form a brain. Incoming information is analyzed within the brain, by passing from interneuron to interneuron, and a decision is made about how the body will respond. Information about how to respond is then passed to motor neurons, which carry commands to either a gland or a muscle—the two kinds of tissue that can bring about a bodily response. The gland releases its hormone, which alters cellular activity, or the muscle contracts, which causes movement (fig. 17.12).

Consider for example the neural network that regulates an animal's response to what it sees. Suppose you are walking along the street and see someone approaching you. How do you respond? Sensory neurons in your eyes send information to your brain about the person's appearance—the body shape, clothing, and pattern of walking. In your brain, this information is evaluated; it is sent from

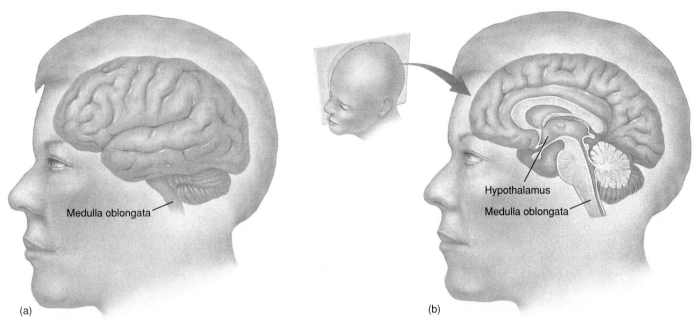

(a)   (b)

Figure 17.11 Two neural centers of homeostasis: the hypothalamus and the medulla oblongata. In vertebrate animals, these two regions of the brain control most aspects of homeostasis. (a) The intact brain, as seen from the left side.

The medulla oblongata is visible but the hypothalamus is hidden by more external regions of the brain. (b) The same brain, cut through the middle and with the left half removed. Both the hypothalamus and the medulla oblongata can be seen.

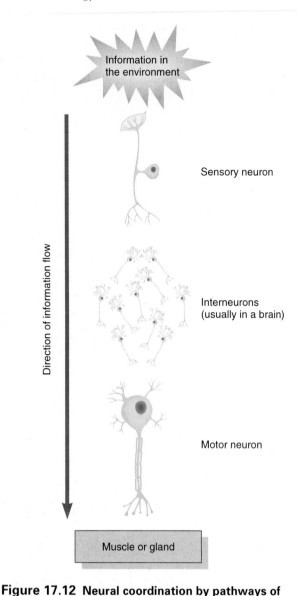

Information in the environment

Direction of information flow

Sensory neuron

Interneurons (usually in a brain)

Motor neuron

Muscle or gland

Figure 17.12 Neural coordination by pathways of neurons. Information enters the nervous system by way of a sensory neuron and then passes to interneurons, which are grouped together to form a brain in most animals. The information then passes from interneuron to interneuron, where it is evaluated by comparison with memories stored within the interneurons. A decision is made and passed to a motor neuron, which passes commands to the appropriate muscle or gland. Information passes from neuron to neuron and from neuron to muscle or gland by way of chemical messengers called neurotransmitters.

interneuron to interneuron and compared with memories—has the person been seen before and, if so, in what context? A decision is made, still within the brain, as to the appropriate response—to look away, to greet the person, or to avoid an interaction by crossing the street. Commands are then sent to your muscles and glands, and they bring about the response.

SUMMARY

1. Animals form the kingdom Animalia. They are multicellular, eukaryotic organisms without cell walls. An animal acquires energy and nutrients by heterotrophy—digesting (internally) the molecules of other organisms. Animals probably evolved from protozoa in response to selection for larger size, which helps them capture food and evade being captured, and division of labor among cells. The first animal must have appeared sometime between 700 million years ago and 1 billion years ago.

2. Three distinctive characteristics of animals are heterotrophy, locomotion, and homeostasis.

3. The many kinds of cells within an animal are organized according to function. A tissue is a group of similar cells, located near one another, that contributes in a similar way to the same task. Animals have epithelial, connective, muscle, and nervous tissues. An organ is a group of different tissues, located near one another, that contribute in different ways to the same task. An organ system is a group of organs, located in different parts of the body, that contribute in different ways to the same bodily function.

4. Cellular activities within an animal are coordinated by the endocrine system and nervous system. Homeostasis in vertebrate animals is governed largely by the medulla oblongata and hypothalamus of the brain.

KEY TERMS

Connective tissue (page 316)
Digestion (page 313)
Endocrine (EN-doh-krin) **gland** (page 326)
Epithelial (EH-peh-THEE-lee-ul) **tissue** (page 316)
Extracellular fluid (page 315)
Heterotroph (HEH-tur-oh-TROHF) (page 313)
Homeostasis (HOH-mee-oh-STAY-sis) (page 313)
Hormone (HOR-mohn) (page 326)
Hypothalamus (HEYE-poh-THAL-ah-mus) (page 327)
Interstitial (IN-tur-STIH-shul) **fluid** (page 315)
Locomotion (loh-koh-MOH-shun) (page 313)
Medulla oblongata (meh-DOO-la AH-blong-GAH-tah) (page 327)
Muscle tissue (page 316)
Negative feedback (page 314)
Nervous tissue (page 317)
Neuron (NUR-ahn) (page 317)
Organ (OR-gun) (page 321)
Organ system (page 322)
Plasma (PLAEH-zmah) (page 315)
Secretion (seh-KREE-shun) (page 316)
Tissue (TIH-shoo) (page 316)

STUDY QUESTIONS

1. How does an animal cell differ from a plant cell? From a fungal cell?
2. What is the advantage of locomotion? What is the advantage of homeostasis?
3. Why is a rabbit able to remain active during the winter whereas a snake is not? Why is a Pacific salmon able to survive in both saltwater and fresh water whereas a rainbow trout is not?
4. What is the role of blood in homeostasis?
5. How can one distinguish among epithelial tissue, connective tissue, muscle tissue, and nervous tissue? Name one part of the body formed by each kind of tissue.
6. Describe the pathway between environmental information and a response triggered by a hormone.
7. Among all the organisms, only animals have nervous tissue. What advantage is this kind of tissue?

CRITICAL THINKING PROBLEMS

1. Suppose you were given a dead, preserved organism and asked to determine whether it was an animal. What would you look for in the organism and why?
2. Describe your own anatomical, physiological, and behavioral adaptations for maintaining body temperature on a cold, windy day.
3. List as many ways as you can in which the skin contributes to homeostasis.
4. Imagine that you are cold-blooded rather than warm-blooded. Describe the ways in which you could alter your daily life in order to adjust for this difference.
5. Describe a situation in your own life that suggests a link between the immune system and the nervous system.

SUGGESTED READINGS

Introductory Level:

Hall, S. S. "A Molecular Code Links Emotions, Mind and Health," *Smithsonian,* June 1989, 62–71. A review of research on interactions between organ systems, with special emphasis on the nervous system and immune system.

Hole, J. W., Jr. *Human Anatomy and Physiology,* 4th ed. Dubuque, IA: William C. Brown Publishers, 1987. Chapter 5 provides a thorough description of the four basic kinds of tissues.

Schmidt-Nielsen, K. "How Are Control Systems Controlled?" *American Scientist,* January–February 1994, 38–44. A review of the mechanisms that govern homeostasis, with special emphasis on salt appetite, thirst, normal body temperature, and fever.

Advanced Level:

Cannon, W. B. *The Wisdom of the Body.* New York: W. W. Norton, 1967 (original copyright 1932). A classic description of homeostasis, including both a general discussion of the phenomenon and detailed discussions of temperature and the various components of extracellular fluids.

Langley, L. L. *Homeostasis.* New York: Von Nostrand Reinhold, 1965. An easy-to-read presentation of the scientific history and general principles of homeostasis, as well as a description of specific mechanisms.

Vander, A. J., J. H. Sherman, and D. S. Luciano. *Human Physiology,* 5th ed. New York: McGraw-Hill, 1990. Chapters 6, 7, and 8 are excellent descriptions of homeostasis and control systems.

Nutrition and Digestion

Objectives

In this chapter, you will learn
–what determines dietary requirements;
–the best kinds of energy foods;
–the need for amino acids, fatty acids, vitamins, minerals, and fiber;
–how the digestive system processes food.

Food supplies body cells with energy and building materials.

We eat for a variety of reasons—to celebrate, to indulge in irresistible foods, to socialize, to counteract stress. The biological reason for eating, however, is to provide body cells with energy and building materials.

An organism's food must provide all the materials its cells need but cannot synthesize. The kinds of materials the cells can synthesize, in turn, depend on the kinds of genes the organism has inherited. Each kind of animal has a different set of requirements. What are the dietary requirements of a human? What should we eat?

The answer to this question used to be simple—eat as much meat as you can afford and avoid starchy foods. Roasts, steaks, and whole milk were in; spaghetti, potatoes, and bread were out. Now we are told to cut back on meat and dairy products and to consume more grains, fruits, and vegetables. Why have nutritional guidelines changed?

One reason for the change in guidelines is that we have changed our eating habits. Prior to the 1950s, many people subsisted mainly on potatoes and bread because that is all they could afford. Thus, nutritionists stressed the need for meat and dairy products. Now most Americans are more affluent and can afford to buy meat and to eat out more often. As a consequence, we eat too much protein and fat, especially in the convenient form of hamburgers, hot dogs, pizzas, and other fast foods. At the same time that our diet came to include more fat, we also became more sedentary. The excess calories in our food become body fat rather than fuel for physical activity.

Another reason the guidelines have changed is that medical researchers continue to discover new relations between diet and health. Fifty years ago, we did not know that a high-fat diet increases our susceptibility to heart attacks, strokes, and colon cancer, nor that a high-protein diet may overwork the kidneys. Medical research proceeds slowly, often taking us down false paths, because we cannot experiment on humans like we can on laboratory animals. Much of our evidence is circumstantial and confounded by aspects of life-style other than diet.

Nutritional advice will continue to change in response to changes in our eating habits and new discoveries by medical researchers. Yet the fundamentals of a good diet, described in this chapter, remain about the same. And a good diet promotes good health, which is much more than just the absence of disease. Good health is a feeling of well-being, with enthusiasm for physical and mental activities.

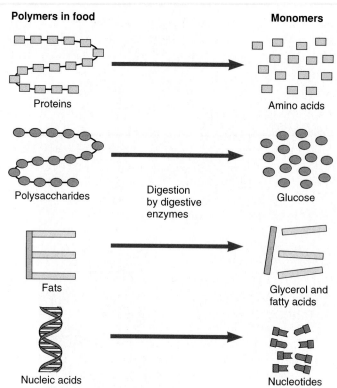

Figure 18.1 Digestion. Digestive enzymes split polymers into their component monomers. Proteins are digested into amino acids, polysaccharides into glucose, fats into glycerol and fatty acids, and nucleic acids into nucleotides.

Nutrition

An animal eats to provide its body cells with energy and building materials. Cells use monomers as direct sources of energy and building materials. **Monomers** (see page 40) are small organic molecules that combine to form **polymers**—the large molecules that characterize life (see page 40). Glucose monomers form polysaccharides; fatty acids and glycerol form fats; amino acids form proteins; and nucleotides form nucleic acids (see chapter 2).

Some organisms can build their own monomers from inorganic materials. Plants, for example, build glucose from carbon dioxide and water. Animal cells cannot synthesize their own monomers. They acquire monomers by eating other organisms.

When an animal consumes a plant or another animal, it acquires a package of polymers that were synthesized by the consumed organism. These polymers are broken apart by digestive enzymes within the animal's digestive tract. Digestion converts polymers into monomers (fig. 18.1). The monomers are moved from the digestive tract into the body cells, where they are used as sources of energy and materials for building the cells' own kinds of polymers.

Dietary Requirements

Each animal chooses only certain foods. One reason, of course, is it can only find and capture certain kinds of organisms. Other reasons, discussed here, are biochemical: the animal's cells require specific raw materials for their activities, and the animal's digestive enzymes are limited in the kinds of materials they can digest.

Nutritional Needs   An animal's diet must furnish the materials that its cells need but cannot synthesize. All cells need glucose, twenty kinds of amino acids, a variety of fatty acids, and many kinds of vitamins and minerals. Each animal can synthesize some but not all of these materials. Those it cannot synthesize are dietary requirements.

The capacity to synthesize a particular molecule is an inherited trait. The earliest animals, which probably resembled sponges, acquired food by filtering unicellular organisms from their surroundings. These animals were unselective, except for size, in the kinds of foods they consumed. They could synthesize everything they needed from whatever raw materials were in the organisms they filtered. Their dietary requirements were minimal.

Later, animals became more selective about what they ate; they chose only certain kinds of organisms to eat. Some, for example, ate only grasses and others only worms. As a consequence of this more specialized diet, some of the organic molecules that the animal needed were always present in its food. The animal's cells no longer had to synthesize these molecules and, over many generations, the ability to synthesize them was lost. Molecules that were supplied regularly by the animal's food became dietary requirements.

Our requirement for vitamin C appears to have developed in this way. All animals must have this organic molecule to build collagen, a protein that supports and binds together the cells of animal tissues. Most animals can synthesize their own vitamin C. Those that cannot include the guinea pigs, some bats, a few types of birds, and some primates (including humans). What the animals on this list have in common is they all eat fruit—a rich source of vitamin C. Over the generations, the enzymes for synthesizing vitamin C became superfluous; cells never built this molecule since it was always supplied by the food. As a consequence, the genes for building these enzymes played no role in survival or reproduction and were lost from the genetic repertoire. Vitamin C then became a dietary requirement.

Digestive Capabilities   An animal's digestive enzymes convert food polymers into monomers that its cells can use. Each type of polymer has unique chemical bonds between its monomers, and each type of digestive enzyme can break only one kind of bond. Thus, the kinds of polymers an animal can digest depend on the kinds of digestive enzymes it can synthesize. Polymers that are not digested pass through the digestive tract and leave the body as feces.

Each type of animal has a unique and limited set of digestive enzymes that it can synthesize, as determined by its genes. For example, animals lack the ability to make cellulase, the enzyme that digests cellulose into glucose. (Strange, since cellulose is the most abundant polymer on earth.) Cellulose within the salad greens, fruits, and vegetables that we eat is not digested; it passes right through our digestive tract and does not nourish our cells. Some animals harvest the glucose within cellulose by harboring in their bodies certain microorganisms that can synthesize cellulase. Cows, deer, and antelope have cellulose-digesting bacteria in their stomachs; horses, rabbits, and beaver have these bacteria in their intestines; termites have cellulose-digesting protozoa (unicellular, eukaryotic organisms) in their stomachs.

Food for Energy

All activities of an animal, from thinking to running, require energy. Each animal has unique energy requirements based on body size, basal metabolic rate (how active the cells are when the body is resting), and degree of physical activity.

Energy for doing biological work is provided by cellular respiration (📖 *see page 88*), a biochemical process that releases chemical energy from the covalent bonds of monomers. While any monomer can be fuel for cellular respiration, the best fuels are glucose and fatty acids.

Energy Sources   The best source of energy for humans is starch. This polysaccharide (or complex carbohydrate) is abundant in foods made from seeds (e.g., pasta and bread) as well as starch-storing regions of stems (e.g., white potatoes) and roots (e.g., carrots). Starch is digested slowly and distributed efficiently to body cells (fig. 18.2a). Sugar, by contrast, is not a good source of energy because it is digested too fast to be distributed efficiently (fig. 18.2b).

Fats contain large quantities of energy—about twice as many calories,[1] per unit weight, as polysaccharides or proteins. Most of us, however, eat far too much fat and should

---

1. Chemical energy is measured in terms of how much heat a molecule releases when it is burned completely. Heat, in turn, is measured in calories. One calorie is the amount of heat that will raise the temperature of 1 gram of water 1° centigrade. Chemical energy in our food is expressed in larger units, called kilocalories (or Calories, with a capital C). One kilocalorie equals 1,000 calories.

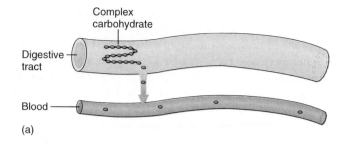

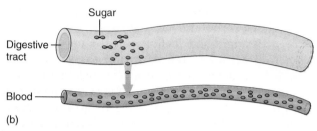

Figure 18.2 Carbohydrate digestion. (a) Complex carbohydrates are digested slowly because of their many bonds. Glucose trickles into the bloodstream and can be efficiently distributed to the liver and other body cells. (b) Sugars are digested rapidly because each has just one bond. Table sugar (sucrose), for example, consists of one glucose bonded to one fructose. The monomers flood the bloodstream and cannot be efficiently distributed. Some may be eliminated in the urine.

cut back on this energy-rich nutrient. No more than 30% of our calories should come from fat, whereas in the average American diet approximately half the calories come from fat. (All fats contain about 110 Calories per tablespoon.)

Of the fats that we do ingest, most should be unsaturated rather than saturated. Saturated fats (📖 *see page 44*) contribute to hardening of the arteries and certain kinds of cancer. They are abundant in the meat of cows, pigs, and chickens, as well as in lard, butter, margarine, whole milk, and cheese. Unsaturated fats (📖 *see page 44*), which are abundant in most vegetable oils, are better sources of energy because they prevent rather than cause hardening of the arteries.

Energy Stores   Animals store energy as glycogen and fat. Glycogen, a polysaccharide made of glucose, is stored in small quantities in the liver and muscles. Larger energy stores typically consist of fat rather than glycogen because fat weighs less than glycogen. (For equivalent amounts of energy, fat weighs half as much as glycogen.) An animal that relies on rapid movement for survival cannot afford to be heavy; it stores energy largely as fat. Only sedentary animals, such as oysters and mussels, store most of their energy as glycogen.

The liver synthesizes fat from excess monomers of any type. An excess of 3,500 Calories (accumulated over any time period) adds 1 pound of body fat; a deficiency of 3,500 Calories removes 1 pound of body fat. (As most adults burn between 1,500 and 2,500 Calories a day, you can see why it is virtually impossible to lose a pound of fat a day on any weight-loss program.)

Starvation is the condition in which fewer calories are consumed than used for a prolonged period of time. During early phases of starvation, the body cells draw mainly on the fatty acids within fat stores for energy. Some muscle protein is digested to provide amino acids for building essential proteins. When the body fat has been used up, the cells obtain energy entirely from amino acids within muscle proteins. Skeletal muscles (the ones that move bones) are used first and then the muscles of the stomach, intestine, and heart. A starving person dies at this point, since the muscles of these organs are essential to life.

It is healthy to be slim, but some people carry dieting too far. Two serious dieting extremes, anorexia and bulimia, are described in the accompanying Wellness Report.

Food for Building Materials

An animal needs materials for building its own kinds of molecules. Animal cells build enzymes, antibodies, chemical messengers, cell membranes, and other structures from monomers, vitamins, and minerals. An animal's diet must provide the building materials it needs but cannot synthesize.

Malnourishment occurs when the diet fails to provide enough building materials for the body cells to function optimally. A person can get plenty to eat yet still be malnourished. Someone who exists on hamburgers, pizza, and pop, for example, receives plenty of calories but suffers from malnourishment. Guidelines for choosing a healthful diet are provided in figure 18.3.

An animal's diet must provide four kinds of building materials: amino acids, fatty acids, vitamins, and minerals. The need for each of these materials is described following.

Amino Acids   An animal's diet must include amino acids; its cells cannot build these monomers yet need them to build proteins. Proteins, in turn, form enzymes, antibodies, muscles, and other essential structures. An animal needs to eat proteins every day; it does not store amino acids yet its cells need a steady supply to rebuild cellular proteins that are broken down and rebuilt on a regular basis.

Proteins are synthesized from twenty kinds of amino acids. While all twenty may not be present in

Figure 18.3 The food guide pyramid. This guide to daily food choices was recommended in 1992 by the U.S. Department of Agriculture. The shape of the diagram reminds us to eat mostly grains, vegetables, and fruits and only small amounts of fats, oils, and sweets.
Source: U.S. Department of Agriculture.

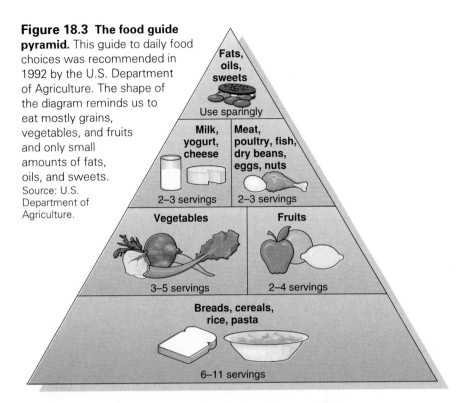

combining certain kinds of plant materials. The proteins in grains (wheat, corn, rice, and oats) complement the proteins in legumes (beans, peanuts, peas, and lentils); a meal that includes both grains and legumes is as good a source of protein as a meal that contains animal products or soybeans. Examples of vegetarian meals that provide all eight essential amino acids in the right proportions are lima beans with corn, pinto beans with rice, and peanut butter on whole wheat bread.

Most of us eat twice as much protein as we need. The excess amino acids are used in cellular respiration and converted into fat. In either case, nitrogen within the amino acids becomes a waste product that is eliminated by the kidneys. A high-protein diet strains the kidneys, which have to work harder to eliminate the excess nitrogen, and the greater volume of urine removes valuable calcium from the body. Our requirement for protein is provided by about half a pound of meat a day.

Protein deficiency occurs most often where wars, crop failures, and overcrowding restrict the diet to just one or a few kinds of high-carbohydrate foods, such as roots or rice. In time, insufficient protein leads to low blood pressure, enzyme deficiencies, and abnormal development. Low blood pressure develops because there are not enough proteins in the blood to hold water within the blood vessels. Consequently, water moves out of the blood by osmosis (*see page 76*):

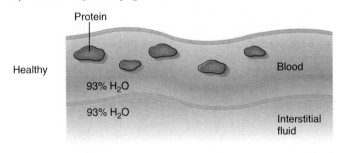

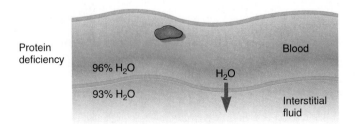

each protein, they must all be available to make the animal's entire set of proteins. If just one amino acid is in short supply, then some of the proteins cannot be synthesized.

An animal can convert some kinds of amino acids into other kinds. **Essential amino acids** are the ones that the animal cannot make; it is essential that they be included in the diet. Nonessential amino acids can be made from the essential ones; it is not essential to include them in the diet. The particular amino acids that are essential vary among the different kinds of animals. In an adult human, eight of the twenty amino acids (nine in young children) are essential; the remaining twelve can be made, by liver cells, from these eight.

Complete proteins have all eight essential amino acids in approximately the right proportions for building human proteins. Meat, milk, eggs, and soybeans are rich in complete proteins, whereas plants are not (except soybeans). Most plant materials provide mainly incomplete proteins; all eight essential amino acids may be present, but some occur in such relatively small amounts that you would have to eat huge amounts of this protein to get enough of all eight kinds.

A strict vegetarian, who eats no animal products, can get all eight essential amino acids in the right proportions (i.e., complete proteins) by eating soybeans or by

Wellness Report

Eating Disorders

For a snack one night, Kim ate two bags of cookies, four candy bars, three sandwiches, a box of crackers with sour-cream dip, a jar of peanut butter, a large pizza, and six pieces of chocolate cake. Then she made herself throw up.

Kim has bulimia, an eating disorder characterized by the overwhelming need to eat massive quantities of food and then to get rid of the food, in order to prevent weight gain, by self-induced vomiting. Some bulimics vomit many times a day. Such regular vomiting has serious consequences, for the highly acidic stomach contents burn the esophagus and dissolve tooth enamel. The vomiting can also lead to high blood pressure, diabetes, respiratory difficulties, and organ damage.

About 90% of all bulimics are women. They tend to be extroverted perfectionists who begin to eat in binges while in their late teens. Bulimia is most common among college freshmen, a time of major changes in life-style when eating can become a way of handling stress, just as an alcoholic uses drinking. To prevent weight gain from the enormous meals, each gorge is followed by a purge. (Bulimia is known among college students as the "scarf and barf" disorder.)

A second type of eating disorder is anorexia, in which there is no appetite for food. This condition begins as an emotional problem (anorexia nervosa) characterized by an overwhelming fear of becoming fat (fig. 18.A). Many anorexics eat very little and exercise compulsively to burn off the few calories they do consume. Anorexia progresses from an emotional problem to a physical problem, in which the appetite control center (in the hypothalamus of the brain) no longer functions as it should. Anorexics become extremely thin, a condition similar to starvation, and may suffer damage to their developing ovaries and uterus as well as other organs. Some starve themselves to death.

Most anorexics come from homes where food is important; where there are power struggles be

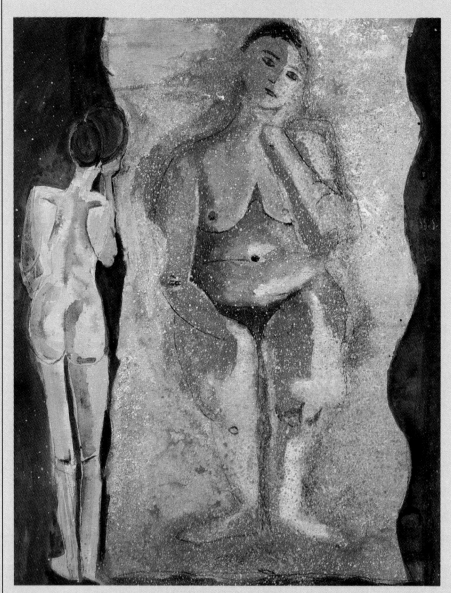

Figure 18.A Anorexia nervosa. Someone suffering from this disorder is so afraid of becoming fat that they eat very little. Shown here is an artist's representation of this irrational fear.

tween the parents and children about eating and jokes and nagging about weight. Anorexics are generally shy, withdrawn girls who develop their symptoms around the onset of puberty. Many of these girls do not want to develop womanly bodies.

Do you think you have bulimic or anorexic tendencies? Take the following test to find out—answer the questions honestly and circle the number of each answer. When you are finished, add all the circled numbers and compare your total with those in the table at the end of the test.

1. I have eating habits that are different from those of my family and friends.
   1) often; 2) sometimes; 3) rarely; 4) never
2. I find myself panicking if I cannot exercise as I planned for fear of gaining weight.
   1) almost always; 2) sometimes; 3) rarely; 4) never
3. My friends tell me I am thin but I don't believe them because I feel fat.
   1) often; 2) sometimes; 3) rarely; 4) never
4. (Females only) My menstrual period has ceased or become irregular due to no known medical reasons.
   1) true; 2) false
5. I have become obsessed with food to the point that I cannot go through a day without worrying about what I will or will not eat.
   1) almost always; 2) sometimes; 3) rarely; 4) never
6. I have lost more than 25% of the normal weight for my height (e.g., 30 pounds from 120 pounds).
   1) true; 2) false
7. I would panic if I got on the scale tomorrow and found out that I had gained 2 pounds.
   1) almost always; 2) sometimes; 3) rarely; 4) never

8. I find that I prefer to eat alone or when I am sure no one will see me; thus, I am making excuses so I can eat less often with others.
   1) often; 2) sometimes; 3) rarely; 4) never
9. I find myself going on uncontrollable eating binges during which I consume large amounts of food to the point that I feel sick and make myself vomit.
   1) three or more times per day;
   2) one or two times per day;
   3) one or two times per week;
   4) rarely; 5) never
10. I use laxatives (medicines to cause bowel movements) as a means of weight control.
   1) on a regular basis;
   2) sometimes; 3) rarely; 4) never
11. I find myself playing games with food (e.g., cutting it into tiny pieces, hiding food so people will think I ate it, chewing it and spitting it out), telling myself certain foods are bad.
   1) often; 2) sometimes; 3) rarely; 4) never
12. People around me have become interested in what I eat and I find myself getting angry at them for pushing food on me.
   1) often; 2) sometimes; 3) rarely; 4) never
13. I have felt more depressed and irritable recently than I used to and/or have been spending increasing amounts of time alone.
   1) true; 2) false
14. I keep a lot of my fears about food and eating to myself because I am afraid no one would understand.
   1) often; 2) sometimes; 3) rarely; 4) never
15. I enjoy making gourmet, high-calorie meals or treats for others as long as I don't have to eat any of it myself.
   1) often; 2) sometimes; 3) rarely; 4) never

16. The most powerful fear in my life is the fear of gaining weight or becoming fat.
   1) often; 2) sometimes; 3) rarely; 4) never
17. I find myself totally absorbed when reading books about dieting, exercising, and calorie-counting to the point that I spend hours studying them.
   1) often; 2) sometimes; 3) rarely; 4) never
18. I tend to be a perfectionist and am not satisfied with myself unless I do things perfectly.
   1) almost always; 2) sometimes; 3) rarely; 4) never
19. I go through long periods of time without eating anything (fasting) as a means of weight control.
   1) often; 2) sometimes; 3) rarely; 4) never
20. It is important to me to try to be thinner than all of my friends.
   1) almost always; 2) sometimes; 3) rarely; 4) never

Compare your score with this table:
   Under 30 points: Strong tendencies toward anorexia nervosa
   30 to 45 points: Strong tendencies toward bulimia
   45 to 55 points: Weight conscious, but not necessarily with anorexic or bulimic tendencies
   Over 55 points: No need for concern

If you scored below 45 it would be wise for you to

1. Seek more information about anorexia and bulimia.
2. Contact an eating disorders program in order to determine what kind of assistance would be of most help to you. (Your student health service may have such a program.)

Reprinted with permission of The Boulder, CO, *Daily Camera.*

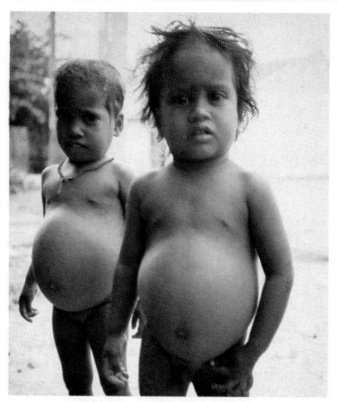

Figure 18.4 Children suffering from a protein-deficient diet. These children get enough calories (they are not thin) but not enough protein in their diet. Their swollen bellies indicate severe edema from not enough protein in their blood. This condition is sometimes known by its African name—kwashiorkor disease.

Water from the blood accumulates in the interstitial fluid (between the cells) where it causes **edema** (swelling), and within the abdominal cavity, where it causes an enlarged abdomen (fig. 18.4). With less water in the blood, the total volume of blood decreases and the blood pressure drops. Abnormally low blood pressure can damage the heart and kidneys.

Enzyme deficiencies develop when there are not enough amino acids to build these important proteins. Normal biochemical pathways cannot proceed and a variety of serious health problems develop.

A child who does not eat enough protein cannot develop normally. The most serious consequence is mental retardation, in which the brain does not develop to its full potential. Protein deficiency in young children typically occurs when a baby is bottle-fed and the milk diluted with water.

Fatty Acids   Fatty acids are monomers of fats, which form membranes, provide energy, and perform many other vital functions. (The other monomer of fats, glycerol, is not a dietary requirement because our cells can build it from glucose.) **Essential fatty acids** are the kinds that cells need but cannot synthesize. They must be provided by the diet. Nonessential fatty acids can be synthesized from the essential ones.

Cells require many kinds of fatty acids, but in humans only three are essential fatty acids: linoleic acid, linolenic acid, and arachidonic acid. All three are particularly abundant in vegetable oils and fish oils.

Vitamins   A **vitamin** is any organic molecule that cells cannot synthesize but need in tiny amounts—even minute deficiencies produce disorders. Many vitamins are coenzymes; they activate enzymes by bonding with them. Vitamin B_1, for example, activates the enzyme that initiates cellular respiration in brain cells. A diet lacking in this vitamin leads to brain damage.

The need for vitamins was discovered long before their chemical structures were known, which is why vitamins are referred to simply by the letters A, B, C, and so forth. The thirteen vitamins required by human cells are listed in table 18.1.

Vitamins A, D, E, and K are fat-soluble. They occur in the fats we eat and can be stored for long periods of time in our liver and fat cells. We should never take more of the fat-soluble vitamins than the body can use, because excess amounts are stored and may reach toxic levels. The other vitamins are water-soluble. They occur in nonfatty foods and are not stored in the human body. When more of these vitamins are taken than used, the excess is eliminated in the urine.

Minerals   A **mineral** is an atom that an animal needs and can use in an inorganic form—as an inorganic ion or part of an inorganic molecule. Calcium and potassium, for example, are minerals because animal cells can use their ions—Ca^{++} and K^+. Carbon and nitrogen, however, are not minerals since animal cells can use these atoms only when they are part of an organic molecule, such as glucose or an amino acid.

Sixty different minerals occur in the human body. Of these, twenty are essential to human health and forty have no known function. Six of the twenty essential minerals are much more abundant in the body than the others. They are calcium, chlorine, magnesium, phosphorus, potassium, and sodium. The other fourteen, called trace minerals, are present in tiny amounts. All 20 essential minerals are listed, along with their dietary sources and cellular functions, in table 18.2. Most of us get too much sodium and not enough calcium or iron in our diets.

The essential minerals readily form ions in body fluids. These ions, also known as electrolytes, tend to leave the body in urine and sweat. For this reason, we need to replenish our supply of these minerals every day.

A mystery of our eating behavior is why we crave table salt, or sodium chloride (NaCl). Most of us eat about ten times more sodium than we need. In some people, excess sodium may produce high blood pressure, a condition that stresses the heart and arteries. The simplest way to reduce sodium intake is to eliminate obvious sources, such as the salt shaker and potato chips, and be wary of foods that have been commercially processed. A single serving of canned soup, for instance, may contain as much sodium as you need for an entire day. With all the natural sources of salt in a varied diet, there is no nutritional reason to add it to food.

Fiber

Polysaccharides that are eaten but not digested are known as **fiber.** These plant materials, largely cellulose, pass directly through the digestive tract. Fiber is neither an energy source nor a building material. It makes the feces more bulky, which prevents constipation, and it removes damaging materials, such as cholesterol and bile. Most of us eat only half as much fiber as we need for optimal health.

Two categories of fiber are recognized: those that are insoluble in water and those that are soluble. Each contributes in a different way toward maintaining health. Most plant materials contain both kinds, but certain foods are particularly rich in one or the other. The sources of these two kinds of fiber and the health advantages of eating them are presented in table 18.3.

The Digestive System

The digestive system consists of the digestive tract, liver, and pancreas. Its function is to transfer monomers, vitamins, ions, and water from the external environment to the body cells. The digestive system contributes to homeostasis (maintaining a constant internal environment) by providing body cells with the raw materials they need to carry out normal activities.

Food is processed for cellular use as it moves through the **digestive tract** (fig. 18.5). This muscular tube, which runs lengthwise through the center of the body, has two openings: the mouth, where food enters, and the anus, where feces leave. Digestive materials are added to the tract by the liver and pancreas.

The walls of the digestive tract contain the four main tissues (see page 316): (1) an inner layer of epithelial tissue, with cells that secrete hormones and digestive materials; (2) several layers of connective tissue; (3) nervous tissue; and (4) two sets of smooth muscles—a set of longitudinal muscles that shorten the tract and a set of circular muscles that narrow the channel of the tract. The arrangement of these tissues is shown in figure 18.6a.

Food processing is coordinated by both the endocrine system and the nervous system. Sensory cells within the lining of the digestive tract respond to the food bulk (distention of the walls) and its particular kinds of molecules. This information is relayed to the nervous system, which sends commands to smooth muscles within the walls of the tract and to secretory cells within the lining of the tract:

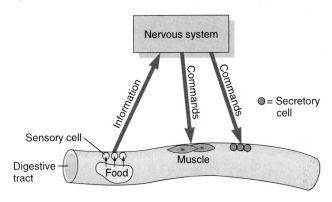

The muscles respond by shortening. Waves of contractions, called **peristalsis,** move food in a single direction through the tract (fig. 18.6b). The secretory cells respond by releasing hormones into the bloodstream, where they travel to their target cells—other kinds of secretory cells within the pancreas and liver. The secretory cells in these organs respond to the hormones by releasing digestive materials into tubes that drain into the tract:

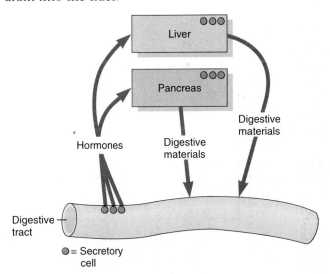

Each region of the digestive tract is specialized for a particular aspect of food processing (fig. 18.7). The different regions can be isolated from one another by **sphincters**—rings of smooth muscles that contract to prevent passage of food. The function of each organ in the digestive system is summarized in table 18.4 and described following.

Table 18.1

Vitamins

Vitamin	Major Food Sources	Adult RDA*	What It Does	Potential Benefits	Supplementation
Fat-soluble					
A	Milk, eggs, liver, cheese, fish oil. Plus fruits and vegetables that contain beta carotene. (You need not consume preformed vitamin A if you eat foods rich in beta carotene.)	8,000 IU,† women; 10,000 IU, men *1 cup milk = 1,400 IU*	Promotes good vision; helps form and maintain skin, teeth, bones, and mucous membranes. Deficiency can increase susceptibility to infectious disease.	May inhibit the development of certain tumors; may increase resistance to infection in children.	Not recommended, since toxic in high doses.
D	Milk, fish oil, fortified margarine; also produced by the body in response to sunlight.	5 mcg (micrograms), or 200 IU; 10 mcg, or 400 IU, before age 25 *1 cup milk = 100 IU*	Promotes strong bones and teeth by aiding the absorption of calcium. Helps maintain blood levels of calcium and phosphorus.	May reduce the risk of osteoporosis.	400 IU for people who do not drink milk or get sun exposure, especially strict vegetarians and elderly. Toxic in high doses.
E	Vegetable oil, nuts, margarine, wheat germ, leafy greens, seeds, almonds, olives, asparagus.	8 mg, women; 10 mg, men (12–15 IU) *1 Tbsp canola oil = 9 mg* *1 Tbsp margarine = 2 mg* *1 oz peanuts = 2 mg* *1 cup kale = 6 mg*	Helps in the formation of red blood cells and the utilization of vitamin K. As an antioxidant, it combats the adverse effects of free radicals.	May reduce the risk of certain cancers, as well as coronary artery disease; may prevent or delay cataracts; may improve immune function in the elderly.	200–800 IU advised for everybody; you can't get that much from food, especially on a low-fat diet. No serious side effects at that level, though diarrhea and headaches have been reported.
K	Intestinal bacteria produce most of the K needed by the body. The rest is supplied by leafy greens, cauliflower, broccoli, cabbage, milk, soybeans, eggs.	60–65 mcg, women; 70–80 mcg, men *1 cup broccoli = 175 mcg* *1 cup milk = 10 mcg*	Essential for normal blood clotting.	May help maintain strong bones in the elderly.	Not necessary, not recommended.
Beta carotene (not a vitamin, but converted to vitamin A in the body)	Carrots, sweet potatoes, cantaloupe, leafy greens, tomatoes, apricots, winter squash, red bell peppers, pink grapefruit, broccoli, mangos, peaches.	No RDA; experts recommend 5–6 mg (milligrams) *1 medium carrot = 12 mg* *1 sweet potato = 15 mg*	Converted into vitamin A in the intestinal wall. As an antioxidant, it combats the adverse effects of free radicals in the body. Best known of a family of substances called carotenoids.	May reduce the risk of certain cancers as well as coronary artery disease.	6–15 mg (equal to 10,000–25,000 IU of vitamin A) a day for anyone not consuming several carotene-rich fruits or vegetables daily. Nontoxic.
Water-soluble					
C (ascorbic acid)	Citrus fruits and juices, strawberries, tomatoes, peppers (especially red), broccoli, potatoes, kale, cauliflower, cantaloupe, Brussels sprouts.	60 mg *1 orange = 70 mg* *1 cup fresh O.J. = 120 mg* *1 cup broccoli = 115 mg*	Helps promote healthy gums and teeth; aids in iron absorption; maintains normal connective tissue; helps in healing of wounds. As an antioxidant, it combats the adverse effects of free radicals.	May reduce the risk of certain cancers, as well as coronary artery disease; may prevent or delay cataracts.	250–500 mg a day for anyone not consuming several fruits or vegetables rich in C daily and for smokers. Larger doses may cause diarrhea.

Vitamin	Best Sources	RDA*	Function	Possible Benefits	Supplement Advice
Thiamin (B₁)	Whole grains, enriched grain products, beans, meats, liver, wheat germ, nuts, fish, brewer's yeast.	1–1.1 mg, women; 1.2–1.5 mg, men 1 pkt oatmeal = 0.5 mg	Helps cells convert carbohydrates into energy. Necessary for healthy brain, nerve cells, and heart function.	Unknown.	Not necessary, not recommended.
Riboflavin (B₂)	Dairy products, liver, meat, chicken, fish, enriched grain products, leafy greens, beans, nuts, eggs, almonds.	1.2–1.3 mg, women; 1.4–1.7 mg, men 1 cup milk = 0.4 mg 3 oz chicken = 0.2 mg	Helps cells convert carbohydrates into energy. Essential for growth, production of red blood cells, and health of skin and eyes.	Unknown.	Not necessary, not recommended.
Niacin (B₃)	Nuts, meat, fish, chicken, liver, enriched grain products, dairy products, peanut butter, brewer's yeast.	13–19 mg 3 oz chicken = 12 mg 1 slice enriched bread = 1 mg	Aids in release of energy from foods. Helps maintain healthy skin, nerves, and digestive system.	Large doses lower elevated blood cholesterol.	Megadoses may be prescribed by doctor to lower blood cholesterol. May cause flushing, liver damage, and irregular heartbeat.
B₆ (pyroxidine)	Whole grains, bananas, meat, beans, nuts, wheat germ, brewer's yeast, chicken, fish, liver.	1.6 mg, women; 2 mg, men 1 banana = 0.7 mg 1 cup lima beans = 0.3 mg	Vital in chemical reactions of proteins and amino acids. Helps maintain brain function and form red blood cells.	May boost immunity in the elderly.	Megadoses can cause numbness and other neurological disorders.
B₁₂	Liver, beef, pork, poultry, eggs, milk, cheese, yogurt, shellfish, fortified cereals, fortified soy products.	2 mcg 1 cup milk = 0.9 mcg 3 oz beef = 2 mcg	Necessary for development of red blood cells. Maintains normal functioning of nervous system.	Unknown.	Strict vegetarians may need supplements. Despite claims, no benefits from megadoses.
Folacin (a B vitamin; also called folate or folic acid)	Leafy greens, wheat germ, liver, beans, whole grains, broccoli, asparagus, citrus fruit and juices.	180 mcg, women; 200 mcg, men 1 cup raw spinach = 110 mcg 1 pkt oatmeal = 150 mcg 1 cup asparagus = 180 mcg	Important in the synthesis of DNA, in normal growth, and in protein metabolism. Adequate intake reduces the risk of certain birth defects, notably spina bifida.	May reduce the risk of cervical cancer.	400 mcg, from food or pills, for all women who may become pregnant, in order to help prevent birth defects.
Biotin (a B vitamin)	Eggs, milk, liver, brewer's yeast, mushrooms, bananas, tomatoes, whole grains.	No RDA; experts recommend 30–100 mcg	Important in metabolism of protein, carbohydrates, and fats.	Unknown.	Not necessary, not recommended.
Pantothenic acid (B₅)	Whole grains, beans, milk, eggs, liver.	No RDA; experts recommend 4–7 mg	Vital for metabolism of food and production of essential body chemicals.	Unknown.	Not necessary, not recommended. May cause diarrhea.

*RDA = Recommended Daily Allowance. These figures are not applicable to pregnant women, who need additional vitamins and should seek professional advice.

†IU = International Units.

Reprinted permission of the University of California at Berkeley *Wellness Letter*, © Health Letter Associates, 1994.

Table 18.2

Essential Minerals

Mineral	Adult RDA*	Sources	What It Does
Abundant:			
Calcium	1,000 mg	Milk, fish bones, broccoli, green leafy vegetables	Builds bones and teeth; regulates heartbeat; helps muscles contract
Chlorine	1,900–5,000 mg	Table salt, all food	Maintains body fluid; forms HCl in stomach
Magnesium	350 mg	Whole grains, salad greens, nuts, apricots	Aids bone growth; helps muscles contract; regulates heartbeat
Phosphorus	800 mg	Meats, milk, beans, nuts	Aids bones and teeth
Potassium	1,500–6,000 mg	Oranges, peanuts, beans, potatoes, cocoa, meat	Regulates heartbeat; helps muscles contract; maintains body fluid
Sodium	2,000 mg	Table salt, all food	Maintains body fluids
Trace:			
Chromium	.05–.20 mg	Meat, cheese, whole grains, beans, peanuts	Helps body use glucose
Copper	2–3 mg	Oysters, nuts, liver, chocolate, corn oil	Forms part of red blood cells; helps release energy
Fluorine	1.5–4 mg	Fluoridated water, fish, tea, gelatin	Forms part of bones and teeth
Iodine	.15 mg	Iodized salt, seafood	Essential for normal cell function
Iron	10 mg	Liver, meat, eggs, nuts, molasses	Forms part of red blood cells
Manganese	2.5–5.0 mg	Nuts, whole grains, vegetables, fruit	Aids bone growth; needed for normal cell function
Molybdenum	.15–.50 mg	Peas, beans, grains, dark green vegetables	Essential for normal cell function
Selenium	.05–.20 mg	Fish, shellfish, red meat, garlic, tomatoes	Helps maintain all molecules
Zinc	15 mg	Shellfish, beef, whole wheat bread, eggs	Needed for growth, reproduction, and wound healing

*Recommended Daily Allowance for an adult male in good health.

Table 18.3

The Two Categories of Fiber

Kind	Composition	Good Sources	Health Benefits
Insoluble fiber	Mostly cellulose	Whole wheat bread; bran cereal; leafy greens	Prevents constipation; helps prevent colon cancer by removing bile salts
Soluble fiber	Mostly pectins	Raw fruit; raw vegetables; brown rice; oats; beans	Helps prevent hardening of the arteries by removing cholesterol; helps control blood sugar by slowing absorption of glucose

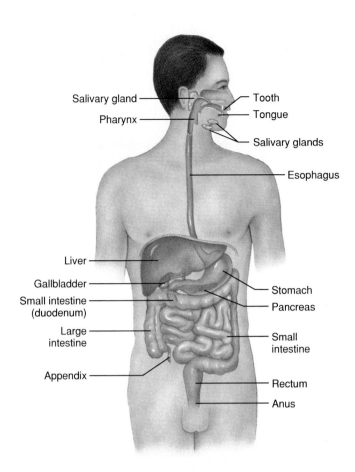

Figure 18.5 **The human digestive system.**

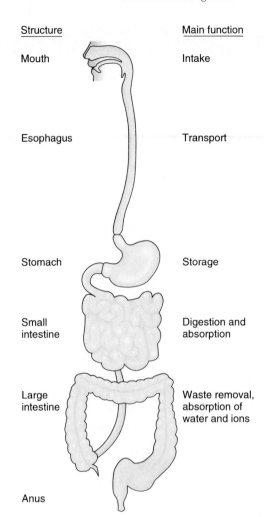

Structure	Main function
Mouth	Intake
Esophagus	Transport
Stomach	Storage
Small intestine	Digestion and absorption
Large intestine	Waste removal, absorption of water and ions
Anus	

Figure 18.7 **The organs of the digestive tract and their specialized functions.** Each region of the tract is a separate organ that contributes in a unique way to food processing. It can be isolated from other organs by sphincters—circular muscles that can contract tightly to close off the flow of food.

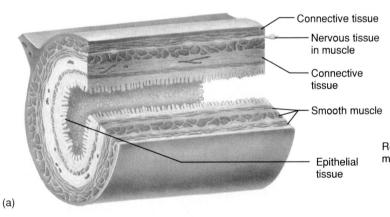

(a)

▣ Figure 18.6 **The wall of the digestive tract.**

(a) Epithelial tissue controls what moves from the tract into the blood and secretes digestive materials into the tract. Smooth muscle contracts to move food through the tract. Nervous tissue picks up information about the food and coordinates the response of the digestive system. Connective tissue holds the other tissues together. (b) Circular muscles in the wall contract to narrow the channel, and longitudinal muscles contract to shorten the channel. They produce waves of contractions, called peristalsis, that push food through the digestive tract.

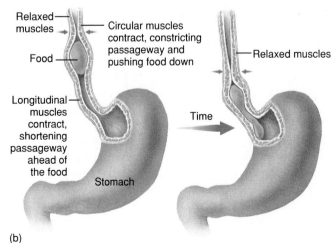

(b)

Table 18.4

Food Processing by Organs of the Digestive System

Organ	Function
Mouth	Tears meat; ruptures cell walls and seed coats; moistens and lubricates food; partially digests starch
Esophagus	Moves food from mouth to stomach
Stomach	Stores food; softens meat and bones; kills bacteria; partially digests proteins
Small intestine	Digests polymers in food; absorbs monomers, vitamins, and minerals into bloodstream
Pancreas	Secretes digestive enzymes and an alkaline solution
Liver	Secretes bile and an alkaline solution; converts and distributes monomers
Appendix	No known function
Large intestine	Removes wastes as feces; absorbs water and ions into bloodstream

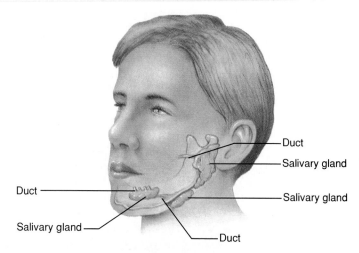

Figure 18.8 The salivary glands. A human has three pairs of salivary glands. Each gland secretes saliva, which flows through a duct (small tube) into the mouth.

Mouth and Esophagus

The mouth is an opening for intake and initial processing of food. The size of the mouth and shape of the teeth vary among animals and are adaptations for handling certain kinds of foods.

Meat does not have to be chewed. The teeth of meat eaters have sharp edges for grabbing and tearing their prey. Plant materials require much more processing than meat because plant cells are enclosed within tough walls of cellulose and seeds are enclosed within tough coats (the bran). These coverings have to be split open before nutrients within the cells or coats become available to an animal. Vertebrate animals that eat plants have short, flat teeth for grinding.

The human mouth is adapted for processing both meat and plant materials. We have a variety of teeth, some for processing each kind of food. (Human teeth have become smaller during the past 10,000 years, apparently an evolutionary response to increased processing of food prior to eating—cooking, chopping, and grinding make chewing less important.) We chew our meat, although it is not necessary, to enhance and prolong its flavor.

Chewing not only grinds food but also mixes it with saliva. Saliva is secreted at all times by three pairs of sali-

vary glands that empty into the mouth (fig. 18.8). Saliva consists chiefly of water but also contains mucus (a slimy fluid), ions, and a digestive enzyme. The water and mucus moisten and lubricate food, holding it together so it can slip easily down the throat when swallowed. The digestive enzyme begins the digestion of starch, breaking the long chains of glucose into much shorter chains. Saliva enables an animal to taste food, since taste buds can detect molecules only when they are dissolved in water.

Swallowing is a muscular activity, involving the tongue and cheek muscles, that pushes food out of the mouth and into the esophagus. The esophagus is a muscular tube, about 25 centimeters (10 in) long in the human (see fig. 18.5), that moves food by peristalsis into the stomach.

Stomach

The stomach is an expanded region of the digestive tract, located within the abdominal cavity. (The human torso has two large cavities: a thoracic cavity and an abdominal cavity, as shown in figs. 18.5 and 20.3.) Its main function is to store food, but it also prepares food for digestion. Secretory cells within the epithelial tissue that lines the stomach secrete mucus, hydrochloric acid, bicarbonate ions, and protein-digesting enzymes. In an adult, the stomach produces between two and three liters of this gastric juice each day.

The walls of an empty stomach are highly folded. As the stomach fills, the walls unfold to expand its capacity—approximately 4 liters of food can be stored in a human stomach. Two sphincters can isolate this part of the digestive tract: one at the entrance to the stomach, which closes to prevent backflow, and another near the entrance to the small intestine, which regulates how fast food is delivered to the small intestine (fig. 18.9).

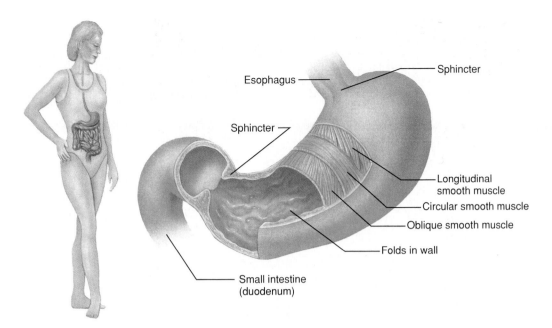

Esophagus

Sphincter

Sphincter

Longitudinal
smooth muscle

Circular smooth muscle

Oblique smooth muscle

Folds in wall

Small intestine
(duodenum)

Figure 18.9 The stomach. Food is churned by rhythmic contractions of three muscle layers: circular, longitudinal, and oblique. A sphincter at the entrance to the stomach prevents acidic food from moving back into the esophagus. When this sphincter fails, we experience a burning sensation known as heartburn. A sphincter at the exit of the stomach prevents large quantities of food from entering the small intestine all at once.

Food remains in the stomach for two to six hours, depending on the size and composition of the meal. While in storage, it is churned, softened, and partially broken down. Hydrochloric acid (HCl) released onto the food softens meat and bones, kills bacteria in the food, and unfolds proteins in the food. It also activates the protein-digesting enzymes, which break protein molecules into shorter chains of amino acids. (Digestive enzymes, which are proteins, do not unfold within the acids because their shape is held by strong disulfide bonds.)

There is always a danger that the protein-digesting materials within the stomach will digest the walls of the digestive tract itself. Normally the walls are protected against digestion, but sometimes this protection breaks down and an ulcer forms. The search for the cause of ulcers is described in the accompanying Research Report.

Small Intestine

Food moves from the stomach to the small intestine, a muscular tube about 4 centimeters (1.6 in) wide and 2 meters (6 ft) long. The small intestine forms more than half the total length of the digestive tract and is folded to fit snugly within the abdominal cavity (see fig. 18.5). Epithelial tissue lining the wall secretes water, ions, mucus, and digestive enzymes. Most digestion of polymers and all absorption of monomers take place within the small intestine.

The first 25 centimeters (10 in) of the small intestine, called the duodenum, receives the highly acidic (📖 *see page 33*) food from the stomach. An alkaline (📖 *see page 33*) solution of bicarbonate ions (HCO₃⁻) is released immediately, from both the liver and pancreas, onto the food. Bicarbonate ions pick up hydrogen ions (H⁺ + HCO₃⁻ ⟶ H₂CO₃), thereby rendering the food less acidic.

Digestion Digestive enzymes are secreted mainly by the pancreas. Each kind of polymer is digested by a different enzyme. Together, the enzymes digest food into monomers.

Before fats can be digested, however, they must be worked on by bile. **Bile** is a complex fluid that is synthesized by liver cells. It consists of water, ions, cholesterol, and bile salts. (A bile salt is a sodium ion bonded to a steroid, which is synthesized from cholesterol in the liver.) Bile salts are emulsifying agents—molecules that break fat into pieces that are small enough to mix with water and associate more closely with the fat-digesting enzymes:

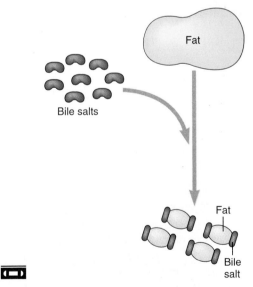

Fat

Bile salts

Fat

Bile
salt

Bile flows from the liver to the duodenum through the bile duct (*duct,* small tube). Some animals, including humans, store and concentrate bile in a gallbladder. This

Research Report

Ulcers: Products of Life-style or Infection?

Stop it—you're giving me an ulcer!" This phrase used to mean, "Don't give me emotional stress because it might produce an ulcer." Now it is more likely to mean, "Don't infect me with the bacteria that are causing your ulcer." Our understanding of ulcers is changing dramatically. For decades, we believed ulcers were the product of life-style—too much stress, too much alcohol, too many aspirins. Now it appears that ulcers are caused by bacteria. What is an ulcer and why do we now think it is an infectious disease?

An ulcer, shown in figure 18.B, is partial digestion of the digestive tract by its own digestive materials (acids and a protein-digesting enzyme). A gastric ulcer develops in the stomach and a duodenal ulcer in the duodenum (the first part of the small intestine). If left untreated, the ulcer may digest the walls of blood vessels, causing bleeding, and eventually eat its way through

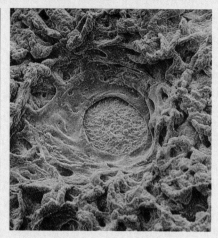

Figure 18.B A stomach ulcer.

the entire wall of the digestive tract. This condition, known as a perforated ulcer, is life-threatening, since contents of the digestive tract then spill out into the abdominal cavity.

The digestive tract is normally protected against self-digestion. The stomach is coated with a layer of mucus and bicarbonate ions, through which digestive materials cannot penetrate. The duodenum contains a solution of bicarbonate ions that neutralize the highly

acidic food as soon as it enters. What goes wrong when an ulcer forms?

Until recently, two major hypotheses (fig. 18.C) about ulcer formation were accepted by most physicians. One states that ulcers develop when certain materials, particularly aspirin, ibuprofen, and alcohol, inhibit secretion of mucus and bicarbonate ions. The other hypothesis states that emotional stress triggers excess production of acids—too many to be neutralized by the bicarbonate ions. Neither hypothesis has been well tested (due largely to the fact that it is unethical to perform the necessary experiments on humans). Nor is either hypothesis supported by the enormous amount of data that has been collected on ulcer patients. Most ulcer patients taking drugs that suppress acid secretion, for example, develop another ulcer.

A third hypothesis states that bacteria cause ulcers. This hypothesis was developed in the 1980s by two Australian physicians, Barry Marshall and J. Robin Warren. They noticed spiral-shaped bacteria (*Helicobacter pylori*) in the ulcers of their patients. About 70%

---

small sac lies next to the liver and is connected to the bile duct (fig. 18.10). The gallbladder can release large amounts of bile all at once when a large amount of fat is consumed. It is not an essential organ and may be surgically removed when obstructed by a gallstone (a hard deposit of cholesterol and/or calcium). In the absence of a gallbladder, bile flows slowly and continuously from the liver through the bile duct and into the duodenum. The only consequence of the operation is that large quantities of fat cannot be digested at once.

Absorption Monomers released during digestion move through the wall of the small intestine into the bloodstream. Vitamins, minerals, and water also enter the blood at this

point. The transfer of materials from the digestive tract to the bloodstream is called **absorption.** It is accomplished by three methods of membrane transport (see page 75): diffusion, facilitated diffusion, and active transport.

The lining of the small intestine is folded extensively and its surface is covered with fingerlike projections known as villi (fig. 18.11). The villi, in turn, are covered with smaller projections called microvilli. The folds, villi, and microvilli greatly enlarge the surface for absorption; if the small intestine were layed out smooth, it would cover an area larger than a tennis court! The larger area provides more space for absorption, so products of digestion can move more rapidly from the digestive tract to the blood.

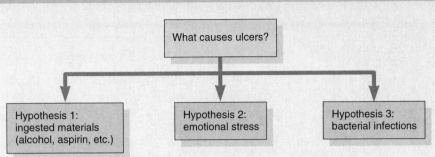

Figure 18.C Hypotheses about the cause of ulcers.

of their patients with stomach ulcers and 90% of their patients with duodenal ulcers had these bacteria growing within the epithelial tissue of their digestive tracts. (As an experiment, Dr. Marshall actually swallowed some of these bacteria. A few days later, the lining of his stomach became inflamed—the first step in development of an ulcer.) The correlation between presence of the bacteria and occurrence of ulcers was so strong that the researchers concluded a cause-and-effect relationship: the bacteria were causing the ulcers.

The bacterial hypothesis was ignored for years by other physicians, in part because everything known about

bacteria indicated they couldn't survive the highly acidic solutions of the stomach. The hypothesis stimulated other researchers to study these bacteria, and soon it was discovered how the bacteria grow in acid—they synthesize an enzyme that forms ammonia (NH_3). Ammonia is a base; it picks up hydrogen ions ($NH_3 + H^+ \longrightarrow NH_4^+$), thereby reducing the acidity of its environs. This work demonstrated that the bacteria associated with ulcers could grow in the highly acidic environment of the stomach. More physicians took notice of the hypothesis.

Every scientist is acutely aware that a correlation does not necessarily indicate cause and effect. The fact that

a particular kind of bacterium is regularly found in ulcered stomachs and duodenums may not mean the bacterium causes the ulcers. It may mean that the ulcers alter the digestive tract in ways that make it easier for the bacteria to invade, or it may mean that both the bacteria and the ulcers are caused by a third factor. How can scientists test for cause and effect?

So far the best test of whether bacteria cause ulcers comes from treatment of ulcer patients with antibiotics, the molecules that kill bacteria. If bacteria cause ulcers, then elimination of the bacteria should prevent recurrence of the ulcers. This turns out to be the case. In one study, involving many patients, only 13% of patients with gastric ulcers and 12% of patients with duodenal ulcers had recurrences after treatment with antibiotics. In contrast, recurrence rates for patients not treated with antibiotics were 74% for gastric ulcers and 95% for duodenal ulcers. Other experiments show similar results. While more research is needed, we can tentatively conclude that bacterial infection plays a critical role in formation of most ulcers.

Fatty acids released by fat digestion are not soluble in water and so are not moved directly into the bloodstream along with the other monomers. Instead, both fatty acids and glycerol (the two components of fats) enter epithelial cells of the small intestine, where they are converted back into fats. The fat molecules are wrapped within a protein coat, which makes them soluble in water so they can be transported in the blood. They move first into lymph vessels (a network of vessels that is part of the immune system) within the villi and later enter the bloodstream by way of a vein near the heart.

Pancreas and Liver

The pancreas and liver contribute to digestion as well as to other bodily functions. Both organs are located in the

upper part of the abdominal cavity. The pancreas is on the left side, behind the stomach, and the liver is on the right side (see fig. 18.5).

The pancreas is both an exocrine gland (secretes materials onto a surface or into a cavity) and an endocrine gland (secretes materials into the bloodstream). The exocrine portion of this organ secretes digestive enzymes and a solution of bicarbonate ions. Both kinds of secretions flow from the pancreas to the duodenum by way of a duct (see fig. 18.10). The endocrine portion of the pancreas secretes two hormones, insulin and glucagon, that regulate the blood-sugar level (described in chapter 17).

The liver is the "biochemical factory" of the body. Its crucial functions, many not related to food processing, are listed in table 18.5. The liver contributes to digestion by secreting bile and bicarbonate ions into the small

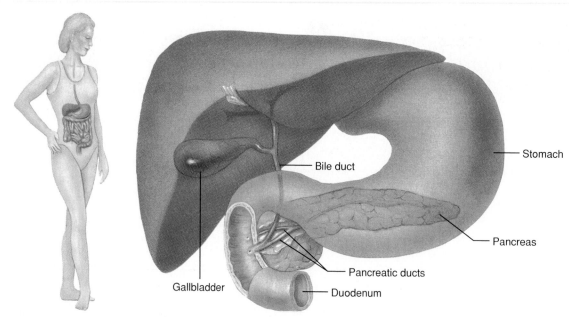

Figure 18.10 Secretions from the liver and pancreas flow into the small intestine. Bile from the liver flows through the bile duct to the gallbladder and the small intestine. When there is no fat in the small intestine, a sphincter in the bile duct shunts bile into the gallbladder, where it is stored. The gallbladder then releases its bile when a large amount of fat enters the duodenum. The pancreas secretes digestive enzymes and bicarbonate ions, which travel to the small intestine by way of the pancreatic duct.

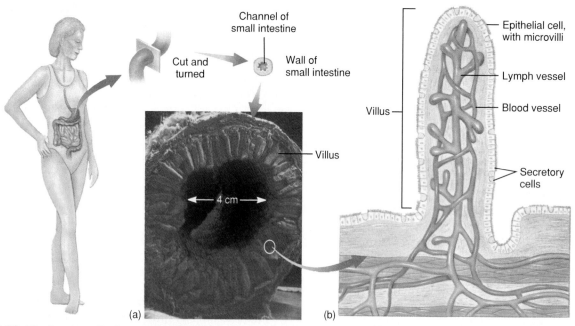

Figure 18.11 The inner wall of the small intestine.
(a) Photograph of the small intestine cut crosswise to reveal its many villi. (b) Each villus is covered with microvilli, which are hairlike extensions of the epithelial cells. Secretory cells, which secrete digestive enzymes and mucus, are interspersed among the epithelial cells. Absorption occurs when monomers, vitamins, and minerals move through the epithelial cells and into blood vessels and lymph vessels.

Table 18.5

Functions of the Liver

What It Does	Significance to the Body
Secretes bile	Emulsifies the fat in food
Secretes a solution of bicarbonate ions	Neutralizes stomach acids
Converts monomers from one form to another	Forms the monomers that are needed from monomers in the food
Stores vitamins and minerals	Prevents deficiencies
Converts glucose to glycogen	Lowers blood sugar when it is too high
Converts glycogen to glucose	Raises blood sugar when it is too low
Builds lipoproteins	Transports fat to fat cells
Builds plasma proteins	Holds water in the blood
Breaks down alcohol, drugs, pesticides, and excess hormones	Protects body cells from toxic materials
Forms blood-clotting materials	Prevents blood loss from wounds
Synthesizes cholesterol	Provides a component of cell membranes
Breaks down old red blood cells	Maintains healthy blood for transporting oxygen gas

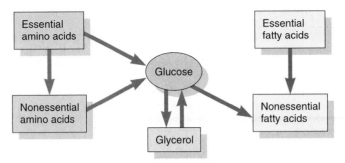

Figure 18.12 The liver converts monomers from one form to another. Monomers carried to the liver from the small intestine are modified, as indicated by the arrows. These conversions reduce our number of dietary requirements.

intestine. It also handles the products of digestion. Vitamins, minerals, and monomers are carried from the small intestine to the liver. There, vitamins and minerals are stored and monomers are converted from one form to another. Liver cells can synthesize all the monomers needed to maintain life as long as the diet includes the essential amino acids, essential fatty acids, vitamins, and minerals (fig. 18.12). The liver releases these nutrients into the bloodstream as they are needed by body cells.

Large Intestine

All food not digested and absorbed in the small intestine passes into the large intestine, or colon (see fig. 18.5). This portion of the digestive tract is approximately 6 centimeters (2.3 in) in diameter and 1.5 meters (5 ft) long. (It is called the large intestine because its channel is broader than the channel of the small intestine.) The large intestine stores wastes and transfers water and ions from the

wastes to the bloodstream. The wastes, called feces, move out of the digestive tract during defecation.

Transport of materials through the large intestine is controlled by two sphincters, one at the end of the small intestine and the other at the anus. The anal sphincter consists of an inner ring of involuntary muscle (cannot be controlled by the mind) and an outer ring of voluntary muscle that we learn, as children, to control.

The appendix is a narrow pouch that dangles from the first part of the large intestine (see fig. 18.5). It is found only in humans, certain apes, and wombats (bearlike mammals of Australia). Its function is not understood. One hypothesis is that the appendix is a vestigial organ—one that used to have a function in some distant ancestor. Another hypothesis, based on its tissues, is that the appendix plays some role in defending the body against infections. Occasionally, the appendix becomes infected from bacteria in the feces, a condition known as appendicitis (*itis*, inflamed). The appendix then swells and must be surgically removed before it bursts and spills feces into the abdominal cavity.

Most of the material that enters the large intestine is fiber. Other materials include water, mucus, bile, and sloughed-off cells from the lining of the digestive tract. Millions of bacteria are normal inhabitants of the large intestine and some are carried out with the feces. These bacteria feed on undigested material (fiber) and release gas as a waste product. They also synthesize vitamin K and certain B vitamins, which are absorbed into the bloodstream and used by the body cells.

The characteristic reddish brown color of feces comes from bilirubin, a breakdown product of hemoglobin that forms during the normal recycling of red blood cells. This red pigment enters the small intestine as part of bile and darkens as it is modified by bacteria in the large intestine. Feces that are not reddish brown can indicate a medical problem, as described in table 18.6.

As undigested food passes through the large intestine, most of its water and ions are absorbed into the bloodstream. The quantity of water removed from the

Table 18.6

Feces Color

Color	What It Indicates
Reddish brown	Good health
Grayish white	Liver damage; gallbladder damage; not enough red blood cells
Bright red	Bleeding in the large intestine
Black, tarlike	Bleeding in the mouth, lungs, stomach, or small intestine
Greenish	Excessive breakdown of red blood cells, probably from strenuous exercise

feces depends on how rapidly they are transported. Bulky feces, which contain a lot of fiber, move rapidly because they stretch the walls and stimulate muscle contractions. They retain sufficient water to form a large, soft stool. Small feces, which are lacking in fiber, move slowly because they do not stretch the walls. Most of their water is absorbed into the bloodstream and they become even smaller, harder, and more difficult to expel.

Diarrhea develops when the walls of the large intestine do not remove enough water from the feces. Many factors can cause diarrhea, including irritating chemicals in certain foods and drugs, emotional stress, and infections of the intestine. Diarrhea in a young child is dangerous; a small loss of water may represent a large portion of the total body fluids.

SUMMARY

1. An animal consumes the polymers of other organisms and digests them into monomers. An animal's diet must include all the materials that its cells need but cannot synthesize. It can only use polymers that its digestive enzymes can split into monomers.

2. An animal's diet provides monomers as sources of energy. Excess monomers are stored as glycogen and fat. When the diet does not provide enough energy, the condition is known as starvation.

3. An animal's diet provides building materials for synthesizing polymers. The term for insufficient building materials is malnourishment. The human diet must provide eight essential amino acids, for building proteins; three essential fatty acids, for building membranes and other structures; thirteen vitamins; and twenty minerals.

4. Fiber is plant material that is eaten but not digested. It prevents constipation and removes damaging materials from the body.

5. Food enters the digestive tract by way of the mouth, where it is torn or ground by teeth and moistened by saliva. An enzyme in saliva initiates starch digestion. Food moves from the mouth to the stomach by way of the esophagus.

6. The stomach is a place of food storage. It also churns food, softens meat and bones, and initiates protein digestion. Cells in the stomach wall secrete mucus, an acid, bicarbonate ions, and a protein-digesting enzyme.

7. The small intestine is where most digestion and all absorption take place. Digestive enzymes are secreted by the pancreas, bile is secreted by the liver, and bicarbonate ions are secreted by the pancreas and liver. The monomers, as well as vitamins and minerals, are absorbed into the bloodstream (or lymph, in the case of fats).

8. The pancreas and liver secrete materials into the small intestine. The pancreas secretes digestive enzymes and bicarbonate ions. The liver secretes bile and bicarbonate ions. The liver also converts monomers (the products of digestion) from one form to another and distributes them as needed.

9. The large intestine eliminates wastes from the digestive tract. Most of the water and ions in the wastes are absorbed into the bloodstream before the feces are eliminated.

KEY TERMS

Absorption (page 346)
Bile (BEYEL) (page 345)
Complete protein (page 335)
Digestive tract (page 339)
Edema (eh-DEE-mah) (page 338)
Essential amino acid (page 335)
Essential fatty acid (page 338)
Fiber (page 339)
Malnourishment (page 334)
Mineral (page 338)
Monomer (MAH-noh-muhr) (page 332)
Peristalsis (PAEHR-ih-STAL-sis) (page 339)
Polymer (PAH-lih-muhr) (page 332)
Sphincter (SFINK-tur) (page 339)
Starvation (page 334)
Vitamin (VEYE-tah-min) (page 338)

STUDY QUESTIONS

1. What is the biological reason for eating? What are two biochemical reasons that an animal consumes only certain kinds of food?
2. Why are complex carbohydrates better sources of energy than sugar?
3. What is the difference between starvation and malnourishment?
4. What are complete proteins? How can a strict vegetarian obtain complete proteins or their equivalent?
5. Describe the symptoms of long-term protein deficiency and explain why these particular symptoms develop.
6. What are the fat-soluble vitamins? Why should you avoid consuming more of these vitamins than your body can use?
7. What is the difference between a vitamin and a mineral?
8. What is fiber? Why is it healthy to eat fiber?
9. Draw the human digestive tract (stretched out rather than packed inside the abdomen). Label each part and give its functions.
10. Describe digestion and absorption.

CRITICAL THINKING PROBLEMS

1. Explain why we store energy as fat rather than glycogen.
2. An advertisement claims a new diet pill enables you to lose at least a pound a day. Is this false advertising? Why or why not?
3. Are you malnourished? Why or why not?
4. Explain why malnourished children often appear fat, with very large bellies.
5. How would you explain, to an inquisitive butcher, the velvety texture of a calf's small intestine?
6. Can you gain weight on a pure protein diet? How?
7. Which one of the following organs would you least like to have removed—stomach, small intestine, or large intestine? Why?

SUGGESTED READINGS

Introductory Level:

Blonz, E. R. *The Really Simple, No Nonsense Nutrition Guide*. Emeryville, CA: Conari Press, 1992. A concise review of contemporary nutritional research, covering issues such as the relation between salt and blood pressure, the effects of sugar on children, the significance of cholesterol, and whether taking vitamins is worthwhile.

Brody, J. *Jane Brody's Good Food Book*. New York: W. W. Norton, 1985. An excellent description of what to eat and why.

Jackson, G., and P. Whitfield. *Digestion: Fueling the System*. New York: Torstar Books, 1984. A medically oriented discussion, written for the layperson, of the human digestive tract and what can go wrong with it.

Lappe, F. M. *Diet for a Small Planet*. New York: Ballantine Books, 1975. An excellent guide for staying healthy on a vegetarian diet.

Thompson, W. G. *Gut Reactions: Understanding Symptoms of the Digestive Tract*. New York: Plenum Publishing, 1989. An explanation, for the layperson, of what can go wrong with the digestive tract and the symptoms of these problems.

Advanced Level:

Eckert, R., D. Randall, and G. Augustine. *Animal Physiology: Mechanisms and Adaptations*, 3d ed. New York: W. H. Freeman and Company, 1988. A good comparison (chapter 15) of feeding and digestion in the different animal forms.

Livingston, E. H., and P. H. Guth. "Peptic Ulcer Disease," *American Scientist* 80 (1992): 592–98. A review of research on the causes of gastric and duodenal ulcers.

Martin, R. J., B. D. White, and M. G. Hulsey. "The Regulation of Body Weight," *American Scientist* 79 (1991): 528–41. A review of research on neural and hormonal controls of appetite.

Willett, W. C. "Diet and Health: What Should We Eat?" *Science* 264 (1994): 532–37. An analysis of the scientific evidence that relates components of the diet to coronary heart disease, cancer, and obesity.

❧ EXPLORATION

Interactive Software: Diet and Weight Loss

This interactive exercise explores how diet and exercise interact to determine whether we gain or lose weight. The program provides a pathway for energy, from food to either the synthesis of ATP or the storage of fat. The student alters the kinds and amounts of foods eaten as well as the amount of energy used in exercise. The effects of these three variables are then determined by measuring changes in the amount of stored fat.

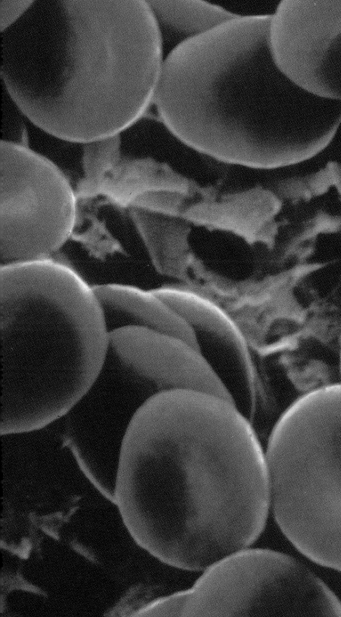

Circulation

Objectives

In this chapter, you will learn
–why animals need internal transport
  systems;
–the kinds of materials carried by
  blood;
–how blood is moved through vessels
  and the heart;
–what determines how much blood the
  heart pumps each minute;
–the factors that govern blood pressure;
–what causes abnormal pressures and
  hardening of the arteries.

Red blood cells are components of blood. Blood, in turn, is part of the circulatory system.

love you with all my heart." "In my heart I know I am right." "The heart has its reasons." "His heart is in the right place." "Her heart has turned to stone." "Eat your heart out." "He is cold-hearted; warm-hearted; kind-hearted; good-hearted; faint-hearted; lion-hearted." "She is pure of heart."

These common sayings imply that the heart is the center of love, reason, morality, courage, compassion, and intellect. We even pledge allegiance to the flag with hand over heart. Just what is this extraordinary organ?

The heart is just a pump. Its extraordinary feature is reliability: your heart contracts about seventy times a minute, every minute of the day and night. (How many times has your heart contracted in your lifetime? In your grandmother's lifetime?) When you sit at your desk, your heart pumps more than a gallon of blood every minute. And when you run, it pumps as many as 10 gallons a minute.

Your heart keeps you alive. If it stopped beating for just a few minutes, some of your brain cells would die from lack of oxygen. If longer, your whole body would die. The heart is so critical that only in the past few decades have surgeons dared to touch it with a scalpel. While open-heart surgery is now routine, it still frightens most of us even more than surgery on the brain—the organ that actually is the center of love, reason, morality, courage, compassion, and intellect.

Internal Transport Systems

Normal cellular activities take place only when conditions are optimal. A cell must have sufficient energy and building materials to do its work. The temperature, concentration of ions, and pH (ratio of acids to bases) must be just right for biochemical reactions. And cellular wastes must be removed as soon as they form. These conditions are maintained by a regular exchange of materials with the surrounding environment.

A simple animal, like the sponge, is built so that each cell is near the surrounding water. (All simple animals live in water.) Their bodies are hollow and their body walls are only a few cells thick. Water flows around and through the animal, enabling a direct exchange of materials with each body cell.

Animals with larger, more complex bodies have cells buried deep within the body, too far from the external environment for a direct exchange of materials. They require special structures for transporting materials between the external environment and the body cells. (Without an in-

Table 19.1

Components of the Circulatory System

Component	Function
Blood	Fluid that carries materials from one part of the body to another
Blood vessels	Tubes through which blood flows
Artery	Carries blood from heart to capillaries
Capillary	Place where materials are exchanged between the blood and the cells
Vein	Carries blood from capillaries to heart
Heart	Pumps blood
Atria	Receive blood from veins and move it into ventricles
Ventricles	Receive blood from atria and pump it into arteries
Valves	Ensure that blood flows in just one direction

ternal transport system, it would take three years for a molecule of oxygen to travel by diffusion from your lungs to your hand.) These structures include a transport fluid, a network of tubes, and at least one heart for pumping the fluid through the tubes.

The Human Circulatory System

A human has two systems of internal transport: a lymphatic system (described in chapter 22) and a circulatory system. The circulatory system consists of blood, a network of blood vessels, and a single heart. It circulates materials, within blood, throughout the body. The major components of the human circulatory system are summarized in table 19.1 and described in this chapter.

Blood contributes to homeostasis (maintaining optimal conditions within the body; see page 313) by keeping conditions optimal within the interstitial fluid around the cells. (Blood does not contact the body cells directly.) Materials move between the blood and the interstitial fluid through the walls of blood vessels. Materials move

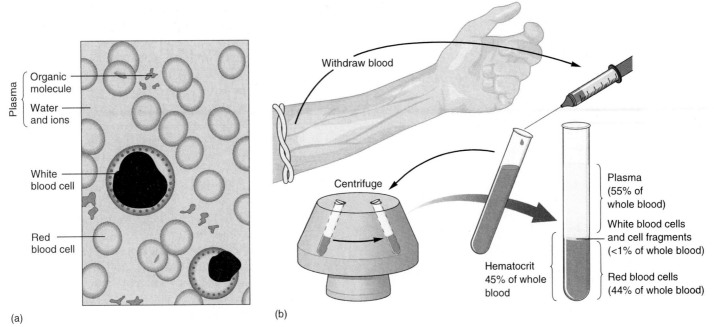

Plasma
- Organic molecule
- Water and ions

White blood cell

Red blood cell

(a)

Withdraw blood

Centrifuge

Plasma (55% of whole blood)

White blood cells and cell fragments (<1% of whole blood)

Hematocrit 45% of whole blood

Red blood cells (44% of whole blood)

(b)

Figure 19.1 Blood. (*a*) Blood is a form of connective tissue, with widely spaced cells immersed within a fluid called plasma. (*b*) Part of a physical examination is a blood test. Blood is removed from a vein in the arm and spun in a centrifuge.

The cells and cell fragments are pulled to the bottom of the tube, where they form a solid—the hematocrit. A normal hematocrit comprises about 45% of the blood.

between the interstitial fluid and the cell interiors through the plasma membranes of the cells:

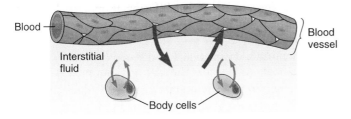

Blood

Interstitial fluid

Blood vessel

Body cells

The movement of materials between the blood and interstitial fluid is chiefly by diffusion. The movement of materials between the interstitial fluid and cell interiors occurs by all four methods of membrane transport (□ *see page 74*): diffusion, facilitated transport, active transport, and bulk transport.

Blood

Blood is a form of connective tissue (□ *see page 316*); it consists of widely spaced cells immersed within a fluid (fig. 19.1). Blood cells form approximately 45% of the blood volume. Most of the cells are red blood cells, which transport oxygen gas, and the rest are white blood cells, which fight infections and cancer, and cell fragments (platelets)

that help form blood clots. The remaining 55% of the blood is plasma, a pale yellow fluid consisting of molecules and ions dissolved in water. The human body contains approximately 5 liters (5.25 qt) of blood.

What the Blood Transports   The blood carries all the materials that cells need and all the wastes that cells produce. It carries eight categories of materials:

1. oxygen (O_2) from the lungs, where it is breathed in, to all the cells of the body, where it is used in cellular respiration;
2. carbon dioxide (CO_2) from the cells, where it is a waste product of cellular respiration, to the lungs, where it is breathed out;
3. food monomers, vitamins, and minerals from the digestive tract to the liver, where they are modified and stored, and from the liver to other body cells as needed;
4. waste products (other than CO_2) from the cells, where they form, to the kidneys, where they are eliminated in the urine;
5. heat from the deeper parts of the body to the skin, where it can be released (heat is a by-product of all chemical reactions);

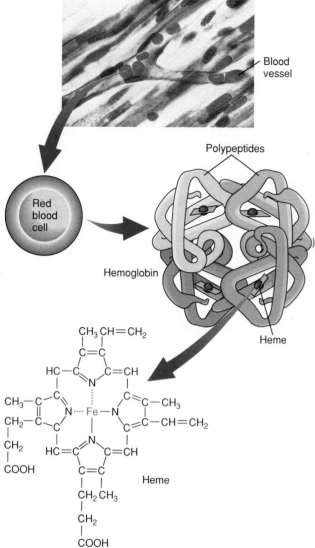

Figure 19.2 Hemoglobin. A red blood cell holds approximately 300 million molecules of hemoglobin. Each molecule is formed of four polypeptides (long chains of amino acids) and four heme groups. Each heme group can carry a molecule of oxygen, which binds with iron (Fe) in the center of the structure. It picks up O_2 where this gas is more concentrated (in the lungs) and releases O_2 where the gas is less concentrated (near the body cells) as well where CO_2 is more concentrated (near more active body cells).

6. materials of the immune system, such as white blood cells and antibodies, which fight infections and cancer;
7. materials that repair broken vessels by forming blood clots;
8. special molecules, such as hormones, that are formed in one part of the body and used in other parts.

Red Blood Cells   The function of a red blood cell is to pick up and carry oxygen gas, which does not dissolve well enough in water to be carried by the plasma alone. A red blood cell is packed with **hemoglobin,** a large protein built around an atom of iron (fig. 19.2). It is the iron in hemoglobin that binds with oxygen. Hemoglobin turns bright red when combined with as much oxygen as it can hold and is bluish when not.

Red blood cells lose their nuclei as they mature, providing more space for hemoglobin. Without a nucleus, however, a cell has no genes to direct protein synthesis and cannot live long. The life span of a red blood cell is about four months.

Dead blood cells are replaced continuously by new ones. The dead ones are broken down in the liver and spleen (a large organ behind the stomach). New ones are constructed in the bone marrow (the soft interior of bones). Every second of your life, approximately 2.5 million new red blood cells are made and an equal number of old ones die within your body.

Anemia   **Anemia** is any condition in which there are not enough red blood cells or molecules of hemoglobin for adequate transport of oxygen. The symptoms of anemia are fatigue, dizziness, and ringing in the ears.

Anemia has many causes. Some common reasons for inadequate oxygen transport are:

1. loss of blood from a wound or from menstruation;
2. inability to build red blood cells because of damage to the bone marrow from radioactive materials or certain drugs;
3. inability to form enough red blood cells or hemoglobin because the diet does not provide enough iron, protein, folic acid (needed for cell division), or vitamin B_{12} (needed to activate folic acid);
4. premature death of red blood cells due to a genetic disorder (e.g., sickle-cell anemia).

Many people suffer unknowingly from anemia. The condition may become apparent only at high altitudes, where oxygen transport is already impaired, or during physical exercise, when there is a heightened demand for oxygen transport.

Network of Vessels

Blood moves throughout the body within a closed network of tubes called the **cardiovascular system** (fig. 19.3). While we give the different parts of this network different names—heart, arteries, capillaries, and veins—they are all part of the same continuous structure.

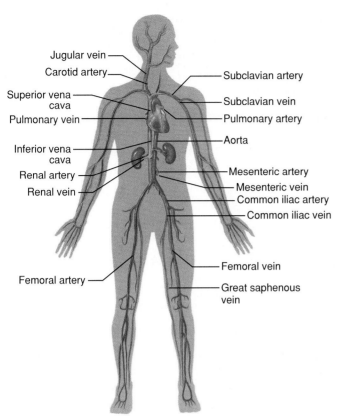

Figure 19.3 Major components of the human cardiovascular system. The cardiovascular system consists of the heart, arteries, capillaries, and veins. Only the heart and the major arteries and veins are shown here. Smaller arteries, veins, and capillaries connect the major arteries with major veins.

Blood flows in a single direction through the cardiovascular system. It circulates throughout the body, traveling from the heart to the arteries to the capillaries to the veins and then back again to the heart:

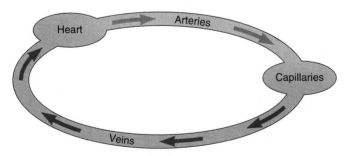

The vessels branch and merge to form a dense network that penetrates every part of the body. If all your blood vessels were lined up end to end, they would extend about 100,000 kilometers—more than twice the distance around the world!

Arteries, Capillaries, and Veins   A blood vessel is a tube formed of three kinds of tissues (□ *see page 316*): epithelial tissue, connective tissue, and smooth muscle. All vessels have an inner layer of epithelial tissue. They differ in how much connective tissue and smooth muscle surround the inner layer.

Arteries carry blood from the heart to the capillaries. Blood flows out of the heart through large arteries, which branch again and again until they form millions of tiny arteries called arterioles.

The wall of an artery has thick layers of connective tissue and smooth muscle around the inner layer of epithelial tissue (fig. 19.4). The connective tissue contains elastic fibers (proteins) that encircle the channel and change length in response to how much blood is within the channel. When more blood enters the artery, the fibers stretch and the channel broadens. When less blood enters the artery, the fibers shorten (by elastic recoil, like a rubber band) and the channel narrows. In this way, the fibers minimize changes in pressure against the artery walls. Smooth muscles also encircle the channel. They change length in response to commands from the endocrine and nervous systems. When the muscles contract, they constrict the artery and narrow its channel. When the muscles relax, they dilate the artery and broaden its channel:

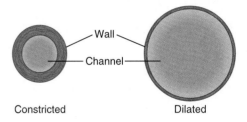

The arterioles connect to capillaries, the only part of the circulatory system where materials can move between the blood and the interstitial fluid. The channel of a capillary is narrow and the wall contains only epithelial tissue—usually just a single layer of flat cells (fig. 19.4). The cells are coated with their own secretions: glycoproteins, which provide adhesion, and protein fibers, which provide support. A capillary has no connective tissue or smooth muscle; it cannot change in diameter. The wall must be very thin in order for materials to move between the blood and the interstitial fluid. The human body has roughly 7 billion capillaries; they cover every region of the body and account for almost half the total length of the vessels. No body cell is farther than 0.1 millimeter from a capillary.

The network of capillaries within a tissue or organ is called a capillary bed. It begins with an arteriole, which

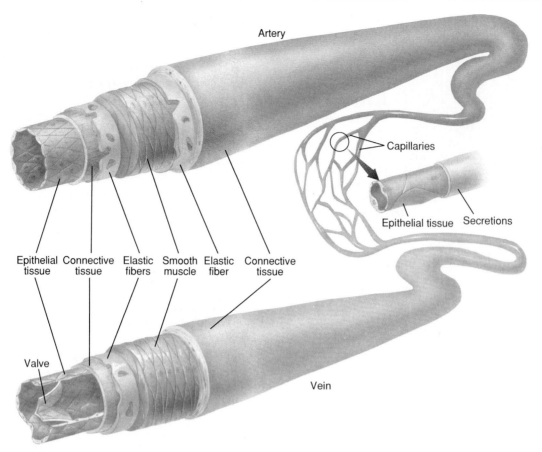

Figure 19.4 Tissues that form blood vessels. The wall of an artery is made of epithelial tissue, connective tissue (with elastic fibers), and smooth muscle. The wall of a capillary consists only of a thin layer of epithelial tissue covered with secretions. Materials move into and out of the blood only through the walls of the capillaries. The wall of a vein resembles the wall of an artery except it contains less smooth muscle. Veins are the only blood vessels with valves.

branches into capillaries, which, in turn, branch into more capillaries. These capillaries then merge to form fewer capillaries, which merge into a tiny vein called a venule:

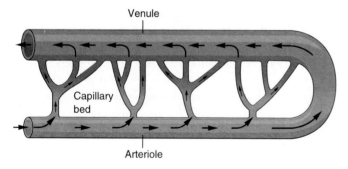

Veins carry blood from capillaries to the heart. The wall of a vein (and venule) has an inner layer of epithelial tissue surrounded by layers of smooth muscle and connective tissue with elastic fibers (fig. 19.4). A vein has less smooth muscle than an artery; its diameter is changed more by elastic fibers than by smooth muscles. Veins also have one-way valves (described later), which are not found in arteries or capillaries.

Millions of venules merge to form veins, and the many veins merge to form a few large veins that connect with the heart (see fig. 19.3).

Shunting the Blood   Sometimes an organ is very active and requires a large flow of blood to provide materials and remove wastes. Other times it is relatively inactive and can be maintained by a small flow of blood. Since the total amount of blood within the body remains fairly constant, the amount in one area of the body can be increased only by decreasing the flow to other parts of the body. Blood is shunted from organs that are less active to organs that are more active.

Blood is shunted by dilating the arterioles in more active organs and constricting the arterioles in other organs (fig. 19.5). After eating a heavy meal, for example,

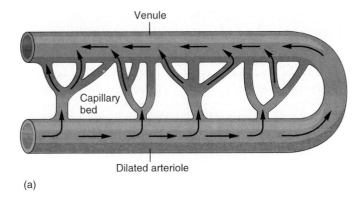

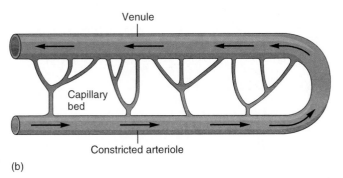

Figure 19.5 A blood shunt. (*a*) Blood is shunted into more active tissues by dilating the arterioles that bring blood into the capillaries. (*b*) Blood is shunted away from less active tissues by constricting the arterioles that bring blood into the capillaries.

less blood flows into your skeletal muscles so that more blood can flow into your digestive tract to aid in digestion and absorption. In this state, you are not inclined to exert yourself and cannot exercise at peak intensity.

Blood shunts are controlled by the nervous system and the endocrine system. They are also controlled by local changes within the tissues themselves in response to changes in cellular activities. Some arterioles, for example, dilate in response to higher concentrations of carbon dioxide, which reflect higher rates of cellular respiration (recall that carbon dioxide is a waste product of this energy-releasing reaction).

The Heart

A heart is an enlarged, highly muscular region of the vessel network. Its muscles, called cardiac muscles, are unique in both structure and function. They contract rhythmically to squeeze blood out of one area and into another. The human heart, about the size of a fist, is located in the center of the chest (see fig. 19.3). The structure, function, and location are the same in all mammals and differ only slightly between mammals and birds.

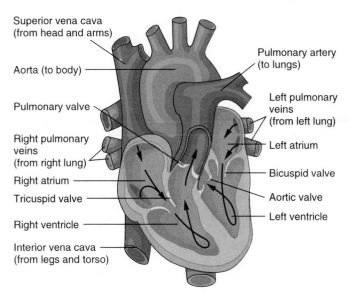

Figure 19.6 The human heart. The heart consists of two atria, two ventricles, and four valves. Blood enters the right atrium by way of the inferior and superior venae cavae. It leaves the right ventricle by way of the pulmonary artery. Blood enters the left atrium through four pulmonary veins and leaves the left ventricle through a single, large artery called the aorta. By convention, a body or organ—like the heart shown here—is illustrated as though we were looking at it, so that what is on your left is actually the right side of the body or organ.

Anatomy   The heart has four chambers (expanded spaces): two atria and two ventricles (fig. 19.6). An atrium (singular of atria) receives blood from veins and passes it to a ventricle. A ventricle receives blood from an atrium and pumps it into an artery:

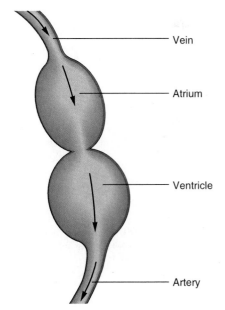

One-way valves between different regions of the heart keep the blood flowing in a single direction. These simple flaps open and close in response to differences in pressure on their two sides:

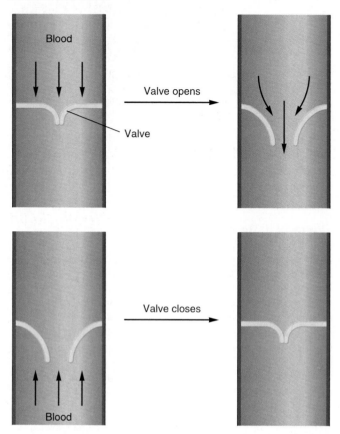

The human heart is really two pumps: the atrium and ventricle on the right side receive blood from all parts of the body except the lungs and pump blood to the lungs. The atrium and ventricle on the left side receive blood from the lungs and pump it to all parts of the body:

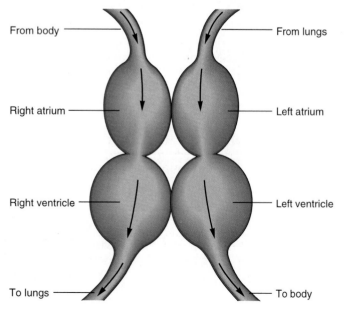

Table 19.2
Heart Disorders

Disorder	Problem	Effect on Circulation
Heart murmur	Defective valve	Edema or heart failure
Heart failure	Heart cannot pump with enough force	Blood flows too slowly or not at all
Heart attack	Damaged heart muscles	Blood flows too slowly or not at all
Heart block	Signals from pacemaker are impeded	Uncoordinated muscle contractions; blood flows too slowly or not at all

This sketch shows the basic design of the heart. Figure 19.6 shows the heart as it really is—with more veins connected to the atria and with the ventricles twisted up and over the atria. The flow of blood through the heart is described following. Various heart disorders that interfere with this flow are defined in table 19.2.

Blood Flow   Blood from all organs of the body (except the lungs) enters the right atrium by way of two large veins: the superior vena cava (*superior*, upper; *vena*, vein; *cava*, hollow), which brings blood from the head and arms, and the inferior vena cava (*inferior*, lower), which brings blood from the torso and legs (see fig. 19.3). This blood is low in oxygen, which cells have used in cellular respiration (see page 88), and high in carbon dioxide, which cells have released as a waste product of cellular respiration.

From the right atrium, blood is sucked into the right ventricle as the muscles of the right ventricle relax and the chamber enlarges (fig. 19.7a). The pressure of blood on the atrium side of the tricuspid valve (*tri*, three; *cuspid*, part), located between the atrium and ventricle, pushes the valve open. The right atrium then contracts to complete its emptying (fig. 19.7b).

After the right ventricle has filled with blood, it contracts and blood is squeezed out of the ventricle and into the pulmonary artery (fig. 19.7c). Blood pressing against the tricuspid valve closes it so there is no backflow into the atrium. Blood pressing against the pulmonary valve, which lies between the right ventricle and the pulmonary artery, opens the valve and allows blood flow into the artery.

When the ventricle relaxes and its chamber enlarges (see fig. 19.7a), blood in the pulmonary artery is pulled backward against the pulmonary valve, closing it to prevent backflow into the ventricle.

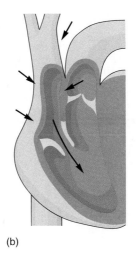

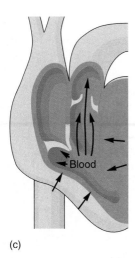

(a)   (b)   (c)

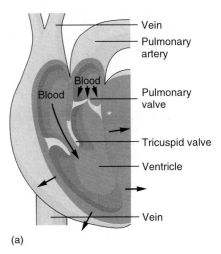 **Figure 19.7 How the heart pumps blood.** Only the right side of the heart is shown; the left side pumps blood in the same manner. (*a*) Blood is sucked out of the atrium and into the ventricle as the ventricle enlarges. Blood flowing from the atrium toward the ventricle pushes the tricuspid valve open. Backflow of blood from the pulmonary artery toward the ventricle pushes the pulmonary valve closed. (*b*) The atrium then contracts to fully empty its chamber. (*c*) Blood is squeezed out of the ventricle into the pulmonary artery when the ventricle contracts. Blood pushed toward the pulmonary artery opens the pulmonary valve. Blood pushed toward the atrium closes the tricuspid valve.

The pulmonary artery carries blood to capillary beds within the lungs. There the blood picks up oxygen from the lungs and releases carbon dioxide. It then returns to the left atrium of the heart by way of four pulmonary veins. Hemoglobin molecules in the red blood cells are now fully saturated with oxygen, and most of the carbon dioxide has been removed through exhalation from the lungs.

Blood moves through the left atrium and ventricle in the same way as it moves through the right atrium and ventricle. It is sucked out of the left atrium by expansion of the left ventricle. The left atrium then contracts to complete the emptying. The left ventricle contracts, forcing blood under considerable pressure out of the heart through the aorta—the largest artery, which carries blood to all parts of the body except the lungs. (The left ventricle is the most muscular part of the heart because it pumps blood the greatest distance.) The bicuspid valve (*bi*, two) between the left atrium and left ventricle and the aortic valve between the left ventricle and aorta prevent backflow of blood. (The bicuspid valve is also known as the mitral valve.)

The aorta curves up and around the heart and then runs downward in front of the backbone. It branches to form more arteries and then branches again and again to form millions of arterioles that carry blood to capillary beds within all the tissues and organs of the body (except the lungs). From the capillaries, the blood returns to the right atrium by way of venules and veins.

Heart Sounds The sounds of a beating heart are vibrations of the valves as they snap shut. A normal heart makes a "lub-dub" sound. While there are four valves, there are only two sounds because the right and left atria contract at the same time as do the right and left ventricles. The first sound ("lub") is the simultaneous closing of the tricuspid and bicuspid valves. The second sound ("dub") is the simultaneous closing of the pulmonary and aortic valves. The heart makes unusual sounds when its valves are not working properly. These sounds are called heart murmurs.

When a valve between an atrium and ventricle does not close completely, due to scar tissue or a hole, the heart makes a "lub-hiss-dub" sound. The "hiss" is the sound of blood flowing back into the atrium through a partially open valve. Blood then backs up all the way from the atrium to the veins and capillaries. Pressure from the accumulated blood forces plasma out of the capillaries, forming edema (excess fluid between the body cells). When it is the tricuspid valve that does not close completely, edema may develop in the abdomen, legs, and feet. When it is the bicuspid valve that is faulty, edema develops in the lungs, where it causes shortness of breath and, in extreme cases, individuals may actually drown in their own plasma.

When the valve between a ventricle and artery does not close completely, the heart sound is a "lub-dub-hiss." The "hiss" is the sound of blood flowing back into the ventricle through a partially open valve. The heart's effectiveness as a pump is diminished. In a young person, the only consequence may be an enlargement of the muscles within the ventricle as they work harder to counteract the backflow. The problem is more serious in a heart that is already weak, as is often the case in older people. In them, the increased workload may be too much for the weak heart

361

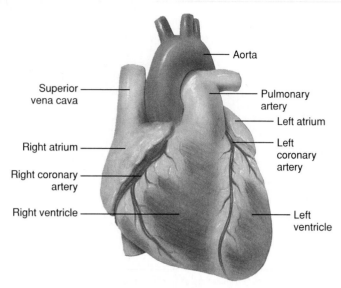

Figure 19.8 The coronary arteries. The coronary arteries are the first to branch off from the aorta. They begin as two large arteries, one on each side of the heart, and branch to form many smaller arteries and then millions of arterioles. Arterioles carry blood to capillary beds within the heart muscles. Blood then returns to the inferior vena cava by way of coronary veins.

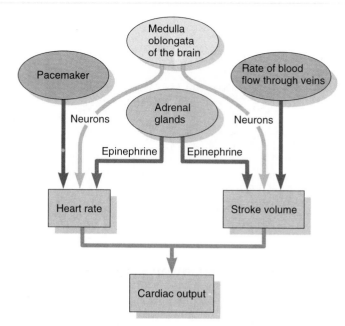

Figure 19.9 Control of cardiac output.

muscles as they struggle to overcome the backflow. This condition and others in which the heart muscles cannot pump blood adequately are called heart failure.

Because the heart valves operate according to simple mechanical principles, they are relatively easy to repair or replace. Defective valves are replaced by artificial valves made of synthetic materials or by valves transplanted from other mammals (usually the pig).

The Coronary Arteries   Blood passing through the chambers of the heart does not nourish the heart muscles. Instead, blood reaches capillaries within the heart muscles by way of **coronary arteries** on the outer surface of the heart (fig. 19.8). They are the first arteries to branch off from the aorta.

A heart attack occurs when a coronary artery does not deliver sufficient blood to the heart muscles. This happens when the artery is blocked by a blood clot or an accumulation of materials on its inner wall. Portions of the heart that are normally nourished by the blocked artery die from insufficient oxygen. How to recognize a heart attack is described in the accompanying Wellness Report.

Cardiac Output

The amount of blood pumped by each heart ventricle each minute is the **cardiac output.** When you are resting, each

ventricle pumps about 5 liters of blood—the total amount in the body—each minute. During physical exercise, your cardiac output may rise as high as 37 liters per minute.

Cardiac output is the product of two factors: (1) the **heart rate,** which is the number of contractions per minute, and (2) the **stroke volume,** which is the amount of blood ejected from each ventricle during a contraction. These factors and how they are regulated are summarized in figure 19.9 and described following.

Heart Rate

The number of times the heart contracts each minute is the heart rate, or pulse rate. It is measured by feeling the impact of blood on the wall of an artery. The pulse can be felt wherever a large artery lies near the skin: at the wrist (on the thumb side) and on each side of the throat. The impact of blood on the arterial wall is strongest when the left ventricle contracts, so that the frequency of the pulse reflects the frequency of contractions of the left ventricle.

The heart contracts as often as needed to provide adequate blood flow. Each person has a characteristic heart rate that quickens during exercise and emotions. Normally, the heart beats between 50 and 80 times a minute when you are resting and increases to as many as 200 beats a minute during strenuous exercise. The heart rate is regulated by the pacemaker, the brain, and the adrenal glands.

Control by the Pacemaker   The main regulator of heart rate is the **pacemaker,** a patch of specialized cells within the wall of the right atrium. The pacemaker sends out electrical signals that stimulate the heart muscles to contract. A signal moves first through the two atria, which re-

Wellness Report

Heart Attack

On Sunday, John planned to celebrate his sixtieth birthday with brunch at his favorite restaurant. When he awoke that morning, however, his driveway was covered with a foot of new snow. He bundled up and went out to shovel. John had almost cleared the driveway when he began to feel a heavy pressure in his chest. He felt nauseated and went inside to lie down. The chest pain continued, now extending down his left arm, and he began to sweat profusely. John was having a heart attack.

A heart attack occurs when muscles of the heart are damaged by an insufficient supply of oxygen (fig. 19.A). Typically it is caused by blockage of a coronary artery or arteriole, when blood cannot flow to the heart muscles because of a blood clot or an accumulation of cholesterol and other fatty deposits on the inner walls. Without oxygen, a heart muscle deteriorates and eventually dies. The damage is permanent. A blocked artery causes more damage, for it feeds more muscles, than a blocked arteriole.

A heart attack is not a sudden event: it develops over four to six hours. As the attack proceeds, more and more heart muscle is damaged. Thus, it is important to recognize the signs of a heart attack and to seek medical treatment as soon as possible. (While you may be too young to be concerned about yourself, it is important that you recognize the signs in order to help someone else.) A person who is having a

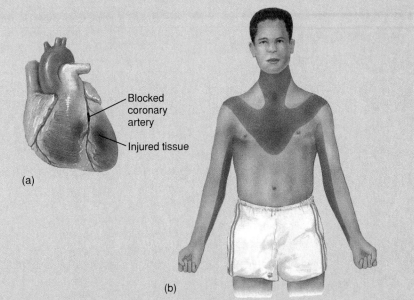

Figure 19.A A heart attack. (*a*) A heart attack occurs when a coronary artery becomes blocked. Muscles nourished by blood in the blocked artery die from lack of oxygen. (*b*) Signs of a heart attack include pressure or pain in the chest, neck, and arms (shown in red).

heart attack will experience some of the following warning signs:

- Pressure or squeezing pain in the chest, spreading to the shoulders, neck, or arms (fig. 19.A);
- profuse sweating;
- shortness of breath;
- nausea;
- dizziness or fainting.

If you suspect someone is having a heart attack, call 911 immediately. (Do not delay calling because you are afraid to risk the embarrassment of a false alarm.) If the person is unconscious, a 911 dispatcher may advise you to begin CPR (cardiopulmonary resuscitation), involving mouth-to-mouth breathing and chest compression (described in chapter 20). Even if you are not trained in this emergency procedure, a dispatcher can instruct you in CPR until help arrives (usually within five minutes).

Half of all people who have heart attacks wait too long before seeking medical help. And half—300,000 Americans each year—die before reaching the hospital. Of those who survive, most of the permanent damage done to the heart occurs during the first hour.

---

spond by contracting, and then to a similar patch of specialized cells at the base of the atrium. This patch relays the signals to the two ventricles and they contract about 0.1 second after the atria.

Irregularities in the heart muscles, such as dead tissue from a heart attack, delay the flow of electricity from the pacemaker through the heart muscles. These disturbances,

called heart blocks, interfere with the coordination of heart muscles and weaken the heart's ability to function as a pump.

Sometimes the pacemaker itself becomes defective and loses its ability to maintain a normal, steady heart beat. It can be replaced by an artificial pacemaker, which is a small battery-operated device that is inserted into the

Research Report

Discovery of How Neurons Control the Heart Rate

In the early part of the nineteenth century, biologists observed that a frog's heart beats more slowly in response to electrical stimulation of neurons that connect the medulla oblongata of the brain to the muscles of the heart. The response occurs even in a heart that has been removed, along with its connecting neurons, and placed in a solution of water and ions that resemble body fluids. A century later, this laboratory preparation of the frog heart was used to demonstrate how neurons stimulate muscles to contract.

Otto Loewi (1873–1961), a German physiologist, wanted to find out how neurons stimulate muscles. He developed, in 1903, two alternative hypotheses regarding neuronal stimulation of muscles: (1) a neuron passes electrical signals directly to a muscle, and (2) a neuron passes chemical messengers to a muscle. His logic is dia-

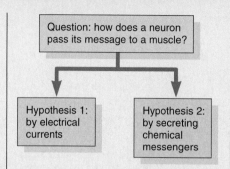

Figure 19.B Loewi's hypotheses.

grammed in figure 19.B. The problem was, Loewi could think of no way to test the hypotheses—to disprove one or both of them. For eighteen years, the problem rested within his mind.

On the night before Easter in 1921, Loewi thought of a way to test his hypotheses. Here, in his own words, is how it happened:

> I awoke, turned on the light, and jotted down a few notes on a tiny slip of paper. Then I fell asleep again. It occurred to me at six o'clock in the morning that during the night I had written down something most important, but I

was unable to decipher the scrawl. That Sunday was the most desperate day in my whole scientific life. During the next night, however, I awoke again, at three o'clock, and I remembered what it was. This time I did not take any risk; I got up immediately, went to the laboratory, made the experiment on the frog's heart . . . and at five o'clock the chemical transmission of nervous impulses was conclusively proved.

His experiment was performed on hearts removed from frogs and immersed within fluids. His predictions from the hypotheses were as follows. If stimulation of the heart muscle is by chemical messengers, then some of these chemicals will leak into the surrounding fluid. Another heart, without attached neurons, will respond when exposed to this same fluid. If stimulation is by electrical signals, then there will be no change in the surrounding fluid. Another heart, without attached neurons, will show no response. These predictions are diagrammed in figure 19.C.

chest and connected to the heart. It sends signals, usually in the form of microwaves, to the heart muscles.

Control by the Brain and Adrenal Glands The heart rate is also under the influence of the brain and the adrenal glands (which lie on top of the kidneys). A cluster of neurons within the medulla oblongata of the brain (see fig. 17.11), called the **cardiovascular center,** regulates heart rate in response to emotions, level of physical activity, and blood pressure. It receives information about these conditions from other parts of the brain as well as from the blood and blood vessels. In response to this information, the cardiovascular center sends commands, via neurons, to the pacemaker of the heart. One set of neurons releases the neurotransmitter (a chemical messenger) acetylcholine, which slows the heart rate. It is this set of neurons that is usually active (when the connection between the brain and heart is severed, the heart beats more rapidly).

The experiment that first showed the effects of acetylcholine is described in the accompanying Research Report. The second set of neurons releases the neurotransmitter epinephrine, which accelerates the heart rate. These neurons become more active during physical exercise, when blood must flow more rapidly.

Epinephrine is secreted by the adrenal glands as well as by neurons. This hormone (also known by its brand name, adrenalin) travels through the bloodstream to the right atrium, where it accelerates the heart rate during physical exercise and emotional stress—fear, anger, and excitement.

Epinephrine from the adrenal glands (rather than from a neuron) travels throughout the body and alters the activities of many cells. It triggers the **fight-or-flight response,** which prepares the entire body for immediate action—to fight or run away—without going through a gradual warm-up phase. The response is initiated by emo-

Loewi removed the hearts from two frogs. The first heart, dissected with its neurons intact, was stimulated with an electric current applied to the neurons. The heart muscles responded by contracting more slowly. Some of the fluid surrounding this heart was then placed next to the second heart, in which the neurons had been removed. The heart muscles responded by contracting more slowly.

Loewi concluded that a chemical released from the first heart-neuron preparation controlled heart contractions in the second preparation, which consisted of a heart but not neurons. The chemical, now known as acetylcholine, was the first neurotransmitter to be discovered.

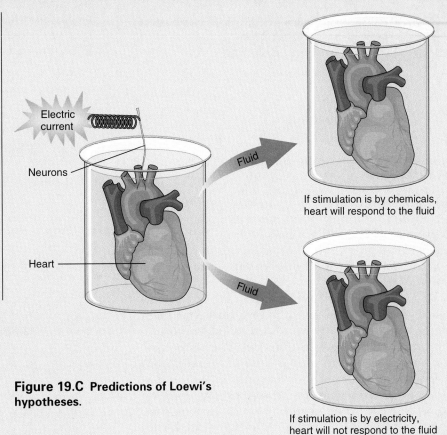

Figure 19.C Predictions of Loewi's hypotheses.

tional centers in the brain, which send messages to the medulla oblongata of the brain and the adrenal glands. These centers of neural and endocrine control then prepare the body for emergency action by raising the cardiac output and shunting the blood to more active organs.

Stroke Volume

As with heart rate, the stroke volume rises during times of strenuous exercise and emotional stress. The amount of blood pumped during each contraction depends on how much blood enters the ventricles and how completely the ventricles empty each time they contract.

The amount of blood entering the ventricles depends on the amount of blood that returns to the heart. When the body is at rest and the blood is moving slowly, approximately 60% of the blood is in the veins. When the body is more active, the veins constrict and blood passes

more rapidly through the veins so that more reaches the heart. The stroke volume then increases.

Heart muscles contract more forcefully, and empty their chambers more completely, when they are stretched. The muscles are stretched more when their chambers receive more blood from the veins, as described previously. In addition, stronger contractions are stimulated by signals from the medulla oblongata of the brain and the hormone epinephrine.

Blood Pressure

Blood pressure is the force exerted by blood on the wall of a blood vessel. The force is greater when more blood flows through the vessel and when its channel is narrower (fig. 19.10). When a person is resting, blood pressure

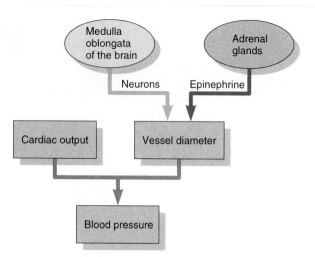

Figure 19.10 Control of blood pressure. For the factors that control cardiac output, see figure 19.9.

reflects the strength of the heart and the elasticity of the vessel walls. The resting blood pressure is kept within a narrow range of values by the cardiovascular center of the medulla oblongata (by way of neural signals to muscles in the heart and vessels) and by the adrenal glands (by way of epinephrine).

Blood pressure is measured from the main artery in the arm and expressed as the height in millimeters (mm) that the pressure will raise a column of mercury (Hg). Two measurements are taken: the systolic pressure, which is the maximum pressure during contraction of the left ventricle, and the diastolic pressure, the minimum pressure during relaxation of the left ventricle. The average blood pressure for men, taken while sitting, is about 120 mm Hg for systolic pressure and 80 mm Hg for diastolic pressure (abbreviated 120/80). The average pressure for women is somewhat lower, 110/70.[1]

Pressure Within the Vessels

Blood pressure is highest when blood leaves the left ventricle and decreases steadily as it travels farther from its pump:

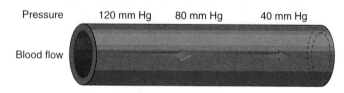

It loses pressure because its channel widens: the total width of all the arteries is smaller than the total width of all the capillaries. It is this difference in pressure, from arteries to capillaries, that propels the blood from the heart to the tissues and organs.

1. These average blood pressures are calculated from a wide range of pressures measured on a large number of people. It is unrealistic to expect your own blood pressure to be exactly the same as the average. What is important is how your normal blood pressure responds to changes in age and life-style.

The Arteries When the left ventricle contracts, blood leaves the heart under considerable pressure. When the left ventricle relaxes, no blood leaves the heart and the pressure is slight. The elasticity of the artery walls offsets these extreme changes in pressure. Elastic proteins within the connective tissue expand to accommodate the greater blood flow when the left ventricle contracts. They return to their original length when the ventricle relaxes and less blood flows out of the heart:

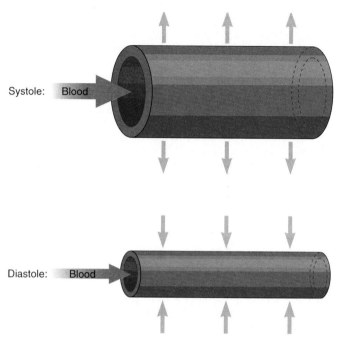

In this way, the arteries smooth out the pulsations of blood coming from the heart.

Blood pressure decreases and varies less as blood moves farther from the ventricles. By the time it reaches the capillaries, the pressure is about 18 mm Hg and the flow of blood through each capillary is sufficiently slow and steady for the exchange of materials with the interstitial fluid.

The Veins When the blood reaches the veins its pressure is only about 7 mm Hg—not high enough to establish a gradient of pressure that could propel blood back to the heart. Other forces are needed.

Blood is moved through veins by a squeeze-hold procedure: it is squeezed along by pressures exerted on the veins and then held in the new position, nearer the heart, by valves within the veins (fig. 19.11). The squeezing actions come from two sources: contractions of skeletal muscles and movements of the chest during breathing.

Skeletal muscles squeeze the veins when they contract. During normal movements of the body, the regular contractions of leg and arm muscles push blood toward the heart. (You feel faint when you stand motionless for a long time because your leg muscles are not contracting enough to squeeze the blood along.) An increase in physical activity, such as running, returns blood more rapidly to the heart and increases the cardiac output.

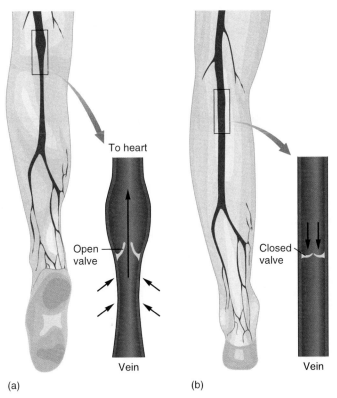

(a)   (b)

Figure 19.11 How blood moves through a vein. Blood is pushed through a vein by intermittent squeezings of the skeletal muscles and by breathing movements. (*a*) In the vein of a leg, blood is squeezed every time the calf muscles contract. Blood squeezed toward the heart presses against a one-way valve and opens it. Blood is allowed to flow in this direction through the valve. (*b*) When the calf muscles relax, backflow of blood against the valve closes it. The blood is held in its new position, nearer the heart.

Movements of the chest during breathing squeeze veins in the abdominal cavity and chest cavity. When air is inhaled, the expanding chest pushes downward against the abdomen, squeezing the veins in the abdominal cavity. When air is exhaled, the chest becomes smaller and squeezes veins in the chest cavity. During strenuous exercise, deeper, more rapid breathing moves blood more rapidly and increases cardiac output.

One-way valves in the veins ensure that blood flows toward the heart (fig. 19.11). When a vein is squeezed, blood presses against the valve and forces it open so that blood moves toward the heart. Between squeezes, blood flows back against the valve and closes it to prevent further backflow.

Varicose veins develop when one of these valves ruptures. This condition develops from poor blood flow, in which blood accumulates in a vein and places so much weight on a closed valve that it ruptures. Varicose veins develop most often in surface veins, which are not surrounded by skeletal muscle, in the legs when a person sits or stands for long periods of time. Similarly, hemorrhoids develop when blood stagnates within a vein near the anus or inside the rectum. As the vein expands with blood, a valve either ruptures or cannot close completely.

Table 19.3

Factors That Cause Abnormal Blood Pressures

Type of Blood Pressure	Causes
Chronic hypotension	Weakened heart muscles; damaged heart valves; malnutrition; slow leakage of blood
Sudden drop in blood pressure (cardiovascular shock)	Sudden blood loss; severe allergic reaction
Sudden rise in blood pressure	Physical exercise; emotions
Chronic hypertension	Unknown
Arteriosclerosis	Rigid or clogged arteries from hypertension, stress, drugs, diet, obesity, or lack of exercise

Low Blood Pressure

Low blood pressure may be normal for a person or it may indicate a health problem. Someone whose blood pressure has always been below average probably has an unusually strong heart and healthy arteries. Someone whose blood pressure has dropped below what is normal for them, however, may have a health problem—either hypotension or cardiovascular shock. These two conditions are summarized, along with the conditions of high blood pressure, in table 19.3 and described following.

Hypotension When blood pressure gradually drops below normal the condition is called **hypotension** (*hypo*, less than; *tension*, pressure). It is caused by a heart that has weakened as a pump and by a decrease in the amount of blood.

A weakened heart develops from deteriorated muscles or faulty valves. Heart muscles deteriorate from partially blocked coronary arteries, heart infections, and the aging process. The weakened muscles can no longer contract with enough force to build a normal blood pressure. A defective heart valve reduces the quantity of blood that is pumped by allowing blood to slosh back and forth within the heart.

The amount of blood within the vessels may be reduced gradually by malnutrition and blood loss. A low-protein diet can lower the concentration of plasma proteins, which normally hold water within the vessels. Water then moves by osmosis out of the blood and into the interstitial fluid, reducing the volume of blood. A slow leakage of blood from the vessels, as from a bleeding ulcer, also lowers the blood volume.

Cardiovascular Shock   A sudden drop in blood pressure, called cardiovascular shock, reflects a rapid loss of blood (from ruptured vessels, as in a wound) or a sudden dilation of all the arteries (from a severe allergic reaction). In either case, there is too little blood for the size of the vessels and there is much less pressure of blood against the vessel walls.

Cardiovascular shock is a serious condition that requires immediate medical care. The danger is that the heart may stop pumping blood. Very low blood pressure slows the circulation of blood so that the ventricles do not fill before they contract. Consequently, the heart muscles may not be stimulated to contract with enough force to pump blood to the capillaries.

(Emotional shock is an entirely different condition. It is part of a normal response in which the body prepares itself for emergency action. The blood pressure rises rather than falls during emotional shock, as described in the following section on high blood pressure.)

High Blood Pressure

High blood pressure may be acute (temporary) or chronic (long term). An acute rise in blood pressure is normal during times of physical exertion or extreme emotion. Chronic high blood pressure, which persists even during times of rest, is abnormal and unhealthy. These two kinds of high blood pressure are summarized in table 19.3 and discussed following.

Physical Exercise and Emotion   During times of physical exertion or emotional stress (the fight-or-flight response), the heart pumps faster and stronger, and blood is shunted from less active organs to the heart, brain, and skeletal muscles (fig. 19.12). More blood surges through narrower channels and the blood pressure rises. This temporary rise in blood pressure reflects an adjustment of the circulatory system for transporting materials more rapidly to the more active cells.

Hypertension   High blood pressure that does not go away when the body is at rest is called **hypertension** (*hyper,* more than; *tension,* pressure). It develops from excessive constriction of the arterioles. Severe hypertension (diastolic pressure greater than 115) is likely to cause heart failure or kidney failure. Mild hypertension, which usually has no symptoms, is likely to lead to arteriosclerosis.

Arteriosclerosis, or hardening of the arteries, develops when the arteries become too rigid and narrow to ease the pressure of blood pumped out of the heart. It may develop from hypertension, prolonged stress (the fight-or-flight response constricts the arteries), or habitual use of a stimulant—nicotine, caffeine, cocaine, or amphetamines. Most often, however, it develops when materials accumulate on the inner walls of the arteries, a particular form of arteriosclerosis known as **atherosclerosis.**

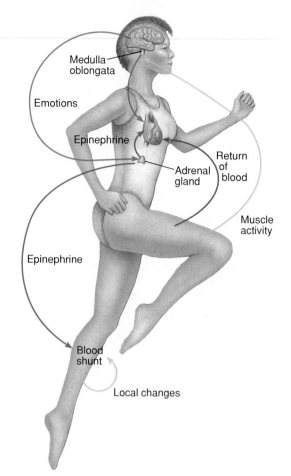

Figure 19.12 Response of the circulatory system to physical exercise. During physical exercise, the heart beats faster and with greater force in response to stimulation by the medulla oblongata (which monitors the products of muscle activity), by epinephrine from the adrenal glands (which react to emotions), and increased return of blood through the veins. Blood is shunted to the working muscles in response to stimulation by epinephrine and to local changes within the working muscles.

Atherosclerosis appears to develop when tiny packets of molecules, called low-density lipoprotein (LDL), deposit cholesterol and fats on the inner wall of an artery. This deposit, called a plaque, interferes with normal blood flow. The LDL may also stimulate cell division in smooth muscles within the artery wall, further blocking the flow of blood. The normal function of LDL is to transport cholesterol from liver cells to other body cells, where it is used in building membranes. (Cholesterol is not soluble in water and can be transported in blood only when wrapped in other materials—proteins and phospholipids in the case of LDL.) Once a plaque forms, it may trigger a blood clot that blocks the flow of blood. Or the plaque itself may grow until it obstructs the flow (fig. 19.13).

When atherosclerosis blocks a major artery, the organ that was nourished by the artery dies. A heart attack occurs when a coronary artery is blocked; it damages heart muscles. A stroke occurs when an artery that carries blood to the brain is blocked or ruptured; it damages neurons within a particular region of the brain. A stroke victim may expe-

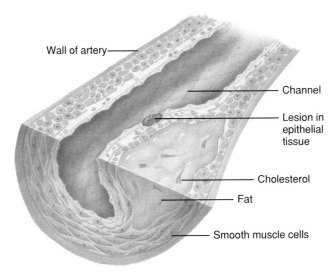

Wall of artery

Channel

Lesion in epithelial tissue

Cholesterol

Fat

Smooth muscle cells

Figure 19.13 Atherosclerosis. This disorder develops when a plaque grows on the inner wall of an artery. The plaque, which consists of cholesterol and fat sandwiched between layers of smooth muscles, grows until it blocks the flow of blood. Lesions may form within the epithelial tissue and trigger blood clots, which also block the flow of blood.

rience numbness or paralysis on just one side of the body, difficulty speaking or understanding speech, impaired vision in one eye, or dizziness. Blockage of arteries to various regions of the digestive tract also cause serious damage.

Arteriosclerosis is the major cause of death in America. A particularly disturbing fact is that three out of every four young American men already have this disease in its early form. (Young women are usually protected by estrogen, the female hormone.) While we do not know precisely what triggers the development of arteriosclerosis, we do know that it is correlated with aspects of our life-style. Relations between life-style and hardening of the arteries are listed in table 19.4.

SUMMARY

1. Most animals have internal transport systems that carry materials between their cells and the external environment. The transport system of humans is the circulatory system, which consists of blood, a heart, and a network of vessels. The blood carries a great variety of materials. Red blood cells transport oxygen. Anemia occurs when there is not enough red blood cells or hemoglobin.

2. The heart pumps blood into arteries and from arteries to capillary beds to veins and then back to the heart. Blood is shunted to areas of greatest need by dilating the arteries there and constricting the arteries in other areas.

3. The heart consists of two muscular pumps: a right side that pumps blood to the lungs and a left side that pumps blood to the rest of the body. Blood flows

---

Table 19.4

Aspects of Life-style and Their Relations to Arteriosclerosis

Aspect of Life-style	Effect on the Arteries
Diet high in saturated fats (e.g., animal fats)	Raises LDL concentration, which initiates plaque formation
Artificial stimulants (especially nicotine)	Constricts arteries until they can no longer dilate
Obesity	Keeps cardiac output high, which strains the arteries
Emotional stress	Constricts arteries until they can no longer dilate; keeps cardiac output high, which strains the arteries; releases chemicals that damage arteries
Salty food	May increase blood volume, which raises blood pressure and strains the arteries
Too little fiber in the diet	Causes plaques by not removing cholesterol from the body
Too little physical exercise	Causes plaques by decreasing HDLs and increasing LDLs;* promotes obesity

*HDLs are high-density lipoproteins, which carry cholesterol away from the artery walls to the liver for disposal. LDLs are low-density lipoproteins, which deposit cholesterol and fats on the artery wall.

through the right heart along the following pathway: inferior and superior venae cavae, right atrium, tricuspid valve, right ventricle, pulmonary valve, pulmonary artery. Blood flows through the left heart along the following pathway: pulmonary veins, left atrium, bicuspid valve, left ventricle, aortic valve, aorta.

4. The sounds of the heart are vibrations from the valves as they close. Heart murmurs are unusual sounds from defective valves.

5. Coronary arteries carry blood to the heart muscles. A coronary heart attack occurs when one of these arteries is blocked.

6. Cardiac output is the quantity of blood pumped by the heart each minute. It depends on the heart rate and stroke volume. Heart rate is controlled mainly by the pacemaker but also by the brain and adrenal glands. Stroke volume depends mainly on how rapidly blood in veins is returned to the heart.

7. Blood pressure decreases as blood moves from the heart to the capillaries. It is too low in the veins to move blood back to the heart. Blood is moved through veins by intermittent squeezings, from skeletal muscles and breathing movements, and is held by one-way valves.

8. Persistent low blood pressure, called hypotension, is caused by heart defects, malnutrition, and bleeding. Sudden low blood pressure, called cardiovascular shock, is caused by rapid loss of blood or extreme dilation of arteries during an allergic reaction.

9. Blood pressure rises during physical exercise and the fight-or-flight reaction as the cardiac output increases and blood is shunted to more active organs. Chronic high blood pressure, called hypertension, may lead to arteriosclerosis, in which the arteries are rigid and narrow. In atherosclerosis, a form of arteriosclerosis, cholesterol and fats are deposited on the inner walls of arteries.

KEY TERMS

Anemia (ah-NEE-mee-ah) (page 356)
Arteriosclerosis (ar-TEER-ee-oh-sklur-OH-sis (page 368)
Atherosclerosis (AEH-thur-oh-sklur-OH-sis) (page 368)
Blood pressure (page 365)
Cardiac output (page 362)
Cardiovascular center (page 364)
Cardiovascular (KAHR-dee-oh-VASS-kyoo-lahr) system (page 356)
Coronary (KOR-oh-nah-ree) artery (page 362)
Fight-or-flight response (page 364)
Heart rate (page 362)
Hemoglobin (HEE-moh-GLOH-bin) (page 356)
Hypertension (HEYE-pur-TEN-shun) (page 368)
Hypotension (HEYE-poh-TEN-shun) (page 367)
Pacemaker (page 362)
Stroke volume (page 362)

STUDY QUESTIONS

1. Why do most animals need a system of internal transport?
2. List eight kinds of materials carried by the human circulatory system.
3. Define anemia. Give four causes of anemia and explain how each gives rise to this disorder.
4. Describe a blood shunt. What is its function?

5. Draw a heart, showing the positions of the left ventricle, right ventricle, left atrium, right atrium, inferior vena cava, superior vena cava, pulmonary veins, aorta, and pulmonary artery. Indicate with arrows the direction of blood flow through each part of the heart.
6. A heart murmur is most serious when it is the bicuspid valve that is defective. Explain why, using a diagram to show blood flow through the heart.
7. Describe the two factors that determine cardiac output.
8. What causes blood to flow from a ventricle to a capillary bed?
9. What causes blood to move from a capillary bed in a foot to the right atrium?
10. What is cardiovascular shock? What are two situations in which this type of shock develops?

CRITICAL THINKING PROBLEMS

1. Estimate the number of times that your heart contracted in the time it took you to read this chapter.
2. How does blood contribute to homeostasis?
3. Explain why an existing anemia might become apparent only after arriving in the mountains for a skiing vacation.
4. Why are veins blue?
5. What can a physician discover about your health by listening to the sounds of your heart?
6. Why do soldiers sometimes faint when standing at attention?
7. Why do the feet and ankles of older people often swell?
8. What changes in life-style could you make to improve your cardiovascular fitness?

SUGGESTED READINGS

Introductory Level:

Duke, M. *The Development of Medical Techniques and Treatments: From Leeches to Heart Surgery.* Madison, CT: International University Press, 1991. A fascinating history of the treatment of heart disease, written for the layperson and well referenced.

Moore, T. J. *Heart Failure.* New York: Simon and Schuster, 1989. A critical analysis of modern treatments of heart failures, including research methods and hospital treatments.

Punchot, R. (ed.). *Blood: The River of Life*. New York: Torstar Books, 1985. A medically oriented description of the blood—its functions, disorders, and capacity to self-heal.

Vogel, S. *Vital Circuits: On Pumps, Pipes, and the Workings of the Circulatory System*. New York: Oxford University Press, 1992. A remarkably clear account of how the circulatory system works.

Advanced Level:

Milnor, W. R. *Cardiovascular Physiology,* New York: Oxford University Press, 1990. A good reference for information on the circulatory system.

Schmidt-Nielsen, K. *Animal Physiology: Adaptation and Environment*. New York: Cambridge University Press, 1990. Good comparisons of circulatory systems in a wide variety of animals.

❋ EXPLORATION

Interactive Software: Evolution of the Human Heart

This interactive exercise explores evolution of the vertebrate heart. The program provides a diagram of a fish heart—a simple structure with one atrium and one ventricle. The students modify this heart by forming partitions within each atrium and ventricle (as occurred during evolution from fishes to mammals) and observe how these partitions affect the animal's blood pressure and oxygen delivery to body cells.

Respiration

Objectives

In this chapter, you will learn
−why your lungs are large and wet;
−why you must breathe continually;
−the structure of lungs;
−what causes inhalation and
  exhalation;
−how breathing rate is regulated;
−the structure of air passages to the
  lungs.

A swimmer is acutely aware of the
need to breathe.

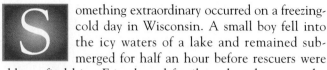omething extraordinary occurred on a freezing-cold day in Wisconsin. A small boy fell into the icy waters of a lake and remained submerged for half an hour before rescuers were able to find him. Friends and family gathered next to the lake, hoping for a miracle yet knowing there was little chance the child was alive. When pulled from the water, the boy was gray and had no detectable heartbeat. Everyone was sure he was dead. But as he warmed up, he began to show signs of life. Miraculously, the long period of time without air had not killed him.

Even more extraordinary, when he regained consciousness it was clear that he had suffered no brain damage. Ordinarily, the brain is permanently damaged when deprived of oxygen for only a few minutes. Yet the boy had not breathed air for thirty minutes. His complete recovery puzzled the medical specialists.

After studying the problem, medical researchers concluded that the boy adjusted to submersion in the same way that a marine mammal (seal, whale, or dolphin) adjusts to its deep dives. These animals hold their breaths during their dives, which typically last between fifteen minutes to an hour. They remain physically and mentally active, hunting for food, even though their lungs do not take in oxygen while they are under water.

A diving mammal experiences what is called a dive reflex: the heart rate slows so that the heart muscles require less oxygen. The blood is redistributed so that the essential organs—the brain and heart—receive most of its oxygen. Blood flow to other parts of the body decreases, causing these parts to cool and become less active, thereby requiring less oxygen.

It turns out that humans, too, have a dive reflex that is especially pronounced in young children. When the face is immersed in cold water, the heart slows down and most of the blood is shunted to the heart and brain. The body then requires very little oxygen. And that is what saved the boy during his long "dive" into the icy lake.

Gas Exchange

Respiration contributes to homeostasis (maintaining a constant internal environment) by bringing oxygen into an animal and moving carbon dioxide out. The movement of oxygen and carbon dioxide through a body surface is known as **gas exchange.**

Every animal needs some method of gas exchange to sustain cellular respiration (see page 88), the biochemical reaction that releases energy from organic molecules. Animal cells get most of their energy from aerobic cellular respiration, which requires oxygen to proceed and forms carbon dioxide as a waste product:

$$\text{Organic molecule} + O_2 \longrightarrow CO_2 + H_2O + \text{Energy}$$

Urgency of Gas Exchange

Life is sustained by energy. A cell dies when it no longer receives energy from cellular respiration. Cellular respiration, in turn, requires a steady inflow of oxygen and outflow of carbon dioxide.

Oxygen cannot be stored within the body; it must be continuously taken into the body from the external environment. A nerve cell, for example, dies within a few minutes when deprived of oxygen.

Animals must eliminate carbon dioxide as rapidly as it forms during cellular respiration. This gas is lethal in high concentrations because it makes the body fluids too acidic. Carbon dioxide combines with water to form carbonic acid:

$$CO_2 + H_2O \longrightarrow \underset{\substack{\text{Carbonic} \\ \text{acid}}}{H_2CO_3}$$

Carbonic acid cannot hold onto both hydrogen ions; it immediately loses one to form a bicarbonate ion:

$$H_2CO_3 \longrightarrow \underset{\substack{\text{Hydrogen} \\ \text{ion}}}{H^+} + \underset{\substack{\text{Bicarbonate} \\ \text{ion}}}{HCO_3^-}$$

The increased concentration of hydrogen ions makes fluids within the cells more acidic, a change that interferes with the activities of enzymes. Biochemical reactions then shut down and the cells die.

Surfaces for Gas Exchange

Carbon dioxide and oxygen move into and out of the body by diffusion, the process that moves material from regions of high concentration to regions of low concentration. Oxygen gas is always less concentrated inside the body than outside, because it is incorporated during cellular respiration into water molecules (H_2O) and thus is no longer in the form of oxygen gas (O_2):

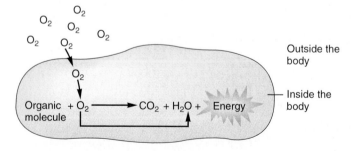

Therefore, oxygen tends to enter the body by diffusion.

Conversely, carbon dioxide is formed during cellular respiration and so is more concentrated inside than outside the body:

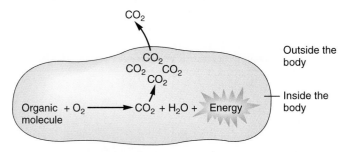

Therefore, carbon dioxide tends to leave the body by diffusion.

Gas exchange is urgent, yet gases cross cellular membranes by the slow process of diffusion. Only certain kinds of surfaces can help gases move into and out of the body fast enough to support cellular activities. Gas-exchange surfaces, whether they are the skins of earthworms, the gills of fish, or the lungs of humans, must be large, moist, and ventilated. And in most animals these surfaces are closely linked with a circulatory system.

Large Surface   Diffusion proceeds slowly. In order to take in large amounts of oxygen and get rid of large amounts of carbon dioxide, diffusion must occur in many places at the same time. Thus, the surface through which gases diffuse must be large.

Most gas-exchange surfaces contain tiny depressions or projections to enlarge the area of diffusion and yet not take up a lot of space within the body. The larger the gas-exchange surface, the greater the volume of gases exchanged and the more active the animal can be.

Wet Surface   Gases diffuse through cellular membranes only when they are dissolved in water. For this reason, the cells that form gas-exchange surfaces must be kept wet at all times.

Animals that live in water have no difficulty maintaining a wet skin or gills for gas exchange. Animals that live on land, however, require special adaptations for maintaining wet surfaces in the dry air. Their gas-exchange surfaces are folded inside the body and connected to the outside air by narrow channels:

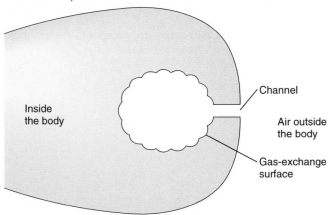

Air within these pouches is kept humid so that a thin film of water covers the gas-exchange surface. A land animal loses (and must replenish) some of this water every time it exhales. You see this water vapor in your own breath on a cold day and when you breathe on a mirror.

Ventilated Surface   Diffusion proceeds more rapidly when the concentration gradient is steeper: when the concentration of molecules is much higher on one side of the gas-exchange surface than on the other side. To maintain steep gradients of oxygen and carbon dioxide, the air or water next to the gas-exchange surface is changed regularly (fig. 20.1). Fresh air or water brought to the gas-exchange surface raises the concentration of oxygen gas, so that oxygen diffuses rapidly into the body. Used air or water carried away from the surface lowers the concentration of carbon dioxide, so that carbon dioxide diffuses rapidly out of the body.

This process of moving air or water over a gas-exchange surface is known as **ventilation.** Without ventilation, the gases would eventually become equally distributed on both sides of the surface and diffusion would cease.

Animals ventilate in two basic ways: (1) by moving their bodies through the surrounding water or air, away from areas where they have picked up oxygen and released carbon dioxide, or (2) by moving the surrounding water or air over their gas-exchange surfaces. Few animals use the first alternative, for to do so requires continuous locomotion. A tuna, for example, cannot pump water over its gills; it ventilates by swimming and will suffocate if forced to remain motionless. Most aquatic animals pump water across gills. Terrestrial vertebrates pump air into and out of lungs.

Linked with a Circulatory System   An animal uses diffusion to move gases only over very short distances—less than 1 millimeter—because the process is too slow over long distances. For longer distances, an animal transports gases with a circulatory system.

A circulatory system consists of blood, one or more hearts, and a network of vessels. A heart pumps blood through vessels that permeate all regions of the body. Some of the vessels lie next to the gas-exchange surface. There, oxygen gas diffuses from the air or water into the blood and carbon dioxide diffuses from the blood into the surrounding air or water. Other vessels lie next to body cells, where oxygen diffuses from the blood into the cells and carbon dioxide diffuses from the cells into the blood. The blood transports the gases rapidly between the gas-exchange surface and the body cells (fig. 20.2).

The Human Respiratory System

The human respiratory system is a group of organs that work together to bring oxygen into the body and carry carbon dioxide out of the body. The system consists of the

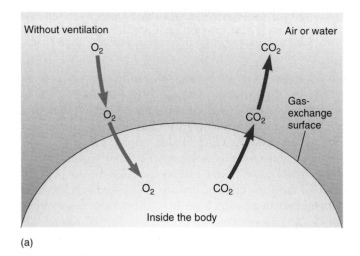

(a)

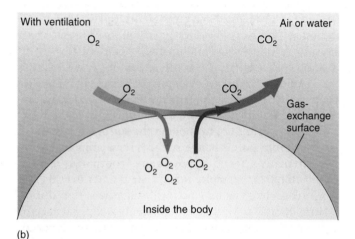

(b)

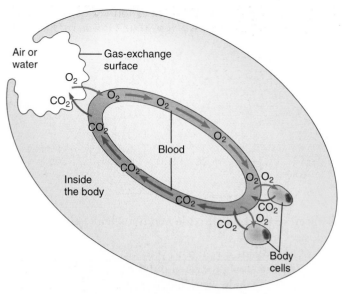

Figure 20.2 Gas-exchange surfaces and the circulatory system. In many animals, the gas-exchange surface is linked with a circulatory system for more rapid transport of the gases. Blood carries oxygen from the gas-exchange surface to the body cells, where it is used in cellular respiration. Blood carries carbon dioxide from the body cells, where it forms during cellular respiration, to the gas-exchange surface.

Figure 20.1 Ventilation. Ventilation maintains steep concentration gradients of gases by moving air or water over the gas-exchange surface. (*a*) Without ventilation, oxygen that diffuses into the body is replaced very slowly, by diffusion from other regions of the air or water to the gas-exchange surface. Also, carbon dioxide that diffuses out of the body leaves the gas-exchange surface very slowly, by diffusion to other regions of the air or water. The concentrations of these two gases are similar on both sides of the gas-exchange surface. (*b*) With ventilation, oxygen moving into the body is replaced rapidly, and carbon dioxide leaving the body is carried away rapidly. Steep gradients of concentration are maintained, and gases diffuse rapidly into and out of the body.

lungs, pleural membranes, diaphragm, rib cage, nose, pharynx, trachea, bronchi, and bronchioles. These organs are shown in figure 20.3 and listed, along with their functions, in table 20.1.

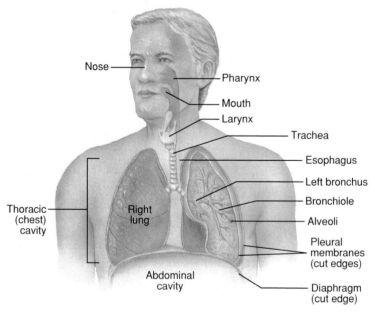

Figure 20.3 The human respiratory system. The left lung is cut away to show internal structures.

Table 20.1

Organs of the Respiratory System

Organ	Function
Lungs (alveoli)	Surfaces for gas exchange
Pleural membranes	Adhere lungs to inner wall of chest and diaphragm
Rib cage and its muscles	Ventilate lungs by making them larger (draws air in) and smaller (pushes air out)
Diaphragm	Ventilates lungs by making them larger (draws air in) and smaller (pushes air out)
Nose	Opening to respiratory tract; cleans, moistens, and warms the air
Pharynx	Channel between the nose and larynx; prevents food from entering the larynx and nose
Larynx	Channel between the pharynx and trachea; specialized for sound production
Trachea	Channel between the larynx and bronchi; cleans and moistens the air
Bronchi	Channels between trachea and bronchioles
Bronchioles	Channels between the bronchi and alveoli of the lungs

The Lungs

The human lungs are gas-exchange surfaces folded into the cavity of the chest. There the moist, delicate tissues are protected from dehydration and injury. Internal gas-exchange surfaces, such as lungs, adapt an animal to life on land.

There are two lungs—two separate "pouches," one on the right and one on the left side of the body. The right lung is partially divided into three sections, called lobes, and the left lung is partially divided into just two lobes to make room for the heart.

The gas-exchange surfaces of the lungs are thin sheets of densely packed cells (epithelial tissue; □ *see page 316*) supported by a network of elastic fibers. The fibers are long protein molecules that shorten after being stretched, enabling the lungs to change size so that air can be drawn in and pushed out. The tissue is indented to form 300 million or so tiny depressions called **alveoli** (fig. 20.4*a*). The total surface area of all the alveoli in both lungs is roughly the size of a tennis court!

On the internal side of the lungs, facing the body cells, the alveoli are covered with a dense network of capillaries (blood vessels; □ *see page 357*), as shown in figure 20.4*a*. These vessels receive blood from the right ventricle of the heart, via the pulmonary artery. After the blood has picked up oxygen and released carbon dioxide at the alveoli, it returns directly to the heart by way of the pulmonary veins (fig. 20.4*b*). The walls of the capillaries and the tissue of the alveoli are so thin that air inside the lungs is separated from blood inside the vessels by a distance less than the width of a cell (fig. 20.5).

The immense surface area of the lungs and the immense network of capillaries in contact with the lungs compensate for the slowness of diffusion. Each gas molecule diffuses slowly, but billions of gas molecules are diffusing at the same time and over tiny distances between the air and the blood.

On the external surface of the lungs, facing the air-filled cavity, the alveoli are moistened by a thin film of liquid. The liquid contains a **surfactant,** which is a material that prevents wet surfaces from sticking together. Without the surfactant, the alveoli stick together and the lung cavity cannot fill with air. This often happens in babies who are born prematurely, before they begin to synthesize surfactant. The lungs of such babies are now sprayed with a synthetic version of surfactant.

Lungs are delicate tissues. While their location protects them from most types of damage, they are vulnerable to airborne materials and disease organisms. The effects of polluted air on our lungs are examined in the accompanying Research Report.

Breathing

Human lungs are ventilated by breathing, in which air flows into and out of the lungs. About ten to fourteen times a minute, the lungs expand and fresh air is drawn in. After each expansion, the lungs shrink and "used" air is

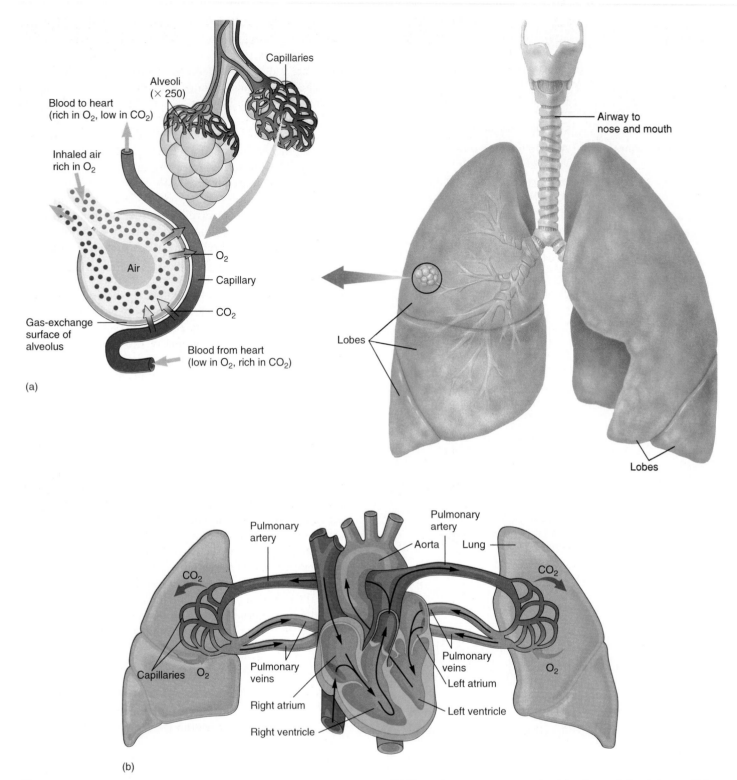

Figure 20.4 Alveoli. (*a*) Lung tissue is shaped into alveoli, which are tiny cup-shaped depressions, and surrounded by a network of capillaries. Gases diffuse across the short distance between air within the alveoli and blood within the vessels.

(*b*) The pulmonary artery carries blood out of the right ventricle. It then bifurcates to form two arteries that carry blood to the two lungs. Blood travels through four pulmonary veins from the lungs to the left atrium of the heart.

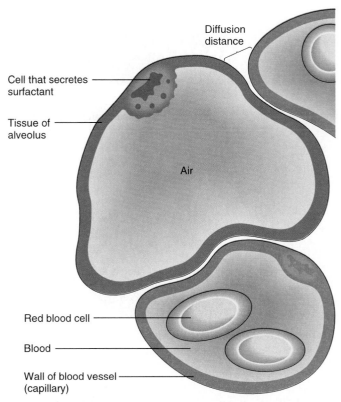

Figure 20.5 Air in the alveoli is closely associated with blood in the blood vessels. The distance between the air and the blood is less than the width of a blood cell. The tissue of an alveolus includes cells that secrete surfactant, a material that prevents the tissues from sticking together.

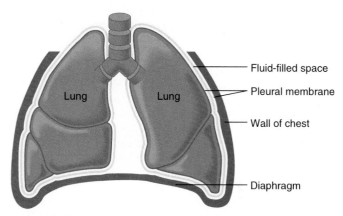

	%O$_2$	%CO$_2$	%N$_2$	%H$_2$O
Inhaled air	20.71	0.04	78.00	1.25
Exhaled air	14.60	4.00	75.50	5.90

Table 20.2

Compositions of Inhaled Air and Exhaled Air

Figure 20.6 The pleural membranes. The pleural membranes are two sheets of cells (shown in yellow), one enclosing each lung. Each membrane is folded back upon itself. Because the membrane is covered on both sides with a thin layer of fluid, it sticks to the lungs, to itself, and to both the diaphragm and the inner wall of the chest. In this way, the lungs are attached to the wall of the chest and the diaphragm. (The two folds of each membrane are pressed tightly against each other as well as to the lungs, the chest wall, and the diaphragm. They do not form such large spaces as shown here.)

pushed out. In this way, air within the lungs is replaced on a regular basis. The compositions of inhaled air and exhaled air are shown in table 20.2.

The lungs expand and shrink in response to changes in the size of the chest cavity. The lungs inflate as the chest enlarges because they adhere to the chest wall, held by a thin, moist membrane called the **pleural membrane.** (This membrane is a sheet of cells, whereas a cellular membrane is a sheet of molecules.) The membrane is folded so that one part covers the internal surface of the chest cavity and the other part covers the internal surface of the lung (the side next to the body cells), as shown in figure 20.6. These two portions of the membrane, both wet and without surfactant, stick together yet slide smoothly over each other as the chest and lungs move during breathing.

Inhalation During inhalation (breathing in), air is drawn into the lungs as the chest cavity enlarges (fig. 20.7a). Two sets of muscles are involved. One set forms the **diaphragm,** a sheet of muscles that separates the chest cavity from the

abdominal cavity. At rest, the diaphragm is dome-shaped and curves upward into the chest cavity. When its muscles contract, the dome moves downward to enlarge the chest from top to bottom. The other set of muscles lies between the ribs. Contractions of these muscles raise the ribs upward and outward, enlarging the chest from front to back. As the chest and lungs expand, air is pulled into the cavities of the lungs.

Exhalation During quiet breathing, exhalation (breathing out) is passive and requires no muscle contractions. The muscles of the diaphragm and ribs simply relax, causing the diaphragm to move upward and the ribs to move downward and inward (fig. 20.7b). The tissues of the

Tobacco Smoke, Smog, and the Lungs

We imagine ourselves growing old with fewer disabilities than our grandparents because of advances in modern medicine. However, we may develop one disability that our grandparents did not—lung damage from smog. Hazy, brownish air is a relatively recent phenomenon and one that will affect our health in ways we do not yet understand.

While everyone believes that smog is bad for us, we still do not know exactly what it does to cells within the lungs or how long the effects last. A working hypothesis is that smog affects our lungs in the same way as tobacco smoke.

Tobacco smoke damages the delicate tissues of the lungs (fig. 20.A). Secondhand smoke, from other smokers, turns out to be almost as deadly as firsthand smoke. Lung disorders, the third leading cause of death in America, include lung cancer as well as bronchitis and emphysema, which are described following.

Tobacco smoke destroys the cilia that move mucus upward through the air passages to the nose and mouth. Without these structures, mucus accumulates within the channels and breathing—especially exhaling—becomes extremely difficult. This condition is called bronchitis (the plugged channels are the bronchioles, shown in fig. 20.3). Air then becomes trapped within the lungs and the alveoli balloon out from the pressure. In time, the alveoli lose their elasticity and may burst. This more advanced condition, in which the alveoli lose their elasticity, is called emphysema. Bronchitis and emphysema generally occur together and the condition worsens as smoking (or inhaling secondhand smoke) continues. Breathing becomes increasingly more difficult, and the brain may be damaged from an inadequate supply of oxygen.

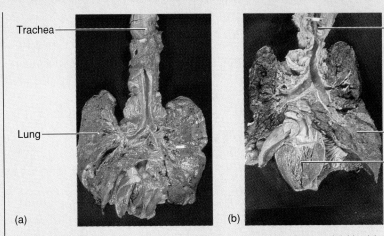

Figure 20.A Effects of tobacco smoke on human lungs. (*a*) Healthy lungs of a nonsmoker. (*b*) Damaged lungs and heart of a heavy smoker.

How does smog affect our lungs? Evidence suggests that at least one ingredient in smog—ozone—acts just like tobacco smoke in causing bronchitis and emphysema. Ozone (O_3) is a major ingredient in photochemical smog, the kind that forms from a combination of automobile exhaust and sunlight. It is especially abundant over cities on the Atlantic and Pacific coasts. (Ozone that forms a layer approximately 24 kilometers (15 mi) above the earth's surface, however, is beneficial: it reduces the amount of ultraviolet radiation that reaches the earth's surface.)

The strongest evidence in support of the hypothesis that smog affects lungs in the same way as tobacco smoke comes from an experiment with rats. In this experiment, conducted by biologists in the Environmental Protection Agency, rats were exposed to conditions that simulated the conditions experienced by people who live most of their lives in areas polluted with ozone. The rats breathed air that contained realistic levels of ozone (the amount present on an average day in Los Angeles) for eighteen months, which is approximately two-thirds of a rat's lifetime. The results of the experiment supported the hypothesis: all the rats developed bronchitis and emphysema, and the damage to their lungs worsened with age. Their old age was not a pleasant one.

Indirect evidence (other than experimental) of lung damage from ozone includes the following:

1. One-fourth of the young people (between fifteen and twenty-five years of age) who live in Los Angeles County and 80% of the children who live in inner regions of the city itself have damaged lungs, a condition that was rare forty years ago.
2. The lungs of people exposed to high levels of ozone retain more tiny particles (the particulates of air pollution) than the lungs of people in cleaner areas.
3. The death rate (largely from respiratory and heart ailments) in the smoggiest city (Steubenville, Ohio) is 26% higher than in the cleanest cities (Topeka, Kansas, and Portage, Wisconsin).

If it is true that smog permanently damages our lungs and that the damage accumulates from year to year, then the consequences are profound. More than half of all Americans live in cities that exceed the health standard for ozone or particulates. These people, like the rats in the experiment, may develop severe and incurable breathing difficulties as they grow older.

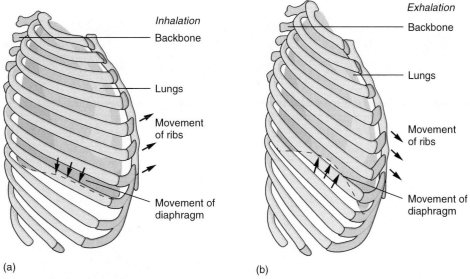

(a)   (b)

Figure 20.7 Breathing movements. Breathing alters the size of the lungs by altering the size of the chest cavity. (*a*) During inhalation, muscles between the ribs contract, lifting the ribs upward and outward. The chest cavity enlarges from front to back. At the same time, the diaphragm contracts, moving downward this partition between the cavities of the chest and abdomen. The chest enlarges from top to bottom. (*b*) During exhalation, muscles between the ribs relax and the ribs drop downward and inward. The chest cavity becomes smaller from front to back. At the same time, muscles of the diaphragm relax and this partition moves upward. The chest cavity becomes smaller from top to bottom.

lungs, made in part of elastic fibers, shrink (like a balloon) when they are no longer stretched. As the lung cavities become smaller, air is forced out (just as air moves out of a balloon).

During vigorous exercise, exhalation is an active process that involves muscle contractions. Contractions of the abdominal muscles compress the abdominal cavity, which forces the diaphragm rapidly upward. Contractions of muscles between the ribs (a different set from the muscles used in inhalation) pull the ribs rapidly downward and inward. The chest is rapidly reduced in size and air is rapidly forced out of the lungs.

Artificial Respiration   Breathing movements are so simple and mechanical that they can be simulated. Artificial respiration is an effective means of ventilating the lungs when natural mechanisms fail. Emergency methods for restoring breathing and for removing obstructions in the air passages are described in the accompanying Wellness Report.

Regulation of Breathing

Normal breathing is an involuntary action, automatically adjusted to suit the body's need for gas exchange. We do not have to remember to inhale and exhale. Our minds can, however, exert control over our breathing. We would not be able to speak, sing, swim, or play a clarinet without this control.

The Basic Rhythm of Breathing

Breathing is governed by clusters of neurons, called the respiratory center, within the **medulla oblongata** of the brain (see fig. 17.11). The center generates the basic rhythm of quiet breathing—about ten to fourteen breaths per minute—by sending commands along neurons to muscles of the diaphragm and rib cage. The muscles respond by contracting, and an inhalation occurs. When the commands cease, the muscles relax and an exhalation occurs.

(The respiratory center is depressed and the breathing rate slowed by barbituates [in some sleeping pills] and morphine. An overdose of these drugs may depress the center to such an extent that breathing stops entirely. There is no antidote to barbituates or morphine. Life can be sustained under these conditions only by artificial respiration, which must be continued for several hours until the drug wears off.)

Adjusting the Breathing Rate and Depth   The respiratory center within the medulla oblongata adjusts breathing to different levels of activity: it lowers the rate during sleep and raises the rate and depth during times of physical exercise, fever, and intense emotions. (The function of a yawn—a very deep inhalation—is not understood. It appears to function as a muscle stretch and may promote arousal in situations where an animal needs to stay alert.)

The respiratory center monitors levels of cellular activity primarily by keeping track of how much carbon dioxide (or H^+, which forms when CO_2 dissolves in water) is in the blood. More active cells release more carbon dioxide because they have higher rates of cellular respiration. During vigorous exercise, the respiratory center receives information from the muscles and joints as well. It also receives information from emotional centers within the brain during the fight-or-flight response. The respiratory

Life-Saving Techniques

Everyone should know how to save the life of a person who suddenly has ceased to breathe or to pump blood. Once the lungs or heart have stopped working, you have just four minutes to get them started again—after four minutes there is likely to be permanent brain damage followed shortly thereafter by death. While almost always, the first thing to do is to call for medical help (by dialing 911), it is unlikely that help will arrive within four minutes. For this reason, you should know how to perform life-saving techniques while you wait for a professional to arrive.

Life-saving techniques are mastered most effectively by taking a course in first aid, and we recommend strongly that you take such a course. The techniques are outlined here because you already have sufficient knowledge, from studying this chapter and the one on circulation, about the mechanics of breathing and heart contractions to prolong life under emergency conditions.

Breathing is stopped by a variety of events: choking, drug overdose, suffocation, drowning, gas poisoning, electric shock, heart attack, and injuries to the head or chest. Three procedures are used, depending on the problem, to initiate breathing: the Heimlich maneuver, mouth-to-mouth resuscitation, and cardiopulmonary resuscitation.

The Heimlich maneuver is applied when someone is *choking*—that is, cannot inhale or exhale because an object, usually a piece of food or vomit, is blocking the windpipe (or trachea). Signs of this condition include sudden inability to speak, blue skin, choking gesture (hand held to the throat), and physical agitation. The Heimlich maneuver uses air already in the lungs to blow the object out through the mouth. Position yourself behind the individual and wrap your arms around him or her above the waist. Place a fist

between the breastbone and navel, with the thumb against the person's body. Grab your fist with your other hand and thrust quickly and firmly inward and upward:

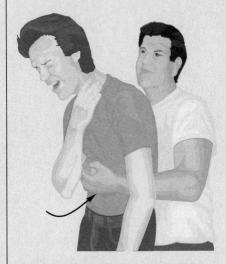

Repeat until the food or object is dislodged. (Do not apply this procedure to children less than one year old, who need different emergency care when choking.) If you are alone and choking, press your abdomen (between the breastbone and navel) into a table or sink.

Simple mouth-to-mouth resuscitation is used when a person has *stopped breathing* but is not choking and still has a heartbeat. (You should be aware that some microbes, including the virus that causes AIDS, can be

transferred in saliva during this procedure.) First, clear the mouth of vomit, blood, loose teeth, food, or chewing gum. Next, straighten the airway by tilting the chin upward while pressing on the forehead:

Then begin mouth-to-mouth resuscitation to force air into the person's lungs. (Your exhaled breath still contains a lot of oxygen gas, as shown in table 20.2, and plenty of carbon dioxide to stimulate the respiratory center.) Pinch the nose closed and cover the mouth with your own:

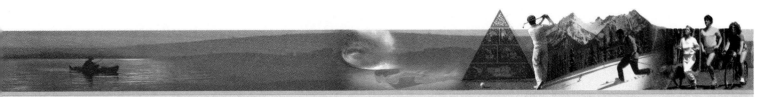

Blow air into the mouth until you see the chest rise. (Blow less air into a child, for a child's lungs are much smaller than your own.) Remove your mouth and allow the victim to exhale passively. Repeat the cycle every five seconds (about twelve times a minute—the same rate as breathing occurs naturally).

Cardiopulmonary resuscitation (CPR) is applied when *both breathing and the heartbeat* have stopped. Feel for a pulse to determine whether or not the heart is beating. The best place to do this is in the carotid artery (the large artery along either side of the windpipe just beneath the jaw). If you feel no pulse, be sure that you call 911 immediately, before beginning CPR, since effective treatment may require special machines.

Cardiopulmonary resuscitation has two components: mouth-to-mouth resuscitation (described previously) to restore breathing and chest compressions to restore the heart contractions. The chest compressions are done as follows:

1. Place the individual on his or her back and kneel alongside. Find the lower tip of the breastbone. Two fingerbreadths above this point, place the heel of one hand, with the other hand on top:

2. With your body directly over your clasped hands, push down 80 to 100 times a minute—slightly faster than the normal heart rate—to force blood through the heart. Press hard enough to depress the breastbone about 3 to 4 centimeters (1.5 to 2 in) (less in a child) (*right*):

If you are alone with the individual, depress the chest fifteen times and then inflate the lungs twice (by mouth-to-mouth resuscitation). If someone else is available to assist you, have that person inflate the lungs once for every eight times you compress the chest. Repeat the sequence until normal breathing is restored or professional help arrives.

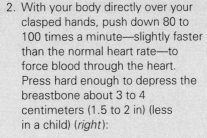

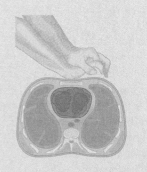

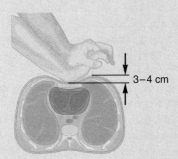

3–4 cm

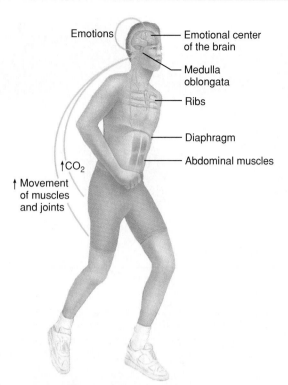

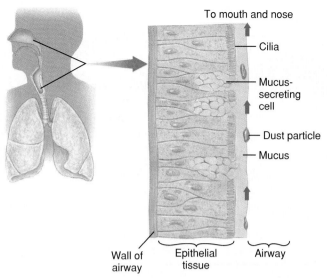

Figure 20.8 Regulation of breathing rate. Breathing is regulated by a respiratory center in the medulla oblongata of the brain. Neurons within this center detect higher levels of physical activity by monitoring concentrations of carbon dioxide, strain on the muscles and joints, and emotions generated elsewhere within the brain. The neurons respond by sending signals to muscles of the rib cage, diaphragm, and abdomen, which initiate faster and deeper breathing.

Figure 20.9 The mucus escalator. The upper respiratory tract is lined with epithelial tissue. Some of the cells within this tissue secrete mucus and some are covered with cilia. Foreign particles that enter the tract stick to the mucus, which is propelled continuously toward the nose and mouth by wavelike movements of the cilia. Together, the mucus and cilia are known as the "mucus escalator."

center responds to all this information by adjusting the rate and depth of breathing (fig. 20.8).

The Upper Respiratory Tract

Air enters and leaves the lungs through a series of channels called the upper respiratory tract (see fig. 20.3). Epithelial tissue, which forms the inner lining of the tract, modifies incoming air to make it more suitable for gas exchange. Some of the cells in this tissue secrete **mucus,** a sticky fluid that traps dust, small insects, and other debris that enters the respiratory tract. Other cells have hairlike projections, called **cilia,** that wave in synchrony and propel the debris-covered mucus toward the nostrils or mouth, where it can be blown out or swallowed. Together the mucus and cilia form a "mucus escalator," as shown in figure 20.9.

The Nose Air enters and leaves the respiratory tract through the nose or mouth. When resting, all mammals (except some humans) breathe through their noses, as this part of the respiratory system modifies air prior to contact with the delicate tissues of the lungs.

The nose cleans, moistens, and warms the incoming air. The nostrils, or openings to the nose, are lined with hairs that trap foreign materials in the air. The cavities inside the nose are covered with mucus and cilia to further cleanse the incoming air. Air passing through the nose is moistened as water evaporates from the mucus. Air is warmed as it passes near warm blood within vessels that line the nasal passages.

The Pharynx The **pharynx** connects the nose and mouth with the channel for air (the larynx and trachea) and with the channel for food (the esophagus), as shown in figure 20.10a. The larynx and trachea lie in front of the esophagus. Air entering the larynx from the nose passes over the opening to the esophagus, and food entering the esophagus from the mouth passes over the entrance to the larynx.

Food does not enter the larynx and trachea because of a flap of cartilage called the **epiglottis** (fig. 20.10b). (Cartilage is like bone but not as hard.) During a swallow, the epiglottis moves over the entrance to the larynx to prevent food from passing into this airway.

Food does not enter the nose because of the **soft palate,** which forms the roof of the mouth behind the bony part. During a swallow, the soft palate closes over the airway from the nose (fig. 20.10c). When you are not swallowing, a portion of the soft palate dangles from the roof of the mouth. Breathing movements during sleep may vibrate this part of the soft palate and cause snoring.

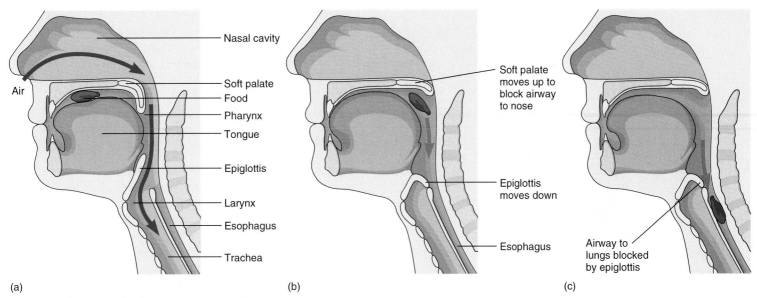

(a)   (b)   (c)

Figure 20.10 The pharynx during swallowing. (a) Air from the nose passes through the pharynx on its way to the larynx and trachea. Food also passes through the pharynx on its way to the esophagus. The soft palate and epiglottis prevent food from entering the air passages. (b) When food is swallowed, the soft palate moves upward to close off the airway between the pharynx and nose, and the epiglottis moves downward to close off the airway between the pharynx and the larynx. (c) Passing the epiglottis, food moves directly into the esophagus.

The Larynx   The **larynx** (or voice box) connects the pharynx with the trachea below. It consists of two elastic tissues, called vocal cords, stretched across the airway:

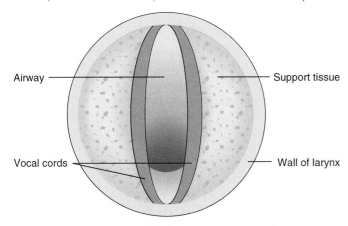

When the cords are shortened during an exhalation, they vibrate and produce a sound. The shorter the cords become, the more rapidly they vibrate and the higher the pitch of the sound. The larynx, also known as the Adam's apple, can be seen and felt as a bulge in the front of the neck.

The Trachea   Warmed, moistened, and cleansed air moves from the larynx to the **trachea** (or windpipe), as shown in figure 20.11. This rigid tube can be felt just beneath the skin in the front of the neck. In addition to epithelial tissue, which forms the inner lining, the wall of the trachea contains smooth muscle and cartilage. The smooth muscle enables slight changes in diameter of the airway and the rigid cartilage holds the airway open.

As in the nose, cilia and mucus in the trachea transport foreign materials toward the mouth. A cough helps remove larger quantities of mucus. During a cough, the epiglottis closes over the trachea and at the same time a rapid, forceful exhalation is initiated by strong contractions of the abdominal muscles. The epiglottis then opens suddenly and a blast of air passes from the lungs up through the trachea and mouth. Frequent coughing accompanies a cold, when excess mucus is secreted to trap microorganisms. Frequent coughing also accompanies heavy smoking. Inhaling tobacco smoke not only adds foreign particles to the respiratory tract, it also paralyzes and eventually destroys the cilia so that the only way to eliminate debris-covered mucus is to cough it up.

The Bronchi and Bronchioles   Below the trachea, the single airway branches into a pair of **bronchi** (the plural of bronchus) (fig. 20.11). The walls of the bronchi are made of epithelial tissue (which forms cilia and mucus), smooth muscle, and cartilage. They contain less cartilage, and are therefore less rigid, than the wall of the trachea.

The bronchi branch repeatedly to form thousands of **bronchioles** (fig. 20.12). These smaller tubes are lined with mucus and cilia for moving foreign materials toward the mouth. The tips of the bronchi are covered with a cluster of alveoli, the tiny gas-exchange surfaces of the lungs.

Asthma is a bronchial disorder in which the epithelial tissue swells and the smooth muscles contract in spasms. It is an inappropriate overreaction of the immune system to particles in the air (such as pollen, cat dander, molds, smoke, and smog), exertion, or emotional stress.

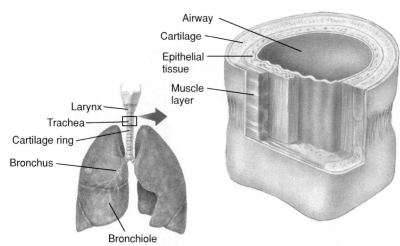

Figure 20.11 The larynx, trachea, and bronchi. The larynx is specialized for sound production. The trachea and bronchi are air passages between the larynx and the bronchioles in the lungs. Their walls are formed of cartilage, a rigid tissue that keeps the airways open; muscles, which alter the diameter of the airways; and epithelial tissue, which contains cells that secrete mucus and cells covered with cilia.

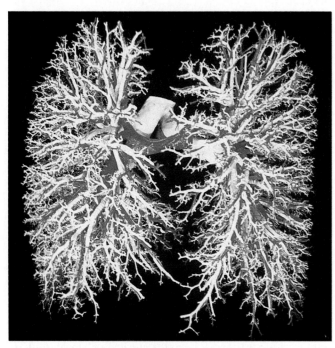

Figure 20.12 The bronchioles. The two bronchi branch repeatedly to form more than 250,000 tiny bronchioles within the lungs. Each bronchiole ends in a cluster of alveoli (not shown here). In this cast, made by filling the lungs and blood vessels of a cadaver with resin, the bronchioles are white and the blood vessels are red.

The airways become narrow and exhalation becomes extremely difficult. (Asthma is characterized by a wheezing sound during exhalation.) In extreme cases, air becomes trapped in alveoli; the continuous pressure of air on these delicate structures causes them to lose their elasticity and even to rupture.

SUMMARY

1. An animal takes in oxygen (for cellular respiration) and releases carbon dioxide (a waste product of cellular respiration). This exchange of gases is urgent, for oxygen cannot be stored and carbon dioxide is toxic.

2. Gas-exchange surfaces are large, to provide sufficient places for diffusion; wet, to enable diffusion through cellular membranes; ventilated, to maintain concentration gradients for diffusion; and usually linked with a circulatory system, for rapid transport of gases to and from body cells.

3. The human respiratory system includes the lungs, muscles of the rib cage and diaphragm, medulla oblongata of the brain, nose, pharynx, larynx, trachea, bronchi, and bronchioles.

4. The lungs consist of alveoli, which are cup-shaped sheets of elastic epithelial tissue. Their external side is covered with a fluid, which contains surfactant to prevent sticking, and their internal side is covered with capillaries.

5. Breathing movements ventilate the lungs. During inhalation, the chest cavity enlarges, by means of contractions of muscles within the diaphragm and between the ribs. The lungs also enlarge, for they adhere by pleural membranes to the chest wall, and air is sucked in. During exhalation, the chest cavity is compressed as the same muscles relax. The lungs shrink and air is pushed out. Breathing can be simulated by artificial respiration.

6. Inhalation and exhalation occur in response to signals from the respiratory center in the medulla oblongata of the brain. The center sets the basic rhythm of quiet breathing, and it also adjusts the rate and depth of breathing to different levels of cellular activity.

7. The upper respiratory tract consists of airways to the lungs. Air entering the nose is cleansed, moistened, and warmed. It then moves to the pharynx, where the epiglottis and soft palate prevent food from entering the air passages, and from the pharynx to the larynx, where sound is produced. From the larynx, inhaled air moves through the trachea to the bronchi and bronchioles. Each bronchiole terminates in a cluster of alveoli.

KEY TERMS

Alveolus (al-VEE-oh-lus) (page 377)
Bronchiole (BRONK-ee-ohl) (page 385)
Bronchus (BRONK-us) (page 385)
Cilium (SILL-ee-uhm) (page 384)
Diaphragm (DEYE-ah-frahm) (page 379)
Epiglottis (EH-pih-GLAH-tus) (page 384)
Gas exchange (page 374)
Larynx (LAEHR-enks) (page 385)
Medulla oblongata (meh-DOO-lah AH-blong-GAH-tah) (page 381)
Mucus (MYOO-kus) (page 384)
Pharynx (FAER-enks) (page 384)
Pleural (PLUR-ahl) **membrane** (page 379)
Respiration (page 374)
Soft palate (PAEL-uht) (page 384)
Surfactant (sur-FAEK-tunt) (page 377)
Trachea (TRAY-kee-ah) (page 385)
Ventilation (page 375)

STUDY QUESTIONS

1. Why is regular breathing crucial to staying alive?
2. Explain why lungs need to be large, wet, ventilated, and very close to capillaries.
3. Why is the concentration of carbon dioxide in the blood a good measure of cellular activity?
4. Describe how each of the following structures contributes to inhalation: pleural membranes, diaphragm, muscles between the ribs, surfactant, and medulla oblongata of the brain.
5. What prevents food from entering the lungs?
6. What are the advantages of breathing through the nose rather than the mouth?

CRITICAL THINKING PROBLEMS

1. Design an animal that would not need a circulatory system to transport gases.
2. Why are internal gas-exchange surfaces crucial to life on land?

3. Why is the water content of exhaled air greater than of inhaled air?
4. You have a friend who smokes cigarettes. Explain to this friend why he coughs so much in the mornings.
5. If you hyperventilate (breathe very rapidly) for several minutes, you then can hold your breath longer than normal. Why is this so? Why is it a bad idea to hyperventilate prior to diving into the water?

SUGGESTED READINGS

Introductory Level:

Huyghe, P. "The Big Yawn," *Discover*, June 1989, 78–81. An interesting description of research dealing with the function of yawning.

Sebel, P., D. M. Stoddart, R. E. Waldhorn, C. S. Waldmann, and P. Whitfield. *Respiration: The Breath of Life*. New York: Torstar Books, 1985. A medically oriented description of respiration, including the history of our knowledge, the embryological development of lungs, and respiratory disorders.

Vogel, S., and D. Manhoff. *Emergency Medical Treatment*. Wilmette, IL: EMT, 1993. An excellent guide for dealing with medical emergencies.

Advanced Level:

Eckert, R., D. Randall, and G. Augustine. *Animal Physiology: Mechanisms and Adaptations*, 3d ed. New York: W. H. Freeman and Company, 1988. An excellent presentation, in chapter 14, of respiration as it occurs in a broad range of animals.

Egginton, S., and H. F. Ross (eds.). *Oxygen Transport in Biological Systems*. New York: Cambridge University Press, 1993. An analysis of optimal ways, both theoretical and real, of transferring oxygen from the external environment to the internal transport fluid.

Randall, D. J., W. W. Burggren, A. P. Farrell, and M. S. Haswell. *The Evolution of Air Breathing in Vertebrates*. New York: Cambridge University Press, 1981. A description of adaptations for air breathing in amphibians, reptiles, birds, and mammals.

Humans are superb distance runners.
We can run far without overheating
because our kidneys conserve water
and our sweat glands use the water
to cool our bodies.

Body Fluids and Temperature

Chapter Outline

Objectives

In this chapter, you will learn
–how the urinary system helps
    maintain an optimal body chemistry;
–how the kidneys form urine from
    blood;
–that an animal's anatomy and
    behavior affect its body temperature;
–that some animals, like yourself,
    maintain a high and fairly constant
    body temperature.

Humans are superb distance runners. Our skeletons and muscles are arranged for efficient movement during a slow run. Our kidneys are equipped to conserve water. And we are exceptionally well-adapted for keeping cool—our skin has between 1,000 and 2,000 sweat glands per square inch, capable of forming as much as a gallon of sweat an hour! Sweat cools by evaporating, and our virtual lack of hair enhances evaporation.

Why are we so well equipped for distance running? One hypothesis is that our ancestors hunted by running down large game. An antelope can run faster than a human but not as far during the heat of the day. It keeps cool mainly by panting rather than sweating. Panting is very effective in cooling the brain, but it is not very effective in cooling the entire body (an antelope's temperature may rise, during flight, to as high as 43° C, or 110° F). Another disadvantage of panting is that rapid, shallow breathing is not compatible with the deep breathing required of endurance running. In addition, panting removes too much carbon dioxide from the blood, making the body fluids too alkaline for normal cellular activities. Early humans, with their superior adaptations for keeping cool, probably chased their prey until these animals collapsed.

Regulation of Body Fluids

An animal consists mostly of water. A crucial aspect of homeostasis (see page 313) is maintaining an optimal amount of water in the body, as well as optimal concentrations of ions in the water.

Cellular activities are extremely sensitive to the composition of fluids within the cell. Animals control the composition of their cellular fluids by controlling the composition of their interstitial fluids. These fluids, in turn, are controlled by the blood. The composition of the blood, in turn, is controlled primarily by excretory organs—kidneys in vertebrate animals (fig. 21.1).

An animal's excretory organs maintain optimal body fluids in spite of enormous variations in input and output. An animal may find water and drink copiously or go for days without a sip; it may find a salt lick and gorge on salt; it may sweat or salivate profusely and lose large quantities of water and ions. Regardless of these activities, an animal's body fluids remain fairly constant and optimal for cellular activitites.

All excretory organs are based on the same basic structure: a tube that is closed on one end and open to the outside on the other end. The closed end and, in most cases, the sides of the tube are formed of epithelial tissue (see page 316), which controls the movement of mate-

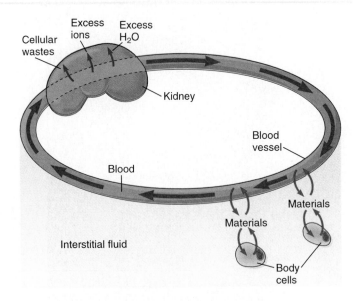

Figure 21.1 Control of cellular fluids. A body cell is surrounded by interstitial fluid. The composition of this fluid, crucial to cellular activities, is maintained by the composition of the blood. The blood carries a great variety of materials, including wastes from cellular activities as well as food monomers, ions, vitamins, and water from the digestive tract. An optimal composition of the blood is maintained by excretory organs, called kidneys in vertebrate animals.

rials between the body fluids and the interior of the tube. Fluid that remains inside the tube becomes urine and drains to the outside of the animal:

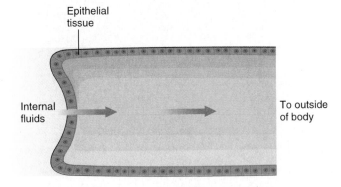

In vertebrate animals, these basic units of excretion form the kidneys, which, together with associated organs, comprise the urinary system.

The Urinary System of a Human

The **urinary system** of a human consists of two kidneys, two ureters, a urinary bladder, and a urethra (fig. 21.2). These organs are listed, together with their functions, in table 21.1.

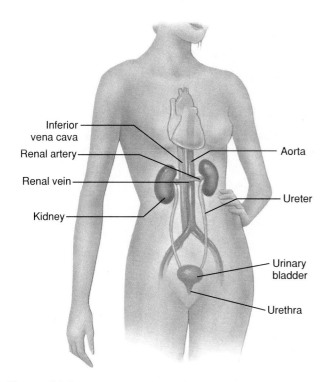

Figure 21.2 The human urinary system.

Inferior vena cava
Renal artery
Renal vein
Kidney
Aorta
Ureter
Urinary bladder
Urethra

Table 21.1

Organs of the Urinary System

Organ	Function
Kidneys	Form urine from components of the blood
Ureters	Transport urine from the kidneys to the urinary bladder
Urinary bladder	Stores urine
Urethra	Transports urine from the bladder to the outside of the body

Organs That Form Urine The **kidneys** form urine from components of the blood. The two kidneys, each about the size of a fist, are embedded within the wall of the abdominal cavity behind the liver and stomach. These organs process 4 or 5 liters (all the blood in the body) every forty minutes. The kidneys perform the following vital functions: (1) remove wastes formed by all the body cells; (2) adjust the amounts of ions and water in the blood; and (3) adjust the acidity of the blood.

The kidneys contain the basic excretory structures, called **nephrons** (fig. 21.3a). Each kidney contains more

than a million of these tiny tubes. The wall of the entire nephron is formed of epithelial tissue, through which materials enter and leave as urine is formed from blood. Filtered blood enters one end of a nephron and is transformed into urine as it passes through the nephron.

Filtered blood enters a nephron through a cuplike depression called the **glomerular capsule** (also known as Bowman's capsule). A bed of capillaries, called the **glomerulus** (*glomus*, mass; *ulus*, small), fills the depression of the capsule. It receives blood from branches of the renal artery (*renal*, pertaining to a kidney) (fig. 21.3b).

Some of the blood passing through the glomerulus is filtered into the nephron, and the rest passes through the capillary bed and into another artery (instead of a vein, as is usually the case after a capillary bed). This artery carries the unfiltered blood to a second capillary bed that surrounds and is intimately associated with the remaining segments of the nephron (fig. 21.3b). In these segments, materials are exchanged between the filtered blood inside the nephron and unfiltered blood inside the capillaries:

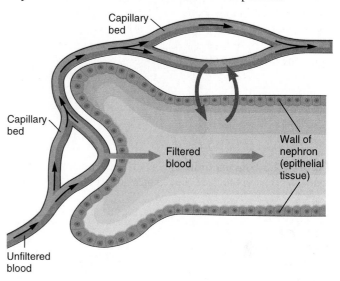

Capillary bed
Capillary bed
Filtered blood
Wall of nephron (epithelial tissue)
Unfiltered blood

The blood then flows from the capillaries to venules and into the renal vein, which empties into the inferior vena cava (see fig. 21.2).

Organs That Transport Urine Urine formed within all the nephrons of each kidney drains into a tube called the **ureter.** There are two ureters, one connected to each kidney. They propel urine, by contractions of smooth muscles within their walls, from the kidneys to the bladder.

The **urinary bladder** is a hollow, muscular organ that stores urine. It holds up to three cups of urine. Emptying of the bladder begins with contractions of the smooth muscles within the walls. The contractions alter the bladder shape, which, in turn, opens the internal sphincter—a ring of muscles that controls an opening at the base of the

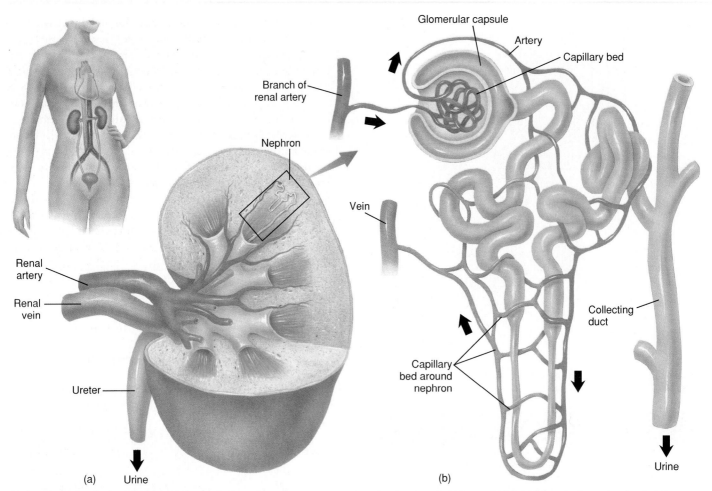

Figure 21.3 The nephron. (*a*) Each kidney contains more than a million nephrons. (*b*) Each nephron has two capillary beds—one bed lies within the glomerular capsule and the other bed surrounds the nephron.

bladder. These involuntary changes are sensed as an urge to urinate. Voluntary relaxation of a second sphincter, called the external sphincter, then results in urination (fig. 21.4).

Urine is squeezed out of the bladder and into a single tube, the **urethra,** that connects to the outside of the body. The urethra of a woman is about 4 centimeters (1.5 in) long and opens to the outside just in front of the vagina. Its only function is to transport urine. The urethra of a man is longer—18 centimeters (7 in) or so—and opens to the outside through the penis. It transports both urine and sperm.

Disease microbes sometimes enter the body through the urethra. There they feed on cells that line the urinary tract: urethra, bladder, and ureters. This painful condition is discussed in the accompanying Wellness Report.

The Formation of Urine

The nephrons form urine from materials in the blood, which is filtered into one end of the nephron (glomerular capsule) and modified as it travels through the remainder of the nephron. The composition of the filtered blood (the so-called filtrate) is quite different from the composition of urine, as shown in table 21.2. Each segment of the nephron

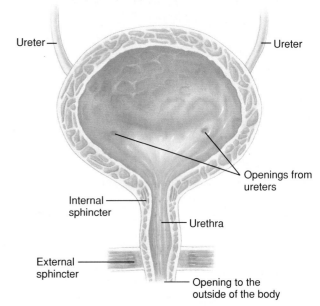

Figure 21.4 The urinary bladder and urethra. Urine enters the bladder by way of two ureters, one connected to each kidney. Urine leaves the bladder by way of the urethra, as controlled by two sphincters: an internal sphincter, which opens involuntarily when the bladder is distended, and an external sphincter, which opens under voluntary control.

Urinary Tract Infections

Infections of the urinary tract are caused by bacteria that feed on epithelial cells that line the urethra, urinary bladder, and ureters. Most often these bacteria come from the intestines, where they normally reside but do not cause infections, and move into the urinary tract by way of feces. Urinary tract infections are more common in women than men because the shorter urethra places the bladder closer to the anus. The bacteria are not usually passed from person to person.

Most often the infection settles in the bladder, where it is known medically as cystitis (*cyst*, fluid-filled sac; *itis*, inflammation, accompanied by pain and swelling). The main symptom of cystitis is a frequent and urgent need to urinate yet great pain in doing so. The pain makes it difficult to get the stream started, and once started the stream is usually stopped before the bladder is emptied. The infection becomes much more serious if it passes to the kidneys, where it causes chills, fever, vomiting, and pain in the abdomen or back.

The growth of bacteria in the bladder is promoted by incomplete emptying. Dehydration is the most common reason that a person retains small amounts of urine in the bladder for long periods of time. Cystitis is also likely to develop in women after frequent sexual intercourse, when bacteria in the urethra are pushed into the bladder. Such a condition is often referred to as "honeymoon cystitis."

To prevent a urinary tract infection, urinate often and empty the bladder completely, drink six to eight glasses of fluids a day, wipe from front to back (women) so that fecal material does not contact the urethra, and keep the area dry by wearing cotton underwear. Honeymoon cystitis can be avoided by urinating right after intercourse, thereby flushing out of the bladder any microorganisms that may have invaded.

Urinary tract infections can be treated successfully with antibiotics. Whenever you experience the symptoms, see a doctor right away. The infection can spread quickly, and a delay may allow the bacteria to invade the kidneys, where they cause more serious damage.

Table 21.2

Differences Between Filtrate and Urine (averages)

Substance	Filtrate (amount formed per day)	Urine (amount formed per day)
Water	180 liters	1.8 liters
Sodium ions	630 grams	3.2 grams
Potassium ions	22 grams	2.1 grams
Glucose	180 grams	0
Urea	475 grams	262 grams

Table 21.3

Segments of the Nephron

Segment	Function
Glomerular capsule	Filters components of the blood into the nephron
Proximal tubule	Returns water and other useful materials to the blood
Loop of Henle	Returns water to the blood
Distal tubule	Moves ions between filtrate and blood as needed
Collecting duct	Returns water to the blood as needed
All segments	Acquire hydrogen ions from the blood as needed

is specialized, by means of specialized cells within its epithelial tissue, for performing a particular aspect of urine formation. These segments and their functions are listed in table 21.3, illustrated in figure 21.5, and described following.

Filtration   Urine formation begins at the glomerular capsule of each nephron. Blood inside the glomerulus is separated from the hollow interior of the nephron by a filter formed of three structures: the capillary wall, a membrane made of cellular secretions (mainly proteins), and the epithelial tissue of the glomerular capsule (fig. 21.6). Blood pressure, which is unusually high within these capillaries (about 55 mm Hg rather than the usual 18 mm for a capillary), forces fluid through these tissues. The filtered blood becomes filtrate within the nephron.

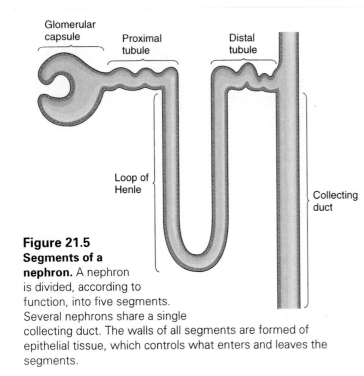

Figure 21.5 Segments of a nephron. A nephron is divided, according to function, into five segments. Several nephrons share a single collecting duct. The walls of all segments are formed of epithelial tissue, which controls what enters and leaves the segments.

Figure 21.6 The glomerular capsule. Blood in the capillaries of the glomerulus is filtered into the glomerular capsule of the nephron. It passes through three layers: the wall of the capillary (epithelial tissue), a noncellular membrane (formed mainly of proteins), and the wall of the glomerular capsule (epithelial tissue).

The filtrate consists of almost everything in the blood except blood cells, fragments of blood cells (platelets), and large proteins. It contains water, ions, glucose, amino acids, vitamins, and urea (a waste product of protein breakdown). Most of these materials are returned to the blood as they pass through other segments of the nephron. The kidney cleans the blood like someone who cleans a room by first removing everything that fits through the door (the filter) and then returning to the room everything they want to keep. The advantage is that the nephron needs to recognize only what it must keep rather than all possible waste materials.

Volume Reduction   Most of the nutrients, water, and ions that enter a nephron return to the blood by passing through the walls of the **proximal tubule** (*proximal*, near the origin; *tubule*, tiny tube), which is the next segment of the nephron after the glomerular capsule (see fig. 21.5). All the amino acids, glucose, vitamins, and most of the ions are pumped (by active transport) out of the tubule, through the interstitial fluid, and into the capillaries surrounding the tubule:

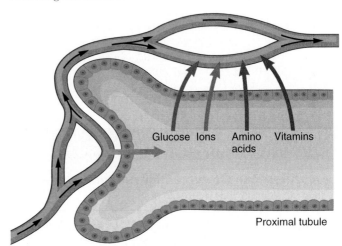

As these materials leave the tubule, water follows by osmosis (see page 76).

An enormous amount of fluid—about 180 liters (45 gal)—enters the nephrons of the two kidneys each day. Of this fluid, however, only 1 to 3 liters become urine for excretion. (The exact amount of urine depends on the needs of the body.) Approximately 99% of the fluid that enters the nephrons returns to the blood, most of it through the walls of the proximal tubule.

After the filtrate is reduced in volume by the proximal tubule, it passes into the next segment of the nephron, which is the **loop of Henle** (see fig. 21.5). Here, more water leaves the nephron by a remarkable system in which sodium ions are concentrated within the interstitial fluid just outside the nephron. The loop of Henle, found only in mammals, is an adaptation of the kidney for conserving water.

The loop has a descending limb and an ascending limb, which lie next to each other. The descending limb is permeable to water but the ascending limb is not:

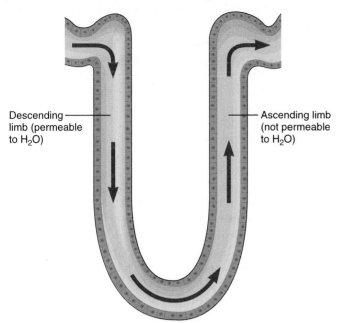

Descending limb (permeable to H₂O)

Ascending limb (not permeable to H₂O)

Sodium ions (Na⁺) are pumped (by active transport) out of the ascending limb, especially near the hairpin turn, and into the interstitial fluid surrounding the loop:

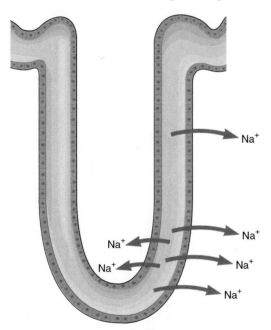

Na⁺

Na⁺
Na⁺

Na⁺
Na⁺

Na⁺

Na⁺

Chloride ions, not dealt with here, passively follow the sodium ions out of the nephron. Water, however, cannot follow because the wall of the ascending limb is impermeable to it.

The higher concentration of ions surrounding the loop of Henle draws water, by osmosis, out of the filtrate in the descending limb and into the interstitial fluid:

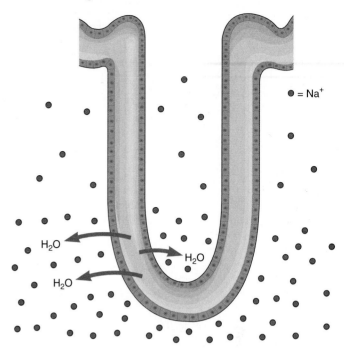

● = Na⁺

H₂O

H₂O

H₂O

The water then moves from the interstitial fluid to the blood.

Some of the sodium ions pumped into the interstitial fluid return to the filtrate by entering the descending limb. Some move into the blood but return eventually to the descending limb when they reenter a nephron at the glomerular capsule. Either route brings the ions back to the descending limb and then on to the ascending limb, where they are pumped back into the interstitial fluid. Thus, sodium ions are recycled:

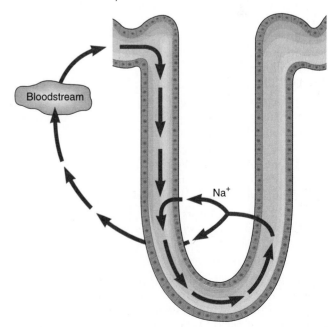

Bloodstream

Na⁺

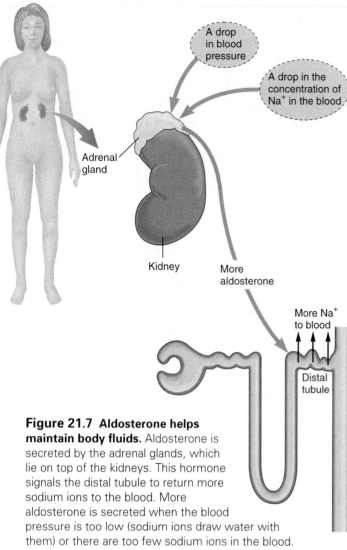

Figure 21.7 Aldosterone helps maintain body fluids. Aldosterone is secreted by the adrenal glands, which lie on top of the kidneys. This hormone signals the distal tubule to return more sodium ions to the blood. More aldosterone is secreted when the blood pressure is too low (sodium ions draw water with them) or there are too few sodium ions in the blood.

Because the sodium ions cycle again and again through the interstitial fluid, they are always highly concentrated around the loop of Henle. These ions become important again in drawing water out of the collecting ducts, as described following.

The filtrate contains a high concentration of ions by the time it reaches the hairpin turn of the loop of Henle. As it moves up the ascending limb, where sodium ions are pumped out, it becomes less concentrated again.

Adjustments of Ions and Water   The filtrate, now greatly reduced in volume, moves from the loop of Henle into the **distal tubule** (*distal*, far from the origin) of a nephron (see fig. 21.5). Here the transfer of ions is adjusted to maintain optimal concentrations within the blood. (Symptoms of abnormal concentrations of ions include food cravings, abnormal blood pressure, abnormal heartbeat, muscle weakness, muscle twitches or cramps, mental confusion, convulsions, and coma.)

The wall of the distal tubule is permeable to ions but not to water. Many kinds of ions (including sodium, potas-

sium, calcium, magnesium, and chloride) move between the nephron and the blood at this segment of the nephron. The transfer of sodium ions out of the nephron and into the blood is governed by **aldosterone,** a hormone secreted by the adrenal glands. (Other ions are governed by other factors.) The more aldosterone present, the more sodium ions return to the blood (fig. 21.7). Aldosterone secretion increases whenever the blood pressure drops below normal, as described following, and whenever the blood contains too few sodium ions in relation to water.

The filtrate then passes from the distal tubule to the **collecting duct** (see fig. 21.5), a small tube that carries the filtrate, now called urine, into the ureters. Several nephrons empty into each collecting duct. Here the amount of water in the urine is adjusted to maintain an optimal concentration in the blood.

Fine adjustments in the amount of water in the blood are controlled by **antidiuretic hormone** (ADH, also known as vasopressin), which is synthesized in the hypothalamus (a major control center for homeostasis in the brain) and stored in the pituitary gland (fig. 21.8). The hypothalamus monitors the concentration of water in the blood directly from blood passing through it and receives information about blood volume from sensory cells in the arteries of the neck. When blood is low in volume or has too low a ratio of water to sodium ions, the hypothalamus signals the pituitary gland to secrete more ADH. (Under these conditions, the hypothalamus also stimulates thirst.)

In the presence of ADH, the wall of the collecting duct becomes more permeable to water. Since this duct lies next to the loop of Henle, which is always surrounded by a high concentration of sodium ions, water leaves the duct by osmosis whenever it is free to do so:

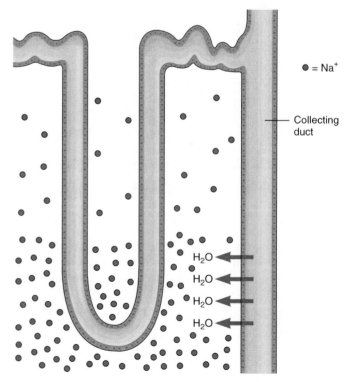

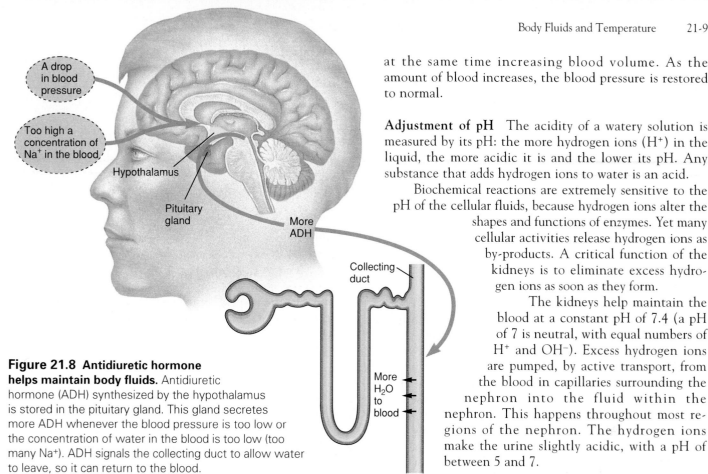

Figure 21.8 Antidiuretic hormone helps maintain body fluids. Antidiuretic hormone (ADH) synthesized by the hypothalamus is stored in the pituitary gland. This gland secretes more ADH whenever the blood pressure is too low or the concentration of water in the blood is too low (too many Na⁺). ADH signals the collecting duct to allow water to leave, so it can return to the blood.

Labels in figure: A drop in blood pressure; Too high a concentration of Na⁺ in the blood; Hypothalamus; Pituitary gland; More ADH; Collecting duct; More H₂O to blood

at the same time increasing blood volume. As the amount of blood increases, the blood pressure is restored to normal.

Adjustment of pH   The acidity of a watery solution is measured by its pH: the more hydrogen ions (H^+) in the liquid, the more acidic it is and the lower its pH. Any substance that adds hydrogen ions to water is an acid.

Biochemical reactions are extremely sensitive to the pH of the cellular fluids, because hydrogen ions alter the shapes and functions of enzymes. Yet many cellular activities release hydrogen ions as by-products. A critical function of the kidneys is to eliminate excess hydrogen ions as soon as they form.

The kidneys help maintain the blood at a constant pH of 7.4 (a pH of 7 is neutral, with equal numbers of H^+ and OH^-). Excess hydrogen ions are pumped, by active transport, from the blood in capillaries surrounding the nephron into the fluid within the nephron. This happens throughout most regions of the nephron. The hydrogen ions make the urine slightly acidic, with a pH of between 5 and 7.

The more ADH present, the more permeable the wall of the collecting duct and the more water removed from the urine. Consequently, the blood acquires more water and the urine becomes more concentrated.

Alcohol suppresses the release of ADH, thereby diminishing the transfer of water from the collecting duct to the blood. When a large amount of alcohol is consumed, a large amount of dilute urine forms and the body becomes dehydrated.

Someone who cannot secrete enough antidiuretic hormone has diabetes insipidus (*diabetes*, constant urination; *insipidus*, tasteless, to distinguish it from a different disease called diabetes mellitus, in which the urine tastes sweet). The collecting ducts of the nephrons remain impermeable to water, and enormous quantities of dilute urine—as much as 25 liters a day—are excreted. A person with diabetes insipidus is always thirsty.

Regulation of Blood Pressure   Low blood pressure triggers the secretion of more aldosterone and more antidiuretic hormone. As more aldosterone is released, more sodium ions are transferred from the distal tubule to the blood. As more antidiuretic hormone is released, more water moves from the collecting duct to the blood. By working together, the two hormones maintain an optimal concentration of water and ions in the blood while

Regulation of Body Temperature

We rarely think about the temperature of our bodies. We have no need to, since homeostatic mechanisms normally keep our internal temperature constant and optimal. Our temperature remains about the same whether we are basking in the sun or walking in the snow.

Animals strive to keep their bodies warm. The ideal temperature for most cells is near 37.8° C (100° F)—warm, but not so warm as to be lethal. Warmer cells are more active, since the rate at which chemical reactions proceed doubles with each 10° C (18° F) rise in temperature. A warmer animal with more active cells is better able to catch food, evade enemies, and attract mates.

The ideal temperature is not, however, much warmer than 37.8° C (100° F). Death becomes imminent, because enzymes can no longer function when the body temperature rises about 6° C above normal—around 43° C (110° F) in most mammals. (Lethal temperatures vary because thermal damage occurs at different temperatures for different enzymes, and cells synthesize "heat-shock" proteins that bind to certain enzymes and help them hold their shape.) Without functional enzymes, there can be no biochemical reactions. The brain is particularly susceptible to high temperatures and produces early danger signs: mental confusion, lack of coordination, and convulsions (involuntary muscle contractions). While these signs do not cause death, they are warnings that the body is nearing a lethal temperature.

The ideal body temperature, then, is the warmest that is compatible with life—relatively near the upper lethal temperature but not so near that a slight fluctuation is fatal. Maintaining this temperature is an important aspect of homeostasis. Animals inherit anatomical, behavioral, and physiological adaptations that promote ideal body temperatures.

Anatomical and Behavioral Adaptations

Most animals have anatomical and behavioral adaptations (see page 17) that help them control how much heat enters and leaves their bodies. An animal that governs its body temperature primarily by anatomical and behavioral adaptations is an **ectotherm** (*ecto*, outside; *therm*, heat). Sometimes ectotherms are called "cold-blooded" animals. Their blood is not, however, necessarily cold because they have ways of acquiring heat from their surroundings. The advantage of being an ectotherm is that acquiring heat from the environment is not costly; an ectotherm does not burn food to produce body heat. All animals except birds and mammals are ectotherms.

An **endotherm** (*endo*, within) is an animal that warms itself from within, by heat generated from cellular respiration (see page 88):

$$\text{Organic molecule} + O_2 \longrightarrow CO_2 + H_2O + \begin{array}{c} \text{Energy} \\ \text{ATP} \quad \text{Heat} \end{array}$$

The advantage of being an endotherm is the ability to remain very active at all times. The disadvantage is the much greater demand for food, particularly in cold environments, to provide fuel for warming the body. Birds and mammals are endotherms.

How an Ectotherm Acquires Heat An ectotherm needs to acquire enough heat from its environment to remain active. It warms its body mainly by moving into sunny places or pressing against warm objects and, while doing so, assuming postures that increase the uptake of heat.

Anatomical adaptations that help an animal acquire heat from the sun involve body size and surface tissues. A smaller animal warms up more rapidly when basking in the sun. In addition to body size, the surface of an ectotherm is not insulated—there is no fat beneath the skin, no fur or feathers to impede the inflow of heat.

How an Endotherm Retains Heat An endotherm warms its body from the inside, with heat from cellular respiration. Behavioral and anatomical adaptations then help it retain this heat when the outside air or water is cooler than the optimal body temperature.

Figure 21.9 An endotherm can retain much of the heat it generates internally. By curling up with their bushy tails over their faces, these sled dogs expose less surface area to the cold environment. Their body surfaces are well insulated with fat and hair.

Behavioral ways of retaining heat include seeking shelter from wind, rain, and extreme cold. Social animals often huddle to keep warm. Anatomical adaptations for preventing heat loss involve the body surface (fig. 21.9). Birds and mammals in cool environments insulate their bodies with layers of fat just beneath the skin. Fat is a superb insulator: heat moves three times more slowly through fat than through any other biological material. The layer of fat thickens just before winter to provide greater protection against heat loss. (An arctic seal, which feeds in cold water, may have half its body mass invested in blubber.) In addition, the bodies of endotherms are insulated with feathers or hair, which trap warm air next to the skin and slow the loss of heat. When a bird or mammal becomes very cold, it raises its feathers or hair to further thicken the layer of trapped air. Humans have very little hair yet have tiny muscles for raising body hairs. When we are cold, these muscles contract to form "goosebumps" on the skin. The muscles are remnants of an adaptation that once kept our more hairy ancestors warm.

How an Animal Eliminates Excess Heat Both ectotherms and endotherms cool themselves by behavioral adjustments. They press their bodies against cool objects and seek cooler places beneath vegetation, rocks, or soil. Many are active only during the cool of the night.

Most animals have some means of evaporative cooling, in which water evaporates from their surfaces. A large amount of heat is needed to convert liquid water to water vapor, which is why evaporative cooling is such an effective way of getting rid of heat. Lizards, birds, and dogs pant; they take rapid, shallow breaths that increase the flow of air over the moist surfaces of the mouth and throat. The more rapidly the moistened air is carried away, by panting, the more rapidly water evaporates from the surface. An insect expands and contracts its abdomen to pump more air into and out of the moist, air-filled tubes that extend throughout its body. Elephants spray themselves with water from a lake or river. In addition, most mammals sweat, a physiological adaptation described below.

Physiological Adaptations

An endotherm, like yourself, maintains a high and fairly constant body temperature by means of physiological adaptations (physiology refers to internal processes involving chemical reactions). Temperature is controlled by a cluster of neurons within the hypothalamus of the brain, a major center of homeostasis in vertebrate animals.

The temperature control center in the hypothalamus receives information about body temperature from three sources: sensory neurons (📖 *see page 321*) in the skin, sensory neurons in the abdominal cavity; and directly from blood as it passes through the hypothalamus. The center compares this information with its "thermostat," which is set at an ideal temperature. On average, the body temperature of a human is maintained at about 37° C (99° F), with variations of around two degrees among different individuals and in daily cycles within each individual.

After evaluating the information, the hypothalamus sends commands via motor neurons (📖 *see page 321*) to glands and to muscles that move the skeleton and walls of blood vessels:

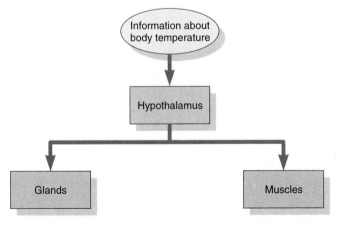

The glands and muscles respond in ways that restore the temperature to its ideal.

Lowering the Body Temperature   Heat moves through the body within the blood and leaves the body through the

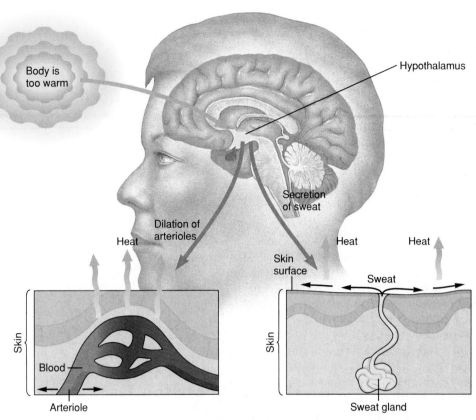

Figure 21.10 An endotherm has internal mechanisms for eliminating excess heat. When a mammal becomes too warm, this information travels to the temperature control center in the hypothalamus. The center responds by commanding arterioles in the skin to dilate, bringing more heat to the body surface, and sweat glands in the skin to release sweat onto the surface, which carries heat away as it evaporates.

skin. Two physiological adjustments increase heat loss: an increase in the amount of blood that reaches the skin, and an increase in the amount of water released onto the skin (fig. 21.10).

A small rise in body temperature, which occurs normally, is corrected by a slight dilation of arterioles (tiny arteries) in the skin. The control center commands the muscles within the walls of these vessels to relax. The channels then enlarge and more blood flows into the skin. More heat (carried in the blood) reaches the body surface, where it can radiate into the surrounding air.

When the control center within the hypothalamus detects a large rise in body temperature (a few degrees or more), it commands two changes: an extreme dilation of arterioles in the skin, bringing as much as 12% of all the blood to the surface; and the release of water, in the form of sweat or saliva, onto the surface of the body.

Most mammals sweat, at least on their foot pads, but only large mammals, including humans, horses, and cattle, release sweat all over their bodies. Humans are particularly good at cooling their bodies this way. Sweat is a dilute solution of water and ions (mainly sodium, potassium, magnesium, and chloride, which are also called electrolytes). It is secreted by **sweat glands** (see fig. 17.8). The 2.5 million sweat glands distributed over the surface of a human body can produce as much as 4 liters (about a gallon) of sweat in an hour provided the water is replaced by drinking.

Sweat must evaporate in order to cool the body. As water molecules within sweat acquire heat from the skin, they move about more rapidly. Eventually, they break free from the liquid and become water vapor in the air, carrying heat with them.

Sweat is secreted whenever the amount of heat generated by the body cells exceeds the amount of heat lost by slight dilations of arterioles in the skin. This happens during exercise, when muscle cells are more active and generate more heat from cellular respiration. When not exercising, sweat is released whenever the air temperature exceeds approximately 22° C (72° F) because above that temperature heat radiates out of the body more slowly than it forms.

As long as there is enough water in the body to produce sweat, a human can tolerate very high environmental temperatures. Experiments show that in rooms as hot as 121° C (250° F), a normal body temperature can be maintained as long as there is plenty of water to drink and the surrounding air is dry enough to promote rapid evaporation of the water in sweat.

Not all sweat functions to cool the body. There is another form of sweat, called emotional sweat, secreted from different kinds of glands in the skin. The functions of this sweat are discussed in the accompanying Research Report.

Raising the Body Temperature   When the body temperature drops slightly below normal, which happens normally, the control center commands the muscles of arterioles in the skin to contract slightly, narrowing the channel and decreasing the flow of blood (and heat) to the body surface.

When the body temperature drops more than a degree below normal, the control center sends commands to the arterioles in the skin, signaling them to contract strongly; and the skeletal muscles, signaling them to contract, which involves an increase in cellular respiration (fig. 21.11).

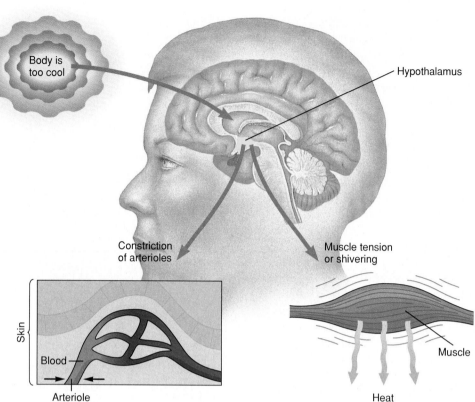

Figure 21.11   An endotherm has internal mechanisms for generating and retaining heat. When a bird or mammal becomes too cool, this information travels to the temperature control center in the hypothalamus. The center responds by commanding arterioles in the skin to constrict, retaining more heat within deeper regions of the body, and muscles to tense up or shiver, generating more heat from cellular respiration.

When muscles in the walls of the arterioles contract, the channel narrows and less blood reaches the skin. Heat is retained deep inside the body. During prolonged exposure to cold, these arterioles occasionally dilate, bringing warm blood into the skin and preventing frostbite.

When the skeletal muscles (the ones that move bones) contract more frequently they generate more heat from cellular respiration. If the temperature drops only a degree or so, the muscles contract slightly and become more tense. If the temperature drops below about 36° C (97° F), the muscles contract more forcefully and shiver. Energy for these contractions comes from fermentation (see page 90), a form of cellular respiration that uses only glucose as a fuel. Shivering releases up to four times as much heat as normal muscle contractions.

Fever   The hypothalamus does not lose control of body temperature during a fever or hibernation. Instead, the "thermostat" is set higher or lower. This new temperature is then maintained by the same mechanisms that maintain the normal body temperature.

A fever is part of the body's defense against infectious diseases; it stimulates the immune system in ways that are described in chapter 22. When disease microorganisms

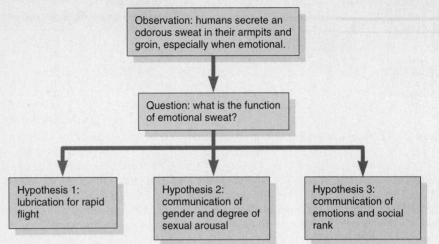

Research Report

Emotional Sweat

A human sweats not only when too warm but also when emotional. Fear, anxiety, embarrassment, and sexual arousal stimulate the release of sweat. Emotional sweat differs from cooling sweat in several ways. First, it is released from anatomically different glands, which are connected to hair follicles. Second, the glands are located only in the armpits, groin, and near the navel. Third, emotional sweat has a distinctive odor. What is the function of this kind of sweat? Three hypotheses are described here and outlined in figure 21.A.

One hypothesis is that the function of emotional sweat is to lubricate the armpits and groin to prevent chaffing while running. The secretion of this kind of sweat is viewed as part of the fight-or-flight response in which the body prepares for intense physical activity. For most of us, such sweating is no longer an appropriate reaction: a lubricant in the armpits and groin does not help us deliver a speech, write an exam, or interview for a job. Emotional sweat may be a relic of our more physically active past. Evidence against this hypothesis is found in the odor. Why would a lubricant have a distinctive odor?

Strangely, the sweat itself has no odor, but it contains an organic molecule that, when acted upon by bacteria, develops an odor. Bacteria live under our arms and near our genitals, where thick hair provides the warm, moist environments that are optimal for bacterial growth. Why do we have thick hair in these areas? Is it so that the bacteria will thrive and give our sweat an odor? If so, what is the advantage of having a distinctive body odor?

Perhaps the function of a person's emotional sweat is to communicate information, by means of scent, to other people. (Other mammals use odors to communicate social rank and sexual receptivity.) Men and women have

Figure 21.A **Alternative hypotheses about emotional sweat.**

very different odors, and within each sex there are individual differences. In one experiment, husbands were able to identify, by smell, the clothing worn by their wives from piles of clothing worn by many people. And the wives could do likewise.

These observations lead to a second hypothesis: the function of emotional sweat is to communicate information about mates—whether the person is male or female and whether or not they are sexually receptive. In support of this hypothesis, researchers note that emotional sweat is not secreted until puberty nor do the armpits and groin develop hair until that time. Moreover, both sweat production and hair density decline with old age. The odor is strongest during the ages of greatest sexual activity. This hypothesis is being tested, largely by evaluating the effects of a particular odor on attitudes toward people of the opposite sex.

A third hypothesis is that emotional sweat communicates information about emotions. It tells us whether another person is frightened, anxious, or embarrassed. This may have helped our ancestors assess one another's social rank, making dangerous fights for dominance unnecessary.

This hypothesis is not easily tested, as it is difficult (and unkind) to make human subjects frightened, anxious, or embarrassed.

Whatever its original role, modern humans try to get rid of their body odors. We bathe and change clothes frequently, apply perfumes to mask the smell, use deodorants to kill the bacteria, and use antiperspirants to plug the sweat glands.

Ironically, many of our perfumes contain the very same odors we are trying to eliminate. One ingredient is andostenone, originally extracted from glands of male deer, beaver, and civets (a kind of cat) but now synthesized in laboratories. Bacteria living on human skin also produce this molecule from testosterone, the male sex hormone. It is andostenone that is responsible for the distinctive odor of a man's underarm sweat, saliva, scalp oil, and urine. (Yet men wash their armpits and then apply andostenone in aftershave lotion, and women wear perfumes formed of this male scent!) Women are much more sensitive than men to this musky odor, and they become 100 times more sensitive during the time of the month when they can get pregnant.

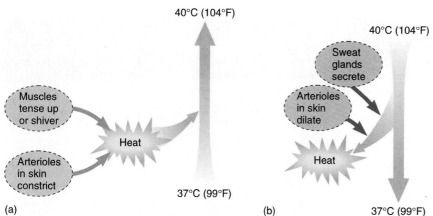

Figure 21.12 How a fever is generated and reduced. (a) A fever develops when the thermostat in the hypothalamus is set at about 40° C, or 104° F. The hypothalamus then commands the muscles to tense up or shiver, generating more heat from cellular respiration, and the arterioles in the skin to constrict, retaining more heat deep inside the body. (b) The fever subsides when the thermostat is reset to normal (about 37° C, or 99° F). The hypothalamus then commands the sweat glands to release sweat, which evaporates and removes heat, and the arterioles in the skin to dilate, which brings more heat to the surface for removal.

invade the bloodstream, white blood cells secrete chemical messengers that signal the temperature control center in the hypothalamus to maintain a body temperature that is several degrees above normal. The control center commands arterioles in the skin to constrict and skeletal muscles to tense up or to shiver. The body temperature then rises to about 40° C (104° F), where it is held by the temperature control center. When the infection subsides, the temperature is returned to normal by dilations of arterioles in the skin and by sweating (fig. 21.12).

When Temperature Control Fails Sometimes the mechanisms that control body temperature fail, and a rise or fall in temperature is not corrected. Such failures can quickly become lethal.

Heat stroke (or sunstroke) occurs when the body cannot eliminate heat by sweating. The body temperature then rises rapidly. One cause is dehydration—there is simply not enough water in the body to produce sweat. This happens during strenuous exercise, such as a prolonged athletic event, when more water is lost in sweat than replaced by drinking. Another cause of heat stroke is humid weather, when there is too much water vapor in the air for sweat to evaporate. Sweat that drips does not cool. Someone with heat stroke should be artificially cooled, by placing cool water on the skin and then fanning the skin with air to hasten evaporation.

Hypothermia occurs when the body can no longer generate heat by shivering. This happens after prolonged exposure to cold, especially when a person falls into cold water or is exposed to cold wind. Muscles can shiver for only about thirty minutes or so, because the muscles run out of glucose. At this point, the body has no way of generating more heat and its temperature drops steadily. At 35° C (95° F), the brain does not function well and the person becomes disoriented, irritable, and uncoordinated. At 32° C (90° F), the hypothalamus stops working and can no longer command arterioles in the skin to constrict. At this point, the only way to keep the person alive is to add heat from the outside. The safest way to do this is to use your own body heat—to press against the person's body while wrapped in a blanket or sleeping bag.

SUMMARY

1. Organs of the urinary system maintain optimal composition of the body fluids. The kidneys form urine from blood. The ureters carry urine from the kidneys to the bladder. The urinary bladder stores urine. The urethra carries urine outside the body.
2. The kidneys consist of nephrons, which form urine from blood. Within each nephron, blood is filtered from capillaries into the glomerular capsule. All components of the blood except cells, cell fragments, and large proteins pass into the nephron.
3. The filtrate is modified as it travels through the nephron. Useful materials are returned to the blood at the proximal tubule. More water returns at the loop of Henle. Ions are returned, as needed, to the blood at the distal tubule. The return of sodium ions is governed by aldosterone. Water is returned, as needed, at the collecting duct as governed by antidiuretic hormone. The pH of the blood is adjusted by transferring excess hydrogen ions into the nephron.

4. Animals have adaptations that help them maintain body temperatures that are warm but not so warm as to be lethal. Ectotherms acquire heat from the environment, whereas endotherms generate heat from cellular respiration.

5. An ectotherm acquires heat by moving into warm places and assuming postures that maximize heat uptake. It has no fat beneath the skin, and no hair or feathers to impede the inflow of heat. An endotherm acquires heat from cellular respiration and retains this heat by seeking shelter, maintaining layers of fat beneath the skin, and having either feathers or hair.

6. Animals eliminate heat by moving into cooler places and releasing fluids onto their surfaces.

7. An endotherm has a control center in the hypothalamus that maintains an optimal body temperature. When the temperature rises too high, the center commands arterioles in the skin to dilate and glands to release fluids onto the body surface. When the body temperature drops too low, the hypothalamus commands arterioles in the skin to constrict and the skeletal muscles to tense up or to shiver. A fever develops when the control center sets its thermostat several degrees higher than normal.

8. Temperature control mechanisms fail during a heat stroke, when sweat cannot get rid of excess heat, and during hypothermia, when the muscles can no longer shiver.

KEY TERMS

Aldosterone (ahl-DAH-stur-ohn) (page 396)
Antidiuretic (AN-teye-deye-yur-EH-tik) **hormone** (page 396)
Collecting duct (page 396)
Distal tubule (page 396)
Ectotherm (EK-toh-thurm) (page 398)
Endotherm (EN-doh-thurm) (page 398)
Glomerular (gloh-MEHR-yoo-lar) **capsule** (page 391)
Glomerulus (gloh-MEHR-yoo-lus) (page 391)
Kidney (page 391)
Loop of Henle (HEN-lee) (page 394)
Nephron (NEH-fron) (page 391)
Proximal tubule (page 394)
Sweat gland (page 400)
Ureter (YUR-eh-tur) (page 391)
Urethra (yur-EE-thrah) (page 392)
Urinary (YUR-in-ahr-ee) **bladder** (page 391)
Urinary system (page 390)

STUDY QUESTIONS

1. What is the role of the kidney in homeostasis?
2. Draw a nephron, indicating the locations of the distal tubule, proximal tubule, collecting duct, loop of Henle, glomerular capsule, and glomerulus. What happens at the proximal tubule?
3. Explain how aldosterone and antidiuretic hormone help maintain optimal blood pressure.
4. What is the difference between an ectotherm and an endotherm? Give three examples of each kind of animal.
5. How does an animal get rid of excess heat? Describe an adaptation that makes humans especially good at cooling down.
6. What is the role of shivering in temperature regulation? What is the role of arterioles?

CRITICAL THINKING PROBLEMS

1. Suppose you were given an unknown animal to dissect and found it had an unusually long loop of Henle. What could you conclude about the animal's life-style?
2. What behavioral and physiological adaptations do you use to keep cool on a hot day? On a cold day?
3. Urine formation often stops when a person goes into cardiovascular shock (sudden drop in blood pressure). Explain why this happens.
4. It is a hot, dry day and you are running with a friend, who begins to act in a strange manner—staggering and talking nonsense. He then collapses to the ground and is unable to tell you how he feels. His skin is warm and dry. What is most likely the cause of your friend's collapse? What should you do to help him recover? In what way will this first-aid help him recover?
5. It is a cold, windy day and you are camping in the mountains with a friend, who begins to act in a strange manner—staggering and talking nonsense. She then collapses and shivers violently. After about half an hour, she stops shivering and wants to sleep. What is most likely the cause of your friend's collapse? What should you do to help her recover? In what way will this first-aid help her recover?

SUGGESTED READINGS

Introductory Level:

"Beating the Heat," *Natural History,* August 1993. The entire issue is devoted to how animals, including humans, survive hot environments.

"Keeping Warm," *Natural History,* October 1981. The entire issue is devoted to how animals, including humans, survive cold environments.

Schmidt-Nielsen, K. "How are Control Systems Controlled?" *American Scientist* 82 (1994): 38–44. A discussion of such topics as how wild mammals obtain ions, camels regulate temperature, ectotherms develop fevers, and marine birds excrete salt.

Wolkomir, R. "Chilling Out for Science," *Discover,* February 1988, 43–51. A description of research, including methods and results, on hypothermia.

Advanced Level:

Cannon, W. B. *The Wisdom of the Body.* New York: W. W. Norton, 1967 (original copyright 1932). A classic description of homeostasis, with emphasis on the body fluids and body temperature.

Crawshaw, L. I., B. P. Moffitt, D. E. Lemons, and J. A. Downey. "The Evolutionary Development of Vertebrate Thermoregulation," *American Scientist* (September–October) 1981: 543–50. A description of how vertebrates, from fish to humans, control their body temperatures.

Kluger, M. J. *Fever: Its Biology, Evolution, and Function.* Princeton, NJ: Princeton University Press, 1979. A description of all aspects of a fever, including how both ectotherms and endotherms raise their body temperatures as part of their fight against diseases.

Smith, H. W. *From Fish to Philosopher.* Boston: Little, Brown, 1953. A classic, easy-to-read book describing the evolution of the vertebrate kidney.

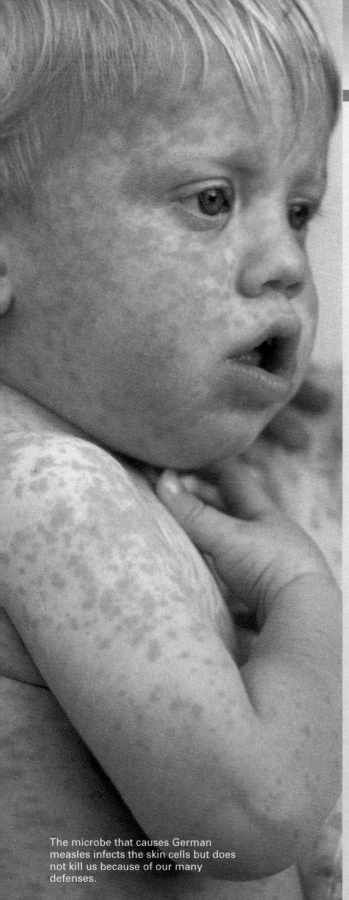

The microbe that causes German measles infects the skin cells but does not kill us because of our many defenses.

Defenses Against Diseases

Objectives

In this chapter, you will learn
–how parasites live with their hosts;
–about the microorganisms that
  parasitize humans;
–how your body prevents invasions by
  microbes;
–how internal defenses localize an
  invasion;
–how internal defenses fight an
  invasion that has spread;
–why you cannot get the same disease
  twice;
–what happens when your internal
  defenses attack your own body cells;
–why the flu and AIDS are hard to
  combat.

An enormous quantity of food was stockpiled within the fort. Outside the wall, the enemy searched relentlessly for an opening into the fort. Inside, guards patrolled back and forth, making sure there were no openings and checking individuals for identity cards to make sure there were no invaders. The standoff continued for weeks without incident.

Then one morning the wall was penetrated: a kitchen knife slipped, the skin was cut, and germs poured into the body. The patrolling guards sounded an alarm, informing other defenders that an invasion had occurred and exactly who the invaders were. Specific weapons for fighting that particular germ were brought to the skin. There the two armies clashed, one trying to penetrate farther into the fort (the body) and the other trying to contain the enemy within a small area (the wound).

Such battles, involving individual cells and molecules, are fought all along the boundaries of your body. No matter how often you wash, microorganisms cover your skin as well as the linings of your respiratory tract, digestive tract, and reproductive tract. Some of these organisms are disease microbes (or germs) that live by invading our bodies and feeding on our molecules. Normally, microbes are prevented from doing so by an arsenal of weapons known as the immune system. Without this remarkable defense, disease microbes would kill you within a week.

Disease Microbes

A great variety of microorganisms live on your skin and internal surfaces. Most of them are harmless, some are beneficial, and some can make you sick. If the ones that can make you sick—the so-called disease microbes—manage to penetrate your body surface, they will feed on your cells and multiply within your body.

Each species of disease microbe typically causes illness in just one species of organism. The microbes that infect dogs, for example, rarely infect humans. Moreover, each species of microbe specializes in just one tissue or organ: the microbe that causes mumps infects salivary glands, and the microbe that causes German measles infects the skin. Disease symptoms that occur elsewhere in the body, such as headaches, fever, and muscle aches, are usually caused by the body's defense rather than by the microbe itself.

The Host-Parasite Relationship

A host-parasite relationship develops when two kinds of organisms live together and one benefits while the other is harmed (but seldom killed) by the relationship. The organism that benefits is the **parasite** and the organism that is harmed is the **host.** Host-parasite relationships between microorganisms and humans lead to infectious diseases—illnesses that spread from one person to another. Examples of infectious diseases are the flu, measles, tuberculosis, and acquired immune deficiency syndrome (AIDS).

Balance of Power   A parasitic microbe rarely kills its host because both species have adaptations for dealing with each other. A parasite and its host evolve together: an improved offense by the parasite selects for an improved defense by the host, and an improved defense by the host selects for an improved offense by the parasite. These opposing selective forces produce an evolutionary "arms race" that maintains a balance of power.

When Parasites Kill   Sometimes the balance of power is upset and the disease microbe kills its host. Generally the microbe gains an upper hand because its host is weakened by malnourishment, age, physical exhaustion, or emotional stress.

Disease and famine go together. For example, more than half the deaths during the severe famine (early 1990s) in Somalia were from measles and dysentery. Age also has its effects: young children are more vulnerable to infectious diseases, for their immune systems are not fully developed, as are older people, for their immune systems are deteriorating. Someone with AIDS also has a deteriorating immune system and is extremely susceptible to infections. Even under normal conditions, a loss of three or more hours of sleep can weaken your immune system by as much as 50%.

An environmental change can upset the balance of power between host and parasite. Modern women inadvertently gave an advantage to the microbe that causes toxic shock syndrome when they began to wear tampons during their menstrual periods. Highly absorbent material inside the vagina concentrates nutrients that the microbes require, enabling the microbes to flourish and overwhelm the body's defenses. Legionnaires' disease struck when new kinds of large air conditioners provided ideal environments for microbes that normally live within moist soil. The microbes multiplied within the air conditioners and were wafted into rooms and inhaled by humans.

Parasites also kill when they contact a previously unexposed population of hosts. The microbes that cause measles and smallpox, for example, were introduced into North America by Europeans in the eighteenth and nineteenth centuries. These microbes, which caused minor illnesses in European children, killed more than half the Native Americans. Even today, diseases that have been restricted to small, isolated populations are now being spread throughout the world as these people establish closer contact with the rest of the world.

Table 22.1

Some Microbes and the Diseases They Cause

Microbe	Some Diseases	Cells Infected
Protozoan	Amoebic dysentery	Large intestine
	Giardia	Small intestine
	Malaria	Red blood cells
	Sleeping sickness	Brain
	Tularemia	Skin
Fungus	Athlete's foot	Skin of the foot
	Ringworm	Skin
	Vaginitis	Vagina
	Valley fever	Lungs
Bacterium	Cholera	Small intestine
	Gonorrhea	Reproductive tract
	Legionnaires' disease	Lungs
	Lyme disease	Joints and heart
	Traveler's diarrhea	Large intestine
	Tuberculosis	Lungs
	Whooping cough	Lungs
Virus	AIDS	White blood cells
	Chicken pox	Skin
	Common cold	Respiratory tract
	Hantavirus	Lungs
	Hepatitis	Liver
	Herpes	Lips or reproductive tract
	Influenza	Respiratory tract
	Mononucleosis	White blood cells
	Mumps	Salivary glands
	Rabies	Brain and spinal cord

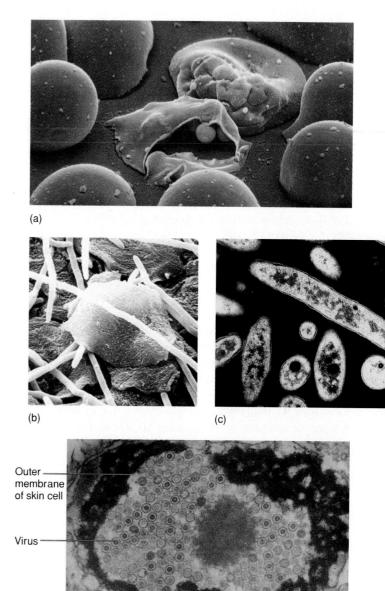

(a)

(b)

(c)

Outer membrane of skin cell

Virus

(d)

Figure 22.1 Microbes that cause human diseases. (a) The protozoan (tiny spheres stained yellow) that causes malaria lives within red blood cells (larger spheres stained red). After reproducing, the new protozoa break out of the cell and the person develops a fever. (b) Threads of a fungus growing over a flat skin cell. The most common fungal infection is athlete's foot. (c) A cluster of the bacteria that cause Legionnaires' disease. (d) Herpesviruses within a skin cell.

Microbes That Cause Human Diseases

Four major categories of disease-causing microorganisms are the protozoa, fungi, bacteria, and viruses. (Disease macroorganisms, not discussed here, are chiefly worms, such as the tapeworms that live within human intestines.) Examples of these microorganisms and their diseases are given in table 22.1. Photographs of microbes from each category are shown in figure 22.1.

All microbes except viruses are cells with the same fundamental processes as our own cells. Like us, the protozoa, fungi, and bacteria use organic monomers (e.g., glucose and amino acids) as sources of energy and as materials for building their own kinds of macromolecules. Microbes that cause human diseases acquire their organic monomers from human cells. Viruses, on the other hand, are not cells. They are clusters of molecules that invade cells and use cellular machinery to reproduce.

Protozoa The protozoa are members of the kingdom Protista (see the inside back cover of this book). A protozoan is a large eukaryotic cell (see page 62), with a nucleus, chromosomes, mitochondria, and other organelles. It has no cell wall.

At least a dozen species of protozoa live on you. One species lives in your mouth, where it eats bacteria and food particles lodged between your teeth. Several species live in your intestines, where they feed on bacteria that live there. Most of these protozoa are harmless and a few even help you by consuming disease microbes. Some protozoa, however, consume human cells (see table 22.1). A protozoan feeds by absorbing nutrients through its outer membrane, by engulfing and digesting cells, or by entering a cell and digesting molecules from within.

Diseases caused by protozoa are difficult to treat because their cells are so similar to our own. Anything that destroys a protozoan is likely to damage us as well.

Fungi   Organisms in the kingdom Fungi (see inside back cover of this book) that cause human diseases are yeasts and molds. They consist of eukaryotic cells with cell walls, and they feed by secreting digestive enzymes onto their food.

The human body provides a favorable environment for many kinds of fungi. They live on surfaces—the skin, nails, vagina, and lungs, where their digestive enzymes produce lesions. Fungi cause only a small percentage of human diseases, some of which are listed in table 22.1.

Fungal infections of the skin and vagina cause itching but are not serious. They can be treated with chemicals called fungicides. Fungal infections of the lungs, however, are very serious because oxygen and carbon dioxide do not diffuse readily through a lung that is coated with mold. Lung infections are difficult to treat because the delicate host cells are readily damaged by fungicides.

Bacteria   Bacteria form the kingdom Monera (see inside back cover). A bacterium is a prokaryotic cell (see page 59) with a cell wall but no nucleus, mitochondria, or other membrane-enclosed organelles.

About 150 human diseases are caused by bacteria, of which a few are listed in table 22.1. Some disease bacteria invade body cells and feed directly on materials there. The bacteria that cause salmonella food poisoning, for example, enter epithelial cells that line the large intestine. Other bacteria, like the ones that cause cholera and tetanus, release toxins that damage body cells.

Bacterial infections can be cured with antibiotics, such as penicillin and tetracycline. These organic molecules kill prokaryotic cells (bacteria) without harming eukaryotic cells (our cells). How they kill is described in the Wellness Report in chapter 11. Since the early 1940s, antibiotics have been our miracle drugs, virtually eliminating bacterial diseases wherever they were applied.

Unfortunately, bacteria evolve resistance to antibiotics. When a bacterial infection is treated with an antibiotic, any bacterium that is less vulnerable to the antibiotic will survive and reproduce with much greater success than bacteria that are more vulnerable to the antibiotic. In a short period of time, the bacterial population consists mainly of resistant individuals. (It is for this reason that physicians are reluctant to prescribe antibiotics unless absolutely necessary.) Tuberculosis, for example, was a relatively common and serious lung disease prior to 1950, but it virtually disappeared in the United States after the discovery of antibiotics. Now, new strains of the tuberculosis bacteria are resistant to antibiotics and this serious disease is making a comeback.

Viruses   A virus (see page 204) is not a cell but rather a cluster of molecules. (We refer to it as a microbe, but it is not really an organism at all.) The basic structure of a virus is a core of nucleic acids (either DNA or RNA) surrounded by a coat of proteins. In some viruses the coat contains other molecules, in addition to proteins, and many viruses have portions of cellular membranes (taken from their host cells) around the coat.

A virus reproduces by invading a cell and then commanding it to cease normal activities and begin making more viruses of its kind. The conversion of a normal cell into a factory for manufacturing viruses typically kills the cell. Viruses are not placed in any kingdom of life.

Many human diseases are caused by viruses (see examples in table 22.1). In the United States, viruses are responsible for 80% of all infectious diseases. In some cases, like measles or mumps, the virus enters a cell, reproduces rapidly, and kills its host cell. In other cases, the virus remains inactive and hidden within a cell for months or years. When activated by certain conditions, the virus suddenly causes its particular illness. The herpesvirus that causes cold sores, for example, rests within nerves of the face until the person experiences intense sunlight or stress. The virus then becomes active and migrates to cells within the lips, where it reproduces. A blister forms as these cells break open and release the newly formed viruses.

Viral infections are difficult to treat because viruses reside within our cells. Anything that destroys the virus is likely to damage our own cells.

Natural Defenses of the Human Body

The natural defenses of the body form the immune system, which is a widely dispersed group of cells, tissues, and organs that work together to destroy disease microbes. The immune system consists of the body surfaces, blood, thymus gland, bone marrow, spleen, lymph, lymph vessels, lymph nodes, tonsils, adenoids, and certain areas

Table 22.2

Major Components of the Immune System

Organ or Tissue	Function
Skin and linings of internal channels	Form barriers to invasions
Lymph	Returns interstitial fluid to blood; transports white blood cells and antibodies; brings microbes to lymph nodes
Lymph vessels	Form channels for lymph flow
Lymph nodes	Filter microbes out of the lymph and then destroy them; store white blood cells
Adenoids and tonsils	Filter microbes out of the lymph and then destroy them; store white blood cells
Peyer's patches on small intestine	Prevent bacteria from entering bloodstream
Blood	Transports cells and molecules that fight microbes
Bone marrow	Produces immature white blood cells; develops mature B cells*
Spleen	Stores white blood cells
Thymus gland	Develops mature T cells*

*B and T cells are forms of white blood cells.

of the small intestine. These components are listed, along with their functions, in table 22.2 and illustrated in figure 22.2. The immune system contributes to homeostasis (see page 313) by protecting the body cells from damage by microbes.

Animals have both nonspecific and specific weapons against microbes. A nonspecific weapon is a cell that devours anything it recognizes as foreign. A specific weapon is a cell or molecule that is specialized for, and therefore very good at, recognizing and attacking just one kind of microbe. Invertebrate animals, like starfish and moths, have mainly nonspecific weapons whereas vertebrate animals have many weapons of both kinds. Microbe-fighting cells are the white blood cells (table 22.3); some are nonspecific and others act specifically against a particular microbe. They secrete chemical messengers (table 22.4) that coordinate the different components of the immune system.

Preventing Invasions

The first line of defense against disease microbes is your body surface—your skin and linings of the tubes that form your digestive tract, respiratory tract, urinary tract, and reproductive tract. The tightly packed cells of epithelial tissues (see page 316), which form all our outer boundaries, are physical barriers to invasions by microbes. A person with severely burned skin has lost this barrier and is extremely vulnerable to infections.

In addition to forming a barrier, cells of the epithelial tissue secrete materials that kill microbes on the body surfaces. The salts in sweat and tears draw water out of microbes, shrinking and killing the ones that live on your skin and in your eyes. The acids in the stomach and vagina burn them. An enzyme (called lysozyme) in sweat, tears, saliva, earwax, and nasal secretions ruptures bacteria by digesting their cell walls. Mucus (a sticky fluid) and cilia (hairlike projections) line our internal tubes, trapping microbes and transporting them out of the body. Respiratory infections usually begin with a sore throat because that is where the mucous layer is thinnest.

The Local Response

When a microbe penetrates the body's outer defenses, it triggers a powerful internal defense at the invasion site. The function of this **local response** is to prevent the microbes from spreading farther into the body. This line of defense, summarized in table 22.5, has two components: the release of interferons and the inflammatory response.

The Release of Interferons An **interferon** is a chemical messenger, built mainly of protein, that defends the body against viruses. Several kinds of interferons are secreted by cells that have been invaded by viruses. They signal other cells to begin synthesizing enzymes that prevent viruses from reproducing.

Interferons work against all kinds of viruses; they are a nonspecific form of defense. Someone already fighting one kind of virus is protected, by the release of interferons, against infections by other kinds of viruses.

The Inflammatory Response Certain kinds of white blood cells, called mast cells, secrete histamine whenever they detect something foreign in the body. Unlike other white blood cells, which travel within the bloodstream, mast cells are embedded within connective tissues (see page 316) throughout the body. They are particularly abundant within epithelial tissue and along blood vessels. **Histamine** is a chemical messenger that

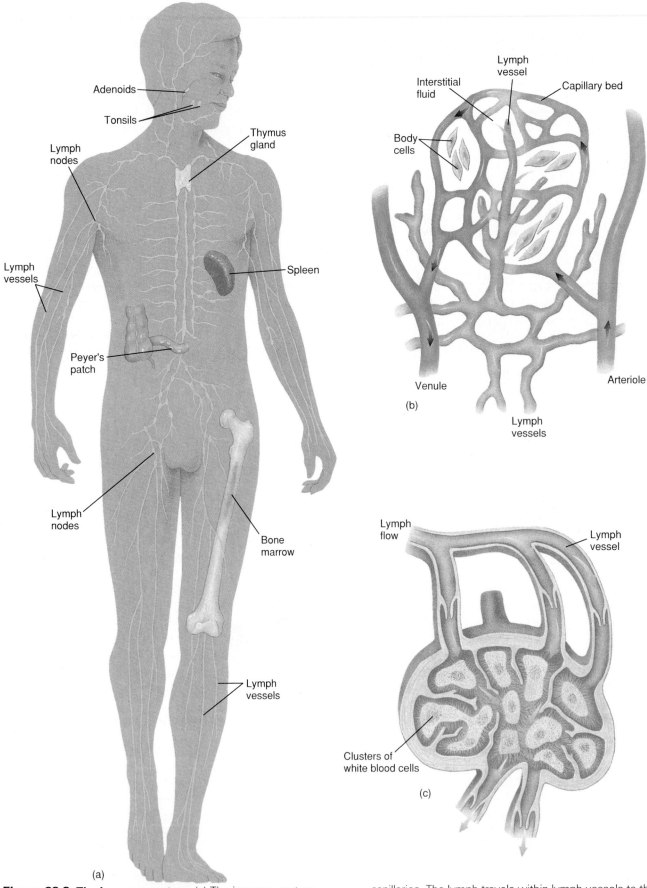

Figure 22.2 The immune system. (*a*) The immune system consists of the skin, blood, thymus gland, spleen, tonsils, adenoids, Peyer's patches on the small intestine, bone marrow, lymph nodes, lymph, and a network of lymph vessels. (*b*) Materials from the interstitial fluid enter the lymph at the lymph capillaries. The lymph travels within lymph vessels to the inferior vena cava (a large vein that carries blood into the heart), where it joins the bloodstream. (*c*) Microbes are filtered out of the lymph at the lymph nodes, which are packed with white blood cells. Each node is between 1 and 12 mm (less than an inch) wide.

Table 22.3

Major Kinds of White Blood Cells

Name of Cells	Function
Macrophages	Devour microbes; identify microbes for helper T cells; secrete chemicals that initiate a fever
Mast cells	Secrete histamine, which triggers the inflammatory response
Lymphocytes	
Helper T cells	Secrete chemicals that coordinate the general response
Killer T cells	Kill virus-infected cells and cancer cells
B cells	Secrete antibodies, which inactivate microbes and toxins

Table 22.4

Some Chemical Messengers of the Immune System

Messenger	Some of Its Functions
Histamine	Initiates inflammatory response
Interferons (at least three forms)	Stimulate body cells to produce materials that fight viruses; make cancer cells more "visible" to killer T cells; enhance effects of white blood cells
Interleukins (at least twelve forms)	
Interleukin-1	Stimulates reproduction of helper T cells; initiates a fever
Interleukin-2	Stimulates reproduction of B cells
Interleukin-8	Stimulates macrophages to move out of capillaries into the infected region
Interleukin-12	Stimulates killer T cells to reproduce, kill, and secrete chemical messengers

Table 22.5

The Local Response

Component	Function
Interferons	Signal cells to make enzymes that block viral reproduction
Inflammatory response	
Blood clots	Barricade the infected site
Macrophages	Devour microbes; identify microbes for helper T cells; secrete chemical messengers that alert other components of immune system
Antibodies	Clump microbes; prevent viruses from entering cells; activate complement
Complement	Digests holes in microbes; guides macrophages to microbes; assists macrophages in devouring microbes

fluid, which leaks out of the blood vessels; and painful from the pressure of the fluid on nerve endings. Histamine also commands the capillaries to become more permeable to materials of the immune system. Four kinds of materials move out of the blood and into the infected tissue: clotting materials, macrophages, antibodies, and complement.

Clotting materials are proteins and cell fragments (called platelets) carried within the blood. They leave the blood and form clots around the infected area, thus sealing off the microbes and preventing them from spreading to other regions of the body.

A **macrophage** (*macro*, large; *phago*, eat) is a white blood cell (fig. 22.4a) that kills anything it recognizes as foreign to the body. It is a nonspecific form of defense. This unusually large cell kills microbes by engulfing and digesting them. It then places a distinctive fragment of the microbe's protein on its outer surface. The "flagged" macrophage enters the bloodstream and travels to a lymph node, where its flag communicates to other white blood cells the kind of microbe that has invaded (fig. 22.4b). The phagocyte also secretes chemical messengers that activate other white blood cells.

As the local battle continues, dead white blood cells, dead microbes, and fragments of damaged body cells accumulate. This debris forms a whitish fluid, called pus, that oozes from the infected area.

An **antibody** is a protein that recognizes and binds with a particular molecule on a particular kind of microbe or its toxin; it is a specific form of defense. The molecules that antibodies bind with are called antigens, or *anti*body *generating* molecules, because their presence stimulates the production of more antibodies of that type.

triggers an **inflammatory response,** characterized by redness, warmth, swelling, and pain in the infected area (fig. 22.3).

Histamine commands the arterioles to dilate and bring more blood to the area. The accumulating blood generates the symptoms of an inflammatory response: the region becomes warm and red from the blood; swollen with

411

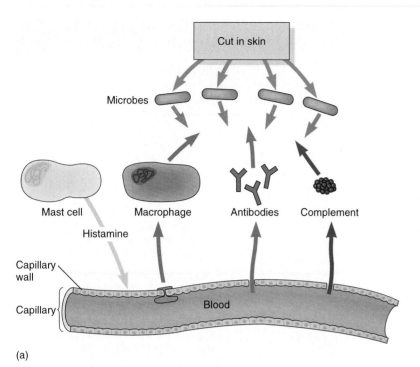

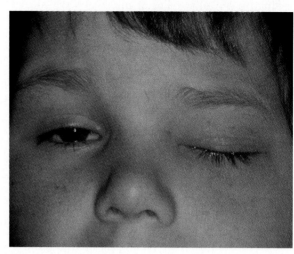

(a)

(b)

Figure 22.3 An inflammatory response. (*a*) Histamine released from mast cells brings more blood to the area and makes the capillaries more permeable to microbe-fighting materials—macrophages, antibodies, and complement. (*b*) It is the inflammatory response that makes a wound warm, red, swollen, and painful.

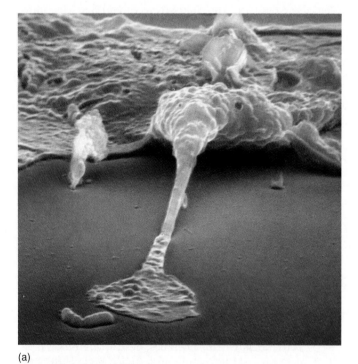

(a)

Figure 22.4 Macrophage. (*a*) A macrophage (stained white) capturing a bacterium (stained orange). (*b*) A macrophage wraps its outer membrane around the microbe and then forms an internal vesicle of the membrane. Inside the vesicle, the microbe is digested by enzymes. Some of the microbe's molecules are then displayed on the macrophage's surface, as a flag that informs other white blood cells what kind of microbe has invaded.

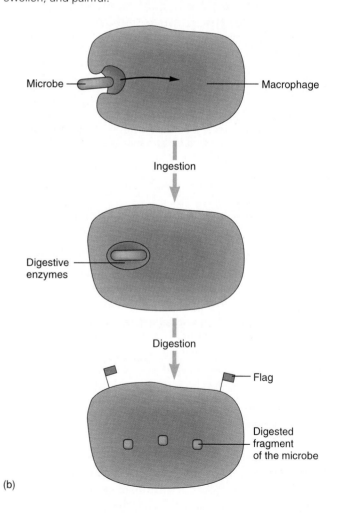

(b)

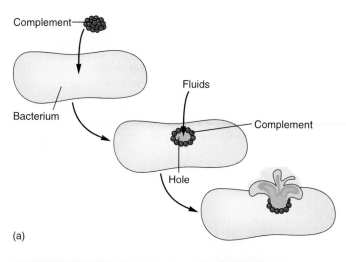

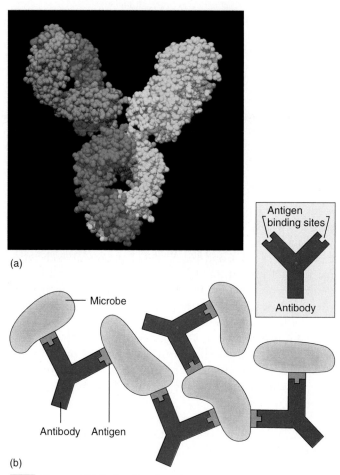

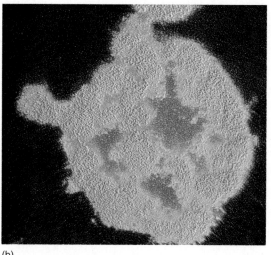

Figure 22.5 Antibodies. (*a*) An antibody is a large Y-shaped protein. The base of the Y contains information that specifies which species it comes from—whether the antibody came from a mouse, for example, or a human. The ends of the arms (top) bind with a particular antigen (kind of foreign molecule). (*b*) An important function of antibodies is to clump microbes, which happens when the two arms of each antibody bind with two microbes.

Figure 22.6 The effects of complement. (*a*) Complement is a cluster of proteins that digests a hole in the wall of a bacterium. Without a wall to resist swelling, fluid flows into the bacterium and it bursts. (*b*) The effects of complement: a bursting bacterium.

The human body can make more than 100 million kinds of antibodies, one for each kind of foreign molecule. We have, for example, one form of antibody that binds only to a particular molecule found on the measles virus and another form that binds only to a molecule found on the tuberculosis bacterium.

An antibody does not harm the microbe directly but brands it for destruction. Its main function is to clump the microbes, making it easier for macrophages and complement (described following) to find and destroy them. Each antibody is Y-shaped, with two identical binding sites so that it can attach to two microbes at the same time (fig. 22.5*a*). A group of antibodies, then, can hold many microbes together, as shown in figure 22.5*b*.

A second function of an antibody is to inactivate viruses and toxins by binding with their active sites. A virus cannot invade a cell when its receptors are covered over by an antibody; a toxin can do no harm when its active site is covered.

A third function of an antibody is to activate complement. **Complement** is a bundle of proteins that travels within the bloodstream at all times. The proteins are activated, by conversion into enzymes, when they contact a bacterium or an antibody that is bound to its microbe. Some of the enzymes then kill the microbe by digesting a hole in its outer membrane (fig. 22.6). Others guide macrophages to the clump of microbes and assist them in other ways. The activities of complement are complementary to the activities of antibodies: complement needs antibodies to identify the microbes and antibodies need complement to kill the microbes.

Table 22.6

The General Response

Component	Function
Macrophages	Devour microbes; identify microbes
Lymph	Carries microbes to lymph nodes
Lymph nodes	Filter microbes out of lymph; localize the fight against microbes
Helper T cells	Coordinate components of the immune system; stimulate the B cells to reproduce and secrete their antibodies
Killer T cells	Destroy abnormal body cells
B cells	Secrete antibodies
Fever	Enhances effects of interferons; increases reproduction of T cells; deprives bacteria of nutrients

The General Response

If the local response does not control the infection, the microbes enter the blood and lymph, where they spread throughout the body. A widespread defense, called the **general response,** is then brought into play. This response, summarized in table 22.6, has three lines of defense: macrophages within the vessels, an inflammatory response within the lymph nodes, and a fever.

Macrophages Within Vessels The first line of defense after the microbes have spread involves macrophages within the blood and lymph. These white blood cells creep along the inner walls of the vessels, searching out and devouring microbes.

The Inflammatory Response Within Lymph Nodes Microbes that survive the macrophages travel within the blood and lymph to the lymph nodes. **Lymph nodes** are organs, located through the body (see fig. 22.2), that filter microbes out of the lymph. They are packed with white blood cells.

When a microbe arrives at a lymph node, mast cells located there release histamine, which triggers an inflammatory response within the node. This inflammatory response is more powerful than the one in the local response, because the lymph nodes contain **lymphocytes**—white blood cells specific to each kind of microbe.

Lymphocytes are concentrated within lymph nodes but not confined to these organs. They circulate continuously within the bloodstream. At the capillaries, some of the lymphocytes leave the blood and enter the interstitial fluid (between the cells). This fluid accumulates as lymph and then enters lymph vessels and returns to a lymph node (see fig. 22.2b and c). The circulation of lymphocytes increases the likelihood that an invading microbe will encounter a lymphocyte programmed to recognize and destroy it.

There are several microbe-fighting lymphocytes. The most important in fighting an infection are the helper T cells, killer T cells, and B cells. T cells form in the bone marrow, like all blood cells, but mature in the thymus gland (see fig. 22.2). B cells form and mature within the bone marrow. As a T or B cell matures, its surface becomes studded with receptors—proteins embedded within the membrane that bind with a particular kind of protein or polysaccharide. Millions of different kinds of helper T cells, killer T cells, and B cells develop, one for each possible kind of protein or polysaccharide. The T and B cells with receptors that bind with the body's own molecules are then destroyed, leaving only the lymphocytes that bind with foreign molecules.

Thus we have at least one kind of helper T cell, killer T cell, and B cell against each kind of microbe and toxin. These cells reproduce, forming many more of their type, whenever they contact the microbe or toxin they are designed to fight.

A **helper T cell** coordinates the general response (fig. 22.7). It detects the presence of its particular microbe by pieces of the microbe's molecules (the flags) carried on macrophages that have devoured it. A receptor on the helper T cell then binds with a flag, an event known as the "kiss," and this stimulates the T cell to reproduce and to secrete chemical messengers that activate microbe-specific killer T cells and B cells.

A **killer T cell** destroys abnormal cells when activated by chemical messengers from helper T cells (fig. 22.8). Abnormal cells include infected cells, cancer cells, and foreign cells. A killer T cell recognizes an infected cell by fragments of the microbe's proteins displayed on the surface of the infected cell. It recognizes a cancer cell by unusual molecules on its surface and cells from someone else (part of a tissue or organ transplant) by foreign molecules on their surfaces. When a killer T cell binds with these foreign or unusual molecules, it is stimulated to reproduce and to secrete an enzyme that rips apart the abnormal cell.

A **B cell** manufactures antibodies specific to a particular foreign molecule. Some of a B cell's antibodies are embedded on its surface, where they serve as receptors for the foreign molecule. A B cell is activated by two events: a foreign molecule binds with its receptor and, at the same time, it receives chemical messengers from a helper T cell. The B cell responds by reproducing many times and secreting millions of copies of its antibody (fig. 22.9). The antibodies enter the bloodstream and travel throughout the body. Typically, it takes the B cells approximately ten days

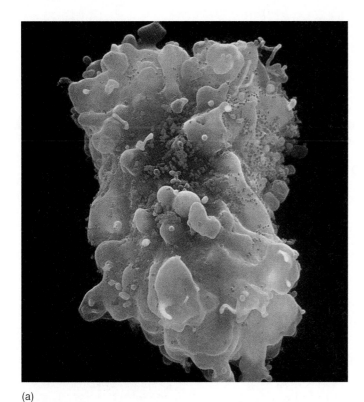

(a)

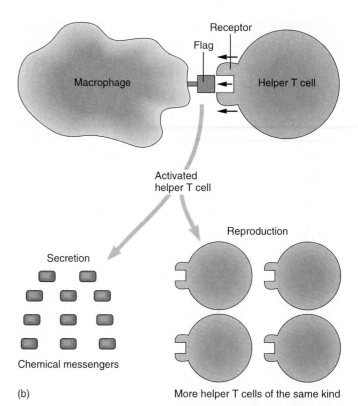

(b)

Chemical messengers

More helper T cells of the same kind

Figure 22.7 Helper T cell. (*a*) This helper T cell (X27,000; stained pink) is infected with HIV (stained blue), the virus that causes AIDS. (*b*) A receptor on the surface of a helper T cell bonds with a flag on a macrophage. The flag is

a piece of the specific microbe that the T cell fights. The binding stimulates the T cell to reproduce and to secrete chemical messengers.

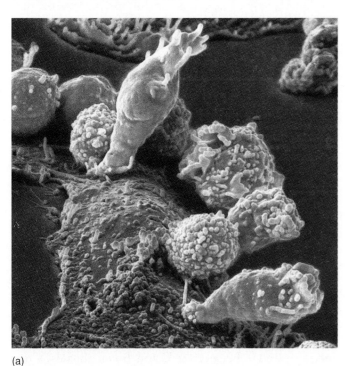

(a)

Figure 22.8 Killer T cells. (*a*) Killer T cells (stained green) attacking a cancer cell (stained orange). (*b*) A receptor on a killer T cell bonds with an abnormal molecule on an infected cell or a cancer cell. The bond stimulates the T cell to reproduce and secrete enzymes that destroy the abnormal cell.

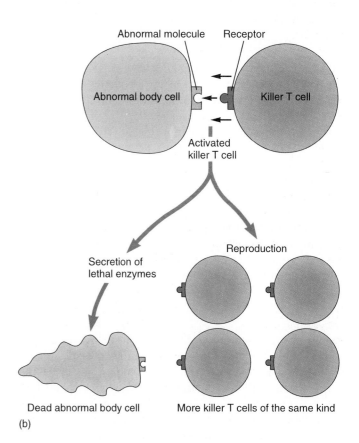

Dead abnormal body cell

More killer T cells of the same kind

(b)

415

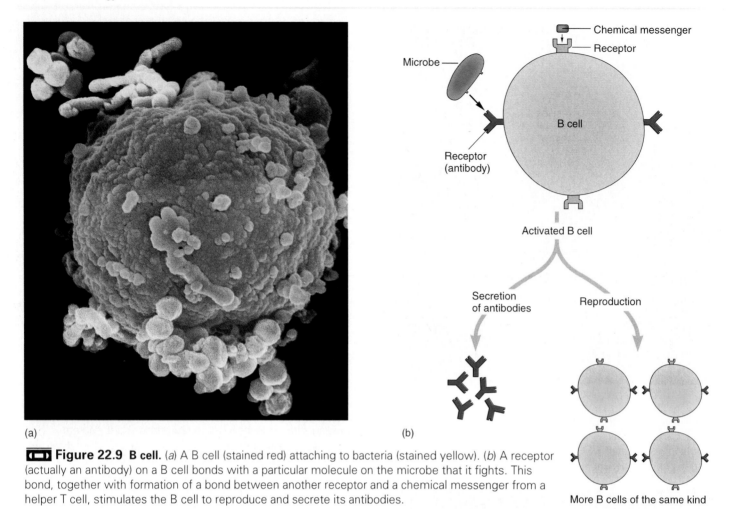

Figure 22.9 B cell. (*a*) A B cell (stained red) attaching to bacteria (stained yellow). (*b*) A receptor (actually an antibody) on a B cell bonds with a particular molecule on the microbe that it fights. This bond, together with formation of a bond between another receptor and a chemical messenger from a helper T cell, stimulates the B cell to reproduce and secrete its antibodies.

to manufacture enough antibodies to completely eliminate their particular microbe from the body. Thus, a serious infection typically lasts a week or so.

Fever During a fever, the body's thermostat is adjusted higher (to around 40° C, or 104° F) and held there until the infection is conquered. It develops in response to a chemical messenger secreted by the T cells and is part of the body's defense against diseases.

The elevated body temperature helps fight microbes by enhancing the effect of interferons, accelerating the reproduction of T cells, and removing minerals from the blood so that bacteria cannot use them as nutrients.

Immunity

Immunity is long-term protection against a particular disease. It develops from the accumulation of many T cells and B cells, as well as circulating antibodies, specific to the particular microbe that causes the disease. After development of immunity, the body contains so many of these specialized weapons that the microbes are destroyed as they attempt to invade.

Naturally Acquired Immunity Immunity acquired from actually having the disease (or being exposed to it) is naturally acquired immunity. During the first infection by a particular microbe, some of the T and B cells specific to that microbe are stored as "memory" cells within the lymph nodes, spleen, and bone marrow. During a second invasion by the same kind of microbe, the stored T and B cells quickly reproduce and become so numerous that they overwhelm the microbe before it can cause its illness (fig. 22.10). Ordinarily, you cannot get the same disease twice.

Artificially Acquired Immunity Immunity acquired from a vaccine is artificial immunity. A **vaccine** is a harmless form of a disease microbe: a dead microbe, a portion of the microbe or its toxin, or a living microbe that has been genetically altered so it can no longer cause its disease. When placed inside the body, a vaccine stimulates reproduction of microbe-specific T and B cells but does not cause the disease. Some of the newly formed T and B cells are stored as memory cells, just as when you actually have the disease. Ordinarily, you cannot get a disease after being vaccinated against it.

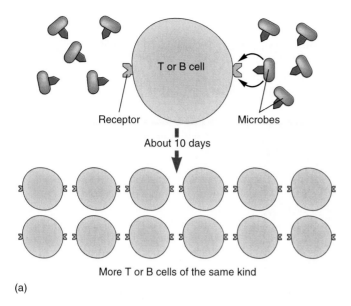

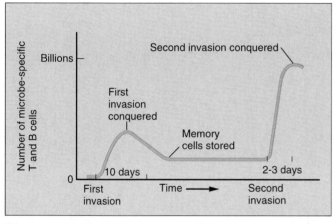

(a)

(b)

Figure 22.10 The development of immunity. (*a*) Contact with a microbe stimulates the microbe-specific T and B cells to reproduce. Enough T and B cells form, usually within ten days, to kill all the microbes. (*b*) The number of T and B cells increases during the first invasion by a microbe. Some of these cells are then stored, as memory cells, in the spleen, lymph nodes, and bone marrow. When the same microbe invades a second time, the T and B cells multiply so rapidly that the microbe is killed before it can cause its disease.

We still need to be vaccinated even though most infectious diseases have become rare. A discussion of vaccinations and the ones you should have is provided in the accompanying Wellness Report.

A less effective way of acquiring immunity is to transfer specific antibodies, which fight the specific disease, into the body. This passive form of immunity lasts only a few months and is less effective than immunity from an infection or a vaccination. It is used to combat unexpected infec-

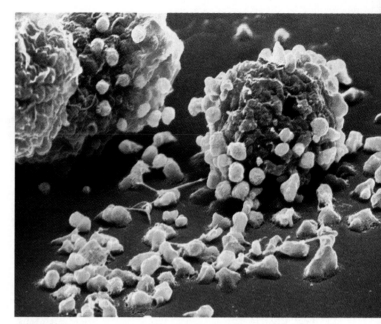

Figure 22.11 A mast cell releasing histamine. This particular kind of white blood cell (stained red) releases vast quantities of histamine (stained white) during an allergic reaction. The histamine triggers an extreme inflammatory response.

tions. Monoclonal antibodies, developed in the laboratory, may be used in novel ways to provide passive immunity, as described in the accompanying Research Report.

Undesirable Side Effects of the Immune System

Sometimes the immune system responds inappropriately and makes us ill. These undesirable side effects of our natural defenses cause allergies and autoimmune disorders.

Allergies   An **allergy** is an inappropriate overreaction of the immune system. The reaction is inappropriate because it develops in response to a foreign material that is harmless—one that is not a microbe. It is an overreaction in that the mast cells, located within connective tissues, release unusually large amounts of histamine (fig. 22.11) and other chemical messengers.

Almost any foreign material can trigger an allergic reaction. Some that commonly do so are dust, pollen, cat dander (discarded skin cells), insect stings, and a chemical in poison ivy. (Nettle actually secretes histamine directly onto its leaves.) Some common allergies and the tissues they affect are hives (skin), hay fever (nasal passages), asthma (bronchi), and eczema (skin).

During an allergic reaction, the large amount of histamine triggers an extreme inflammatory response. In addition to the normal symptoms of this response (redness,

Vaccinations

The scene is familiar to physicians: an infectious disease spreads rapidly across campus, characterized by a runny nose, watery eyes, cough, and fever—as high as 40.6° C (105° F). After a few days, a skin rash appears. Some students develop pneumonia, have seizures, or become deaf or blind. A few enter into comas and die. Is this disease caused by some new kind of virus? No, the disease is measles.

No American should get measles, for we have an effective vaccine against this disease (fig. 22.A). The vaccine was introduced in 1963 and

Figure 22.A A vaccination against measles.

since then the incidence of measles has declined rapidly, as shown in figure 22.B. As measles became rare, parents began to view it as no longer a threat to their children's health. Consequently, many children did not get vaccinated and this has caused an increase in the incidence of measles, as shown in figure 22.B. What many parents have failed to understand is that measles is rare *because* most children are vaccinated. All viruses (except smallpox) that cause infectious diseases, however rare, are still around—many are on you right now. If your immune system has never encountered one of these microbes, you are likely to get the disease.

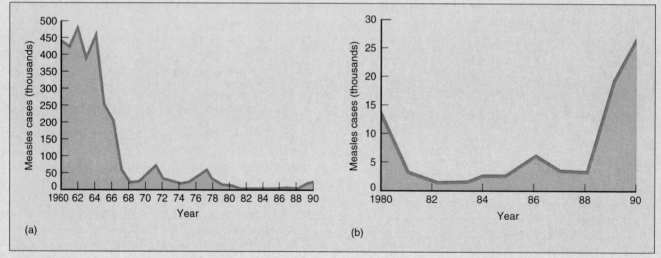

(a)

(b)

Figure 22.B The effects of vaccinations on the incidence of measles in the United States. (a) Since introduction of the measles vaccine in 1963, the incidence of this disease has declined dramatically. (b) Using a larger scale on the Y axis, the recent increase in measles can be seen more clearly. Measles is becoming more common because some children are not vaccinated and some received an ineffective vaccine.

warmth, swelling, and pain), there is sneezing, coughing, itching, labored breathing, or rashes, depending on what part of the body is affected. In a severe reaction, called anaphylactic shock (pronounced AN-ah-fil-AK-tik), the airways constrict and most of the arterioles dilate. There is difficulty breathing and the blood pressure drops so precipitously that the individual goes into cardiovascular shock (see page 368).

All these symptoms are side effects of the immune system's attack on what is otherwise a harmless material. An allergic reaction is treated with antihistamines, which are drugs that block the release of histamine, and with cortisone, a hormone that reduces swelling. Anaphylactic shock is treated with epinephrine, a hormone that expands airways and constricts the arterioles.

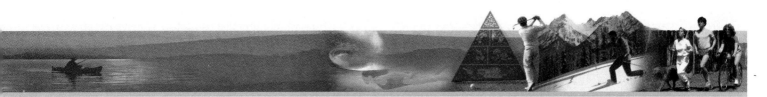

Measles and other childhood diseases are making a comeback among adults, primarily because we were not adequately vaccinated as children. And these diseases are much more serious in adults than in children. For some reason, not fully understood, the adult immune system is less effective in fighting the microbes that cause childhood diseases. The microbes tend to spread beyond their normal tissues, often invading the nervous system. Brain damage, paralysis, blindness, and deafness are not unusual. You should take measures to protect yourself from these diseases.

Most common diseases can be avoided by getting the nine vaccinations listed in table 22.A. To know which vaccinations you need, you have to find out your vaccination history. Ask your parents and your childhood doctor; consult school records. If you cannot find evidence of a vaccination but think you had either the vaccination or the disease, you can have your blood tested to find out whether you are carrying antibodies to that particular disease.

Research on vaccines is a very active field of medicine, due to the importance of vaccines in preventing illness and to new approaches provided by genetic engineering. Vaccines for genital herpes, meningitis, cholera, chicken pox, and Lyme disease are being developed. Find out about new vaccines as they become available and then have the vaccinations you need. It's not so bad—kids do it all the time.

Table 22.A

Vaccinations Every American Should Have

Disease	Vaccination Procedure	Effects of the Disease in Adults
Diphtheria	An initial series of three injections followed by a booster at age six or every ten years	Brain and heart damage
Hepatitis B	A series of three injections	Liver damage or cancer
Measles (rubeola)	A series of two injections; some of us were given an ineffective vaccine*	Brain damage; blindness; deafness
Mumps	A series of two injections; some of us were given an ineffective vaccine*	Deafness; sterility in men; miscarriages
Polio	An initial injection or liquid swallow followed by boosters; some of us were given an ineffective vaccine*	Permanent paralysis
German measles (rubella)	A series of two injections	Severe defects in the unborn
Tetanus	An initial series of three injections followed by a booster every 10 years	Painful, prolonged muscle contractions
Whooping cough (pertussis)	An initial series of three to five injections	Violent cough; pneumonia
Hemophilus influenzae type B	A series of at least two injections	Hearing loss; seizures; brain damage

*Vaccines used in the mid-1960s used dead viruses and were not very effective.

Autoimmune Disorders Sometimes the immune system attacks the very cells it is supposed to defend. Whenever the immune system attacks body cells, the condition is referred to as **autoimmunity.** Why this happens is not clear. Nor do we understand why many of these disorders, such as rheumatoid arthritis and lupus, are much more common in women than in men.

Autoimmune disorders were once considered part of the aging process, since they are characterized by a gradual deterioration of certain tissues and organs. Now autoimmunity is viewed as more than forty separate diseases that can, perhaps, be treated and even cured. Some of these diseases are listed in table 22.7.

Research Report

Monoclonal Antibodies

One of the triumphs of medical science is the mass production of antibodies. These weapons against disease are used to treat people who have been infected unexpectedly, as in a severe burn, animal bite, food poisoning (by hepatitis A), or open wound, when it is too late to be vaccinated. Injections of antibodies that bind with and inactivate viruses or bacterial toxins can be lifesaving.

Antibodies for use in humans were once made in other animals. To get antibodies against rabies (a fatal disease transferred through the bite of a dog, raccoon, skunk, bat, or fox), for example, the rabies virus was injected into a horse or rabbit. The animal responded by developing specific T cells, B cells, and antibodies against the virus. Blood was removed from the animal and the portion that contains antibodies, called gamma globulin, was isolated and stored in a refrigerator.

Someone bitten by a rabid animal was then injected with these antibodies.

This technique of manufacturing antibodies was slow and the injections were impure—other kinds of antibodies as well as other components of the blood were injected along with the microbe-specific antibodies. To really fight an infection, massive quantities of pure antibodies were needed. Standard techniques of recombinant DNA, in which a gene is placed inside a bacterium for synthesis of its protein, have proven difficult. The instructions for building an antibody are not located within a single gene but rather within a combination of many genes. Finding the right combination of genes for each antibody is difficult.

Researchers have tried to grow B cells in the laboratory, hoping they could develop large populations that synthesize vast quantities of specific microbes. Unfortunately, they cannot get human B cells to divide in a laboratory dish.

A breakthrough took place in 1971, when César Milstein and George Kohler (of Cambridge University, En-

gland) discovered they could fuse a B cell with a cancer cell, thereby taking advantage of the cancer cell's unusual capacity for cell division. The fused cell multiplied to form a clone (a group of identical cells), and all the cells in the clone secreted the same kind of antibody. Huge amounts of the specific antibody could be harvested. Such antibodies are called monoclonal antibodies because they are synthesized by cells of the same clone.

Unfortunately, Milstein and Kohler had to use mouse B cells because they couldn't get human B cells to fuse with cancer cells. (The cancer cells were human cells that cause melanoma.) To obtain mouse B cells against a particular disease, they injected a mouse with the microbe and then isolated the B cells it formed against that particular microbe.

While mouse antibodies have proven helpful in fighting diseases, they are not nearly as effective as hoped. The problem is that the immune system recognizes mouse antibodies as foreign proteins and de-

Table 22.7

Some Autoimmune Disorders

Disorder	Cause
Graves' disease	Antibodies attack the thyroid gland
Juvenile diabetes	T cells attack the insulin-secreting cells in the pancreas
Multiple sclerosis	T cells attack nerve cells
Myasthenia gravis	Antibodies attack the synapses between nerves and muscles
Psoriasis	T cells attack skin cells
Rheumatic fever	Antibodies attack heart muscles
Rheumatoid arthritis	T cells attack the linings of joints
Systemic lupus erythematosus (lupus)	Antibodies attack connective tissue in various parts of the body

Some Unconquered Diseases

For most infectious diseases, we can assist our own immune systems by treating the infection with antibiotics (against bacteria) or by preventing the illness with vaccinations (against all kinds of microbes). Some diseases, however, cannot be treated or prevented by these artificial means, and we must rely on our natural defenses. Two such diseases are influenza (the flu) and acquired immune deficiency syndrome (AIDS).

Influenza

One of the more common diseases of modern times is influenza, and it is one of the few illnesses that a person can get over and over again. This disease, also known as the flu, is a viral infection of the respiratory tract. It begins abruptly with these familiar symptoms: sore throat, runny nose, headache, fever, aching muscles, coughing, and extreme fatigue. The disease typically runs its course in three to seven days, although the fatigue may last much

stroys them within a few days. More-over, mouse antibodies do not recruit macrophages and T cells to the invasion site like human antibodies do.

Researchers are now developing ways of making mouse antibodies more acceptable to the human immune system. They have discovered that the base of the Y-shaped protein signals the species—whether the antibody is built by a human cell or a mouse cell. The two arms, by contrast, are the functional parts, which bind the antibody to its particular foreign molecule. If the base can be removed, or replaced by a base from a human antibody, the antibody may bind with its foreign molecule but not itself be recognized as foreign by the human immune system. Genetic engineers have developed more than a dozen such antibodies, some of which are shown in figure 22.C. These antibodies remain in the bloodstream much longer than mouse antibodies, and they recruit other components of the immune system to the invasion site. So far, "humanized" mouse antibodies show real promise.

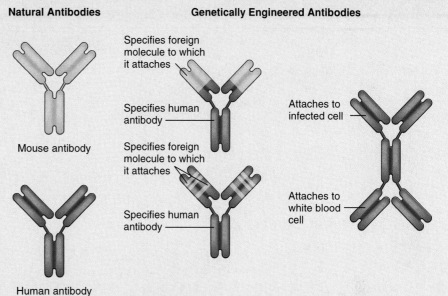

Natural Antibodies

Mouse antibody

Human antibody

Genetically Engineered Antibodies

Specifies foreign molecule to which it attaches

Specifies human antibody

Specifies foreign molecule to which it attaches

Specifies human antibody

Attaches to infected cell

Attaches to white blood cell

Figure 22.C Designer antibodies. Researchers use mice to manufacture antibodies from mouse genes, human genes, and various combinations of the two kinds. The base of the Y-shaped protein specifies whether it is formed from a mouse gene or a human gene. The two arms specify the molecule to which they bind.

longer. The virus passes readily from person to person through the air, especially in winter when people are crowded within closed rooms.

The influenza virus (see page 205) was one of the first viruses to be identified, in 1933. The molecular structure of this virus, as well as its mode of infection, is described in chapter 11. Our antibodies recognize the virus by two proteins—the H protein (hemagglutinin) and the N protein (neuraminidase)—that protrude from its surface.

Our immune systems fail to prevent repeated attacks of influenza because the H and N proteins on the virus keep changing. Antibodies developed against last year's H and N proteins may not bind with this year's proteins. Minor changes in viral proteins occur every year as the viral genes undergo small mutations. These new strains produce minor illnesses, since antibodies from last year's infection provide some protection. Major changes in viral proteins occur less often (every ten to twenty years) as the viruses acquire one or more entirely new RNA molecules. These major changes produce serious outbreaks of the illness.

What causes these genetic changes in the flu virus? The genes are distributed among eight molecules of RNA.

Minor changes are characteristic of all genetic material, but RNA mutates more often than DNA. Major changes, involving entire molecules of RNA, may be peculiar to the flu virus. They develop when flu viruses from a human and flu viruses from a bird come together within the cells of a pig. There, RNA molecules from the two viruses mix, and the newly formed viruses receive some molecules from the human viruses and some molecules from the bird viruses (fig. 22.12). Many new kinds of viruses can form from different combinations of RNA molecules from the human and bird viruses. This mixing of RNA molecules happens most often in Asia, where efficient farming methods bring farmers, pigs, and ducks in close proximity. Strains of the flu virus, named after their place of origin, usually have Asian names.

The H and N proteins on new forms of flu virus are not recognized by the stockpile of antibodies developed against previous infections with the flu virus. You can get the flu every year. To be protected by a vaccine, you have to be vaccinated every year with new vaccines developed from the new strains of virus.

(a)

(b)

Figure 22.12 The formation of new strains of the flu virus. (*a*) Many Asian farmers house their domestic animals so that the feces and urine of one kind of animal are used by another kind. Here, wastes from ducks are consumed by pigs, and wastes from pigs fertilize a fish pond. The close association among ducks, pigs, and farmers brings together different kinds of flu viruses. (*b*) The human virus (red) is exhaled by the farmer and inhaled by the pig. The bird virus (blue) leaves the duck by way of feces, which are eaten by the pig. Both viruses infect the same pig cells, where their RNA molecules become mixed. Newly formed viruses contain some RNA molecules from the farmer and some from the duck.

Acquired Immune Deficiency Syndrome

Acquired immune deficiency syndrome, or AIDS, is a viral infection of white blood cells, mainly the helper T cells that coordinate the general response. As the virus destroys these cells, the body becomes vulnerable to other infectious diseases and to cancer. The virus that causes AIDS is named human immunodeficiency virus (HIV; 📖 *see page 208*). Its structure and mode of infection are described in chapter 11.

Illness from an HIV infection may not develop for many years after the virus enters the body. Early symptoms include weight loss, fatigue, excessive sweating, and swollen lymph nodes. As the disease progresses, these symptoms are accompanied by persistent infections (fig. 22.13). Some patients also develop Kaposi's sarcoma, a cancer of the blood vessels that forms purple lesions in the skin. Persons with AIDS die of some other infectious disease, such as tuberculosis or pneumonia, due to their weakened immune systems.

AIDS gradually erodes the immune system. Within a few days after infection, B cells secrete HIV-specific antibodies. As the antibodies become numerous in the bloodstream, however, the viruses migrate out of the bloodstream and into the lymph nodes. There the viruses insert their genes into the chromosomes of helper T cells, where they hide from the immune system. Over the next few years, the viruses multiply, killing helper T cells in the process, and mutate into thousands of new genetic strains. Eventually the lymph nodes deteriorate and the newly formed viruses spill into the bloodstream. The existing antibodies, formed early in the infection, are not effective against the mutated strains. The immune system then attempts to develop new T and B cells to fight all the new

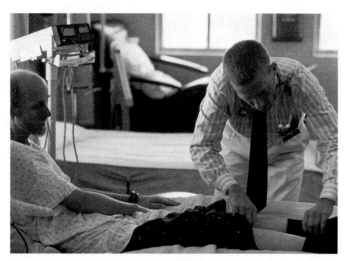

Figure 22.13 A person with AIDS.

strains. Odds are stacked in favor of the viruses, however, since most of the helper T cells, which coordinate the defense, were killed in the lymph nodes. The immune system is overwhelmed.

The HIV-specific antibodies form the basis of a blood test that determines whether or not a person has been infected with HIV. This test is not accurate until about six months after infection, for there are too few antibodies to detect (someone who is HIV positive may at first test HIV negative). Almost everyone who tests HIV positive eventually develops AIDS.

The AIDS virus spreads from one person to another when body fluids, especially blood and semen, of an infected person get into the body fluids of another person. The virus spreads readily among drug users who share needles. It also spreads by sexual intercourse, when the semen or vaginal fluids of an infected person enter the bloodstream of another person through a cut in the mouth, vagina, penis, or rectum. Someone with lesions from other sexually transmitted diseases (e.g., herpes or syphilis) is particularly susceptible. The virus also passes readily from an infected mother to her child during pregnancy.

Why have we failed to develop a cure or vaccine for this horrible disease? It is not for lack of research: the virus is the most intensely studied virus in history. One reason the research has not been fruitful, as yet, is that the viral genes are RNA molecules with unusually high rates of mutation; the virus changes rapidly. A second reason is the genetic material hides within the chromosomes of our own cells, where it cannot be distinguished from our own genes. Finally, AIDS research is impeded by ethical considerations—researchers cannot expose human subjects to this lethal virus, even in a dead or altered version.

At least for now, our best hope in fighting AIDS is to prevent infection. While illicit drug use is not encouraged, some communities are making an effort to supply drug users with disposable needles. Infection by way of sexual intercourse can be prevented by abstinence, few sexual partners, and the use of condoms.

SUMMARY

1. Infectious diseases result from host-parasite relations, in which two species live together and one (the parasite) benefits at the expense of the other (the host). The two species have adaptations that prevent them from killing each other. A disease microbe kills its host only when some change weakens the host's defenses.

2. Four categories of microorganisms cause human diseases: protozoa, fungi, bacteria, and viruses. Infections by protozoa and fungi are difficult to treat, for they have cells similar to our own. Bacteria are vulnerable to antibiotics. Viruses are difficult to treat for they live within our own cells.

3. The immune system's first line of defense includes barriers (the skin and internal linings) as well as poisons (salts, acids, and enzymes), mucus, and cilia on these barriers.

4. The second line of defense is the local response, which consists of interferons and the inflammatory response. Interferons prevent viruses from reproducing. The inflammatory response, triggered by histamine, brings clotting materials, macrophages, antibodies, and complement to the infected area.

5. The third line of defense is the general response, which consists of macrophages in the vessels and an inflammatory response within the lymph nodes. This response includes the activities of T and B cells specific to the particular microbe. Helper T cells coordinate the response, killer T cells destroy abnormal cells, and B cells produce microbe-specific antibodies. A fever enhances the effects of the immune system and deprives bacteria of nutrients.

6. Immunity is the buildup of so many microbe-specific T and B cells that the microbe cannot cause an illness. Naturally acquired immunity develops from being infected. Artificially acquired immunity develops from a vaccine or antibodies transferred into the body.

7. Some illnesses are caused by side effects of the immune system. An allergy develops when the immune system overreacts to a harmless foreign material. An autoimmune disorder develops when the immune system attacks the body's own cells.

8. Some diseases continue to plague us because we have no treatments or vaccines to assist our natural defenses. Our immune systems do not develop lifelong immunity to influenza because the virus regularly changes its surface proteins. Our immune systems do not win battles with the AIDS virus because the virus attacks the immune system itself, hides its genes within our own chromosomes, and regularly changes its surface proteins.

KEY TERMS

Allergy (page 417)
Antibody (ANN-tih-bah-dee) (page 411)
Autoimmunity (page 419)
B cell (page 414)
Complement (page 413)
General response (page 414)
Helper T cell (page 414)
Histamine (HISS-tah-meen) (page 409)
Host (page 406)
Immunity (ih-MYOO-nih-tee) (page 416)
Inflammatory (in-FLAH-mah-toh-ree) **response** (page 411)
Interferon (IN-tur-FEER-ahn) (page 409)
Killer T cell (page 414)
Local response (page 409)
Lymph node (limf nohd) (page 414)
Lymphocyte (LIM-foh-seyet) (page 414)
Macrophage (MAEH-kroh-fahje) (page 411)
Parasite (page 406)
Vaccine (page 416)

STUDY QUESTIONS

1. Discuss why a parasite does not normally kill its host.
2. Explain why antibiotics are not used to treat malaria or the flu.
3. Describe the local response to an infection. Why does the region become warm, red, swollen, and painful?
4. Describe how macrophages, helper T cells, killer T cells, and B cells fight an infection.
5. Describe how naturally acquired immunity develops.
6. What is a vaccine? How does it provide immunity?
7. Describe what happens during an allergic reaction.
8. Explain why we can only get most diseases once. Explain why we can get the flu almost every year.
9. Why is it so difficult to develop vaccines for the flu and AIDS?

CRITICAL THINKING PROBLEMS

1. Why have some bacterial infections, such as tuberculosis, become more common in the past ten years?
2. Why would someone with the flu probably not get a cold at the same time?
3. What is the pus that oozes from a wound?
4. Describe how a sore throat differs from swollen "glands" (which are really swollen lymph nodes).
5. Why do we have to get a flu vaccination every year to be protected from this disease?
6. Suppose you are hiking with a friend who is stung by a bee. She collapses. You feel her pulse and note that it is rapid and weak. Describe what has most likely happened within her body. What should you do to save her life?
7. Explain why antihistamines are used to treat allergies.

SUGGESTED READINGS

Introductory Level:

Levins, R., T. Awerbuch, U. Brinkmann, I. Eckardt, P. Eptstein, N. Makhoul, C. Albuquerque de Possas, C. Puccia, A. Spielman, and M. E. Wilson. "The Emergence of New Diseases," *American Scientist* 82 (January–February 1994): 52–60. A discussion of why new diseases are appearing and old ones are reappearing.

McNeill, W. H. *Plagues and People.* New York: Anchor Press/Doubleday, 1977. Written by a historian, this book traces the origins of infectious diseases and their role in human affairs throughout history.

Radetsky, P. *The Invisible Invaders: The Story of the Emerging Age of Viruses.* Boston: Little, Brown, 1991. An entertaining and informative book about viruses: their structure, activities, history, and the development of vaccines against them.

Rogers, N. *Dirt and Disease: Polio Before FDR.* New Brunswick, NJ: Rutgers University Press, 1992. A social and medical history of polio, focusing on the 1916 epidemic.

Scholtissek, C. "Cultivating a Killer Virus," *Natural History* (January 1992): 2–6. A description of how efficient farming methods give rise to major changes in the flu virus.

Advanced Level:

Eigen, M. "Viral Quasispecies," *Scientific American* (July 1993): 42–49. A description of the different kinds of viruses.

Goodenough, U. S. "Deception by Pathogens," *American Scientist* (July–August 1991): 344–55. An explanation of what the lymphocytes do and how they are fooled by microbes.

Johnson, H. M., F. W. Bazer, B. E. Szente, and M. A. Jarpe. "How Interferons Fight Disease," *Scientific American* (May 1994): 68–75. A description of what interferons do and how they may be useful in treating infectious diseases and cancer.

"Life, Death, and the Immune System," *Scientific American* (September 1993). The entire issue is devoted to immunity, including such topics as tissue transplants, cancer, AIDS, allergies, autoimmune diseases, and the future of the microbe-human relationship.

Rietschel, E. R., and H. Brade. "Bacterial Endotoxins," *Scientific American* (August 1992): 54–61. A description of endotoxins and how they affect the human body.

Travis, J. "Tracing the Immune System's Evolutionary History," *Science* (July 9, 1993): 164–65. A review of research on animals other than mammals, including an attempt to construct an evolutionary history.

🌑 EXPLORATION

Interactive Software: Immune Response

This interactive exercise allows students to explore the way the human body defends itself from cancer and against invasion by viruses and microbes by varying the effectiveness of the body's defenses. The exercise presents a cross section of the human body, with a variety of white blood cells circulating, whose numbers change in response to an infection. The "infection" may be a virus (HIV), a microbe (*Pneumocystis carinii*), or a cancer (Kaposi's sarcoma), each of which elicits a different response from the immune system. Students can investigate how the defense works by altering the numbers of particular cell types, raising or lowering the number of macrophages and lymphocytes.

23

Hormonal Control

Objectives

In this chapter, you will learn
–how hormones maintain optimal
  conditions within the body;
–how the secretion of each hormone is
  regulated;
–how a hormone affects what a cell
  does;
–the organization of the endocrine
  system;
–how hormonal control is linked with
  neural control.

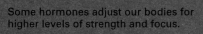

Some hormones adjust our bodies for
higher levels of strength and focus.

S ara was busy preparing dinner when she glanced out the window to check on Mark, her toddler, who was in his playpen. What she saw terrified her. A mountain lion was stalking her baby! As Sara ran outside she grabbed a heavy brass candlestick—the only object she saw that might serve as a weapon. When she reached the yard, the lion already had Mark in its jaws and was dragging him toward the woods. Sara lunged at the animal and beat on its head with the candlestick. The wounded animal eventually dropped the child and slouched away. Later, the lion was captured and examined by a veterinarian, who was amazed at the severe damage to its skull. How could a small woman like Sara wield such heavy blows?

Sara's mind and body were not functioning normally during the attack. When she first saw the lion, a surge of hormones flooded her body and commanded profound changes in cellular activity. The so-called fight-or-flight response prepared her for intense physical exertion. Her mind became exceptionally alert and focused on how to deal with the emergency situation. Her heart beat faster and her arteries shunted blood to the heart, brain, and skeletal muscles—the organs that brought about her remarkable performance. Her liver poured glucose into the bloodstream and her windpipe expanded to allow more air into the lungs. Pain-killing molecules flooded her body.

Unlike our more primitive ancestors, who dealt with life-threatening situations on a regular basis, we seldom need our fight-or-flight responses. But when we do face danger, our hormones can still rouse our bodies to extraordinary levels of strength and focus.

Hormonal Control

The activities of animal cells are controlled mainly by hormones and neurons. **Hormones,** the subject of this chapter, are chemical messengers secreted into the bloodstream by specialized organs called endocrine glands. A **chemical messenger,** in turn, is any molecule that is secreted (synthesized and released) by one kind of cell and that alters the activities of another kind of cell, called its **target cell.** Hormones are a category of chemical messengers, as shown in figure 23.1.

Endocrine Glands

An **endocrine gland** is a cluster of cells that are specialized for secreting a particular hormone. Most endocrine glands develop from epithelial tissues (see page 316), as described in chapter 17.

Endocrine glands are located in various regions of an animal's body. Their particular locations are not important because their products travel within the bloodstream to all cells of the body. The amount of hormone secreted by each gland, however, is important because it determines the activity of other cells.

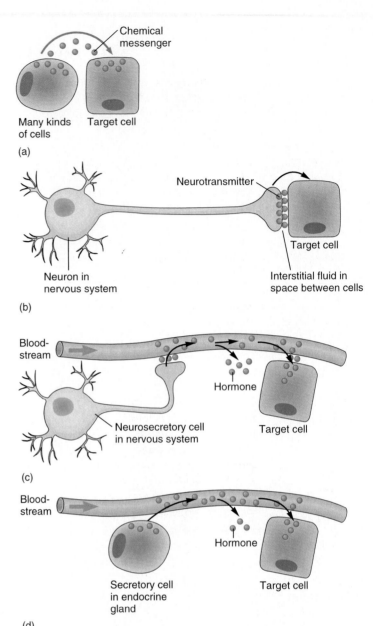

Figure 23.1 Categories of chemical messengers. (*a*) Most cells secrete chemical messengers that affect adjacent cells. (*b*) Neurons secrete neurotransmitters, which pass information to the next cell in the pathway. (*c*) Neurosecretory cells, which develop from the nervous system, secrete hormones into the bloodstream. (*d*) Secretory cells, which develop from epithelial tissue, also secrete hormones into the bloodstream.

The Amount of Hormone Secreted Each endocrine gland secretes more or less of its hormone in accordance with the body's need for the hormone. Some hormones maintain homeostasis, and the need for them depends on how far internal conditions have deviated from optimal. Other hormones adjust the body to unusual states, such as an emergency or pregnancy, and the need for these hormones is sporadic.

A gland that secretes a hormone involved in homeostasis (see page 313) operates by **negative feedback,** in which an increase in the consequence of the hormone's activity causes a decrease in secretion of the hormone, and vice versa. Glucagon, for example, stimulates liver cells to export glucose to the blood, increasing the concentration of glucose there. When the concentration of glucose is high, less glucagon is secreted; when the concentration of glucose is low, more glucagon is secreted.

Glands that coordinate activities other than homeostasis do not use negative feedback to govern their secretion. Their hormones adjust the body to different conditions; they bring about change rather than maintain a steady state. Epinephrine, for example, prepares the body for extreme physical exertion. Large amounts of this hormone are secreted when we are frightened or angry, and very little amounts are secreted when we are calm.

Stimulation of Secretory Cells Hormone secretion by a gland is controlled in one of three ways: by the concentration of a material in the bloodstream, by another hormone, and by the nervous system.

Some glands secrete more or less of their hormone in response to materials within the blood. Too high or too low a concentration of water, a particular inorganic ion, or a particular organic molecule in the bloodstream indicates the need for a particular hormone. Appropriate amounts of the hormone are then secreted to restore optimal concentrations of the material. The secretion of glucagon, for example, is governed by the concentration of glucose in the blood.

Other glands are controlled by hormones secreted by other glands. The secretion of sex hormones by the ovaries and testes (which are glands as well as organs that form eggs and sperm), for example, is triggered by hormones from the pituitary gland, which is located just beneath the brain.

Few glands are controlled by the nervous system. The brain receives information about what is happening inside and outside the body, evaluates the information, and then commands the gland to secrete more or less of its hormone. The secretion of epinephrine, for example, is controlled by emotional centers within the brain.

How Hormones Affect Their Target Cells

A hormone in the bloodstream reaches all the body cells but affects only its target cells. These cells have specific **receptors**—protein molecules to which the hormone binds. Some receptors are located on the cell surface, whereas others are inside the cell. Only the hormone's target cells have its specific receptor. Binding of the hormone to its receptor triggers a change in the cell's activity.

A hormone alters its target cells by either altering the rate of normal cellular activities or altering the permeability of the cell membrane to certain materials. Exactly how a hormone accomplishes this depends in part on its molecular form.

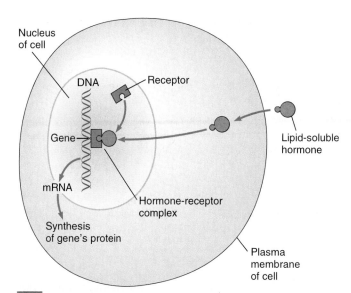

Figure 23.2 A lipid-soluble hormone directly affects gene activity. A steroid hormone (or thyroid hormone) enters its target cell and joins with its specific receptor. The hormone-receptor complex then binds with a particular gene, stimulating it to begin synthesizing its protein.

Lipid-Soluble Hormones Lipid-soluble hormones exert control from within their target cells. Because they dissolve in lipids, these hormones pass directly through the lipid framework of plasma membranes (see page 60) and bind with receptors located inside the cells. Lipid-soluble hormones enter all cells, but only their target cells contain their particular receptors.

A lipid-soluble hormone commands its target cell to step up production of a particular protein (fig. 23.2). It does this by binding with its receptor (a transcription factor; see page 200), forming a complex that attaches to a specific gene in a chromosome. Attachment of the hormone-receptor complex activates transcription—stimulates the gene to initiate synthesis of its protein. Most often this protein is an enzyme that accelerates a particular biochemical reaction. Testosterone, for example, enters cells of the testes, where it triggers the synthesis of enzymes that control sperm development.

There are two categories of lipid-soluble hormones: steroid hormones and thyroid hormones. The main steroid hormones are cortisol, aldosterone, estradiol (a form of estrogen), progesterone, and testosterone (see fig. 2.5). The thyroid hormones, named thyroxine and triiodothyronine, are modified amino acids. Iodine atoms within these molecules make them soluble in lipids.

Water-Soluble Hormones Water-soluble hormones exert control from the outer surface of their target cells. All hormones except the steroid and thyroid hormones described previously are soluble in water. Water-soluble hormones are formed of molecules—typically modified amino acids (but without iodine) or small proteins—that cannot pass directly through the lipid framework of a cellular membrane.

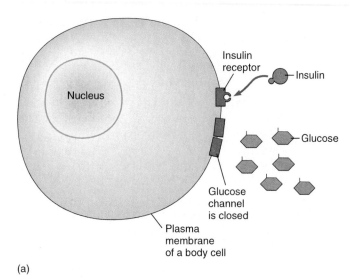

(a)

Figure 23.3 Some hormones alter the permeability of the plasma membrane. (a) Insulin is a water-soluble hormone that controls its target cells by altering what enters and leaves the cell. Insulin binds with its receptor, embedded within the

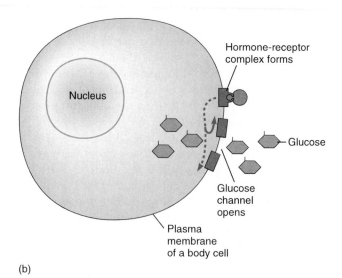

(b)

plasma membrane, to form a hormone-receptor complex. (b) Formation of this complex opens channels that allow glucose to enter the cell.

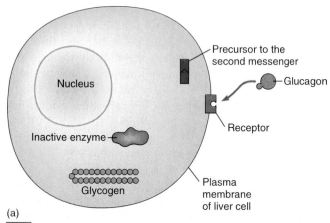

(a)

Figure 23.4 Some hormones stimulate production of intracellular messengers. (a) Glucagon is a water-soluble hormone that binds with receptors on liver cells. (b) Binding of

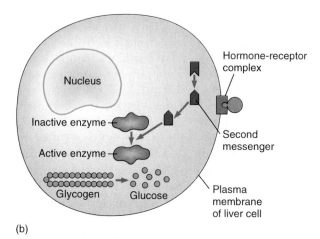

(b)

the hormone with its receptor triggers formation of a second messenger, which activates the enzymes that convert glycogen to glucose.

A water-soluble hormone binds with its specific receptor on the surface of its target cell. The binding triggers either a change in the permeability of the plasma membrane or activation of other messengers inside the cell.

Hormones that change the permeability of the plasma membrane alter the kinds of materials that enter or leave the cell. Insulin, for example, is a protein hormone that binds with a receptor on the cell surface. The hormone-receptor complex then opens channels that allow entry of glucose (fig. 23.3).

Hormones that use intracellular messengers also bind with receptors on the cell surface. The hormone-receptor complex activates messengers that carry the signal from the cell surface to either a gene, where it initi-

ates protein synthesis, or a biochemical pathway, where it alters enzyme activity. When just one messenger carries the signal, it is called a second messenger. Most often this messenger is a calcium ion or a molecule, called cyclic AMP, formed from ATP. When a series of messengers carries the signal, it is called a signal pathway. Most often these messengers are proteins.

Glucagon, for example, is a protein hormone that binds with a receptor on the surface of liver cells. Formation of the hormone-receptor complex stimulates production of a second messenger (most likely cyclic AMP), which activates the enzymes that convert glycogen to glucose (fig. 23.4). The liver cells then release glucose into the bloodstream, raising the blood-glucose level.

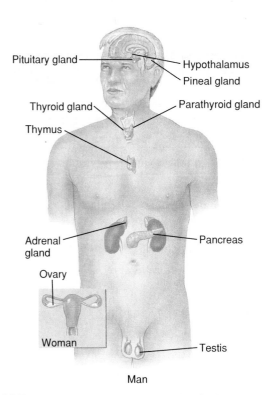

Pituitary gland — — Hypothalamus
— Pineal gland
Thyroid gland — — Parathyroid gland
Thymus —
Adrenal gland — — Pancreas
Ovary
Woman
— Testis
Man

Figure 23.5 The major endocrine glands of a human.

The Endocrine System

The human endocrine system consists of about a dozen glands that together secrete more than twenty-five hormones. Just how many glands there are depends on how a gland is defined. Some so-called glands are actually two glands in one, and some tissues that secrete hormones are not always called glands. Locations of the major glands in a human are shown in figure 23.5. A list of these glands and their hormones is provided in table 23.1.

The Hypothalamus and the Pituitary

The brain and pituitary gland are intimately associated. A region of the brain, the **hypothalamus,** lies at the base of the brain and the pituitary gland extends just beneath it (fig. 23.6a). The two structures are connected by a narrow stalk that contains blood vessels and extensions of neurons.

The hypothalamus is both a region of the brain and an endocrine gland. Some of its cells are typical neurons and others are secretory cells, called **neurosecretory cells** to indicate they are neurons that secrete hormones. It is a major center of homeostasis, maintaining internal conditions by commands sent via neurons and hormones.

One function of the hypothalamus is to control the pituitary gland, which, in turn, controls several other glands. Thus, the hypothalamus forms a direct link between the nervous system and the endocrine system.

The **pituitary** is really two distinct structures: a posterior pituitary, which develops from nervous tissue, and an anterior pituitary, which develops from epithelial tissue (fig. 23.6b).

Posterior Pituitary The posterior pituitary is formed of neurosecretory cells that extend from the hypothalamus. One end of each neurosecretory cell lies within the hypothalamus and synthesizes the hormone; the other end lies within the posterior pituitary and releases the hormone into the bloodstream (fig. 23.6b). Together, the hypothalamus and posterior pituitary form a complete gland.

Two hormones, antidiuretic hormone (ADH, also known as vasopressin; see page 396) and oxytocin, are synthesized in the hypothalamus and released by the posterior pituitary. Antidiuretic hormone affects kidney cells, altering their permeability to water. In the presence of this hormone, more water is returned to the blood rather than eliminated in the urine. Oxytocin stimulates muscle cells in a woman's uterus to contract during childbirth and secretory cells in her breasts to release milk when the baby suckles.

Anterior Pituitary The anterior pituitary synthesizes and releases seven hormones: three hormones that directly affect cells outside the endocrine system and four hormones that regulate other glands. Hormone secretion by the anterior pituitary, in turn, is controlled by hormones from the hypothalamus. The hypothalamus secretes these hormones, called hypothalamic-releasing hormones, into the bloodstream, where they travel directly to the anterior pituitary (fig. 23.6b).

Three hormones of the anterior pituitary gland—prolactin, melanocyte-stimulating hormone, and growth hormone—directly affect the activities of cells outside the endocrine system. Prolactin stimulates the synthesis of milk by target cells within a woman's breasts after childbirth. (It appears to have no function in a man.) Melanocyte-stimulating hormone apparently increases skin pigmentation by dispersing melanin (a brown pigment) within melanocytes of the skin. Growth hormone stimulates body growth and increases the ratio of muscle to fat by accelerating cell division in muscle, bone, and cartilage. A child with too much growth hormone may become a giant (fig. 23.7a). A child with too little may become a dwarf. Children with too little growth hormone can grow to normal size when treated with growth hormone manufactured by recombinant DNA technology, a form of genetic engineering.

What about Pygmies? No one in a Pygmy population (fig. 23.7b) is taller than 1.5 meters (4 ft, 11 in). Initially, researchers hypothesized that these people secrete unusually small amounts of growth hormone. Surprisingly, the hypothesis turns out to be false—a Pygmy secretes the

Table 23.1

Major Glands and Hormones of the Human Endocrine System

Gland	Hormone	Effects
Hypothalamus	Releasing hormones	Secretion of hormones by anterior pituitary
Hypothalamus and posterior pituitary	Antidiuretic (or vasopressin)	Excretion of less water
	Oxytocin	Contractions of uterus; release of milk
Anterior pituitary	Prolactin	Synthesis of milk
	Melanocyte-stimulating hormone	Increased pigmentation of the skin
	Growth hormone	Body growth; increased ratio of muscle to fat
	Adrenocorticotropic hormone	Secretion of hormones by adrenal cortex
	Thyroid-stimulating hormone	Secretion of hormones by thyroid
	Follicle-stimulating hormone	Secretion of hormones by ovaries or testes
	Luteinizing hormone	Secretion of hormones by ovaries or testes
Thyroid	Thyroxine/triiodothyronine	Acceleration of cellular activities
	Calcitonin	Removal of calcium ions from the blood
Parathyroids	Parathyroid	Addition of calcium ions to the blood
Adrenal cortex	Mineralocorticoids	Regulation of ions and water
	Glucocorticoids	Fight-or-flight response
	Gonadocorticoids	May contribute to development of male or female traits
Adrenal medulla	Epinephrine	Fight-or-flight response
	Norepinephrine	Fight-or-flight response
Pancreas	Insulin	Removal of glucose from blood
	Glucagon	Addition of glucose to blood
	Somatostatin	Inhibition of the secretion of insulin and glucagon
Thymus	Thymosin	Maturation of T cells
Pineal	Melatonin	Regulation of daily and seasonal rhythms
Ovaries	Estradiol (a form of estrogen)	Development of female traits
	Progesterone	Maintenance of pregnancy
Testes	Testosterone	Development of male traits

same amount of growth hormone as a taller person. It seems that the shorter stature is caused by fewer receptors on the target cells. Thus, the same amount of hormone has less effect on muscle, bone, and cartilage cells. The short stature apparently adapts these people to life in a jungle.

The remaining four hormones of the anterior pituitary govern the activities of other endocrine glands. They are part of a hierarchy of control: from hypothalamus to anterior pituitary to other glands to target cells outside the endocrine system (fig. 23.8). The target cells, in turn, provide information to the hypothalamus about the need for each hormone. Adrenocorticotropic hormone (ACTH) stimulates the adrenal cortex (part of the adrenal gland) and thyroid-stimulating hormone acts on the thyroid gland. Follicle-stimulating hormone and luteinizing hormone influence secretion of sex hormones by the ovaries and testes. (They also have direct effects on the development of eggs and sperm, as described in chapter 28.)

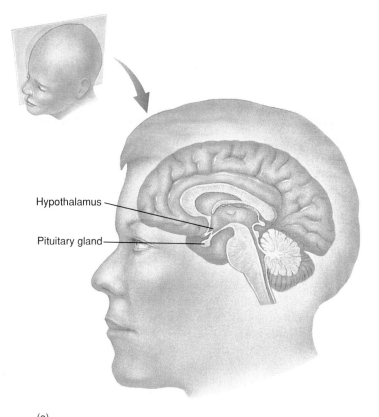

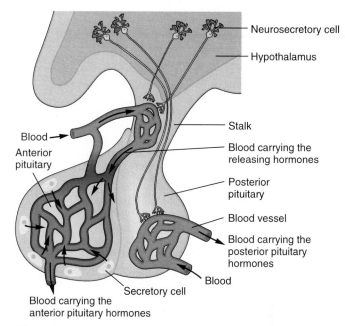

(a)

(b)

Figure 23.6 The hypothalamus and the pituitary. (*a*) The hypothalamus is located in the base of the brain (above the roof of the mouth). The pituitary gland is attached to the underside of the hypothalamus. (*b*) The pituitary gland consists of an anterior pituitary and a posterior pituitary. The posterior pituitary is formed of neurosecretory cells that extend from the hypothalamus. The anterior pituitary is formed of secretory cells, which develop from epithelial tissue, that release hormones into the bloodstream. This gland is controlled by hormones secreted into the bloodstream by neurosecretory cells in the hypothalamus.

(a)

(b)

Figure 23.7 The effects of growth hormone. (*a*) This boy, only seven years old, is already 1.8 meters (6 ft) tall and weighs 90 kilograms (200 lb). A tumor in his anterior pituitary has generated too many secretory cells and, consequently, too much growth hormone. (*b*) Pygmies are short because their bone, muscle, and cartilage cells have about half as many receptors for growth hormone as the cells of taller people in other populations.

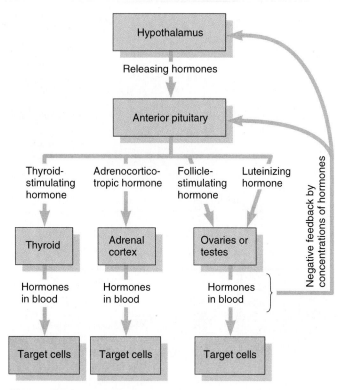

Figure 23.8 Hierarchy of control: from hypothalamus to glands. The hypothalamus secretes releasing hormones, which control hormone secretion by the anterior pituitary. The anterior pituitary responds to these hormones by secreting thyroid-stimulating hormone, adrenocorticotropic hormone, follicle-stimulating hormone, and luteinizing hormone. These hormones, in turn, stimulate secretion by the thyroid, adrenal cortex, and ovaries or testes. Hormones from these glands then control their target cells. They also control, by negative feedback, secretion by the anterior pituitary and hypothalamus: when the thyroid, adrenal cortex, and ovaries or testes secrete too much of their hormones, the anterior pituitary and hypothalamus secrete less of their hormones.

Thyroid Gland

The thyroid gland is located within the neck, alongside the trachea (fig. 23.9). It secretes three hormones: thyroxine, triiodothyronine, and calcitonin.

Thyroxine and triiodothyronine are secreted in response to thyroid-stimulating hormone from the anterior pituitary gland (see fig. 23.8). Their molecular structures are similar, built in part from iodine atoms, and their target cells include most body cells. In a child, these two thyroid hormones are essential for normal physical and mental development. In an adult, they accelerate the activities of almost every body cell.

Adults with too little thyroid hormone have abnormally low rates of cellular activity. They feel tired and depressed, are easily chilled, and tend to gain weight. Most often this condition develops because the thyroid gland does not receive enough iodine from the diet. The thyroid gland responds by enlarging, which helps it build thyroxine and triiodothyronine from what little iodine it has. An enlarged thyroid, known as a goiter (fig. 23.10a), used to be common wherever the soil was deficient in iodine. Nowadays most people get enough iodine from eating iodized salt and dairy products from cows that have been fed iodized grain.

Someone with too much thyroid hormone has abnormally high rates of cellular activity. They feel nervous and irritable, are easily overheated, tend to lose weight, and have bulging eyes (fig. 23.10b). This condition may be caused by a tumor, in which there are many more secretory cells than normal, or autoimmunity (Graves' disease), in which the secretory cells are overstimulated by antibodies that mimic the receptors for thyroid-stimulating hormone.

Calcitonin, the third hormone of the thyroid gland, removes calcium ions from the blood. Calcitonin-secreting cells in the thyroid monitor the body fluids and secrete more of their hormone when they detect too many calcium ions. The hormone stimulates bone cells to take in and store more calcium ions.

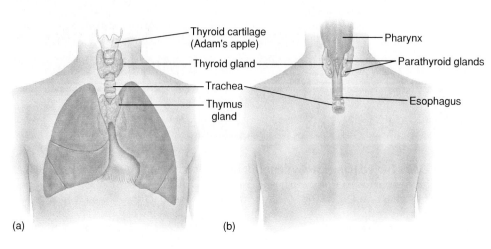

(a)   (b)

Figure 23.9 The thyroid, parathyroid, and thymus glands. (a) In this front view of a human neck and chest, the thyroid gland and thymus gland can be seen next to the trachea (air channel). (b) In this rear view of a human neck, portions of the thyroid gland can be seen next to the trachea and esophagus (food channel). The four parathyroid glands lie on the back side of the thyroid gland.

(a)

Figure 23.10 Disorders of the thyroid gland. (*a*) A goiter is an enlarged thyroid gland, caused typically by an iodine-

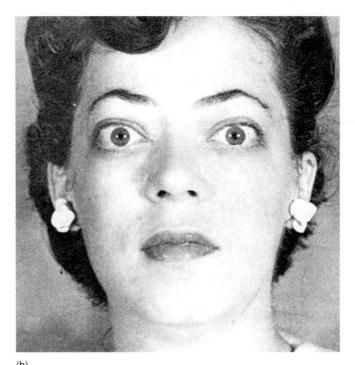

(b)

deficient diet. (*b*) Protruding eyes are symptomatic of an overactive thyroid gland.

Parathyroid Glands

Four tiny glands, the parathyroids, lie close to the thyroid gland (see fig. 23.9). They secrete parathyroid hormone, which helps regulate the concentrations of ions in the blood. Parathyroid hormone stimulates cells within the bones, kidneys, and small intestine to transfer more calcium and magnesium ions into the blood, and it stimulates cells within the kidneys to excrete more phosphate ions. Together, parathyroid hormone and calcitonin (from the thyroid) maintain appropriate amounts of calcium ions in the blood.

Adrenal Glands

There are two adrenal glands, one on top of each kidney. Each gland is actually two glands in one—an outer layer, called the cortex, and a center, called the medulla (fig. 23.11).

Adrenal Cortex Hormones of the adrenal cortex are secreted in response to adrenocorticotropic hormone from the anterior pituitary (see fig. 23.8). These hormones are steroids, called corticoids, of two kinds: the mineralocorticoids and the glucocorticoids. (A third kind, the gonadocorticoids, can influence sexual development but typically are secreted in such small amounts they have no noticeable effect.)

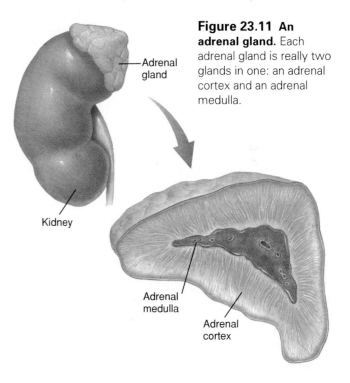

Figure 23.11 An adrenal gland. Each adrenal gland is really two glands in one: an adrenal cortex and an adrenal medulla.

Adrenal gland

Kidney

Adrenal medulla

Adrenal cortex

Mineralocorticoids regulate the amounts of inorganic ions (minerals) and water in the body fluids. The main hormone of this kind is aldosterone (*see page 396*), which alters the permeability of kidney cells to sodium so that fewer of these ions are excreted in the urine. When too many mineralocorticoids are secreted, the body fluids retain too many ions and too much water. Edema develops as fluid accumulates within the interstitial fluids, and high blood pressure develops as fluid accumulates within the blood.

Glucocorticoids help develop the fight-or-flight response during times of physical stress. The main form of this hormone is cortisol, which has many target cells. Cortisol suppresses the activity of white blood cells, thereby reducing the inflammatory response (swelling and pain) associated with injuries during strenuous exercise. It also provides more fuel for cellular respiration by stimulating fat cells to release fatty acids and other kinds of cells to convert fats and proteins into glucose. Cortisol also works with epinephrine to help raise the blood pressure.

Someone with a glucocorticoid deficiency tends to be weak (from low blood sugar and blood pressure) and to lose weight. Someone with an excess is susceptible to infections and tends to gain weight, especially as fat on the back and shoulders.

Physicians and athletic trainers administer cortisone (a molecule similar to cortisol) to reduce the swelling from arthritis, allergic reactions, autoimmune disorders, and injuries. These treatments, when prolonged, produce the same health problems as when the adrenals secrete excess amounts of glucocorticoids (infections and weight gain). A serious problem with cortisone is addiction—within a week of treatment, the adrenal glands stop secreting cortisol and the body needs the cortisone medication to replace it. Withdrawal of cortisone is followed by a period of cortisol deficiency (with weakness and weight loss) that may last many weeks before the adrenal cortex returns to its normal levels of secretion.

Adrenal Medulla   The adrenal medulla contains neurosecretory cells that release two hormones, epinephrine and norepinephrine (also known by their brand names, adrenalin and noradrenalin) into the bloodstream. The hormones are secreted in response to neural commands from the hypothalamus, which reach the adrenal medulla by way of neurons.

Epinephrine and norepinephrine help the body deal with stress by developing, along with the adrenal cortex and the nervous system, the fight-or-flight response, in which the whole body is geared up for emergency action. Hormones involved in developing this response are diagrammed in figure 23.12. During prolonged stress, the continued presence of these hormones can damage the heart, brain, and blood vessels. Such stress can also overwork the adrenal glands, resulting in a decreased ability to secrete cortisol, epinephrine, and norepinephrine. Without these hormones, the body cannot adjust to physical exertion, extreme cold, infections, or intense emotions.

Pancreas

The pancreas is located within the abdominal cavity near the stomach (see fig. 23.5). It is both an exocrine gland, releasing digestive materials into the small intestine, and an endocrine gland, releasing hormones into the bloodstream (fig. 23.13). The pancreas secretes three hormones: insulin, glucagon, and somatostatin.

Insulin and glucagon govern the concentration of glucose in the blood. Insulin removes glucose from the blood by making the plasma membranes of muscle, fat, and liver cells more permeable to glucose, enabling glucose to enter, and by stimulating liver cells to convert more glucose into glycogen. Glucagon adds glucose to the blood by stimulating liver cells to convert glycogen back into glucose. In this way, the two pancreatic hormones maintain an optimal blood-sugar level. The secretion of insulin and glucagon, in turn, is inhibited by somatostatin.

When the pancreas does not secrete enough insulin or the body cells do not respond to insulin, the body cells do not receive enough glucose and cannot use this high-quality fuel. This disorder is described in the accompanying Wellness Report.

Thymus Gland

The thymus gland, located within the upper chest (see fig. 23.9), secretes thymosin. This hormone controls maturation of certain white blood cells called T cells. The thymus gland is most active during infancy, the time in life when T cells develop their specific microbe-fighting properties. As a child grows older, the gland becomes smaller and secretes less thymosin. The thymus gland of an adult is barely visible.

Pineal Gland

The pineal gland is a tiny structure located near the center of the brain (fig. 23.14). It secretes melatonin in response to neural commands from the hypothalamus, which receives information about light by way of neurons from the eyes.

The function of melatonin is to inform body cells about the time of the day and season of the year. This information, in turn, is used to govern daily and seasonal cycles of cellular activity.

Body cells monitor the amount of melatonin in the body. This hormone is secreted only at night, when the hypothalamus receives no information about light from the eyes. It accumulates during the night and reaches a peak concentration at dawn. No melatonin is secreted during

Figure 23.12 Some of the ways in which stress triggers the fight-or-flight response.

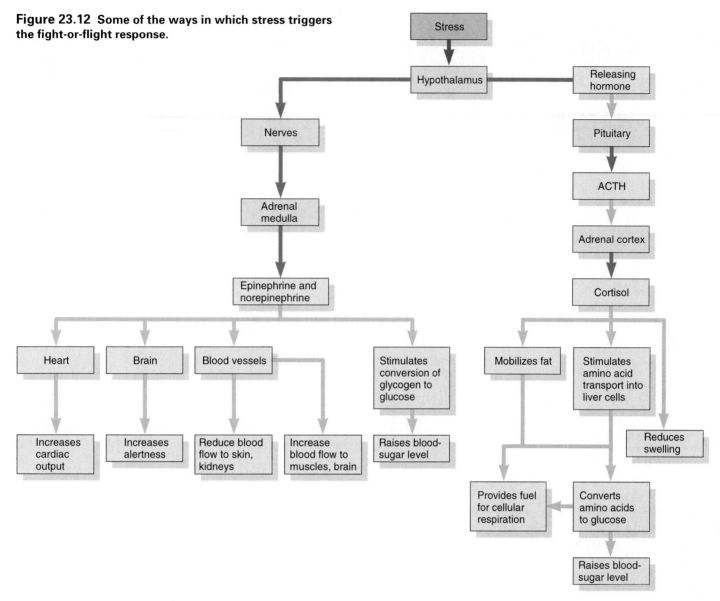

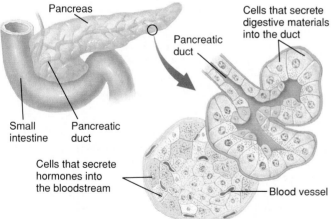

Figure 23.13 The pancreas is both an exocrine gland and an endocrine gland. As an exocrine gland, the pancreas secretes digestive materials into a duct that empties into the small intestine. As an endocrine gland, the pancreas secretes hormones into the bloodstream.

the day, when the hypothalamus receives information about light. The concentration of this hormone declines steadily, reaching a low at dusk (fig. 23.15).

In this simple way, the amount of melatonin in the body reflects the time of day and night. On a yearly basis, more melatonin is secreted during the long winter nights than during the short summer nights. Thus, differences in the total amount of melatonin that accumulates during a night reflect the season of the year. Body cells respond to these differences by initiating seasonal activities.

In birds and in mammals other than humans, the amount of melatonin is known to govern such seasonal activities as reproduction, migration, and hibernation. In us, it appears to govern daily cycles of activity, such as sleep cycles, as well as jet lag and seasonal disorders, such as winter depression. These cycles and the factors that govern melatonin concentration are described in the accompanying Research Report.

Wellness Report

Diabetes Mellitus

Diabetes mellitus is a serious disorder that affects one in every twenty Americans. It develops when the pancreas secretes too little insulin or the body cells cannot use it effectively.

The function of insulin is to open channels in a cell membrane so that glucose can enter the cell. It does this by binding with its specific receptor in the plasma membrane of its target cells, which include muscle, liver, and fat cells. The hormone-receptor complex then triggers nearby proteins in the membrane to change shape, forming a channel through which glucose can pass. In the absence of insulin or its receptor, glucose cannot enter a cell. This important fuel normally provides energy for cellular activities in muscle cells and is stored in liver and fat cells. If glucose cannot enter these cells, it accumulates in the bloodstream (fig. 23.A) and produces a variety of health problems. There are two kinds of diabetes mellitus: type I diabetes and type II diabetes.

Type I diabetes, also known as juvenile-onset diabetes or insulin-dependent diabetes, is an autoimmune disorder in which antibodies and killer T cells attack the insulin-secreting cells of the pancreas. As these cells die, less and less insulin is secreted. This type of diabetes, which accounts for about 10% of all diabetes cases, develops during childhood or adolescence.

Type II diabetes, also known as adult-onset diabetes or noninsulin-dependent diabetes, typically develops after age thirty. In most cases, it is caused by a decline in the number of insulin receptors on muscle, liver, and fat cells. Most diabetics are considerably overweight. A diet high in sugar and fat leads to obesity and appears to

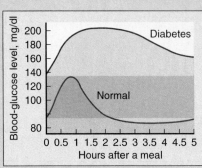

Figure 23.A Blood-sugar levels as a function of time after a meal.
Large amounts of glucose are absorbed into the blood as a meal is digested. Normally, enough insulin is present to transfer most of the excess glucose out of the blood and into body cells within an hour or so. In untreated diabetes, there is not enough insulin to remove excess glucose from the blood. Over a period of several hours, the excess glucose is gradually excreted in the urine.

eliminate insulin receptors from their target cells. A cell with fewer receptors takes in less glucose. Normally, the pancreas compensates for fewer receptors by secreting more insulin. In some people, however, the pancreas cannot step up its insulin secretion and diabetes develops.

Symptoms of diabetes include extreme fatigue, frequent urination, increased thirst, numbness in the feet, frequent skin infections, and blurred vision. Except for fatigue, which develops because cells lack glucose, the symptoms of diabetes are caused by too much glucose in the blood.

The kidneys work to eliminate excess glucose in the blood. Glucose filtered from the blood is diluted with water to form urine. The more glucose in the blood, the more water required to dilute it. Consequently, a diabetic forms large amounts of urine each day

and is almost always thirsty. Many diabetics suffer from kidney damage.

Excess blood glucose thickens capillary walls, impeding flow, and damages nerves. The feet in particular suffer nerve damage and become numb, causing the diabetic to stumble frequently and suffer cuts and bruises. These wounds do not heal well, because of poor blood flow, and may develop into gangrene (tissue death from insufficient blood). The feet may have to be amputated to save the person's life.

Blurred vision develops from damage to cells in the eyes. Cells in the retina and lens absorb glucose without the help of insulin. They take in too much glucose when there is too much glucose in the blood. The excess glucose is converted into an alcohol, which draws water into the cells by osmosis. The cells swell and die. Diabetes is the leading cause of blindness in the United States.

Diabetes is treated in several ways. Type I diabetics must be given insulin to live. And it must be administered by injection: insulin is a protein that, when taken by mouth, is broken apart by digestive enzymes. New treatments being developed include transplants of pancreatic tissues enclosed within membranes that are not permeable to the immune system and techniques for a continuous delivery of insulin without injection. Type II diabetes can be controlled, typically, by changes in life-style that help maintain a normal blood-sugar level. Most important is to adopt a diet high in fiber and low in both sugar and fat. Next in importance is weight loss, which apparently restores the normal density of insulin receptors on body cells. Exercise also helps. Some type II diabetics require insulin injections and some take oral medications that reduce the amount of sugar in the blood or stimulate the pancreas to secrete more insulin.

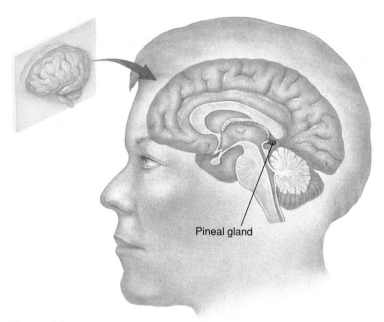

Figure 23.14 The pineal gland. This relatively obscure gland, located deep within the brain, secretes melatonin.

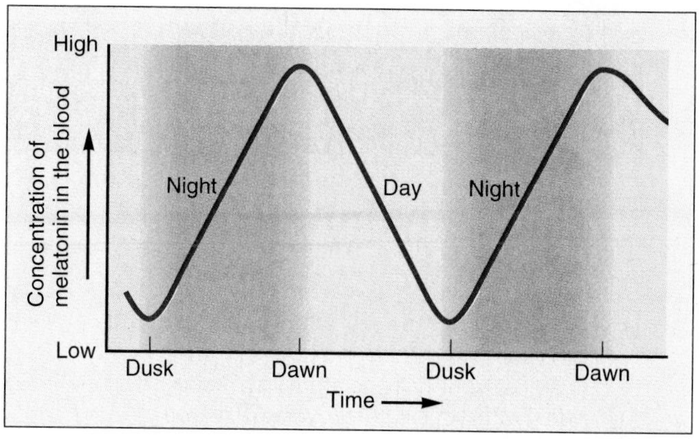

Figure 23.15 Concentrations of melatonin vary with time of day. Melatonin is secreted only when no light reaches the eyes. It accumulates in the blood during the night, reaching a peak at dawn. It then disintegrates during the day, reaching a low at dusk. These daily cycles in melatonin concentration regulate our cycles of activity.

Ovaries and Testes

The ovaries of a woman and the testes of a man secrete the sex hormones, estrogen and testosterone. The ovaries also secrete progesterone, a hormone that maintains pregnancy. These reproductive glands secrete their hormones in response to follicle-stimulating and luteinizing hormones from the anterior pituitary (see fig. 23.8). The ovaries, testes, and their hormones are described in chapters 28 and 29, where reproduction and development are discussed.

Links Between Hormonal and Neural Control

The hormonal and neural systems of control are linked in their task of coordinating the activities of the body cells. They are connected in the following ways: (1) some neurons secrete hormones; (2) most glands are controlled, directly or indirectly, by the nervous system; and (3) many bodily responses require commands from both the endocrine and nervous systems.

The distinction between neuron and secretory cell is blurred in certain cells of the hypothalamus and adrenal medulla. These cells, called neurosecretory cells, develop in nervous tissue but later secrete hormones. Neurosecretory cells in the hypothalamus secrete antidiuretic hormone, oxytocin, and releasing hormones, which travel through the bloodstream to their target cells. Neurosecretory cells in the adrenal medulla secrete epinephrine and norepinephrine, which also reach their target cells by way of the bloodstream.

The nervous system controls four endocrine glands by the link between the hypothalamus and the anterior pituitary gland. The hypothalamus directly controls the anterior pituitary, which, in turn, controls the thyroid, adrenal cortex, testes, and ovaries. The hypothalamus also has direct control over the adrenal medulla and the pineal gland. The advantage of neural control over hormone secretion is that it allows a more specific response to an environmental change. The sense organs can receive information and pass it to the brain, which can store and evaluate the information. The brain can then decide on the best response to this particular information and command appropriate glands to secrete their hormones.

Many responses of the body are coordinated by both the endocrine and nervous systems, working together to bring about the overall response. The fight-or-flight response, for example, is rapidly initiated by the nervous system, via neurons from the brain, and then maintained by the slower-acting hormones of the adrenal glands.

Research Report

Daily Rhythms

Are you aware of your daily rhythms? Do you know how to take advantage of them—to adjust your daily schedule so it coincides with your natural cycles? Do you know what time of day is best for you to cram for a quiz, solve thought problems, and compete in sports?

Almost every bodily function follows a daily rhythm—heart rate, blood pressure, body temperature, blood-sugar level, brain activity, physical strength, and hormone secretion. These rhythms wake us in the morning and put us to sleep at night; they make us feel energetic at certain times of the day and run down at other times. A typical activity schedule is shown in figure 23.B.

What causes these daily cycles of activity? Two hypotheses have been studied (fig. 23.C). The first hypothesis was developed in response to the observation that cells in laboratory dishes maintain daily rhythms of activity. According to this hypothesis, every cell has an internal clock that regulates its daily cycles of activity. The second hypothesis was developed in response to the observation that animals also have seasonal cycles of activity (e.g., reproductive cycles), and their bodies recognize seasons by day length. According to this hypothesis, daily cycles of sunlight and darkness generate daily cycles of cellular activity.

The two hypotheses have been tested by keeping animals, including

7-9 A.M. Levels of sexual hormones reach a peak, triggering a desire for sexual intercourse.

8-10 A.M. Time of least sensitivity to physical pain.

9-10 A.M. Best time for using short-term memory, such as cramming for a quiz that will be taken within ten minutes.

9 A.M.-12 P.M. Best time for using analytical and reasoning skills, such as solving problems and developing strategies.

10 A.M.-12 P.M. Mental alertness and speaking skills reach a peak.

1-3 P.M. Time of lowest mental alertness. Best use of the time: take a nap.

3-4 P.M. Mental alertness rises again and long-term memory skills reach a peak. This is the best time to study for an exam.

3-5 P.M. Mood and emotional outlook improves during these hours.

4-5 P.M. Peak time for handling confrontations with other people.

4-6 P.M. Physical dexterity and strength reach a peak. Best time to play a musical instrument or engage in a sport.

6-9 P.M. Thinking skills and physical reflexes begin to decline in preparation for sleep. Best time to relax; worst time to exercise, since it will interfere with sleep.

7-9 P.M. Peak time of sensory awareness; best time to experience tastes, smells, sights, sounds, and touch.

11 P.M.-1 A.M. Most creative time of the day, best-suited for artistic endeavors.

Figure 23.B A schedule of daily rhythms. This schedule reflects average times during the day in which young adults experience highs and lows of various activities. It assumes that you go to sleep at about 11 P.M., sleep soundly, and waken at about 6 A.M. If your day runs earlier or later, the schedule is shifted by a corresponding amount.

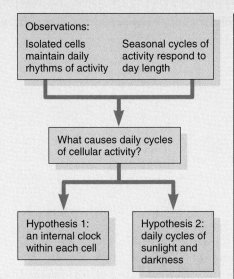

Observations:

| Isolated cells maintain daily rhythms of activity | Seasonal cycles of activity respond to day length |

What causes daily cycles of cellular activity?

Hypothesis 1: an internal clock within each cell

Hypothesis 2: daily cycles of sunlight and darkness

Figure 23.C Two hypotheses that explain daily cycles of cellular activity.

people, in continuous darkness. If the first hypothesis is true, then the elimination of light should have no effect on the internal clock. The daily cycles of activity should persist without change. If the second hypothesis is true, then the elimination of light should eliminate the internal clock. The daily cycles of activity should disappear.

Surprisingly, the daily cycles of activity persisted but were no longer twenty-four-hour cycles. Most of the animals (including the human subjects) settled into twenty-five-hour cycles; some developed forty-eight to fifty-

hour cycles, staying awake for thirty-three to thirty-six hours and sleeping for twelve to fifteen hours. Researchers conclude that both hypotheses are correct: cells have internal clocks that regulate daily cycles, but light is needed to keep these clocks on a twenty-four-hour cycle.

Further research revealed that mammals have a master clock located within the hypothalamus. Neurons that form this clock receive information about light from the eyes and pass it to the pineal gland. This gland responds by secreting melatonin only when it is dark, so that concentrations of this hormone rise during the night and fall during the day. Other body cells use these concentrations of melatonin to govern their daily cycles of activity.

Our bodies rely on natural cycles of sunlight and darkness to synchronize internal activities. Aspects of modern life, however, interfere with compliance with these natural cycles. Artificial light allows us to rise before dawn and stay awake long after sunset, but it is not strong enough to reset our internal clocks. Consequently, our behavioral activities are governed by artificial cycles of light, whereas our physiological activities are governed by natural cycles. We may stay up much later on Saturday night, sleep much later on Sunday morning, but suffer the consequences on Monday morning when our internal rhythms are out of synchrony.

Rapid travel by jet also interferes with natural cycles. If you fly east or west, the normal day-night cycle is shifted forward or backward so that the internal rhythms get out of synchrony, like an orchestra without a conductor. Symptoms of this internal confusion, called jet lag, include headaches, irritability, insomnia, indigestion, joint pains, and slow reactions.

Can we adjust our internal clocks, setting them forward or backward at will? Initial research on jet lag indicates that exposure to sunlight can be manipulated in ways that reset the master clock. When traveling westward by jet, and crossing fewer than seven time zones (e.g., flying from New York to Los Angeles), the internal clock can be set forward by spending a few hours in the sun late in the afternoon. When traveling eastward and crossing fewer than seven time zones (e.g., flying from Los Angeles to New York), the internal clock can be set backward by spending a few hours in the sun early in the morning. Traveling greater distances requires at least two full days of sunshine.

These instructions for combating jet lag, as well as for dealing with Monday morning blues, winter depression, and night jobs, will become more explicit as we learn more about the effects of light on our internal rhythms. If we cannot fit our behavioral schedules to our internal rhythms, perhaps we can adjust our internal rhythms to fit our behavioral schedules.

SUMMARY

1. Hormonal control of cellular activities is organized into an endocrine system consisting of more than a dozen glands, which consist of secretory cells that release hormones into the bloodstream. Each gland releases as much hormone as is needed to maintain homeostasis or to adjust the body to unusual conditions. Hormone secretion is regulated by materials in the blood, hormones from other glands, or the nervous system.

2. A hormone alters its target cells by binding with specific receptors. This binding alters either the permeability of the plasma membrane or the rate of some cellular activity. Lipid-soluble hormones enter their target cells and attach to receptors there. Water-soluble hormones attach to receptors on the cell surface, altering membrane permeability or activating other messengers inside the cell.

3. The hypothalamus and pituitary gland form a direct link between the nervous system and the endocrine system. The posterior pituitary is an extension of neurosecretory cells within the hypothalamus. These cells secrete antidiuretic hormone and oxytocin. The anterior pituitary is governed by hypothalamic-releasing hormones. It secretes prolactin, melanocyte-stimulating hormone, growth hormone, adrenocorticotropic hormone, thyroid-stimulating hormone, follicle-stimulating hormone, and luteinizing hormone.

4. The thyroid gland secretes thyroxine, triiodothyronine, and calcitonin. The parathyroid glands secrete parathyroid hormone.

5. Each adrenal gland consists of a cortex and a medulla. The adrenal cortex secretes mineralocorticoids and glucocorticoids. The adrenal medulla secretes epinephrine and norepinephrine.

6. The pancreas maintains the blood-sugar level by secreting insulin, glucagon, and somatostatin.

7. Other glands include the thymus, which secretes thymosin; the pineal, which secretes melatonin; the ovaries, which secrete estrogen and progesterone; and the testes, which secrete testosterone.

8. Hormonal control and neural control are linked in several ways: some neurons are also secretory cells; the brain governs secretion by most glands; and certain bodily responses are coordinated by both the nervous system and the endocrine system.

KEY TERMS

Chemical messenger (page 428)
Endocrine (EN-doh-krin) **gland** (page 428)
Hormone (HOR-mohn) (page 428)
Hypothalamus (HEYE-poh-THAL-ah-mus) (page 431)
Negative feedback (page 428)
Neurosecretory (NUR-oh-SEE-krih-tohr-ee) **cell** (page 431)
Pituitary (pih-TOO-ih-tahr-ee) (page 431)
Receptor (ree-SEP-tohr) (page 429)
Target cell (page 428)

STUDY QUESTIONS

1. Define chemical messenger, hormone, and endocrine gland.
2. Describe how the secretion of some hormones is controlled by negative feedback.
3. How does a hormone alter the activity of certain cells and not others?
4. Explain how the hypothalamus is both a part of the brain and an endocrine gland. How does it form a link between the nervous and endocrine systems?
5. What controls secretion by the anterior pituitary gland?
6. Why is the pituitary gland considered to be two glands in one? Why is an adrenal gland considered to be two glands in one?
7. Describe the fight-or-flight response, including the three main hormones that help develop and maintain it. How was this response an adaptation to the lifestyle of our distant ancestors?
8. What is the advantage of an endocrine system that is controlled in part by the brain?

CRITICAL THINKING PROBLEMS

1. What does the location of an endocrine gland tell you about its function?
2. Why is iodine added to table salt?
3. There is some evidence that more thyroxine is secreted during the winter than the summer. What would be the advantage of having more of this hormone in the body when the weather is cold?

4. People often describe their reaction to a dangerous situation by saying, "the adrenalin was really flowing." Explain in biological terms what is meant by this phrase.

5. Design an experiment to test the hypothesis that the length of night governs mating behavior in canaries.

6. The adrenal cortex secretes less and less cortisone as a person takes more and more of this hormone as a medication. Explain why the adrenal cortex responds in this way.

7. Discuss the ethical issues that might arise if growth hormone is available, in drug stores, to all children regardless of their height.

SUGGESTED READINGS

Introductory Level:

Dolnick, E. "Snap Out of It," *Health* (February/March 1992): 86–90. A description of daily cycles, especially those related to mental activities.

The Endocrine System: Miraculous Messengers. New York: Torstar Books, 1985. A medically oriented treatment, written for the layperson, of glands and hormones.

Fellman, B. "A Clockwork Gland," *Science* 85 (May 1985): 76–83. An easy-to-read discussion of how the pineal gland marks time and how it affects sleep cycles and mood.

Orlock, C. *Inner Time: The Science of Body Clocks and What Makes Us Tick.* New York: Birch Lane Press, 1993. An explanation of inner cycles and clocks, as well as suggestions of ways to control or take advantage of them.

Advanced Level:

Lienhard, G. E., J. W. Slot, D. E. James, and M. M. Mueckler. "How Cells Absorb Glucose," *Scientific American* (January 1992): 86–91. A description of the transporter that pumps glucose into a cell and its activation by the hormone-receptor complex.

Linder, M. E., and A. G. Gilman. "G Proteins," *Scientific American* (July 1992): 56–65. A discussion of the proteins involved in calling up second messengers.

Sapolsky, R. M. "Stress in the Wild," *Scientific American* (January 1990): 116–23. A description of research on the effects of stress on baboons of different social rank.

Wetterberg, L. (ed.). *Light and Biological Rhythms in Man.* Tarrytown, NY: Pergamon (Elsevier Science), 1994. A compilation of papers that summarize what we know about light as a regulator of biological rhythms in relation to health.

❂ EXPLORATIONS

Interactive Software: Hormone Action

This interactive exercise allows students to explore how the hormone insulin regulates levels of sugar in the blood. The exercise presents a diagram of a human body with liver and muscles highlighted and levels of circulating blood glucose indicated. An insert shows a section of plasma membrane with insulin receptors in cross section. By varying diet and amount of exercise, students can investigate how the interaction of insulin and glucagon keeps levels of blood glucose constant despite wide fluctuations in dietary intake of calories and utilization of calories for exercise.

Interactive Software: Cell-Cell Interactions

The key role of cell surface receptors in communication between cells is explored in this interactive exercise. The user alters the design of a hypothetical receptor and its intracellular signalling system, and assesses the consequences on intercellular communication. The user explores the critical roles of receptor design, G-proteins, and phosphorylation cascades in amplifying a signal, evaluating how changes in these elements alter the communication between two cells.

The appreciation and performance of music are based on neurons and their chemical messengers.

Neural Control

Chapter Outline

Objectives

In this chapter, you will learn
–the difference between neural control
  and hormonal control;
–how nerve cells are organized for rapid
  communication;
–the way in which signals pass through
  a nerve cell;
–how signals pass from a nerve cell to
  another cell;
–the organization of the human
  nervous system;
–how emotions and thoughts are
  interwoven within the human brain.

ourteen-year-old Melissa lay on the operating table. She was alert while the famous neurosurgeon, Wilder Penfield, was treating her for epilepsy. Suddenly, without warning, she flashed back to an afternoon when she was seven years old. It was summer and she was walking through a grassy field with her brothers. A man with a large sack appeared behind her and said, "How would you like to get into this bag with the snakes?" The teenager on the operating table screamed out in terror.

This disturbing memory, long forgotten, was brought to Melissa's conscious mind by an instrument probing her brain. Electricity from the instrument stimulated a particular cluster of cells in her brain, evoking the same images and emotions that she experienced many years before. The recall was so vivid that she believed it was really happening.

How can the brain preserve in such detail the experience of a grassy field in the summer sun and a strange man with a bag of snakes? What happens to brain cells when they store memories and when they return these memories to the conscious mind? How do brain cells link memories with emotions? Neurobiologists are still searching for complete answers to these questions. In this chapter, we explore what is known about how the brain works.

Organization of the Nervous System

Neural control of cellular activities is accomplished by a nervous system—a network of neurons (nerve cells) and chemical messengers through which information travels rapidly and precisely from one part of the body to another. This remarkable system, found only in animals, receives information from the environment, evaluates the information, and then commands muscles and glands to bring about appropriate responses.

The adaptive value of a nervous system is threefold: (1) it contributes to homeostasis (see page 313) by coordinating an animal's internal functions; (2) it enables an animal to move rapidly through its environment; and (3) it enables an animal to learn appropriate responses to food, shelter, mates, and dangerous situations.

The advantage of neural control over hormonal control is that messages, even complex ones, travel more rapidly. Messages flow through neurons as fast as 100 meters per second (approximately 200 miles an hour); an animal can respond instantaneously to an event in its environment. Messages carried by hormones travel more slowly, requiring several minutes to reach their target cells. Almost everything an animal does is controlled at least in part by its nervous system. An animal could not even

stand in place without instantaneous communication between its brain and the muscles that position its skeleton.

Pathways for the Flow of Information

Most of an animal's neurons are clustered within a **central nervous system,** where interactions among neurons evaluate incoming information and decide how the body should respond. In vertebrate animals (fish, amphibians, reptiles, birds, and mammals), the central nervous system consists of a brain and spinal cord. The rest of an animal's neurons form a **peripheral nervous system,** which connects the central nervous system with the rest of the body. Some neurons of the peripheral nervous system connect with sense organs, such as eyes and ears, and bring information into the central nervous system. Other neurons connect the central nervous system with muscles and glands—the organs that bring about responses to information. Thus, the basic organization of the nervous system is as follows:

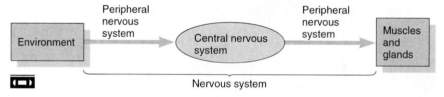

Take, for example, a situation in which you are walking down the street and see someone approaching you. Information about that person travels to your brain by way of neurons in your peripheral nervous system. In your brain, the information is evaluated by comparing it with previous experiences, stored as memories. A decision is then made about how to respond to the situation. Commands are sent, by way of motor neurons in your peripheral nervous system, to the muscles that produce facial expressions and move your body. You smile or frown; approach or withdraw. If the person is someone you fear, commands are sent also to your adrenal glands, signaling them to secrete hormones that contribute to the fight-or-flight response (see page 364).

Neural Tissue

Neural tissue consists of neurons, glial cells, and, in vertebrate animals, a blood-brain barrier. These three components are illustrated in figure 24.1 and described briefly in table 24.1. Most neurons are so specialized for carrying messages that they cannot maintain themselves without help. They are kept alive by the glial cells. On the average, each neuron in the human brain requires the assistance of ten glial cells.

Neurons   A neuron is a cell that receives information and passes it to other cells. It uses mainly glucose in cellular respiration, yet it stores very little glycogen. Neurons rely on a steady supply of glucose from the blood, which is

Figure 24.1 Neural tissue. Neural tissue consists of neurons, glial cells, and a blood-brain barrier. Neurons pass messages. Glial cells maintain and assist the neurons. They also wrap themselves around portions of neurons (the axons), speeding up the flow of information, and around capillaries, contributing to the blood-brain barrier.

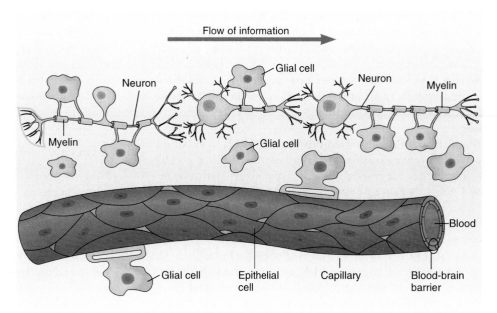

Table 24.1

Components of Neural Tissue

Structure	Main Function
Neuron	Carries messages
Sensory neuron	Carries messages from environment to the central nervous system
Interneuron	Carries messages within the central nervous system
Motor neuron	Carries messages from the central nervous system to muscles and glands
Glial cell	Maintains neurons; contributes to blood-brain barrier
Blood-brain barrier	Controls what moves from the blood to the fluid around neurons

tem. The environment includes everything outside the body (the external environment) as well as everything inside the body (the internal environment) that is outside the nervous system. Sensory neurons in your body, for example, carry information about the color of your notebook as well as about whether your stomach is empty or full.

An **interneuron** receives information from one neuron (a sensory neuron or another interneuron) and passes it to another neuron (an interneuron or a motor neuron). Interneurons and glial cells comprise most of the central nervous system, which evaluates incoming information by passing it from interneuron to interneuron, comparing it with previous information (memory) stored within the pathways. By this process, the interneurons decide how the body will respond to the information and send appropriate commands to the motor neurons.

A **motor neuron** carries commands from another neuron to organs that can bring about the response: either a muscle or a gland. A muscle responds to the information by shortening in a contraction. A gland responds by secreting more or less of its hormone.

The three kinds of neurons fit into the basic organization of the nervous system as follows:

why a drop in the blood-glucose level leads quickly to mental confusion, dizziness, convulsions, and coma. There are three categories of neurons: sensory neurons, interneurons, and motor neurons. They form pathways for the flow of information within a nervous system (fig. 24.1).

A **sensory neuron** carries information from the environment to interneurons within the central nervous sys-

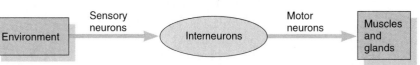

Glial Cells Neurons are kept alive and healthy by glial cells. Several kinds of glial cells assist neurons in various ways. Most glial cells provide neurons with nutrients, dispose of wastes, maintain appropriate concentrations of ions, and hold the neurons in position within the tissue. Some glial cells wrap their plasma membranes (see page 60) around neurons (fig. 24.1), forming a fatty layer called **myelin** that insulates the neurons and speeds up the flow of information. Still others play roles in the blood-brain barrier, as described following.

Blood-Brain Barrier   Neurons are extremely specialized cells that function only within a narrow range of conditions. If the interstitial fluid deviates even slightly from optimal, a neuron may lose its ability to transmit messages. Neurons in the brain are shielded from fluctuations in body chemistry by a **blood-brain barrier** within the capillary walls.

All capillary walls are formed of epithelial cells (see page 316). In tissues and organs other than the brain, materials move from the blood to the interstitial fluid through small gaps between the epithelial cells. In all but a few regions of the brain, however, the cells are so tightly packed that most materials can leave the blood only by passing through the plasma membrane of an epithelial cell. These membranes are selectively permeable; they allow certain materials to pass through channels. Glial cells wrapped around the capillaries (fig. 24.1) contribute to the barrier by stimulating the epithelial cells to press against one another.

The movement of lipid-soluble materials through the capillary walls cannot, however, be controlled by epithelial cells. These materials pass directly through the lipid framework of the plasma membranes. Such materials include ether, alcohol, nicotine, caffeine, heroin, and barbiturates.

The hypothalamus has no blood-brain barrier. Its neurons need to maintain contact with the blood in order to monitor its contents. The hypothalamus is a major center of homeostasis; it responds to deviations in blood chemistry by commanding changes that return the blood to an optimal state. The vomiting center (a cluster of neurons within the brain stem) also has no blood-brain barrier. It monitors the blood for poisons. When poisons are detected, this part of the brain induces vomiting, which rids the stomach of poisons.

The Neuron

A neuron is built to receive and pass messages. Information enters one end of the neuron and passes rapidly to the other end, where it is relayed to another cell. If you touch a hot stove, messages travel through a series of neurons to your spinal cord and brain. Almost immediately, you feel pain and pull your hand away. How can cells pass messages so rapidly?

A neuron transmits information in two forms: electrical signals and chemical messengers. Information re-

ceived by one end of a neuron travels as an electrical signal along the cell's plasma membrane until it reaches the other end of the neuron. There the signal triggers secretion of a chemical messenger, which carries the message to the next cell in the pathway. The flow of information from one neuron to the next is called an electrochemical impulse to indicate its two components.

Structure

Neurons appear in a variety of shapes, depending on where in the nervous system they relay messages. Typically, however, a neuron has four distinct regions: dendrites, cell body, axon, and terminal buttons. These structures are shown, for a motor neuron, in figure 24.2.

The **dendrites** are short projections, shaped like a tree (*dendron*, tree), that function like antennae. They are located at one end of the neuron. The plasma membrane of a dendrite receives information from the environment or another neuron and passes it to the axon. The branching shape of a dendrite greatly increases the surface area for receiving messages.

The **cell body** is a rounded portion of the neuron that holds the nucleus, mitochondria, endoplasmic reticulum, and other organelles characteristic of animal cells. It performs basic activities, such as protein synthesis and cellular respiration. The plasma membrane around the cell body may be formed into dendrites and receive information.

The **axon** is a slender extension of the cell body on the side opposite the dendrites. The first part of an axon, adjacent to the cell body, is the **axon hillock.** (Its significance will become clear in the following section on electrical flow.) The axon carries electrical signals from the cell body to the terminal buttons. Axons vary greatly in length. The axon of an interneuron may be shorter than a millimeter, whereas the axon of a motor neuron may extend all the way from the brain to a toe.

Most axons branch several times, especially near the end opposite the cell body. The ends of the axon are shaped into little knobs, called **terminal buttons,** where chemical messengers are stored. When an electrical message reaches the end of an axon, it stimulates the terminal buttons to release chemical messengers.

The narrow, fluid-filled space between the terminal button and the next cell is the synaptic cleft. The plasma membrane of the terminal button plus the synaptic cleft plus the plasma membrane of the next cell in the pathway form the **synapse** (fig. 24.2).

Electrical Flow of Information

An electrical signal travels along a neuron when ions enter and leave the neuron through its plasma membrane. The cell's plasma membrane separates two solutions of ions, with more negatively charged ions inside the cell than outside. This difference in electric charge is called the mem-

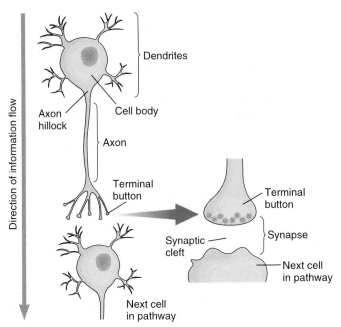

Figure 24.2 The structure of a neuron. A neuron (motor neuron shown here) receives information through its dendrites. The information then flows, as an electrical signal, from dendrite to cell body to axon hillock to axon and finally to a terminal button. There, the electrical signal becomes a chemical message that passes through the synapse—from the terminal button through the synaptic cleft to the next cell in the pathway.

brane potential ($\Box$ *see page 80*), because it is potential (stored) energy. If free to move, some ions would enter the cell and others would leave.

Ions are not, however, free to enter or leave a cell. The passage of an ion through the plasma membrane is controlled by a specific protein channel within the membrane. It is the opening and closing of these channels, permitting an inflow and outflow of certain ions, that forms an electrical signal. It is the frequency and pattern of the signals that form the message. The development of an electrical signal, described following, may appear at first complex. Yet, when you consider some of its results—an appreciation of spring flowers, a symphony, or a mathematical formula—the process is remarkably simple. The electrical signal develops from a particular concentration of ions within the resting neuron.

The Resting Neuron When a neuron is not transmitting a message, its membrane potential is −70 millivolts (1 millivolt = 1/1,000 volt), which means that the inside of the neuron is 70 millivolts more negative than the outside. This electric charge, called the resting potential, is the consequence of a particular distribution of ions, shown in figure 24.3*a*, that is maintained by channels within the membrane. It "cocks" the neuron for firing.

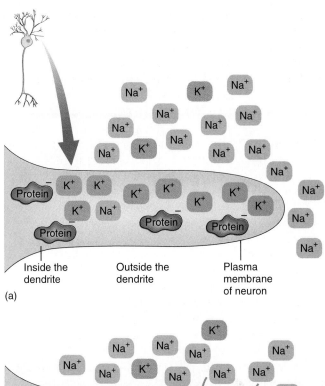

Figure 24.3 What happens when a neuron receives information. (*a*) A resting neuron (its dendrite shown here) is 70 millivolts more negative than the fluid around it, mainly because of its negatively charged protein ions. There are more potassium ions (K⁺) and fewer sodium ions (Na⁺) inside the neuron than outside. (*b*) When information is received, sodium channels within the dendrite open. Sodium ions enter the dendrite, moving by diffusion and by attraction to the more negatively charged interior.

The most important aspect of this distribution of ions is that sodium ions (Na⁺) would enter the neuron if they could do so. They would be moved by two forces: by diffusion (sodium ions are ten times more concentrated outside than inside) and by attractions of opposite charges (the interior has an overall negative charge). As long as the neuron is resting, however, channels within the plasma membrane remain closed and sodium ions cannot enter the neuron.

Receiving Information A neuron receives information when something outside the neuron, such as a chemical

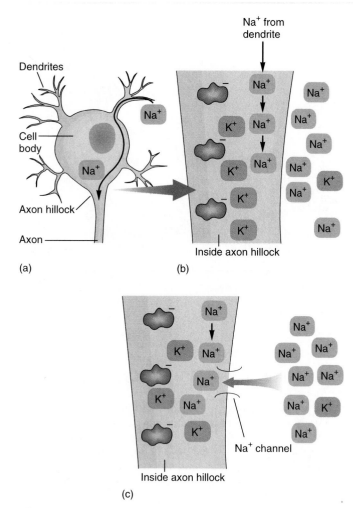

Figure 24.4 Development of an action potential.
(*a*) Sodium ions that enter a dendrite (when information is received, see fig. 24.3*b*) flow along the inner surface of the plasma membrane to the axon hillock. (*b*) At the axon hillock, the additional sodium ions make the inside of the neuron less negative than when at rest. (*c*) If enough sodium ions flow into the axon hillock to change the electric charge from −70 to −55 millivolts, then sodium channels open and sodium flows into the axon. So many sodium ions enter that the electric charge becomes +40 millivolts.

messenger, opens sodium channels within the plasma membrane of a dendrite. These channels are narrow, and only a few sodium ions trickle in (fig. 24.3*b*). Some of the sodium ions that enter the dendrite then diffuse along the inner side of the plasma membrane toward the axon hillock (fig. 24.4*a*), which is the site of the first action potential.

The Action Potential   If enough sodium ions accumulate at the axon hillock to change the membrane potential from −70 to −55 millivolts (fig. 24.4*b*), then an action potential is generated. An **action potential** is a change in electric charge, from −55 to +40 millivolts, followed by a return to the resting potential of −70 millivolts. It occurs within a small region of the plasma membrane and takes less than 2 milliseconds to complete.

An action potential develops whenever the membrane potential reaches −55 millivolts, because this particular charge opens sodium channels in the plasma membrane of an axon. These channels open wide and many sodium ions rush into the axon. So many of these positive ions enter (fig. 24.4*c*) that the charge inside the axon reaches +40 millivolts.

A charge of +40 millivolts triggers the sodium channels to close and potassium channels to open. Sodium ions are held inside the axon while potassium ions pour out, moving down a concentration gradient and away from the now positively charged interior. So many potassium ions leave the axon that its electric charge changes from +40 to −80 millivolts, becoming somewhat more negative than when the neuron is resting.

This immediate return to a negative charge is accomplished by a temporary redistribution of ions—more sodium ions inside the axon and more potassium ions outside the axon than in the resting condition. The original distribution of ions, with most of the potassium ions inside and most of the sodium ions outside, is established more slowly. Transport proteins, called sodium-potassium pumps (*see page 78*), within the membrane carry sodium ions out and bring potassium ions into the neuron until the original distribution of ions is restored. The plasma membrane is now "cocked" for another action potential. Every action potential is exactly the same, as shown graphically in figure 24.5.

Passing Information Along the Axon   Each action potential occurs within a tiny region of the axon. An electrical signal travels along the axon in the form of a series of action potentials that develop in one tiny region after another. Each action potential triggers an identical action potential farther down the axon.

One action potential triggers another when some of the sodium ions that rush into the axon diffuse away from their place of entry. When sufficient sodium ions accumulate at a new site to change the electric charge from −70 to −55 millivolts, then sodium channels at the new site open and an action potential develops (fig. 24.6*a*).

The series of action potentials travels only in one direction, from axon hillock to terminal button, because a second action potential cannot recur at the same site until the resting potential has been restored. New action potentials develop in new sites, farther along the axon, rather than in previous sites.

An animal reacts more rapidly to life-threatening situations when messages flow more rapidly along its axons. A faster flow can be accomplished by warmer temperatures, wider axons, or insulated axons. Most animals have adaptations for warming their bodies, enabling more rapid responses to their environments. In addition, invertebrate animals have thick axons, some as much as a hundred times wider than ours, that rapidly convey commands to their muscles. The jet-propelled escape of a squid, for ex-

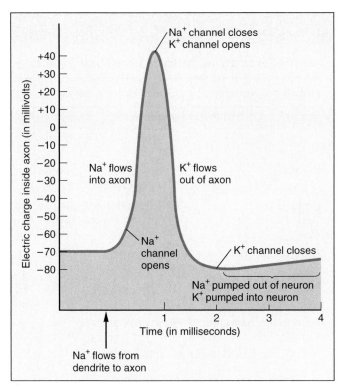

Figure 24.5 Graph of the electrical changes during an **action potential.** An action potential is an electrical change at a particular site within an axon. At rest, the charge is −70 millivolts. As sodium ions enter the axon from the dendrite, the charge becomes less negative. If the charge reaches −55 millivolts, then it triggers the sodium channels to open. Sodium ions flow into the axon and the electric charge becomes +40 millivolts. This particular charge closes the sodium channels and opens the potassium channels. Potassium ions then flow out of the axon until the electric charge becomes −80 millivolts, a charge that closes the potassium channels. Transport proteins within the plasma membrane then pump sodium ions out of the axon and potassium ions back in until the original distribution of ions is restored.

ample, is triggered by a broad axon (more than .5 mm in diameter) that carries messages at 25 meters per second. Vertebrate animals have too many axons to use this method: if our axons were as broad as a squid's, our spinal cord would be more than 30 centimeters (1 ft) in diameter! Instead, vertebrate animals gain fast responses by insulating their narrow axons with myelin:

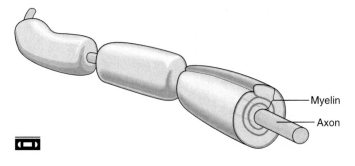

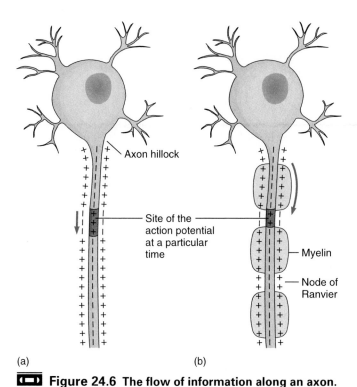

Figure 24.6 The flow of information along an axon.
(a) Information flows as action potentials develop, one after the other, in a continuous series along the axon. The first action potential is always at the axon hillock and the last one at a terminal button. (b) The axons of vertebrate neurons are coated with myelin, which prevents action potentials. An action potential can develop only in an unmyelinated region, called a node of Ranvier. Consequently, the electrical signal jumps from node to node and flows more rapidly.

This fatty material, formed of glial cells wrapped around the axon, prevents action potentials from occurring. Action potentials develop only in a few regions of the axon, called nodes of Ranvier, where myelin is absent (fig. 24.6b). Consequently the electrical signal jumps from one node to the next and travels down the axon fifty times more rapidly than in an axon that has no myelin.

Multiple sclerosis is an autoimmune disease in which the body's own white blood cells attack myelin. As myelin degenerates, an axon develops multiple scars (sclerosis) that impede the flow of action potentials. Symptoms of multiple sclerosis include diminished muscular coordination, caused by scars on motor neurons, and impaired vision, caused by scars on sensory neurons.

What happens when the electrical signal reaches the end of the axon? A few neurons have electrical synapses, in which the plasma membrane of the neuron makes direct contact with the plasma membrane of the next cell in the pathway. Signals pass from one cell to the next in the same way they pass down an axon; an action potential in the end of the axon triggers another action potential in the adjoining cell. Some of the neurons that control muscles of the heart and digestive tract pass signals in this way. Most

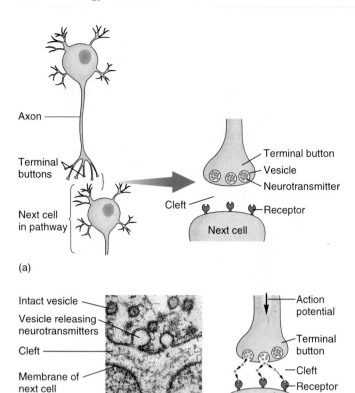

(a)

(b)   (c)

Figure 24.7 A synapse. (a) In a resting synapse, chemical messengers called neurotransmitters are stored inside vesicles within the terminal button. (b) Photograph of an active synapse. Vesicles have fused with the plasma membrane of a terminal button and are releasing their neurotransmitters into the cleft. (The neurotransmitters are too small to be seen as distinct objects.) (c) Diagram of an active synapse. An action potential in the terminal button stimulates migration of vesicles to the plasma membrane. There, the vesicle fuses with the plasma membrane, and its neurotransmitters are released into the cleft. The neurotransmitters diffuse through the cleft and latch onto receptors within the plasma membrane of the next cell.

neurons, however, have chemical synapses, in which the plasma membranes do not make direct contact. Instead, the neurons secrete chemicals that carry messages to the next cell in the pathway. The advantage of a chemical synapse is flexibility of response, as described following.

Chemical Flow of Information

Most neurons relay messages by secreting chemical messengers, called **neurotransmitters,** into the synaptic cleft. These messengers are typically amino acids, modified amino acids, or small proteins. A neuron synthesizes neurotransmitters in its cell body and stores them inside vesicles (membrane-enclosed sacs) within its terminal buttons (fig. 24.7a).

How the Message Is Relayed   A series of action potentials travels along an axon until it reaches the terminal

Table 24.2

Some Neurotransmitters and Their Effects

Neurotransmitter	Some of Its Effects
Acetylcholine	Stimulates contractions of skeletal muscles; contributes to memory
Serotonin	Calms emotions; promotes sleep and dreaming
Dopamine	Excites emotions; elevates mood; enhances muscle coordination
Norepinephrine	Excites emotions; elevates mood; increases alertness
Endorphin	Reduces pain; promotes euphoria
Gamma-aminobutyric acid (GABA)	Alleviates anxiety

button. There, the action potential stimulates the vesicles to fuse with the plasma membrane and release neurotransmitters into the synaptic cleft. The neurotransmitters diffuse through the cleft to the plasma membrane of the adjacent cell (another neuron, a muscle cell, or a secretory cell within a gland), where they bind with special proteins called receptors (fig. 24.7c). The brain has more than sixty kinds of neurotransmitters (a few are listed in table 24.2), and each binds with a particular kind of receptor.

A synapse may be excitatory or inhibitory, depending on what happens to channels for ion flow, located within the plasma membrane of the next cell. In an excitatory synapse, binding of the neurotransmitter with its receptor opens channels for a flow of ions (e.g., Na^+) that makes the cell interior less negative. An action potential is then more likely to develop. In an inhibitory synapse, the binding opens channels for a flow of ions (e.g., Cl^-) that makes the interior more negative. An action potential is then less likely to develop.

Some neurotransmitters do not cause immediate changes in the ion channels. Instead, they activate intracellular messengers (second messengers or signal relays) that initiate basic changes in the kinds and amounts of receptors, channels, and neurotransmitters that are synthesized. These basic changes are associated with such processes as thinking, feeling, and remembering.

Neurons or glial cells remove neurotransmitters from the synapse as soon as the message has been relayed. Some neurotransmitters are broken down by enzymes and others are picked up by carrier molecules and returned to the terminal button. If these messengers were not removed from the synapse, they would accumulate and overstimulate the next cell in the pathway. Some mind-altering drugs, such as cocaine, interfere with this removal process; they overstimulate neurons by keeping neurotransmitters within the synapse.

The Role of Chemical Synapses The advantage of a chemical synapse is flexibility of response. A typical interneuron (within the brain) secretes more than one kind of neurotransmitter, and each kind of neurotransmitter may have more than one kind of receptor. Moreover, a typical interneuron receives information from thousands of other interneurons. Whether or not the neuron develops a series of action potentials, and so relays a message, depends on the balance of excitatory and inhibitory synapses that have been activated. The interplay among neurons, each capable of a variety of responses, provides great flexibility of response.

Connections between neurons are the products of both genes and environment. Synapses that form during early development can be modified by experiences. Laboratory animals raised in enriched environments, with toys and mazes, develop more synapses than animals in deprived environments. Modification of synapses appears to be the physical basis of learning.

We are just beginning to understand how synapses influence mood, thought, and behavior. Most of our clues come from the study of mental disorders, in which symptoms can often be alleviated by drugs that resemble neurotransmitters. Excessive anxiety, for example, can be alleviated by Valium, a drug that mimics the neurotransmitter GABA (gamma-aminobutyric acid). Depression can be alleviated by Prozac, a drug that keeps more of the neurotransmitter serotonin in the synapse. Parkinson's disease can be controlled by L-dopa, a chemical that neurons convert into the neurotransmitter dopamine. Many mental illnesses develop from imbalances among neurotransmitters or their receptors within the brain.

Since prehistoric times, humans have used drugs to alter their minds. Our distant ancestors drank alcohol brewed from fruit, chewed coca leaves and peyote buttons, drank juice from the opium poppy, and consumed sacred mushrooms to enhance certain moods. Why are we now so concerned about drug use in our society? Because we have learned to concentrate these drugs into extremely potent doses. LSD, for example, affects the mind in the same way as mescaline (in peyote) but is 10,000 times more active. Mind-altering drugs interfere with neurotransmitters in ways that are described in the accompanying Wellness Report.

The Human Nervous System

The nervous systems of all vertebrate animals, including ourselves, are organized in about the same way. The part

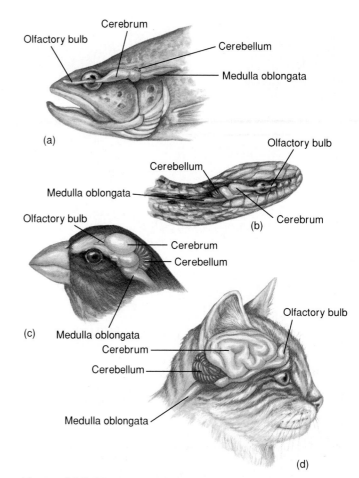

Figure 24.8 The brains of some vertebrate animals.
The most striking difference in the brains of different vertebrates is the size of the cerebrum (or cerebral cortex). (*a*) A fish has the simplest brain of all vertebrate animals. Note the large olfactory bulbs—smelling is the main way a fish learns about its environment. (*b*) The olfactory bulbs of a reptile are smaller, but the cerebrum is larger, than in a fish. (*c*) The brain of a bird is slightly larger but otherwise similar to the brain of a reptile. (*d*) The cerebrum of a mammal has many folds, which increase the surface area, and is so large that it covers other regions of the brain.

that varies most is the cerebrum, a region of the brain that governs intelligent behavior. The cerebrum becomes progressively larger and more intricate from the lower vertebrates (fishes and amphibians) to the higher vertebrates (reptiles, birds, and mammals). Some of these differences are shown in figure 24.8.

In mammals, the cerebrum is so large that it extends backward over other regions of the brain, and neurons that govern learned behaviors are clustered mainly within an outer layer called the cerebral cortex (cortex = bark). In more advanced mammals, such as dogs, cats, and humans, the cortex is large and deeply folded to fit inside the skull. The larger cerebrum of advanced mammals is the physical basis of more intelligent behavior.

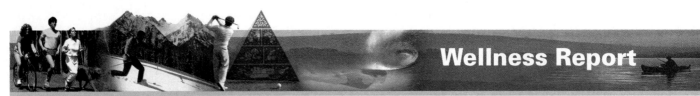

Wellness Report

Mind-Altering Drugs and Addictions

Mind-altering drugs influence the brain by modifying synapses. Some drugs, like heroin, bind with the same receptors as neurotransmitters and stimulate the same mental responses. Other drugs, like cocaine, prevent removal of neurotransmitters from the synapse. In most cases, we do not know enough about neural networks and neurotransmitters to identify exactly what a drug does to the brain.

Heroin is the best-understood of the mind-altering drugs. It comes from the opium poppy (*Papaver somniferum*, which grows in regions of the Middle East), where its function is to discourage insects from feeding on the developing seeds. Heroin binds with receptors on neurons in the thalamus, where deep burning pain is generated, and in the limbic system, where emotions are generated. These are the same receptors used by endorphins, the neurotransmitters that normally act to reduce pain and enhance pleasure. Small quantities of endorphins, secreted at all times, counteract normal tendencies toward depression and eliminate the continuous pain associated with routine internal processes, such as contractions of the digestive tract. Larger quantities

are secreted during physical exertion, emotional stress, childbirth, and injury. (It is endorphins that give rise to "runner's high.")

Neurons monitor the amount of neurotransmitter present in the brain and adjust their secretion accordingly. A problem with drug use is that neurons may not recognize the difference between a drug and a natural neurotransmitter. When someone takes heroin on a regular basis, their neurons interpret the presence of this drug as too many of their own endorphins. The neurons gradually stop making endorphins so that more and more heroin is needed to substitute for less and less neurotransmitter. This situation is diagrammed in figure 24.A.

Figure 24.A The physical basis of an addiction to heroin. (*a*) A normally functioning neuron releases neurotransmitters called endorphins, which latch onto receptors in the plasma membranes of other neurons. Joining of an endorphin with its receptor alters the neuron's activity, reducing pain and increasing pleasure. (*b*) When someone takes heroin, their neurons mistake it for endorphins and stop producing this neurotransmitter. The person must then have heroin to feel good. (*c*) During withdrawal from heroin, the addict has neither endorphins nor heroin to reduce pain and depression.

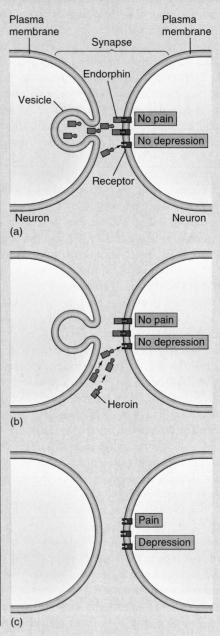

The vertebrate nervous system consists of a peripheral nervous system and a central nervous system, as shown in figure 24.9. The structures within each system are listed, along with their functions, in table 24.3.

The Peripheral Nervous System

The peripheral nervous system consists of all neurons or parts of neurons that lie outside the brain and spinal cord.

It consists chiefly of sensory neurons, which carry information from the environment (external and internal) to the central nervous system, and motor neurons, which carry information from the central nervous system to muscles and glands.

The neurons carrying information to and from the central nervous system are arranged in bundles called **nerves.** Each nerve may consist of only sensory neurons, only motor neurons, or a combination of both. There are

A certain amount of either endorphins or heroin is needed just to feel okay—to be without constant pain and depression. The longer the person takes heroin, the fewer endorphins their neurons produce and the more heroin that is needed. A habitual user of heroin who tries to stop "cold turkey" will at first have neither heroin nor endorphins in the synapses and will experience severe nausea, abdominal cramps, and depression. This chemical imbalance is the physical basis of addiction.

The withdrawal phase may last many weeks, during which time the neurons synthesize more and more endorphins until the chemistry of the brain is restored. Addiction apparently involves an emotional need as well as a chemical imbalance. Stimulation of pleasure centers in the limbic system gives such an emotional high that an addict finds it almost impossible to return to normal levels of emotion.

Addictions to other mind-altering drugs develop in a similar way. Cocaine occurs within the leaves of a shrub (*Erythroxylon coca*) that grows on the eastern slopes of the Andes Mountains. Its function is to protect the leaves of this plant from caterpillars. In humans, cocaine affects synapses between neurons in the midbrain, which secrete dopamine, and neurons in the limbic system, which give rise to pleasure. Cocaine binds with a carrier that normally picks up

dopamine and returns it to the terminal buttons for recycling:

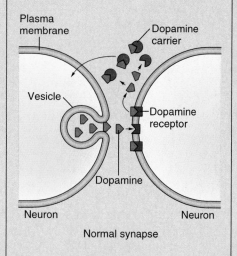

Normal synapse

Synapse with cocaine

Thus, cocaine holds dopamine in the synaptic cleft and overstimulates neurons in pleasure centers of the limbic system. Addiction develops as the

neurons secrete less and less dopamine (since so much remains in the synapse) and reduce the number of receptors. The addict then needs cocaine to enhance the effects of the small amounts of this neurotransmitter that their neurons receive. Without cocaine, the addict does not have enough dopamine and feels depressed and lethargic.

Less is known about the actions of marijuana, found in the leaves and flowers of the hemp plant (*Cannabis sativa*). This tall, annual herb is native to Asia but now grows in many parts of the world. Its active ingredients (more than sixty molecules, called cannabinoids) produce a variety of responses, depending on the person and dosage. The most common responses are euphoria, alterations of time and space, and cravings for sweets. In the late 1980s, researchers discovered a marijuana receptor on neurons within the limbic system and cerebellum. These neurons play important roles in emotions, perception, memory, and muscle coordination. In 1992, a neurotransmitter that binds with the same receptors was discovered. The function of this neurotransmitter, named amandamide (after a Sanskrit word meaning "internal bliss"), is not yet understood. Marijuana can be addictive. Withdrawal symptoms include irritability, headaches, and dietary cravings.

forty-three pairs of nerves: twelve pairs of cranial nerves that connect with the brain and thirty-one pairs of spinal nerves that connect with the spinal cord (fig. 24.9). For each pair, one nerve connects with the left side and the other connects with the right side of the brain or spinal cord.

The peripheral nervous system is subdivided into a somatic system and an autonomic system, described following and summarized in table 24.3. This subdivision is based

solely on the functions of the neurons; it is in no way related to where the neurons are located within cranial and spinal nerves.

Somatic Nervous System   The **somatic nervous system** consists of sensory neurons and the motor neurons that govern skeletal muscles (the ones that move the skeleton, face, and tongue). The sensory neurons bring information into the central nervous system and the motor neurons

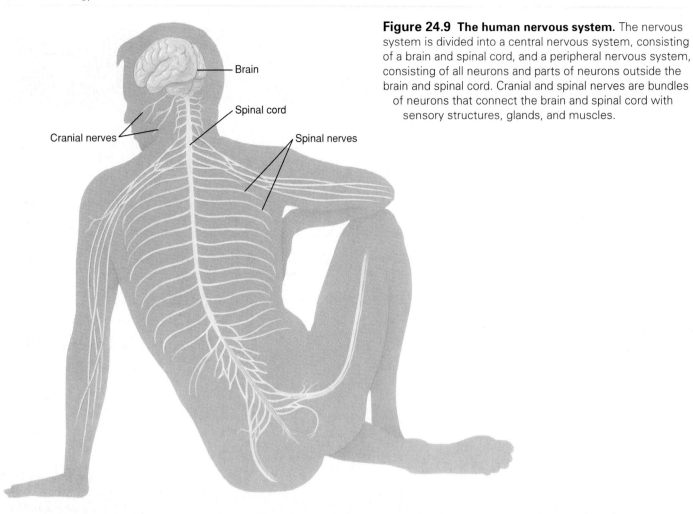

Figure 24.9 The human nervous system. The nervous system is divided into a central nervous system, consisting of a brain and spinal cord, and a peripheral nervous system, consisting of all neurons and parts of neurons outside the brain and spinal cord. Cranial and spinal nerves are bundles of neurons that connect the brain and spinal cord with sensory structures, glands, and muscles.

Table 24.3

Organization of the Human Nervous System

Division	Main Functions
Peripheral nervous system	Carries information between central nervous system and other parts of the body
Somatic nervous system	
Sensory neurons	Carry information from the external environment and parts of the body to the central nervous system
Motor neurons	Carry information from the central nervous system to voluntary muscles
Autonomic division	
Motor neurons	Carry information from the central nervous system to glands and involuntary muscles
Sympathetic system	Excites or inhibits glands and internal organs
Parasympathetic system	Excites or inhibits glands and internal organs
Central nervous system	Evaluates information; makes decisions; controls responses by muscles and glands
Spinal cord	Carries information between brain and other body parts; effects spinal reflexes
Brain	Evaluates information; makes decisions; produces cranial reflexes (reflex reactions to sights and sounds)

control voluntary activities—movements and behaviors that a person consciously decides to perform. Walking, smiling, and chewing, for example, are controlled by neurons within the somatic system.

Autonomic Nervous System The **autonomic nervous system** consists of motor neurons that control glands and involuntary muscles (the cardiac muscles of the heart and the smooth muscles of other internal organs). It governs the activities of exocrine glands (e.g., salivary glands), the contractions of heart muscles, peristalsis of the digestive tract, and many other events that occur without conscious control. The autonomic nervous system also participates in the fight-or-flight response by controlling secretion of epinephrine and norepinephrine by the adrenal medulla (see fig. 23.12). Meditation and other forms of stress management aim at controlling the autonomic nervous system.

The autonomic nervous system, in turn, is divided into a sympathetic division and a parasympathetic division. Motor neurons from both divisions connect to the muscles of most internal organs. The two divisions work in opposition to each other: neurons from one division excite the organ and neurons from the other inhibit it. Heart rate, for example, is accelerated by motor neurons in the sympathetic division and depressed by motor neurons in the parasympathetic division.

The Spinal Cord

The spinal cord is a long rod of nervous tissue, about as thick as your little finger, that extends downward from the brain (fig. 24.9). It is enclosed within the backbone. The spinal cord consists of interneurons as well as portions of sensory neurons and motor neurons. Its functions are to carry information to and from the brain and to bring about spinal reflexes.

Damage to the spinal cord may block the flow of information to and from the brain, causing numbness (no information from part of the body to the brain) or paralysis (no information from the brain to the muscles). The extent of the injury depends on where the cord is damaged: a broken neck affects most regions of the torso and limbs, whereas a broken back may affect only the legs, bowels, and bladder.

A spinal reflex happens automatically, without evaluation by the brain. Most reflexes involve just a few neurons. The simplest reflex—the knee jerk—involves only one sensory neuron and one motor neuron, which form a

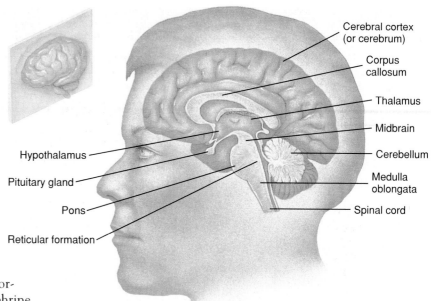

Figure 24.10 The human brain. This side view of the brain interior shows the major structures.

synapse within the spinal cord. Other reflexes have at least one interneuron between the sensory and motor neurons.

The Brain

The human brain is by far the most remarkable organ of the body. Weighing around 1.5 kilograms (3 lb), it contains approximately 200 billion neurons interconnected by more than 100 trillion synapses. The brain forms the mind—the unique personality, talents, attitudes, moods, desires, and memories of a person.

Major regions of the brain form visibly distinct structures (fig. 24.10) that specialize in different activities, as summarized in table 24.4. A modern technique for viewing brain function, called the PET scan, reveals higher levels of cellular activity in different regions when the brain is processing different kinds of information (fig. 24.11).

Almost every mental process involves an interplay among various regions of the brain. As your eyes move along the lines of this page, for example, information enters your brain at the visual cortex—a region of the brain that is specialized for receiving and evaluating all visual information. From the visual cortex, the information travels through billions of other neurons, in various regions, where it is further evaluated with regard to language, memory, imagination, and curiosity.

Moreover, the brain is shaped by experiences. The same activity may be performed by different regions in

Table 24.4

Organization of the Human Brain

Area	Major Functions
Brain stem	
Cerebellum	Coordinates skeletal muscles; memory of motor skills
Medulla oblongata	Governs heart rate, breathing rate, and artery diameter
Reticular formation	Governs state of awareness; screens incoming information
Pons	Governs facial expressions and aspects of sleep
Midbrain	Controls cranial reflexes; relays messages; coordinates posture
Olfactory bulbs	Relay information from nose to cerebrum
Limbic system (includes hypothalamus)	Generates emotions and sexual responses; contributes to memory; governs sleep; regulates hormone levels, water balance, appetite, and body temperature
Thalamus	Relays information from senses (except smell) to limbic system and cerebral cortex
Cerebrum (or cerebral cortex)	Initiates appropriate behaviors; suppresses intense emotions
Corpus callosum	Carries information between left and right hemispheres of cerebral cortex

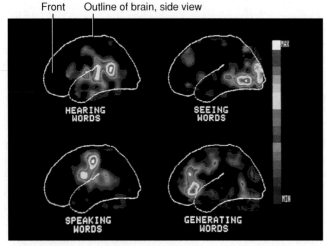

Front   Outline of brain, side view

HEARING WORDS

SEEING WORDS

SPEAKING WORDS

GENERATING WORDS

MAX

MIN

Figure 24.11 Images of more active regions of the brain during various mental processes. The technique used to produce these images, called a PET scan (for positron emission tomography), involves injecting radioactively labeled glucose into the bloodstream. More of this glucose enters brain cells that are utilizing glucose more rapidly in cellular respiration. The rate of cellular activity, reflected in the amount of radioactivity in the cell, generates a color-coded map of brain activity.

different people. When listening to music, for example, the brain of a novice who appreciates music processes the information in different regions from the brain of a professional musician.

Two regions of the brain are particularly important in governing behavior. These regions, the limbic system and cortex, are closely interconnected and maintain a delicate balance between emotions and thoughts.

The Limbic System   The **limbic system** is a ring (*limbus*, border) of interneurons located just beneath the cerebral cortex (fig. 24.12). The neurons are arranged in clusters, each specialized to some extent for a different function. A major function of the limbic system is to generate such emotions as pleasure, fear, rage, and lust. Mood-altering drugs, such as Valium, Prozac, cocaine, and heroin, affect this part of the brain.

Emotions are linked with thoughts. The way we feel influences the things we think about and vice versa. This interplay between emotions and thoughts is the consequence of extensive connections between the cortex and limbic system. We are not mere logic machines nor are we simply bundles of unrestrained emotions.

Together, the limbic system and cortex influence the hypothalamus, which commands the fight-or-flight response. Imagine you are home alone and hear the stairs creak. The information is relayed to your cerebral cortex, where it is interpreted, and then passed to your limbic system, where fear is generated. Interplay between these two regions connects the presence of an intruder with fear of physical pain. This emotional message is passed to your hypothalamus, which sends commands via neurons and hormones to parts of the body that participate in the

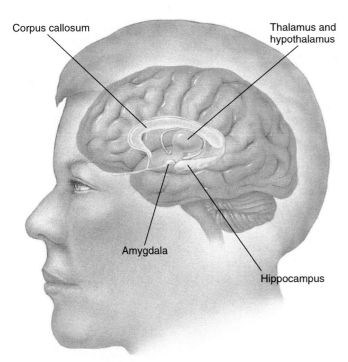

Figure 24.12 The limbic system. This region of the brain (tinted orange) consists of many interconnected clusters of neurons, including the thalamus, hypothalamus, amygdala, and hippocampus.

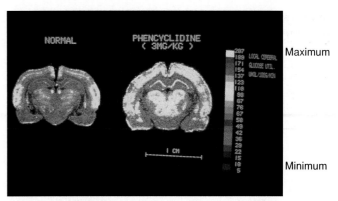

Figure 24.13 Effects of "angel dust" (phencyclidine) on the limbic system. PET scans of a normal limbic system (left) compared with a limbic system under the influence of the drug called "angel dust" (right). More active neurons appear in yellow, orange, and red. "Angel dust" intensifies emotions by overstimulating neurons in the limbic system.

fight-or-flight response. Your body prepares for emergency action—to fight or to run away.

The limbic system also stores memories, particularly the ones associated with strong emotions. Strangely, these memories can be evoked by sights, sounds, tastes, and smells. The reason for this strange connection is that information from our sense organs passes to the limbic system before traveling to the cortex. Odors are particularly powerful stimulants for recalling memories, because neurons from the nose pass messages directly to the limbic system. (Information from other sense organs is modified by passing through neurons within the thalamus before reaching the limbic system.)

Emotions generated in the limbic system are usually inhibited by the cerebral cortex. Some mind-altering drugs remove this inhibition, releasing intense emotions. The drug known as "angel dust" (phencyclidine, or PCP), for example, appears to excite the limbic system in this way (fig. 24.13).

The Cerebral Cortex The **cerebral cortex** is a layer of interneurons (fig. 24.14a), about 3 millimeters deep, that overlays other regions of the human brain. It is highly folded, with many fissures that serve to enlarge its surface

area. Approximately 70% of all neurons in the brain are within this remarkable part of the brain.

The cortex is divided into halves, with a left hemisphere and a right hemisphere (fig. 24.14b). Beneath the cortex are axons that connect neurons in the cortex with other regions of the brain. A large bundle of axons, called the corpus callosum (fig. 24.14c), connects the left hemisphere with the right hemisphere.

Interneurons of the cortex can be grouped, according to primary function, into three categories: a sensory cortex, which receives information from the senses (via sensory neurons); a motor cortex, which sends commands to voluntary muscles (via motor neurons); and an association cortex, which evaluates information from the sensory cortex, decides how the body will respond, and informs the motor cortex what the response should be:

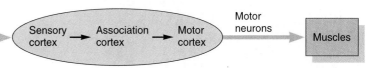

The association cortex is where higher mental processes occur. Locations of these regions are shown in figure 24.15.

The sensory cortex is subdivided, according to where the information comes from, into a visual cortex, auditory cortex, olfactory cortex, taste cortex, and somatosensory cortex. The somatosensory cortex receives information about pressure on the skin (i.e., touch), stretching of the muscles, and positions of bones at the joints. Neurons in the right hemisphere receive information from the left side of the body, and neurons in the left hemisphere receive information from the right side. Moreover, each region within the somatosensory cortex receives information

Figure 24.14 The cerebral cortex of a human. (*a*) The cortex forms a thin outer layer of the cerebrum. (*b*) It is divided into a left hemisphere and a right hemisphere. (*c*) The two hemispheres are connected by a bundle of axons called the corpus callosum.

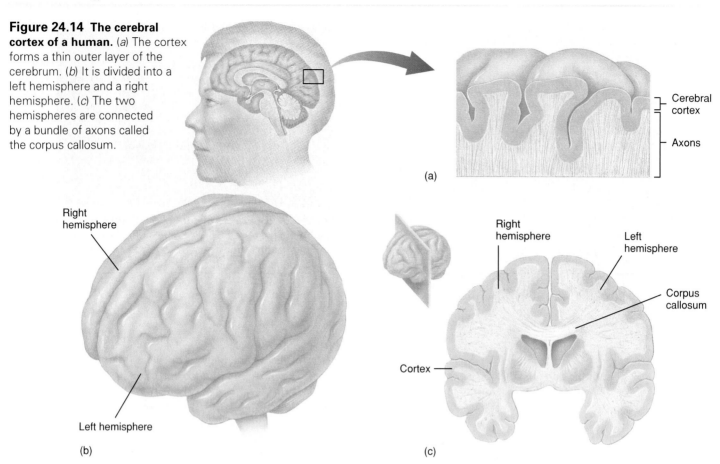

(a)

Cerebral cortex

Axons

Right hemisphere

Left hemisphere

Corpus callosum

Cortex

(b)

(c)

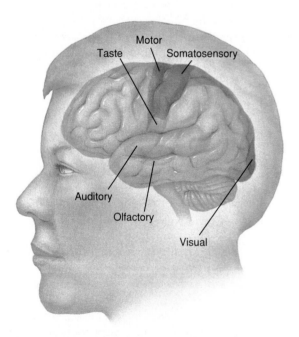

Motor
Taste
Somatosensory
Auditory
Olfactory
Visual

Figure 24.15 Map of the cerebral cortex. The cortex consists of a sensory cortex, motor cortex, and association cortex. The sensory cortex is further divided into a somatosensory, visual, olfactory, auditory, and taste cortex. Everything that is not sensory cortex or motor cortex is association cortex.

about a particular part of the body. A map of these regions is shown in figure 24.16. A stroke (neuron damage caused by a blocked artery) in this region often causes numbness on one side of the face or body.

The motor cortex receives information from the association cortex (where responses are determined) and sends commands along motor neurons to skeletal muscles, which move the skeleton, face, and tongue. The motor cortex of the left hemisphere sends commands to the right side of the body and the motor cortex of the right hemisphere to the left side. And within each hemisphere, specific regions of the motor cortex are responsible for contracting the muscles of specific parts of the body. A map of the motor cortex is shown in figure 24.17. A stroke in this region often paralyzes one side of the face or body.

The association cortex consists of all neurons within the cortex that are not part of the motor cortex or sensory cortex. It is involved with perception (interpreting information from the senses), abstract thought, planning for the future, memory, and language (fig. 24.18). A stroke in the association cortex often affects the ability to understand language or to speak language.

Association cortex within the right hemisphere has somewhat different functions from association cortex in the left hemisphere. For reasons not yet understood, which side performs which tasks depends on whether you are

Figure 24.16 Map of the somatosensory cortex. The somatosensory cortex of the right hemisphere receives information from the skin, muscles, and joints of the left side of the body, and the somatosensory cortex of the left hemisphere receives similar information from the right side of the body. Each cluster of neurons receives information from a particular part of the body. Regions represented by larger areas of the cortex (e.g., the lips) are more sensitive than regions represented by smaller areas (e.g., the hips).

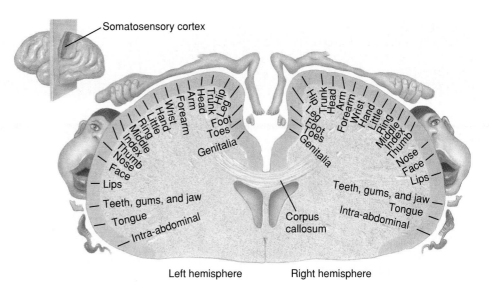

Figure 24.17 Map of the motor cortex. The motor cortex of the right hemisphere sends information to voluntary muscles on the left side of the body, and the left hemisphere sends information to voluntary muscles on the right side. Areas of the body that are more dexterous (e.g., the hands) are controlled by larger portions of the motor cortex than areas that are less dexterous (e.g., the legs).

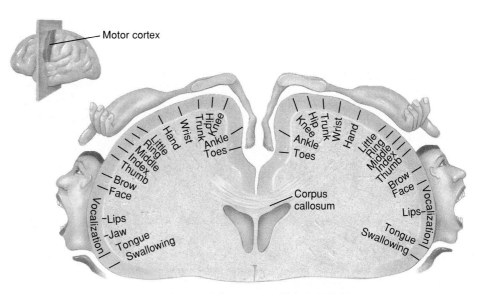

Figure 24.18 Regions of the left hemisphere that deal with language. Different regions of the left hemisphere deal with the mechanics of speaking and writing. Language concepts, on the other hand, involve large regions of both hemispheres.

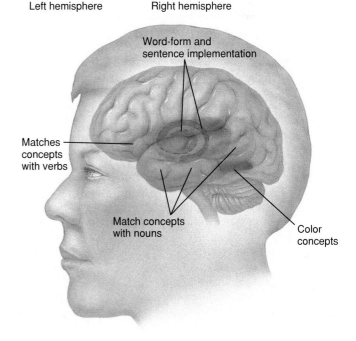

right-handed or left-handed. In right-handers, the right association cortex deals more with concrete phenomena, such as navigation, recognizing faces, music appreciation, art awareness, and simple arithmetic. The left association cortex deals more with symbols and concepts, such as language, logic, and mathematics. In left-handers, the tasks are distributed more evenly between the two hemispheres.

The human mind is an amazing biological phenomenon. It is difficult to imagine how a geometric arrangement of neurons and chemicals can produce a poem, a symphony, or a mathematical formula. An information gap remains between knowledge about the properties of individual neurons and about the properties of the human mind. A clue to the physical basis of memory is found in people with extraordinary powers of memory, described in the accompanying Research Report.

Research Report

The Savant Syndrome

A person with savant syndrome is mentally retarded yet brilliant beyond belief. Leslie, for example, is so handicapped in daily tasks that he cannot manage a knife and fork. Yet on the piano he can play any musical composition after hearing it just once. George, on the other hand, is a calendar calculator. He has difficulty adding two plus two, but he can name the day of the week for any date (taking leap years into consideration) within a span of 40,000 years. He can tell you within a few seconds on what day of the week your birthday will fall in the year 3314. Other savants create sculptures and memorize telephone books.

How does the intellect of a person become so focused on a single skill? Are we all capable of such intellectual feats but simply lack concentration? Or do these savants have unusual brains?

Savant syndrome was first diagnosed by J. Langdon Down (the same man who described Down syndrome) about a century ago. We still do not know the cause of this condition, primarily because we still do not know enough about the brain to explain normal thinking processes. What we do know about the syndrome comes from observing the savants.

A typical savant is a male who was born prematurely and has a below-average IQ. (A few savants are of average intelligence, but they develop the syndrome after a brain injury or infection.) The phenomenal memory of a savant is associated with certain skills, such as music, art, and arithmetic, generated by the right hemisphere of the cerebral cortex. At the same time, there are deficiencies in left-hemisphere functions, such as language and abstract concepts. When brain scans are done, they show abnormal tissue in the left hemisphere. A savant's particular talent often occurs in an area where other members of the

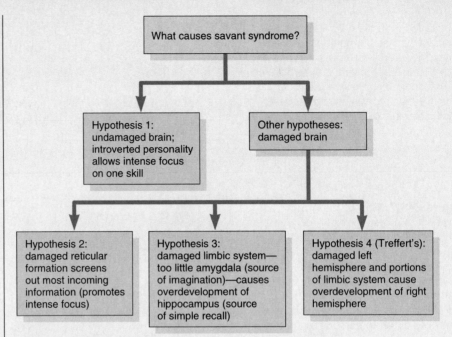

Figure 24.B Hypotheses that explain savant syndrome.

family also exhibit talent; there appears to be a genetic predisposition for the particular skill that develops.

The seemingly limitless memory of a savant is not associated with emotions. Instead, it works like a tape recorder. Musical savants do not sit down at the piano to enjoy the pleasures of music; they play when someone asks them and continue to play until the piece is finished. And if interrupted during the performance, they become confused and have to start at the beginning again.

Several hypotheses have been offered to explain the savant syndrome, as outlined in figure 24.B. One hypothesis is that everyone has the capacity to develop extraordinary skill in a single task but most of us do not focus our attention enough on just one aspect of life. Other hypotheses suggest damage of various sorts to the brain. One of these hypotheses, developed by the psychiatrist Darold Treffert, is that savant syndrome develops from damage to the left hemisphere of the cortex and portions of the limbic system. Damage

to the neurons occurs during the eighth or ninth month of fetal development, due to a lack of oxygen during premature birth or excess testosterone (normally secreted by males even before they are born). The right hemisphere is not affected because it develops earlier than the left hemisphere. Billions of developing neurons in the fetal brain find only dead cells where they were programmed to make connections. As a consequence, they grow extensions into the right hemisphere and connect with neurons there. Thus, a portion of the right hemisphere (perhaps an area already overdeveloped for genetic reasons) connects with an unusually large number of neurons and becomes a supercomputer with regard to such right-side functions as music, art, and arithmetic. And damage to the limbic system enhances rote memory in part by freeing the mind from distracting emotions.

Treffert's hypothesis remains controversial and the subject of much debate. It is one of several best guesses about the origin of the savant syndrome.

SUMMARY

1. Neural control is produced by a network of neurons and chemical messengers. Most nervous systems are organized into a central system (aggregates of neurons) and a peripheral system.

2. Neural tissue consists of neurons, which transmit signals; glial cells, which assist neurons; and a blood-brain barrier, which controls what materials enter neural tissue. There are three kinds of neurons: sensory neurons, interneurons, and motor neurons.

3. A neuron consists of dendrites, which receive information; a cell body, which performs basic activities; an axon, which carries signals; and a terminal button, which passes signals through a synapse to the next cell.

4. A neuron receives information when sodium channels in its dendrite open slightly to let in sodium ions. Some of the incoming sodium ions travel to the axon hillock, where they generate an action potential—a change in membrane potential from −70 to −55 to +40 to −70 millivolts. The signal then flows along the axon as a series of action potentials. The flow is more rapid in broader axons and in myelinated axons.

5. An action potential in the terminal button releases neurotransmitters into the synapse. There they bind with receptors in the plasma membrane of the next cell, triggering a change that may initiate an action potential.

6. The nervous systems of all vertebrate animals are similar, differing mainly in the size of the cerebrum. The peripheral nervous system consists of all neurons and parts of neurons outside the brain and spinal cord. It is subdivided into a somatic system, which receives sensory information and controls voluntary muscles; and an autonomic system, which controls glands and involuntary muscles. The autonomic system is further divided into sympathetic and parasympathetic systems, which have opposite effects on glands and internal organs.

7. The central nervous system consists of the spinal cord and brain. The spinal cord relays information between the brain and the body and controls spinal reflexes. The brain consists of several major structures, including the limbic system and cerebral cortex. The limbic system deals primarily with emotions and aspects of memory. The cerebral cortex consists of two hemispheres, subdivided into a sensory cortex, which receives information; a motor cortex, which sends commands to muscles and glands; and an association cortex, which deals with higher mental processes. The two hemispheres are connected by the corpus callosum.

KEY TERMS

Action potential (page 450)
Autonomic (aw-toh-NAH-mik) **nervous system** (page 457)
Axon (AK-sohn) (page 448)
Axon hillock (page 448)
Blood-brain barrier (page 448)
Cell body (page 448)
Central nervous system (page 446)
Cerebral (sur-EE-bruhl) **cortex** (page 459)
Dendrite (DEN-dreyet) (page 448)
Interneuron (page 447)
Limbic system (LIM-bik) (page 458)
Motor neuron (page 447)
Myelin (MEYE-uh-lin) (page 448)
Nerve (page 454)
Neurotransmitter (NUR-oh-trans-MIH-tuhr) (page 452)
Peripheral nervous system (page 446)
Sensory neuron (page 447)
Somatic (soh-MAH-tik) **nervous system** (page 455)
Synapse (SIN-aps) (page 448)
Terminal button (page 448)

STUDY QUESTIONS

1. Draw a neuron, identifying the dendrites, cell body, axon, and terminal buttons. Show the direction of information flow and explain why it is called an electrochemical impulse.

2. The electric charge is −70 millivolts inside a resting neuron and +40 millivolts at a region of an axon where information is being passed. Describe what happens to the axon to bring about this change in electric charge. After a region of the axon has passed information, its electric charge returns immediately to −80 millivolts and then more slowly to −70 millivolts. Describe what happens in the axon to bring about this return to a resting state.

3. Describe how the nervous system is organized. Include in your description the following terms: interneuron, motor neuron, sensory neuron, central nervous system, muscles or glands, and environment.

4. A poem conveys both thoughts and emotions. What is the physical basis for these phenomena and how do they become linked?

5. Draw the left cerebral cortex of a human, showing the positions of the motor cortex, taste cortex, auditory cortex, olfactory cortex, visual cortex, and somatosensory cortex.

6. Explain how the two hemispheres of the cortex differ in function. How does one hemisphere know what is happening in the other hemisphere?

CRITICAL THINKING PROBLEMS

1. What types of neurons are present in your thumb?
2. Explain why someone with untreated diabetes may go into a coma.
3. In a resting neuron, there are many more sodium ions outside the neuron than inside. Explain why these sodium ions do not enter the neuron.
4. Explain why someone with multiple sclerosis has difficulty walking.
5. How do mind-altering drugs affect the brain?
6. Suppose your grandfather had a stroke and can no longer move the muscles on the left side of his face. Explain why this particular symptom has developed.
7. Suppose a girlfriend has suffered brain damage from a climbing accident. Her brain stem and hypothalamus continue to function, but most of her limbic system and cerebral cortex are irreversibly damaged. Would you recommend that she be taken off all medication and tube feeding, in order to die with dignity? Why or why not?

SUGGESTED READINGS

Introductory Level:

Maguire, J. *Care and Feeding of the Brain: A Guide to Your Gray Matter.* New York: Doubleday, 1990. A book written for laypersons that examines the human mind and answers questions about consciousness, memory, intelligence, and emotions.

McCrone, J. *The Ape That Spoke: Language and the Evolution of the Human Mind.* New York: Avon, 1991. A delightful book that explores such topics as how we form sentences, why we have a conscience, the physical basis of "deja vu," and how our mind differs from the minds of other animals.

Restak, R. M. *The Brain Has a Mind of Its Own: Insights from a Practicing Neurologist.* New York: Crown Publishers, 1991. A collection of essays that deal with the author's personal experiences with patients, speculations about mental disorders, and the effects of modern society on the mind.

Restak, R. M. *Receptors.* New York: Bantam Books, 1994. A description of research on neurotransmitters, their receptors, mental illness, and mind-altering drugs.

Advanced Level:

Barondes, S. H. *Molecules and Mental Illness.* New York: Scientific American Library, 1993. An introduction to biological psychiatry, with explanations of how molecules (especially neurotransmitters) generate behaviors and what goes wrong when the behaviors become abnormal.

Edelman, G. M. *On the Matter of the Mind.* New York: Basic Books, 1992. An argument, based mainly on new information about synapses, that the brain is not "hard-wired" like a computer.

Gazzaniga, M. S. "Organization of the Human Brain," *Science* 245 (September 1, 1989): 947–51. An examination of the relation between structure and function in the human brain with particular emphasis on the cerebral hemispheres.

Hellig, J. B. *Hemispheric Asymmetry: What's Right and What's Left.* Cambridge, MA: Harvard University Press, 1993. An examination of the evidence that the left and right hemispheres of the cerebral cortex are specialized for different functions.

Korenman, S. G., and J. D. Barchas (eds.). *Biological Basis of Substance Abuse.* New York: Oxford University Press, 1993. A collection of papers describing current research on drug abuse, including such topics as drug-induced euphoria, neural adaptations to repeated exposure, and genetic vulnerability to addiction.

LeDoux, J. E. "Emotion, Memory, and the Brain," *Scientific American,* June 1994, 50–57. A description of how different animals form memories about emotional experiences, especially fear.

Tulving, E. "Remembering and Knowing the Past," *American Scientist* (July–August 1989): 361–67. A discussion of what scientists have learned about memory.

✷ EXPLORATIONS

Interactive Software: Nerve Conduction

This interactive exercise allows students to explore how sodium and potassium channels cause a electrochemical impulse to pass down a motor axon by enabling them to alter the architecture of the neuron. The exercise presents a diagram of a motor neuron axon, showing the series of ion channels in the membrane. Students can investigate the consequences of extending or reducing the zones covered by the myelin sheath, measuring the speed of conduction along the axon. By altering the diameter of the axon, students can explore the surprisingly great influence of axon diameter upon the speed with which the impulse travels down the axon.

Interactive Software: Synaptic Transmission

This interactive exercise allows students to explore the ways neurons employ chemicals to pass electrochemical impulses from one cell to another by enabling them to vary the nature of the chemicals. The exercise presents a diagram of a synapse between a neuron and a target cell, containing a variety of receptors and channels. Students can investigate the difference between excitatory and inhibitory synapses by varying the nature of the chemical released into the synapse and seeing what kind of channels are opened in response. Students will also observe the consequences of the channel opening to transmission of the impulse.

Interactive Software: Drug Addiction

This interactive exercise allows students to learn about the physical basis of drug addiction by exploring the direct consequences of the addictive drug cocaine on a neuron. The exercise presents an animated diagram of a synapse within the limbic system of the human brain. The neurotransmitter dopamine crosses this synapse to produce feelings of pleasure in the mind. The student can explore the consequences of introducing cocaine into the synapse, watching it bind up the transporter that normally removes dopamine from the synapse. The result is that dopamine levels stay high and stimulate the synapse repeatedly, producing euphoria. Students can then watch the synapse as it adjusts to this higher neurotransmitter level by lowering the number of its postsynaptic receptor channels. Now there is no pleasure without the drug, which is one way in which addiction starts. By varying the amount and frequency of cocaine use, the student can explore how patterns of drug use reinforce addiction.

The Senses

Objectives

In this chapter, you will learn
–how senses adapt an animal to its
  particular way of life;
–the way in which an eye detects
  objects;
–how cells in the nose and mouth
  detect chemicals;
–how an ear detects sounds and
  changes in body position;
–how the skin, muscles, and joints alert
  an animal to touch, pain,
  temperature, and position of body
  parts.

**An animal's senses provide
information about the environment.**

hat do we know about the environment around us? Only what our senses tell us. Each animal has senses that respond only to certain aspects of the environment. A human, dog, canary, and fly in the same room will perceive it differently, because each animal has different sensory capacities.

Our own senses are limited compared with the senses of other animals. Our eyes tell us about light, yet they are blind to patterns of ultraviolet light on flowers, which bees use to locate nectar. Our ears tell us about sounds, but they are deaf to the low murmurs of elephants and the high-pitched screeches of bats. We taste our food but not nearly as well as a catfish tastes its surroundings. We can smell airborne molecules, but a dog does it fifty times better. Even our sense of touch is poor compared to the mole, which lives underground.

As you can see, there is no "real" world. The world of each animal, including yourself, is an impression developed within its brain in response to the information it receives through its senses. And this information is only a fraction of all the information that is present in the environment. Each animal receives information about aspects of its environment that help it to survive and reproduce.

From Environment to Perception

An animal is aware of events that occur in its environment. Information that it receives about its surroundings help it to locate food and shelter, detect predators, recognize mates, and navigate from one place to another. An animal has adaptations, called senses, that bring certain kinds of information into its nervous system.

What is information and how does it enter a nervous system? Information occurs in two forms: environmental and biological. Environmental information occurs in the form of energy changes—a change in light energy as it strikes objects, a rearrangement of air molecules, a mechanical pressure, or a chemical reaction. Biological infor-

mation occurs in the form of electrical signals within neurons. The function of an animal's senses is to convert environmental information into biological information. How do an animal's cells accomplish this task? How, for example, do cells in the eye convert reflections of light into the image of a friend? How do cells in the ear convert rearrangements of air molecules into a melody?

Receiving Information

An environmental event must be converted into electrical signals within a neuron for it to be processed by the nervous system. This conversion, called transduction, is performed by specialized cells within the eyes, ears, mouth, nose, skin, muscles, and joints.

Transduction   Every change in the environment is accompanied by a change in energy. The conversion of energy changes in the environment into electrical signals within neurons is known as **transduction** (*to transduce*, to transfer).

Exactly how transduction occurs depends on the kind of energy associated with the environmental event—whether it is light, heat, pressure, or chemical changes. The crucial stage of transduction, however, always involves a change in permeability of a cell's plasma membrane, allowing certain ions to move into or out of the cell. This change in the distribution of ions alters the electrical charge of the cell and triggers electrical signals within the nervous system.

Receptor Cells   A cell that transduces environmental energy into electrical signals is a **receptor cell.** The human body has four kinds of receptor cells: photoreceptors, which transduce light energy; thermoreceptors, which transduce heat energy; mechanoreceptors, which transduce the energy of mechanical force; and chemoreceptors, which transduce the presence of certain molecules. These receptor cells are described in table 25.1. A receptor cell may be a sensory neuron (see page 447) or a specialized cell within a sense organ.

Table 25.1		
Receptor Cells and Their Functions		
Type of Receptor Cell	**Type of Information Transduced**	**How Transduction Begins**
Photoreceptor	Changes in light	Absorption of light changes a pigment to an enzyme that affects the plasma membrane
Thermoreceptor	Changes in heat	Not fully understood
Mechanoreceptor	Changes in mechanical force	Distorts the plasma membrane
Chemoreceptor	Changes in the presence of molecules	The molecules bind to receptor molecules in the plasma membrane

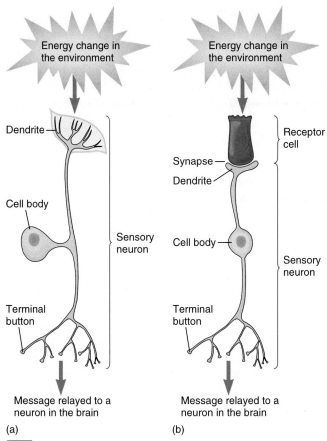

◀▶ Figure 25.1 Transduction by receptor cells.
(*a*) When the receptor cell is a sensory neuron, energy changes within the environment open ion channels in the plasma membrane of a dendrite. Incoming ions trigger electrical signals that travel along the sensory neuron and then to neurons in the brain. (*b*) When the receptor cell is not a sensory neuron, energy changes within the environment open ion channels in the plasma membrane of the cell. As ions enter or leave the cell, they trigger the release of neurotransmitters, which generate electrical signals in an adjacent sensory neuron.

When a receptor cell is a sensory neuron, transduction generates electrical signals within it. The change in energy that accompanies an environmental event opens ion channels within the plasma membrane, altering the distribution of ions and initiating an electrical signal that travels along the axon of the neuron toward the brain (fig. 25.1*a*). Odors, for example, are transduced by sensory neurons within the nose. A receptor cell that is a sensory neuron has two functions: to transduce energy and to carry electrical signals toward the brain.

A receptor cell that is not a sensory neuron is part of a sensory organ, like an eye or an ear. Other cells within the organ assist the receptor cell by amplifying and focusing the energy. This kind of receptor cell, as opposed to a neuron, is a modified epithelial cell (📖 *see page 316*) that secretes chemical messengers called neurotransmitters in response to environmental events. A change in environmental energy opens ion channels in the plasma membrane of the receptor

cell and alters the distribution of ions. The resulting change in electric charge triggers the secretion of a neurotransmitter, which diffuses through a synapse (📖 *see page 448*) to an adjacent sensory neuron, where it generates electrical signals that travel to the brain (fig. 25.1*b*). Light, for example, is transduced by receptor cells in the retina of the eye. Receptor cells within sense organs have just one function: to transduce electrical signals. They do not themselves carry electrical signals toward the brain.

Decoding Information

The sensation of an event, such as a visual image, does not develop until neurons within the brain receive electrical signals and decode them with regard to the type of event, its intensity, and its variation through time. Neurons decode signals by the pathway of neurons through which electrical signals enter the brain and by the pattern of signals within the neurons.

Pathway of Neurons   Once a receptor cell has transduced environmental energy into electrical signals within a neuron, information about the type of event (e.g., whether it involved light or sound) is lost, since electrical signals always take the same form as they travel through neurons. Electrical signals reaching the brain from the various kinds of receptor cells are decoded by the "wiring" of the brain—the pathway of information into the brain.

Electrical signals generated by different types of events travel to different parts of the brain. Signals that reach the visual cortex (📖 *see page 459*) are decoded as visual images; signals that reach the auditory cortex are decoded as sounds; and so on. Moreover, signals that reach certain regions within the visual cortex are decoded as red, green, or blue light; signals that reach certain regions within the auditory cortex are decoded as high-pitched sounds or low-pitched sounds.

Pattern of Electrical Signals   The pattern of electrical signals as they flow into the brain conveys information about the intensity of the event as well as its variation through time.

The intensity of the event, such as the brightness of light or the warmth of heat, is encoded within the rate at which electrical signals reach the brain. A more intense event generates more electrical signals, either by generating a higher frequency of signals within each sensory neuron, as shown in figure 25.2*a*, or by generating electrical signals in a larger number of sensory neurons.

The temporal pattern of the event—how it varies through time—is encoded within the pattern of electrical signals as they travel through a neuron. Patterns of speech or of moving visual images stimulate neurons in an irregular manner—at times many signals appear in a cluster, one right after another, and at other times they appear infrequently. Examples of these patterns are shown in figure 25.2*b*.

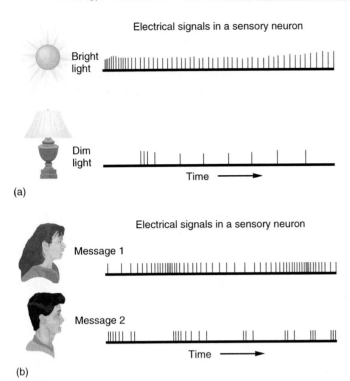

Electrical signals in a sensory neuron

Bright light

Dim light

Time

(a)

Electrical signals in a sensory neuron

Message 1

Message 2

Time

(b)

Figure 25.2 How the brain decodes electrical signals. The two diagrams show electrical signals over comparable time periods. Each vertical line represents an electrical signal. (*a*) Intensity of an event is decoded in terms of the number of signals per second. Bright light generates more frequent electrical signals than dim light. (*b*) Temporal pattern of an event is decoded in terms of how the signals are grouped as they move through neurons. Human speech, for example, is transduced into electrical signals of irregular frequency. Each sentence generates a unique pattern of electrical signals.

Perceiving Information

Electrical signals pass from sensory receptors to the brain, traveling through the thalamus, limbic system, and then on to the cerebral cortex. In the brain, the signals are transformed into meaningful information by means of a complex process known as **perception.** The electrical signals are decoded, as described previously, and then made meaningful by comparing their information with memories and linking it with emotions.

Electrical signals from your eyes, for example, may be decoded by your visual cortex as a red object, approximately your size, moving toward you. You then perceive the image as being your mother approaching in her red coat. You give meaning to the size, color, and movement of the image by comparing it with memories stored within your brain. At the same time, you develop emotions by linking the image with emotional centers in the brain.

Perception involves many areas of the brain, including the reticular formation, limbic system, thalamus, and association cortex. Electrical signals travel rapidly through these areas. As shown in figure 25.3, information about a pitched baseball travels through much of the batter's brain before the swing, just a fraction of a second after the ball is thrown.

Under unusual circumstances, the brain misinterprets the source of its information: sensations that originate within the brain are interpreted as sensations that origi-

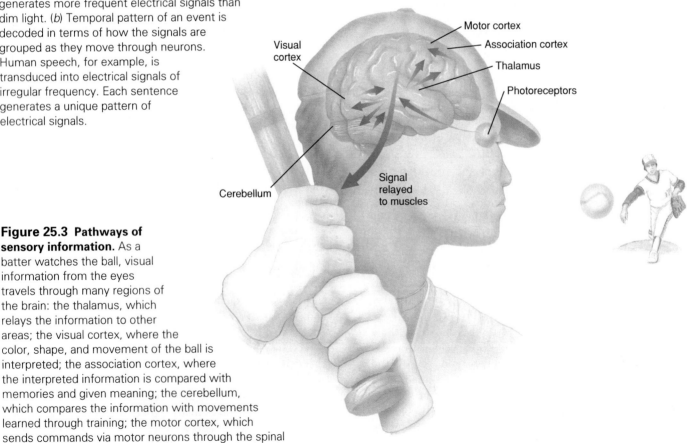

Figure 25.3 Pathways of sensory information. As a batter watches the ball, visual information from the eyes travels through many regions of the brain: the thalamus, which relays the information to other areas; the visual cortex, where the color, shape, and movement of the ball is interpreted; the association cortex, where the interpreted information is compared with memories and given meaning; the cerebellum, which compares the information with movements learned through training; the motor cortex, which sends commands via motor neurons through the spinal cord to muscles of the arm and other parts of the body.

Visual cortex

Motor cortex

Association cortex

Thalamus

Photoreceptors

Signal relayed to muscles

Cerebellum

Figure 25.4 The eyes of a spider. This wolf spider has eight eyes (only six can be seen in this front view) that detect moving objects. Each eye is a long tube ending in a lens, which magnifies the image, and a retina, which distinguishes green and ultraviolet light.

nate in the outside environment. The biological basis of these strange experiences is described in the accompanying Research Report.

The Special Senses

Vision, taste, smell, and hearing are known as the special senses, with receptor cells that are highly localized within sensory organs of the head. (Equilibrium, described later, is also considered a special sense.) Vision provides a mental picture of the surroundings. Taste and smell help an animal find appropriate food and mates. Hearing is used to locate the positions of moving objects and to receive messages from other animals.

Vision

Vision is based on changes in light, a form of energy emitted by the sun. Animals have receptor cells, called **photoreceptors,** that transduce light. These cells provide images of surrounding objects and inform an animal about the time of day and season of year. In most animals, these receptor cells are grouped with other kinds of cells to form sense organs called eyes (fig. 25.4).

In addition to photoreceptors, an eye contains other kinds of cells with other functions. Some of the cells control how much light reaches the photoreceptors and others coordinate the passage of electrical signals to the brain. These "accessory" cells and structures of the vertebrate eye are shown in figure 25.5 and summarized in table 25.2. They are described in more detail following.

Photoreceptors: Rods and Cones   Photoreceptors within an eye transduce light into electrical signals that travel to the brain, where they produce an image of the outside world. There are two kinds of photoreceptors: rods and cones, named for their shapes (fig. 25.6a).

The **rods** detect small differences in light intensity, seen as shades of gray. They provide images in dim light and are particularly sensitive to moving objects. The photoreceptors of nocturnal animals (active at night), such as mice and raccoons, are mostly rods.

Rods respond to light because they contain a light-sensitive pigment (fig. 25.6b) called rhodopsin. When rhodopsin absorbs light, it breaks into two smaller molecules: opsin (an enzyme) and retinal (a lipid derived from vitamin A):

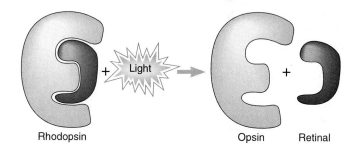

Rhodopsin + Light → Opsin + Retinal

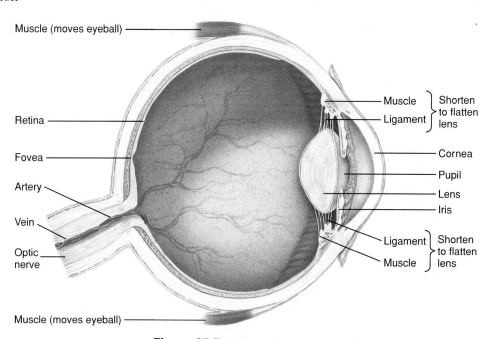

Muscle (moves eyeball)

Retina

Fovea

Artery

Vein

Optic nerve

Muscle (moves eyeball)

Muscle ⎫ Shorten to flatten lens
Ligament ⎭

Cornea

Pupil

Lens

Iris

Ligament ⎫ Shorten to flatten lens
Muscle ⎭

Figure 25.5 The vertebrate eye.

Research Report

Hallucinations

A long-distance truck driver on a lonely stretch of highway at night sees a wall rise in front of him. Later he is startled by a huge spider on the windshield. A lone explorer crossing the North Pole sees her family in bright red robes, flowing together and then apart like the closing and opening of a flower. A prisoner in a dark cell sees cartoon characters against a background of exploding colors and flashes of light.

These visions are hallucinations—false sensations that originate within the brain rather than in response to events in the outside world. Not limited to visual images, a hallucination may occur as sounds, tastes, and smells. It may even involve several senses at once. These bizarre experiences develop under conditions of unusual physical or emotional stress—extreme hunger or thirst, lack of sleep, severe pain, fever, or sensory deprivation. They are also a symptom of a mental disorder, known as psychosis (fig. 25.A). Most of what we know about the phenomenon comes from giving people hallucinogenic drugs (usually mescaline or LSD) and having them describe their sensations.

Drug-induced hallucinations are remarkably similar from one person to another. At first, most people have visions in which they see geometric forms—cobwebs, tunnels, and spirals. The forms, which are extremely bright and intensely colored, move in pulses or rotations. Later sensations are more complex and realistic, drawing more on childhood memories and scenes associated with strong emotional experiences from the past. Yet, almost all the subjects reported seeing the same kinds of images—moving scenes (like movies) of landscapes, small animals, and people (most of them friendly and many in the form of cartoons), usually projected on a background of geometric forms. During the peak hallucinatory periods, the subjects see themselves within the moving scenes and often feel detached from their bodies.

Hallucinations develop when a person is awake and the brain is unusually active (as indicated by measurements of electrical activity) while at the same time the flow of incoming sensory information is impeded. As a result, sensations that originate within the brain, from memory and imagination, are interpreted as sensations that originate in the outside environment.

What happens inside the brain during a hallucination? It is difficult to say because the sensation occurs within the privacy of a person's own mind. Several hypotheses have been developed. One attributes hallucinations to "background noise" in the brain, caused by random movements of molecules. These disturbances, which resemble the hiss in a radio signal when a radio is tuned between stations, stimulate neurons by accident. Another hypothesis attributes hallucinations to an insufficient flow of information into the brain. Normally, certain memories and sensations are suppressed by new information. When this inflow is blocked by drugs, stress, or impairment of the reticular system (a part of the brain stem that governs awareness), the memories and sensations become expressed as sensory experiences.

Figure 25.A *Hallucinations.* **A series of paintings of the same cat as seen by an artist during the development of a psychosis.**

Opsin then opens ion channels of the plasma membrane, which alters the electric charge of the rod cell and triggers the release of its neurotransmitter. The neurotransmitter diffuses through a synapse and activates the next cell in the pathway (described following).

The **cones** detect color and provide fine details in bright light. The photoreceptors of most diurnal animals (active during the day) include cones as well as rods. Each kind of cone contains a pigment that absorbs certain wavelengths of light. The more kinds of cones, the greater the

Table 25.2

The Vertebrate Eye

Structure	Function
Photoreceptors	
Rod cells	Transduce light into black-and-white images
Cone cells	Transduce light into colored images
Accessory cells of the retina	
Cells of the middle layer	Carry information from rods and cones to sensory neurons and from one pathway to another within the retina
Sensory neurons	Carry information from the retina to the brain
Accessory structures of the eye	
Pupil	Lets light into the eye
Iris	Controls size of pupil
Cornea	Bends light to focus it on the retina
Lens	Bends light to focus it on the retina
Eyeball	Moves continuously to keep the rods and cones stimulated

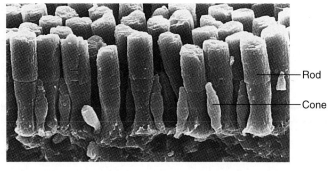

(a)

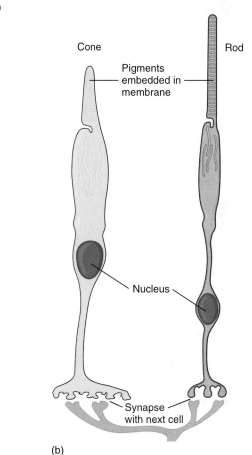

(b)

Figure 25.6 Photoreceptors: rods and cones. (*a*) Highly magnified photograph of the rods and cones in a retina. The rods are larger than the cones. (*b*) Sketch of a rod and a cone. The pigments are located in one end of each cell and the synapse in the other end.

animal's ability to distinguish among colors. Our own eyes have three kinds of cones. Dogs have just two kinds; they can distinguish blue and red but not green. Certain flies and fishes have five kinds of cones and much better color vision than we have. A mantis shrimp, with ten kinds of cones, uses its superb color vision to recognize subtle differences in the colors of other shrimp—differences that our own eyes cannot detect.

The light-sensitive pigments of a cone (fig. 25.6*b*) are formed of retinal and opsin, as in a rod, but in a cone the opsin comes in various forms. In us, each of the three kinds of cones contains a different kind of opsin: one kind absorbs red light, a second absorbs green, and a third absorbs blue. (The opsin in rhodopsin, within the rods, is yet a fourth kind in our eyes.) The three kinds of cones, with their three pigments, combine in various ways to detect a spectrum of colors.

As in a rod, light absorption by a cone breaks the pigment molecule into two parts. One part (opsin) becomes an enzyme that opens ion channels in the plasma membrane of the cone. The change in electric charge stimulates the cone to release its neurotransmitter, which generates electrical signals in adjacent cells.

The Retina The **retina** is a multilayered tissue in the back of the eye, as shown in figure 25.7. It contains several kinds

of cells, including photoreceptors and sensory neurons. The cells are interconnected by synapses to form pathways for the flow of electrical signals generated by light.

The rods and cones form the back layer of the retina; light passes through the other layers before reaching these light-sensitive cells. Most of the cones are in the center of this sheet of photoreceptors, an area known as the fovea (see fig. 25.5), so that colored images are sharpest when the eye looks directly at an object. Most of the rods are in the periphery, so that images in dim light are seen most clearly "out of the corner of the eye."

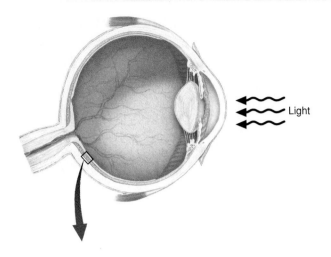

the sensory neurons, and they carry the information to the brain. All the sensory neurons from each eye form a bundle called an optic nerve.

Accessory Structures   The vertebrate eye has many accessory structures that control how light reaches the retina (see fig. 25.5). The amount of light that enters the eye depends on the size of the opening, or **pupil,** through which light passes. Pupil size is controlled by the **iris** (the colored part of the eye), which consists of two rings of muscles—one ring widens and the other narrows the pupil.

The cornea and lens bend incoming light so that it converges on the retina to give a sharp image. The **cornea** is a transparent, convex tissue that forms the front of the eyeball. It bends light but cannot itself change shape. The **lens** is a transparent disc located just behind the iris and pupil. Its shape is adjusted, by small muscles, for the distance of the object being viewed. The nearer an object is to the eye, the more the light must be bent in order to see it clearly:

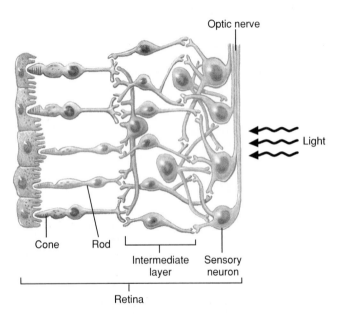

Figure 25.7 Cells of the retina. The human retina contains more than 100 million cells of several types. The back of the retina is a layer of photoreceptors (rods and cones) and the front is a layer of sensory neurons. In between are several kinds of cells that form synapses with one another as well as with the photoreceptors and sensory neurons, integrating visual information before it travels to the brain.

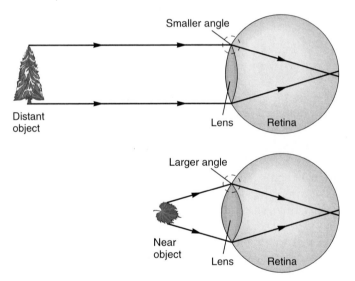

Muscles attached to the lens contract to flatten the lens (bending the light only a small amount) when viewing distant objects. These same muscles relax and the lens becomes more convex (bending the light more) when viewing near objects.

As a person ages, the lens becomes less flexible and incoming light converges either behind or in front of the retina. Images then become blurred. Glasses or contact lenses help focus the light on the retina.

Another accessory structure is the eyeball itself, which is attached to muscles that move it continuously to shift the image from one set of photoreceptors to another. If this movement did not occur, the image of a stationary object would gradually fade from view, since photoreceptor cells respond only to changes in light.

The rods and cones connect by synapses to an intermediate layer of cells (accessory cells) and then to sensory neurons in the front of the retina. When stimulated by light, the rods and cones secrete neurotransmitters that trigger electrical changes in the intermediate layer of cells, which pass the information to one another, integrating visual information from different photoreceptors. Some of these cells then secrete neurotransmitters that stimulate

Table 25.3

Taste Buds of Different Animals

Animal	Number of Taste Buds
Catfish	100,000
Cow	35,000
Rabbit	17,000
Pig	15,000
Goat	15,000
Human	9,000
Bat	800
Bird	200 or fewer

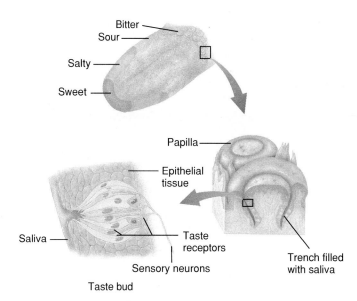

Figure 25.8 Taste buds. (*a*) In vertebrate animals, most taste receptors are located within taste buds on the tongue. Each region of the tongue responds to a particular taste: bitter, sour, salty, or sweet. (*b*) The taste buds are located within papillae, which are surrounded by trenches that accumulate saliva. (*c*) A taste bud consists of taste receptors (a form of chemoreceptor), sensory neurons that form synapses with the taste receptors, and epithelial tissue that surrounds and supports the taste receptors.

Taste

Taste is information about dissolved molecules in the environment. An animal uses this sense to distinguish foods that are good to eat from foods that are spoiled or poisonous. Taste also helps animals choose high-energy foods, enabling them to survive when food is scarce. We may have inherited, from ancestors that lived on the threshold of starvation, this craving for such high-energy foods as cookies, candy, and ice cream.

Cells that transduce the presence of molecules into electrical signals within neurons are **chemoreceptors.** The molecules must be dissolved in fluid (saliva or the surrounding water) to be detected. They attach to receptor molecules within the plasma membrane of the chemoreceptor, and this attachment triggers a change in the membrane's permeability to ions. The resulting change in electric charge triggers the chemoreceptor to secrete its neurotransmitter, which initiates electrical signals in an adjacent sensory neuron. The neuron then carries this message to the brain.

Taste receptors are located within **taste buds,** which are sensory organs formed of several kinds of cells that work together to pick up and relay information about dissolved molecules. The taste buds of most animals are in or near their mouths, but some animals taste food with other parts of their bodies. A butterfly, for example, tastes the nectar in flowers with its feet.

A human has approximately 9,000 taste buds, each with about fifty taste receptors, located mostly on the tongue but also on the throat and roof of the mouth. Some animals have many more taste buds than we have, as shown in table 25.3. A catfish, for example, has 100,000 taste buds distributed all over its body but particularly concentrated on its whiskers. Taste provides the fish with information about its location within the water, identifies edible objects, and tells the fish whether another animal is a friend, foe, or potential mate.

Taste buds on our tongues form tiny bumps, called papillae, that are surrounded by trenches (fig. 25.8). A bud contains approximately 100 taste receptors, each with cilia (hairlike extensions) that project into the trench, where saliva accumulates and brings dissolved molecules in close contact with the plasma membranes of the cilia.

It is hard to know what other animals taste. Vertebrate animals appear to have just four kinds of chemoreceptors: sweet, salty, sour, and bitter. Sweet receptors identify sugar, a high-energy food that is good to eat; salt receptors identify sodium chloride, which is required every day; sour receptors identify unripe foods; and bitter receptors identify plant poisons and spoiled meat. In us, each kind of chemoreceptor is clustered within a particular region of the tongue, as shown in figure 25.8*a*. The number of each kind varies among humans. Some of us, for example, have twice as many sweet receptors, and thus require less sugar to satisfy our "sweet tooth," than other people.

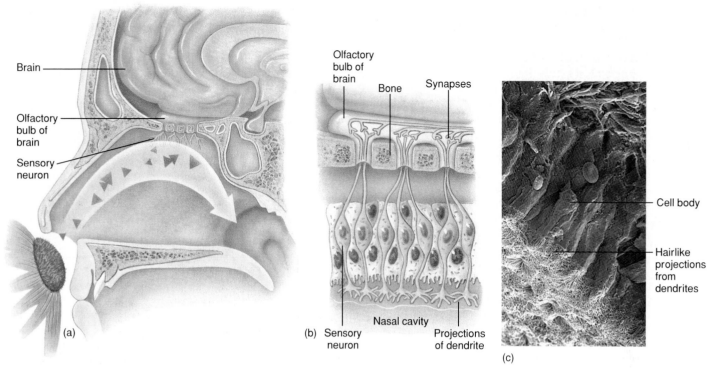

Figure 25.9 Chemoreceptors within a human nose.
(*a*) Chemoreceptors that detect scents are sensory neurons with dendrites that extend into the fluid that coats the roof of the nose. The dendrites contact airborne molecules that become dissolved in the fluid. A human nose has approximately 5 million chemoreceptors. (*b*) A closer view of the chemoreceptors. The dendrites of each sensory neuron contact molecules within the nasal cavity. The opposite end of each neuron forms a synapse with neurons in the olfactory bulb of the brain. (*c*) Highly magnified photograph of the chemoreceptor.

Smell

Smell provides information about airborne molecules in the environment. Most animals rely more on smell than on any other sense. The sense of smell is used chiefly to locate food and to govern social and sexual life. A dog tracks rabbits with its nose and "reads the social news" from scent posts marked with the urine of other dogs. (A bloodhound can recognize a human footprint that is six weeks old.) A male moth has extraordinary organs of smell, on its antennae, that can detect a female moth several miles away.

Our own sense of smell is poor, as is that of most birds and primates. An animal that flies, lives in the trees, or walks upright on two legs has its nose far from the ground, away from the scents of footprints and urine. While we do not rely on scents to survive or reproduce, they do give us sensual pleasure. We delight in certain odors—the smell of warm bread, flowers, and newly mown grass pleases us and evokes memories. Sensory neurons from the nose pass directly to the limbic system, a region of the brain that deals with emotions and memories.

Smell receptors can occur almost anywhere on an animal's body: in ants or snails they are on the antennae; in flies they may be also on the wings. In mammals, smell receptors are sensory neurons located within the nose. At one end of each neuron, dendrites project into the nasal cavity (fig. 25.9). At the other end, the neuron forms synapses with other neurons inside the brain. The nose is the only place in the body where there are direct connections, via single neurons, between the outside world and the brain.

Airborne molecules drawn into the nose dissolve in the mucus that lines the cavities. There they latch onto receptor molecules in plasma membranes of the dendrites of a sensory neuron. Joining of these two molecules opens sodium channels in the membrane and generates electrical signals within the neuron. The neuron carries olfactory signals to the brain, where they are perceived as the smell of lilacs, pizza, sweat, or whatever.

A human has more than 1,000 kinds of chemoreceptors that recognize more than 10,000 distinct odors! In fact, we identify foods more from their odors than from their tastes. The unique smell of chocolate comes from

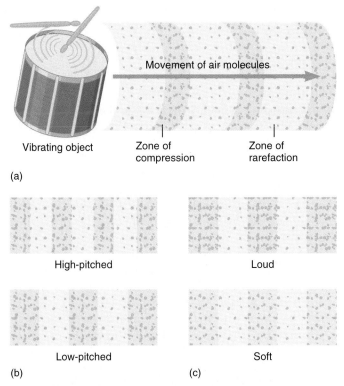

Vibrating object   Zone of   Zone of
                    compression   rarefaction

(a)

High-pitched   Loud

Low-pitched   Soft

(b)   (c)

Figure 25.10 The physical basis of sound. (*a*) Sound develops when a vibrating object disturbs the surrounding air, rearranging the air molecules into zones of compression and zones of rarefaction. These two zones of disturbed air move away from the vibrating object. (*b*) The zones are more narrow in a high-pitched sound than in a low-pitched sound. (*c*) A louder sound has more molecules in the compressed zones and fewer molecules in the rarefied zones than a softer sound.

more than 500 different odors, and the smell of coffee from 800. Food that cannot be smelled has little flavor, as can be demonstrated by eating different flavors of jelly beans while holding your nose closed. Without their odors, all the jelly beans taste the same. Surprisingly, many centers of the brain respond to odors. Sniffs of spiced apple have been shown to reduce stress, and sniffs of peppermint increase mental activity almost as much as drinking coffee or cola.

Hearing

Sound develops when an object, such as a voice box or a slammed door, vibrates. Small movements of the vibrating object rearrange air molecules around it to form zones of compression, where the molecules are densely packed, and zones of rarefaction, where the molecules are far apart. These two zones travel away from the object in alternate waves (fig. 25.10*a*).

The fundamental characteristics of sound are pitch and loudness. Pitch depends on frequency of the zones: a sound is higher-pitched when the zones are more narrow

(fig. 25.10*b*). Loudness depends on the density of molecules in the zones of compression as compared with the zones of rarefaction: a louder sound has more molecules in the compressed zones and fewer molecules in the rarefied zones than a softer sound (fig. 25.10*c*).

Sound traveling through water exhibits the same characteristics, except that water molecules rather than air molecules are displaced. While we do not hear well underwater, a lake or ocean is a noisy place, filled with the grunts and growls of fish as well as the songs of dolphins and whales.

Most animals have sensory cells that detect sounds. A vertebrate animal hears through cells located within ears on the head. A cicada listens through ears on its abdomen, while a cricket has ears on its knees. Animals recognize one another by the sounds they make. They also locate objects by sound, an ability that requires at least two ears; the brain assesses the difference in time the sound takes to reach each of the ears.

Sound is particularly important to animals that hunt in the dark. An owl, for example, uses the rustling sounds of a mouse to pinpoint its precise location. A bat uses sound in a different way; it emits high-pitched sounds as it hunts moths. The sounds bounce off a moth, forming echoes that reveal its location.

Mechanoreceptors: Hair Cells   Sound is transduced by **mechanoreceptors,** which are cells that convert mechanical forces into electrical signals in neurons. The mechanoreceptors of sound are **hair cells.** One end of a hair cell has cilia and the other end forms a synapse with a sensory neuron:

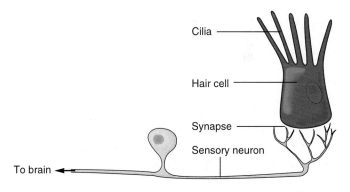

Cilia

Hair cell

Synapse

Sensory neuron

To brain

When a mechanical force bends the cilia, it distorts the plasma membrane and opens ion channels. The resulting change in electric charge stimulates the hair cell to release neurotransmitters into the synapse. The neurotransmitters generate electrical signals in the adjacent sensory neuron.

In vertebrate animals, sound is transduced by hair cells within the ears in the head. In mammals such as ourselves, many accessory structures help bring sound onto these receptor cells, as summarized in table 25.4. The anatomy of the human ear is illustrated in figure 25.11 and described following.

Table 25.4

The Human Ear: Structures That Provide Hearing

Structure	Function
Outer ear	Channels sound onto tympanic membrane
Tympanic membrane	Vibrates in response to sound
Middle ear	
Malleus	Transfers vibrations from tympanic membrane to incus
Incus	Transfers vibrations from malleus to stapes
Stapes	Transfers vibrations from incus to oval window
Cochlea of inner ear	
Oval window	Transfers vibrations of stapes to waves of fluid within the cochlea
Fluid	Transfers vibrations to up-and-down movements of the basilar membrane
Basilar membrane	Transfers movements to hair cells
Hair cells	Transduce sound

The Outer and Middle Ear   The outer and middle ear, shown in figure 25.11a, contain accessory structures that channel, amplify, and transfer sound to hair cells within the inner ear.

The **outer ear** is a funnel that channels sound waves onto the **tympanic membrane,** or eardrum. This membrane vibrates as it is displaced by alternating waves of compressed and rarefied air.

Vibrations of the tympanic membrane are amplified thirtyfold as they travel through a series of small bones known as the **middle ear.** Each bone in the series is smaller, more delicate, and consequently more easily displaced than the previous one. The first bone, called the malleus (or hammer), presses against the tympanic membrane. It moves back and forth with vibrations of this membrane. The malleus amplifies the vibrations and passes them to the second bone, called the incus (or anvil),

which further amplifies the vibrations and passes them to the third bone in the series. The third bone, called the stapes (or stirrup), further amplifies the vibrations and passes them to the inner ear.

The Inner Ear   The **inner ear** consists of the cochlea, which transduces sound, as well as the semicircular canals, utricle, and saccule, which provide the sense of equilibrium (described later).

The **cochlea** receives sound vibrations from the middle ear. It is a long, bony tube, about the size of a pea and shaped like the shell of a snail. The tube is filled with fluid. A small region in the wall of the cochlea consists of a membrane, called the **oval window,** rather than bone. The stapes of the middle ear presses against this membrane. Vibrations of the stapes move the oval window in and out, and these movements generate waves within the fluid of the cochlea.

Inside the cochlea (fig. 25.11b), hair cells are positioned between two parallel membranes, the basilar membrane and the tectorial membrane. One end of each hair cell is attached to the basilar membrane. The other end, from which the cilia extend, is embedded within the tectorial membrane. Waves within the cochlear fluid move the basilar membrane up and down (fig. 25.11c). As a wave lifts this membrane upward, cilia on the hair cells press against the tectorial membrane and their plasma membranes become distorted. Ion channels in the membranes open and the electric charge within the hair cell changes. A hair cell responds to the change in electric charge by releasing neurotransmitters, which generate electrical signals in the sensory neuron. All the sensory neurons from a cochlea form a bundle called the auditory nerve. In the brain, information from the two ears is compared and the direction of sound pinpointed.

Each frequency of sound excites only a fraction of the approximately 15,000 hair cells on the basilar membrane. Higher-frequency vibrations displace the region nearest the oval window and are perceived by the brain as high-pitched sounds. Lower-frequency vibrations displace regions farther from the oval window and are perceived by the brain as low-pitched sounds.

The louder the sound, the stronger the waves that move the hair cells and the greater the number of electrical signals generated in the sensory neurons. The ear is protected against very loud sounds by tiny muscles that stretch the tympanic membrane to prevent extreme vibrations. Prolonged exposure to loud sounds, however, damages hair cells and causes hearing loss, as described in the accompanying Wellness Report.

■□ Figure 25.11 The human ear: structures involved with hearing.

(*a*) The outer ear funnels sound onto the tympanic membrane. Sound then passes, as vibrations, from the tympanic membrane to the three bones of the middle ear (the malleus, incus, stapes) and then to the oval window, which forms part of the cochlea in the inner ear. The cochlea holds a fluid that develops waves in response to vibrations of the oval window. (*b*) Cross-section of the cochlea, showing the structures that transduce sound. (*c*) Close-up of the structures that transduce sound. Waves of fluid move the basilar membrane up and down, pushing the hair cells against the tectorial membrane and distorting their plasma membranes. Electrical changes resulting from the distortions cause the hair cells to release neurotransmitters, which generate electrical signals in adjacent sensory neurons. All the sensory neurons of each ear form the auditory nerve.

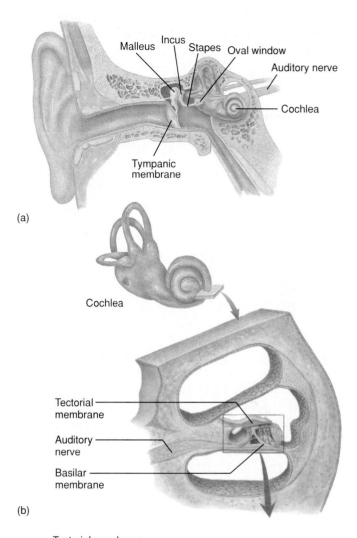

(a)

(b)

The Sense of Equilibrium

A sense of equilibrium provides information about movements and positions of the body. It tells an animal when its body is falling or turning, which way is up, and how the body is tilted with regard to gravity (fig. 25.12). This sense is essential for locomotion: it enables an animal to maintain balance by compensating for displacements of the body. In a vertebrate animal, sense organs dealing with equilibrium are located within the inner ear.

A common equilibrium disorder, called motion sickness, develops when information from the inner ear contradicts information from the eyes. When inside a ship, for example, the inner ears detect displacements of the body in relation to gravity, whereas the eyes detect a fixed position of the body in relation to the interior of the cabin. The brain receives conflicting information, and for reasons not yet clear it sends messages to the vomiting center.

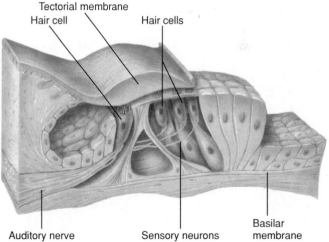

(c)

Figure 25.12 A cat has a superb sense of equilibrium.
Urban cats often fall from the balconies of high-rise
apartments but are rarely injured. A falling cat instinctively
twists its body so that its legs point downward and the impact
is spread over all four legs. It also flexes its legs as it lands,
further distributing the force of the impact through many
muscles and joints.

Table 25.5	
The Human Ear:	
Structures That Provide Equilibrium	
Structure	**Function**
Semicircular canals	
Bony walls	Hold hair cells in place; hold fluid
Fluid	Pulls on the cilia of hair cells
Hair cells	Transduce rotations of the head
Utricle and saccule	
Bony walls	Hold hair cells in place; hold fluid
Fluid	Pulls on the cilia of hair cells
Hair cells	Transduce tilts of the head

Receptor Cells

Equilibrium is sensed by hair cells (mechanoreceptors),
usually within the head of an animal. One end of each hair
cell is attached to a bone. The other end is covered with
cilia and immersed within a thick, heavy fluid. When the
head moves, the fluid slides over the stationary hair cells.
The cilia, which move with the fluid, become bent and
their plasma membranes distorted:

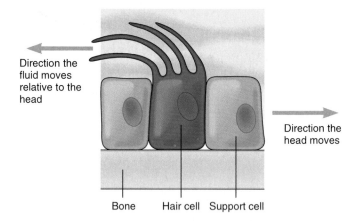

Direction the fluid moves relative to the head

Direction the head moves

Bone   Hair cell   Support cell

The distortion opens ion channels in the membrane and
changes the hair cell's electric charge, which triggers the
release of neurotransmitters. These chemical messengers
generate electrical signals in an adjacent neuron, which
carries the information to the brain.

The Sense Organs

In vertebrate animals, the organs that sense equilibrium
are located within the inner ear. Hair cells are attached to
the inner walls of five bony structures: three curved tubes,
called the semicircular canals, and two cavities, called the
utricle and saccule. Their functions are listed in table 25.5.

Wellness Report

Music That Deafens

Millions of young people are sacrificing their hearing to the pleasures of music. Loud sounds, particularly from amplified music (fig. 25.B) and earphones, destroy hair cells in the inner ear—the cells that convert sounds into electrical signals within neurons. Loud sounds generate such strong waves within the fluid of the cochlea that the delicate hair cells (fig. 25.C) are slammed into the tectorial membrane and ripped apart. If someone else can hear sounds from the earphones on your personal stereo, then your hair cells are being damaged. The hearing loss is permanent. In time, the earphones will have to be swapped for a hearing aid.

How do you know if your hair cells are damaged? The ability to hear high-pitched sounds is lost first. In particular, you will have trouble distinguishing consonants within human speech, because the sounds of these letters are higher in frequency than the sounds of vowels. Words like "thin" and "fin" or "sheep" and "cheap" will sound the same. You will also have difficulty separating voices near you from background noises, making it difficult to carry on a conversation during a party or to follow a television program in noisy surroundings. Hearing loss affects every facet of life: social, cultural, educational, vocational, and recreational. Not surprisingly, people who

Figure 25.B Amplified music can cause permanent loss of hearing.

do not hear well tend to become depressed and to withdraw from social interactions.

Eighteen-year-olds used to have the sharpest ears of anyone. Now most of them hear no better than their grandparents, whose hearing has declined because of aging. What will happen as these young people grow older? Only time will reveal what problems today's youth will experience when the normal hearing loss associated with the aging process is combined with their already impaired hearing.

Approximately 80% of all deafness in the United States is caused by damaged hair cells. Researchers believe they will eventually find a way to regenerate the damaged hair cells. They are optimistic because they

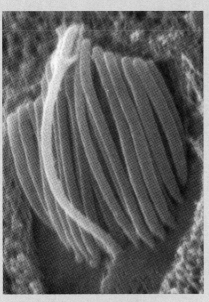

Figure 25.C Cilia projecting from a hair cell in the inner ear.

know it happens in other kinds of vertebrate animals: fishes, amphibians, and reptiles form new hair cells (just like ours) throughout life, and birds can replace damaged hair cells with new ones. While mammals lack this ability, there is some indication they could do so if given the appropriate chemical signal. A promising avenue of research is to find the molecule that stimulates regeneration in birds and then use it on mammals. If this molecule stimulates regeneration of mammalian hair cells, then we will have a cure for most cases of deafness. In the meantime, turn down the sound.

The Semicircular Canals   The **semicircular canals** inform the brain about sudden movements and rotations of the head. They help an animal maintain balance while accelerating, decelerating, falling, and turning. The three canals are oriented along the three planes of movement: forward-backward, up-down, and left-right (fig. 25.13*a*). Each canal is filled with a thick fluid and lined at one end with hair cells (fig. 25.13*b*). When the head moves rapidly, the semicircular canals and the hair cells move with the

head, but the heavy fluid in one of the canals (the one oriented in the same direction as the movement) lags behind (just as when you move a bowl of water quickly, the water moves more slowly than the bowl). Cilia of the hair cells, which are immersed in the lagging fluid, become bent and distorted, opening channels in their plasma membranes. In response to these electrical changes, the hair cells release neurotransmitters into synapses between the hair cells and sensory neurons.

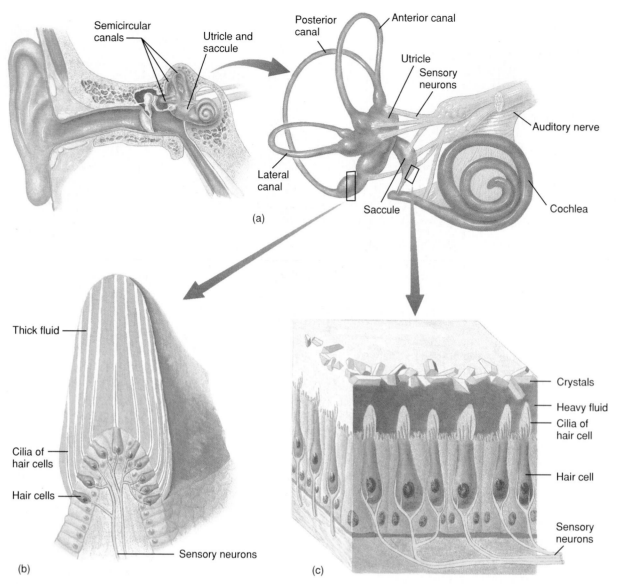

Figure 25.13 The human ear: structures that provide equilibrium. (*a*) The semicircular canals, utricle, and saccule provide information about equilibrium. The vestibular nerve consists of all sensory neurons carrying information about equilibrium. (*b*) The enlarged end of a semicircular canal contains the hair cells that transduce movements of the head. When the head is moved suddenly or rotated, the hair cells move with the head but their cilia, which are immersed in a thick fluid, move more slowly. As the cilia bend, their plasma membranes become distorted and the hair cells release neurotransmitters that generate electrical signals in adjacent sensory neurons. (*c*) The utricle and saccule are lined with hair cells and filled with a fluid made heavy with crystals of calcium carbonate. When the head is tilted, the heavy fluid flows downward and pulls the hair cells, distorting their plasma membranes and triggering the release of neurotransmitters. The neurotransmitters generate electrical signals in adjacent sensory neurons.

The Utricle and Saccule The **utricle** and **saccule** inform the brain about the tilt of the head—its position with regard to gravity. These two interconnected cavities are lined with hair cells and filled with a fluid (fig. 25.13c).

The hair cells are attached to the inner walls of the cavities but their cilia extend into the fluid, which is studded with crystals of calcium carbonate (limestone). The crystals make the fluid heavier and more responsive to gravity. When the head tilts, gravity pulls the fluid downward and bends the cilia, opening ion channels in their membranes. The hair cells release neurotransmitters that generate electrical signals in adjacent sensory neurons.

Figure 25.14 The touch receptors of animals reflect their way of life. A pig, which feeds by digging in the soil, senses its environment largely through touch receptors on its large nose.

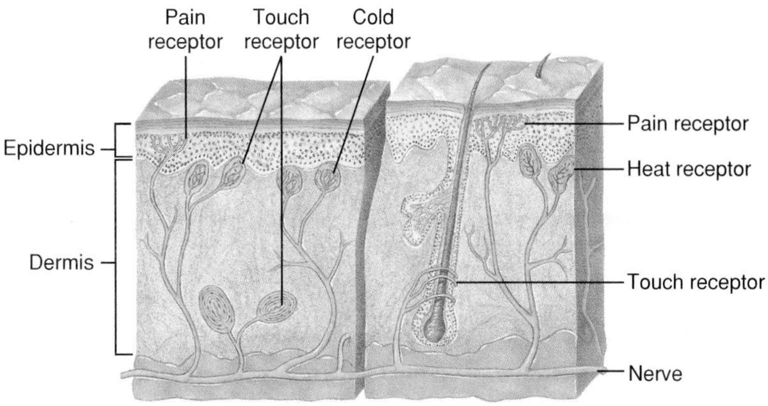

Figure 25.15 Receptor cells of human skin. The skin contains touch receptors, which respond to pressure; pain receptors, which respond to the chemicals released from damaged or stretched cells; and thermoreceptors, which respond to changes in temperature.

The Somatic Senses

If you close your eyes and plug your ears, you still receive a great deal of information about events taking place on your skin and inside your body. You know when part of your body is in contact with a solid object, how fast the wind is blowing, the temperature of the air, and the positions of your legs and arms. The somatic senses inform your brain about what is happening on the surface of your body and inside it. They deal with touch, pain, skin temperature, and body position.

Touch

A sense of touch provides information about contact with solid objects and about movements of air and water across the body surface. It is a property of the skin. Each animal has more touch receptors in certain regions of the skin than in others (fig. 25.14), depending on its way of life. Our own skin has more touch receptors in the fingertips, lips, tongue, and genitalia than elsewhere.

Touch is one of our first senses to develop and is usually the last to diminish; newborns receive most of their information through the skin, and the elderly remain sensitive to touch long after their sight, hearing, and sense of smell have faded.

A touch receptor is a sensory neuron that generates electrical signals when its plasma membrane is distorted by pressure. Touch receptors located within different layers of the skin respond to varying degrees of pressure (fig. 25.15). Additional touch receptors are wrapped around the roots of body hairs and respond to slight air currents and light pressures, such as the landing of a mosquito. Touch receptors are a form of mechanoreceptor, which transduces the energy of mechanical force.

Pain

Pain protects an animal by evoking behaviors that prevent further injury. If we touch a hot stove, for instance, pain from our fingertips stimulates us to pull away. People who cannot feel pain in their legs, because of strokes or spinal cord injuries, experience frequent cuts, bruises, burns, and sores from not moving their legs away from sources of injury.

Pain receptors are sensory neurons (fig. 25.15) that send electrical signals to the brain, where pain is perceived. (They also carry information about itching.) Pain receptors are located within the skin, internal organs, cornea of the eye, and tissues surrounding the muscles and bones. Surprisingly, the brain itself has no pain receptors; brain surgery can be performed without anesthetics.

The sensation of pain develops in response to a variety of stimuli, such as a pinch, a cut, or intestinal gas. The damaged or stretched cells trigger the formation of chemical messengers, bradykinin and prostaglandins, within the tissues. These messengers, in turn, stimulate pain receptors directly by binding with molecules in their plasma membranes. They also generate pain indirectly, by initiating swelling that presses on and distorts the plasma membranes of pain receptors. In response to these membranes changes, pain receptors develop electrical signals that travel to the brain, where they are perceived as pain.

Pain-killing drugs act in different ways. The opiates (heroin, morphine, and codeine) act on pain centers within the brain. They relieve deep, cramping types of pain by generating a feeling of detachment, in which pain sensations are disconnected from emotions. Local anesthetics, such as novocaine, act directly on pain receptors in the skin and gums. Aspirin, ibuprofen, and naproxen block the formation of prostaglandins, thereby preventing inflammation and the pressure of fluids on pain receptors.

They alleviate pain from sprains, toothaches, and headaches. Ibuprofen is also effective against the cramping pain associated with menstruation and intestinal gas.

Skin Temperature

Temperature receptors in the skin help an animal regulate its body temperature, either by internal mechanisms (homeostasis) or by behavior (changing posture; moving to a warmer or cooler place).

Thermoreceptors in the skin transduce changes in heat energy into electrical signals within neurons. How this happens is not well understood. The human skin contains two types of thermoreceptors: one type responds to temperatures warmer than 25° C (77° F) and the other responds to temperatures cooler than 20° C (68° F). These receptors, which are sensory neurons (fig. 25.15), send the information to the temperature control center within the hypothalamus (see page 327).

Kinesthesia

Kinesthesia informs the brain about position and movement of body parts. This sense is provided by sensory neurons that are mechanoreceptors. Located within muscles, tendons, and ligaments, they generate electrical signals in response to stretch and pressure. The signals travel to the brain and initiate movements that help the animal to maintain posture, coordinate locomotion, and avoid overstretching its tissues.

SUMMARY

1. An event outside the nervous system is perceived only after transduction, in which energy of the event is transformed into electrical signals within neurons. Transduction is performed by receptor cells, which may be sensory neurons or specialized cells within sense organs.

2. Electrical signals generated by receptor cells are carried to the brain for initial decoding. Sensory pathways reveal the type of event. Patterns of electrical signals reveal the intensity and temporal pattern of the event. The event is perceived by linking its information with memories and emotions.

3. An animal sees when photoreceptors (rods and cones) within its eyes transduce light. Light absorption alters pigments within these receptors in ways that generate electrical signals within the nervous system. Rods provide black-and-white images; cones provide colored images. Other cells in the retina assist the rods and cones by relaying and integrating visual information. Accessory structures of the eye include the iris and pupil, which control how much light enters; the cornea and lens, which focus the light; and the eye-

ball, which, by moving, locates objects and maintains stimulation of the photoreceptors.

4. An animal tastes when dissolved molecules are transduced by chemoreceptors, which, in vertebrates, are located within taste buds. Transduction occurs as molecules latch onto receptor molecules within the plasma membranes.

5. An animal smells when airborne molecules within the nose are transduced by chemoreceptors. When the molecules bind to the plasma membrane of a chemoreceptor, electrical signals are generated within the nervous system.

6. An animal hears sounds, which are rearrangements of air molecules in response to vibrations of moving objects. The outer ear funnels sound into the middle ear, which amplifies the sound. The inner ear contains the hair cells that transduce sound.

7. The sense of equilibrium is provided by mechanoreceptors, called hair cells, within the head (the inner ear of a vertebrate). Movements of the head distort cilia on the hair cells, which results in a release of neurotransmitters that generate electrical signals. Hair cells within the semicircular canals inform the brain about rotations of the head. Hair cells within the utricle and saccule inform the brain about tilts of the head.

8. The somatic senses inform the brain about touch, pressure, temperature, and position of the body parts. Touch is transduced by several kinds of mechanoreceptors in the skin; they respond to distortions of their plasma membranes. Pain is transduced by receptor cells located throughout the body. Temperature of the skin is transduced by two kinds of thermoreceptors: one that responds to warmer and another that responds to cooler temperatures. Movements of body parts, a sensation known as kinesthesia, are transduced by mechanoreceptors within the muscles, tendons, and joints.

KEY TERMS

Chemoreceptor (KEE-moh-ree-SEP-tuhr) (page 475)
Cochlea (KOH-klee-ah) (page 478)
Cone (page 472)
Cornea (KOHR-nee-ah) (page 474)
Hair cell (page 477)
Inner ear (page 478)
Iris (EYE-ris) (page 474)
Kinesthesia (KIH-ness-THEE-zee-ah) (page 484)
Lens (page 474)
Mechanoreceptor (page 477)
Middle ear (page 478)
Outer ear (page 478)
Oval window (page 478)
Perception (pur-SEP-shun) (page 470)

Photoreceptor (page 471)
Pupil (PYOO-pil) (page 474)
Receptor (ree-SEP-tuhr) cell (page 468)
Retina (REH-tih-nuh) (page 473)
Rod (page 471)
Saccule (SAK-yool) (page 482)
Semicircular canals (page 481)
Taste bud (page 475)
Thermoreceptor (page 484)
Transduction (trans-DUK-shun) (page 468)
Tympanic (tim-PAEH-nik) membrane (page 478)
Utricle (YOO-trih-kuhl) (page 482)

STUDY QUESTIONS

1. Define transduction. Explain how transduction by a sensory neuron differs from transduction by a cell that is not a neuron.
2. Describe how light energy entering the eye becomes electrical signals in the optic nerve.
3. Describe how light is focused onto the retina.
4. Describe how airborne molecules that enter the nose become electrical signals in the olfactory nerve.
5. Trace the pathway of information from the time a friend says "Hello" to the time it enters the auditory nerve.
6. Explain how you know the position of your body (both head and torso) even when your eyes are closed.

CRITICAL THINKING PROBLEMS

1. What are some events that may be occurring around you, of which you are not aware? What kinds of animals could detect these events?
2. Why would it be difficult for a diurnal predator, who locates prey by sight, to become a nocturnal predator?
3. Explain why, in terms of adaptation, some animals have no cones and others have many cones in their retinas.
4. Evaluate the saying that "eating a lot of carrots will help you see better at night." (Hint: carrots are rich in beta carotene, which the body converts into vitamin A.)
5. Suppose that you dive off a pier into a lake on a dark night. Describe how your brain would receive information about your environment during the dive.

SUGGESTED READINGS

Introductory Level:

Ackerman, D. *Natural History of the Senses*. New York: Random House, 1990. Everything you might want to know about your own senses and those of other animals, written in a style that is scientific and yet entertaining and personal.

Kern, G. "Good Vibrations," *Discover* (June 1993): 45–54. A description of what loud sounds are doing to the hair cells of our inner ears.

Kern, G. "In the Realm of the Chemical," *Discover* (June 1993): 69–76. A description of our senses of smell and taste and how they helped our ancestors survive.

Montgomery, G. "Color Perception: Seeing with the Brain," *Discover* (December 1988): 52–59. A description of how we see relative, rather than absolute, colors and how this discovery was made.

Advanced Level:

Carlson, N. R. *The Physiology of Behavior*, 4th ed. Boston: Allyn and Bacon, 1991. An advanced textbook dealing with the physiological basis of behavior. Chapters 6 and 7, which deal with the senses, are exceptionally good.

Dusenbery, D. B. *Sensory Ecology: How Animals Acquire and Respond to Information*. New York: W. H. Freeman and Company, 1992. An analysis of the various ways in which animals receive and use information about their environments.

Farbman, A. I. *Cell Biology of Olfaction*. New York: Cambridge University Press, 1992. A detailed discussion of what is known about the mechanisms underlying odor detection, transduction, and perception.

Schnapf, J. L., and D. A. Bayler. "How Photoreceptor Cells Respond to Light," *Scientific American* (April 1987): 40–47. A description of how absorption of light alters the electrical charges of rods and cones.

Stein, B. E., and M. A. Meredith. *The Merging of the Senses*. Cambridge, MA: MIT Press, 1993. An analysis of the ways in which different kinds of sensory information become integrated within the brain.

Locomotion

Objectives

In this chapter, you will learn
–how animals use muscles to move
  their entire bodies from one place to
  another;
–what makes a bone strong;
–how bones are joined to form a
  skeleton;
–how muscles move bones;
–what causes a muscle contraction;
–how muscle cells influence athletic
  talent.

Locomotion in a barn owl.

ne of the few certainties of life is that regular exercise improves physical fitness. Running, bicycling, swimming, skiing, and other forms of aerobic exercise are guaranteed to improve endurance, physical appearance, and mood. A workout program always pays off.

Endurance reflects the body's capacity to acquire oxygen and to use it in cellular respiration, the biochemical process that releases energy from organic molecules. Aerobic exercise improves the ability of the lungs to bring oxygen into the blood. It strengthens the heart muscles, so that more blood is pumped with each contraction. It makes the blood vessels more flexible, so they can adjust more readily to changes in blood pressure. Heart and skeletal muscles develop whole new networks of capillaries, which transport more blood to the muscle cells. And muscle cells develop more mitochondria, the organelles that perform cellular respiration.

Physical appearance benefits from regular exercise. The body becomes trim as excess fat disappears. The skeletal muscles become firm, improving posture and giving a spring to the step. A physically fit person appears alert, vigorous, and energetic.

Exercise also elevates the mood. It not only reduces tensions but alters the chemistry of the brain in ways that generate feelings of well-being. Long aerobic workouts apparently stimulate the release of endorphins. These neurotransmitters, known as the body's own opiates, are responsible for "runner's high," characterized by feelings of euphoria, tranquility, and inner strength.

These desirable states of health were natural by-products of the great physical effort our ancestors had to make just to stay alive. Although our life-style no longer demands such effort, exercise remains essential for a healthy and vigorous life.

Animal Movement

Nonrandom movement is a striking characteristic of life. Every cell has internal movements that transport materials and organelles from one place to another. Every cell moves with precision during cell division, when chromosomes line up in pairs and then move apart, and when the whole cell pinches in two.

In addition to cellular movements associated with internal function and division, some organisms move in a controlled fashion from one place to another. This type of movement, characteristic of animals, is called **locomotion.**

Animals crawl, walk, jump, fly, swim, slither, and burrow. An animal must move from place to place to capture food and to avoid being eaten. Animal movement is accomplished by muscles—bundles of cells specialized for changing their lengths. Muscles move an animal by squeezing or pulling more rigid parts of the body.

An earthworm uses both squeezing and pulling muscles to crawl along surfaces. Each segment of its body has two sets of muscles—a set of longitudinal muscles that pull and a set of circular muscles that squeeze, as shown in figure 26.1a. When the circular muscles contract, they squeeze the front of the body into a longer, thinner shape. Body fluids are pumped into the front segments, making them rigid and extending them forward. Hairs along the front segments grasp the ground. The rest of the body then is pulled forward as the longitudinal muscles contract to shorten the body, as shown in figure 26.1b.

Locomotion in most animals is accomplished by muscles that pull components of a skeleton rather than body fluids. The exoskeleton of an invertebrate animal, such as a lobster or an insect, consists of many rigid plates that lie external to the muscles that move them (fig. 26.2a and b). The endoskeleton of a vertebrate animal, such as a fish or a mouse, consists of bones that lie internal to the muscles that move them (fig. 26.2c and d).

Vertebrate locomotion is the main topic of this chapter. It is accomplished by four kinds of structures: bones, joints, tendons, and muscles. These structures are described briefly in table 26.1 and in detail following.

The Skeleton

A vertebrate animal accomplishes locomotion by moving its skeleton. The bones of a skeleton provide a rigid frame (contractions of a soft blob would not produce effective locomotion). In many animals the bones also support the body above the ground to facilitate locomotion. Each bone is a living organ that changes in response to pressure.

All vertebrate skeletons are similar in design (fig. 26.3), since all vertebrate animals have inherited genes for developing this skeleton from the same ancestral vertebrate—a primitive form of fish that lived approximately 540 million years ago. Skeletal differences between major groups of animals reflect differences in their ways of life. A fish, for example, does not have ribs or leg bones. All mammals have the same number of bones (206) interconnected by the same kinds of joints. Differences in the shapes of mammalian skeletons are due to differences in the lengths and widths of their bones. The neck bones of a giraffe, for example, are much longer than the neck bones of a whale. The bones of a human skeleton are listed in table 26.2 and illustrated in figure 26.4.

Bones

We tend to think of bones as dry and brittle because that is how we usually see them—the skull of a cow in a field or the skeleton of a human in a biology laboratory. Living bones are not at all like this. They are moist, pink, flexible tissues made in part of living cells. The shape of a

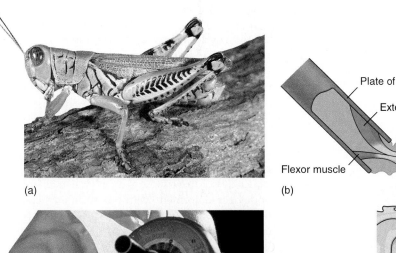

Figure 26.1 Locomotion in an earthworm. (*a*) An earthworm has many body segments. The wall of each segment contains a sheet of longitudinal muscles and a sheet of circular muscles. (*b*) Contractions of the circular muscles make the body narrow and rigid, extending the front segments forward (*l and 2*). The worm then grasps the surface with its front segments (*3*). Contractions of the longitudinal muscles pull the rest of the body forward (*4 and 5*). The earthworm has moved one "step" forward.

(a)

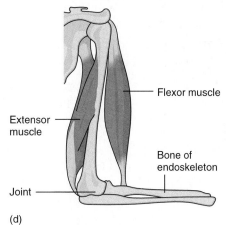

(c)

(b)

Plate of exoskeleton

Extensor muscle

Joint

Flexor muscle

Figure 26.2 How muscles move animal skeletons. (*a*) Insects and other kinds of invertebrates have exoskeletons—hard plates on the outside of their bodies. (*b*) An insect moves its plates by contracting muscles that lie beneath its skeleton. (*c*) Mammals and other vertebrates have endoskeletons—bones that lie beneath the muscles. (*d*) A vertebrate moves its bones by contracting muscles that lie outside its skeleton.

Flexor muscle

Extensor muscle

Bone of endoskeleton

Joint

(d)

489

Table 26.1

Vertebrate Locomotion: The Structures and Their Functions

Structure	Function
Bone	Supports and protects soft tissues; forms blood cells; stores minerals and fats
Joint	Enables movement of bones
Ligament	Holds bones together
Cartilage	Cushions the ends of bones
Fluid	Lubricates places where bones rub together
Tendon	Attaches a muscle to a bone
Muscle	Moves a bone

weight-bearing bone, such as a leg bone or backbone, is continually modified in response to pressures placed upon it.

Bone Structure   Bone is a type of connective tissue (📖 *see page 316*) formed of widely spaced bone cells immersed within a hard material. Portions of each bone cell extend outward to connect with other bone cells and with blood vessels (fig. 26.5). The cells are arranged in concentric circles around the blood vessels, which supply them with nutrients and dispose of their wastes.

About 45% of a bone consists of hard material that surrounds the cells and vessels. This material, secreted by the bone cells, consists mainly of collagen and minerals.

Collagen is a remarkable protein that is abundant in all animals; it forms one-third of all the protein in your body. A collagen molecule consists of three intertwined polypeptides (chains of amino acids):

Bundles of collagen molecules form very strong fibers, which give strength to all components of the skeleton. The minerals, which make a bone firm, are chiefly calcium and phosphorus but also magnesium, manganese, boron, and others. Strong bones are maintained by dietary minerals, certain steroids, and regular pressure.

Strong bones contain high concentrations of minerals. These nutrients, which come from our food, are especially abundant in milk, fish bones (e.g., in sardines and canned salmon), and dark green leafy vegetables.

Strong bones also require certain steroids (📖 *see page 45*)—vitamin D and the sex hormones (either estrogen or testosterone). These steroids help transport minerals from the digestive tract into the bloodstream, where they can be taken up by bone cells, and stimulate the bone cells to secrete collagen. Without sufficient vitamin D, a child develops rickets, in which the leg bones are deformed. Without sufficient sex hormones, an adult develops osteoporosis (*osteo*, bone; *porus*, pore), in which the bones are easily broken (fig. 26.6). This condition develops most often in older women, owing to a dramatic reduction in estrogen after menopause, but also occurs in men as their testosterone levels gradually decline with age. It is aggravated, among the elderly,

Pelvic girdle

Scapula (shoulder blade)

Femur

Humerus

Ulna

Fibula

Tibia

Radius

(a)

(b)

(c)

Figure 26.3 All vertebrate animals have similar skeletons. (*a*) Cat. (*b*) Lizard. (*c*) Fish. The same colors in the different animals represent the same kinds of bones.

by poor eating habits (insufficient minerals), insufficient sunlight (insufficient vitamin D), and not enough exercise (insufficient pressure on the bones).

Strong bones require regular pressure. The more pressure placed on a bone, the more minerals the bone cells secrete and the stronger the bone becomes. A heavier person has stronger bones, as does someone who lifts weights or runs regularly. The weakest bones are found in people confined to bed and in astronauts who have experienced long periods of weightlessness.

Each human skeleton is unique; it reflects not only the sex, age, and race of the person but also the life-style, illnesses, and injuries. These distinctive features enable experts to identify humans from remains of their skeletons, as described in the accompanying Research Report.

Bone Functions Bones have four functions: to provide a rigid frame, to protect the softer tissues, to form blood cells, and to store materials.

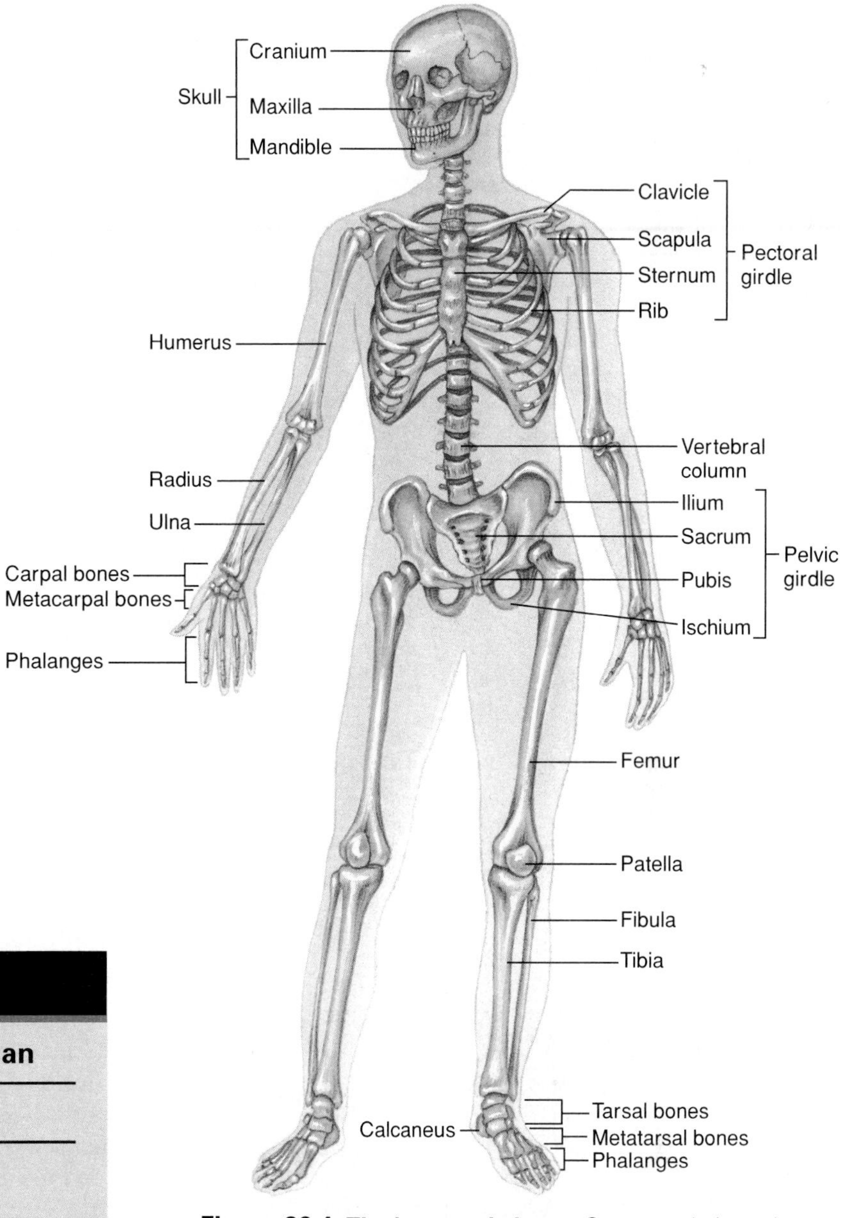

Figure 26.4 The human skeleton. Our own skeleton is almost the same as a cat skeleton (see fig. 26.3a) except that we stand upright on two legs.

Table 26.2

The Distribution of Bones in a Human

Part of the Body	Number of Bones
Skull	22
Ears	6 (3 each)
Backbone	26
Sternum	1
Throat	1
Shoulders	4
Arms	6 (3 each)
Hands	56 (28 each)
Hips	2
Legs	6 (3 each)
Feet	52 (26 each)
Ribs	24 (12 on each side)*

* One out of every twenty people has an extra pair of ribs. This trait is more common in men than women.

Most bones contribute to a rigid frame for locomotion. The bones of terrestrial vertebrates (except snakes and some amphibians) also hold the body aboveground to facilitate locomotion.

Some bones protect the softer tissues of the body from injury. Bones that form the skull, vertebrae (backbone), and ribs protect the brain, spinal cord, heart, and lungs. In fact, protection was the original function of bones—in the first vertebrate animals (the early fishes), the bones were on the outside of the body, where they formed heavy plates that protected softer tissues from attacks by predators.

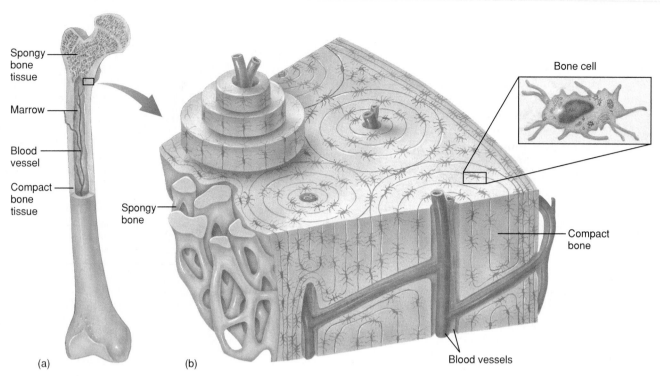

Spongy bone tissue

Marrow

Blood vessel

Compact bone tissue

(a)

Spongy bone

(b)

Bone cell

Compact bone

Blood vessels

Figure 26.5 Bone tissue. (*a*) A bone, such as the femur shown, has two kinds of tissues, spongy and compact. Spongy bone tissue is porous and lightweight; it forms the ends of some bones as well as the marrow. Compact bone tissue, which forms the "walls," is denser and stronger than the spongy bone tissue. (*b*) In compact bone tissue, bone cells are arranged around blood vessels. The cells are far apart and surrounded by a hard material formed of collagen and minerals. Each cell has extensions that reach out toward other cells and regions near blood vessels.

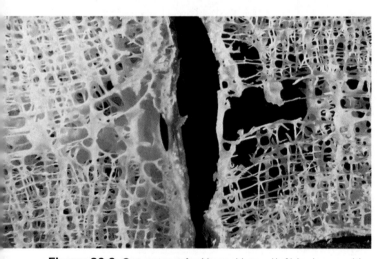

Figure 26.6 Osteoporosis. Normal bone (*left*) is dense with mineral deposits. Osteoporotic bone (*right*) has lost its minerals and become porous and easily broken.

The interior of a bone is soft tissue called **marrow** (see fig. 26.5). In the shorter bones—the skull, vertebrae, ribs, and pelvis—the marrow is red and its function is to form both red blood cells and white blood cells. In the longer bones of the arms and legs, the marrow is yellow and its function is to store excess minerals and fat.

Joints

A **joint** is a connection between two or more bones. Most joints are movable; they allow the bones to change position in relation to one another. A movable joint is made of bones, ligaments, cartilage, and fluid (see table 26.1). The largest joint of the body—the knee—is shown in figure 26.7.

Ligaments A **ligament** is a rope of collagen that holds together the bones of a joint (fig. 26.7). It is very strong and somewhat elastic, resembling a powerful rubber band.

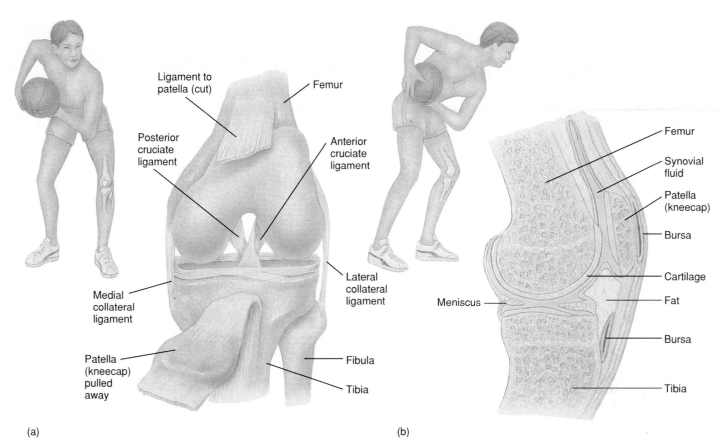

(a)

(b)

Figure 26.7 Anatomy of the knee. (*a*) The right knee as seen from the front, with the kneecap removed (whether it is right or left can be distinguished by the position of the fibula, as can be seen in fig. 26.4). Four major ligaments hold the femur and tibia together—the lateral collateral, medial collateral (most often injured in football from blows hitting the outside of the knee), anterior cruciate (most often injured in basketball and skiing from extending the lower leg too far forward), and posterior cruciate. (*b*) The knee as seen from the side, showing the cavity of the joint.

Although ligaments are strong, they are the weakest parts of the skeleton. Violent movements that stretch a ligament more than about 15% of its normal length cause small tears, a condition known as a sprain. Severe stress on a joint may stretch the ligaments so much that one bone is moved out of alignment with another, a condition known as a dislocation.

Cartilage Cartilage is a connective tissue made of cartilage cells and collagen. It resembles bone but lacks mineral deposits and blood vessels. Cartilage is softer and more flexible than bone and does not heal as well as bone.

Cartilage forms cushions at the ends of bones (fig. 26.7b), protecting them from damage as they press against each other. In the knee, cartilage also forms meniscuses, which are wedges that stabilize the joint. Cartilage substitutes for bone within the outer ears and the tip of the nose.

Cartilage is damaged when too much stress is placed on the joints. The knee is particularly vulnerable to this kind of injury. Runners and basketball players in particular suffer from stress-related damage to the cartilage within their knees. Small pieces of cartilage sometimes break off from a meniscus and, unless surgically removed, can lock the knee in one position. Bone spurs may develop, especially in the foot, as damaged cartilage inappropriately acquires mineral deposits. Almost everyone suffers some degeneration of cartilage after age forty-five. This condition, called osteoarthritis or "wear-and-tear" arthritis, stiffens the hips, knees, feet, and backbone.

What Bones Can Tell Us

Two kids walking through a field in early spring found the remains of a human. The body was badly decomposed—basically only a skeleton with some hair. There was no longer a face or fingerprints. There were no clothes, wallet, or other personal items. What could be done to identify the body?

A great deal. Each one of the 206 bones and thirty-two teeth has a tale to tell. People who are trained to read clues from human remains are called forensic anthropologists (*forensic*, pertaining to courts of law; *anthropology*, the study of human populations). They help detectives and pathologists identify victims of crime, airplane crashes, and other tragedies. Each nameless skeleton poses a unique mystery to these professionals, and it is their job to solve it.

The first question: What was the sex of the victim? This question is the easiest to answer. A woman's pelvis is broader and shallower than a man's pelvis, her breastbone is wider and shorter, her wrists are slimmer, and her lower arms join her upper arms at a greater angle than a man's. Even when only a skull is available, the sex can be identified by differences in the size of the mastoid processes (the bony knobs at the base of the skull) and the ridges behind the eyebrows. After measuring the skeleton of the body found in the field, the forensic anthropologist concluded it was of a female.

Next, can anything be concluded about race? Was the victim of African, European, or Asian descent? Many traits provide clues, but the size and shape of the facial bones and the relative lengths of various other bones in the skeleton often fall into one of the three categories. After measuring all the bones and comparing them with standard measurements for each race, forensic experts concluded that the body was of European ancestry.

Approximately how old was the female when she died? A young child has more bones (350 as compared with 206) than an adult, because some bones fuse during development. The skull bones in a child, for example, are not yet fused; certain ones fuse in the late teens and others not until the mid-twenties. The collarbone, too, does not become a single bone until a person is between eighteen and twenty-five years of age. Age can also be estimated by the amount of cartilage at the ends of the leg bones: the amount declines as the legs grow. After adult stature is reached, the cartilage remains the same until about thirty, after which it again declines with age. All the skull bones in the victim were fused, as was the collarbone. Cartilage at the ends of the leg bones indicated the woman had stopped growing but her cartilage had not begun to degenerate. Experts estimated she was between twenty-five and thirty years old.

How tall was the woman? Was she athletic? Height is estimated by the length of the leg bones. Muscle strength is estimated by measuring the size of places on the bones where the muscles attached; larger muscles leave larger "scars" where they were joined to bones. These sites of muscle attachment also indicate whether the person was right-handed or left-handed, since they are larger on the arm that was used more. Experts concluded the dead woman was about 1.7 meters (5 ft, 6 in) tall, left-handed, and not particularly athletic.

What else can we surmise from a skeleton? Diseases leave their marks: arthritis leaves bony growths in the joints, tuberculosis breaks down bones of the back, and syphilis attacks bones of the skull, lower leg, and lower arm. Injuries also leave marks: knife wounds, bullet holes, and fractures, even when the injury occurred a long time ago and has completely healed. Forensic experts found no indications of diseases on the woman's bones, but they noted that she had broken her ankle when a young child. A more dramatic finding was a bullet hole through the rib that covered her heart. From the angle of the bullet and the fact that she was left-handed, detectives concluded she was murdered.

Who was the murdered woman? How can she be identified? X rays of the bones and teeth were compared with X rays in the medical records of missing persons. Certain patterns inside the bones, such as the shapes of sinuses, are unique to each person, as are the number and location of dental cavities and fillings. In this case, however, X rays of the victim could not be matched with those of any missing person.

As a final attempt at identification, the woman's face was reconstructed from her skull. Each person's facial bones—cheekbones, jawbones, nasal openings, eye sockets, and brow ridges—are unique and become transformed into a distinctive face when covered with clay of appropriate thickness (to simulate fat, muscle, and skin) and when eyelids, teeth, ears, lips, and a wig are added. This technique, shown in figure 26.A for the fossil of an early human, produced a distinctive face, which was printed on posters and in newspapers.

The murdered woman's face was recognized immediately by employees of a restaurant near the field where she was found. They recalled how she had stopped coming to work, more than a year ago, but at the time they thought she had simply moved on to another job. The murderer has not yet been found.

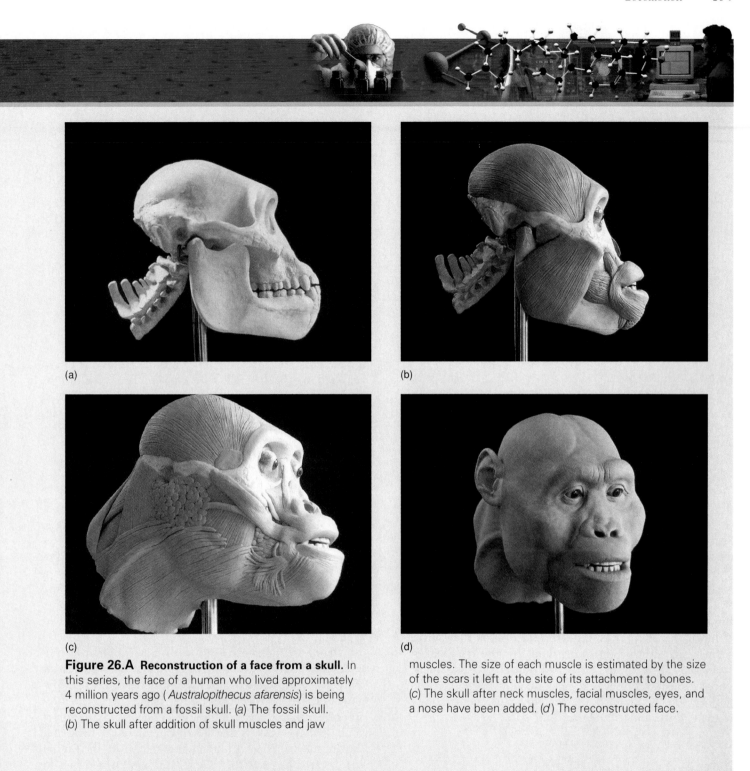

(a)

(b)

(c)

(d)

Figure 26.A Reconstruction of a face from a skull. In this series, the face of a human who lived approximately 4 million years ago (*Australopithecus afarensis*) is being reconstructed from a fossil skull. (*a*) The fossil skull. (*b*) The skull after addition of skull muscles and jaw muscles. The size of each muscle is estimated by the size of the scars it left at the site of its attachment to bones. (*c*) The skull after neck muscles, facial muscles, eyes, and a nose have been added. (*d*) The reconstructed face.

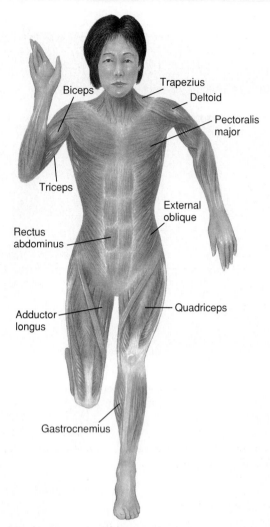

Figure 26.8 Muscles of the human body. More than 600 muscles move the bones of the skeleton. Only the larger muscles are shown here.

Synovial Fluid   The cavity of a joint is filled with **synovial fluid**—a thick material with the consistency of egg white. It lubricates the bones, reducing friction as they move against each other. Lubricating fluids are also held within small sacs called bursae (fig. 26.7) in the elbows, shoulders, knees, heels, and elsewhere. The amount of fluid in the joints gradually diminishes as you grow older, contributing to the stiffness and pain of the joints.

Muscles

A vertebrate animal moves its body by moving its bones. The bones, in turn, are moved by skeletal muscles—organs made of cells that specialize in changing length. A vertebrate animal has hundreds of skeletal muscles; half your body weight is skeletal muscle (fig. 26.8).

Each muscle is attached to two or more bones, which are connected by a joint. A muscle has power only when it shortens—not when it lengthens; it contracts to pull a

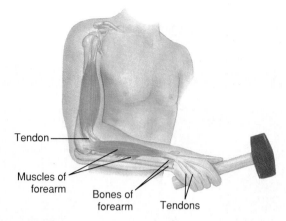

Figure 26.9 Tendons of the fingers. Bones in the fingers are moved by muscles in the forearm. Long tendons connect these bones and muscles, freeing the hand of bulky muscles.

bone into a new position but cannot then push the bone back to its original position. It takes two muscles to move a bone back and forth: one to pull the bone in one direction and another to pull it back again. Two muscles that work together in this way are called an **antagonistic pair.** The biceps and triceps of the arm (fig. 26.8), for example, are an antagonistic pair of muscles—the biceps raises the forearm and the triceps lowers it.

Attachments to Bones

A **tendon** is a rope of collagen that attaches a muscle to a bone. Of all the structures involved in movement—bone, ligament, tendon, and muscle—tendons are the strongest and least likely to break. Severe stress may rip a tendon from its attachment to a bone but almost never rips a tendon from its attachment to a muscle. Small tears, however, develop from repetitive activities like hitting a tennis ball or distance running. This condition is known as tendinitis.

Some tendons are long, with muscles far from the bones they move. This arrangement allows certain parts of the body to be slim and not encumbered by muscles. The muscles that raise the fingers, for example, are in the forearm rather than the hand; they are connected to the finger bones by long tendons that lie on the top of the hand and the fingers (fig. 26.9). If the muscles that move the fingers were near the bones they move, the hand would be too bulky to manipulate small objects.

Muscle Structure

A single muscle, such as the biceps of the arm, consists of many **muscle bundles** (fig. 26.10a). Each muscle bundle, in turn, consists of hundreds of muscle cells arranged in parallel and wrapped within a **muscle sheath,** a tough covering made of collagen. The sheaths of all the muscle bun-

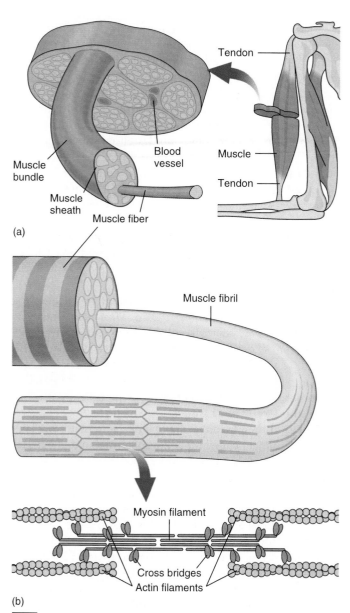

(a)

(b)

Figure 26.10 Anatomy of a muscle. (a) A muscle is made of many muscle bundles—each holding hundreds of muscle fibers and enclosed within a sheath. (b) Each muscle fiber (cell), in turn, contains hundreds of muscle fibrils. Each fibril is made of overlapping filaments of actin and myosin.

dles merge at the end of the muscle to form the tendon. Because a tendon is a continuation of many muscle sheaths, it is unusually strong and unlikely to tear at the end, where it joins with the muscle.

A **muscle fiber** is a muscle cell—long and cylindrical in shape, with mitochondria, ribosomes, and other structures typical of animal cells. A muscle fiber is packed with long structures, called fibrils, which are arranged in parallel and run the length of the cell (fig. 26.10b).

A **muscle fibril** consists of two kinds of proteins—actin and myosin. **Actin** is a globular (spherical) protein that joins with other actin molecules to form two intertwined chains called the actin filament:

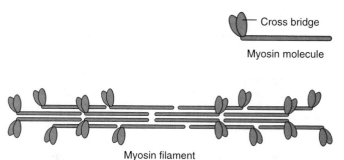

Myosin is a long protein that terminates in two knobs that are called a **cross bridge.** Myosin molecules bind with one another to form a myosin filament:

The cross bridges on the filament play important roles in muscle contractions. In a muscle fibril, the actin and myosin filaments are arranged one after another in an overlapping pattern (see fig. 26.10b).

Muscle Function

A muscle shortens, to about half its resting length, when its muscle cells (fibers) contract. The strength of the contraction depends on how many muscle cells contract. To lift a pencil, only a few muscle cells in the arm contract. To lift a child, many muscle cells in the arm contract.

Contraction of a Muscle Cell   When a muscle cell contracts, all its fibrils shorten; when a muscle cell relaxes, all its fibrils lengthen.

The length of a muscle cell depends on the degree of overlap between the actin and myosin filaments within its fibrils. When a fibril is relaxed, there is little overlap because the actin filaments are relatively far apart:

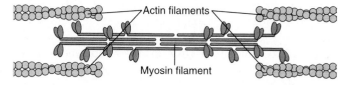

When a fibril is contracted, there is a lot of overlap because the actin filaments are closer together:

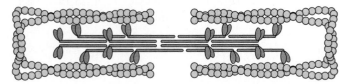

During a contraction, the actin filaments are pulled toward each other by the myosin filaments. Each cross bridge on the myosin filament attaches to one actin filament and pulls it toward another actin filament:

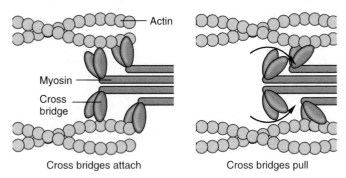

Cross bridges attach   Cross bridges pull

The attachment is a chemical bond that, once formed, alters the shape of the cross bridge in such a way that it pulls the actin filament. Each cross bridge pulls and releases the actin filament many times during a contraction.

When the contraction is complete, all the cross bridges detach from the actin filaments. The actin filaments then slide away from each other and the fibril lengthens.

Energy of a Contraction   As within every other cellular process, the energy of a contraction comes from ATP (adenosine triphosphate). A resting fibril is already "loaded" with energy from ATP (see page 87); it does not need an input of ATP to contract. Each cross bridge is like a mousetrap that has been set and only needs the right stimulus to activate it. When stimulated, the cross bridge uses its stored energy to pull the actin filament (fig. 26.11a–c).

After a cross bridge has moved an actin filament, it must be reloaded with energy from a molecule of ATP. The ATP is used to (1) break the chemical bond between each cross bridge and actin molecule (fig. 26.11d), so that the cross bridge can detach; and (2) recharge the cross bridge in preparation for the next contraction (fig. 26.11a). Most muscular action involves continuous work, rather than a single contraction, and requires a continuous supply of ATP.

Thus, an input of ATP is needed for a muscle to relax but not for a muscle to contract. This surprising fact explains rigor mortis—sustained contractions of the skeletal muscles after death. When muscle cells are no longer active, there is no ATP for detaching the cross bridges from the actin. The fibrils remain in a contracted position for a day or so after death, until the actin and myosin filaments degenerate and can no longer hold their positions.

Stimulation of a Contraction   When a muscle cell is relaxed, bonding sites on the actin filaments are covered with proteins (the tropomyosin-troponin complex) so that the myosin cross bridges cannot attach. When a muscle cell is stimulated, these proteins change shape (in response to the presence of calcium ions) and the bonding sites become exposed. The cross bridges then bond to the actin filaments and pull them closer together.

A muscle contraction is stimulated by an electrochemical impulse in a motor neuron. Most motor neurons (see page 447) that control skeletal muscles carry commands from the motor cortex (see page 460) of the brain, as described in chapter 24. Each motor neuron forms synapses with one or more muscle cells. Electrical signals in the motor neuron trigger the release of its neurotransmitter (acetylcholine; see page 452), which generates electrical signals in the plasma membrane of the muscle cell. This electrical change, called an action potential and generated (as in a neuron) by movements of sodium and potassium ions, stimulates the release of calcium ions (Ca^{++}) from small sacs that lie next to the muscle fibrils (fig. 26.12).

The calcium ions then trigger two reactions within the muscle fibrils: (1) a change in the proteins that cover actin, exposing bonding sites and allowing attachment of the cross bridges; and (2) conversion of myosin into an enzyme that releases energy from ATP.

After the contraction, the calcium ions are returned to their sacs, a process that requires energy from ATP. There the ions remain until the muscle cell is stimulated again. In the absence of calcium ions, proteins on the actin filaments move back over the bonding sites and prevent the cross bridges from attaching.

The entire sequence of events, from electrical signals within the motor neuron to contraction of the muscle cell, is diagrammed in figure 26.13.

Types of Muscles

Not all muscles work in the same way. The leg muscles of a cottontail rabbit are specialized for sprinting, so the animal can quickly reach shelter; the leg muscles of a jackrabbit are specialized for distance running, so the animal can out-

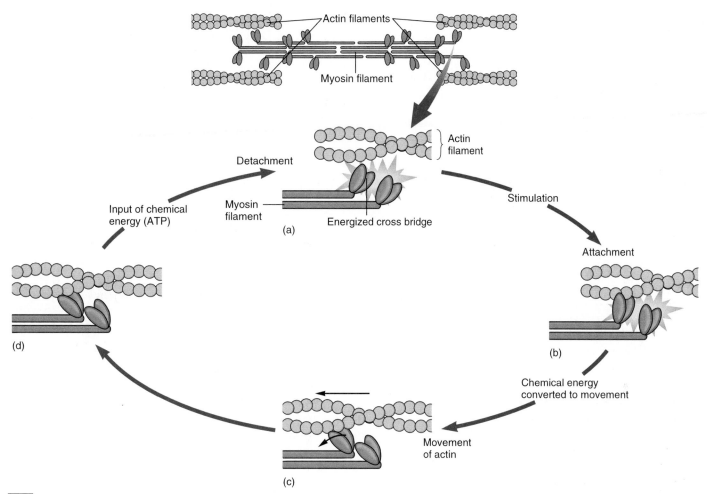

Figure 26.11 Contraction of a muscle fibril. A fibril contracts when cross bridges on the myosin filament pull the actin filaments closer together. (*a*) When a muscle fibril is relaxed, each cross bridge is energized (has received energy from ATP) but is not attached to the actin filament. (*b*) When a muscle fibril is stimulated, the cross bridge attaches to the actin filament. (*c*) Attachment of the cross bridge changes its shape, causing it to pull the actin filament. This movement converts chemical energy within the cross bridge into the energy of movement. The cross bridge is no longer energized. (*d*) Energy from ATP is added to the cross bridge, causing it to detach from the actin and also recharging it for the next pulling movement.

last its enemies. Similarly, the leg muscles of a chicken are built for slow walking, whereas the breast muscles are built for rapid flying. Humans, too, exhibit these differences: the leg muscles of world-class sprinters are not the same as the leg muscles of world-class marathoners.

Muscles differ in how they generate ATP, the immediate source of energy for muscle contractions. Some muscles are specialized for rapid but inefficient ATP production. Others are specialized for efficient but slow ATP production. Before examining these specializations, we need to review the ways in which cells generate ATP.

ATP Production by Muscle Cells Muscle cells produce ATP in three ways: from creatine phosphate, by aerobic cellular respiration, and by anaerobic cellular respiration (fermentation). These three sources of ATP are compared in table 26.3 and described following.

While ATP cannot be stored, a similar molecule called **creatine phosphate** is stored right next to the muscle fibrils. Creatine phosphate is synthesized in a resting muscle from ATP and creatine (a small molecule made of carbon, hydrogen, oxygen, and nitrogen). The terminal phosphate group (ⓅP; 📖 *see page 87*)

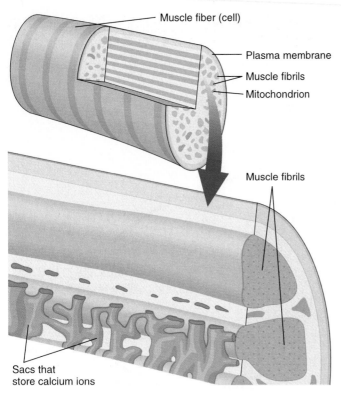

Figure 26.12 Calcium stores in a muscle fiber. Calcium ions are stored within sacs that surround the fibrils of a muscle fiber (cell). When the muscle cell receives a signal from a neuron, the sacs release their calcium ions.

within ATP is transferred, along with energy from the chemical bond, to the creatine molecule:

$$\text{Adenosine-}\textcircled{P}\text{-}\textcircled{P}\text{-}\textcircled{P} + \text{Creatine} \longrightarrow \text{Adenosine-}\textcircled{P}\text{-}\textcircled{P} + \text{Creatine-}\textcircled{P}$$
$$\text{ATP} \qquad\qquad\qquad\qquad\qquad\qquad\qquad \text{Creatine phosphate}$$

When energy is needed quickly, creatine phosphate is used to form ATP, which provides energy for muscle contractions:

$$\text{Creatine-}\textcircled{P} + \text{Adenosine-}\textcircled{P}\text{-}\textcircled{P} \longrightarrow \text{Creatine} + \text{Adenosine-}\textcircled{P}\text{-}\textcircled{P}\text{-}\textcircled{P}$$

The energy from creatine phosphate is only enough to fuel a few seconds of intense exertion. But a few seconds of sprinting can make a difference to a predator or its prey (as well as to a competitor in the 100-meter dash). The energy available from both ATP and creatine phosphate comes ultimately from cellular respiration (☐ *see page 88*). This energy-releasing reaction has two forms: aerobic and anaerobic.

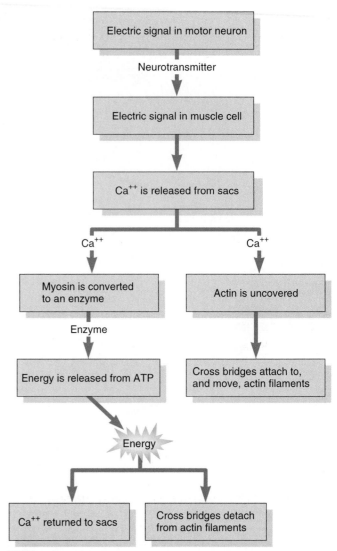

Figure 26.13 Sequence of events during a muscle contraction.

Table 26.3

Sources of ATP for Muscle Contractions

Source	Description
Creatine phosphate	Small amounts stored next to the fibrils and used to quickly build ATP
Aerobic cellular respiration	Uses oxygen to build large amounts of ATP from either a glucose or a fatty acid molecule
Anaerobic cellular respiration (or fermentation)	Requires no oxygen to rapidly build small amounts of ATP from a glucose molecule

Table 26.4

Characteristics of Slow Versus Fast Fibers

Characteristic	Slow Fibers	Fast Fibers
Main source of ATP	Aerobic cellular respiration	Anaerobic cellular respiration (fermentation)
Speed of contraction	Slow	Fast
Rate of fatigue	Slow	Fast
Myoglobin content	High	Low
Mitochondria	Many	Few
Glycogen content	High	High
Fat content	High	Low
Capillaries	Many	Few
Reaction to work	Builds capillaries	Builds fibrils

Aerobic cellular respiration generates ATP efficiently but slowly. The overall equation of this biochemical reaction is:

$$\text{Organic molecule} + O_2 \longrightarrow CO_2 + H_2O + \begin{array}{c} \text{Energy} \\ \text{ATP} \quad \text{Heat} \end{array}$$

The organic molecule is usually glucose or fatty acid. Most steps in the reaction take place within the mitochondria of the muscle cell. Aerobic cellular respiration fuels moderate levels of muscular work (less than 70% maximum exertion). This work, called aerobic exercise, uses a combination of fuels—typically 30% glucose and 70% fatty acids.

Anaerobic cellular respiration is fermentation, the biochemical process that generates ATP without oxygen gas. It forms lactic acid as a waste product:

$$\text{Glucose} \longrightarrow \text{Lactic acid} + \begin{array}{c} \text{Energy} \\ \text{ATP} \quad \text{Heat} \end{array}$$

Fermentation takes place outside the mitochondria, within the fluids of the muscle cell. It is inefficient, generating much less ATP per glucose molecule than aerobic cellular respiration, but it is fast—a great deal of ATP can be generated (from a lot of glucose) in a short time. Fermentation provides ATP for sudden bursts of power. Muscle contractions fueled by ATP from fermentation are called anaerobic work.

Anaerobic work builds up an **oxygen debt,** which is the amount of oxygen needed to eliminate the lactic acid that has formed. Lactic acid is toxic to cells. The oxygen debt is paid off after the muscles have stopped working and the body is able to acquire excess oxygen. In paying off the debt, oxygen is combined with some of the lactic acid (about 20%) to release energy:

$$\text{Lactic acid} + O_2 \longrightarrow CO_2 + H_2O + \text{Energy}$$

The energy is used to convert the remainder of the lactic acid (about 80%) back into glucose and then to join the glucose molecules into glycogen (the large molecule used for storing glucose):

$$\text{Lactic acid} + \text{Energy} \longrightarrow \text{Glucose}$$

$$\text{Many glucose} + \text{Energy} \longrightarrow \text{Glycogen}$$

Muscle cells use fermentation whenever there is not enough oxygen for aerobic cellular respiration. This situation develops whenever the contractions are rapid, as in a sprint, or sustained, as when lifting a heavy weight and the bulging muscles cut off the blood supply. Fermentation cannot, however, continue for long (less than a minute in an all-out sprint) because it uses only glucose, which is quickly used up, and it accumulates lactic acid, which poisons the muscle cells.

Fermentation is important during the warm-up phase of aerobic exercise, when the body is adjusting to a higher level of activity. This interesting period of transition is described in the accompanying Wellness Report.

Specialized Muscle Fibers Muscle fibers vary with regard to the kind of cellular respiration they are specialized for using. While there are several types of fibers, the two extreme types are slow fibers, with adaptations for aerobic work, and fast fibers, with adaptations for anaerobic work. The features of these two types of muscle cells are listed in table 26.4 and described following.

Muscle cells specialized for aerobic work contract slowly and are called **slow fibers.** They are dark in color because they contain myoglobin, a dark pigment similar to hemoglobin. (The dark meat of a turkey consists mainly of

The Warm-Up Phase of Aerobic Exercise

Humans are exceptionally good at sustained exercise. We are well adapted for distance running, perhaps because our distant ancestors chased their prey to exhaustion. The warm-up phase of aerobic exercise is a transition period between the resting state and a much higher level of physical activity.

The warm-up phase lasts between five and ten minutes. It occurs at the onset of running, swimming, bicycling, or some other form of aerobic exercise, when the muscles are working harder but the circulatory and respiratory systems are not yet delivering oxygen more rapidly. Some of the ATP used to power the muscle contractions comes, therefore, from anaerobic cellular respiration (fermentation). How long you can exercise, as well as the intensity of the workout, depends largely on how much of the energy used during the warm-up phase comes from anaerobic cellular respiration.

Anaerobic exercise uses only glucose, and only small quantities of glucose are stored in the muscles. The more glucose used in the warm-up phase, the less remains as fuel for the aerobic work and the more slowly this phase of the workout must proceed. In addition to using a lot of glucose, anaerobic cellular respiration produces lactic acid, which makes the muscles sluggish and muscle contractions painful; lactic acid is not eliminated until after the exercise period is over, when the oxygen debt is paid off. A good workout should begin gradually, to conserve glucose and minimize lactic acid.

As the warm-up phase proceeds, the amount of carbon dioxide (from aerobic work) in the blood increases, and it is detected by neurons within the medulla oblongata of the brain. These neurons respond by stimulating an increase in heart rate as well as the breathing rate and depth. The blood carries more oxygen and circulates more rapidly.

During the warm-up phase, the adrenal glands secrete more epinephrine—the fight-or-flight hormone that prepares the body for strenuous work. Epinephrine improves blood flow to the working muscles by stimulating the spleen to release stored blood and by initiating blood shunts: arteries in nonworking parts of the body constrict while arteries in the working muscles dilate. In addition, epinephrine dilates the air passages, increases the strength of heart contractions, and stimulates the breakdown of glycogen and fats into glucose and fatty acids so they can be used as fuels.

At first, blood is shunted to the working muscles and away from the skin, digestive tract, and kidneys. With less blood flowing to the skin, more heat is retained in the body and the temperature of the muscles rises to about 39° C (102° F), which accelerates the diffusion of oxygen into the cells as well as the rate of all chemical reactions. The warmer temperature also prevents injuries: it softens the ligaments and cartilage to make the joints more flexible. Cartilage within the joints absorbs water, like a sponge, expanding by about 12% and providing a softer cushion.

At the end of the warm-up, all the ATP comes from aerobic cellular respiration. The circulatory and respiratory systems are working hard to move blood and oxygen rapidly to the working muscles (table 26.A). More blood is shunted to the skin so that the additional heat, generated from cellular respiration in the muscles, is released and the body temperature stabilized. Homeostasis is established at a new and higher level of activity.

Table 26.A

Physiological Measures (Averages) During Rest and During Strenuous Aerobic Exercise

Activity	Resting State	Aerobic Exercise
Cardiac output (liters of blood pumped per minute)	5	30
Pulse (number of heart contractions per minute)	70	180
Blood pressure (systolic)	120	175
Blood flow to working skeletal muscles (liters per minute)	1.2	12.5
Blood flow to heart muscles (liters per minute)	0.3	0.8
Blood flow to skin (liters per minute)	0.5	1.9
Liters of air moved into the lungs per minute	7	140

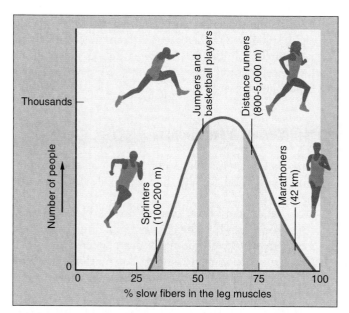

Figure 26.14 Percentages of slow fibers in the leg muscles of world-class athletes. The leg muscles of most people (peak of curve) are about 60% slow fibers and 40% fast fibers. World-class sprinters have the smallest percentage of slow fibers (and, consequently, the highest percentage of fast fibers) in their leg muscles. World-class marathoners have mostly slow fibers. Jumpers and basketball players have equal numbers of both kinds of fibers. While these two sports are largely anaerobic (sprinting) and use fast fibers, jumping requires good coordination, which is accomplished by slow fibers.

slow fibers, which the bird uses to stand and to walk.) The main function of myoglobin is to facilitate diffusion of oxygen from the plasma membrane to the mitochondria within the muscle cell.

In addition, slow fibers have large stores of both glycogen and fats. They are not easily fatigued because they seldom run out of fuel. The fibers contain many mitochondria (the site of aerobic cellular respiration) and are surrounded by dense networks of capillaries, which supply them with oxygen and remove carbon dioxide. When exercised, slow fibers develop more capillaries but do not become larger.

Muscle cells specialized for anaerobic work contract rapidly and are called **fast fibers.** These fibers are easily fatigued as they run out of glucose and accumulate lactic acid. Fast fibers have little or no myoglobin and are light in color (the breast muscles of a turkey are mainly fast fibers used for rapid wing beats). They have few mitochondria (fermentation occurs outside the mitochondria), store glucose but not much fat, and are surrounded by sparse networks of capillaries. When exercised, fast fibers become larger (by building more fibrils) but do not develop more capillaries.

Athletic Performance   A pronghorn antelope is an endurance athlete with mainly slow fibers in its leg muscles. It can run, at speeds of around 65 kilometers an hour, for more than thirty minutes. A cheetah, by contrast, is a sprinter with mainly fast fibers in its leg muscles. It can sprint faster than 110 kilometers an hour, but only for a minute or so. The muscles of each animal are adapted to its particular way of life.

All muscles in the human body have both slow and fast fibers, but the percentage varies from muscle to muscle. Muscles of the back, for example, are used for posture and have more slow fibers than muscles of the arms, which are used for lifting.

Within each muscle, the ratio of slow fibers to fast fibers varies from person to person. The ratio is a genetic trait that cannot be altered by training. The leg muscles of Olympic marathoners are about 95% slow fibers and 5% fast fibers, whereas the leg muscles of Olympic sprinters are about 35% slow fibers and 65% fast fibers. The distribution of world-class athletes, with regard to these two kinds of muscle cells, is shown in figure 26.14.

SUMMARY

1. Locomotion characterizes animal life. An animal moves its entire body as muscle contractions alter the positions of rigid body parts—compressed body fluids, an exoskeleton, or an endoskeleton.

2. A vertebrate animal has an endoskeleton, formed of bones, which consist of bone cells, blood vessels, collagen, and minerals. Bones provide a rigid frame, protect softer tissues, form blood cells, and store minerals and fats.

3. Joints are places where movable bones come together. Ligaments hold the bones together. Cartilage cushions the bones. Synovial fluid lubricates the bones.

4. Muscles that pull bones occur in antagonistic pairs. They are connected to bones by tendons. A muscle consists of muscle bundles, a bundle consists of muscle cells wrapped in a sheath, and a muscle cell (or fiber) consists of muscle fibrils. A muscle fibril, formed of actin and myosin filaments, shortens to produce a contraction. The stronger a muscle contraction the more fibers contracting. Within each fiber, all the fibrils contract or none of the fibrils contract.

5. A fibril contracts when cross bridges on the myosin filaments latch onto and pull the actin filaments closer together. Energy for a contraction comes from ATP, which is added to the cross bridges right after a contraction. ATP enables a cross bridge to detach from the actin and prepares the cross bridge for the next contraction.

6. A fibril is stimulated to contract when an electrical signal within the muscle cell (generated by a motor neuron) releases calcium ions. The calcium ions allow attachment of myosin to actin and convert myosin to an enzyme that releases energy from ATP.

7. Muscle cells generate ATP in three ways: from creatine phosphate (very rapidly), by aerobic cellular respiration (efficiently but slowly), and by anaerobic cellular respiration (fermentation) (rapidly but inefficiently). Muscle cells are specialized for either aerobic work (slow fibers) or anaerobic work (fast fibers). Animal muscles are adapted, by their proportions of slow and fast fibers, for sprinting or endurance.

KEY TERMS

Actin (AEK-tin) (page 497)
Antagonistic pair (page 496)
Bone (page 490)
Cartilage (KAHR-tah-lahj) (page 493)
Collagen (KAHL-ah-jen) (page 490)
Creatine (KREE-ah-tin) **phosphate** (page 499)
Cross bridge (page 497)
Fast fiber (page 503)
Joint (page 492)
Ligament (LIG-ah-munt) (page 492)
Locomotion (loh-koh-MOH-shun) (page 488)
Marrow (MAEH-roh) (page 492)
Muscle bundle (page 496)
Muscle fiber (FEYE-bur) (page 497)
Muscle fibril (feye-BRIL) (page 497)
Muscle sheath (page 496)
Myosin (MEYE-oh-sin) (page 497)
Oxygen debt (page 501)
Slow fiber (page 501)
Synovial (sih-NOH-vee-ahl) **fluid** (page 496)
Tendon (page 496)

STUDY QUESTIONS

1. Describe the effects on a bone of dietary minerals and steroid molecules (vitamin D, testosterone, and estrogen).
2. Draw a front view of the knee, showing the three bones, six ligaments, and meniscuses. Identify whether it is the left or the right knee.
3. Draw a muscle that has been cut crosswise, showing the muscle bundles, sheaths, fibers, and fibrils.
4. Draw a muscle fibril in its relaxed and contracted phases. Include in your drawing the actin and myosin filaments as well as the cross bridges.

5. Outline the sequence of events from the time a muscle cell is stimulated to the time it shortens in a contraction.
6. Explain why a chicken breast is lighter in color and has fewer calories than an equal-sized serving of meat from a chicken leg.

CRITICAL THINKING PROBLEMS

1. The skeleton of a whale comprises a smaller proportion of its total weight than the skeleton of an elephant. Why?
2. Abnormal amounts of ions within the blood, brought about for example by heavy sweating, often cause muscle cramps. Why do you think this happens?
3. What proportion of slow fibers do you think you have in your leg muscles? Explain how you came to this conclusion.
4. List the occasions, during a typical day, in which your muscles use anaerobic cellular respiration (fermentation) to generate ATP.
5. Different people inherit different proportions of fast and slow fibers. Account for these differences in terms of adaptations to environments in distant ancestors.
6. Curare is a molecule extracted from certain South American trees. In animals, curare prevents acetylcholine from binding with its receptor on muscle cells. What noticeable effect would this have on an animal?

SUGGESTED READINGS

Introductory Level:

Joyce, C., and E. Stover. *Witnesses from the Grave: The Story Bones Tell*. Boston: Little, Brown, 1991. A fascinating description of forensic anthropology (the identification of dead people from their bones) and of its leading authority, Clyde Snow. Chapters 8, 9, and 10 deal specifically with the identification of Joseph Mengele, the infamous Nazi doctor who disappeared in South America.

Shepart, R. J. *Exercise Physiology*. Philadelphia: Decker Publishing, 1987. A basic textbook describing what happens to the body during exercise and how to improve fitness.

Ubelaker, D., and H. Scammell. *Bones: A Forensic Detective's Casebook*. New York: Burlingame (HarperCollins), 1992. A very readable and personal account of the technical procedures used to identify human skeletons or decomposed remains.

Advanced Level:

Gray, J. *How Animals Move*. Cambridge, England: Cambridge University Press, 1960. A classic book on how various animals move from place to place, including how each type of locomotion obeys the laws of mechanics and reflects the animal's life-style and evolutionary history.

Huxley, A. F. *Reflections on Muscle*. Princeton, NJ: Princeton University Press, 1980. History of research on muscle by a foremost scientist in the area.

Margaria, R. "The Sources of Muscular Energy," *Scientific American* 226, no. 3 (1972): 84–91. A description of how energy becomes available for muscular work, beginning with food and ending with ATP, and suggestions on how to improve energy stores.

Nadel, E. R. "Physiological Adaptations to Aerobic Training," *American Scientist* 73 (1985): 334–43. An interesting discussion of what happens to the body, including muscle efficiency, oxygen transport, and temperature control, during a fitness program.

Shipman, P., A. Walker, and D. Bichell. *The Human Skeleton*. Cambridge, MA: Harvard University Press, 1986. A description of the nature of bones and bone function as well as the diagnosis of age, sex, race, and stature from the evidence of bones.

✪ EXPLORATION

Interactive Software: Muscle Contraction

This interactive exercise allows students to explore how the proteins in muscle cells interact with one another to produce muscle contraction. The exercise presents a diagram of a fibril, with myosin "walking" along actin. Students can explore how changing the ATP concentration affects the speed of muscle contraction, learning that diminished ATP slows rather than weakens the force of an individual fibril's contraction. Students can then investigate a muscle fiber made of many fibrils, and explore how the force of the fiber's contraction depends upon changes in calcium ion levels, and why a more forceful muscle contraction results from repeated firing of the motor neuron.

Behavior

Objectives

In this chapter, you will learn
–how a nervous system produces a
  behavior;
–how animals choose a place to live,
  find food to eat, and evade predators;
–why animals migrate;
–how animals recognize suitable mates
  and cooperate in raising their young;
–how animals divide resources without
  continual squabbling;
–why some animals live in social
  groups.

Animals have contests to determine first access to resources. Here, two polar bears fight to determine dominance.

Do you ever wonder what other animals think about? Do they make plans? Do they think about the consequences of their actions? What emotions do they feel? What do they communicate to one another?

We know a great deal about how other animals behave but almost nothing about their thoughts. The only way to find out what another animal is thinking is to have it tell us, which animals of other species cannot do. Our best clues come perhaps from watching our pets, since we observe so much of what they do.

Dogs and cats appear to think about more complex things than where to find food, water, warmth, and sex. They seem to have mental images of certain goals and to choose behaviors that realize the goals. A dog who trots directly to the place where the school bus stops each afternoon must have a plan to achieve a goal—an image of the children and the rewards of interacting with them. A tomcat that lies concealed in a flower bed until a female cat walks by, and then jumps out to startle her, seems to have anticipated her arrival and the reward he would get from seeing her reaction. We think we understand these behaviors, but do we really? Are they really goal oriented or have we trained our pets to behave as we would? How much of our interpretation of animal behavior is based on what we ourselves would be thinking? How can we test our hypotheses?

The study of wild animals is less tainted by our own emotions but more difficult than the study of pets. Laboratory experiments are designed to see how animals react to different situations and problems. They provide unbiased information but tend to underestimate an animal's intelligence and array of behaviors. It is difficult to simulate, in the laboratory, the complex problems and strong motivations of animals in their natural settings. Field studies, by contrast, present other difficulties: wild animals are timid and do not follow normal behavioral patterns when they are aware of humans in their surroundings. Moreover, both the condition of the animal and its environment vary so much from time to time that it is hard to identify cause-and-effect relations.

In spite of these difficulties, biologists have learned a great deal about how animals search for food, build shelters, avoid predators, and solve problems, as well as how they communicate with one another, choose mates, and teach their young. Their thoughts, however, are beyond the realm of science.

Mechanisms of Behavior

A behavior is an observable activity that an animal performs, ranging from a simple movement to solving a complex problem. It is a form of adaptation (📖 *see page 17*),

in which an animal inherits certain ways of responding or the ability to learn certain tasks that improve its chances of surviving and reproducing. An animal in its natural setting finds food, moves into favorable places, and communicates with others of its kind.

A behavior is the product of an animal's sense organs, nervous system, and muscles. It is triggered by a **stimulus,** which is information that enters the nervous system through the sense organs. The stimulus may come from inside the animal, as in the case of hunger, or outside the animal, as when a predator is seen. Except for very simple behaviors, the incoming information is evaluated by the brain and a decision is made about how to respond. The brain then commands certain muscles to contract in ways that carry out the decision. The resulting action is the animal's **behavioral response** to the stimulus:

$$\text{Stimulus} \rightarrow \left(\begin{array}{c}\text{Sense}\\\text{organ}\end{array} \rightarrow \text{Brain} \rightarrow \text{Muscles}\right) \rightarrow \text{Response}$$

Hormones often play roles in behavior. An environmental event may not be a stimulus unless the level of a certain hormone is high. The sight of a female, for example, has little effect on the male unless his blood carries a certain level of male sex hormones. Hormones also help carry out responses. A stimulus that demands a fight-or-flight response, for example, triggers the release of epinephrine, which brings about physiological changes that prepare the body for a higher level of physical activity.

How does the brain evaluate a stimulus and decide what to do about it? The physical processes, called the mechanisms of behavior, consist of neural pathways through which information flows. These processes form in two ways: (1) by genetic programming, in which pathways are layed down according to instructions by an animal's genes during early development; and (2) by learning, in which pathways are modified in response to experiences.

The relative contributions of genes and experiences to behavior vary from one group of animals to another and among the many behaviors exhibited by each animal. They also vary in the amount each kind of pathway contributes to a particular behavior: some behaviors are the result of neural pathways established by genes and then modified by learning. Behaviors are classified into many forms, depending on their complexity and whether they are based mostly on genetic programs or mostly on learning. These forms of behavior are summarized in table 27.1 and described following.

Genetic Programming

Some behaviors are controlled mainly by neural pathways layed down according to instructions from genes. These genetically programmed behaviors are called **innate behaviors,** or instinctive behaviors.

Table 27.1

Basic Categories of Behavior

Form	Description
Innate behaviors	Pathways in the brain are developed by the animal's genes
Kinesis	One stimulus causes one response; the response is not oriented with regard to location of the stimulus
Taxis	One stimulus causes one response; the response is oriented with regard to location of the stimulus
Fixed-action pattern	One stimulus causes a sequence of behaviors
Learned behaviors	Pathways in the brain are developed and strengthened by experiences
Trial-and-error	An association between a behavior and its consequences is made after direct experience
Insight	An association between a behavior and its consequences is made without direct experience
Imprinted behaviors	A particular task is learned only at a genetically programmed time

A female graylag goose, for example, is programmed to rotate her eggs several times a day to prevent the embryos from sticking to the shells. As she does this, an egg sometimes rolls out of the nest. She retrieves the misplaced egg by a sequence of innate movements, rolling the egg with her bill as she backs toward her nest (fig. 27.1). Once the behavioral sequence begins, it continues even when inappropriate—when the egg has been taken away, for example, or the object is a baseball rather than an egg. The goose is programmed to retrieve anything near her nest that looks remotely like an egg.

An innate behavior occurs automatically in response to a stimulus and is performed perfectly the first time. Such behaviors are adaptive in situations where the correct response is always the same, as in the case of egg retrieving, and when a wrong response could have serious consequences, as in responses to predators. Innate behaviors range from simple changes in locomotion to complex responses involving long sequences of behaviors.

Simple Innate Behaviors   Animals move themselves, or at least parts of their bodies, toward or away from certain objects or conditions. These simple behaviors involve a single stimulus and a single, genetically programmed response. There are two categories of simple innate behaviors: kinesis and taxis.

A **kinesis** is a change in locomotion in response to a stimulus but without regard to location of the stimulus. The animal responds by changing its rate of movement or turning. A wood louse, for example, crawls faster when in a dry area than when in a humid area of its environment. As a consequence, the louse spends more time in humid areas, where it survives better.

Figure 27.1 Egg retrieval by a graylag goose. The goose retrieves misplaced eggs by a sequence of stereotyped behaviors triggered by sight of the egg. She stares intently at the egg and then stretches her neck toward it. She rolls the egg toward the nest by using a side-to-side motion of her head, pulling first on one side of the egg and then on the other side.

A **taxis** is a change in locomotion that moves an animal directly toward or away from a stimulus. For example, when a male moth detects the scent of a female moth, he flies directly toward her.

Fixed-Action Patterns A **fixed-action pattern** is a sequence of behaviors that occurs in response to a single stimulus. The behaviors are said to be fixed because they occur in nearly the same way each time. In the egg-rolling behavior just described, the sight of a misplaced egg triggers a sequence of behaviors.

Another example is cocoon building by female tropical wandering spiders. The sight of a suitable place for a cocoon triggers a sequence of approximately 6,400 movements in which silken threads are woven into a cocoon. The spider first builds a floor and then constructs a wall around it. She then lays eggs on the floor and covers the top of the cocoon. If construction is interrupted (by removing the floor, for example), she does not adjust her behaviors to compensate for the change; she simply continues the fixed sequence of behaviors. Once the first behavior occurs, the others follow automatically.

Most fixed-action patterns occur only when the animal is in a particular physiological state. The sight of a round object near the nest triggers egg-retrieving behavior only while a goose has eggs in her nest; the sight of a suitable place for a cocoon triggers cocoon building only when a spider is ready to lay eggs.

Learning

A learned behavior is a response that has been modified by experience. We do not know exactly what happens inside the nervous system when an animal learns. The process certainly requires memory, in which information associated with a previous experience is stored and compared with new information. Memory, in turn, requires neural pathways that can be modified in response to the new information. While all animals learn, the behavior of animals with simpler nervous systems (e.g., worms) is less readily modified by learning than the behavior of animals with more complex nervous systems (e.g., mammals).

An animal does not learn all things equally well; it inherits tendencies to learn tasks that contributed to the survival and reproduction of its ancestors. Even among closely related animals, the tendencies to learn may be quite different. Some breeds of dogs, for example, inherit the tendency to point at their prey (fig. 27.2), and can be easily trained to do so, whereas other breeds inherit the tendency to protect their masters. Similarly, herring gulls and kittiwake gulls are closely related species of birds, but herring gulls learn to recognize their chicks and kittiwake gulls do not. Herring gulls nest on the ground in dense colonies, where the young are apt to wander from the nest and mingle with other chicks. The parents learn to identify their own chicks, a behavior that is essential for parental care. Kittiwake gulls nest on the narrow ledges of cliffs, where there is no room for chicks to wander. A kittiwake gull parent learns the location of its nest but not how to identify its own chicks. It cares for the chicks in its nest, which are always its own.

Figure 27.2 Some dogs inherit the tendency to point. This English pointer has been trained to remain motionless while pointing toward a quail in the grass. Some breeds of dogs, like this one, learn to point much more readily than others. Human hunters use pointers to locate game.

Trial-and-Error Learning In **trial-and-error learning,** an animal makes the connection between a behavior and its consequences. A behavior followed by something pleasant is repeated; a behavior followed by something unpleasant is not repeated. (Trial-and-error learning is called operant conditioning when it occurs under controlled conditions within the laboratory.)

Consider a blue jay that eats a yellow butterfly. If the butterfly tastes good, then the jay will continue to feed on yellow butterflies; if it tastes bad, then the jay will avoid yellow butterflies. The animal learns by doing.

Trial-and-error learning is fundamental to the behavior of almost all animals. By trying a behavior and seeing its consequences, an animal adjusts to changes in its environment. It learns the foods that can be eaten, the materials that can be used to build a nest, and the sources of danger. It is by trial-and-error learning that some animals adjust to environments dominated by humans. British tits (relatives of chickadees) learn to open milk bottles left on doorsteps, raccoons learn to open cat doors, and sparrows learn to open the automatic doors of supermarkets by flying slowly past electronic sensors. Crows learn to follow corn planters, eating the seeds as they are planted, and to distinguish a gun from a hoe. Often these new behaviors pass from animal to animal by observational learning, in which one animal learns by watching another.

Insight **Insight learning** occurs when an animal makes new associations between previously learned tasks in order to solve a new problem. It is an advanced form of trial-and-error learning in which the trials and errors occur mentally rather than physically: the animal considers a variety of possible behaviors and their consequences before any action is taken.

Figure 27.3 Insight learning. A chimpanzee can figure out how to use a pole and a few boxes to reach bananas suspended from the ceiling.

Insight learning in animals other than humans was first demonstrated in chimpanzees. In a classic series of experiments, the chimps learned to join several short poles to make one long pole and use this tool to knock down bananas suspended from the ceiling. They also learned to stack boxes and climb the stack to reach the bananas. As a final problem, the bananas were suspended too high to be reached by either the long pole or the stacked boxes. The chimps solved this problem by using both the poles and the boxes: they connected the poles, stacked the boxes, and then climbed the stack and knocked the bananas down with the pole (fig. 27.3).

How widespread is insight learning? Many biologists believe this advanced form of learning occurs only in primates, whereas others believe it is widespread but difficult to demonstrate. Consider the following actual observations of four female lions hunting antelopes. Two of the lions sit in full view of the antelopes, who see the pair but know they are too far away to pose a threat. Meanwhile, a third lion creeps along a ditch on the opposite side of the herd and a fourth circles around through some trees. The fourth bursts out of the woods and stampedes the herd across the ditch where her partner is crouched. The lion in the ditch leaps up and makes a kill. All four lions then share the feast. Did the first two lions position themselves with the intent of preventing a stampede in their direction, forcing the herd over the ditch? Did the fourth lion dart from the woods with the intent of frightening the herd toward the ditch? In other words, did each lion consider the conse-

Figure 27.4 Imprinting. Konrad Lorenz, who discovered imprinting, is followed by goslings that have imprinted on him rather than their mother.

quences of her own behavior as well as the behavior of the other lions before she acted? If so, then her behavior was governed by insight learning.

Imprinting

Imprinting occurs when an animal learns to respond in a particular way to just one type of animal or object. It is genetically programmed to take place at a specific age and under a particular set of conditions.

Imprinting was first identified by the German biologist Konrad Lorenz, who showed that a young goose, within a day after hatching, forms a strong attachment to the first slowly moving object it sees. The gosling becomes imprinted on the moving object. Normally this object is the mother, and the imprinted behavior functions to keep young geese near their mothers. Goslings raised without a mother will imprint on almost anything that moves. They readily imprint on a human and will try desperately to follow the person everywhere (fig. 27.4). Once a gosling has imprinted on an object other than its mother, it cannot be trained later to follow its own mother; the sensitive period for learning this behavior has passed.

Imprinting occurs in all vertebrate animals. Horses imprint on their mothers during the first and second day after they are born. Dogs form lasting attachments to other dogs (or humans) during the fourth to sixth week after birth. A white-crowned sparrow learns the song of its species, from listening to its parent, when it is between ten and fifty days old. A jackdaw (a crowlike bird) learns early in life, by imprinting on its parent, what its future mate should look like.

Behavioral Ecology

Most of an animal's behavior is related to its ecology—its interactions with other species and nonliving aspects of its environment. An animal evades predators and finds a place to live, food to eat, a shelter for sleeping, and a nest site for raising young. These behaviors have been strongly selected for in the past, for they have a direct effect on survival and are appropriate for the particular environment in which the species evolved.

Acquiring Resources

Each animal selects certain resources from its environment. Some animals are specialists and select only a few kinds of foods, nesting sites, and sleeping places. Other animals are generalists and make use of a wide variety of resources. The degree to which an animal is a generalist depends largely on the degree to which it can modify its choices by learning.

Habitat Selection   A **habitat** is the kind of environment in which an individual lives. It provides food, shelter, nesting sites, and specific conditions of temperature and moisture suited to the animal's needs. On land, an animal's habitat is described in terms of the landscape—a meadow, for example, or pine forest. In water, the habitat is described in terms of water depth, rate of flow, or whether the bottom consists of sand, mud, pebbles, or bedrock.

Choosing an appropriate place to live is extremely important. An animal's ancestors lived in a particular habitat, and it has inherited traits for dealing with that kind of habitat. Not surprisingly, habitat selection typically involves behaviors that are innate or imprinted shortly after birth. Young deer mice, for example, inherit a preference for grasslands that cannot be altered by early experience.

Animals typically choose their habitat at a young age, when they are forced by crowded conditions to leave their place of birth. They travel through unfamiliar places in search of a place to settle down and raise their own young. These wanderers apparently have a mental image of how the habitat they seek should look or smell. Young

Figure 27.5 A red fox. The foraging behavior and body of this fox are adapted for hunting mice.

ground squirrels, for example, wander through forests and bogs until they find a grassy meadow with soft, pliable soil in which to dig burrows.

An interesting consequence of innate choice of habitat is seen in house rats. Two species of rats invade our houses, but each chooses a different part of the house. The black rat is descended from populations that nested in trees. It chooses to live in attics. The brown rat is descended from populations that nested in river banks. It chooses to live in damp cellars.

Foraging Behavior   Behavior associated with acquiring food is called foraging behavior: the kind of food an animal chooses, the time of day or night when it looks for food, the places it searches, and the method of search. In some animals these behaviors are largely innate, whereas in others they are modified by learning. Many aspects of foraging, such as whether an animal is a long-distance chaser (like a wolf) or a sit-and-wait sprinter (like a cat), are linked with physical and biochemical traits that cannot be altered. The kinds of sense organs an animal inherits, and whether its muscles are built for endurance or sprinting, control many aspects of its foraging behavior.

An animal expends time while acquiring food, and during this time it is more vulnerable to predators. Its foraging behavior, therefore, is a compromise between the need to acquire food and the need to minimize exposure time. A deer, for example, browses mainly at dawn and dusk, when the conditions of light make it difficult for wolves and mountain lions to see.

An animal expends energy while acquiring food, and its foraging behavior is adapted so that more energy is acquired than expended. An animal searches where food is abundant, pursues prey it is likely to catch, and chooses food that is nourishing. Consider the red fox (fig. 27.5), which hunts chiefly for mice along the edges of forest trails. These openings in the forest receive more sunlight and so have more grasses than the dark understory of the forest. More grasses mean more mice, and so a fox is likely to find more food along trails. It also chooses trails because they are quieter than a forest floor; they have no leaves to rustle or twigs to snap. As the fox travels, it watches and listens for mice and then creeps close enough to pounce. A fox is built for catching mice. Its movable ears detect faint rustlings. It moves quietly on soft foot pads with retractable claws. Its leg muscles are built for pouncing onto its prey.

Some animals change foraging behaviors when the rate at which they find food drops below a certain level. They switch to another kind of food or another place to forage. Hunting flexibility appears to be facilitated by a search image—a picture in the animal's mind of what it is looking for. The image helps them find a certain kind of prey and can be replaced by another image when they switch prey. When a coyote hunts rabbits, for example, it has a search image for rabbits and considers no other kind of prey. When rabbits are hard to find, the coyote switches to mice and develops a new search image.

Evading Predators

Appropriate responses to predators are essential to survival. These behaviors are governed by genetic programming, since errors are usually lethal. Antipredator behaviors are postures and movements that make it difficult for a predator to find and catch the prey.

Flash behavior is a sudden movement or sound that focuses the predator's attention. Grasshoppers, for example, make clicking sounds and expose their brightly colored wings when flying. After landing, however, they become silent and fold their wings. A predator that focuses on the sound and color of the flying grasshopper cannot locate the insect after it lands. The white underside of a deer's tail (fig. 27.6) may serve the same function—a predator focuses on the white color of the fleeing deer but then cannot locate the prey when it stops fleeing and lowers its tail.

Startle behaviors are sudden sights, sounds, or smells that surprise a predator, giving the prey a few moments to flee. Many insects have false eyespots that are hidden except when a predator is nearby. When the eyespots are suddenly exposed, the predator thinks they are the eyes of a larger animal and is startled (fig. 27.7).

Another way in which animals evade predators is by unpredictable flight patterns. Rabbits, for example, turn left and right at unpredictable times and angles as they flee toward their burrows. This type of flight is even more difficult to follow when performed in the three dimensions of air or water, as in the erratic flights of butterflies and fish. Prey that move in groups, such as bird flocks, fish schools, and mammal herds, evade predators by scattering in many directions. The predator becomes distracted, unable to concentrate on just one member of the group.

Figure 27.6 Flash behavior. A fleeing white-tailed deer raises its tail to expose a patch of white hair. The function of this behavior may be to focus the attention of a predator on the white color, so that when the deer stops and lowers its tail the predator will continue to look for a white color and not be able to find the deer.

Figure 27.7 Startle behavior. This insect larva has colored patterns on its surface that resemble the eyes of a vertebrate. Normally, the patterns are concealed within folds. When the larva detects a predator, it suddenly exposes the eyespots and startles the predator. In this way, the larva gains time to escape.

Figure 27.8 Camouflage requires a freezing posture. This bird, an American bittern, must remain still if it is to resemble the grasses around it.

The most common antipredator adaptation, camouflage, combines physical traits with behavioral traits that conceal the prey from the predator. An animal that looks like its background must not move when a predator is near, or it will be detected. Thus, camouflage goes hand in hand with a freezing posture (fig. 27.8).

Migration

Some animals change habitats, either with the seasons or at different times of their lives. They may travel great distances between their two "homes." Regular journeys between two specific places of residence are called **migrations.**

Seasonal Migrations

An animal migrates with the seasons in order to live in an environment that is warmer or offers more food. Astonishing migrations are undertaken by Monarch butterflies, which seem too fragile for traveling long distances. Each autumn, they leave the cold of northern United States and Canada and fly to Mexico and southern California, a distance of more than 3,200 kilometers (2,000 mi). Each spring they fly back again. The route must be programmed in their tiny brains, since the butterflies that fly north in the spring die during the summer; it is their progeny that fly south in autumn.

Most seasonal migrants are birds, migrating south in the autumn and north in the spring. In North America alone, more than 5 billion birds migrate back and forth each year. Birds that feed in lakes (e.g., geese and ducks) or that eat insects (e.g., flycatchers and warblers) or nectar (hummingbirds) cannot find food during the cold winter months. They leave their frozen habitats and spend the winter in warmer places, where they join communities of tropical birds. (Many seed-eating birds do not migrate, for seeds can be found during the winter.) In the spring, when more food is available in the north than the south, the birds return to their summer residences.

Some birds migrate great distances and have superb navigational skills. They make use of landmarks, such as coastlines, mountain ranges, and rivers, as well as positions of the sun, stars, and the earth's magnetic fields. These skills are typically programmed in their genes, but some birds learn routes from their parents, a topic of the accompanying Environment Report.

Reproductive Migrations

Some animals migrate great distances to and from their breeding grounds. These journeys coincide with the age of the animal rather than with the seasons of the year. Animals with reproductive migrations include sea turtles, salmon, and eels.

Freshwater eels are known for their long migrations. They grow up in lakes and rivers throughout western Europe and eastern North America. When about nine years old, the eels travel downstream to the Atlantic Ocean and swim to the Sargasso Sea, southwest of Bermuda (fig. 27.9a). The journey takes about six months. There the eels spawn (release eggs and sperm that form fertilized eggs) and then die.

When the young eels hatch out of their eggs, they leave the Sargasso Sea and return to the same lakes and rivers where their parents grew up. Only a centimeter or so long when they begin their journey, they grow to roughly 8 centimeters by the time they reach "home" several years later.

The ability of adult eels to find the Sargasso Sea and of young eels to find their parental home is inherited. Their cues are the odors of water (from chemicals and organisms) and the type of current.

Why do eels undergo such a strenuous migration? Probably because Europe and North America were once close together and the Atlantic coasts of both continents were not far from the Sargasso Sea, as illustrated in figure 27.9b. As the continents drifted apart, the eels continued to breed in the same place as their ancestors despite the ever-increasing distances they had to travel.

Social Behavior

Social behavior occurs when an animal interacts with other animals of its kind. The study of the biological basis of social behaviors—their advantages in terms of natural selection—is called **sociobiology.** These behaviors range from simple mating behaviors to helping others raise their young.

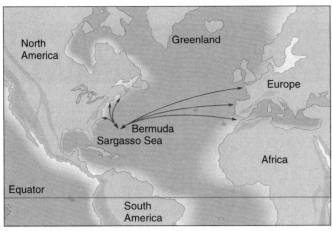

(a)

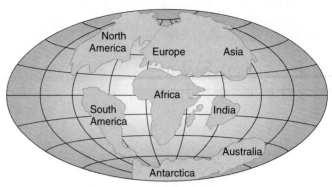

(b)

Figure 27.9 Migration routes of freshwater eels. (*a*) The eels live in lakes and rivers in eastern North America and western Europe but migrate to the Sargasso Sea to breed. The adult eels then die and the young migrate back to the freshwater homes of their parents. (*b*) This peculiar migratory behavior may have evolved millions of years ago when Europe and North America were near each other and the Sargasso Sea. The positions of the continents 65 million years ago are shown.

Reproductive Behavior

All sexually reproducing animals inherit behaviors that result in mating. For a mating to be successful, a male and a female must coordinate their release of eggs and sperm as well as make sure their partner is of their own species. After mating, some pairs form long-lasting bonds that promote cooperation in raising their offspring.

Mating Behavior Every animal knows instinctively how to copulate. These mechanical movements are fixed-action patterns, triggered by high levels of sex hormones and mating signals that are unique to each species. Mating may occur in response to a simple stimulus, such as a particular scent, color, or gesture, or it may require a prolonged period of courtship.

Courtship behaviors are fixed-action patterns unique to each species. They guarantee that mates are of the same species and so can have healthy offspring. Courtship behavior also reduces aggression and synchronizes the release of eggs and sperm.

Elaborate dances and songs characterize courtship in many species of fish, frogs, and birds. One of the best-known dances is performed by the three-spine stickleback fish. At the onset of the reproductive season, the male stickleback constructs a tube-shaped nest from plant material and sediment that he glues together with secretions from his kidneys. He then waits, near the nest, for a female to arrive. The sight of her abdomen, swollen with eggs, stimulates the male to begin the dance, as diagrammed in figure 27.10. Each action by the male triggers a behavioral response by the female, and vice versa. If the dance is not exactly correct, mating behavior is terminated.

Courtship behaviors are by no means limited to vertebrate animals. A male spider, for example, dances for his mate, setting up a specific pattern of vibrations on her web. A male fruit fly dances, by whirling about, and sings, by vibrating his wings and abdomen. Both the dance and the song occur at a frequency that is specific to his species.

Pair-Bonding A pair-bond is a stable relationship between a mated male and female. It develops during mating and is reinforced afterward by greeting rituals, such as birds touching their bills or mammals sniffing and licking. Such rituals promote continuance of the relationship and suppress aggressive tendencies.

Pair-bonds develop in **monogamous** species, where each animal spends most of its time with one mate. Monogamy is an adaptation to conditions in which both parents are needed to raise the young. It is common among birds, where a pair-bond helps the parents cooperate in building a nest and feeding the young. Most mammals are not monogamous, because the mother feeds the young with her own milk and the father, freed of this task, typically mates with more than one female. Exceptions include foxes, weasels, and beavers, which form lasting bonds with their mates. Surprisingly, behaviors that promote pair-bonding and parental care in male prairie voles are triggered by a particular chemical in the brain, as described in the accompanying Research Report.

515

Environment Report

Teaching Swans Where to Migrate

Trumpeter swans, once abundant in North America, were nearly wiped out during the eighteenth and nineteenth centuries. Millions were killed for their large feathers, which were used to make quill pens. In 1932, only seventy trumpeters were left in the United States.

Today the quill pen is no longer used and the trumpeter swans are making a comeback in Canada. Unfortunately, they no longer know how to reach the winter home of their ancestors. Normally, young swans learn the route as they travel with their parents, but so many adults were killed that the swans stopped migrating. They now spend the winter in Canada, near ponds that are kept warm by power plants.

Swans that have been raised in captivity and then released into the wild attempt to migrate in the autumn, but they fly in all directions—some even migrate north for the winter. They are in danger of extinction if they do not take up their ancestral habit of migration.

An extraordinary man, Bill Lishman, has taken on the task of teaching these swans the route between their summer and winter residences (fig. 27.A). He plans to train trumpeter swans to follow his ultralight plane from Ontario, their summer home, to Chesapeake Bay, the winter home of their ancestors.

Bill has developed a training procedure with Canada geese. Each spring, goslings are separated from their mothers and imprinted on Bill instead. They live in shelters near the runway of an airport, where they learn to follow him rather than their moth-

ers. Three times a day, he runs up and down the runway, playing the sound of an ultralight engine on his tape recorder. The fluffy goslings run after him, desperately trying to keep up. In this way, the young birds learn to follow the sound of the plane. At night, Bill beds the goslings down under the wing of his parked ultralight so they learn to stay near the plane.

When the young geese are old enough to fly, they follow Bill and his ultralight in the air. On warm summer days, they can be seen in the Canadian skies: Bill in his ultralight followed by a V-shaped flock of honking geese. His neighbors call him Father Goose.

Bill's training procedure was tested for the first time in October of 1993. Eighteen of his trained geese followed two ultralight planes, piloted by Bill and a friend, from their home near Toronto to Fauquier County, Virginia. It took them seven days, but every goose made the trip. There the geese overwintered. Now the question was: Would they find their way back to the summer grounds in the spring? In April of 1994, ten of these birds appeared on the grassy landing strip outside Bill's home. "They look in great shape," Lishman said, "I took them all kinds of goodies to eat, but they'd rather root in the pond."

Now Bill is transferring his training techniques from Canada geese to trumpeter swans. If he can teach just a few of the swans how to find Chesapeake Bay, then this information may be passed on to future generations. He may be able to save this rare and beautiful bird from extinction.

Figure 27.A Bill Lishman and his geese.

Parental Care   Parental care is the helpful behavior of parents toward their offspring (fig. 27.11). These behaviors include feeding, sheltering, grooming, protecting, and teaching the young.

Parental care requires a sacrifice on the part of the parent. The minimum sacrifice is the time and energy spent on the young that could otherwise be used in improving the parent's own chances of survival and future reproduction. The maximum sacrifice is when a parent gives up its own life to ensure survival of its offspring.

Predators are a major threat to most young animals. Some parents protect against predators by carrying their young with them. Opossums and certain aquatic bugs carry their young on their backs, and some frogs carry them in their mouths. Other parents keep their young in concealed nests and, if necessary, fight off a predator. The females of some ground-nesting birds, such as sandpipers, lure foxes and other nonflying predators away from their chicks by pretending to be easy prey. When a fox approaches the nest, the hen flies off and then drops to the ground nearby and pretends to be injured, dragging a wing and crying out. Sensing an easy kill, the fox stalks the "injured" bird. She lures the fox farther and farther from her chicks and, when they are a long way from the nest, she then flies away.

Dividing a Resource

An animal competes with other members of its population for the same kinds of food, shelter, nesting sites, and mates. Many species have systems of

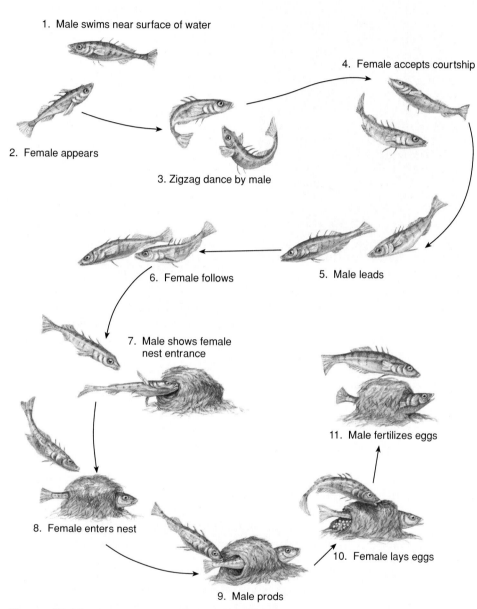

Figure 27.10 The courtship dance of sticklebacks. Attracted by the nest and swimming movements of the male (*1*), a female approaches him (*2*) and swims in a head-up posture. The sight of her enlarged belly, swollen with eggs, stimulates the male to dance. He performs a zigzag dance while moving between his nest and the female (*3*). She accepts courtship (*4*) and follows him to the nest (*5* and *6*), where he makes a series of quick thrusts with his nose to indicate the entrance (*7*). As the female enters the nest (*8*), the male prods his head against her tail and quivers (*9*). In response, the female releases her eggs (*10*). She then leaves and the male enters to release his sperm onto the eggs (*11*).

Figure 27.11 Parental care. This emperor penguin feeds her young, grooms it, and keeps it warm by holding it near her body as well as on her feet—out of contact with the frozen ground.

Research Report

The Chemical Basis of Monogamy

What makes some males stay with one female while others play the field? In voles, the key seems to be a molecule, called vasopressin, secreted by neurons in the brain.

Field biologists were surprised to discover that prairie voles are monogamous: each male remains with a single female, apparently for life, and prevents other males from getting near her (fig. 27.B). Once he becomes a father, a male prairie vole defends and helps care for his offspring. Other types of voles behave quite differently: the male mates with as many females as possible and never even knows his offspring. How can the prairie vole behave so differently from its close relatives? How do their brains differ?

Neurobiologists examined the brains of prairie voles and closely related species, trying to find some physical difference that might explain the vast difference in behavior. In 1992, they discovered something re-

Figure 27.B Prairie voles.
Prolonged periods of mating in the prairie vole help facilitate the formation of pair-bonds.

markable: the brains of prairie voles have a unique distribution of receptors for the neurotransmitter vasopressin. (Vasopressin is both a neurotransmitter in the brain and a hormone, also

Observations:
• Male prairie voles exhibit monogamous behaviors, whereas the males of other voles do not.
• Prairie voles have a unique distribution of receptors, which bind with vasopressin, on neurons in their brains.

Question:
Is the unique distribution of receptors related to monogamous behavior?

Hypothesis:
Mating in prairie voles increases the secretion of vasopressin, which binds with its receptors and induces monogamous behaviors.

Figure 27.C Scientific logic with regard to monogamy in prairie voles.

known as antidiuretic hormone, that travels in the bloodstream.)

From these observations, researchers developed the following hypothesis (fig. 27.C) to explain pair-

---

dividing up the resources in ways that minimize squabbling. These systems include home ranges, territories, and dominance hierarchies.

Home Ranges A **home range** is a space used predominantly by an individual, pair, or group of animals but not actively defended. It is the area where an animal normally lives and often surrounds a smaller defended space, such as a nest, burrow, or den. A mountain lion, for example, hunts within its own home range.

A system of home ranges disperses the population, so that each animal has an uncrowded place to search for food. It also reduces social interactions, thereby eliminating the time, energy, and risk of injury associated with aggression. A home range helps an animal evade its predators, for an animal can learn escape routes and hiding places within the confines of a familiar place much better than if it roamed from place to place.

Home ranges are essential when there are young in the nest. The parent needs to forage near the nest, for it feeds the young at regular intervals and cannot leave them unprotected for long periods of time. A home range system of resource division is particularly common in mammals.

Territories A space that is defended by an individual, pair, or group of animals is a **territory.** An animal within its own territory has a monopoly on the resources in that space—food, shelter, nesting sites, mates, or places for courtship. Territorial behavior occurs in almost every major group of animals, both invertebrate and vertebrate.

The defense of a territory occurs in a sequence of rituals: advertisement of ownership, then threat, and finally fighting. Advertisements or threats usually resolve the issue and eliminate actual combat.

An advertisement is a signal that the area is owned and will be defended (fig. 27.12). Bird songs and cricket

bonding (which includes defense of a female) and parental care in the male prairie vole: The act of mating stimulates neurons within the brain to secrete larger amounts of vasopressin, which bind with receptors on neurons within the limbic system (the emotional center of the brain). This binding activates neural pathways that were established, according to genetic programs, during development of the animal. The activated pathways generate monogamous behaviors: pair-bonding and parental care.

A prediction of the hypothesis is that if these receptors are blocked, so that vasopressin cannot stimulate neurons in the limbic system, then monogamous behaviors will not be exhibited. The prediction has been tested with regard to both pair-bonding and parental care. These experiments were carried out in 1993.

In one experiment, researchers tested the prediction that if the actions of vasopressin are blocked, then a male prairie vole will not form a pair-bond. Male prairie voles were divided into two groups: one group received injections of a chemical that prevents vasopressin from binding with its receptor, and the other group received injections of a salt solution, which should have no effect on the body. The voles were then put in cages with females and given twenty-four hours to mate. After a period of time, during which a pair-bond would normally form, foreign males were placed in the cages with the mates. Males that had received the chemical that blocked vasopressin showed no response to the strangers. Males that had received the salt solution attacked the foreign males, a normal pair-bonding behavior. The results of this experiment confirmed the hypothesis: pair-bonding is based on the effects of vasopressin on neurons in the brain.

In another experiment, researchers tested the prediction that if vasopressin activity is prevented, then male prairie voles will not exhibit parental care. Virgin males were divided into three groups: one group received injections of vasopressin (directly into their limbic systems), another group received injections of the chemical that prevents vasopressin from attaching to its receptors, and a third group received injections of a salt solution. Each male was then placed in a cage with a very young vole pup. Males in the group that received vasopressin spent significantly more time grooming and cuddling the pups than males in other groups. Males in the other groups acted like bachelors: they paid little attention to the pups and in several cases actually attacked them. (The virgin males that received salt solutions have low levels of vasopressin, since its secretion is stimulated by mating.)

Researchers caution that vasopressin may not transform men into ideal husbands and fathers. Our own sexual and parental care behaviors are influenced to a large degree by environmental and cultural factors. What is important, and astonishing, about this research is the revelation that changes in the concentration of a single molecule can cause profound changes in complex behaviors.

Figure 27.12 Territorial advertisement. (*a*) A ring-necked pheasant advertises his territory with a harsh, crowing call while stretching his neck and flapping his wings. (*b*) A cheetah advertises his territory by urinating on it.

(a)

(b)

Figure 27.13 Threat behavior. The open mouth and piercing stare of a rhesus monkey serve to intimidate would-be attackers.

Figure 27.14 Submissive behavior. A wolf of lower status licks the mouth of a wolf of higher status. It communicates submission.

chirps are advertisements, as are the scent marks of a mammal's urine, musk, or scented oils. A threat is a particular posture (fig. 27.13), often combined with behaviors that make an animal look larger than it really is. A bird raises is feathers; a mammal raises its fur; a frog inflates its lungs with air. A threat may be accompanied by vocalizations and movement toward the intruder.

If advertisements and threats fail to intimidate the intruder, aggression escalates to a fight. This, too, is ritualized; as soon as the outcome of the fight is clear, the loser exhibits submissive behavior—exposes a vulnerable part of its body, such as its neck, or displays juvenile behaviors. These behaviors appease the winner, enabling the loser to leave the area without further injury. Ritualized aggression saves energy and reduces the chance of serious injury or death.

Dominance Hierarchies   A **dominance hierarchy** is a set of dominant-submissive relations among members of a group. It is a ranking of social status that controls access to resources.

A dominance hierarchy reduces strife under crowded conditions, when a group of animals is taking advantage of a locally abundant resource, such as a fruit-laden tree, cluster of nesting sites on a cliff, or a group of sexually receptive females. The more dominant animals have first access to the resource and the less dominant animals get what is left over.

Hierarchies form through repeated threats and fights among members of the group. Once established, each ani-

mal communicates its status by appearance or signals rather than by fighting. Factors that determine the outcomes of these interactions include size, age, strength, general health, previous experience, and social status of the parents. After the hierarchy is established, each animal exhibits submissive behaviors to its superiors (fig. 27.14) and aggressive encounters subside.

For animals that live in groups, a dominance hierarchy functions to eliminate continuous fighting. Thus, each animal has more time and energy for foraging, reproducing, and watching for predators. Individuals with higher status gain the most, usually surviving and reproducing better than others. Individuals with lower status may emigrate, seeking higher status in another group, or they may remain and try to move up in the hierarchy when they are older.

Cooperative Living

Many animals live with other members of their own species and cooperate in various ways. The most common advantage of cooperative living is group defense against predators. A group has more eyes, ears, and noses for detecting a predator and can divide labor so that some watch while others feed. In addition, a group can defend their young more effectively than a single parent. Musk oxen, for example, form a circle when wolves are near, with the young ones inside and the horned adults facing outward.

Figure 27.15 Altruistic behavior. A jackal caring for his younger brother. By helping his parents raise another litter, the older brother greatly increases his brother's chance of survival. Without him, the parents can only raise one pup at a time. With his help, they can raise three or four pups at a time.
© Patricia D. Moehlman 1980.

Some animals, such as wolves, African lions, and hyenas, form hunting parties that can capture larger prey than an individual on its own. Others, like beavers and rabbits, cooperate in building communal shelters. Another advantage of group living is communication of where rich food sources are located. When a raven or vulture, for example, finds a large carcass it communicates the location to other members of its flock. A bird that provides the information one day benefits by receiving information on other days. Honeybees also communicate the location of food, apparently by dances, to other members of the hive.

Some social animals exhibit **altruistic behaviors,** which are acts that increase the chances of another animal's survival or reproduction but decrease the chances of their own survival or reproduction. In most cases, the two animals are related and selection for the altruistic behavior can be explained by **kin selection.** In this kind of selection, an animal who cannot reproduce on its own leaves more copies of its genes when it promotes survival and reproduction of its kin, who share many of the same genes.

An unmated jay who helps her sister to raise nestlings leaves more copies of her own genes—the same genes in her sister and, therefore, in her sister's offspring—in the next generation than if she lived on her own and did not find a nesting site. A young jackal that stays with his parents after he is weaned and helps them raise another

Figure 27.16 Warning signal. This meerkat is a sentry, positioned where he can scan the landscape for enemies. When he sees a predator, he gives an alarm call that warns other meerkats to run for cover. He is roughly 60 centimeters (2 ft) long, including his tail. Meerkats, which are related to weasels, live in social groups on the grasslands of South Africa.

litter (fig. 27.15) leaves more copies of his genes in the next generation (in his brothers and sisters) than if he left his parents but did not mate. A young jackal is unlikely to compete successfully with older males for mates.

Some altruistic behaviors may evolve by **reciprocal altruism,** in which an animal that exhibits altruism is likely to benefit from another animal's altruism at another time. An animal who gives a warning signal, for example, informs other members of its group that a predator is nearby. The signal draws attention to the individual who gives it but protects others because they can seek shelter. Many birds give warning signals, as do prairie dogs, pikas, beavers, chimpanzees, and meerkats (fig. 27.16).

SUMMARY

1. Behavior is the product of an animal's sense organs, nervous system, muscles, and sometimes hormones. It involves neural pathways that are genetically programmed and that are modified by experience.

2. Innate behaviors are governed by neural pathways that are genetically programmed. Simple innate behaviors include kineses and taxes; more complex innate behaviors include fixed-action patterns.

3. Learned behaviors include trial-and-error learning, in which an action becomes associated with its consequences; insight learning, in which problems are solved by thinking; and imprinting, in which a particular task is learned during a genetically programmed time.

4. Most of an animal's behavior relates to its ecology. Habitat selection is usually innate or imprinted. Foraging behavior includes food choice as well as the time, place, and method of search. Antipredator behaviors include flash behavior, startle behavior, unpredictable flight patterns, and freezing postures.

5. Migration is regular travel between two places of residence. Many animals migrate with the seasons to places that are warmer or have more food. Some animals migrate to special places for the purposes of reproduction.

6. Social behavior occurs when an animal interacts with others of its species. Sociobiology is the study of the biological basis of social behaviors.

7. Reproductive behaviors are social behaviors that facilitate mating and raising of young. Mating behaviors range from the simple movements of copulation to elaborate courtship behaviors. Pair-bonds develop in monogamous species. Parental care is exhibited by most females and monogamous males.

8. Many animals have social behaviors that divide resources, thereby reducing continual aggression. A home range is a space that is used predominantly by one individual or a group but not defended. A territory is a defended space that is used exclusively by one or a few individuals. It is defended by advertisements, threats, and fights. A dominance hierarchy is a social ranking in which higher ranked individuals get first access to resources.

9. Animals that live in groups have social behaviors that enhance cooperation. Some of these behaviors are altruistic; they evolve by kin selection or reciprocal altruism.

KEY TERMS

Altruistic (AL-troo-ISS-tik) **behavior** (page 521)
Behavioral response (page 508)
Dominance hierarchy (page 520)
Fixed-action pattern (page 510)
Habitat (page 512)
Home range (page 518)
Imprinting (page 511)
Innate (ih-NAYT) **behavior** (page 508)
Insight learning (page 510)
Kinesis (kih-NEE-sis) (page 509)
Kin selection (page 521)
Migration (page 514)
Monogamous (moh-NAH-gah-muss) (page 515)
Reciprocal altruism (page 521)
Sociobiology (page 514)
Stimulus (page 508)
Taxis (TAK-sis) (page 509)
Territory (page 518)
Trial-and-error learning (page 510)

STUDY QUESTIONS

1. Describe a situation in which an innate behavior is more advantageous than a learned behavior. Describe a situation in which a learned behavior is better than an innate behavior.

2. Explain why an animal needs to minimize the time and energy it spends foraging.

3. Why does a bird migrate south in the autumn? Why does it migrate north in the spring?

4. Why are courtship behaviors so elaborate?

5. What is the function of a pair-bond? Why do some animals form pair-bonds and others do not?

6. Describe the three ways in which members of a group divide resources. Describe a condition in which a dominance hierarchy would work better than a system of territories.

7. Describe three situations in which animals are better off living in a group than alone.

CRITICAL THINKING PROBLEMS

1. State a hypothesis to explain the observations that ducklings imprint on their mothers within a day of hatching, whereas young owls do not imprint until they are much older.

2. List two hypotheses that explain why freshwater eels migrate to the Sargasso Sea in order to spawn.
3. What is the advantage, to a hummingbird, of flying north in the springtime?
4. Most male mammals are not monogamous because they cannot feed their young directly. Yet only slight changes in hormones can trigger milk production in males. Provide a hypothesis that explains why male mammals do not breast-feed.
5. Can you think of any examples of the following social behaviors in humans: territoriality, dominance hierarchies, submissive behaviors, altruism?

SUGGESTED READINGS

Introductory Level:

Alcock, J. *Animal Behavior: An Evolutionary Approach.* 4th ed. Sunderland, MA: Sinauer Associates, 1989. An introduction to behavior, covering a broad range of animals and emphasizing the evolution of behavior.

Benjus, J. M. *Beastly Behavior: A Watcher's Guide to How Animals Act and Why.* Reading, MA: Addison-Wesley, 1992. A collection of entertaining, scientific essays (with drawings) on the behaviors of vertebrate animals.

Gould, J. L., and C. G. Gould. *Animal Minds.* New York: Scientific American Library, 1994. A exploration of animal minds, from insects to humans, with regard to the hypothesis that animal behavior is more flexible than previously believed. Good descriptions of research.

Morris, D. *Animal Watching.* New York: Crown Publishers, 1990. Fascinating explanations of why animals behave as they do. Topics include aggression, territories, dominance hierarchies, and cannibalism.

Windybank, S. *Wild Sex: Way Beyond the Birds and the Bees.* New York: St. Martin's Press, 1992. Descriptions of courtship and mating in a wide variety of animals.

Advanced Level:

Dusenbery, D. B. *Sensory Ecology: How Organisms Acquire and Respond to Information.* New York: W. H. Freeman and Company, 1992. A new approach to behavior that bridges sensory biology and ecology, with sections on stimulus properties, stimulus generation, and how animals locate resources, communicate, and navigate.

Griffin, D. R. *Animal Minds.* Chicago: University of Chicago Press, 1992. A compilation of the evidence that many animals have conscious awareness.

Kamil, A. C., J. R. Krebs, and H. R. Pulliam. *Foraging Behavior.* New York: Plenum Publishing, 1987. Proceedings of a symposium, with sections on foraging theory, patch utilization, reproductive consequences, and learning.

Tinbergen, N. *The Animal and Its Work: Explorations of an Ethologist.* Cambridge, MA: Harvard University Press, 1973. A delightful classic by a researcher who contributed substantially to our knowledge of imprinting and aggression.

Wilson, E. O. *Sociobiology: The New Synthesis.* Cambridge, MA: Belknap/Harvard University Press, 1975. A comprehensive review, with references, of social behavior and its evolution.

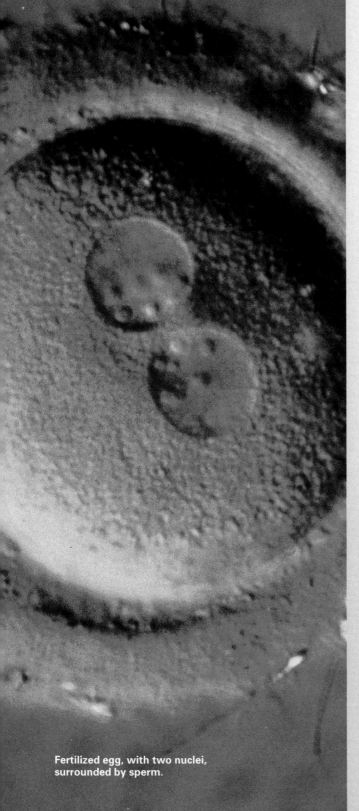

Fertilized egg, with two nuclei, surrounded by sperm.

Reproduction

Chapter Outline

Objectives

In this chapter, you will learn
–why some animals have external
  fertilization and others have internal
  fertilization;
–how eggs and sperm, formed in
  females and males, reach the site of
  fertilization;
–how the genes of a sperm get inside an
  egg;
–about methods of birth control.

here does a baby come from? Is it delivered by a stork? Does it dwell inside a fruit? These and other fanciful notions were imagined by our ancestors because they did not know how a baby forms, yet they needed an explanation for this prominent aspect of their lives.

Our ancestors knew, of course, that a baby is born of a woman. That part of reproduction they could see. But the unseen processes—fertilization and the first nine months of development—were a mystery. In the absence of real information, people of the Middle Ages (from A.D. 500 to A.D. 1500) invented myths to explain the origin of children.

A common myth was that birds delivered children. In northern Europe, for example, the stork was believed to retrieve the souls of babies from marshes, where they were stored, and carry them to people's homes. Couples who wanted children encouraged the storks to build nests on their roofs and placed food on windowsills to attract the birds. Couples who did not want children destroyed the stork nests and chased the birds away. An unwanted pregnancy happened, they believed, when the stork slipped in to deliver a child while the woman was sleeping.

Other cultures had other myths. On some islands of the South Pacific, spirit children were thought to dwell inside the fruit of trees. A woman who ate one of these fruits might swallow the soul of a child and become pregnant. On other islands, the spirits were believed to live within aquatic animals, so that a woman who did not want a child avoided the eyes of crabs as she bathed in the river and did not venture into deep water, where fish and snakes waited to impregnate her.

The connection between the sex act and a newborn was not understood until sperm and eggs were seen, in the late seventeenth century, through a microscope. (The sperm were at first assumed to be some dreadful parasite that infected only men.) Even then, the contribution of the sperm and the egg to formation of a child was not understood until the late nineteenth century, when chromosomes and their function were discovered. Surprisingly, it was not until the 1930s that we learned which days of the menstrual cycle a woman was most likely to become pregnant.

Animal Reproduction

All forms of life have the capacity to reproduce themselves. Most animals reproduce sexually, a process in which the offspring receives genes from its father, via his sperm, and from its mother, via her egg.

Figure 28.1 External fertilization. In frogs, the sperm and eggs unite within the surrounding water. The male "hugs" the female, forcing the eggs out of her body. He then sprays them with sperm.

Sperm are small, mobile cells. They swim through water or body fluids to reach and penetrate the eggs. Eggs are larger and less mobile cells, laden with food for the young embryo. An animal that produces sperm is a male; an animal that produces eggs is a female. Joining of a sperm with an egg is called fertilization.

Most aquatic animals have external fertilization, in which the sperm and egg join in the surrounding environment. The male and female come together, during the reproductive season, and release their sperm and eggs into the water. Mating behaviors ensure that both sperm and eggs are released at the same time and place (fig. 28.1). The sperm then swim to the eggs and fertilization takes place. Marine invertebrates, fishes, and frogs have external fertilization.

Most terrestrial animals have internal fertilization, in which the sperm and egg join within the female. The female retains her eggs and the male places his sperm inside her body. There, the sperm swim through body fluids to meet and fertilize the eggs. Internal fertilization protects the sperm and eggs from drying out and provides body fluids for the sperm to swim through. A male's penis, which places the sperm inside a female, is an adaptation to life on land. Insects, spiders, reptiles, birds, and mammals have internal fertilization.

Formation of Eggs

A female's function in reproduction is to develop eggs and place them where they can be reached by a male's sperm. Eggs are produced by meiosis (📖 *see page 144*), a form of cell division that halves the number of chromosomes. In addition to these basic reproductive functions, many females nourish and shelter their offspring until they are able to fend for themselves (fig. 28.2).

Female Reproductive Systems

Every female has special cells that produce eggs. In simpler animals, such as sponges and sea anemones, certain body cells develop (by meiosis) into eggs during the reproductive season. In more complex animals, eggs are produced within special organs, called **ovaries,** and transported through tubes to the site of fertilization. The arrangement of these tubes depends largely on whether fertilization is external or internal.

In animals with external fertilization, eggs released from an ovary travel through tubes to the outside environment (fig. 28.3a). In animals with internal fertilization, eggs from an ovary travel to a tube that connects with the outside environment. The male then deposits his sperm into this tube and they swim to meet the eggs (fig. 28.3b). The fertilized eggs then either travel out of the female (via the same tube as the sperm entered) or remain inside for development into embryos.

Figure 28.2 Parental care. Some parents shelter and nourish their young after they are born or hatched.

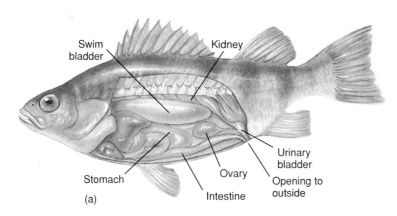

(a)

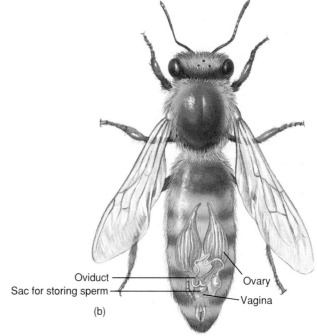

Oviduct

Sac for storing sperm

Ovary

Vagina

(b)

Figure 28.3 Female reproductive systems. (a) In animals with external fertilization, the female reproductive system typically consists of one or more ovaries connected to a tube that carries unfertilized eggs out of the body. (b) In animals with internal fertilization, the female reproductive system typically consists of one or more ovaries connected by tubes to a vagina. A male places sperm in the vagina, and fertilization occurs there or within the tubes. In insects, the fertilized eggs are covered with protective materials and then released through the vagina.

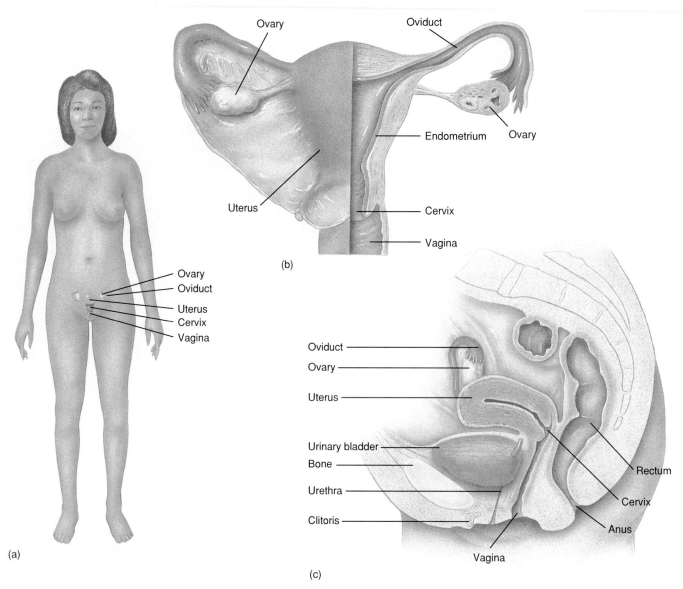

Figure 28.4 A woman's reproductive system. (*a*) The organs are located within the lower abdominal cavity. (*b*) Front view of the organs, with one side cut open to show the internal structure. (*c*) Section through the center of the body, with a side view of the organs.

The Human System

The organs of a woman's reproductive system are located within the lower part of her abdominal cavity, between the urinary bladder and the rectum (fig. 28.4). They include the ovaries, oviducts, uterus, and vagina. These organs and their functions are listed in table 28.1 and described following.

Ovaries   A woman has two ovaries, one on each side of her lower abdominal cavity. They have two functions: to develop mature eggs and to secrete the female sex hormones.

An ovary contains eggs in various stages of development (fig. 28.5). A woman is born with some 2 million immature eggs, of which only 400 or so will develop into eggs that can be fertilized. Each egg is enclosed within a **follicle,** consisting of a single layer of cells when the egg is immature and several layers when the egg is mature:

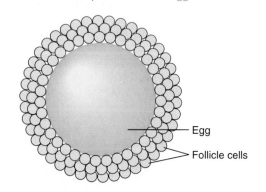

Table 28.1

A Woman's Reproductive System

Organ	Function
Ovaries	Develop mature eggs; secrete estrogen and progesterone
Oviducts	Receive mature eggs from the ovaries; receive sperm placed in the vagina; sites of fertilization
Uterus	Shelters and nourishes the embryo as it develops into a baby
Vagina	Receives sperm during intercourse; serves as a birth canal; drains menstrual fluids
Clitoris	Sexual arousal

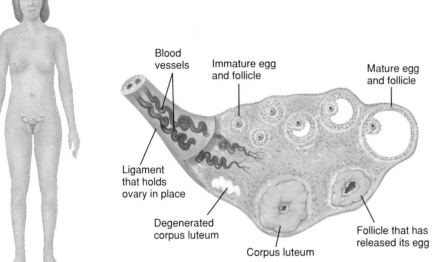

Figure 28.5 Anatomy of an ovary. An ovary contains eggs in various stages of development as well as ruptured follicles and corpora luteum in various stages of degeneration.

When an egg nears maturity, it leaves its follicle and the ovary. The empty follicle then becomes a **corpus luteum** (*corpus*, body; *luteum*, yellow). Both the follicle and corpus luteum are endocrine glands that secrete female sex hormones.

Oviducts   When an egg leaves an ovary, it moves into an **oviduct** (*ovi*, egg; *duct*, tube), a tube that connects with the uterus. (The oviduct is also called a fallopian tube or a uterine tube.) There are two oviducts, one on each side of the uterus. They receive eggs from the two ovaries.

One end of each oviduct lies next to an ovary. This end, which is not directly connected to the ovary, is ex-panded to form a funnel (see fig. 28.4) that captures eggs as they emerge from the ovary.

The other end of each oviduct opens into the uterus. The inner linings of the oviducts are covered with cilia (☐ *see page 66*) that wave back and forth, conveying the egg toward the uterus.

Uterus and Vagina   The **uterus** is a muscular chamber that shelters and nourishes the developing baby. Its internal lining, called the **endometrium,** is richly supplied with blood vessels for nourishing the embryo. (Endometriosis is a disorder in which the endometrium grows through the oviducts and into the abdominal cavity.) The uterus connects with the vagina through the **cervix.** Except during childbirth, the opening in the cervix is very small.

The **vagina** is a muscular tube that extends from the cervix to the outside of the body. The vaginal opening lies behind the opening to the urethra (a much narrower tube for urine) and in front of the anus. The vagina has three functions: to drain menstrual fluids, to receive sperm during sexual intercourse, and to provide an exit route for the baby at the time of birth.

The **clitoris** is a small structure in front of the opening to the urethra (see fig. 28.4). This sensitive organ is the female counterpart to the penis. Like the penis, it enlarges during sexual arousal. (In some cultures, the clitoris is surgically removed in order to prevent sexual pleasure.)

The Menstrual Cycle

From puberty until menopause, a woman's reproductive organs go through regular periods of change known as menstrual cycles. During each cycle, an egg matures and leaves its ovary, while at the same time the endometrium of the uterus is first built up and then torn down. These events are illustrated in figure 28.6. Each cycle lasts an average of twenty-eight days and is marked by two events: menstruation and ovulation.

Menstruation   A menstrual cycle begins with menstruation, a period of several days during which the endometrium degenerates and is expelled through the vagina. The first day of uterine bleeding is considered the first day of the menstrual cycle.

Menstruation is initiated by an extreme constriction of the arteries that transport blood to the endometrium. With their blood supply cut off, the cells of this inner lining die. The dead layer of cells peels off the inner wall, tearing capillaries so they bleed. Contractions of muscles within the uterine wall help push the dead cells and blood out of the uterus and into the vagina, where they drain out of the body. A woman typically loses about half a cup of blood during menstruation.

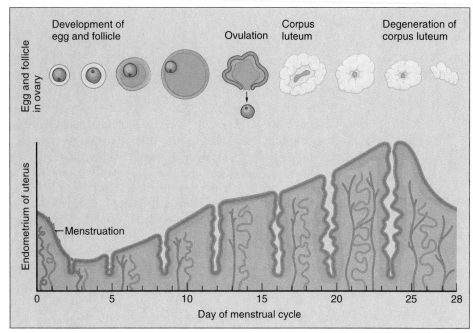

Figure 28.6 The menstrual cycle. The cycle involves changes in the ovaries and the endometrium of the uterus. Within the ovaries, an egg develops within its follicle from day 1 to day 14 of the cycle. On day 14, it breaks out of its follicle and leaves the ovary, an event known as ovulation. Its follicle becomes a corpus luteum. Meanwhile, the endometrium is shed between days 1 and 5 of the cycle. It then thickens between days 5 and 28 in preparation for pregnancy. If pregnancy does not occur, menstruation occurs and another cycle begins.

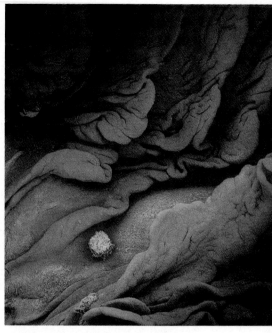

Figure 28.8 A mature egg within an oviduct. The egg, still covered with smaller cells from the follicle, drifts slowly down the oviduct toward the uterus.

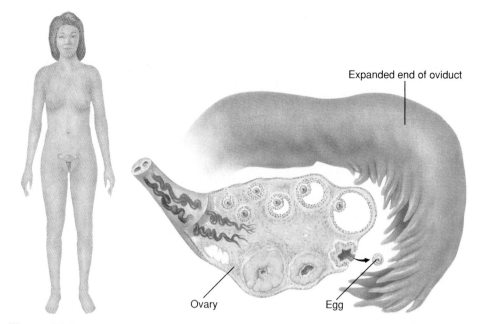

Figure 28.7 Ovulation. Once a month, an egg emerges from an ovary. The walls of the follicle and ovary break apart, releasing the egg into the oviduct. Some cells from the follicle adhere to the released egg. The rest of the cells become the corpus luteum.

After menstruation (about day 5 of the cycle), the endometrium begins once again to thicken as new cells and capillaries form in preparation for pregnancy.

Ovulation On day 14 of a typical cycle, a mature egg is expelled from its follicle and ovary (fig. 28.7). While the egg is mature, it has not yet completed meiosis: cell division is held at metaphase of meiosis II until the egg is fertilized. The two ovaries alternate, with each ovary releasing one egg every other month. The egg moves from the ovary and into the oviduct, an event known as **ovulation.** It is at this time of the month that a woman can become pregnant, because the egg in her oviduct (fig. 28.8) can be reached by a sperm.

After ovulation, the endometrium of the uterus continues to thicken in preparation for receiving an embryo. If the egg is not fertilized by a sperm, then the endometrium degenerates about fourteen days after ovulation (the twenty-eighth day of the cycle). This onset of menstruation marks the first day of the next menstrual cycle.

Hormonal Control of the Menstrual Cycle

Four hormones regulate the menstrual cycle: follicle-stimulating hormone, luteinizing hormone, estrogen, and progesterone. Follicle-stimulating hormone and luteinizing hormone are secreted by the anterior pituitary gland (see page 431). Estrogen is secreted by both the follicle and the corpus luteum within the ovaries. Progesterone is secreted only by the corpus luteum. The levels of these hormones are illustrated in figure 28.9 and their roles are listed in table 28.2.

Follicle-stimulating hormone (FSH) promotes growth of a follicle and maturation of its egg. Luteinizing hormone (LH) stimulates the follicle to secrete estrogen:

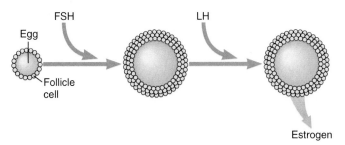

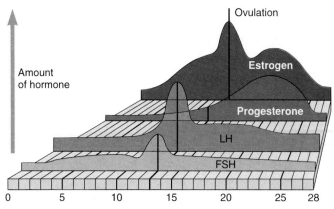

Figure 28.9 Hormone levels change during the menstrual cycle. Amounts of follicle-stimulating hormone (FSH) and luteinizing hormone (LH) in the blood reach a peak just prior to day 14, triggering ovulation. Estrogen, secreted first by the follicle and then by the corpus luteum, is also most abundant just prior to ovulation. Progesterone, secreted by the corpus luteum, becomes abundant only after ovulation.

The follicle secretes more and more estrogen as it grows. The high levels of estrogen, in turn, stimulate secretion of more LH. Both hormones reach peak concentrations just prior to day 14. This surge in hormones triggers ovulation: the follicle ruptures and the mature egg passes out of the ovary and into the oviduct.

After ovulation, the concentrations of both FSH and LH decline. The ruptured follicle becomes a corpus luteum, which secretes both estrogen and progesterone.

Table 28.2

Hormonal Regulation of the Menstrual Cycle

Gland	Hormone	Function
Anterior pituitary	Follicle-stimulating hormone (FSH)	Growth of follicle; maturation of egg
	Luteinizing hormone (LH)	Converts follicle into a gland; triggers ovulation; converts ruptured follicle into corpus luteum
Follicle	Estrogen	Thickens endometrium; high levels stimulate LH secretion
Corpus luteum	Estrogen	Thickens endometrium
	Progesterone	Thickens endometrium; inhibits FSH and LH secretion

531

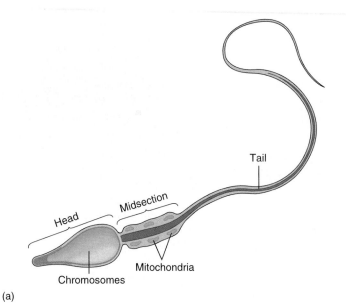

(a)

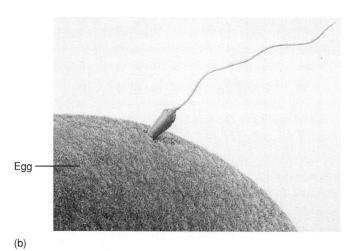

(b)

Figure 28.10 The sperm. (*a*) A sperm consists of a head, midsection, and tail. The head contains the nucleus with a set of chromosomes. The midsection is packed with mitochondria, which provide energy for locomotion. The tail lashes back and forth, propelling the sperm forward. (*b*) Photograph of a sperm fertilizing an egg.

These two hormones act together to prepare the endometrium for pregnancy—to make it soft, moist, and richly supplied with blood.

If the egg is fertilized and pregnancy occurs, then the corpus luteum continues to secrete both estrogen and progesterone. Both hormones cause the endometrium to thicken. Progesterone also inhibits the secretion of FSH and LH so that no new follicles and eggs mature during pregnancy.

If the egg is not fertilized within twenty-four hours after ovulation, then it deteriorates within the oviduct. The corpus luteum gradually degenerates and stops secreting estrogen and progesterone. (The dramatic drop in progesterone level produces, in some women, a condition known as premenstrual syndrome, or PMS, characterized by mood swings, food cravings, and fatigue.) Fourteen days after ovulation, the low levels of estrogen and progesterone (see fig. 28.9) trigger menstruation, which eliminates the endometrium, and increased secretion of FSH, which initiates development of a new follicle and egg.

In addition to its roles in the menstrual cycle, estrogen stimulates development of the female reproductive system and secondary sex characteristics—fat deposits in breasts, hips, and thighs; narrow shoulders and broad hips; hair in the armpits and pubic area. Progesterone maintains pregnancy and stimulates enlargement of the breasts during pregnancy.

Formation of Sperm

A sperm is tiny compared with the egg it fertilizes (see chapter opening photograph). It is a cell specialized for transporting a male's genes to a female's egg. A sperm consists of a head, a midsection, and a tail (fig. 28.10). The head contains the nucleus, with half the number of chromosomes as other cells of the body. (The chromosome number is halved during meiosis, an early stage of sperm development.) The midsection of a sperm is packed with mitochondria (see page 64) that provide energy for swimming. The tail lashes back and forth, propelling the sperm through water or body fluids.

Male Reproductive Systems

Specialized cells produce sperm. In the simpler animals, certain cells within the body develop into sperm during the reproductive season. In more advanced animals, sperm develop in specialized organs called **testes.**

Mature sperm travel through tubes from the testes to the place where fertilization takes place (fig. 28.11). Glands secrete fluids, through which the sperm swim, into these tubes. In animals with internal fertilization, the male usually has a penis for placing sperm inside a female's body. A few animals use, instead, a modified leg, fin, tentacle, or some other appendage. A spider, for example, transfers his sperm onto a palp (a leglike appendage near the mouth), which he then inserts into the female's vagina.

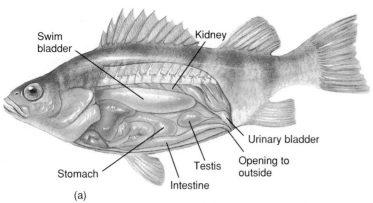

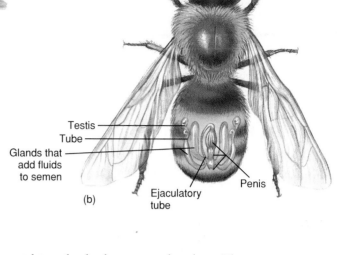

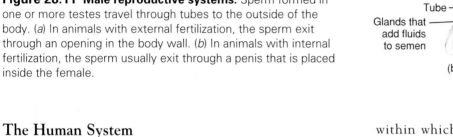

Figure 28.11 Male reproductive systems. Sperm formed in one or more testes travel through tubes to the outside of the body. (*a*) In animals with external fertilization, the sperm exit through an opening in the body wall. (*b*) In animals with internal fertilization, the sperm usually exit through a penis that is placed inside the female.

The Human System

A man's reproductive function is to form sperm and to place them inside the woman's vagina. His reproductive system includes testes, a system of tubes that store and transport sperm, and glands that contribute fluids through which the sperm swim. These organs are illustrated in figure 28.12. Their functions are described briefly in table 28.3 and in greater detail below.

Testes   The testes (sing., testis) are two glands suspended outside the abdominal cavity. They have two functions: to produce sperm and to secrete the male sex hormone, testosterone.

The testes develop, in a male embryo, within the abdominal cavity near the kidneys. Two months before birth, they migrate out of the abdominal cavity and into the **scrotum,** a pouch of skin suspended behind the penis. The reason for this migration is to leave the interior of the body, which is too warm for the sperm. The cooler temperature (about 34° C, or 94° F) within the scrotum is optimal for sperm development.

The testes are packed with tiny tubes, the **seminiferous tubules** (*seminiferous,* semen-carrying; *tubule,* little tube),

within which the sperm develop. There are so many tubules that if they were placed end to end they would be nearly as long as three football fields! Between the tubules are cells that secrete testosterone.

Epididymis, Vas Deferens, and Urethra   The seminiferous tubules connect with another tube, the **epididymis,** which is coiled on top of the testis (fig. 28.12). There are two epididymides, one on each testis. As sperm mature within the seminiferous tubules, they move into the epididymis for storage.

Each epididymis connects, in turn, to yet another tube, the **vas deferens,** that carries sperm into the abdominal cavity. The two vasa deferentia (the plural of vas deferens), one from each testis, extend around the urinary bladder and then join with the **urethra,** which carries urine from the urinary bladder. The urethra extends through the penis. It carries both sperm and urine (though not at the same time) out of the body.

Three types of glands—a pair of seminal vesicles, a single prostate, and a pair of bulbourethral glands—connect with the vasa deferentia and urethra by way of small tubes through which their secretions flow (fig. 28.12). Their functions are described in a later section.

533

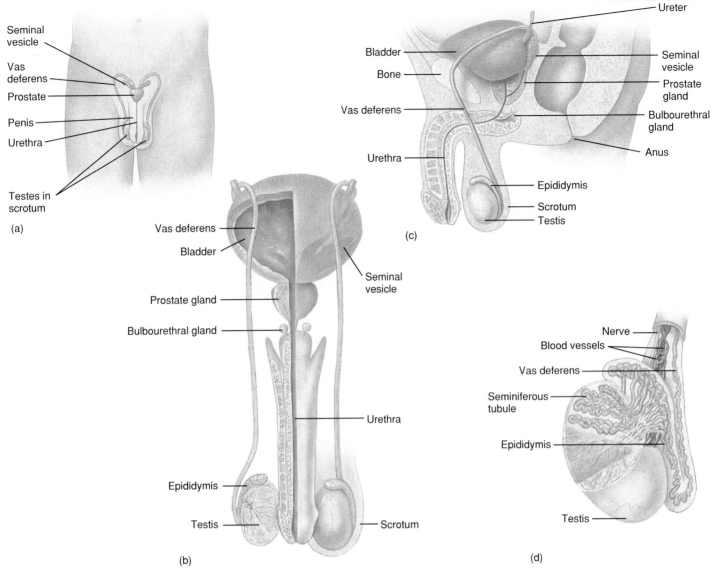

Figure 28.12 A man's reproductive system. (a) Most of the organs are located inside the abdominal cavity. The testes, however, are suspended outside the body where temperatures are cooler. (b) Front view of the organs, with one side cut open to show the internal structures. (c) Side view of the organs. (d) Close-up of the interior of a testis, showing the seminiferous tubes and epididymis.

Sperm Development

The mature testes produce sperm continuously. Each sperm takes about sixty-four days to develop, and all stages of sperm development are present in the testes at the same time. Approximately 300 million sperm mature each day. Over a lifetime, a man produces roughly 8 trillion sperm.

Mature sperm are stored in the epididymides. Sperm that are stored for several weeks (i.e., not ejaculated) degenerate and their materials are recycled. Sperm ejaculated into a woman live approximately three days.

Something strange has been happening to sperm production in men all over the world. Between 1940 and 1990, the average concentration of sperm in semen has dropped from 113 million to 66 million per milliliter. A current hypothesis to explain this change is that sperm-producing cells are suppressed by environmental pollutants that mimic estrogen. These pollutants, which include pesticides and industrial by-products, may also be responsible for another phenomenon: the feminization of male reproductive organs, as observed in male sturgeon, alligators, gulls, Florida panthers, and other wildlife.

Hormonal Control of Male Reproduction

Three hormones govern the development of sperm: follicle-stimulating hormone (FSH), luteinizing hormone (LH), and testosterone (table 28.4). The concentrations of these hormones in a man remain fairly constant from day to day.

Follicle-stimulating hormone and luteinizing hormone are named for their female functions, which were discovered first (FSH stimulates the follicle; LH stimulates the corpus luteum). They are secreted, in both sexes, by the anterior pituitary gland. In a man, FSH stimulates the development of sperm within the testes and LH stimulates the production of testosterone by the testes.

Testosterone, in turn, has many functions. It works with FSH in the formation of sperm and is necessary for both an erection of the penis and ejaculation of the semen. Testosterone also governs development of the male reproductive organs and male secondary sex characteristics—a deep voice; hair on the face, chest, underarms, and pubic area; the masculine distribution of muscle and fat; and development of the sex drive.

Fertilization

During sexual intercourse, the man delivers sperm inside the woman's vagina. The sperm then travel through the woman's reproductive tract to the oviducts. Fertilization occurs if one of the sperm reaches an egg and fuses with it, bringing the two sets of chromosomes together within a single, nutrient-rich cell. The newly formed cell is the fertilized egg.

Journey of the Sperm

Delivery of sperm requires an erection, in which the penis becomes sufficiently rigid to enter the vagina. The penis engorges with blood in response to physical or emotional stimulation. Arteries dilate, bringing more blood to the organ and squeezing the veins so that less blood leaves. The pressure of the accumulated blood makes the penis rigid and erect.

Pathway Out of the Man   An ejaculation propels sperm out of a man. Strong contractions of muscles within the walls of the epididymides and vasa deferentia move the sperm to the penis. As sperm travel through the vasa deferentia, they acquire fluids from the seminal vesicles; as they travel through the urethra, they acquire fluids from the prostate and bulbourethral glands. These glands and their secretions are summarized in table 28.3 and described following.

The **seminal vesicles** secrete a fluid rich in fructose and prostaglandins. Fructose is a sugar used as fuel in cellular respiration within the mitochondria of the sperm, providing energy (as ATP) for swimming. Prostaglandins are chemical messengers that stimulate muscle contractions in the uterus, creating currents of fluid that carry the sperm into the oviducts. (Prostaglandins are named for the prostate gland, which early investigators believed, mistakenly, secreted them.) The **prostate** secretes a fluid rich in enzymes that stimulate sperm locomotion. The **bulbourethral glands** secrete a lubricant that coats the tip of the penis just prior to ejaculation. All these fluids are alkaline; they neutralize the fluids of the urethra and vagina, which are too acidic for optimal sperm activity. (The function of these acids is to kill bacteria that cause diseases.)

Table 28.3

A Man's Reproductive System

Organ	Function
Testes	Form sperm (inside seminiferous tubules); secrete testosterone
Scrotum	Holds testes outside body, in a cooler environment
Epididymides	Store mature sperm
Vasa deferentia	Transport sperm to urethra
Urethra	Transports sperm outside penis
Seminal vesicles	Secrete alkaline fluid that contains sugars and prostaglandins
Prostate gland	Secretes alkaline fluid with an enzyme that facilitates sperm movement
Bulbourethral glands	Secrete alkaline lubricant

Table 28.4

Hormonal Regulation of Male Reproduction

Gland	Hormone	Function
Anterior pituitary	Follicle-stimulating hormone (FSH)	Sperm development
	Luteinizing hormone (LH)	Testosterone production
Testes	Testosterone	Sperm development; erection; ejaculation

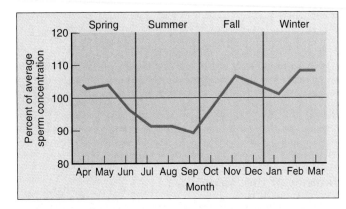

Figure 28.13 Sperm production varies with the season of the year. Sperm production peaks in late winter, at nearly 10% above average, and drops at the end of summer.

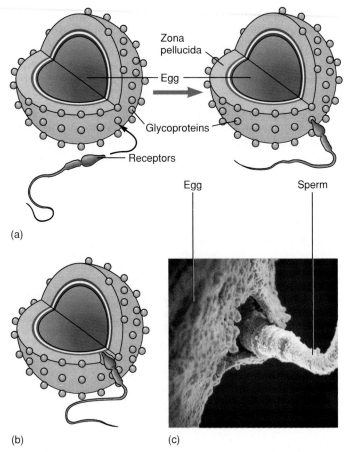

Figure 28.14 Fertilization. (*a*) The head of a sperm contains protein receptors that bind with glycoproteins in the zona pellucida that surrounds an egg. (*b*) Binding of the two kinds of molecules triggers the release of enzymes from the sperm head. These enzymes digest a path for the sperm through the zona pellucida. (*c*) Photograph of a sperm entering the zona pellucida of an egg.

The semen, consisting of sperm and the secretions just described, is ejaculated from the penis through the urethra. Each ejaculate consists of about a teaspoonful or so of fluid and approximately 300 million sperm. For reasons not yet understood, the sperm concentration varies with the season of the year (fig. 28.13).

Pathway Within the Woman Typically, the sperm are deposited far into the vagina, near the cervix. They travel through the cervix, uterus, and oviducts (see fig. 28.4). Their journey to the fertilization site, approximately 13 centimeters (5 in) from the cervix, takes the sperm from thirty minutes to several hours.

The sperm swim through fluids within the female reproductive tract. They are also swept along by cilia that line the tract and by contractions of the uterus. Only a few hundred sperm reach the place where an egg can be fertilized. As an egg travels along an oviduct, from ovary to uterus, it is receptive to fertilization only when passing through the outer third of the oviduct. After that, it is too old to be fertilized.

Joining of an Egg and a Sperm

As an egg travels through the oviduct, it releases a chemical that signals its location to the sperm. The sperm follows a concentration gradient of this chemical toward the egg. Fertilization takes place if both sperm and egg are together within the outer third of an oviduct.

To enter the egg, a sperm has to work its way first through a layer of cells (originally part of the follicle) and then through the zona pellucida, an outer coat made of glycoproteins (chains of amino acids to which sugars are attached). Some of the glycoproteins fit, like keys in locks, into receptor proteins on the sperm. Binding of these two molecules releases digestive enzymes from the head of the sperm. (Many sperm may contribute these enzymes.) The enzymes digest the zona pellucida and one of the sperm quickly reaches the egg surface (fig. 28.14).

Only sperm of the same species as the egg have receptor proteins that bind with the egg's glycoproteins. Sperm of another species will not bind and so do not release digestive enzymes. They cannot approach the egg surface. This species-specific barrier prevents reproduction between different species.

When a sperm contacts the egg, the two plasma membranes fuse to form a single cell, the fertilized egg. The egg then completes meiosis, and its nucleus contains just one set of chromosomes. All parts of the sperm degenerate except the nucleus, which merges with the egg nucleus. The two sets of chromosomes—one from the father, via his sperm, and the other from the mother, via her egg—are brought together. The fertilized egg contains all the necessary information to build a human.

The egg reacts instantaneously to penetration by a sperm, altering its surface so that other sperm cannot enter. The fertilized egg then begins to synthesize proteins and DNA in preparation for transformation into a multicellular embryo. As these changes occur, the egg continues to drift slowly down the oviduct toward the uterus.

Fertilization does not guarantee pregnancy. A woman is not pregnant until the fertilized egg attaches to the endometrium of her uterus. Only about 40% of all fertilized eggs survive the journey to the uterus and implant within its inner wall.

Sometimes a woman ovulates two eggs at the same time. If both eggs are fertilized, then two babies develop simultaneously. Such fraternal twins are not identical. (Identical twins develop when a young embryo splits in two, as described in the next chapter.) Fraternal twins are as similar, genetically, as other siblings, since each child receives its genes from a different egg and a different sperm. The tendency to release two eggs at the same time rises after the age of twenty-five. It also seems to be inherited, since fraternal twins are more common in some family lines than in others. Fertility drugs also stimulate ovulation of more than one egg (sometimes many) at the same time.

Preventing Fertilization

Our knowledge of human reproduction enables us to separate the sexual act from reproduction; there can be sexual intercourse without pregnancy. Methods of birth control that prevent fertilization are described here and in figure 28.15. (Methods that act after fertilization are described in the next chapter.)

Rhythm Method The rhythm method of contraception was developed in the 1930s, soon after the menstrual cycle was described. With this method, a couple does not have intercourse when an egg is likely to be present within an oviduct. This so-called unsafe time of the month is estimated as follows.

An egg is released from an ovary midway through a menstrual cycle. The egg survives about twenty-four hours within the oviduct. A sperm can fertilize an egg during a three-day period after entering the woman. Thus, if intercourse is avoided for three days prior to ovulation (the life span of a sperm) and for a day after ovulation (the life span of an egg), intercourse will probably not lead to pregnancy (fig. 28.16a). The problem lies in knowing exactly when ovulation is occurring.

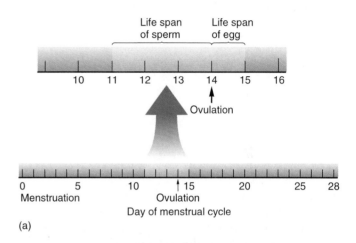

(a)

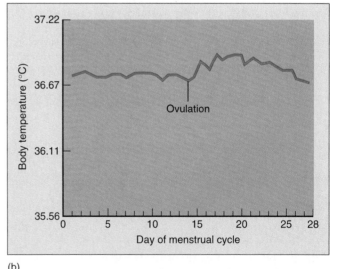

(b)

Figure 28.16 The unsafe period for the rhythm method.
(a) The unsafe period (shown in red), when a woman can become pregnant, occurs near the time of ovulation. She can get pregnant from sperm acquired during intercourse three days prior to ovulation, because sperm can live for three days. She can get pregnant from intercourse one day after ovulation, because an egg lives that long in the oviduct. (b) A woman's body temperature rises slightly just after ovulation. By graphing her body temperature for many months, a woman can predict her time of ovulation in a typical cycle.

Effectiveness

Extremely | Highly | Effective | Moderately | Fairly | Unreliable

100% | 99.6% | 98% | 98% | 95% | 95% | 90% | 87% | 83% | 82% | 76% | 75% | 74% | 70% | 40% | 10%

Tubal ligation | Vasectomy | The Pill | Intrauterine device (IUD) plus spermicide | IUD alone | Condom (good brand) plus spermicide | Condom (good brand) | Diaphragm plus spermicide | Vaginal sponge with spermicide | Spermicide (foam) | Rhythm method (based on daily temperature readings) | Spermicide (creams, jellies, suppositories) | Withdrawal of penis before ejaculation | Condom (cheap brand) | Douche | Chance (no method)

Figure 28.15 Effectiveness of birth-control methods.
Effectiveness is the percentage of sexually active women who did not get pregnant when using the method for one year.

Wellness Report

Sexually Transmitted Diseases

Diseases that spread from person to person during sexual intercourse are called sexually transmitted diseases. Six such diseases are common: syphilis, gonorrhea, chlamydia, herpes, genital warts, and acquired immune deficiency syndrome (AIDS).

Twelve million new cases of sexually transmitted diseases are reported each year in the United States. Approximately two-thirds of these infections are in people between the ages of fifteen and twenty-five. The chances of contracting all these diseases, which spread through direct contact between sexual organs, can be reduced by using condoms.

Bacterial Infections

Three common diseases caused by bacteria are syphilis, gonorrhea, and chlamydia. They are diagnosed by laboratory analysis through detection of the bacteria in urine and fluids from sores, as well as detection of specific antibodies in the blood. The main defense against infections is the immune system, but bacteria can also be killed by antibiotics. Unfortunately, many strains of bacterial infections are now resistant to all known antibiotics.

Figure 28.A Syphilis sores on a penis.

Syphilis The first symptoms of syphilis appear ten to ninety days after sexual contact with an infected person. Sores develop on the penis (fig. 28.A), vagina, mouth, or anus where there has been contact between the tissues. The sores are painless, do not itch, and may be so small that they are unnoticed, ignored, or mistaken for some other skin infection. They disappear in roughly two weeks, even without medical treatment, and are typically forgotten. But the syphilitic bacteria have not gone away; they have moved from the skin into the bloodstream.

The second stage of infection occurs two to six months after the sores have disappeared. By that time, a huge population of the bacteria lives within the bloodstream, causing a fever, the loss of small patches of hair, a sore throat, fatigue, and a skin rash (typically on the soles of the feet and the palms of the hands). During the third and final stage of a syphilitic infection, which may occur years later, the bacteria invade the internal organs. The eyes, heart, bones, liver, and brain may become severely damaged.

Gonorrhea and Chlamydia Gonorrhea and chlamydia are similar diseases. They are common diseases because infected people (especially women) often exhibit no symptoms and unknowingly pass the disease bacteria to others. Both gonorrhea and chlamydia can cause sterility in both men and women.

In men, the bacteria infect cells that line the urethra, causing a painful, burning sensation during urination and a discharge from the penis—yellow in the case of gonorrhea and clear in the case of chlamydia. In women, the bacteria usually infect cells that line the cervix, but they can spread to the uterus and oviducts (a condition known as pelvic inflammatory disease, or PID), and sometimes to the urethra. Symptoms in women include a thick, white discharge from the vagina; aches or pains in the lower abdomen; and discomfort when urinating—the same symptoms of other,

---

A woman estimates her ovulation date by keeping records of her body temperature, cervical mucus, or saliva. Body temperature rises right after ovulation (fig. 28.16*b*) while at the same time the cervical mucus and saliva become less thick. In this way, a woman knows her ovulation date *after* it has occurred. By keeping records she can estimate when during the succeeding months she is likely to ovulate.

The estimated date of ovulation is not a sure thing, however, since time of ovulation is influenced by emotions and illnesses. The only true safe period is from two days after ovulation (estimated from previous records of temperature, mucus, or saliva) until the next menstrual period. Home kits that measure levels of estrogen and progesterone (see fig. 28.9) in the urine are being developed to make the rhythm method more reliable.

Barrier Methods Several methods of birth control use barriers that prevent sperm from entering the cervix. Examples of barriers are condoms, diaphragms, and cervical caps.

less serious conditions. The infection often leaves scars in the oviducts, which cause sterility by blocking passage of the egg. The scars also lead to tubal pregnancies (a form of ectopic pregnancy), in which an embryo develops in the oviduct because a fertilized egg cannot pass through the oviduct to the uterus. Many women do not know they have had gonorrhea or chlamydia until they try unsuccessfully to get pregnant.

Viral Infections

Three sexually transmitted diseases caused by viruses are herpes, genital warts, and AIDS. They are diagnosed by laboratory analysis of fluids from the infected cells and blood. There are no medical cures for these viral diseases; recovery depends entirely on the immune system.

Herpes Genital herpes are painful blisters that eventually break open and spread the infection. They resemble cold sores but are caused by different viruses: herpes simplex virus-1 generally produces cold sores around the mouth, while herpes simplex virus-2 generally causes sores on the genitals. Neither type of virus always confines itself to one place; oral sex can spread the virus from one place to another. Genital herpes spreads readily.

A first attack of genital herpes usually occurs four to eight days after exposure. Early symptoms include itching and burning at the site of in-fection. In men, painful sores develop on the penis and lining of the urethra; in women, they develop on the vagina and surrounding skin. The sores may last a week or more. The virus may invade the bloodstream and spread throughout the body, causing fever, muscle aches, and swollen lymph nodes.

Once a person has been infected with herpes, the virus never leaves the body. It remains dormant in the nerves, erupting and spreading to new cells from time to time. It is mainly during these active periods that the virus spreads from one person to another. Anyone with active sores should refrain from sex until the sores have healed. Women with a history of this disease need to be carefully monitored during pregnancy, since the baby can become infected during passage through the vagina at the time of birth. Herpes in babies is serious; it causes blindness, mental retardation, and even death. If active sores are present at the time of childbirth, the baby may be delivered by Caesarean section (through an incision in the abdomen).

Warts Genital warts, caused by the human papillomavirus (HPV), are so small and flat they are nearly invisible. For reasons not yet understood, the warts are particularly common in young women. An infected person may have no visible signs of the disease but can pass it to their sexual partners. People infected with certain strains of this virus are more likely to develop cancer of the cervix or penis.

AIDS The most deadly disease transmitted by sexual contact is acquired immune deficiency syndrome (AIDS). In this disease, human immunodeficiency virus (HIV) attacks the white blood cells, interfering with normal defenses against infectious disease (described in chapter 22).

HIV spreads from person to person through body fluids, mainly semen and blood but also vaginal secretions and perhaps saliva. Infection requires direct contact between the body fluids, as when the semen of an infected person contacts a cut or sore (from another sexually transmitted disease) in the vagina, mouth, or rectum of another person.

You cannot tell by looking at someone whether they are carrying HIV; a person may have it for as long as ten years before symptoms of AIDS appear. When symptoms do occur, they include swollen lymph nodes, persistent fever, weight loss, and infections (often seen as white patches) of the skin, mouth, and vagina. As the immune system weakens, the person with AIDS becomes increasingly more susceptible to all kinds of infectious diseases and to cancer.

The standard condom is a latex sheath placed over the erect penis prior to intercourse. A new kind of condom, made of plastic, fits inside a woman's vagina. In either case, ejaculation can occur within a woman's vagina but sperm cannot enter her cervix. A tremendous advantage of the condom is that it also prevents the spread of sexually transmitted diseases, by preventing direct contact between the man's penis (and semen) and the woman's vagina and cervix. These diseases are described in the accompanying Wellness Report.

A diaphragm is a rubber cap placed over the cervix to block entry of sperm into the uterus. It is inserted into the vagina prior to intercourse and kept in place for at least eight hours afterward. A cervical cap is like a diaphragm but smaller and stronger. Both must be used with a spermicide (a chemical that kills sperm). A woman needs to be fitted by a physician for either of these devices and to be refitted whenever she gains or loses 4.5 kilograms (10 lb) or more.

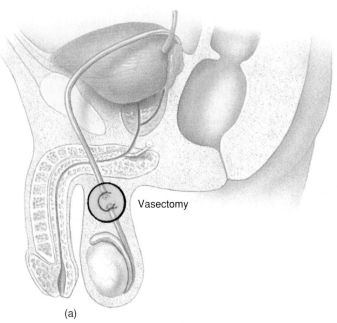

(a)

(b)

Figure 28.17 Sterility: methods of permanent birth control. (*a*) A vasectomy severs the vasa deferentia on both sides of the scrotum. The cut ends are tied, clipped, or burned shut. (*b*) A tubal ligation severs the oviducts on both sides. The cut ends are tied off. The operation prevents sperm from traveling to the fertilization site and eggs from traveling to the uterus.

Sterilization A person who no longer wants to procreate may have a vasectomy or a tubal ligation. Both procedures are considered permanent and are difficult to reverse.

A vasectomy is a surgical operation, shown in figure 28.17*a*, in which the vasa deferentia of a man are blocked. Surgical incisions are made in the vasa deferentia near the epididymides, and the severed ends are tied, cut, or burned shut. This simple operation, under local anesthetic, is performed in a doctor's office. A man who has had a vasectomy continues to secrete testosterone, and therefore to maintain his masculinity, as well as the fluid portion of semen. He looks the same and has both erections and ejaculations. The only difference is that his semen no longer contains sperm. If a man wants a vasectomy but thinks he may want to father a child later in life, he may have some of his sperm frozen and stored before the operation.

A tubal ligation is a surgical operation, shown in figure 28.17*b*, in which the oviducts are cut and tied. It severs the connection between the uterus and the fertilization site; sperm cannot reach the unfertilized eggs, and eggs cannot reach the uterus. The surgery is performed through incisions in the abdominal wall or vagina. A woman who

has had a tubal ligation can still bear children by using a technique that bypasses the blocked oviducts. This procedure is described in the accompanying Research Report.

Birth-Control Pill Birth-control pills, or oral contraceptives, were developed in the late 1950s. They provide a woman with a simple way of controlling her own reproduction without devices, such as condoms and diaphragms, that interfere with spontaneity.

The birth-control pill contains synthetic versions of the female sex hormones, estrogen and progesterone. The dosages are not particularly large, but they maintain fairly constant levels of these two hormones during the menstrual cycle. Without hormonal fluctuations, ovulation does not take place. Without ovulation, there are no eggs in the oviducts that can be fertilized.

A new way of administering oral contraceptives is an implant, called Norplant, inserted beneath the skin of the upper arm. It slowly releases a synthetic version of progesterone, which blocks ovulation and thickens the cervical mucus so that sperm cannot penetrate. The implant prevents pregnancy for five years but can be removed any time the woman wants to become pregnant.

Test-Tube Babies

Many women cannot become pregnant because their oviducts never developed correctly or have become blocked by tubal ligations or scars. In vitro fertilization (*in vitro*, in glass) is a technique that enables these women to become pregnant by bypassing the blocked oviducts: the man's sperm and the woman's egg are united in a shallow dish rather than in the oviduct. After fertilization, the tiny embryo is placed inside the woman's uterus, where it develops normally. Babies developed from in vitro fertilization are known as test-tube babies. The procedure, which has produced more than 20,000 babies, is diagrammed in figure 28.B.

In vitro fertilization was made possible by the invention of a medical instrument called the laparoscope. It is a long metal tube, about as thick as a pencil, with a light and a tiny telescopic lens connected to a television screen. Inserted through a small incision in the abdominal wall, the laparoscope gives a magnified video picture of a woman's ovaries so that mature eggs can be located.

The procedure usually begins by administering a fertility drug to the woman so that several eggs mature simultaneously. Hormone levels in the blood are then charted to predict when she will ovulate. The tiny eggs, each no larger than a pencil dot, are located through the laparoscope. They are extracted from the ovary through a hollow needle and stored in a fluid that mimics conditions inside the oviducts.

Sperm are collected from the man (by masturbation) and mixed with the eggs in the dish. Within two days, the sperm fertilize the eggs. The fertilized eggs develop into embryos of approximately eight cells each.

One or more of the tiny embryos is sucked into a plastic tube (fig. 28.C), which is inserted into the woman through her vagina. The embryos are deposited in the uterus or in part of an oviduct near the uterus. If all goes well, an embryo embeds in the uterus and develops in a normal fashion. The other embryos in the dish are frozen

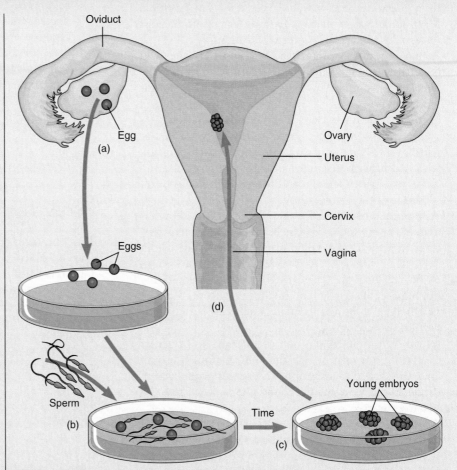

Figure 28.B Preparing a test-tube embryo. Before this procedure, the woman is treated with hormones to stimulate maturation of eggs in her ovaries. (*a*) Mature eggs are removed from the ovary (through a small incision in the abdomen) and placed in a dish. (*b*) Sperm taken from a man (by masturbation) are added to the dish. (*c*) The fertilized eggs develop into embryos. (*d*) An embryo (or several) is placed in the uterus (or an oviduct near the uterus) where it continues its development.

for later use, in case the embryo in the uterus fails to develop.

Surprisingly, the woman who donates the egg does not have to be the same as the woman who receives the embryo. Women well beyond menopause, who can no longer produce eggs, have received embryos. They have given birth to babies developed from their husband's sperm but another woman's eggs. The procedure has also been used by women who can produce eggs but miscarry early in their pregnancies. These women have had their eggs removed and fertilized by their husband's sperm in a dish. The embryo was then implanted in another woman who eventually gave birth.

In vitro fertilization will probably never become common, since it is ex-

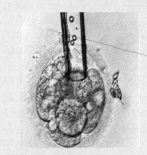

Figure 28.C An eight-celled embryo being transported by a small tube.

pensive and does not guarantee success. The chance of producing a baby on the first try is between 30% to 60%, depending on the age and condition of the mother as well as the number of embryos implanted.

SUMMARY

1. In sexual reproduction, a sperm and an egg join to form a fertilized egg, which develops into an offspring. Most aquatic animals have external fertilization; most terrestrial animals have internal fertilization.

2. A female forms eggs and places them where they can be fertilized. Some females also nourish and shelter the developing young. A woman's reproductive system includes ovaries, which develop eggs and secrete sex hormones; oviducts, which serve as fertilization sites and pathways from the ovaries to the uterus; a uterus, which nourishes and shelters the fetus; a vagina, which drains menstrual fluid, receives sperm, and serves as birth canal; and a clitoris, which swells in response to sexual stimulation.

3. A woman's menstrual cycle, which lasts approximately twenty-eight days, is highlighted by menstruation, in which the endometrium is shed, and ovulation, when a mature egg moves from an ovary to an oviduct. The cycle is governed by follicle-stimulating hormone (FSH), which stimulates growth of a follicle and its egg; luteinizing hormone (LH), which converts the growing follicle into a gland, converts the ruptured follicle into a corpus luteum, and triggers ovulation; estrogen, which thickens the endometrium and stimulates LH secretion; and progesterone, which thickens the endometrium and suppresses FSH and LH secretion.

4. A male forms sperm and delivers them to the site of fertilization. A man's reproductive system includes two testes and a system of tubes. The testes produce sperm and secrete testosterone. The tubes include the epididymides, which store sperm; and the vasa deferentia and urethra, which transport sperm out of the body. Three glands—the seminal vesicles, prostate, and bulbourethral glands—contribute fluids to the semen. Mature sperm form every day within a man.

5. Male reproduction is governed by follicle-stimulating hormone, which governs sperm development; luteinizing hormone, which governs testosterone production; and testosterone, which governs sperm development, erection, and ejaculation.

6. Sperm travel through both male and female reproductive tracts to reach an egg. Within the man, sperm travel from the epididymides through the vasa deferentia and urethra and into the vagina. The seminal vesicles, prostate, and bulbourethral glands add secretions along the way. Within the woman, sperm swim from the vagina through the cervix and uterus and on into the oviducts.

7. Fertilization occurs within the outer third of an oviduct. The sperm attaches to molecules in the zona pellucida, triggering the release of enzymes that digest this layer. When the first sperm contacts the egg, its nucleus merges with the egg nucleus. The two sets of chromosomes come together within the fertilized egg.

8. Most methods of birth control prevent fertilization. The rhythm method prevents sperm from being in the oviduct at the same time as an egg. Barrier methods (condoms, diaphragms, and cervical caps) prevent sperm from reaching the cervix. Sterilization methods (vasectomies and tubal ligations) prevent sperm from reaching the egg. The birth-control pill prevents ovulation.

KEY TERMS

Bulbourethral (BUL-boh-yur-EE-thral) **glands** (page 535)
Cervix (SUR-viks) (page 529)
Clitoris (KLIH-tor-us) (page 529)
Corpus luteum (KOR-pus LOO-tee-um) (page 529)
Endometrium (En-doh-MEE-tree-um) (page 529)
Epididymis (EH-pih-DIH-dih-mus) (page 533)
Follicle (FAH-leh-kul) (page 528)
Ovary (OH-vur-ee) (page 527)
Oviduct (OH-veh-dukt) (page 529)
Ovulation (AH-vyoo-LAY-shun) (page 530)
Prostate (PRAH-stayt) (page 535)
Scrotum (SKROH-tum) (page 533)
Seminal vesicle (SEH-mih-nul VEH-sih-kul) (page 535)
Seminiferous (SEH-min-IH-fur-us) **tubule** (page 533)
Testes (TES-teez) (page 532)
Urethra (yur-EE-thrah) (page 533)
Uterus (YOO-tur-us) (page 529)
Vagina (vah-JEYE-nah) (page 529)
Vas deferens (vaz DEH-fur-ens) (page 533)

STUDY QUESTIONS

1. Draw the lower abdominal region of a woman, using a front view, showing the positions of the cervix, ovaries, uterus, vagina, and oviducts. Mark with an X the place where fertilization takes place.

2. Draw the lower abdominal region of a man, using a side view, showing the positions of the testis, bladder, prostate, urethra, vas deferens, and epididymis.

3. Describe what happens in a woman during menstruation. Describe what happens during ovulation.

4. Where are the eggs of a woman produced? How do they get to the place where sperm can fertilize them? Where are the sperm of a man produced? Trace the pathway taken by the sperm on their way to the site of fertilization.

5. List two functions of the testes. List two functions of the ovaries.

6. What is the male sex hormone? List four of its effects on the body. What are the two female sex hormones? For each hormone, list three of its effects on the body.

CRITICAL THINKING PROBLEMS

1. Compare the investments, in terms of energy and nutrients, of a woman and a man in their fertilized egg.

2. Present a hypothesis to explain the fact that a woman stops producing eggs when she is about fifty years old, whereas a man produces sperm all his life.

3. Modern technology allows women who are beyond menopause to have children. What do you think about older women having children?

4. New techniques using in vitro fertilization allow a couple to choose their baby's sex. Present arguments, both for and against, the widespread use of this technique.

5. A thirty-year-old mother of two children is considering having her tubes tied (a tubal ligation) to prevent further pregnancies. She is worried, however, that the procedure will induce menopause. Will it? Why or why not?

6. Some animals are hermaphrodites, with each individual forming both eggs and sperm. A hermaphrodite does not mate with itself. Instead, two hermaphrodites mate and each receives sperm from the other. Under what conditions would natural selection favor such a mating system? (Hint: snails are hermaphrodites.)

SUGGESTED READINGS

Introductory Level:

Discover, June 1992. An issue devoted mainly to sexual reproduction—origin, sexual development, traits of sperm and egg, orgasm, the role of love, and the paradox of hermaphrodites.

Maxwell, K. *The Sex Imperative: An Evolutionary Tale of Sexual Survival*. New York: Plenum Publishing, 1994. A discussion of the wide variety of mating systems among organisms and how each system adapts the particular organism to its environment.

"The Other Sexual Diseases," *University of California, Berkeley Wellness Letter* 6, no. 7 (April 1990): 4–5. A review of what we know about sexually transmitted diseases other than AIDS.

Riddle, J. M., and J. W. Estes. "Oral Contraceptives in Ancient and Medieval Times," *American Scientist* 80 (May–June 1992): 226–33. A fascinating account of herbal medicines used by women in ancient societies to prevent pregnancy.

Ridley, M. *The Red Queen: Sex and the Evolution of Human Nature*. New York: Macmillan, 1994. A discussion of why humans reproduce sexually, with interesting ideas about the advantages of having genetically variable offspring.

Advanced Level:

Austin, C. R., and R. V. Short (eds.). *Reproduction in Mammals I: Germ Cells and Fertilization*. New York: Cambridge University Press, 1972. A detailed description of the formation of eggs and sperm, female cycles, and fertilization in various mammals.

Daly, M., and M. Wilson. *Sex and Behavior: Adapting for Reproduction*. Belmont, CA: Wadsworth Publishing, 1978. A comparison of how different kinds of animals reproduce.

Wasserman, P. M. "Fertilization in Mammals," *Scientific American*, December 1988, 78–84. A description of the research that led to the discovery of species-specific receptors in the zona pellucida and sperm.

❉ EXPLORATION

Interactive Software: AIDS

This interactive exercise allows students to explore the risk factors associated with the AIDS epidemic by varying key aspects of behavior and policy. The exercise presents a world map on which appear charts of the incidence of AIDS in that geographical region by year and projections of the future state of the epidemic. Students can investigate the past history of the epidemic and explore the potential future consequences of altering patterns of sexual behavior and drug use, use of safe sex, and level of public education. The potential for explosive future growth of the epidemic in Asia will become evident as future projections are explored in different world regions.

The miracle of development: two fertilized eggs have developed into two unique individuals.

Development

Chapter Outline

Objectives

In this chapter, you will learn
–how a fertilized egg becomes a mass of
  many cells;
–how the mass of cells forms tissues
  and organs;
–how each cell becomes specialized in
  structure and function;
–the way in which a human embryo
  and fetus develop;
–the methods for testing a fetus for
  abnormalities;
–how a baby is born.

Every human begins life as a single cell no larger than a pencil dot. How does this cell, the fertilized egg, transform itself into a child of some 200 billion cells? How do the cells organize themselves into a brain, eyes, heart, arms, fingernails, and other perfectly formed structures?

We cannot actually watch the development of a child, since it happens inside the mother. We can, however, follow the development of other animals, such as frogs and sea urchins, where development proceeds outside the mother's body. To watch a single cell transform itself into a perfectly formed organism is an awesome experience. First, the fertilized egg divides many times to form a hollow ball of cells. Some of the cells then stream into the hollow interior and begin to form their particular organs. Cells at one end form a brain and sense organs, and other cells become heart, liver, lungs, and digestive tract. Every organ appears in the right place at the right time.

Each animal builds itself according to instructions within its genes. The fertilized egg of a frog contains recipes for building another frog; the fertilized egg of a human contains recipes for building another human. As the egg divides to form many cells, each cell selects only a few of its recipes to use—a neuron uses one set of recipes and a muscle cell uses another set. The mystery is how each cell chooses which recipes to use—how it "knows" what it is supposed to be.

Animal Development

Development transforms a fertilized egg into an adult organism. Most animals release nutrient-rich eggs into the surrounding environment and the young develop outside the mother's body. Sea urchins, frogs, and most fishes release unfertilized eggs into the water, where they are fertilized and then develop into self-feeding young. Insects lay fertilized eggs in the soil, on leaves, and under bark, where they develop into immature forms that can feed on their own. The young of reptiles and birds develop inside large eggs enclosed within protective shells.

The females of some animals retain their eggs inside their bodies, where the eggs are fertilized and the offspring develop. Internal development occurs in a few kinds of fishes and insects as well as most mammals. When a certain level of development has been reached, the young are born. The length of time that an offspring develops in its mother, called the gestation time, is shown for various mammals in table 29.1.

Table 29.1

Gestation Times for Some Mammals

Animal	Gestation Time* (days)
Elephant	625
Giraffe	410
Horse	340
Human	266
Chimpanzee	237
Dog	63
Cat	60
Rabbit	30
Mouse	19

*Gestation time is the number of days from fertilization to birth.

The early stages of animal development are remarkably similar. These stages and how they form are described following.

Formation of Many Cells

Fertilization initiates development. The egg begins to divide as soon as it is penetrated by a sperm, forming many smaller cells that become building blocks for tissues and organs.

At first the cells divide by a special form of mitosis (see page 128) known as **cleavage,** which is cell division without cell growth. Cleavage divides the fertilized egg into a mass of smaller cells, the early embryo, which is no larger than the original egg. Cleavage of a frog egg is illustrated in figure 29.1.

The pattern of cleavage depends on how much yolk (stored nutrients) the egg contains. Where there is yolk, it is unevenly distributed within the early embryo so that portions with more yolk divide more slowly than portions with little or no yolk. The amount of yolk is related to how long the embryo relies on these stored nutrients. Reptiles and birds, which develop completely within shelled eggs, lay eggs with lots of yolk. Fishes and amphibians, which quickly develop into self-feeding young, produce eggs with moderate amounts of yolk. Mammals, which develop inside their mother and receive nutrients from her, form eggs with no yolk.

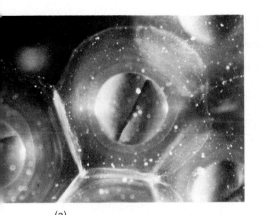

(a)

(b)

(c)

Figure 29.1 Cleavage of the fertilized egg. A frog egg, shown here, has small amounts of yolk, and each cleavage nearly cuts the cell in half. (*a*) The first cleavage divides the fertilized egg into two cells. (*b*) The second cleavage divides the egg into four cells. (*c*) The third cleavage results in an eight-celled embryo. (*d*) The fourth cleavage yields sixteen cells. (*e*) The fifth cleavage results in thirty-two cells and the sixth cleavage, shown here, produces a sixty-four-celled embryo.

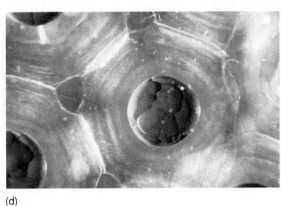

(d)

(e)

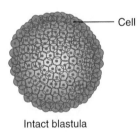

Cell

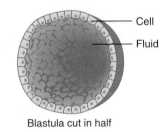

Cell

Fluid

Intact blastula   Blastula cut in half

(a)

Figure 29.2 The blastula. (*a*) A sea urchin egg has hardly any yolk and its early blastula consists of sixty-four cells all about the same size. The interior of the blastula contains fluid but no cells. (*b*) A chicken egg contains large amounts of yolk. Its blastula forms as a tiny structure that rests on top of the huge yolk supply. The chick blastula is a hollow disc rather than a sphere.

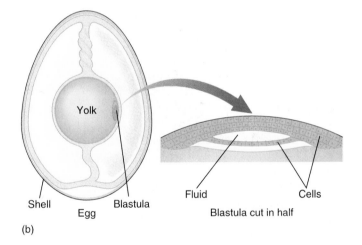

Yolk

Shell   Blastula
        Egg

Fluid   Cells

Blastula cut in half

(b)

Cleavage transforms the egg into a hollow ball of cells called the **blastula** (see fig. 29.1*e*). This early form of embryo, with between 64 and 1,000 cells, is no larger than the original egg. In some animals, the blastula is a sphere, whereas in others it is shaped more like a disc (fig. 29.2). Except for differences in shape, the blastula gives no indication of what kind of animal it will become.

Formation of Tissues

Cells of the blastula are all the same. The next step in development is to form a **gastrula,** which is an embryo with three layers of cells. These cells then become specialized in structure and function as they form a **neurula,** which is an embryo with a rudimentary nervous system.

The Gastrula   The blastula develops into a gastrula, with three layers of cells as well as a top, bottom, front, and rear. The transformation is brought about chiefly by migrations of cells from the outer surface of the blastula into its hollow interior:

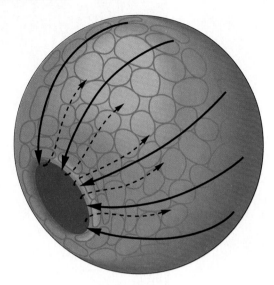

The particular pattern of cell migration depends on how much yolk the embryo carries. It is described here for an embryo with little yolk.

Formation of a gastrula begins when cells that form one side of the blastula migrate into the hollow interior. The new structure resembles a ball with one end pressed inward:

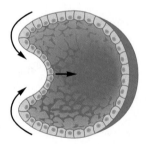

There are now two layers of cells: an outer layer, called the **ectoderm** (*ecto*, outer; *derm*, skin), and an inner layer, called the **endoderm** (*endo*, inner). The indented region is called the **blastopore:**

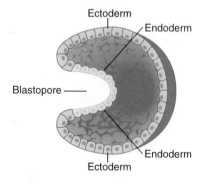

Table 29.2
Fates of the Embryonic Tissues in a Mammal

Tissue	What It Forms in the Adult
Ectoderm	Nervous system, lens of eye, inner ear, epidermis of the skin, hair, fingernails, toenails, adrenal medulla
Mesoderm	Skeletal muscles, skeleton, dermis of the skin, heart, blood vessels, outer coverings of internal organs, kidneys, ovaries or testes, adrenal cortex
Endoderm	Linings of internal channels (digestive, urinary, reproductive, respiratory), liver, pancreas, glands

Then more cells stream inward, between the ectoderm and endoderm near the blastopore, to form a middle layer called the **mesoderm** (*meso*, middle):

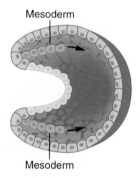

The embryo is now a gastrula with three layers of cells. It also has a rough body plan with a front, rear, top, and bottom. In vertebrate animals, the blastopore will become the rear end. In most other animals, the blastopore will become the head end.

Each cell layer will go on to form only certain kinds of structures. The ectoderm will form, among other things, the nervous system and skin. The mesoderm will develop into muscles and bones. The endoderm will form the lining of the digestive tract. A more complete list of the structures that develop from these three embryonic tissues is provided in table 29.2.

The three layers of cells then develop into organs characteristic of the species. In vertebrate animals, the first organs that form are the brain and spinal cord; the gastrula develops into a neurula.

The Neurula   In a vertebrate animal, the gastrula develops into a neurula, which is an embryo with a rudimentary

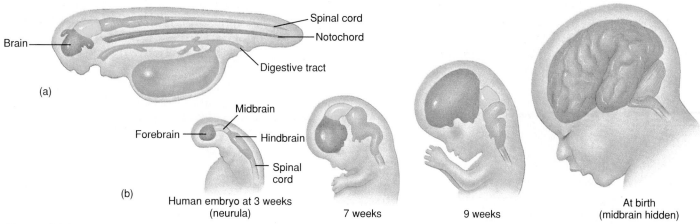

Figure 29.3 **The neurula.** All vertebrate animals form the same kind of neurula. (*a*) An early neurula, showing the rudimentary brain, spinal cord, notochord, and digestive tract. (*b*) Development of the human brain from the neurula stage to birth.

brain and spinal cord (fig. 29.3*a*). The transformation involves additional migrations of cells, as described following.

Development of a neurula begins when some of the mesodermal cells come together to form a **notochord**—a flexible rod that runs the length of the embryo and provides support:

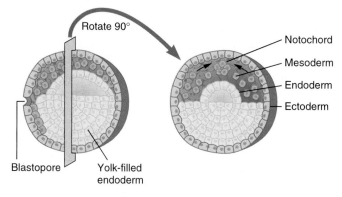

(The notochord is replaced, later in development, by the vertebrae.)

Next, a flat sheet of ectodermal cells on the upper surface of the embryo sinks downward to form a groove that runs the length of the embryo above the notochord:

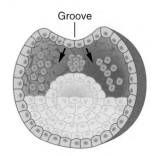

The groove becomes a tube as its edges migrate toward each other and fuse:

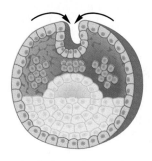

The tube then sinks below the surface of the embryo, where it lies beneath a sheet of cells that eventually form the skin on the back of the animal.

One end of the tube is broader than the rest: it develops into the brain. The narrow portions form the spinal cord (see fig. 29.3*a*). (Failure of the groove to form a tube results in spina bifida, in which the vertebrae do not completely enclose the spinal cord.) Development of the human brain is shown in figure 29.3*b*.

The remaining mesodermal cells (those not involved in formation of the notochord) organize themselves into blocks of tissue called **somites.** There are two rows of somites, one on each side of the notochord (fig. 29.4). They will form the bones and muscles of the back as well as other internal structures.

As the neurula develops, cells become committed to form particular parts of the body. The body elongates and develops a head, trunk, tail, limbs, and the initial forms of internal organs.

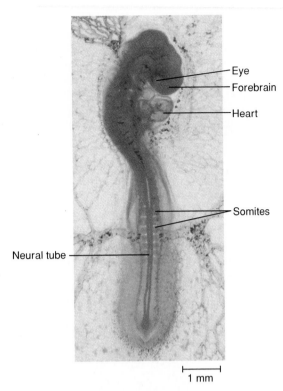

Eye
Forebrain
Heart
Somites
Neural tube

1 mm

Figure 29.4 Somites. A vertebrate animal develops somites while it is a neurula. These blocks of mesoderm occur in pairs, one on each side of the developing spinal cord. They will eventually form the muscles and bones of the back as well as certain other internal structures.

Development of Adult Form   In fishes and amphibians, the embryo hatches out of its membranes shortly after becoming a neurula. A fish then gradually grows into its mature form. A frog, however, remains fishlike for a short period of time and then changes dramatically, a process called metamorphosis, into the adult form.

Other vertebrate embryos (reptiles, birds, and mammals) continue to develop within the protected environment of a shelled egg or a mother's uterus. When the adult pattern of tissues and organs is established, a young bird or reptile hatches and a young mammal is born.

Differentiation of Cells

All cells within an embryo carry the same set of genes. As development proceeds, these genetically identical cells become different in structure and function. This process of cell specialization, called **differentiation,** involves turning on some of the genes and turning off others. Genes contain instructions for building proteins, so that genes that are turned on have their proteins synthesized. The particular proteins of a cell—especially its enzymes—govern its particular activities.

Cellular differentiation gives rise to the many different kinds of cells: skin cells, muscle cells, blood cells, nerve cells, and others. All cells in a human, for example, contain genes that direct the synthesis of antibodies, but these genes are only active within certain kinds of white blood cells. In all other cells of the body, the genes for antibodies remain inactive.

Embryonic Induction   What causes a cell to turn on certain genes and leave others turned off? How does it "know" what it is to become? We do not know the answers to these questions. We know only that the way a cell differentiates depends on where it is located within the embryo. Apparently, the cells in one part of the embryo differentiate first and then control how adjacent cells differentiate. This process, in which one group of cells governs differentiation of an adjacent group, is known as **embryonic induction.** It apparently involves cell-to-cell communication by means of chemical messengers. The classic experiment that revealed this process is described in the accompanying Research Report.

Each cell is not committed to a particular structure until sometime during the gastrula stage. Cells in an early gastrula are not differentiated. When cells that normally form a leg, because of their position in the embryo, are transferred to the interior of the early gastrula, they will develop into part of the digestive tract rather than a leg. In an older gastrula, the cells are differentiated. The same group of cells will form a leg no matter where they are placed in the embryo.

The Homeobox   The basic plan of an animal is laid out early in the blastula stage. A set of genes, known as the **homeobox** (*homeo,* alike) **genes,** controls which part of the body becomes the front, middle, and rear. We are just learning about this fascinating set of genes. In flies, homeobox genes tell the cells in various parts of the embryo what kind of structure to make. With regard to appendages, for example, the genes tell cells in the head to form antennae and cells in the thorax to form legs.

The homeobox genes control development in an indirect fashion: they contain instructions for a set of proteins, called the **homeobox,** that control which genes are turned on and which genes remain turned off in each cell. In the head of a fly, for example, the homeobox turns on genes that develop antennae; in the thorax, it turns on genes that develop legs.

Researchers have found, to their astonishment, that the homeobox appears the same in all animals with a head-to-tail body plan. The sequence of bases within the DNA of homeobox genes, as well as their arrangement within chromosomes, is virtually identical whether the animal is an insect, a worm, or a person. (The genes they turn on, of course, are different: the heads of an insect, a worm, and a person have different features.) The fundamental head-to-tail organization, which is laid down very early in embryonic development, is governed by the same set of genes.

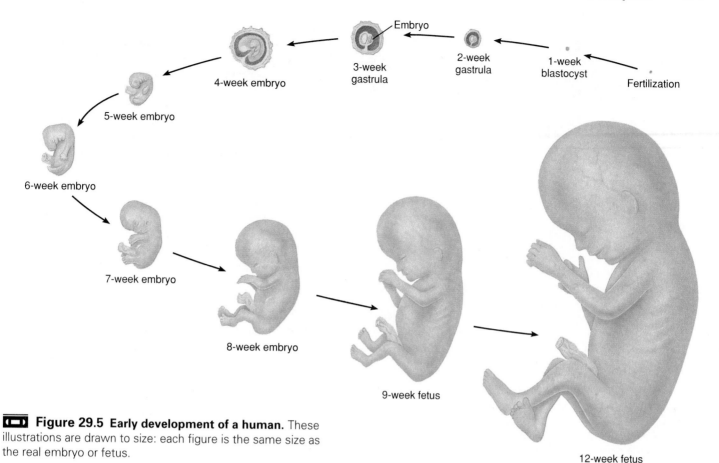

Figure 29.5 Early development of a human. These illustrations are drawn to size: each figure is the same size as the real embryo or fetus.

This genetic similarity implies that homeobox genes are ancient, having appeared in some of the earliest animals (approximately 700,000 years ago) and been passed from generation to generation since that time. It is strong evidence that animals evolved from a common ancestor.

Human Development

As development proceeds, it becomes more and more specific to the kind of animal being formed. All animals have similar development through the gastrula. All vertebrate animals have similar development through the neurula. After that, developmental patterns diverge as they form organs unique to each kind of animal. We describe here the specific development of a human.

It takes nine months to develop a human baby. All major organs form during the first two months, a period of time when the developing child is known as an **embryo.** The following seven months, during which the developing child is called a **fetus,** are mainly a time of growth of already formed organs. Major events in human development are shown in figure 29.5 and listed in table 29.3. They are described in more detail following.

Formation of the Blastocyst

The egg begins to cleave as soon as it is fertilized, forming many smaller cells from the single large one. As it cleaves, the embryo travels along the oviduct toward the uterus (fig. 29.6). It receives nutrients from cells that adhere to it (originally part of the follicle in the ovary) and fluid within the oviduct. By the fourth day the embryo has sixteen cells and enters the uterus. There it drifts freely for several days as it continues to divide and begins to grow.

At the sixty-four-cell stage, the embryo becomes a blastula, which in mammals is called a **blastocyst.** It consists of a fluid-filled cavity and two groups of cells: an inner cell mass and a trophoblast:

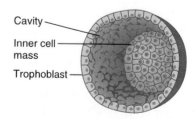

The **inner cell mass** is the young embryo; it will go on to become a gastrula and neurula. It will also form some of the extraembryonic membranes, described following,

551

Research Report

The Discovery of Induction

One of the first scientists to probe the miracle of development was Hans Spemann. This German embryologist, who received a Nobel Prize in 1935, was particularly interested in finding out when and how the cells of a young embryo become committed to becoming a certain type of specialized cell. At what point is the developmental pathway of cells fixed?

Working with salamanders, Spemann observed that when he separated a blastula into two fragments, both fragments went on to form two complete salamanders. The developmental pathways were apparently not yet fixed; half of a blastula can form a complete salamander. When he separated a gastrula, however, neither fragment formed anything at all. The developmental pathways were fixed; half a gastrula can no longer form a salamander. Some irreversible change occurs within embryonic cells between the time they are part of a blastula and the time they are part of a gastrula.

To pinpoint the exact time that cells become set on a particular pathway, Spemann divided embryos into fragments at various times as the blastula becomes a gastrula. He was surprised to find that one particular fragment was crucial to development: a fragment located just above the blastopore (the indented area where cells migrate inward) always formed a complete embryo, whereas other fragments never formed embryos:

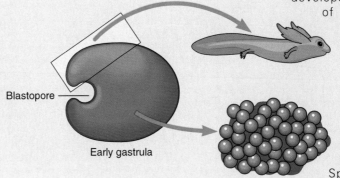

Blastopore

Early gastrula

Spemann named this special region of the embryo the dorsal lip of the blastopore. From these observations, he developed the following hypothesis: Cells within the dorsal lip of the blastopore organize the development of the embryo; they initiate a sequence of changes that transforms unspecialized cells into specific tissues and organs. He predicted that if a gastrula were given a second dorsal lip of the blastopore, then it would develop into two animals rather than one.

To test his hypothesis, Spemann and his student, Hilde Mangold, transferred a dorsal lip of the blastopore from one early gastrula to another (fig. 29.A). (They used gastrulas from two species of salamander—one with dark cells and the other with light cells—so they could distinguish between the transplanted cells and the host cells.) The host embryo, with both its own dorsal lip and the transplanted one, developed into a sort of Siamese twin with two brains, spinal cords, and associated organs. The experiment confirmed the hypothesis. The dorsal lip of the blastopore is now known as Spemann's organizer and the phenomenon, in which one group of cells influences how another group differentiates, is known as induction. Since Spemann's work, induction has been found to govern many other aspects of development.

What is the physical basis for induction? Many hypotheses have been offered, but most biologists now believe it must be by means of chemical messengers. Cells in the dorsal lip of the blastopore, for example, must secrete chemical messengers that attach to receptors on nearby cells, thereby turning on certain genes that guide the cells along a particular pathway of specialization. Such messengers, however, have been hard to find. In the

that surround the embryo. The **trophoblast** forms only extraembryonic membranes.

The cells of the inner cell mass are loosely held together. If they should, by accident, separate into two groups, then identical twins may develop. They carry the same genes, since they are descendants of the same fertilized egg. (Fraternal twins, by contrast, develop from two fertilized eggs; see page 537.) The chance of having identical twins is about 1 in 200.

While identical twins form by accident, we now have the technical expertise to produce such multiple births deliberately. In this procedure, known as cloning, the cells of an early embryo (at the thirty-two-cell stage or earlier) are separated from one another and allowed to develop on their own. At this early stage of development, each cell has the potential to form a complete individual. Such cloning has been carried out with various mammals, including sheep, but has never been carried to completion in a human.

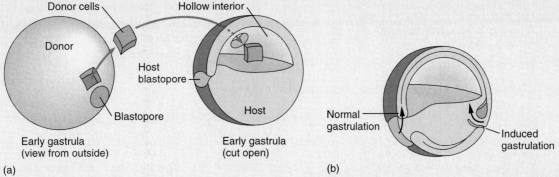

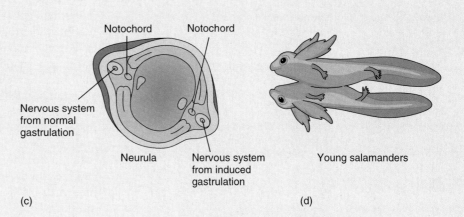

Figure 29.A The experiment that first demonstrated induction.
(a) Cells just above the blastopore were cut out of one early gastrula (the donor, shown in red) and placed in the hollow interior of another early gastrula (the host). (b) Cells near the host's own blastopore and its donor blastopore (shown in red) migrated inward to form two gastrulas. (c) Both gastrulas developed into neurulas: the blastopore cells in both regions induced formation of notochords, which induced formation of nervous systems. (d) The host embryo developed into two young salamanders attached to each other.

early 1990s, researchers isolated a protein, named noggin, that when applied in large doses is capable of inducing the ectoderm of a gastrula to develop into part of the brain. This particular induction occurs naturally during formation of a neurula. (The protein was named *noggin* because very high doses produce embryos with exceptionally large heads, or "noggins.") While it remains to be shown conclusively that noggin is an inducer, the question still remains: Why, in spite of so much research effort, are these chemical messengers so hard to find?

Implantation

While the fertilized egg develops into a blastocyst, the uterus prepares to receive it. In the ovary that produced the egg, the corpus luteum (see page 529) secretes estrogen and progesterone. (A corpus luteum forms from a follicle after its egg has been released.) Estrogen and progesterone stimulate the endometrium (the inner lining of the uterus) to thicken, to secrete mucus (a sticky fluid), and to form more capillaries in preparation for attachment of the embryo.

One week after fertilization, the blastocyst begins to burrow into the thickened endometrium. This process, known as **implantation,** takes about a week to complete (see fig. 29.6). The embedded embryo then receives nutrients from the endometrium.

The Extraembryonic Membranes The blastocyst develops four **extraembryonic membranes** (*extraembryonic,* outside the embryo) as it implants into the endometrium. They are the amnion, yolk sac, allantois,

Table 29.3

Major Events of Human Development

Time after Fertilization	Event
First week	Cleavage of the fertilized egg forms a blastocyst
Second week	Implantation into endometrium; formation of extraembryonic membranes; formation of gastrula
Third week	Formation of extraembryonic membranes; secretion of human chorionic gonadotropin; formation of neurula; formation of placenta
Weeks 4 through 8	Formation of all organs, including the umbilical cord
Months 3 and 4	Growth; formation of ears and eyelids; movements of arms, legs, and head; development of facial features; secretion of oily coat onto skin
Months 5, 6, and 7	Growth to about 1 kilogram (2 lb); development of body hair; perfection of organ function
Months 8 and 9	Growth to about 3 kilograms (6.5 lb); development of sensory responses; deposition of fat beneath skin; swallowing and sucking movements

and chorion (fig. 29.7). These membranes are described briefly in table 29.4 and here in detail.

The **amnion** surrounds the embryo. Its function is to hold a clear fluid, called the amniotic fluid, next to the embryo. The fluid absorbs shocks, protecting the embryo against jolts (a fall rarely causes a miscarriage).

The **yolk sac** gives rise to certain cells that become part of the embryo. It is named for its function, which is to hold yolk in the eggs of reptiles and birds. In mammals, the yolk sac does not contain yolk. Instead, it forms embryonic blood cells as well as the cells that eventually develop into eggs or sperm.

The **allantois** extends between the embryo and the endometrium. It forms blood vessels within the umbilical cord.

The **chorion** encloses everything that develops from the fertilized egg: embryo, amnion, yolk sac, and allantois. It forms part of the placenta, described following, and secretes **human chorionic gonadotropin** (HCG), a hormone that stimulates the corpus luteum to continue secreting estrogen and progesterone, which, in turn, maintain the thickened endometrium. Without HCG, secretion of these hormones would cease and menstruation would begin, shedding the embryo along with the endometrium. (The abortion pill, RU-486, works by blocking progesterone.)

HCG produces the unpleasant symptoms of early pregnancy: nausea, vomiting, food cravings, and mood swings. Some HCG is excreted in the urine and can be used as a test of pregnancy as early as two weeks after conception. Home pregnancy kits consist of antibodies that bind with HCG in the urine. If a woman is pregnant, the antibody-HCG complex appears as small clumps or a color change in the test solution.

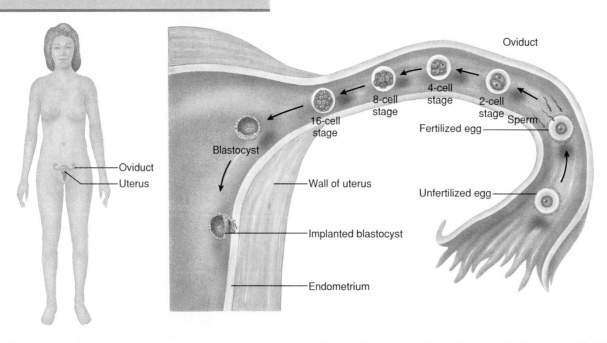

Figure 29.6 Journey of the young embryo. The egg is fertilized in the outer third of the oviduct. It then begins to cleave and to move down the oviduct toward the uterus. It reaches the uterus when it has sixteen cells, roughly four days after fertilization, and then floats within the uterine fluid for three days as it develops into a blastocyst. Seven days after fertilization, the blastocyst burrows into the inner lining (endometrium) of the uterus.

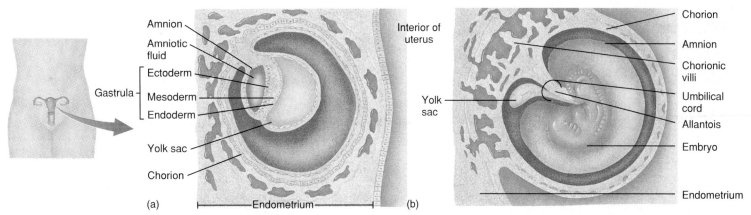

Figure 29.7 The extraembryonic membranes. (*a*) Three weeks after fertilization, the blastocyst has become a gastrula (consisting of ectoderm, mesoderm, and endoderm) with four extraembryonic membranes: the amnion, yolk sac, allantois, and chorion. (*b*) Four weeks after fertilization, the embryo has rudimentary organs. The region of the chorion next to the uterus has formed fingerlike projections, called chorionic villi, into the endometrium.

Table 29.4

The Extraembryonic Membranes

Membrane	Function
Amnion	Holds fluid that cushions embryo
Yolk sac	Produces the first blood cells; forms cells that later become sperm or eggs
Allantois	Forms blood vessels of umbilical cord
Chorion	Encloses embryo and its other membranes; forms part of placenta; secretes human chorionic gonadotropin

The portion of chorion next to the endometrium will form part of the placenta and umbilical cord (described following). The rest fuses with the amnion to form a single membrane around the amniotic fluid.

The Placenta   Soon after implantation, a network of blood vessels forms between the embryo and the mother's circulatory system. Projections of the chorion (called villi) extend into the endometrium (fig. 29.7b). Blood vessels from the embryo then grow into the projections, where they contact pools of blood from maternal blood vessels. The two sources of blood do not mix: they are kept separate by the walls of the embryonic blood vessels. Materials diffuse between the mother's blood and the embryo's blood through the walls of the embryonic blood vessels. This system of embryonic blood vessels, pools of blood, and maternal blood vessels becomes the **placenta** (fig. 29.8).

The placenta serves as a lung, intestine, and kidney for the developing embryo. Oxygen, glucose, amino acids, and other nutrients move from the mother's blood into the embryo's blood. Carbon dioxide and other wastes move from the embryo's blood into the mother's blood.

Only certain materials can pass through the walls of the embryonic blood vessels. Small materials, such as amino acids, glucose, fatty acids, oxygen, and antibodies pass freely from the mother's blood to the embryo. Larger materials, such as red blood cells and bacteria, do not. Unfortunately, alcohol, drugs, and viruses also move from the mother's blood into her developing child. When a pregnant woman drinks alcohol, for example, the drug travels through her bloodstream to the placenta. There it diffuses into the embryo's cells and interferes with normal cellular activities. The resulting deformities, known as fetal alcohol syndrome, include mental retardation, small head, and deformed face. Cocaine also has serious effects, as described in the accompanying Wellness Report.

In addition to connecting the embryo with its mother's bloodstream, the placenta functions as a gland. At first, the portion of the placenta formed from the chorion secretes human chorionic gonadotropin, which stimulates the corpus luteum to continue its secretion of estrogen and progesterone. Later, during the third month of pregnancy, the placenta stops secreting human chorionic gonadotropin (causing degeneration of the corpus luteum) and begins to secrete estrogen and progesterone itself.

Development of the Embryo

The embryo changes dramatically as it implants into the endometrium. The blastocyst, which is a mass of identical cells, develops into a gastrula with three distinct layers of

Wellness Report

Cocaine Babies

Jonathan is almost a year old but appears only six months old. He is small, immature, and irritable and has not yet learned to sit up on his own. His parents are not even sure he can see. Jonathan is a cocaine baby.

Cocaine babies began to appear in large numbers in the 1980s and now, in some cities, represent one out of every ten newborns. The number of affected embryos is even greater, since coke users have an extraordinarily high incidence of miscarriage.

How does cocaine affect the developing baby? It triggers an extreme constriction of certain arteries. In a pregnant woman, the arteries that carry blood to the placenta become so narrow that very little blood flows to the unborn child. Cells in the baby's brain may die from lack of oxygen, and tiny arteries in its brain may burst (a stroke). These changes can leave the infant with permanent physical and mental damage. In at least one case, a

child was born completely paralyzed on one side (as in a stroke in an older person); the child's mother admitted that while she was pregnant, she had celebrated her wedding anniversary with 5 grams of coke (an average-sized snort is less than 0.1 gram).

At least a third of the cocaine circulating in a pregnant woman's blood gets into the bloodstream of her unborn child. There it affects physical development of the baby. Cocaine babies have smaller heads than normal, as well as malformed kidneys, urinary tracts, and lungs. They develop more slowly. Exposure to cocaine also affects the chemistry of the brain. It seems to interfere with the biological clock, generating a sort of "jet lag" that makes the newborn restless and irritable. And, of course, cocaine babies are cocaine addicts. They suffer agonizing withdrawal symptoms after they are born.

Cocaine babies are difficult to care for. They cry incessantly and do not like to be touched. Nor can they tolerate bright light or loud sounds. These

responses interfere with mother-child bonding and may eventually lead to problems adjusting to family and social life. As toddlers, they have shorter attention spans and are more impulsive than normal.

What happens to these children as they grow older depends in part on their childhood environment. If they are neglected, as is likely if they remain with addicted parents, they may never learn simple tasks or develop normal coordination. They will form a "lost generation." If they grow up in a nourishing environment, with proper food and loving attention, they may overcome most of their physical and mental problems.

The terrible injustice done to these babies has generated public anger at women who use drugs while pregnant. In some states, these women can be convicted and sent to prison for passing drugs to their babies through the umbilical cord. Regardless of the law, a pregnant woman should act in a responsible manner toward her unborn child.

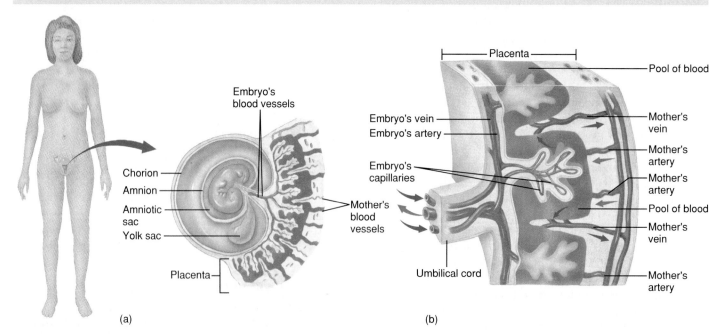

(a)   (b)

Figure 29.8 The placenta. (a) The placenta forms within the endometrium, where the embryo is implanted. (b) Close-up of the placenta at a later stage of development. Arteries and veins within the umbilical cord connect the embryo's circulatory system with the endometrium. Arteries and veins within the uterus connect the mother's circulatory system with the endometrium. In the endometrium, maternal blood leaves the blood vessels and forms pools around the embryonic capillaries. Materials diffuse between these pools of blood and the embryo's blood through the walls of the embryo's capillaries.

cells (see fig. 29.7a) during the second week of development (before a pregnant woman has missed her first period). Then, during the third week, the gastrula becomes a neurula with a rudimentary brain and spinal cord. The basic pattern of a human gradually emerges.

Organ Formation   All major organs form during the first two months of development. They include the nervous system, sense organs, skeleton, muscles, heart, testes or ovaries, liver, kidneys, lungs, and digestive tract.

During the fifth week, the embryo begins to form a pair of organs that later become either testes or ovaries. If the embryo has inherited a Y chromosome (see page 167) from its father, then the organs develop into testes. If it has not inherited a Y chromosome (i.e., has inherited two X chromosomes), then the organs develop into ovaries.

As the sixth week begins, the embryo is still so tiny it could fit on a quarter. Its head comprises half its length (fig. 29.9). The face begins to emerge, as do the fingers and toes. Portions of the extraembryonic membranes and placenta form the **umbilical cord,** which contains blood vessels that carry materials between the embryo and its mother. Another function of the cord is to suspend the embryo away from the placenta, where there is more room to grow.

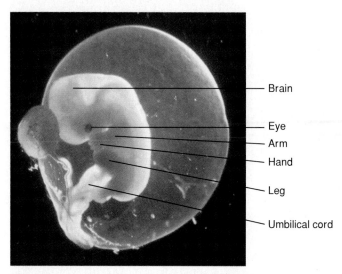

Figure 29.9 A human embryo at six weeks. The entire embryo is less than 3 centimeters long. Its face, hands, and feet are already forming.

During the eighth week, a male's testes begin to secrete testosterone, which causes development of a male reproductive tract. In the absence of this hormone, a female reproductive tract develops. These alternative developmental patterns are illustrated in figure 29.10.

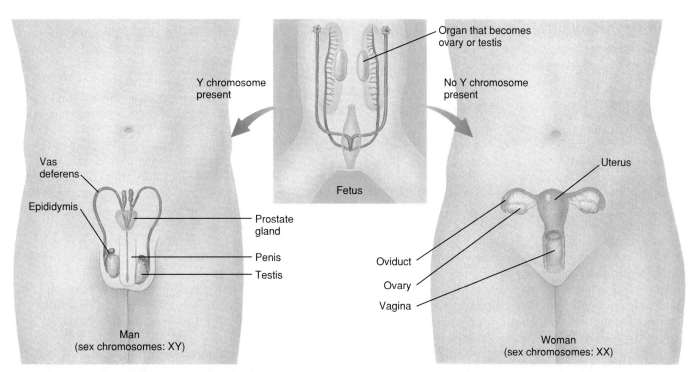

Figure 29.10 Development of male and female reproductive organs. During the fifth week, the embryo forms a pair of organs that will develop into either testes or ovaries. If the embryo is a male, genes within its Y chromosome convert the organs into testes. If the embryo is a female, with no Y chromosome, the organs develop into ovaries. A few weeks after the testes have formed, they begin to secrete testosterone, which governs development of the male reproductive system. In the absence of testosterone, a female reproductive system develops.

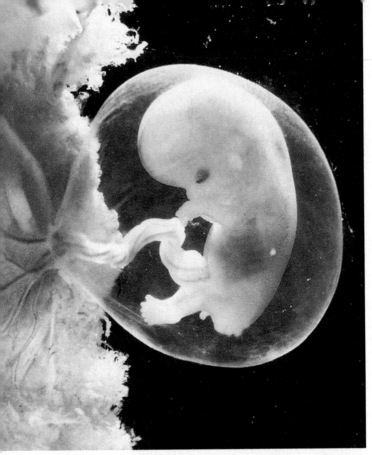

Figure 29.11 A human fetus, nine weeks after conception. Still tiny—less than 4 centimeters long—the fetus is attached to its mother's circulatory system by an umbilical cord.

By the end of the eighth week (second month) of pregnancy, the most significant aspects of development have taken place and the embryo looks like a human (fig. 29.11). It is still very small—less than 4 centimeters in length.

Protecting the Embryo   The first two months of pregnancy are the most crucial period, when any disturbance may deform the embryo (fig. 29.12). Virtually all birth defects arise during this early stage of pregnancy. Many women, however, do not know they are pregnant at this time and so do not adjust their life-style to ensure normal development.

If the embryo does not receive all the nutrients essential for development, then certain organs will fail to develop. If the embryo receives toxic materials through its mother's blood, then certain organs may be damaged. Approximately half of all pregnancies end in miscarriages (spontaneous abortions) during the first two months of development. Most miscarried embryos have gross abnormalities caused by genetic defects, drugs, or alcohol.

Development of the Fetus

By the end of the second month, most of the organs have formed but the embryo is still tiny. The remaining seven months are devoted to growth and perfection of details. After the second month, the embryo is called a fetus.

By the end of the third month, the fetus weighs about 28 grams (an ounce) and is 8 centimeters long. It has well-formed ears and eyelids and begins to move its arms, kick its legs, and nod its head. The face develops during the fourth month. An oily coating covers the skin, protecting the fetus from abrasions and helping to maintain a constant body temperature. This coating remains on the skin until birth, when it is washed off.

During the fifth and sixth months, the fetus acquires antibodies against infections to which its mother has developed immunity. Its skin becomes covered with fine hair, which disappears before birth. By the end of the sixth month, the fetus reacts to sound; some music, for example, has a calming effect and other music stimulates kicking.

At seven months, the fetus weighs about 1 kilogram (2 lb) and has a fair chance of surviving outside the uterus. By eight months, it develops a layer of fat beneath the skin and can perceive light (if born prematurely). It swallows about half a liter of the amniotic fluid every day and may suck its thumb, activities that prepare it for feeding after birth. Most of the swallowed fluid is absorbed into the fetal bloodstream, but some accumulates in the intestines and is defecated shortly after birth.

Examining the Fetus

Will the baby be healthy? Every parent worries that something may have gone wrong—that the baby inherited a disorder or was damaged during development. The concern is not unrealistic: nearly one child in twenty is born with some abnormality. These worries can be dealt with, in part, by examining the fetus.

Tests for Genetic Defects   More than 300 genetic disorders can be detected before a child is born. Some of them are listed in table 29.5. Samples of fetal cells are removed, as described following, and grown in the laboratory. Too many or too few chromosomes, which cause such debilitating conditions as Down syndrome, are easily spotted. More subtle genetic defects, involving individual genes, are detected by examining the DNA directly or the proteins it codes for. There is always a risk, however, that the technique used to sample fetal cells will cause a miscarriage.

Genes of a fetus can be examined by chorionic villi sampling as early as the ninth week of development. In this procedure, which is performed in a physician's office, a long tube is inserted through the vagina and into the uterus (fig. 29.13a). Position of the tube is guided by images formed by ultrasound. A sample of cells from the chorion, where it forms part of the placenta, is suctioned into the tube. These cells, genetically identical to the cells within the fetus, are examined in the laboratory. The risk of miscarriage from the procedure is about 1%. There is also an increased risk (one chance in 2,900 procedures) that the procedure will cause deformities of the fingers or toes.

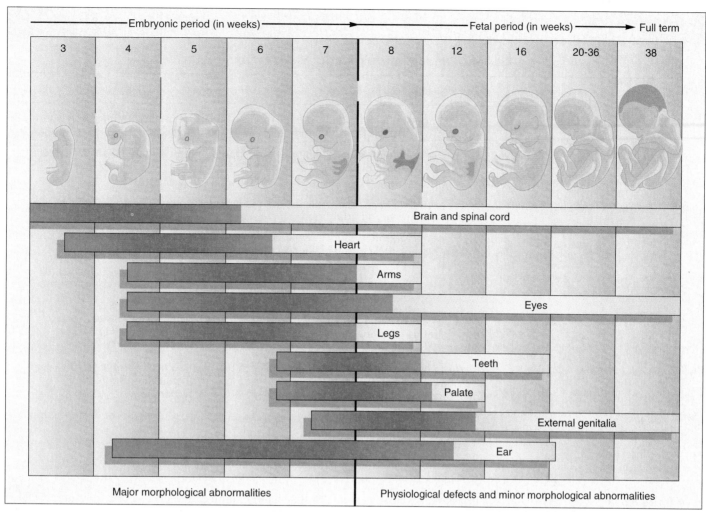

Figure 29.12 Vulnerability of the embryo and fetus to toxic materials. The developing baby is most vulnerable to alcohol, drugs, viruses, and other toxic materials during its first eight weeks (two months) of development. The red bars indicate highly sensitive periods; the yellow bars indicate less sensitive periods.

Another technique, called amniocentesis, can be performed in a doctor's office after the twelfth week of pregnancy (fig. 29.13*b*). A sample of the amniotic fluid is withdrawn through a needle inserted into the mother's abdomen. The fluid contains cells that have sloughed off the fetus. The cells are grown in the laboratory and examined for abnormal chromosomes and gene products. The risk of a miscarriage from this procedure is less than 1%.

Amniocentesis can be combined with in vitro fertilization (see page 541) to test for genetic defects *before* pregnancy. In this relatively expensive procedure, an egg is removed from the ovary and fertilized by a sperm within a laboratory dish. When the young embryo consists of eight cells, one of the cells is removed and analyzed for genetic defects. If no defect is found, the embryo is transferred to the uterus or oviduct for further development.

Tests for Physical Defects   After the fifth month of development, the fetus can be viewed through an instrument called a fetoscope. A thin tube is inserted through the abdominal wall and into the uterus. Details of the fetus can be seen through optical fibers within the tube. External features, such as fingers, eyes, ears, mouth, and genitals, can be examined directly and samples of the skin and blood can be removed for laboratory analysis. The risk of a miscarriage from this procedure is about 5%.

Ultrasound imaging, a very safe way to look at the fetus, can be done after the third month. Sound waves are passed into the uterus, where they bounce off solid objects and form an outline of the fetus. Gross abnormalities can be detected, as well as whether it is a boy or a girl and the presence of more than one fetus. If the fetus is sucking its thumb, then the image will show whether it is left-handed or right-handed.

Correcting Defects   What happens if the tests reveal serious abnormalities? A few conditions, such as urinary tract obstructions and hydrocephalus (accumulation of fluids inside the skull) can be surgically corrected while the child is still in the womb. Some genetic disorders may eventually

Table 29.5

Genetic Defects and Procedures for Detection

The Defect	The Effects
Alpha$_1$ anti-trypsin deficiency	Enzyme deficiency that can lead to cirrhosis of the liver in early infancy and pulmonary emphysema and degenerative lung disease in middle age.
Alpha thalassemia	Severe anemia that reduces ability of the blood to carry oxygen. Nearly all affected infants are stillborn or die soon after birth.
Beta thalassemia (Cooley's anemia)	Severe anemia resulting in weakness, fatigue, and frequent illness. Usually fatal in adolescence or young adulthood.
Cystic fibrosis	Body makes too much mucus, which collects in the lungs and digestive tract. Children don't grow normally and usually don't live beyond age twenty.
Down syndrome	Minor to severe mental retardation caused by an extra twenty-first chromosome.
Duchenne's muscular dystrophy	Fatal disease found only in males, marked by muscle weakness. Minor mental retardation is common. Respiratory failure and death usually occur in young adulthood.
Fragile-X syndrome	Minor to severe mental retardation. Symptoms, which are more severe in males, include delayed speech and motor development, speech impairments, and hyperactivity. Considered one of the main causes of autism.
Hemophilia	Excessive bleeding affecting only males. In its most severe form, can lead to crippling arthritis in adulthood.
Neural tube defects Anencephaly	Absence of brain tissue. Infants are stillborn or die soon after birth.
Spina bifida	Incompletely closed spinal canal, resulting in muscle weakness or paralysis and loss of bladder and bowel control. Often accompanied by hydrocephalus, an accumulation of spinal fluid in the brain, which can lead to mental retardation.
Polycystic kidney disease	Infantile form: enlarged kidneys, leading to respiratory problems and congestive heart failure. Adult form: kidney pain, kidney stones, and hypertension resulting in chronic kidney failure. Symptoms usually begin around age thirty.
Sex chromosome abnormality	Minor to severe developmental and learning disabilities, caused by missing X or extra X or Y chromosome.
Sickle-cell anemia	Deformed, fragile red blood cells that can clog the blood vessels, depriving the body of oxygen. Symptoms include severe pain, stunted growth, frequent infections, leg ulcers, gallstones, susceptibility to pneumonia, and stroke.
Tay-Sachs disease	Degenerative disease of the brain and nerve cells, resulting in death before the age of five.
Trisomy 13 (Patau syndrome)	Severe mental retardation and heart, kidney, and other organ defects, caused by the presence of an extra thirteenth chromosome. Usually fatal soon after birth.
Trisomy 18 (Edwards syndrome)	Severe mental retardation and heart defects, caused by presence of an extra eighteenth chromosome. Usually fatal soon after birth.

*Cannot predict severity unless otherwise indicated.
†Chorionic villi sampling
Compiled by Valerie Fahey, *In: Hippocrates,* May/June 1988, 68–69.

Who's at Risk	Tests and Their Accuracy*	What Can Be Done
1 in 1,000 Caucasians	Amniocentesis, CVS† accuracy varies, but can sometimes predict severity	No treatment
Primarily families of Malayan, African, and Southeast Asian descent	Amniocentesis, CVS accuracy varies; more accurate if other family members tested for gene	Frequent blood transfusions
Primarily families of Mediterranean descent	Amniocentesis, CVS 95% accurate	Frequent blood transfusions
1 in 2,000 Caucasians	Amniocentesis, CVS accuracy varies; more accurate if other family members tested for gene	Daily physical therapy to loosen mucus
1 in 350 women over age thirty-five; 1 in 800, all women	Amniocentesis, CVS nearly 100% accurate	No treatment
1 in 7,000 male births	Amniocentesis, CVS 95% accurate	No treatment
1 in 1,200 male births; 1 in 2,000 female births	Amniocentesis, CVS 95% accurate	No treatment
1 in 10,000 families with a history of hemophilia	Amniocentesis, CVS 95% accurate	Frequent transfusions of blood with clotting factors
1 in 1,000	Ultrasound, amniocentesis 100% accurate	No treatment
1 in 1,000	Ultrasound, amniocentesis test works only if the spinal cord is leaking fluid into the amnion or is exposed and visible during ultrasound	Surgery to close spinal canal prevents further injury; shunt placed in brain drains excess fluid and prevents mental retardation
1 in 1,000	Infantile form: ultrasound 100% accurate Adult form: amniocentesis 95% accurate	Kidney transplants
1 in 500	Amniocentesis, CVS nearly 100% accurate	Hormonal treatments to trigger puberty
1 in 500 blacks	Amniocentesis, CVS 95% accurate	Painkillers, transfusions for anemia, antibiotics for infections
1 in 3,000 Eastern European Jews	Amniocentesis, CVS 100% accurate	No treatment
1 in 20,000; women over age thirty-five have increased risk	Amniocentesis, CVS, ultrasound nearly 100% accurate	No treatment
1 in 8,000; women over age thirty-five have increased risk	Amniocentesis, CVS, ultrasound nearly 100% accurate	No treatment

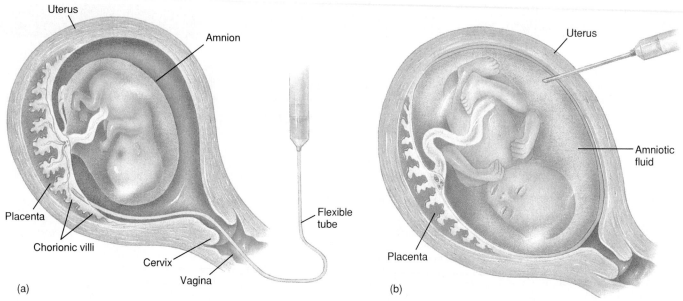

Figure 29.13 Taking a peek at the fetus. Cells from the fetus can be sampled and analyzed for genetic defects. (*a*) In chorionic villi sampling, cells from the chorion are removed by a tube inserted through the vagina. The procedure is guided by images from an ultrasound machine. (*b*) In amniocentesis, sloughed-off cells in the amniotic fluid are removed by a needle inserted through the abdominal wall.

Table 29.6		
Modern Methods of Abortion		
Name of Method	**Time after Conception**	**Technique**
RU-486 pill followed by prostaglandin* pill	First seven weeks	Menstruation is induced by blocking progesterone and stimulating uterine contractions
Menstrual evacuation	First three months	Endometrium is removed by a vacuum instrument
Induced labor	After three months	Premature labor is induced by prostaglandins

*Prostaglandins are chemical messengers that stimulate contractions of the uterus.

be corrected by gene therapy (📖 *see page 223*) on the fetus. Even when treatment of the fetus is not possible, it is useful to be aware of a defect so that treatment can begin as soon as the baby is born.

Unfortunately, most abnormalities cannot be treated. If the embryo is severely defective, the prospective parents may decide to abort it (table 29.6). Procedures for induced abortions within the first two months of pregnancy are relatively simple, safe, and inexpensive.

Childbirth

Approximately nine months (266 days) after conception, the baby emerges from the mother. Strangely, it is the baby who initiates this procedure. When ready to enter the outside world, the baby's brain signals its pituitary gland to secrete a hormone (oxytocin) that acts on the placenta and uterus in ways that initiate birth.

Emergence of the Baby When the baby is ready to be born, it moves downward within the uterus (fig. 29.14*a*), while at the same time the cervix moves upward. The opening in the cervix begins to widen in preparation for passage of the baby.

As the cervical opening enlarges, a plug of mucus that is tinged with blood is expelled through the vagina. This plug has blocked the opening during pregnancy, preventing disease microbes in the vagina from traveling through the cervix and reaching the fetus in the uterus.

Labor typically begins a few days after expulsion of the plug. It consists of rhythmic contractions of muscles within the uterine wall. Labor lasts an average of fourteen hours in a woman giving birth for the first time and less for a woman who has previously had a baby. The contractions occur once every twenty minutes at first and become more frequent as the time of birth approaches. Contractions serve two functions: to widen the opening

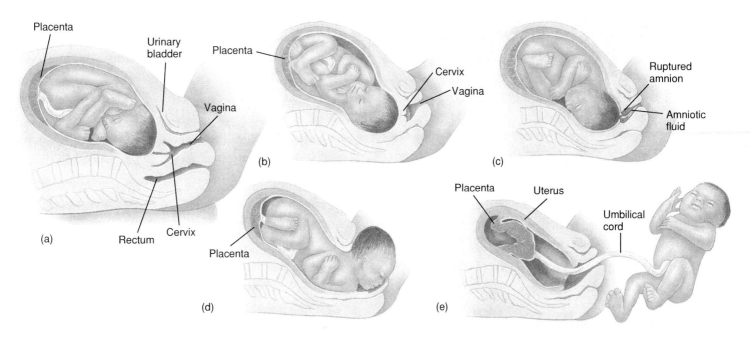

Figure 29.14 Childbirth. (*a*) In most pregnancies, the baby's head comes to rest on the cervix when it is ready to be born. (*b*) As the cervix widens, the baby's head acts as a wedge to hold it open. (*c*) As the time of birth nears, the amnion ruptures and amniotic fluid drains out through the vagina.

(*d*) The baby turns its head as it moves through the cervix and vagina. (*e*) The newborn remains connected, by its umbilical cord, to the placenta within the uterus. The cord is cut and the placenta is shed shortly after the baby is born.

of the cervix and to push the baby out. The baby's head usually presses against the cervix, acting as a wedge to help widen its exit route (fig. 29.14*b*). At some point during labor, the amnion ruptures, an event known as "breaking water," and the amniotic fluid drains out through the vagina (fig. 29.14*c*).

When a contraction occurs every minute, the baby is about to be born. Its head, which normally comes first, takes the longest to pass through the cervix and vagina (often taking as long as half an hour). Once the head has come out (fig. 29.14*d*), the rest of the body slips out easily.

During birth, the baby continues to receive oxygen through the umbilical cord, which is now about 60 centimeters (24 in) long and still attached to the placenta within the uterus. If the head does not come first (a breech delivery), the umbilical cord becomes compressed between the baby's head and the mother's cervix and vagina. The compression blocks the flow of blood, thereby depriving the baby of oxygen and possibly causing brain damage.

Emergence of the Placenta   Approximately fifteen minutes after the baby is born, the placenta is expelled (fig. 29.14*e*). This blood-filled organ is about 23 centimeters (9 in) in diameter and 2.5 centimeters (1 in) thick—about the size and shape of a pie. After the placenta has detached, its hormones (progesterone and estrogen) no

longer enter the mother's bloodstream. The sudden drop in these hormones stimulates her breasts to produce milk.

Detachment of the placenta leaves a raw, bleeding wound on the inner wall of the uterus. The danger of an infection just after childbirth is considerable. Sterile procedures during childbirth and antibiotics afterward greatly reduce this danger. The wound heals within several weeks.

The Newborn   Immediately after birth, blood vessels within the umbilical cord constrict to prevent blood flow between the infant and mother. The cord is cut at this time (fig. 29.15) and the ends tied off. This severs the connection with the mother and forces the newborn to adjust to life on its own.

While developing in the uterus, the baby received oxygen and eliminated carbon dioxide through its mother's blood rather than through its own lungs. Once the umbilical cord is cut, the baby is no longer connected to its mother's bloodstream and must adjust immediately to breathing air. When the newborn takes its first breath, which happens when it begins to cry, its lungs inflate for the first time. The circulatory system then adjusts so that blood flows from the heart to the lungs and then back to the heart, as it does in an adult.

The baby is born with a sucking response, and the natural way for it to receive nourishment is to suckle milk

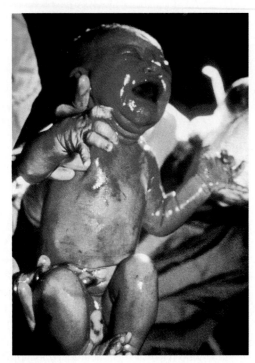

Figure 29.15 The newborn. The newborn infant is coated with an oily material, which is washed off, and attached to a long umbilical cord, which is cut and tied.

from its mother's breasts. Breast-feeding is a characteristic of all mammals, and the milk of each species is adapted to the nutritional needs of its own young. The composition of human milk is uniquely suited to the needs of the growing human infant. Human milk is superior to cow's milk, which is used in a formula to make "milk" for bottle-feeding. Moreover, human milk is more easily digested than cow's milk and contains antibodies, white blood cells, and other materials that protect against diarrhea and other human diseases. In addition, mother's milk is inexpensive, warm, sterile, and conveniently packaged.

The mother benefits from breast-feeding as well. Her uterus shrinks more rapidly to its normal size and she can return to her normal weight more rapidly, since she expends more than 1,000 calories a day in milk production. Suckling relaxes both the mother and the baby and is important in developing a bond between them. In addition, it suppresses menstruation and ovulation, acting as a natural (although not a highly reliable) form of birth control.

SUMMARY

1. Most animals develop in the external environment but some develop inside their mothers. The early stages of development are similar in all animals. The fertilized egg divides by cleavage, forming many smaller cells that become a blastula. Cells of the blastula migrate inward to form a gastrula with three tissues: ectoderm, mesoderm, and endoderm.

2. In vertebrate animals, a gastrula becomes a neurula when ectodermal cells migrate inward to form a neural tube. After the neurula, development varies from one kind of vertebrate to another.

3. As the animal develops, its cells become differentiated—specialized in structure and function. The process occurs as a series of inductions. Homeobox genes, which govern the basic body plan, are very similar in all animals with a head-to-tail orientation.

4. In a human, the blastula (called a blastocyst) forms during the first week as the egg travels from oviduct to uterus. It will develop into an embryo and extraembryonic membranes.

5. Implantation of the blastocyst into the endometrium begins during the second week. As the blastocyst implants, it forms extraembryonic membranes: the amnion, chorion, yolk sac, and allantois. The chorion secretes a hormone that prevents menstruation. After implantation, a placenta forms from the chorion, embryonic blood vessels, and maternal blood vessels. Materials pass through the placenta from the mother's blood to the embryo's blood. The placenta also secretes hormones that maintain pregnancy.

6. The embryo develops from a blastocyst to a gastrula to a neurula and then forms major organs. All major organs have formed by the end of the second month, at which time the embryo becomes a fetus. Months three through nine are times of growth and perfection of details.

7. The fetus can be examined for defects prior to birth. Genetic defects are detected by chorionic villi sampling, amniocentesis, or a procedure used in combination with in vitro fertilization. Physical defects are detected by a fetoscope or ultrasound.

8. Birth of the child occurs after approximately nine months of development. Uterine contractions dilate the cervix and push the baby out. The placenta is then expelled, leaving a raw wound in the wall of the uterus. The newborn undergoes three major adjustments: its lungs open, its blood is shunted to the lungs, and it suckles milk.

KEY TERMS

Allantois (AL-an-TOH-iss) (page 554)
Amnion (AEM-nee-on) (page 554)
Blastocyst (BLAEH-stoh-sist) (page 551)
Blastopore (BLAEH-stoh-pohr) (page 548)
Blastula (BLAEH-stoo-lah) (page 547)
Chorion (KOR-ee-on) (page 554)
Cleavage (KLEE-vahj) (page 546)
Differentiation (DIH-fur-en-shee-AY-shun) (page 550)
Ectoderm (EK-toh-durm) (page 548)
Embryo (EM-bree-oh) (page 551)
Embryonic induction (in-DUK-shun) (page 550)

Endoderm (EN-doh-durm) (page 548)
Extraembryonic membranes (page 553)
Fetus (FEE-tus) (page 551)
Gastrula (GAEH-stroo-lah) (page 547)
Homeobox (HOH-mee-oh-box) (page 550)
Homeobox genes (page 550)
Human chorionic gonadotropin (kor-ee-ON-ik
    go-NAEH-doh-TROH-pin) (page 554)
Implantation (page 553)
Inner cell mass (page 551)
Mesoderm (MEH-zoh-durm) (page 548)
Neurula (NUR-yoo-lah) (page 547)
Notochord (NOH-toh-kord) (page 549)
Placenta (plah-SEN-tah) (page 555)
Somite (SOH-meyet) (page 549)
Trophoblast (TROH-foh-blast) (page 552)
Umbilical (um-BILL-IH-kul) **cord** (page 557)
Yolk sac (page 554)

STUDY QUESTIONS

1. Draw a gastrula and label the three embryonic tissues. List three structures that develop from each tissue.
2. Draw a neurula (side view) and label the brain, spinal cord, notochord, and digestive tract.
3. Discuss the kinds of materials a woman should and should not take into her body during pregnancy. Why is the embryo particularly vulnerable during the first two months?
4. Describe two techniques used to check on the genes of a fetus. How soon can they be performed?
5. Why should sterile materials be used, if possible, when assisting with childbirth? (Hint: where is an infection likely to occur?)

CRITICAL THINKING PROBLEMS

1. Explain why all animals have similar development through the gastrula and all vertebrate animals have similar development through the neurula stage.
2. Explain why, in terms of adaptations to the environment, a frog undergoes metamorphosis but a fish does not.
3. Suppose your wife is about nine months pregnant. How will you know when it is time to take her to a hospital? Suppose you live 100 miles from the nearest hospital. How will you know when it is too late to drive to the hospital? Outline what you would do to assist in the birth.
4. Parkinson's disease and juvenile-onset diabetes could be treated by fetal tissue transplants. In this procedure, tissues taken from an aborted embryo or fetus are transplanted into the brain or pancreas of the patient. Do you think fetal tissue transplants should be permitted? Why or why not?

SUGGESTED READINGS

Introductory Level:

Gilbert, S. F. (ed.). *A Conceptual History of Modern Embryology.* New York: Plenum Publishing, 1991. Easy-to-read descriptions of some of the most fascinating experiments in biology, with special emphasis on the discovery of induction and the roles played by genes in development.

Graham, J. *Your Pregnancy Companion: A Month-by-Month Guide to All You Need to Know Before, During and After Pregnancy.* Buffalo, NY: Prometheus Books, 1991. A guide to the physiological changes in each month of pregnancy. Includes advice about diet, exercise, and handling emotional changes.

Nilsson, L., A. Ingleman-Sundberg, and C. Wirsen. *A Child is Born: The Drama of Life before Birth.* New York: Dell Publishing, 1966. Description, with extraordinary color photographs, of human development.

Wolpert, L. *The Triumph of the Embryo.* New York: Oxford University Press, 1991. A detailed description, written for the nonbiologist, of development with special chapters on fingers and toes, the brain, aging, regeneration, and evolution.

Advanced Level:

Gilbert, S. F. *Developmental Biology.* New York: Plenum Publishing, 1991. A good reference on all aspects of development.

McGinnis, W., and M. Kuziora. "The Molecular Architects of Body Design," *Scientific American* (February 1994): 58–66. Description of research on homeobox genes.

Rossomando, E. F., and S. Alexander (eds.). *Morphogenesis: An Analysis of the Development of Biological Form.* New York: Dekker, 1992. Descriptions of what is known about how differentiated cells become organized into tissues and organs.

Shostak, S. *Embryology: An Introduction to Developmental Biology.* Cambridge, MA: HarperCollins Publishers, 1991. A textbook for advanced biology students that describes both classical and modern experiments and gives a modern outlook on the concepts and unsolved problems of embryology.

Wall, R. *This Side Up: Spatial Determination in the Early Development of Animals.* New York: Cambridge University Press, 1990. A detailed description of what is known about early development in all animals.

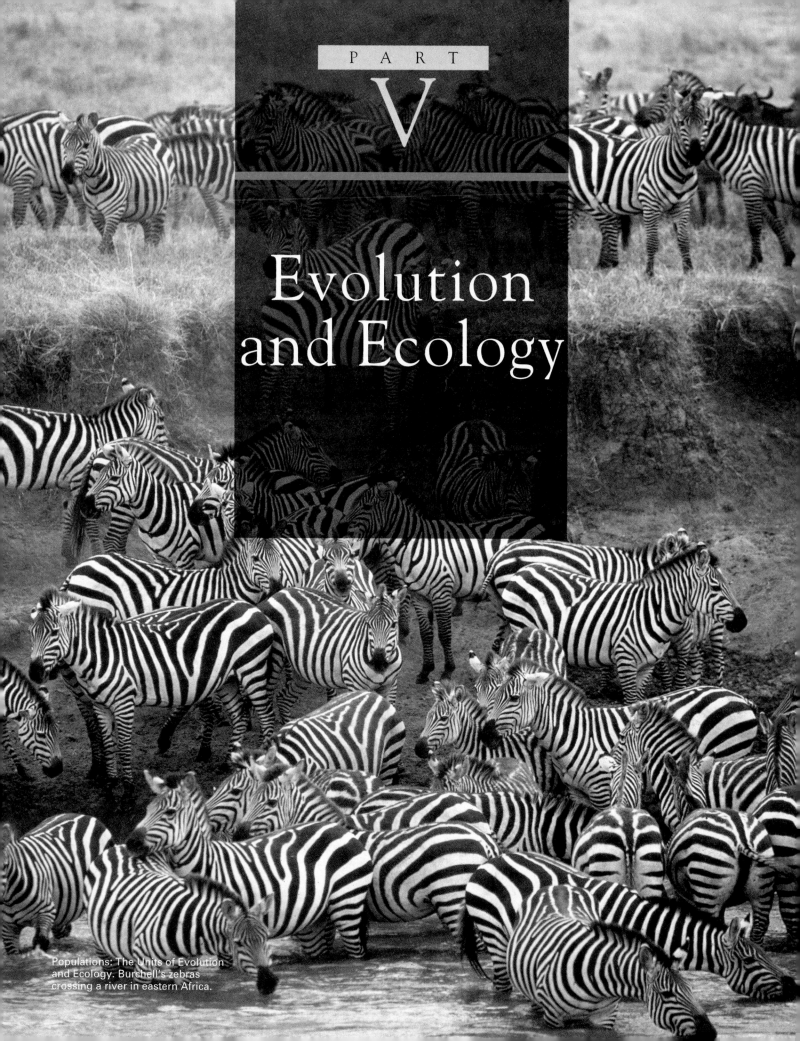

PART

V

Evolution and Ecology

Populations: The Units of Evolution and Ecology. Burchell's zebras crossing a river in eastern Africa.

Exploring Themes

Theme 1

The fundamental unit of life is the cell.

All cells are descendants of an original cell, which formed more than 3.5 billion years ago (chapter 31).

Theme 2

All cells carry out the same basic processes.

The first cells synthesized molecules and acquired energy from organic molecules in their environment. They probably built proteins according to instructions within RNA molecules and reproduced by mitosis (chapter 31).

Theme 3

The cells of a multicellular organism produce adaptations: anatomical, physiological, and behavioral traits that promote survival and reproduction.

Organisms differ because their cells carry out different specialized processes, as governed by the genes they carry. The adaptations of some individuals are better, in terms of survival and reproduction, than of others in the population (chapter 30).

Theme 4

Adaptations evolve in response to interactions with other organisms and with the physical environment.

Individuals that survive better and leave more offspring than others in the population leave more copies of their genes in the next generation; their traits become more common in the population (chapter 30). Each species has a unique pool of genes, maintained by reproductive isolation, and a unique evolutionary history (chapter 31). Interactions among organisms include competition, plant-herbivore interactions, predator-prey interactions, and parasite-host interactions (chapter 32). Interactions between organisms and their physical environment include energy flow and nutrient cycling (chapter 33). Humans dominate other organisms and alter the physical environment (chapter 34).

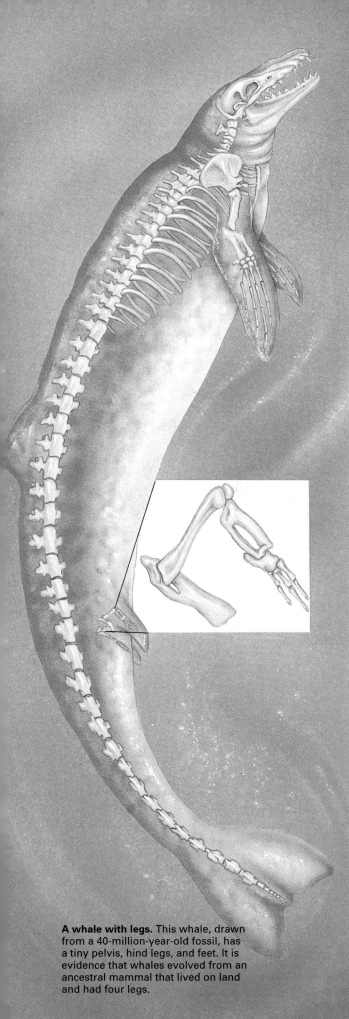

A whale with legs. This whale, drawn from a 40-million-year-old fossil, has a tiny pelvis, hind legs, and feet. It is evidence that whales evolved from an ancestral mammal that lived on land and had four legs.

The Process of Evolution

Objectives

In this chapter, you will learn
–how Darwin's theory has been
  modified by new information;
–how natural selection alters a
  population;
–the evidence that natural selection
  occurs;
–the evidence that modern species
  evolved from ancestral species;
–how new species form.

How could a whale have evolved from a mammal that lived on land? How could an animal that walked on four legs become an animal that swims with its flippers and tail? How could that animal have survived while making the transition from land to sea?

Biologists are delighted to report they now have answers to these questions. Within the past fifteen years, they have found many fossils that represent intermediate stages in the evolution of whales from their terrestrial ancestors. Four major types are described here.

The first type of fossil is of whales that lived approximately 40 million years ago (shown in the opening drawing). These animals were already quite advanced as whales: the spine was long and flexible, for undulating movements during swimming; the forelegs were modified as flippers, for steering and stabilizing the body; and the tail was modified as a fluke, for propelling the body through the water. Bones of the hind legs are present but very tiny—much too small to have supported the weight of the animal. Yet these bones are the same bones as found in terrestrial mammals. These mammals were already adapted for life in water.

The second type of fossil is of an earlier form of whale that lived about 45 million years ago. The leg bones are large enough to have supported the body on land. These whales probably spent most of their time in water, feeding on fishes, but reproduced on land.

The third type of fossil is of an even earlier whale that lived approximately 50 million years ago. This fossil, which consists of a skull, ribs, vertebrae, and limbs, is of a whale adapted to life both on land and in water. Its spine was long and flexible, with a tail not yet modified for propulsion. Its short, stubby forelimbs had limited range of motion; they probably held the body aboveground on land and functioned to steer and stabilize the body in water. The hind legs were large, similar to those of a terrestrial mammal, and could have functioned both in walking and swimming. The most extraordinary feature of this animal is the enormous hind feet, which must have provided the major propulsive force in swimming.

The fourth type of fossil is of a very primitive whale that lived 52 million years ago. Most of its fossil remains are fragments, but a well-preserved skull has been uncovered. The teeth and hearing structures within this skull are similar to those of terrestrial mammals, yet other features of the skull indicate it was a whale. The animal apparently fed on fishes in the sea yet could neither dive deeply nor hear well in water. It apparently spent much of its time on land.

These fossils are evidence that whales evolved from a terrestrial mammal. From them, we can piece together the following history. A terrestrial form of mammal began to feed on fishes in a shallow sea approximately 57 million years ago. Through the generations, the front legs became smaller and functioned as flippers while the hind feet be-

came larger and functioned to propel the animal through water. These semiaquatic mammals then evolved adaptations for diving and for hearing in water. Eventually, the hind legs and feet were replaced by a horizontal extension of the tail, known as a fluke. In this way, a terrestrial mammal became a whale.

Modern Theory of Evolution

The theory of evolution explains how the features of organisms change, through generations, in response to their environments. The process that drives evolution is **natural selection,** in which individuals with certain traits leave more offspring than individuals without these traits. According to the theory, all modern species have descended from one or a few ancestral forms. Evolutionary pathways resemble branches on a tree:

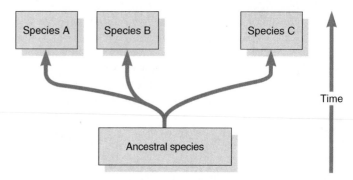

The modern theory of evolution consists of the theory published by Charles Darwin (see page 13), in 1859, as well as all relevant information and ideas generated since that time. Evolutionists continue to refine and clarify the theory but otherwise have found Darwin's ideas to be fundamentally correct.

Darwin's Theory

The theory proposed by Charles Darwin consists of four hypotheses and two predictions. Each hypothesis is a generalization based on many observations. Each prediction follows logically from the hypotheses. The theory, described more thoroughly in the introductory chapter, is outlined as follows:

Hypothesis 1: Every organism has the potential to leave more than one offspring during its lifetime. Consequently, every species has the potential to multiply in number.

Hypothesis 2: The number of individuals within each species usually remains fairly constant over time.

Prediction A: If hypotheses 1 and 2 are true, then not all individuals realize their reproductive potential. There is a struggle for existence (or reproduction).

Hypothesis 3: Individuals of a species vary in terms of their traits.

Hypothesis 4: Some of these variations are inherited.

Prediction B: If prediction A is true and hypotheses 3 and 4 are true, then individuals with certain traits are better suited to their environments and so leave more offspring than individuals with other traits. Inherited traits that promote successful reproduction become more common in future generations of the species. This process is called natural selection. Whether a particular trait promotes success or not depends on local conditions of the environment.

Modern Genetics

Darwin knew a great deal about inheritance but nothing about genes. Everything we now know about the physical basis of inheritance has been discovered since Darwin's time.

The Genetics of Populations In Darwin's theory the unit of evolution is the species, whereas in modern theory it is the population. The shift from species to populations occurred as geneticists extended their studies from organisms in the laboratory to organisms in the wild. They found that a species is not a homogenous group of individuals, with regard to their genes, but rather a mosaic of subgroups. Each subgroup, called a population, has a unique genetic composition.

A **population** is a group of individuals that can mate with one another and *are likely to do so* because they are likely to encounter one another during the reproductive season. Members of different populations of the same species, by contrast, can mate with one another but are unlikely to do so because they rarely encounter one another. Different populations may be separated by a geographic barrier, such as a mountain or river, or they may reproduce at different times or places. Figure 30.1 shows the many populations of the North American song sparrow.

New versions of a gene, called alleles (see page 158), that appear by mutation in one population tend to remain in that population, since matings between populations are rare. In addition, natural selection affects the populations in different ways, since they live in different environments. Thus, each population has a unique collection of genes, known as its **gene pool.**

This new unit of evolution, combined with a greater understanding of population genetics, led to a change in the definition of evolution. Darwin defined evolution as descent with modification, meaning a change in the traits of a species. **Evolution** is now defined as a change in the frequency of an allele within a population in response to local environmental conditions. The frequency of an allele is expressed as a percentage—its abundance in comparison with other versions of the gene in the population. In humans, for example, the frequency of the albino allele, which prevents the development of skin pigments, varies from population to population. In Norwegians it is 1%, whereas in Hopi Indians it is 7%.

The Physical Basis of Variability Since Darwin's time we have learned that inherited traits vary because individuals inherit different alleles. This genetic variability is generated by mutations, sexual reproduction, and immigration.

Mutations are accidental changes in the DNA that produce alleles. They are the original source of genetic variability. Sexual reproduction, involving meiosis (see page 144) and fertilization, then combines the alleles in a variety of ways to generate a variety of offspring. Immigration contributes to the genetic variability of a population when an immigrant brings into the population a novel allele that formed by mutation in another population. The movement of alleles, within immigrants, from one population to another, is known as **gene flow.**

Genetic variability is the raw material of natural selection. The more kinds of individuals in the population, the more likely that some will be better suited than others to changes in the environment. The opposite is also true: when all the individuals are more or less the same, the population has a poor chance of surviving an environmental change. A population loses its genetic variability when it becomes very small. When a few individuals colonize an island, for example, the population has very little genetic variability. This phenomenon is known as the "founder effect." Similarly, an adverse environmental change may decimate a population, as described in the accompanying Environment Report.

Coevolution

An important component of a population's environment is the other species with which it interacts. How successful an individual is, in terms of survival and reproduction, depends at least in part on how it relates with these species. The evolution of species in response to evolutionary changes in each other is known as **coevolution.**

Every organism lives with other kinds of organisms and interacts with at least some of them. Plants interact with the animals that eat them; predators interact with their prey; parasites interact with their hosts; competitors interact with each other. These interactions between species often have a greater influence on survival and reproduction than interactions with the nonliving environment. Coevolution is responsible for major trends in which a species changes in a consistent way and appears to be aiming toward perfection. A steady increase in animal speed is an example of this kind of evolution.

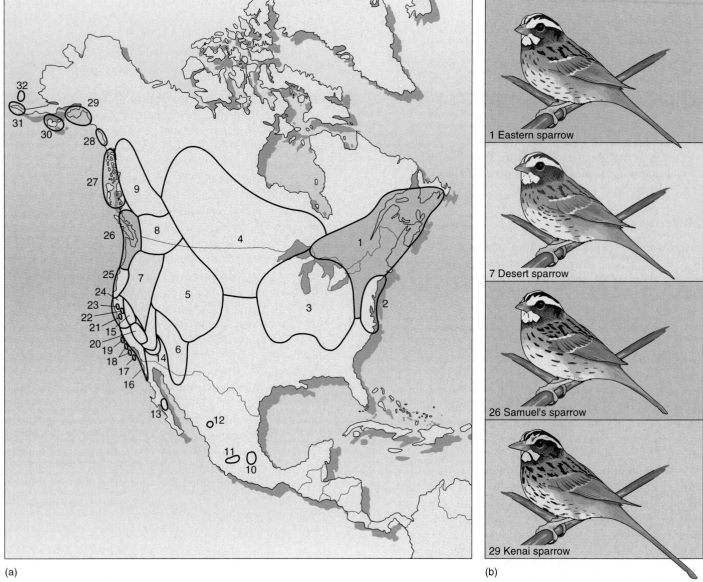

Figure 30.1 Populations of the song sparrow. (*a*) The song sparrow (*Melospiza melodia*) is subdivided into thirty-two populations (or subspecies). On this map, the geographic distribution of each population is enclosed within a solid line. Four of the populations are colored to match the drawings of individual birds shown in (*b*). (*b*) The members of each population are slightly different in size, color pattern, and song from the members of other populations. They all belong to the same species, however, because the sparrows of one population can mate with sparrows of other populations and leave fertile offspring.

Animal speed is the outcome of coevolution between predators and their prey. The gazelle (a form of antelope), for example, eats plants and has no need to run fast except to escape predators. The faster gazelles are more likely to survive and reproduce than the slower gazelles, which are more likely to be caught and eaten. Cheetahs eat gazelles, and the faster cheetahs are more likely to survive and reproduce than the slower cheetahs, which are likely to be undernourished. As the gazelles evolve the ability to run faster, they in turn select for faster-running cheetahs. As the cheetahs evolve the ability to run faster, they in turn select for faster-running gazelles. Coevolution between the gazelles and cheetahs explains the blazing speeds of these two kinds of animals.

An interesting consequence of coevolution is that as long as one species continues to evolve, so also must all species with which it interacts. This consequence is aptly called the "Red Queen effect," named after the character in Lewis Carroll's *Through the Looking-Glass*, who said "It takes all the running you can do, to keep in the same place." In evolution, a species that does not keep pace with its coexisting species becomes extinct.

Cheetahs

The cheetah is the fastest animal alive. It can accelerate to 110 kilometers (70 mi) an hour in less time than a sports car. The cheetah's body is built for sprinting: its lungs and heart are huge, to deliver oxygen rapidly to the working muscles; its claws are long and curved, to grasp the ground for traction; its leg muscles are arranged so that a small amount of contraction produces a large amount of leg movement; its long, slender legs and flexible backbone enable the animal to cover more than 6 meters (20 ft) in a single stride (fig. 30.A).

The awesome beauty, power, and grace of the cheetah conceal its vulnerability to extinction. According to the fossil record, cheetahs first appeared around 7 million years ago and soon became successful predators in the grasslands of Africa. According to genetic analyses, modern cheetahs have very little genetic variability. Geneticists estimate that some catastrophe must have decimated cheetah populations approximately 10,000 years ago. Such a population "bottleneck" usually produces unhealthy individuals and an inability of the population to adapt to change.

Bottlenecks produce unhealthy individuals because of inbreeding, in which matings occur between close

Figure 30.A The cheetah is built for speed.

relatives. Small populations suffer from inbreeding because there are so few members of the opposite sex from which to choose. The only available mate is likely to be a brother or sister, uncle or aunt, cousin, or even a parent. Such matings tend to be barren or to produce unhealthy offspring. Modern cheetahs may suffer from the effects of inbreeding in the past: they are unusually susceptible to disease, the males produce defective sperm that cannot fertilize eggs; many cubs are born dead, and the live ones tend to be sickly.

The second effect of a bottleneck is an inability of the population to adapt to change, owing to insufficient

genetic variability upon which natural selection can act. When most of a population is wiped out, most of its alleles disappear. Even if the population becomes large again, it cannot recover its genetic variability for a long time. It would take millions of years for the cheetah to accumulate (by mutations) the variety of alleles it had prior to the bottleneck.

Today, all cheetahs have nearly the same genes. It seems unlikely they can evolve fast enough to keep up with changes in their environment, particularly the rapid changes brought about by human activities. Only 20,000 of these magnificent animals remain today. The cheetah may soon be extinct.

Kinds of Natural Selection

The essence of natural selection is differential reproduction—some individuals leave more copies of their genes in the next generation than other individuals. In time, the alleles that promote more successful reproduction become more frequent at the expense of other alleles. Natural selection does not simply eliminate defective individuals. It also eliminates adequate individuals who do not win in the competition for resources and mates.

Natural selection occurs in several ways, as described in table 30.1 and following. We look first at the different

effects selection can have on a population and then at two special kinds of selection, sexual and kin.

Effects on the Population

Natural selection controls the composition of a population by controlling which individuals leave more offspring. Over the generations, differential reproduction may change the traits of a population. With regard to the effect on a population, biologists recognize three categories of natural selection: directional selection, stabilizing selection, and disruptive selection.

Table 30.1

The Kinds of Natural Selection

Name	Description
Grouped by effect:	
Directional	Favors individuals with one extreme form of the trait
Stabilizing	Favors individuals with the common form of the trait
Disruptive	Favors individuals with extreme forms of the trait
Sexual	Favors traits that make an individual more attractive as a mate
Kin	Favors traits that increase the reproductive success of a relative

Directional Selection In **directional selection,** the environment selects for an extreme form of a trait (fig. 30.2a). As individuals with this trait reproduce more successfully, the population shifts in one direction. The evolution of horses, for example, shows consistent selection for traits that increase speed: larger bodies, longer legs, and fewer toes, as shown in figure 30.3.

Stabilizing Selection In **stabilizing selection,** individuals with extreme versions of the trait are selected against; they reproduce less well than other individuals (fig. 30.2b). The population becomes less variable but otherwise does not change over the generations.

An example of stabilizing selection is the weight of human babies at the time of birth. In the United States, newborns who weigh between 3.4 and 3.6 kilograms (7 1/2 and 8 lb) survive better than others. Smaller babies are immature or undernourished. Larger babies experience difficulty at birth and are more likely to be injured at that time. (Modern medical procedures have weakened this selection for optimal newborn size.)

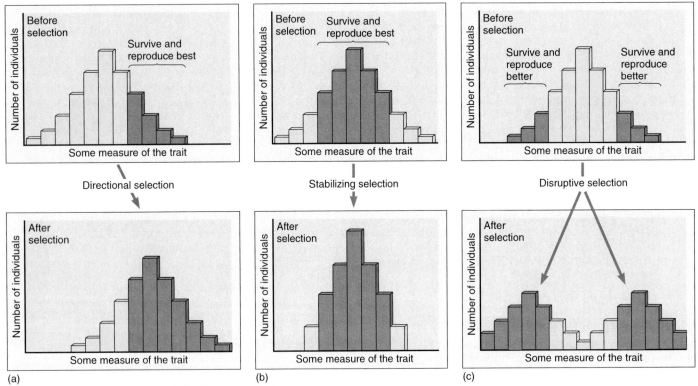

Figure 30.2 The ways that natural selection can affect a population. (a) In directional selection, individuals with an extreme form of the trait (shown in red) survive and reproduce better than others. The population changes in a particular direction. (The new versions of the trait, shown on the extreme right in the bottom graph, form not from selection but rather from new alleles and new combinations of alleles.) (b) In stabilizing selection, individuals with an intermediate form of the trait (shown in red) survive and reproduce better than others. The population becomes less variable after selection. (c) In disruptive selection, individuals with extreme forms of the trait (shown in red) survive and reproduce better than individuals with intermediate forms. After selection, the population exhibits two distinct forms of the trait.

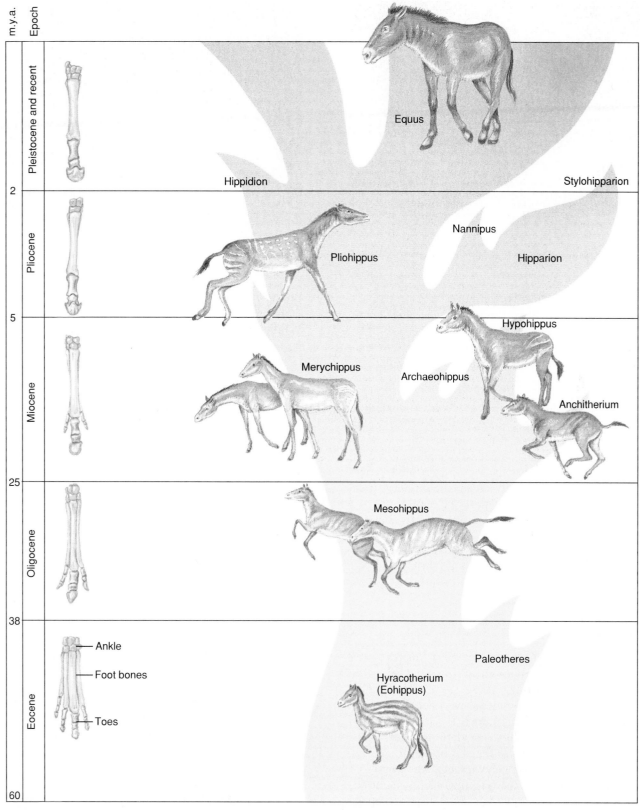

Figure 30.3 Directional selection in horses. Evolutionary changes in horses, based on a series of fossils beginning 60 million years ago, appear to reflect consistent selection for traits that confer speed: larger bodies, longer legs, fewer toes, and the conversion of a toenail into a hoof.

(a)

(b)

Figure 30.4 Sexual dimorphism. (a) Turkeys. (b) African lions.

Stabilizing selection is the most common form of natural selection. Even when it removes a significant fraction of the population each generation, stabilizing selection produces no change in the population. The average (or median) individual is optimal, with regard to the particular trait, for existing conditions.

Disruptive Selection   Selection that favors extreme forms of a trait is called **disruptive selection** (fig. 30.2c). It maintains more than one distinct form of a trait.

Disruptive selection has been documented in land snails. The shells of these snails, which occur in a variety of colors, determine how easily bird predators find them. In early spring, snails with brown shells are concealed against the brown, postwinter vegetation. In summer, snails with green shells are hard to see against the yellowish green vegetation. Disruptive selection by bird predators, favoring first one color and then another (but not intermediates), maintains both brown shells and yellow shells in the snail populations.

Sexual Selection

Natural selection also occurs in response to mates rather than other factors in the environment. In **sexual selection,** individuals with certain traits attract more mates and so leave more offspring than individuals without these traits. Sexual selection is based on competition for mates. It is particularly strong in polygynous populations, where a single male can mate with many females and reproductive success among males is extremely variable. The privileged males, who father most of the offspring each generation, are selected in one of two ways: female choice or competition among males.

Female Choice   When females choose their mates, they often select what appear to be frivolous traits. Female birds, for example, often prefer flashy males with large,

brightly colored feathers. The most colorful cock then mates with many hens and all their sons are colorful. While his brighter, longer feathers increase his vulnerability to predators, the advantages they give him in reproduction counteract the increased risk of dying. The hens, who usually get mated regardless of physical appearance, are drab and well-concealed from predators.

Why do females choose mates with flamboyant traits? Several hypotheses have been offered. One is that the female chooses a mate on the basis of species-recognition signals (colors, songs, and dances that ensure the male is a member of her species), and the more intense the signal, the more attractive it is to the female. Another hypothesis is that the female chooses a mate with superior capacities for survival, and a male with a handicap, such as an extremely long tail or bright colors, must be superior since he has survived in spite of the handicap.

Competition Among Males   When males compete with one another for mating rights, there is some sort of contest that determines who gets first access to the females. The contests may be ritualized fights or they may be struggles to monopolize the food, shelter, or nesting sites that females must have in order to breed. Winners of these contests are typically the larger, stronger animals with more impressive weapons, such as horns, antlers, and spurs. Their sons, in turn, inherit these same traits. The contests are exhausting and often result in injuries, rendering the males more vulnerable to predators and disease.

Sexual Dimorphism   Where sexual selection is strong, the appearances of males and females are strikingly different because they experience different selection. Examples of these differences, known as sexual dimorphisms (*di*, two; *morphism*, appearance), are shown in figure 30.4. In species without strong sexual selection, where each male mates with just one female, the two sexes look alike.

Kin Selection

So far, we have discussed selection for or against the individual, in which certain traits influence how well an individual survives and reproduces. **Kin selection,** by contrast, involves traits that promote the reproductive success of relatives (kin) rather than of the individual itself. This type of selection explains the evolution of altruistic behaviors (see page 521), in which an individual puts itself in danger or gives up the opportunity to reproduce in order to help another individual. By helping a relative to survive or reproduce, an individual increases the number of copies of its own genes in the next generation, since relatives share many of the same genes.

Altruistic behavior is especially common in animals that live in groups where most members are closely related. A dominant baboon acts as sentry for his troop, barking when an enemy approaches, attempting to scare it off, and then protecting the troop as it retreats. In Florida jays, some birds help their sisters to raise nestlings; they sacrifice their own reproductive possibilities (at least for a season) in order to increase the reproductive success of a relative. Nestlings tended by these helpers (aunts and uncles) survive better than nestlings tended only by their parents. By caring for their relatives, the helpers increase the number of copies of their genes in the next generation.

Table 30.2

Artificial Selection

Organism	Years of Selection by Humans	Number of Current Varieties*
Dog	12,000	200
Sheep	10,000	50
Goat	10,000	20
Pig	8,000	35
Cattle	8,000	60
Horse	6,000	60
Cat	5,000	25
Chicken	4,000	125
Duck	3,000	30
Rabbit	1,500	20
Canary	500	20
Parakeet	150	5

*A variety is a distinct type within the species; a poodle, for example, is a variety of dog.

Evidence of Evolution

Is the theory of evolution correct? To answer that question, we divide the theory into two major components: natural selection and the history of life. The validity of each component is then examined by testing its predictions, which are expressed as "if, then" statements. We look first at natural selection, the process by which evolution is proposed to occur, and then at the evidence that this process did in fact produce the species we see today.

Evidence of Natural Selection

According to the theory of evolution, the traits of individuals within a population are changed by natural selection. The evidence in support of natural selection comes in three forms: artificial selection, direct observations, and environmental patterns of adaptation.

Artificial Selection If the environment can change a population by determining which individuals leave more offspring, then humans should be able to simulate this simple process. Selection by humans rather than by nature is called artificial selection.

Humans have used artificial selection to improve and diversify their domestic plants and animals (table 30.2). Our ancestors developed dairy cows from wild cattle by allowing only the cows that gave the most milk to reproduce. They bred chickens that laid the most eggs, pigs that developed the most muscle (meat), and grasses that yielded the most seeds. We still cull from our herds and fields the individuals that do not exhibit the traits we desire. Similarly, we have used artificial selection to develop many varieties of dogs, cats, horses, roses, and other organisms that we have found useful or interesting.

Biologists have demonstrated the effects of artificial selection in laboratory populations. They used fruit flies and other organisms with short generation times, so that many generations could be followed within a few months or years. By removing individuals with certain traits, generation after generation, the researchers caused evolutionary changes in the populations. The density of bristles on a fruit fly, for example, has been increased by removing, generation after generation, individuals with fewer bristles before they reproduce.

Artificial selection differs from natural selection only in that humans, rather than the local environment, determine which individuals leave more copies of their genes (as offspring) to the next generation.

Direct Observations If natural selection happened in the past, then it should be happening now as well. We should be able to find populations that are changing in response to their environments. This prediction holds true: biologists have documented more than 200 cases of natural selection in action.

Figure 30.5 The peppered moth. (a) The dark form is conspicuous against the lichen-covered tree trunk, whereas the light form can barely be seen. (b) On a trunk darkened with soot, the light form is more conspicuous than the dark form.

(a)

(b)

A well-documented example of natural selection is the peppered moths (*Biston betularia*) in central England. The moths appear in two forms: a light form with dark spots (peppered) and a uniformly dark form. Both forms feed at night, like other moths, and rest during the day on the trunks of trees, where many are seen and eaten by birds.

In the mid-nineteenth century, the tree bark in this part of England was covered with gray lichens. The dark moths were much easier for the birds to see (fig. 30.5a) against the light background, and so more of them were eaten than the light moths. In 1848, when the study began, almost all the moths were light (only 1 in every 100,000 was dark). Factories then began to burn coal, which polluted the air and killed the lichens. The tree bark darkened as its natural brown was exposed and covered with soot. The light moths then became more conspicuous to bird predators (fig. 30.5b). In 1898, after fifty years of selection (by birds) against the light forms, 99% of the moths were dark and 1% were light. Natural selection had produced an evolutionary change in the population.

A sequel to this story is that in the 1950s and 1960s, England began to burn less coal and the air became cleaner. Once again, lichens grew on the trees and light moths were harder to see than dark moths. Natural selection reversed the direction of evolution. By 1990, 67% of the moths were light and 33% were dark.

Many examples of natural selection in action involve responses to human activities. When we drastically alter the environment of a population, any individual that by chance can deal with these changes will leave offspring while the others will not. Many bacteria have evolved resistance to antibiotics (📖 *see page 217*), and many insects have evolved resistance to pesticides.

An unusual example of natural selection in response to human activities involves the cockroaches that infest our kitchens and bathrooms. These pests are no longer eliminated by our roach traps—little black trays that contain a poison mixed with glucose. Surprisingly, the cockroaches are still vulnerable to the poison. It is the glucose they will not touch. Over the past decade or so, the rare individuals with an aversion to glucose have survived and multiplied while the once common individuals with a taste for glucose have been poisoned. To combat roaches, we need to keep changing the type of food in the baits.

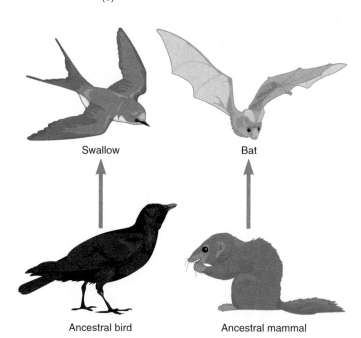

Swallow

Bat

Ancestral bird

Ancestral mammal

Figure 30.6 Convergent evolution. A swallow and a bat look alike because of convergent evolution rather than common ancestry. They do not have a common ancestor that flew. The swallow inherited its adaptations for flight from ancestors that flew. The bat evolved from mammalian ancestors that did not fly.

Patterns of Adaptation   An adaptation is a product of evolution. It is a trait (anatomical, physiological, or behavioral) that promotes survival and reproduction of an organism within a particular environment. It follows, then, that similar environments should give rise to similar adaptations and different environments should give rise to different adaptations.

Convergent evolution occurs when similar environments produce similar adaptations in organisms of very different ancestry; the adaptations were not inherited from a common ancestor. Bats and swallows are not close relatives, since one is a mammal and the other is a bird, yet both have wings and fly. The distant ancestors of bats (the first mammals) did not have wings, whereas the distant ancestors of swallows (the first birds) did have wings (fig. 30.6). Thus, wings evolved independently in the line of mammals leading to bats. The evolutionary paths of bats and birds converged because they lived in the same kind of environment, where selection has promoted the same kinds of traits.

Divergent evolution is when different environments produce different adaptations in closely related organisms. Whales and elephants, for example, are both mammals and must have descended from the same ancestral mammal (fig. 30.7). Their evolutionary pathways diverged because one experienced natural selection in water and the other on land.

Evidence of Common Ancestry

It is clear from the evidence described that natural selection can and does take place. But did it give rise to the various kinds of organisms we see today? Are all organisms descended from the same original form of life? Unfortunately, hypotheses about the origin of existing species cannot be tested directly, since these events took place in the past. The hypotheses can only be tested indirectly, by seeing if their predictions hold true. This type of indirect evidence comes from fossils, comparative anatomy, comparative embryology, and comparative biochemistry.

Fossils If species change through time, and if their remains or imprints appear as fossils that can be dated (fig. 30.8), then we can reconstruct ancestral lines of species.

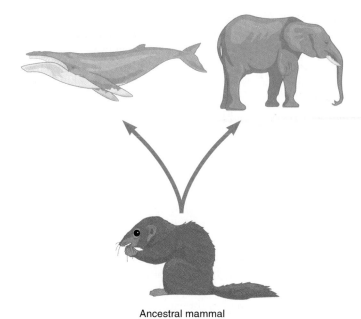

Ancestral mammal

Figure 30.7 Divergent evolution. The whale and the elephant are examples of divergent evolution. They are closely related (both are mammals), but the whale has adaptations for life in the ocean and the elephant has adaptations for life on land.

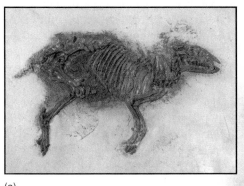

(a)

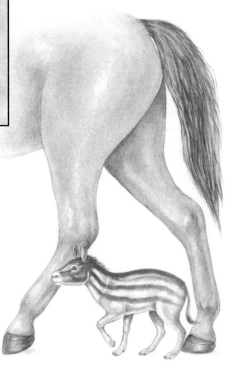

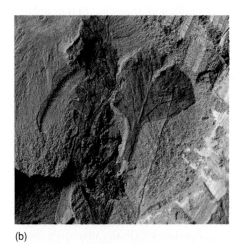

(b)

(c)

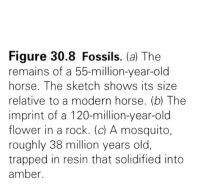

Figure 30.8 Fossils. (*a*) The remains of a 55-million-year-old horse. The sketch shows its size relative to a modern horse. (*b*) The imprint of a 120-million-year-old flower in a rock. (*c*) A mosquito, roughly 38 million years old, trapped in resin that solidified into amber.

Fossils provide our only direct information about ancient organisms. Most fossils form when rock solidifies around an organism and preserves its skeleton or imprint. Some fossils, especially those of insects, form when resin (a sticky plant fluid) traps an organism and then hardens into amber. The age of a fossil can be determined, so we know approximately when the organism died.

An evolutionist attempts to assemble a fossil sequence representing organisms within an ancestral line. Such sequences are necessarily incomplete, since a fossil forms only under extraordinary circumstances. First, the organism must die without being eaten. Second, its body must lie buried in mud or some other environment where there is no oxygen to support the bacteria that would otherwise decompose the tissues. Third, the body must be preserved in sand or mud that becomes compressed into solid rock or in resin that hardens into amber. Once a fossil forms, it still has to be exposed (by rock uplift or soil erosion) and then found by a human.

The geologic record, shown on the inside back cover of this book, is the history of life based on information from rocks and the fossils they hold. It is divided into major blocks of time based on differences in the kinds of organisms found.

Comparative Anatomy If living organisms are descended from a common ancestor, then closely related species should have more similar structures than distantly related species. Comparative anatomy is a field of biology that compares the structures of different kinds of living organisms.

Structures in different species that are similar because they have been inherited from a common ancestor are termed **homologous structures.** Bones within the limbs of amphibians, reptiles, birds, and mammals are considered homologous. The bones in a dog's paw, for example, are almost the same as the bones in your own hand (fig. 30.9). Evolutionists believe the genes for developing the bones of all terrestrial vertebrates have been inherited from a common ancestor, a species of fish that walked on land millions of years ago (fig. 30.10).

Further evidence of common ancestry comes from vestigial structures. A vestigial structure has no function in the organism where it is found but has been inherited from an ancestor, in which it did serve a function. Typically, the structure persists in a rudimentary form. The tiny eyes of a cave-dwelling fish, for example, are vestigial structures, as are the rudimentary leg bones of a whale (see page 569). Some vestigial structures of our own bodies are shown in figure 30.11.

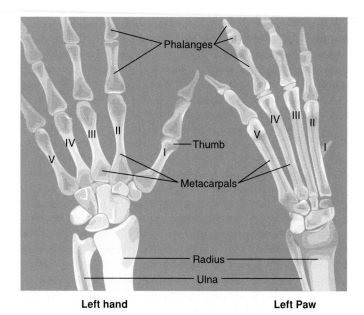

Figure 30.9 The bones in a human hand and in a dog paw. These drawings from X rays show how similar the bones of a dog are to the bones of a human.

Comparative Embryology If living organisms are descended from a common ancestor, then the embryos of closely related species should be more similar than the embryos of distantly related species. The study of different kinds of embryos is a field of biology known as comparative embryology.

The prediction turns out to be true: embryos of closely related species are strikingly similar. The embryos of all vertebrate animals, for example, are indistinguishable at certain stages; they become less similar as development proceeds (fig. 30.12). An embryo cannot withstand, apparently, much deviation from a standard form while establishing its body plan. Stabilizing selection eliminates variations in the early stages of development. Limbs, for example, develop in the same place and at the same stage of development in all vertebrate animals, regardless of whether they eventually form legs, flippers, or wings.

Comparative Biochemistry If all organisms evolved from a common ancestor, then (1) many of the fundamental biochemical pathways should be the same in all organisms, and (2) the molecules of closely related organisms should be more similar than the molecules of more distantly re-

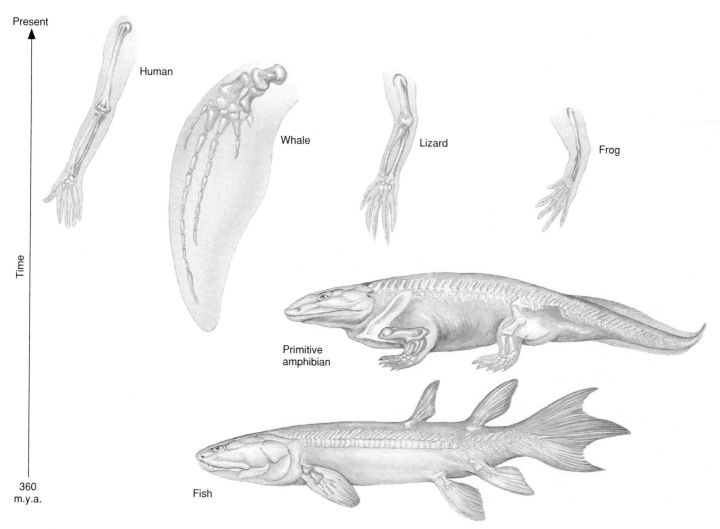

Figure 30.10 Homologous structures in the vertebrate forelimb. All terrestrial vertebrates appear to have inherited genes, from the first fish to walk on land, for developing the same bones in their forelimbs. These bones in the fossil fish are seen in fossil amphibians and in modern amphibians (frogs), reptiles (lizards), and mammals (whales and humans).

Figure 30.11 Vestigial structures of a human. Vestigial structures of humans, which had important functions in our mammalian ancestors, include (1) a nictitating membrane, which when fully developed covered and protected the eye; (2) muscles that move the ears, which located sounds; (3) pointed canines, which tore meat; (4) wisdom teeth, which ground plant materials; (5) hairs on the body, which conserved heat; (6) an appendix, which digested plant materials; (7) a tailbone, which supported a tail that functioned in balance and locomotion; (8) goose bumps from contractions of tiny muscles, which raised the fur when the animal was cold or frightened.

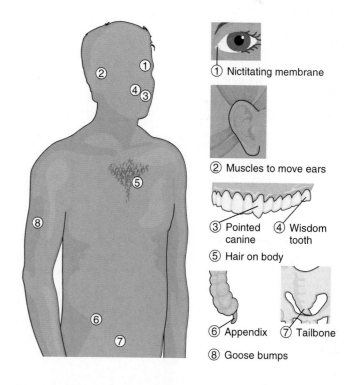

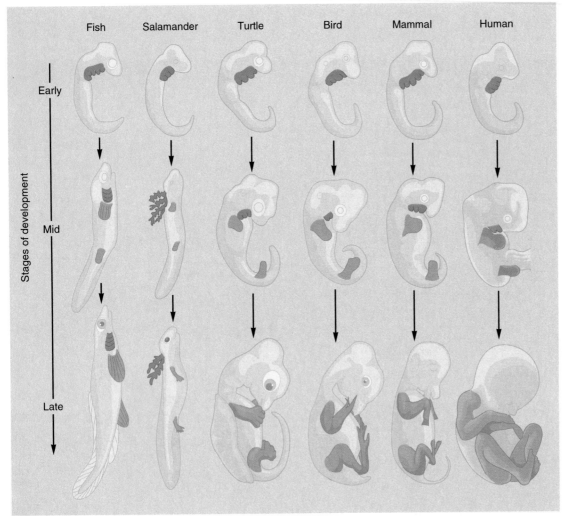

Figure 30.12 Vertebrate embryos. The early stages of development are more similar than the later stages. All young embryos of vertebrates have gills (shown in green) and a tail. All mid-stages develop limbs, whether they become fins, wings, or legs (shown in blue), from the same region of the body.

lated organisms. The study of biochemical pathways and organic molecules in different organisms is known as comparative biochemistry.

The first prediction holds true: many biochemical pathways are the same in all organisms. All cells, from bacteria to humans, synthesize proteins in the same way. And almost all cells use the same enzymes in anaerobic cellular respiration.

The second prediction also is supported by evidence: organisms believed to be closely related (as classified by other techniques) have more similar DNA and more similar proteins than organisms believed to be more distantly related. An example of such biochemical evidence of common ancestry is shown in table 30.3.

Surprisingly, fossils often contain intact DNA, which turns out to be a remarkably stable molecule as long as it is not exposed to water or oxygen gas. Fossil DNA has been recovered from bodies preserved in amber, tar, ice, and mummies. These ancient genes can be replicated in a test tube (the polymerase chain reaction) to form millions of

Table 30.3

Biochemical Similarities Between Humans and Other Animals in Cytochrome C, a Protein Involved in Aerobic Respiration

Animal	Number of Amino Acids That Are Different in the Human Molecule
Monkey	1
Dog	13
Horse	17
Chicken	18
Rattlesnake	20
Tuna	31
Moth	35

copies that can be analyzed in various ways. The sequence of bases can be compared with DNA from other organisms, to test for relatedness, and the genes can be stimulated to direct the synthesis of their proteins. Fossil DNA is a relatively new, and extremely valuable, source of evidence.

The DNA in mitochondria is used to estimate the amount of time that has passed since different organisms had a common ancestor. Mitochondrial DNA, unlike DNA in the nucleus, does not become mixed with other DNA during sexual reproduction. It is inherited only from the mother, within mitochondria of the egg. (Mitochondria in a sperm do not enter the egg at the time of fertilization.) Changes in mitochondrial DNA occur only by mutations, and we know that between 2% and 4% of mitochondrial DNA mutates every million years. Thus, for example, if the mitochondrial DNA of two organisms differs by 1%, they had a common ancestor between 250,000 and 500,000 years ago. This remarkable technique has been used to estimate how long it has been since all humans had the same ancestor. A tentative estimate is 200,000 years ago.

Origin of Species

The living world is organized into discrete groups of individuals called species. The earth supports millions of species. Familiar species are easily recognized as distinct forms of life. A wolf is clearly different from a fox, a rainbow trout from a catfish, and a raspberry from a rose.

Defining a Species

A **species** is a group of organisms with an isolated collection of genes. Although there may be gene flow between different populations within a species, there is no gene flow between species:

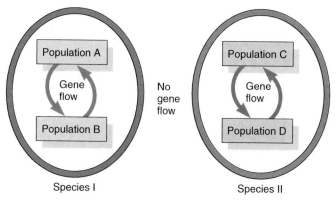

Whether two groups of organisms are indeed two species is determined in various ways. Most often, the different collections of genes are recognized by the different traits they generated. The traits of dandelions and tulips are clearly produced by different sets of genes. Sometimes, when two species are very similar, biologists look to see if there is gene flow—whether individuals in one group reproduce with individuals of the other group.

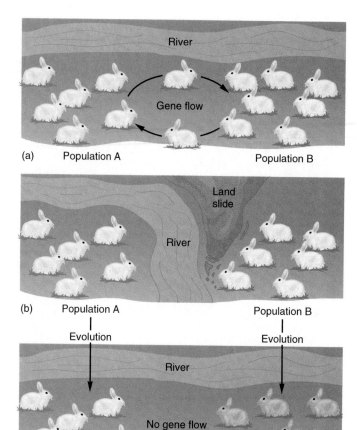

Figure 30.13 Allopatric speciation. This form of speciation begins when two populations of a species become separated by a geographic barrier, such as a river. (*a*) Prior to separation, migrants move between the populations and interbreeding is successful (i.e., there is gene flow). (*b*) A geographic barrier, such as a change in the course of a river, prevents migrations. (*c*) After the two populations have been separated for a long time, they become so different that even when the barrier is removed, individuals from the two populations will no longer mate with each other (i.e., there is no gene flow). Reproductive isolation of the two populations makes them two species.

The formation of a new species is called **speciation.** One way this happens is through time: when a species accumulates new features, at some point biologists recognize it as a new species, distinct from its ancestors. Another way speciation happens is when one species splits into two or more species. The formation of species by splitting, described following, occurs in two ways: allopatric and sympatric speciation.

Allopatric Speciation

When two or more species evolve from two or more populations living in different geographic regions, the process is known as **allopatric speciation** (*allo*, different; *patric*, region). This type of speciation begins with geographic isolation and ends with reproductive isolation (fig. 30.13).

Darwin's Finches

Darwin's finches are thirteen species of sparrowlike birds found on the Galápagos Islands and nearby Cocos Island (fig. 30.B). These Pacific islands are located near the equator, 965 kilometers (600 mi) west of Ecuador. Charles Darwin visited the islands in 1835, and the birds became central to his ideas about species change. They remain a classic example of adaptive radiation, in which many species evolve from a single ancestral species.

Each island has between three and ten finch species, which vary most conspicuously in beak shape. Some have large, thick beaks for opening hard fruits; some have short, thick beaks for crushing seeds; one species uses twigs to probe for insects beneath the tree bark; and several species have long, thin beaks for catching insects. The finch species forage in various places—in trees, on cacti, or on the ground, depending on the kinds of foods they seek. Except for traits related to feeding, the thirteen species are remarkably alike; they are more similar to one another in internal anatomy and biochemistry than to other birds.

It occurred to Darwin that all the different kinds of finches on the islands might be descended from a sin-

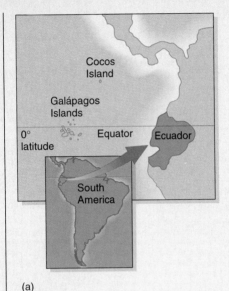

(a)

Figure 30.B Darwin's finches. (*a*) Darwin's finches live on the Galápagos Islands and Cocos Island. (*b*) The thirteen species of finches appear to have evolved from an ancestral species in South America. Each species (only ten shown here) has a unique beak that matches its unique feeding habits.

gle ancestral species in South America. Their adaptive radiation may have proceeded as follows:

1. Some birds from the mainland flew or were blown across the ocean to one of the islands.
2. These colonizers or their offspring then spread to all the islands and established local populations. The

finches on each island became geographically isolated from the finches on other islands; they rarely flew across the ocean to other islands.
3. Over several million years, each isolated population evolved in its own way, accumulating a unique set of alleles (from mutations) and experiencing natural selection in

Geographic Isolation Populations become geographically isolated from each other when something in the landscape prevents individuals from moving between populations. A geographic barrier may develop when a river takes a new course, a glacier forms, or a mountain uplifts. A barrier may develop also when individuals of a species colonize an isolated place, as when a small flock of birds is blown onto an island and forms a new population there.

When gene flow is cut off by a geographic barrier, the gene pools of the isolated populations diverge. Each population accumulates a unique collection of alleles (from mutations) and is molded in unique ways by natural selection in its local environment. The populations evolve in different ways. When they have developed distinctive features, they are called subspecies (or races). When the individuals of two subspecies can no longer reproduce with each other, they are two species.

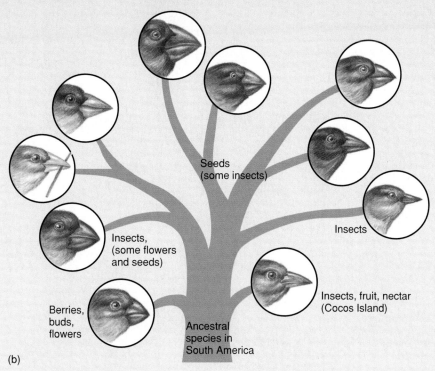

Seeds
(some insects)

Insects,
(some flowers
and seeds)

Insects

Berries,
buds,
flowers

Insects, fruit, nectar
(Cocos Island)

Ancestral
species in
South America

(b)

response to unique conditions on its island. In this way, each population evolved a unique set of adaptations, principally to its local food supply, and became a separate species. Adaptive radiation took place.

4. On rare occasions, some finches flew from their home island to adjacent islands, where they were able to survive and reproduce but were not able to breed successfully with resident species. By a series of such invasions, each island acquired more than one finch species.

Biologists have continued to study these remarkable birds. In 1938, an English ornithologist, David Lack, began to document the feeding habits of Darwin's finches, particularly the species that eat seeds. He observed that when two species share an island, one has a beak that is markedly smaller than the other species. But when only one species of seed-eating finch occurs on an island, its beak is intermediate in size. Lack hypothesized that competition between two species forces each to specialize, one on small seeds and the other on large seeds. In the absence of this competition, the birds evolve beaks that can handle seeds of various sizes.

In 1973, a group of American ecologists, headed by Peter Grant, came to the Galápagos Islands to test Lack's hypothesis. Their approach was to measure all the birds, as well as the available seeds, on a regular basis. While doing so, they documented an unusual event. A severe drought in 1977 greatly reduced the abundance of seeds, particularly the small seeds. The finch species, *Geospiza fortis*, that specializes in small seeds was subject to severe selection. By the end of the year, 85% of the birds had died and no eggs were laid. The few birds of this species that survived the drought were the ones with larger beaks (and, consequently, larger bodies), since they could crack open the larger, harder seeds. In the following years, their offspring inherited these same traits. Within just a few years, *Geospiza fortis* evolved into a larger bird with a larger beak.

Reproductive Isolation The test of whether a subspecies has reached full status as a species is whether it can reproduce successfully with other subspecies. The test occurs naturally when the geographic barrier is eliminated and the two groups contact each other. Three outcomes are possible:

1. Members of the two subspecies mate with one another and their offspring are just as healthy and fertile as the offspring of matings between members of the same subspecies. The two subspecies are not reproductively isolated; they are populations of the same species.

2. Members of the two subspecies do not mate with one another due to behavioral, anatomical, or ecological differences. They are reproductively isolated and have become two species. Occasionally, many species evolve from one species, a phenomenon known as adaptive radiation. A

Figure 30.14 Red-shafted and yellow-shafted flickers.
These two subspecies of flickers are in the process of
becoming two species. They interbreed where their
geographic ranges overlap. Hybrids of these matings have
characteristics of both parental species and are less successful
at reproduction.

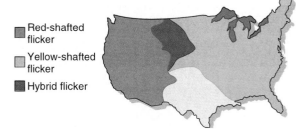

■ Red-shafted
  flicker

□ Yellow-shafted
  flicker

■ Hybrid flicker

Red-shafted

Yellow-shafted

Hybrid

classic example of adaptive radiation is described in
the accompanying Research Report.

3. Members of the two subspecies mate with one
   another but their offspring are less healthy or less
   fertile than the offspring of matings between
   members of the same subspecies. The two
   subspecies are partially isolated, reproductively,
   from each other. Speciation is taking place.
   Reproductive isolation evolves as individuals that
   choose mates from another subspecies are selected
   against: they leave fewer genes (because of
   unhealthy or infertile offspring) to the next
   generation than individuals that choose mates from
   their own subspecies. In time, individuals that tend
   to mate with the wrong subspecies are eliminated

from the population. Reproductive isolation is then
complete and the two subspecies are two species.
The red-shafted and yellow-shafted flickers (fig.
30.14), for example, are subspecies; they interbreed
where their ranges overlap, but the offspring of
these matings do not survive or reproduce as well as
the offspring of matings within each subspecies.

Sympatric Speciation

The formation of two species from a single population
located within the same geographic region is known as
sympatric speciation (*sym*, same; *patric*, region). Under
normal conditions, it is unlikely that reproductive isolation will develop between two groups of the same popula-

Figure 30.15 A hawthorn fly. This male is waiting on an apple for a female to arrive. When she does, they will mate and she will lay an egg inside the apple.

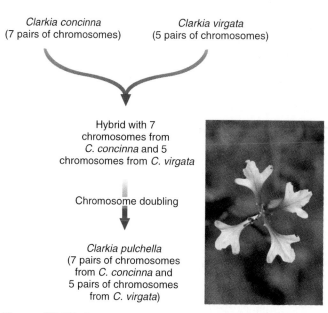

Figure 30.16 Sympatric speciation by allopolyploidy. *Clarkia pulchella*, a wildflower in the Evening Primrose family, arose by allopolyploidy. Speciation began with a mating between *Clarkia concinna*, which has seven pairs of chromosomes, and *Clarkia virgata* which has five pairs of chromosomes. The hybrid offspring, with seven chromosomes of one type and five of another, could not form sperm or eggs because each chromosome had no similar chromosome with which to pair during meiosis I. Then by accident the hybrid doubled its chromosomes so that each chromosome had an identical chromosome with which to pair. The hybrid formed eggs and sperm. It could not reproduce successfully with individuals from either parental species, because of differences in chromosome number, but could mate with itself. Its eggs were fertilized by its sperm, and the offspring became a new species, named *Clarkia pulchella*.

tion, for as long as the individuals live together there is invariably some interbreeding between the two groups. Sympatric speciation happens under two unusual conditions: ecological isolation and chromosomal changes.

Ecological Isolation Sympatric speciation may happen when an abrupt change in the environment separates a population into two ecological groups that no longer meet during the mating season.

Speciation by ecological isolation has been demonstrated in the hawthorn fly (fig. 30.15), a species of fruit fly that lives in eastern and midwestern regions of the United States and Canada. The ecology of this fly was based, originally, on the thorn apple, which is the fruit of the hawthorn tree. Males and females courted and mated only on the thorn apple, and the eggs hatched into larvae that fed on the thorn apple. About 150 years ago, cultivated apple trees were introduced into regions where there had been only hawthorns. Some hawthorn flies began to make use of the cultivated apples instead of the thorn apples, and the population diverged into two ecological groups—one that preferred thorn apples and another that preferred cultivated apples. Since mating takes place on the preferred fruit, members of the two ecological groups have stopped intermating. They are evolving into two species.

Chromosomal Changes A change in the arrangement or number of chromosomes can isolate an individual, repro-

ductively, from other members of its species. The individual may inherit two unmatched sets of chromosomes, from parents of different species (which happens sometimes in plants) or chromosomes that have broken and reattached in an unusual way. If such an individual can form eggs and sperm and mate with itself, it immediately forms a new species. To see how this relatively complex process works, we will examine just one such situation—allopolyploidy, a form of speciation common in plants.

An **allopolyploid** (*allo*, different; *polyploid*, more than two copies of each chromosome) is a plant with more than two sets of chromosomes, inherited from more than one species. Formation of an allopolyploid, shown in figure 30.16, begins with a mating between two species of plants.

The hybrid offspring inherits two sets of chromosomes, one from each parental species, that are so dissimilar that they do not pair during meiosis I (see page 146). Such a hybrid is sterile; it cannot undergo meiosis to form eggs or sperm.

The only way such a hybrid can reproduce is to duplicate all its chromosomes, so that each chromosome has an identical one with which to pair during meiosis. This sometimes happens, by accident, and the hybrid then forms sperm and eggs.

The next problem of an allopolyploid is to find a mate and produce fertile offspring. It is unlikely to find another individual with compatible chromosomes, but if it can mate with itself (a characteristic of many plants), then it can reproduce successfully. The hybrid and its offspring, with their unique chromosomes, immediately form a new species. Almost half the 235,000 existing species of flowering plants, as well as most ferns, appear to have formed by allopolyploid speciation.

SUMMARY

1. The modern theory of evolution consists of Darwin's theory, published in 1859, plus new ideas and information gained since then. Major additions are as follows: recognition of the population as the unit of evolution; definition of evolution as a change in allelic frequency in response to local conditions; discovery of genes and the causes of genetic variability; and recognition of the importance of coevolution.

2. Natural selection is differential reproduction. It has three forms with regard to the effect on the population: directional, stabilizing, and disruptive. Two special forms of selection are sexual selection, which favors traits that are attractive to the opposite sex, and kin selection, which favors traits that promote reproductive success in relatives.

3. The theory of evolution has two components: the process, which is natural selection, and the consequence, which is common ancestry. Direct evidence of natural selection includes artificial selection and observations of wild populations; indirect evidence appears as convergent and divergent patterns of adaptation. Indirect evidence of common ancestry comes from fossils as well as comparative studies of anatomy, embryology, and biochemistry.

4. The living world is subdivided into species that are reproductively isolated from one another. A species may be defined as a group of individuals that can mate with one another and produce fertile offspring. Allopatric speciation occurs when geographic isolation leads to reproductive isolation. Sympatric speciation occurs when ecological isolation or abrupt changes in chromosomes lead to reproductive isolation.

KEY TERMS

Allopatric (AEH-loh-PAEH-trik) **speciation** (page 583)
Allopolyploid (AEH-loh-POL-uh-ployd) (page 587)
Coevolution (page 571)
Convergent evolution (page 578)
Directional selection (page 574)
Disruptive selection (page 576)
Divergent evolution (page 579)
Evolution (page 571)
Gene flow (page 571)
Gene pool (page 571)
Homologous (hoh-MAHL-uh-gus) **structures** (page 580)
Kin selection (page 577)
Natural selection (page 570)
Population (page 571)
Sexual selection (page 576)
Speciation (SPEE-see-AY-shun) (page 583)
Species (SPEE-seez) (page 583)
Stabilizing selection (page 574)
Sympatric (sim-PAEH-trik) **speciation** (page 586)

STUDY QUESTIONS

1. Outline Darwin's theory of evolution. Why have modern biologists changed the unit of evolution from the species, used by Darwin, to the population? What have we learned, since Darwin's time, about the origins of genetic variability?

2. Describe, using an imaginary example, how natural selection works.

3. Why can natural selection work only on traits that are determined by genes and not by traits, such as enlarged muscles or pierced ears, acquired after birth?

4. Define adaptation. Choose a familiar organism and describe two of its adaptations. How does each adaptation help the organism to survive or to reproduce?

5. Consider two populations of prairie dogs that have been separated from each other by a river for a long time. When the river dries up and the two populations interact with each other, what are the three possible outcomes with regard to their ability to mate (and, therefore, their status as separate species)?

CRITICAL THINKING PROBLEMS

1. Is natural selection acting on human populations today? Why or why not?

2. The maximum speeds of a wolf and a moose are approximately the same. Explain the process that led to this similarity.

3. Often a fossil sequence shows no visible change in an organism over millions of years. Does this mean the organism did not evolve? Why or why not?

4. Are humans sexually dimorphic? What does this imply about our past?

5. The mating of a horse and a donkey produces a mule, which is sterile. Does this mean that horses and donkeys are of the same species? Why or why not?

SUGGESTED READINGS

Introductory Level:

Dawkins, R. *The Blind Watchmaker*. New York: W. W. Norton, 1986. An entertaining and well-written description of Darwin's theory and discussion of the evidence of evolution.

Gould, S. J. "What Is a Species?" *Discover* (December 1992): 40–44. A brief history of the species concept and its relevance to laws governing the preservation of species.

Weiner, J. *The Beak of the Finch: A Story of Evolution in Our Time*. New York: Alfred A. Knopf, 1994. An absorbing account of modern research on Darwin's finches, particularly the work by the team led by Peter and Rosemary Grant.

Wills, C. *The Wisdom of the Genes*. Cambridge, MA: Basic Books, 1989. A well-written description, for the layperson, of the history of evolutionary thought and new findings, including the roles of jumping genes and supergenes in the evolutionary process.

Advanced Level:

Endler, J. A. *Natural Selection in the Wild*. Princeton, NJ: Princeton University Press, 1986. A clear account of natural selection theory and how to detect natural selection, examples of natural selection in nature, and a discussion of the generalizations that can be drawn from the examples.

Hall, B. K. (ed.). *Homology: The Hierarchical Basis of Comparative Biology*. San Diego: Academic Press, 1994. A collection of essays, by outstanding comparative biologists, on the history of the homology concept and its philosophical foundations.

Mayr, E., and W. B. Provine (eds.). *Perspectives on the Unification of Biology*. Cambridge, MA: Harvard University Press, 1980. A history of evolutionary theory, with special concentration on the years from 1937 to 1950.

Ott, D., and J. A. Endler (eds.). *Speciation and Its Consequences*. Sunderland, MA: Sinauer Associates, 1989. A collection of research papers dealing with speciation, including the processes that generate species, the definition of a species, and why some genera have many species while others have few.

Smith, J. M. *Essays on Genes, Sex, and Evolution*. New York: Chapman and Hall, 1988. Essays by one of the best biological writers and thinkers on such topics as punctuated equilibrium, sociobiology, game theory, and the evolution of sex and intelligence.

Thompson, J. N. *Interaction and Coevolution*. New York: Wiley-Interscience, 1982. A synthesis of general patterns in the evolution of species interactions and the conditions likely to favor coevolution.

Williams, G. C. *Natural Selection: Domains, Levels, and Challenges*. New York: Oxford University Press, 1992. A clear review of recent advances in evolutionary biology, as well as discussions of paradoxes like the evolution of leks, the absence of viviparous birds, and species as units of selection.

Some marine invertebrates, representatives of the enormous variety of species on earth.

Biodiversity

Chapter Outline

Objectives

In this chapter, you will learn
–how the first cells may have formed;
–about the variety of unicellular
  organisms;
–how the first plants, fungi, and
  animals may have formed;
–about the variety of modern plants,
  fungi, and animals.

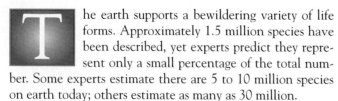

he earth supports a bewildering variety of life forms. Approximately 1.5 million species have been described, yet experts predict they represent only a small percentage of the total number. Some experts estimate there are 5 to 10 million species on earth today; others estimate as many as 30 million.

New species are discovered every year. Eleven new species of whales and porpoises have been discovered in this century. Three new families of flowering plants have been found recently. Within the past decade, 600 new species of beetles were discovered in just one species of tropical tree. Roughly 10,000 species of bacteria are recognized, yet a recent study found thousands of species, most not yet described, in just a single gram of soil.

Tropical rain forests and coral reefs are remarkably diverse. As much as 80% of all terrestrial species reside in the tropics. In some regions, just 7 square kilometers (4 square mi) of tropical rain forest contain as many as 1,500 species of flowering plants, 750 species of trees, 125 species of mammals, 400 species of birds, 100 species of reptiles, 60 species of amphibians, and 150 species of butterflies. Coral reefs are the marine equivalents of rain forests: the 423,000 species of coral reef organisms that have been described represent perhaps 10% of all coral reef species. Unfortunately, the two richest regions of the world—tropical rain forests and coral reefs—are disappearing most rapidly. The majority of their species will go extinct before they are described.

Why are there so many species? Species diversity is the result of two ongoing processes: speciation and extinction. Speciation occurs largely by accident, when two or more populations of a species become reproductively isolated from one another by geographic, ecological, or chromosomal changes. Extinction, by contrast, is the consequence of biological processes. A species becomes extinct when its individuals can no longer produce enough offspring to replace themselves, owing to changes in climate or pressures from competitors, predators, parasites, or human activities.

In this chapter, we describe the major groups of organisms and what is known about their evolution. The geologic record, a history of life based on fossils, is outlined on the inside back cover of the book.

Origin of Cells

The first form of life on earth must have been a single cell. All organisms appear to have descended from this ancestral cell, since all organisms consist of cells and all cells are fundamentally alike. After formation of this first cell, most of its characteristics must have passed, through reproduction, to subsequent cells. How did the first cell form?

The origin of cells happened so long ago that most evidence has disappeared. The oldest fossil cells are in rocks about 3.5 billion years old. The first cell must have arisen earlier, however, since these fossil cells are already complex and diverse. It is unlikely that traces of earlier cells will be found, since rocks older than 3.5 billion years have been altered too much by heat and pressure to retain imprints.

Proteins, nucleic acids, and other organic molecules presumably formed between the time the earth solidified, roughly 4.6 billion years ago, and the time the first cell appeared. We can only speculate on how the transition from inorganic molecules to organic molecules to cells may have occurred. If we assume that complex structures arose from simple structures, then chemical evolution must have proceeded in three major steps: (1) the formation of organic monomers from inorganic molecules; (2) the joining of organic monomers to form polymers, including genes; and (3) the formation of membranes around polymers that formed genes, either DNA or RNA. These steps are illustrated in figure 31.1 and described following.

Formation of Organic Monomers

The building blocks of biological molecules are organic **monomers:** glucose (the building block of complex carbohydrates), amino acids (building blocks of proteins), fatty acids and glycerol (building blocks of lipids), and nucleotides (building blocks of DNA and RNA). The first step in the origin of life had to have been the formation of these monomers from inorganic molecules.

Three main hypotheses about the origin of organic monomers are currently under study. One hypothesis is that organic monomers formed elsewhere in the solar system and arrived on earth via meteorites. A second hypothesis states that monomers formed from inorganic gases in the atmosphere and entered the oceans in rainwater. The third hypothesis is that monomers formed from inorganic gases emitted from vents beneath the oceans.

The first hypothesis is supported by the fact that many meteorites do contain amino acids and other organic monomers. The second and third hypotheses are supported by the fact that organic monomers form readily from inorganic gases in the laboratory. When an electric spark is passed through a mixture of appropriate gases, a variety of organic monomers are formed. Thus, lightning or heat acting upon a mixture of gases in the atmosphere or a deep-sea vent could have produced all the monomers now found in living things.

From Monomers to Polymers

The next step in chemical evolution was formation of polymers from monomers. This may have been the most difficult step. A **polymer** is a large molecule made of monomers. Complex carbohydrates, proteins, lipids, DNA, and RNA are polymers built from monomers of glucose, amino acids, fatty acids, glycerol, and nucleotides. The

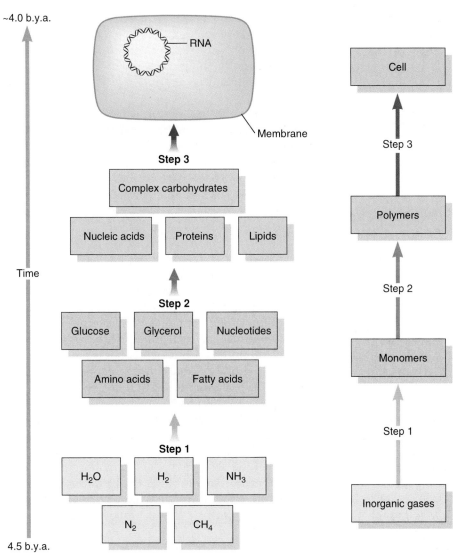

Figure 31.1 Biochemical evolution from inorganic gases to cells. Biochemical evolution involved three major steps: the formation of monomers from inorganic gases; the formation of polymers from monomers; and the enclosure of polymers, including RNA, within a cellular membrane.

monomers are bonded to one another by a chemical reaction called **dehydration synthesis** (see page 41), in which a molecule of water is removed wherever two monomers are joined:

Dehydration synthesis occurs readily inside a cell, where the monomers are highly concentrated and the formation of bonds is facilitated by enzymes and the energy of ATP. The reaction would not have occurred, however, in the open waters of an ocean because (1) the monomers would have been too far apart to be joined and (2) dehydration synthesis does not proceed spontaneously in water: molecules of water will not move spontaneously out of the monomers and into the watery surroundings.

How then could polymers have formed? Possibly on a clay surface. Clay contains ions, which can hold monomers near one another. It also contains iron and zinc, which act like enzymes, twisting the monomers into shapes that promote the formation of bonds. Clay also captures and stores high-energy electrons, which could have provided energy for removing water molecules from the monomers.

However the polymers formed, they could only have accumulated in places that were protected from the sun's rays. At this time in the earth's history, there was no ozone (O_3) layer to block out ultraviolet radiation, which disintegrates large molecules as soon as they form. (O_3 did not develop until photosynthesis evolved and released O_2 into the atmosphere.) Polymers could have accumulated only in places like cracks in rocks, warm vents in the ocean floor, or the bottoms of tide pools.

The first genes were probably in the form of RNA molecules. Of all the polymers, only RNA can perform all three tasks needed for life. Molecules of RNA can (1) replicate themselves without help from protein enzymes or DNA; (2) act as enzymes themselves, promoting protein synthesis and their own replication; and (3) provide instructions for the synthesis of proteins (particularly enzymes, which in turn govern biochemical reactions).

Once RNA molecules formed on the primitive earth, they could have controlled the synthesis of other large molecules—ATP, proteins, lipids, and polysaccharides. These early polymers were probably crude and inefficient versions of the ones found in modern cells. Each enzyme, for example, probably promoted several reactions in a clumsy way. But they worked. All that was needed, then, to make primitive cells was to package the RNA molecules inside membranes.

Assemblage of Membranes

Membranes form in a remarkably simple manner. When phospholipids and proteins are placed in water, they assemble themselves into membranes (see page 72). Each phospholipid has a water-seeking end and a water-avoiding end, so that when many of them are placed in water, they arrange themselves into sheets of molecules:

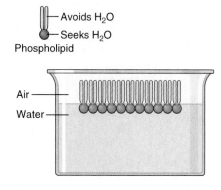

Proteins added to the water become embedded within the lipid sheets to complete the structure of a membrane.

Surprisingly, membranes assembled in this simple way behave like plasma membranes: the phospholipids form a fluid barrier and the proteins transport materials across the barrier. When the water is agitated, the phospholipid molecules form double-layered sheets that round up into hollow spheres, just like the plasma membranes of cells.

Membranes formed in this way tend to capture RNA molecules. Thus it is easy to see how a crude cell, made of RNA molecules enclosed within a membrane, could have formed in the primitive oceans. Such a cell may have acquired organic monomers from its surroundings and used them to build its own kinds of molecules: it would have built protein enzymes according to instructions within its RNA molecules and then used these enzymes to promote the synthesis of other kinds of polymers. The cell would have grown and then reproduced, by replicating its RNA molecule and pinching in two.

At this point in the origin of life, natural selection would have come into play. More efficient versions of RNA, which made more copies of themselves, would have become more abundant at the expense of less efficient versions. In time, the function of gene storage must have been transferred to DNA, which is a more stable molecule, and the function of enzymes must have been transferred to proteins.

Unicellular Organisms

Unicellular organisms are extremely abundant and diverse. Their total weight is twenty times greater than all other forms of life on earth. Each drop of water and every pinch of soil holds thousands of unicellular organisms, as does every speck of your skin. There are more microscopic organisms in your intestines than cells in your body.

Unicellular organisms occur in two forms: the bacteria, with prokaryotic cells, and the protists, with eukaryotic cells.

Bacteria

A bacterium is a **prokaryotic cell** (see page 59), without a nucleus or other membrane-enclosed organelles. The oldest fossil cell resembles a bacterium; it is 3.5 billion years old. While all modern bacteria are structurally similar, they are chemically diverse. The more than 10,000 species of modern bacteria form two kingdoms of life: the kingdom Archaea and the kingdom Eubacteria (see inside back cover of this book).

Archaea   Bacteria in the kingdom Archaea are unusual organisms; their cell walls, ribosomes, lipids, enzymes, and biochemical pathways are quite different from other bacteria. Their distinct chemistry is associated with life under conditions of extreme heat and pressure as well as darkness. Known as "extremophiles," these extreme forms of life thrive in hot springs, volcanoes, ice, very salty lakes, very alkaline or acidic lakes, vents beneath the ocean, and oil reservoirs several kilometers beneath the earth's surface.

The most familiar bacteria in this kingdom are the methanogens, which release a gas called methane (CH_4). They feed on organic molecules and acquire energy by fermentation (in the absence of O_2). Methanogens thrive in decaying vegetation within the stagnant waters of swamps and marshes as well as the digestive tracts of humans, cattle, and other mammals.

Eubacteria   The majority of bacteria within typical environments are eubacteria. They acquire energy and building materials in a great variety of ways. Some feed on organic molecules, and others synthesize their own organic molecules from inorganic materials. Many kinds form mutualistic relations with other organisms.

Some of the bacteria that feed on organic molecules are disease microbes, obtaining energy and building materials from living organisms. Others feed on dead organisms and are crucial to the continuance of life because they break organic molecules into inorganic molecules, which return to the environment for recycling. Every kind of organic molecule is consumed by some kind of bacterium. For many years, we have used bacteria to break down organic wastes at sewage treatment plants. We are now beginning to use them to digest environmental pollutants, including dioxin, pesticides, detergents, and oil.

Cyanobacteria are a particularly common form of life. These photosynthetic bacteria have light-absorbing pigments that reflect blue (*cyano*, bluish), green, and reddish light. The pigments are embedded, along with the enzymes of photosynthesis, on infoldings of their plasma membrane. Cyanobacteria live almost everywhere and are most conspicuous in lakes and ponds, where they form vast populations seen as a bluish green scum.

Chemosynthetic bacteria harvest energy from the chemical bonds within organic molecules, such as hydrogen gas (H_2), sulfite ions (SO_3^-), and ammonia (NH_3). They then use this energy to build sugars and other organic molecules.

Protists

Unicellular organisms with eukaryotic cells form the kingdom Protista. They first appear in the fossil record about 2.1 billion years ago—more than a billion years after the first cells. A **eukaryotic cell** (📖 *see page 62*) is larger and structurally more complex than a prokaryotic cell.

We do not know how eukaryotic cells arose. Considerable evidence from fossils and comparative biochemistry supports the idea that eukaryotes evolved from prokaryotes. There are, however, no fossils as yet of intermediate forms. Several hypotheses have been proposed to explain the origin of these complex cells, as described in the accompanying Research Report.

Protists live wherever there is water. Their cells are the most complex of any organism, with a variety of organelles that resembles the variety of organs in an animal (fig. 31.2).

Classification of the protists remains controversial. Modern systems, based on DNA analyses, indicate there should be as many as a dozen or so kingdoms of unicellular eukaryotes. The simplest system, used here, places all unicellular eukaryotes within a single kingdom and subdivides the kingdom according to how they acquire energy and building materials. According to this system, the kingdom Protista consists of three major groups: the algae, molds, and protozoa.

Unicellular algae live in water wherever there is sunlight. They manufacture their own sugars by photosynthesis within chloroplasts. Most photosynthesis in oceans and lakes is by these invisible organisms.

Molds in the kingdom Protista are slime molds and water molds, which secrete digestive enzymes onto food outside their cells and then absorb the digested monomers. Their food consists of organic molecules within dead plants and animals. Many are multicellular during at least some stage of their life.

Protozoa take organic molecules into their cells for digestion. Some are predators with adaptations for attacking, capturing, and ingesting tiny prey organisms. Some are par-

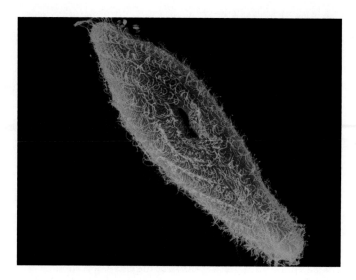

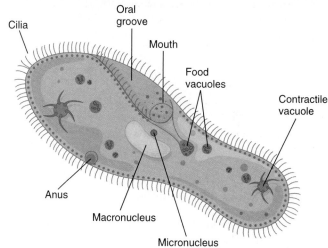

Figure 31.2 A protist is a very complex cell. This paramecium has many organelles, including an oral groove and mouth for ingestion of food, a food vacuole for digestion of food, an anus for removal of feces, a contractile vacuole for pumping out excess water, a micronucleus for cell division, a macronucleus for protein synthesis, and cilia for locomotion.

asites that live on or in other organisms, causing illnesses but typically not killing their host. Malaria and dysentery, for example, are caused by protozoa. Still other protozoa feed on dead organisms, breaking down organic molecules and releasing the inorganic components for recycling.

Plants

A plant is a multicellular organism, with eukaryotic cells, that builds its own sugars by photosynthesis. The cells contain chloroplasts (for photosynthesis) and are enclosed within cell walls. All plants belong to the kingdom Plantae.

Research Report

Origin of Eukaryotic Cells

We do not know how eukaryotic cells arose. We presume they evolved in some way from prokaryotic cells, since they appear much later in the fossil record and have many of the same biochemical pathways as prokaryotic cells. There are, however, no fossils as yet of intermediate forms. Two main hypotheses have been proposed to explain the origin of eukaryotic cells: the endocytosis hypothesis and the endosymbiotic hypothesis (fig. 31.A).

According to the endocytosis hypothesis, an early form of prokaryote lost its ability to synthesize muramic acid, a sugar that formed part of its cell wall. (Other forms of prokaryotes retained muramic acid and became bacteria in the kingdom Eubacteria.) In the absence of this sugar, the wall disintegrated and the cell became vulnerable to damage. Strong selection for cells with some kind of support led to the evolution of two new groups of organisms. One group evolved a new kind of cell wall, built without muramic acid, and eventually became bacteria in the kingdom Archaea. The other group developed a cytoskeleton, formed of protein fibers, and eventually became eukaryotic cells:

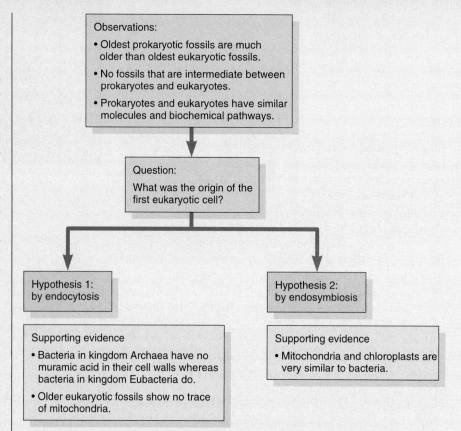

Figure 31.A   Scientific logic about the origin of eukaryotic cells.

Observations:
- Oldest prokaryotic fossils are much older than oldest eukaryotic fossils.
- No fossils that are intermediate between prokaryotes and eukaryotes.
- Prokaryotes and eukaryotes have similar molecules and biochemical pathways.

Question:
What was the origin of the first eukaryotic cell?

Hypothesis 1: by endocytosis

Supporting evidence
- Bacteria in kingdom Archaea have no muramic acid in their cell walls whereas bacteria in kingdom Eubacteria do.
- Older eukaryotic fossils show no trace of mitochondria.

Hypothesis 2: by endosymbiosis

Supporting evidence
- Mitochondria and chloroplasts are very similar to bacteria.

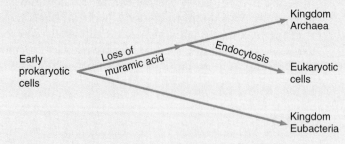

A cell with a cytoskeleton is enclosed within a fluid plasma membrane rather than a rigid cell wall. According to the endocytosis hypothesis, loss of the cell wall enabled these bacteria to evolve endocytosis, in which materials are brought into the cell by indentations of the plasma membrane. Fragments of the membrane involved in the transport then form vesicles within the cell:

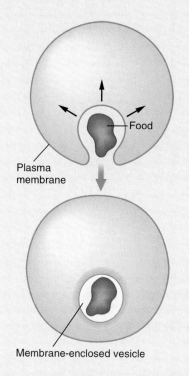

In this way, indentations of the plasma membrane surrounded portions of the cell interior, such as molecules of DNA, and became organelles, such as the nucleus:

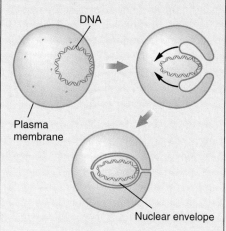

According to the endosymbiotic hypothesis, prokaryotic cells merged to form eukaryotic cells. Each merger occurred as a smaller prokaryote came to live inside a larger prokaryote; it either entered the cell as a parasite or was engulfed as food but not digested. Such mergers have been observed in modern bacteria (fig. 31.B). The two cells then lived together for many generations and developed specializations that made them dependent on each other. Such a relation is called endosymbiosis (*endo,* within; *symbiosis,* close association of individuals from two species).

According to the endosymbiosis hypothesis, mitochondria evolved when a large prokaryote that acquired energy by fermentation (in the absence of oxygen gas) engulfed a smaller prokaryote that acquired energy by the more efficient pathway of aerobic respiration. After many generations the larger cell specialized in acquiring food, while the smaller cell specialized in aerobic respiration. In time, the smaller cell became a mitochondrion within a eukaryotic cell:

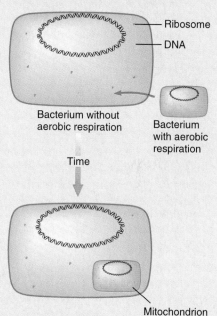

Similarly, chloroplasts evolved from the merger of a large cell, which fed on organic materials, and a smaller cell, which carried out photosynthesis. In time, the larger cell specialized in acquiring inorganic materials and the smaller cell specialized in photosynthesis. The smaller cell became a chloroplast within a eukaryotic cell.

Continued

Figure 31.B Endosymbiosis in modern bacteria. A small bacterium invading a larger one.

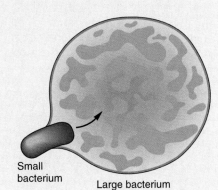

Small bacterium

Large bacterium

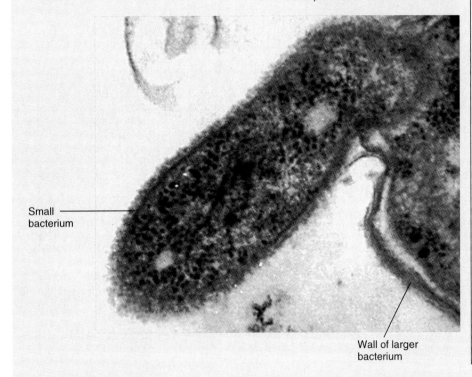

Small bacterium

Wall of larger bacterium

Comparisons between modern bacteria and eukaryotic organelles favor the endosymbiosis hypothesis. Mitochondria and chloroplasts resemble bacteria in several ways. They are about the same size, they have their own ribosomes (which are more similar to the ribosomes of bacteria than of eukaryotic cells), and they have their own DNA, which appears in circular strands that resemble the DNA in bacteria. Both organelles are enclosed within two membranes: an inner membrane, which is structurally similar to the plasma membrane of a bacterium, and an outer membrane, which is structurally similar to the plasma membrane of a eukaryotic cell. These two membranes could have developed when two ancestral cells merged:

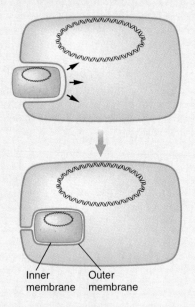

Inner membrane   Outer membrane

Moreover, mitochondria and chloroplasts grow, replicate their DNA, and divide independently of the rest of the cell.

The endocytosis hypothesis is favored by the fossil record. Cells with nuclei appeared approximately 2.1 billion years ago, yet cells with mitochondria did not appear until 850 million years ago. This suggests that cells evolved nuclei and other internal organelles long before they evolved mitochondria and chloroplasts.

It is of course possible that both hypotheses are correct: some organelles, such as mitochondria and chloroplasts, may have arisen by endosymbiosis whereas others, such as the nucleus, endoplasmic reticulum, and Golgi apparatus, may have evolved by endocytosis.

Plants originated in the ocean. The first plant was probably a green filament—a column of photosynthetic cells that remained together after cell division rather than separating to lead independent lives as unicellular algae. They appear in the fossil record approximately 1.4 billion years ago—almost a billion years after the first eukaryotic cell.

The more than 450,000 species of living plants are grouped into thirteen divisions (botanists use the term *division* in place of phylum), as listed in table 31.1. Four of the divisions are **nonvascular plants** (*vascular,* fluid-transporting vessels); they have no xylem or phloem for transporting fluids. The remaining nine divisions are **vascular plants,** which have xylem and phloem. An evolu-

Table 31.1

Classification of the Kingdom Plantae

Category	Division	Examples
Nonvascular Aquatic Plants (multicellular algae)		
	Chlorophyta (green)	Sea lettuce
	Phaeophyta (brown)	Kelp
	Rhodophyta (red)	Coral-forming algae
Nonvascular Land Plants		
	Bryophyta	Common mosses
Vascular Plants Without Seeds		
	Psilophyta	Whisk ferns
	Lycophyta	Club mosses
	Sphenophyta	Horsetails
	Pterophyta	Common ferns
Vascular Plants With Seeds		
Gymnosperms:	Cycadophyta	Cycad
	Ginkgophyta	Maidenhair tree
	Gnetophyta	Mormon tea
	Coniferophyta	Pine tree
Angiosperms:	Anthophyta	Dandelion

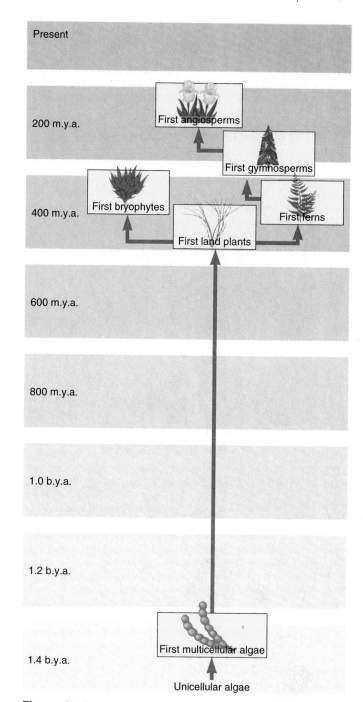

Figure 31.3 An evolutionary history of plants. The dates indicate the times when each major form of plant is believed to have appeared.

tionary history of the major groups, based largely on fossils, is diagrammed in figure 31.3 and described following.

Nonvascular Plants

Nonvascular plants are the multicellular algae and the bryophytes. (Some systems of classification place some or all multicellular algae in the kingdom Protista.) These plants are simple in structure and live either in water or within moist habitats.

Multicellular Algae Multicellular algae live submerged in water, and some probably resemble the very first plants. They vary from long filaments to flat sheets of cells, from tiny seaweeds to giant kelps. Algae have no cuticle, stomata, leaves, roots, xylem, or phloem. Reproduction occurs in water, with a sperm that swims to meet the egg.

The various kinds of multicellular algae are classified chiefly according to their pigments (table 31.1), which absorb different wavelengths of light. The green algae (division Chlorophyta) are an extremely diverse group (fig. 31.4) that probably gave rise to the land plants approxi-

mately 430 million years ago. Their main pigments are chlorophylls that capture light near the water surface. Most green algae live in lakes, ponds, rivers, and streams. The brown algae (division Phaeophyta) have brownish yellow pigments, in addition to chlorophylls, that capture wavelengths beneath the surface waters. Some, like the kelp, have gas-filled bladders that hold them afloat. Brown algae are abundant along rocky shores of oceans where the water is cool. The red algae (division Rhodophyta) have

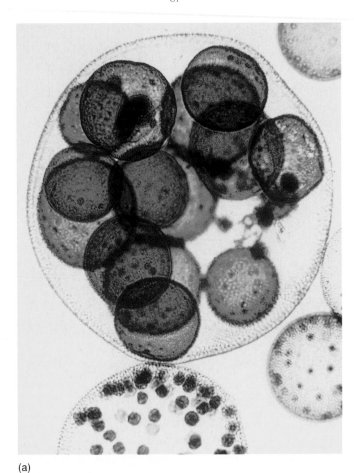

(a)

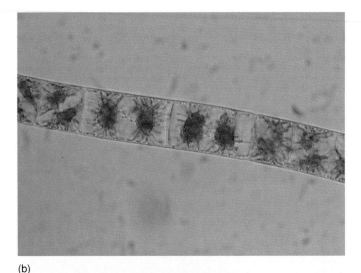

(b)

(c)

Figure 31.4 Green algae. (*a*) *Volvox,* a hollow ball of cells, is intermediate between a colony of individuals and a multicellular organism. (*b*) *Zygnema* is a filament of cells. (*c*) *Ulva,* or sea lettuce, consists of sheets of cells.

reddish pigments, in addition to chlorophylls, that absorb wavelengths in deeper waters. They are abundant in the deep, clear waters of tropical oceans.

Bryophytes   The bryophytes, division Bryophyta, are plants that live on land but only in moist places: marshes, swamps, damp logs, and along the banks of streams. Reproduction requires at least a film of water, since the sperm must swim to reach the egg. The main adaptations of bryophytes to life on land are a cuticle and stomata (📖 *see page 246*), which protect against water loss. They have no roots, leaves, xylem, or phloem. Modern forms are divided into three groups: the mosses, liverworts (fig. 31.5), and hornworts, which differ mainly in body form. They apparently evolved, approximately 380 million years ago, from the first land plants, which appear in the fossil record about 430 million years ago.

The body of a bryophyte consists of a small mass of photosynthetic cells anchored to the ground by hairlike

Figure 31.5 A liverwort. The conspicuous, umbrella-shaped structures are female reproductive organs. Sperm swim up the stalks through a film of water to fertilize the eggs.

structures. The upper surface is coated with a waxy cuticle and dotted with stomata, but the openings cannot be closed to prevent evaporation of water. The lower, uncoated surface absorbs water and ions from the ground and moist air:

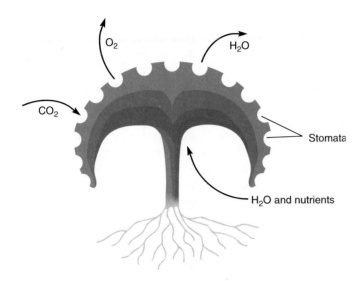

Vascular Plants

Vascular plants are adapted to life on land. In addition to a cuticle and stomata, adaptations of vascular plants include (1) xylem for transporting water from the soil to the leaves, and phloem for transporting sugars and other materials from one part of the plant to another; (2) roots, which absorb water and nutrients from the soil; and (3) leaves, which specialize in photosynthesis. Major forms of vascular plants are the ferns, gymnosperms, and angiosperms.

Ferns   The ancestors of modern ferns appeared roughly 400 million years ago—about 30 million years after the first land plant. They apparently evolved directly from the first land plants and not from the bryophytes. While the body of a fern is adapted to life on land, sexual reproduction is confined to moist habitats. It has a swimming sperm, which requires moist soil to reach the egg, and it has no seeds to protect the embryo from drying out.

Ferns were the first large plants to live on land. Most ferns we see today are short, with their stems underneath the ground. Ferns similar to the early forms—as tall as trees, with aboveground stems—still grow in the tropics.

Gymnosperms   Gymnosperms first appeared around 350 million years ago. They flourished when the climate became drier, approximately 300 million years ago, as all the continents merged to form a single continent known as Pangaea. The interior of this vast continent was very dry. The gymnosperms evolved from ferns that became adapted to dry environments as they spread throughout Pangaea.

Modern gymnosperms are adapted for reproduction on dry land. A major breakthrough was modification of the sperm so they no longer swim to reach the egg. Instead, the sperm develop within pollen grains that are dispersed by the wind. Another important adaptation is development of the embryo within a seed, which has a waterproof coat that protects against desiccation.

Of the four divisions of gymnosperms (see table 31.1), the most common today is the Coniferophyta. These are the conifers: pines, spruces, firs, junipers, larches, yews, cedars, cypresses, and redwoods. Their needle-shaped leaves conserve water by promoting air turbulence, which carries away heat that otherwise would have to be removed by the evaporation of water. Nearly all conifers are evergreen, retaining their leaves for photosynthesis all year round.

Angiosperms   Angiosperms, or flowering plants, evolved from some form of gymnosperm approximately 130 million years ago. They now comprise 96% of all modern species of plants. Angiosperms form a single large division called the Anthophyta (see table 31.1).

Angiosperms differ from gymnosperms in aspects of sexual reproduction (see page 272). First, the flower promotes transport of pollen by animals, which carry pollen more directly to the eggs than wind. Because animal transport conserves pollen, an angiosperm can allocate smaller amounts of nutrients than a gymnosperm to the production of pollen. Second, a flowering plant has double fertilization, in which one sperm nucleus fertilizes the egg and the other fertilizes a cell that develops into the endosperm, a starchy food for the embryo. Double fertilization conserves nutrients, since a food supply does not develop unless the egg is fertilized. A conifer, by contrast, provides food for every egg prior to fertilization. Third, the ovary of a flower develops into a fruit, which promotes dispersal of the seeds.

Fungi

A fungus is a multicellular organism, with eukaryotic cells, that feeds by releasing digestive enzymes onto the organic molecules of other organisms and then absorbing the digested materials. The cells are enclosed within cell walls made of polysaccharides and chitin, a different composition from the cell walls of plants. While fungi are considered multicellular organisms, some (like the yeasts) live most of the time in the unicellular phase of their life cycle. There is

Figure 31.6 The reproductive body of a mushroom.
Shown here is a common woodland species, *Mycena leaiana.*

little division of labor among the cells of a fungus. The kingdom Fungi includes yeasts, molds, and mushrooms. Classification is based mainly on how they reproduce.

The evolutionary origin of fungi remains a mystery. The fossil record indicates they arose more than 400 million years ago. Many experts believe, however, that fungi evolved soon after the first eukaryotic cell, as long ago as 2 billion years. Surprisingly, analyses of DNA indicate the fungi are more closely related to animals than to plants. According to these studies, the line leading to animals and fungi split from the line leading to plants approximately 1.1 billion years ago.

The body of a fungus is built of filaments called **hyphae.** Each hypha is a chain of cells linked end to end. The hyphae are interwoven to form a mesh, called a **mycelium.** All the cells of the mycelium have about the same structure and function.

An unusual feature of the mycelium is that connections between adjoining cells permit cellular materials to flow from one cell to the next. New molecules that are synthesized in older cells flow to the growing tips of the hyphae, so that the mycelium can grow rapidly outward and take advantage of nearby sources of food.

What most people think of as a mushroom is only the reproductive body of a fungus (fig. 31.6). It is connected to, and nourished by, an underground mycelium. (The mycelium of a monster mushroom, found in Wash-

ington State, is estimated to cover more than 4 square kilometers beneath the forest.) The extensive underground body of a fungus enables it to rapidly form reproductive bodies whenever environmental conditions are favorable.

Fungi feed on a variety of organic materials. Some species specialize on dead organisms while others are active predators or parasites that cause diseases in plants and animals. A few form mutualistic relations, in which two or more species live in close association and both benefit. A **lichen** is a mutualism between a fungus and a unicellular alga or cyanobacterium. The fungus obtains sugars from the photosynthesizing algae or bacteria; the algae or bacteria obtain water, inorganic ions, and protection against desiccation. A **mycorrhiza** (☐ *see page 264*) is a mutualism between a fungus and a plant. The fungus grows either around or within the plant root, greatly increasing the surface area for absorption of soil nutrients. In return, the fungus receives sugars synthesized by the plant. The fungi that form mycorrhizae appeared roughly 400 million years ago and may have facilitated the colonization of land by plants.

Animals

An animal is a multicellular organism, with eukaryotic cells, that feeds on organic molecules. All animals belong to the kingdom Animalia. Unlike a fungus, which digests food outside its body, a typical animal digests food inside a body cavity.

Animals arose in the ocean, sometime between 700 million and 1 billion years ago. The first animals apparently had soft bodies, since their only fossils are imprints of tracks and burrows on the ocean floor. The oldest fossils of animal bodies, around 700 million years old, are already quite advanced and diverse.

A huge variety of animals came on the scene near the beginning of the Cambrian period, roughly 530 million years ago. They appeared suddenly, within a period of 10 million years, in a wild profusion of different kinds. Every major form of animal, including sponges, sea anemones, worms of various kinds, sea urchins, clams, octopuses, and fishes, appeared during this period. The reason for this evolutionary frenzy, known as the Cambrian explosion, remains a mystery.

Invertebrate Animals

Invertebrate animals, which have no backbones, constitute 90% of all animal species. Sexual reproduction involves a swimming sperm and, in most invertebrates, a **larva**—a developmental stage that lives on its own and has a different body form and life-style from the adult.

Table 31.2

Major Phyla of the Animal Kingdom

Phylum	Distinguishing Features
Porifera (sponges)	Radial symmetry; intracellular digestion; specialized cells
Cnidaria (e.g., sea anemones)	Radial symmetry; digestive cavity; tissues
Ctenophora (comb jellies)	Radial symmetry; digestive cavity; tissues
Platyhelminthes (flatworms)	Bilateral symmetry; digestive cavity; organs but no circulatory system
Nematoda (round worms)	Bilateral symmetry; digestive tract; organs but no circulatory system
Mollusca (e.g., snails)	Bilateral symmetry; digestive tract; organs; coelom; gills; no segments
Annelida (segmented worms)	Bilateral symmetry; digestive tract; organs and circulatory system with hearts; coelom; segments
Arthropoda (e.g., insects)	Bilateral symmetry; digestive tract; coelom; exoskeleton; segments; gills, lungs, or trachea
Echinodermata (e.g., starfish)	Radial symmetry; digestive tract; coelom; no kidneys; no segments
Chordata (sea squirts; lancelets; vertebrates)	Bilateral symmetry; digestive tract; coelom; dorsal hollow nerve cord; notochord; pharynx with gill slits

Invertebrate animals constitute all but one of the more than thirty animal phyla (only the chordate phylum contains vertebrate animals). Distinguishing features of some of the phyla are listed in table 31.2. Relationships among many of the phyla are not well known, for we do not have sufficient fossils to show the transition from one kind of animal to another. Their likely relationships, based on the body forms and developmental patterns of living animals, are shown in figure 31.7.

The Simple Invertebrates The relatively simple animals are sponges, sea anemones, corals, jellyfish, hydra, and comb jellies. They are marine invertebrates with **radial symmetry,** in which there are no distinct sides of the animal (as opposed to **bilateral symmetry,** in which there are distinct left and right sides), as shown in figure 31.8. These animals have specialized cells and tissues but no organization of tissues into organs. They reproduce asexually most of the time. When sexual reproduction does occur, certain cells in the body develop into sperm or eggs. Fertilization is external; the sperm swims through water to meet the egg.

The simple invertebrates belong to three phyla: Porifera, Cnidaria, and Ctenophora. The phylum Porifera consists of sponges—the simplest animals alive today. Nearly all sponges live in marine waters, attached to the ocean floor, where they filter small particles of organic material from water that flows through their bodies (fig. 31.9). Digestion takes place inside the body cells rather than within a digestive cavity.

Animals in the phylum Cnidaria (nee-DAEHR-ee-ah; *cnide,* stinging) include the sea anemones, corals, hydra, and jellyfish. Some are sessile, in which the adult remains attached to the ocean floor. Others creep along the ocean floor, by muscular action, or drift in the currents. The most significant adaptation seen in a cnidarian but not in a sponge is a digestive cavity (fig. 31.10). A cnidarian can store and digest large pieces of food (an entire fish, for example) within its digestive cavity. A sponge, by contrast, has no place to store food and can only ingest food that is small enough to fit inside its cells, where digestion takes place.

The phylum Ctenophora consists of comb jellies, which are similar to the cnidarians but have no muscles. The outer surface of a comb jelly is covered with eight rows of cilia, resembling combs, that are used in locomotion.

Platyhelminthes Platyhelminthes are the flatworms, which are elongated animals with bilateral symmetry and organ systems (fig. 31.11). They are one of several phyla that represent intermediate stages in animal complexity, being more complex than the sponges, cnidarians, and ctenophores but less complex than the mollusks, annelids, arthropods, echinoderms, and chordates (see fig. 31.7).

The advantage of bilateral symmetry, with right and left sides, is that it gives the animal a front end. This end of the animal can then become specialized for moving first through the environment. It can become a head, with sensory receptors for receiving information about what lies ahead and a brain for evaluating this information and organizing responses. Organ systems of the flatworm include a nervous system, excretory system, digestive system with digestive cavity (as in the simple invertebrates), and reproductive system. Fertilization is external; the sperm swims through water to meet the egg.

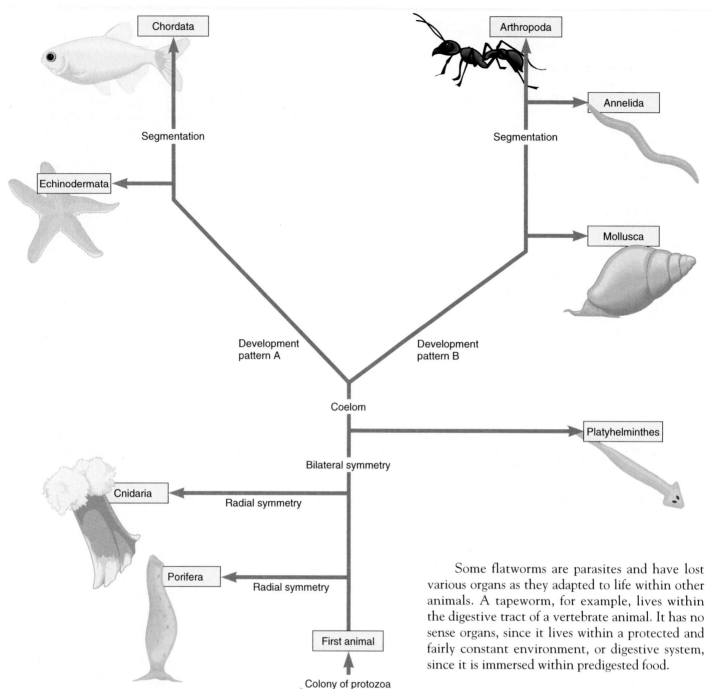

Figure 31.7 Evolutionary relations among the major animal phyla. The evolution of animals begins with a common ancestor, the first animal, which may have formed from a colony of protozoa. Animals in all these phyla are still abundant today.

A flatworm moves from place to place by both cilia and muscles. The cilia, which cover its underside, wave back and forth to "row" the animal through water. The muscles, located within the body wall, provide a creeping type of locomotion.

Some flatworms are parasites and have lost various organs as they adapted to life within other animals. A tapeworm, for example, lives within the digestive tract of a vertebrate animal. It has no sense organs, since it lives within a protected and fairly constant environment, or digestive system, since it is immersed within predigested food.

The Complex Invertebrates More complex invertebrates are in the phyla Mollusca, Annelida, Arthropoda, and Echinodermata. Some have external fertilization, in which fertilization occurs in the surrounding water, and others have internal fertilization, in which the male inserts his sperm into the female.

All complex animals have a **coelom,** or body cavity, within which organs are suspended. (In us, the coelom is divided into a chest cavity and an abdominal cavity.) The coelom enables internal organs to enlarge at certain times—the heart enlarges when filled with blood, the digestive tract enlarges when filled with food, and the

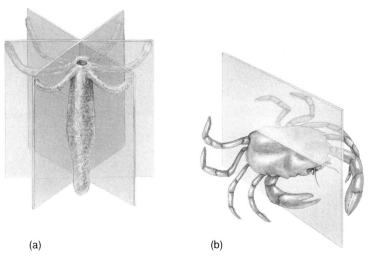

(a)   (b)

Figure 31.8 The two kinds of body symmetry. (*a*) In a radially symmetrical animal, any plane that passes through the center divides the body into halves that are mirror images of one another. (*b*) In a bilaterally symmetrical animal, only one plane that passes through the center divides the body into halves that are mirror images of each other. The animal has a definite left and right side.

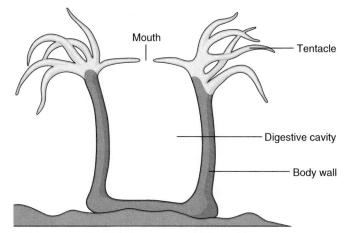

Figure 31.10 A cnidarian. One end of this sea anemone is attached to the ocean floor. The other end is adapted for feeding, with tentacles that capture prey and a mouth that moves prey into the digestive cavity. After digestion, waste materials leave the cavity by way of the mouth.

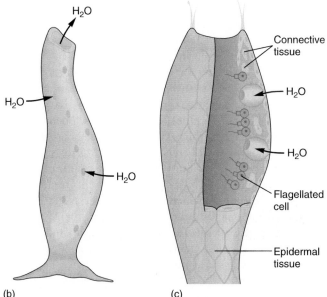

(a)

(b)   (c)

Figure 31.9 The sponge. (*a*) A tube sponge (genus *Verongia*). (*b*) Seawater enters a sponge through pores in the body wall and leaves through an opening at the top. Food is filtered out of the water as it passes through the hollow interior. (*c*) The body wall consists of epidermal tissue, which forms a protective covering; connective tissue, formed of migrating cells that perform a variety of functions; and an inner tissue formed of flagellated cells that filter food from the water. The cells of a sponge secrete a protein that forms a soft skeleton. Some sponges add minerals to the protein to form a hard skeleton.

ovaries or testes enlarge when filled with mature eggs or sperm. It also splits the muscles of the body into two sets, one set around the digestive tract and the other set within the body wall:

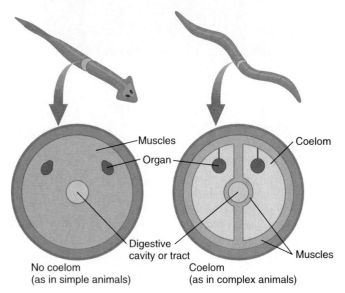

No coelom
(as in simple animals)

Coelom
(as in complex animals)

The phylum Mollusca includes snails, clams, mussels, oysters, squid, and octopuses. Most of these animals have a body plan based on a muscular foot, which they use in burrowing or creeping along surfaces (fig. 31.12). A mollusk is more complex than a flatworm in that it has a digestive tract, circulatory system, and gills. The digestive tract is a long tube with a mouth at one end and an anus at the other. The advantage of a tract over a cavity is that the animal can spend more time feeding, since it can take in food and eliminate wastes at the same time.

The phylum Annelida consists of segmented worms, such as polychaete worms and earthworms (fig. 31.13). Like the mollusks, these animals have a circulatory system and a digestive tract with both a mouth and an anus. An annelid is more complex than a mollusk in that it has a segmented body. The chief advantage of **segmentation** is that different parts of the body can move separately during locomotion. An annelid creeps, swims, or burrows by alternately lengthening and shortening each segment in turn.

The phylum Arthropoda includes all animals with jointed exoskeletons, which are hard outer skeletons that consist of many pieces connected by joints. Eighty percent of all living animal species are arthropods, including lobsters, crabs, scorpions, ticks, mites, spiders, copepods, centipedes, millipedes, and insects. Arthropods were the first animals to colonize the land, roughly 400 million years ago. Adaptations of arthropods to life on land include (1) an exoskeleton, which supports the body in air and protects against evaporation; (2) either lungs or trachea (rigid tubes), which are gas exchange surfaces inverted into the body to reduce evaporation; and (3) a penis, for delivery of sperm into the female so that fertilization can take place on dry land.

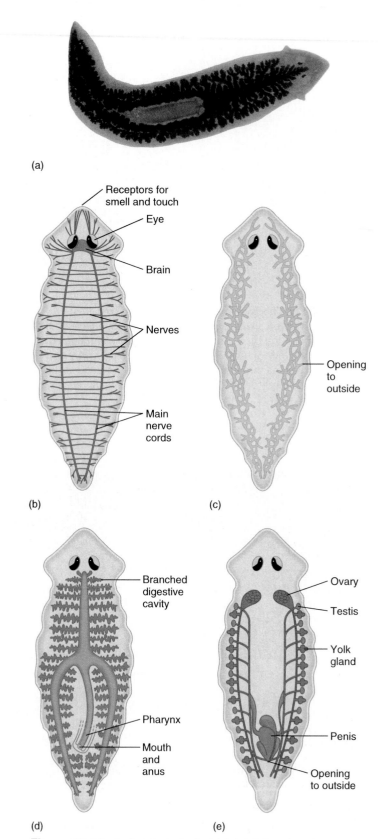

Figure 31.11 A platyhelminth. (*a*) A brightly colored planarian, which is a form of flatworm in the phylum Platyhelminthes. Its head is on the right. (*b*) The nervous system. (*c*) The excretory system. (*d*) The digestive system. (*e*) The reproductive system.

(a)

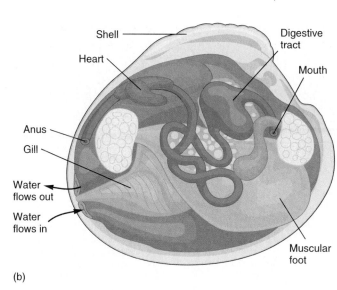

(b)

Figure 31.12 A mollusk. (*a*) A land snail. (*b*) The body plan of most mollusks is based on locomotion by means of a muscular foot. The interior contains many organ systems.

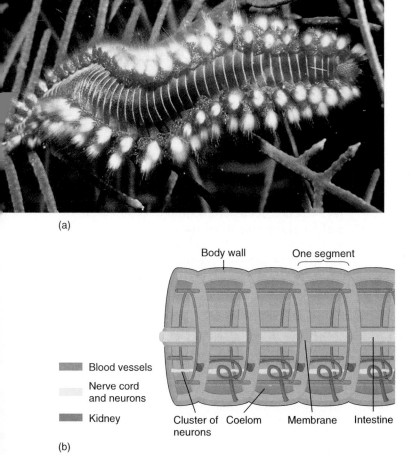

(a)

(b)

Figure 31.13 Annelids. (*a*) A polychaete worm. (*b*) An earthworm, showing the nervous system, circulatory system, digestive system, and excretory system. The body consists of many segments, each with its own coelom, kidneys, and cluster of neurons.

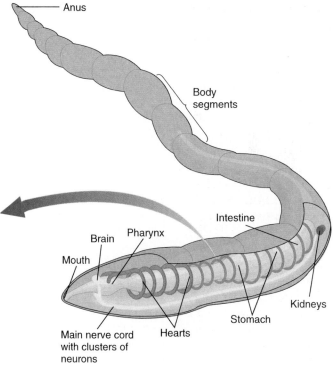

The phylum Echinodermata includes starfish, sea urchins, sea cucumbers, and sand dollars. They are marine animals that move about on the bottom of the ocean. Some graze on algae and others are predators.

Most echinoderms have a rigid skeleton, consisting of hard plates that often bear spines.

Our own phylum (Chordata) is closer to the Echinodermata than to the Mollusca, Annelida, or Arthropoda,

607

(a)

based largely on the pattern of early development. Superficially, the echinoderms look like simple invertebrates. Their internal anatomy, however, is very complex: they have a nervous system, circulatory system, reproductive system, digestive tract with both mouth and anus, muscles for locomotion, and coelom. On the other hand, they apparently lost many features of their wormlike ancestors: they have no kidneys or body segments, and their bodies are radially symmetrical (usually arranged in five parts). The larval form, however, is bilaterally symmetrical; it may represent a clue to the relation between echinoderms and vertebrates, as described following.

Vertebrate Animals

The phylum Chordata has three subphyla: Urochordata, Cephalochordata, and Vertebrata. The subphylum Urochordata consists of tunicates, or sea squirts (fig. 31.14). The adults of these marine animals have cylindrical bodies attached to the ocean floor. Their larvae look like tiny tadpoles, only a millimeter in length. The subphylum Cephalochordata consists of the lancelet (also known by its genus name, *Amphioxus*) and its relatives. These fishlike animals, just 5 centimeters or so in length, have bilateral symmetry and segmented muscles. They spend most of their time in a burrow, filtering food from water they pump into their mouths and out their gills. The subphylum Vertebrata consists of fishes, amphibians, reptiles, birds, and mammals (table 31.3).

Tunicates, lancelets, and vertebrates are placed in the same phylum because they all exhibit, at some time in their lives, the same three features:

1. A **notochord,** which is a stiff but flexible rod (made of cartilage) that runs the length of the body. The presence of a notochord helps the body to bend without shortening when the muscles contract. In most vertebrates, the notochord appears early in development and later is replaced by a stronger,

(b)

Figure 31.14 A urochordate. The tunicate, or sea squirt, is in the subphylum Urochordata. (*a*) The adult tunicate lives attached to the ocean floor. (*b*) The interior of the adult reveals the tunicate's complexity. The adult feeds by pulling water into

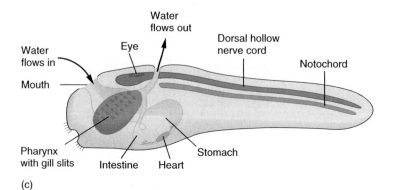

(c)

its body, filtering food through its gill slits, and then ejecting water out another opening. (*c*) The larva has chordate characteristics: a notochord, dorsal hollow nerve cord, and pharynx with gill slits.

Table 31.3

Classes in the Subphylum Vertebrata

Class	Distinguishing Features
Agnatha (hagfish and lampreys)	Fish without jaws; skeletons of cartilage
Chondrichthyes (e.g., sharks)	Fish with jaws; skeletons of cartilage
Osteichthyes (e.g., salmon)	Fish with jaws; skeletons of bone
Amphibia (e.g., frogs)	Semiaquatic vertebrates with lungs, four legs, and muscular tongues; external fertilization and larval development in water
Reptilia (e.g., lizards)	Terrestrial vertebrates with scales, efficient lungs, and water-conserving kidneys; penis and amniotic egg
Aves (e.g., sparrows)	Terrestrial vertebrates with feathers, hollow bones, acute distance vision, efficient lungs, and endothermy; penis and amniotic egg
Mammalia (e.g., squirrel)	Terrestrial vertebrates with hair, endothermy, mammary glands, diaphragm, and large brains; penis; development either in an amniotic egg or a uterus with placenta

more flexible structure—the backbone, which is a row of interconnected bones called vertebrae.

2. A **dorsal hollow nerve cord,** or spinal cord, which lies just above the notochord.

3. A **pharynx** with **gill slits.** The pharynx is the throat, or first part of the digestive tract after the mouth. Gill slits are openings to the outside of the body. A pharynx with gill slits is used for filter feeding (the slits are filters) in some chordates and for gas exchange in others. In reptiles, birds, and mammals, gill slits appear briefly and develop only partially in young embryos.

Origin of Vertebrates   What kind of animal was ancestral to the first vertebrate? No one knows for sure. The oldest fossil vertebrate is a fish, found in rock that is 540 million years old. No fossils have been found of animals intermediate between this fish and its invertebrate ancestor.

The tunicate larva hypothesis about the origin of vertebrates is that the first fish evolved from a tunicate. It is supported by evidence from comparative anatomy, embryology, and biochemistry. While the adult tunicate looks

nothing like a vertebrate, the larval tunicate looks a great deal like a fish (see fig. 31.14c). It has an elongated body with bilateral symmetry as well as a notochord, dorsal hollow nerve cord, and pharynx with gill slits.

According to the hypothesis, tunicate larvae colonized a river, where selection favored individuals who excelled at swimming and feeding in rapidly flowing water. The radially symmetrical, sessile adult form was totally unsuited to this new environment. The larval form, however, survived well; it could swim, by bending its fishlike body, and it could feed, by filtering particles of food from water that flowed into its mouth and out its gills. The poorly adapted adult form was gradually eliminated from the life cycle. The well-adapted larval form began to reproduce before it turned into the adult form. (Reproduction by larvae, a phenomenon known as paedogenesis, is not a far-fetched idea; it has been observed in salamanders as well as a variety of other animals.)

A fish differs from a tunicate larva mainly in its superior adaptations for swimming in swiftly flowing water. A fish is larger and has segmented muscles along each side of a segmented backbone (vertebrae), which generate undulations of the body:

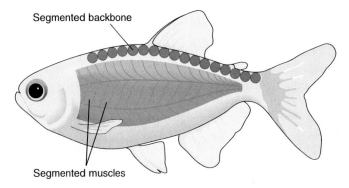

Segmented backbone

Segmented muscles

Unlike a tunicate larva, a fish has a large brain and well-developed sense organs within the head. It is much better adapted for swimming in a particular direction and seeking out food and shelter within the varied environments of a river.

The Fishes   Fossils of early fishes show two striking differences from most modern fishes: an external skeleton and no jaws. The skeleton consisted of heavy plates, made of bones, that cover the body. Most likely it evolved in response to predation by giant arthropods. The head and mouth were large but jawless. The opening of the mouth was on the underside of the head, where it sucked in water (like a vacuum cleaner) from the river bottom. Food was filtered out of the water as it passed through the gills and out of the body.

The jawless, heavily armored fishes that appeared 540 million years ago were the ancestors of all modern fishes (most with jaws) as well as terrestrial vertebrates.

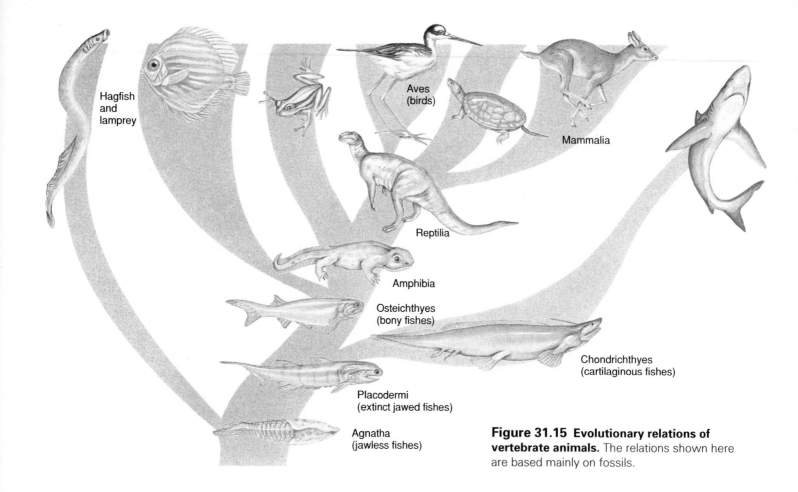

Figure 31.15 Evolutionary relations of vertebrate animals. The relations shown here are based mainly on fossils.

This major pathway of evolution is diagrammed in figure 31.15. Approximately half of all vertebrate species alive today are fish.

Amphibians   An amphibian begins life in water but typically moves onto land as an adult. Amphibians form the class Amphibia, which includes frogs, toads, caecilians (legless forms), and salamanders.

The oldest fossil amphibians are from rocks about 365 million years old. They appeared after the ferns but before the gymnosperms. The first amphibians were large animals, a meter or so in length, with strong legs. Transition from water to land was easier at that time, because the land was subdivided into many smaller continents so that much of the land was near a warm, humid coast. Swampy vegetation, formed of ferns and bryophytes, was inhabited by insects and other invertebrates. Life on land offered an abundance of food as well as shelter from aquatic predators.

We have many fossils that show a transition from fishes to amphibians. The fishes that colonized land had lungs and fleshy fins supported by small bones (see fig. 30.10). They lived in shallow water and used their lungs to occasionally gulp air. The front pair of fins propped up the head so the animal could breathe as well as see, hear, and smell its surroundings. Later, both pairs of fins, front and back, were used for walking on land—short forays to feed and to escape predators. Over the generations, these fishes spent more time on land and less in the water. They became the first amphibians. The adult amphibian is adapted to land. It breathes air through its lungs as well as its skin when moist. It has a muscular tongue for capturing insects and four legs for holding the body off the ground during locomotion. Its mode of reproduction, however, confines it to life near water. Fertilization is external, and the sperm swims through water to meet the egg. Development also requires water, since the eggs hatch into fishlike larvae (tadpoles) that swim, feed, and breathe within water.

Reptiles   Reptiles form the class Reptilia. Modern groups are the snakes, lizards, turtles, alligators, and crocodiles. The first reptiles appeared some 300 million years ago (about 50 million years after the first gymnosperms). Their evolution from amphibians was driven by a drier climate, which developed when all the smaller continents came together to form a single continent known as Pangaea.

Adult adaptations to life on land include (1) a waterproof skin with scales to prevent water loss (a reptile does not breathe through its skin); (2) large lungs that are ventilated by movements of the ribs; and (3) kidneys that conserve water by forming uric acid, which is a crystal rather than a fluid form of urine.

Adaptations for reproduction on land include internal fertilization and development within an egg. The fe-

610

(a)

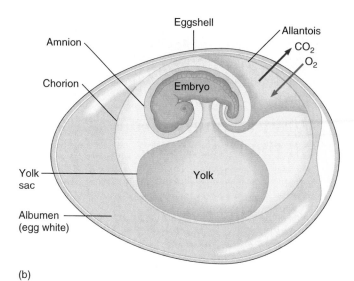

(b)

Figure 31.16 The amniotic egg. (a) Young snakes hatching from amniotic eggs. (b) The egg has an amnion, which encloses the embryo in its own "pond water"; an allantois, which exchanges oxygen and carbon dioxide with the environment and accumulates wastes in the form of uric acid crystals; a chorion, which encloses the embryo and its membranes; a yolk sac, which holds the embryo's food supply; albumen, a solution of proteins that controls osmosis, thereby preventing too much water from entering the embryo; and a shell, which protects against desiccation and predators.

male reptile retains her eggs until they are fertilized by sperm from the male. The male has a penis for depositing sperm directly inside the female so they do not dry out. The sperm swim through internal fluids, produced by both the male and the female, to reach the eggs. The female lays an **amniotic egg** (fig. 31.16), which is almost the same as the egg of a chicken. The egg provides the embryo with a waterproof covering (the shell), a supply of "pond water," plenty of food, and a place for storing wastes. The embryo converts its urine into uric acid crystals, which do not dissolve and pollute the egg. It remains within the egg until ready to fend for itself. The young reptile then hatches out, looking like a small version of the adult.

A spectacular group of reptiles, the dinosaurs, dominated the land from 225 to 65 million years ago. More than 300 species of dinosaur have been described, and they exhibited every life-style available to vertebrate animals. Some species returned to water and lived like whales; others lived like lizards, birds, or mammals. Some were tiny and others were gigantic. All the dinosaurs went extinct 65 million years ago, along with approximately two-thirds of the plants and other animals. The cause of this massive die-off is discussed in the accompanying Research Report.

Birds Birds form the class Aves. They evolved from a reptile, most likely one of the smaller dinosaurs. The oldest fossil bird, *Archaeopteryx*, lived 150 million years ago. A blend of reptilian and bird features, this crow-sized animal had feathers like a bird but small teeth and a long tail like a reptile. Its claws were curved, like those of modern perching birds, apparently for grasping the branches of trees. *Archaeopteryx* had sturdy collarbones (the wish-

bone) but no sternum (breastbone) to anchor flight muscles. It must have been a clumsy flier.

Modern birds and reptiles have similar body forms and reproduce in almost the same way, with internal fertilization and an amniotic egg. Many biologists think of birds as "glorified reptiles" rather than a unique form of vertebrate animal. Nearly all the differences between birds and reptiles involve adaptations for flight.

A bird is adapted for flight by having (1) feathers (modified scales), which control the flow of air; (2) hollow bones, which lighten the load; (3) a modified skeleton, which stabilizes the chest, shifts the weight beneath the wings, and anchors the wing muscles; and (4) superb distance vision.

Flight requires more energy than walking or running. To provide this energy, birds have adaptations that increase the rate of aerobic cellular respiration. They have more efficient lungs than any other animal, for more rapid uptake of oxygen. They are also warm-blooded (endothermic), with a high body temperature that speeds up cellular reactions, including aerobic respiration.

Mammals Mammals form the class Mammalia. Hair and mammary glands (which secrete milk) are two traits that distinguish mammals from other forms of animals. Other features include endothermy, a diaphragm for ventilating the lungs; a middle ear, formed of three tiny bones, for amplifying sound; and a large cerebral cortex for intelligent behavior.

We have good fossil evidence of the evolution of mammals. The transition began with an early form of reptile, the therapsids (fig. 31.17), approximately 225 million

What Happened to All the Dinosaurs?

The oldest dinosaur fossil, discovered in 1991, is 225 million years old. This animal, named *Eoraptor,* was a bipedal predator about the size of a greyhound (fig. 31.C). As a dinosaur, it was quite primitive: the jaws have no hinge to help grasp prey and the pelvis did not allow the animal to rest on its haunches. The descendants of this reptile and its relatives evolved into a great variety of forms, from bipedal predators the size of a turkey to quadripedal herbivores that weighed more than 100 tons.

Dinosaurs dominated the landscape for 110 million years. They spread across the supercontinent of Pangaea, adapting to almost every habitat and way of animal life. Then, 65 million years ago, they vanished. What happened to all the dinosaurs?

Something dramatic must have occurred on earth about 65 million years ago, since dinosaurs were not the only organisms that went extinct at this time. Between 60% and 80% of all existing species disappeared at the same time. This massive die-off is explained by two hypotheses: the impact hypothesis and the climatic change hypothesis.

The impact hypothesis states that a giant comet slammed into the earth, pulverizing rock and spewing dust and sulfur dioxide gas into the air. The dust and gas blocked incoming solar radiation, causing the surface of the earth to become cold and dark. The sulfur dioxide formed sulfuric acid, which returned to earth as acid rain. Photosynthesis was greatly impeded. Plants died, followed by the herbivores, and then the carnivores. The dinosaurs were completely wiped out.

What is the evidence in support of this hypothesis? High concentrations of iridium have been found within 65-million-year-old clay. Iridium is a fairly heavy atom that is abundant in comets but exceedingly rare in the earth's

Figure 31.C The oldest known dinosaur, *Eoraptor,* attacking its prey.

crust. Then, in 1991, a huge crater, 180 kilometers (108 mi) in diameter, was discovered just off the coast of the Yucatan Peninsula in Mexico. The crater, apparently formed by a large comet, has been aged and found to be 65 million years old. The iridium and crater are strong evidence that an enormous comet did crash into the earth about the same time that the dinosaurs and other organisms disappeared.

The climatic change hypothesis is that a gradual change in climate produced a gradual disappearance of species over hundreds of thousands of years. A comet may have crashed to the earth, but its effects were merely the final blow. Evidence in support of this hypothesis lies in the fossil record, which shows many extinctions during the 100,000 years or so prior to 65 million years ago.

Further testing of these two hypotheses will involve more careful analysis of the fossil record. Recently, proponents of both hypotheses analyzed fossils of foraminifera (marine protozoa), which are considered the most reliable measure of the pace of extinction. Results support the impact hypothesis: many species of foraminifera went extinct within a short period of time approximately 65 million years ago.

years ago. The therapsids were enormously successful reptiles that dominated the land for 50 million years until replaced by dinosaurs. They were unusual in having larger brains than other reptiles and in having mammal-like teeth, consisting of incisors, canines, and molars rather than the uniform, cone-shaped teeth of other reptiles. Some therapsids may have been warm-blooded and probably had hair, since traces of whiskers can be seen on some of the fossils.

Figure 31.17 A therapsid. According to the fossil record, a reptile like this (but much smaller) evolved into the first mammal around 225 million years ago.

The first mammals resembled shrews and opossums (fig. 31.18). They were small animals that fed on insects and were active mainly at night, perhaps to avoid the dinosaurs. They layed eggs like their reptilian ancestors but suckled the hatchlings with milk.

Mammals remained small, nocturnal insectivores until approximately 65 million years ago, when the dinosaurs disappeared. They then rapidly evolved into a great variety of forms. The three major lines of mammalian evolution are the monotremes, marsupials, and placentals. They differ chiefly in mode of reproduction.

Monotremes are the most reptilian of mammals. They were once abundant and diverse, but only two kinds remain today: the duck-billed platypus and the spiny anteater. These unusual mammals live only in the Australian region (Australia, New Guinea, Tasmania, and adjacent islands). The female lays eggs, like a reptile. When the young hatch, they lick milk that oozes from her mammary glands; there are no nipples.

Marsupials occur mainly in the Australian region. Examples include the kangaroo, wallaby, koala bear, and opossum. The female does not lay eggs. She has a placenta for nourishing the young inside her body. The marsupial placenta, however, does not grow large enough to support the embryo for very long. The young are still embryos when born (an opossum newborn is the size of a honeybee). The tiny newborns crawl into a pouch on their mother for further development. There they attach to nipples on the mammary glands.

Throughout the world, except for the Australian region, almost all mammals are placentals. The female has a placenta that grows large enough to support the embryo within the uterus for a long period of development. The newborns are well developed; some can stand and run within hours of birth. They suckle milk from nipples on their mother's mammary glands.

Figure 31.18 An opossum. The first mammals probably resembled opossums. Fossils indicate they were small insect-eaters, apparently active only at night.

The unusual distribution of mammals, with monotremes and marsupials restricted largely to the Australian region and placentals everywhere else, can be explained by the breakup of the supercontinent Pangaea. Monotremes and marsupials, which evolved first, had already spread throughout the supercontinent before it broke into the separate continents we see today. Placentals, however, had only just begun to spread and not yet reached lands that eventually became Antarctica and Australia (fig. 31.19). After the breakup, all the continents had all three kinds of mammals except Australia and Antarctica, which had only monotremes and marsupials. Antarctica drifted southward and soon became too cold for terrestrial mammals.

Why are there so few marsupials and no monotremes outside the Australian region? We do not know, for sure, what happened to these mammals. Most likely they could not survive the competition with placentals for food, shelter, and living space.

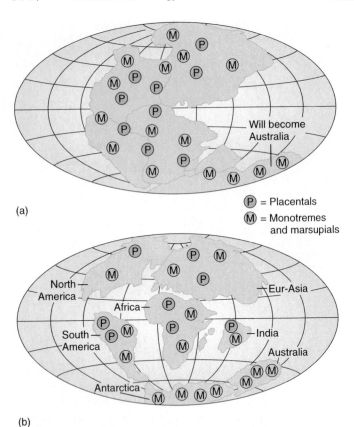

P = Placentals

M = Monotremes and marsupials

(a)

(b)

Figure 31.19 Pangaea and the distribution of mammals. (a) Before the breakup of Pangaea, monotremes and marsupials were found throughout the supercontinent, but placentals had not yet spread to the land that would become Australia. (b) After the breakup, Australia and Antarctica had monotremes and marsupials but no placentals. Elsewhere, all three groups occurred but in time the placentals outcompeted all the monotremes and most of the marsupials except in the Australian region.

SUMMARY

1. All forms of life appear to be descended from the same ancestral cell. The oldest fossil cells are 3.5 billion years old. A long period of chemical evolution, which formed first monomers and then polymers and membranes, must have preceded appearance of the first cell.

2. The first fossil cells are prokaryotic cells. Modern forms of prokaryotes, the bacteria, form the kingdoms Archaea and Eubacteria. The Archaea are a chemically distinct group adapted to extreme environments. The Eubacteria include a great variety of bacteria; some feed on living organisms, some on dead organisms, and some carry out photosynthesis.

3. The first eukaryotic cells appeared 2.1 billion years ago. Modern unicellular eukaryotes form the kingdom Protista, which can be divided into three groups based on how energy and building materials are acquired.

The algae are photosynthetic autotrophs. The slime molds and water molds feed by secreting digestion enzymes onto organic materials. The protozoa feed by ingesting organic materials.

4. A plant is a multicellular organism, with eukaryotic cells (enclosed within cell walls), that carries out photosynthesis. The first plant, a multicellular alga, evolved in coastal waters about 1.4 billion years ago. The kingdom Plantae includes nonvascular and vascular plants.

5. Nonvascular plants include the multicellular algae, which live submerged in water, and the bryophytes, which live on land in moist habitats. They have no xylem, phloem, roots, or leaves. The bryophytes, however, have stomata and a cuticle. Reproduction requires water because of a swimming sperm.

6. Vascular plants include the ferns, gymnosperms, and angiosperms. All have a cuticle, stomata, xylem, phloem, roots, and leaves. The ferns can live in dry habitats but their swimming sperm and unprotected embryos require water. Most modern gymnosperms have pollen grains (rather than swimming sperm) and seeds with waterproof coats; they can reproduce in dry places. Angiosperms have the same adaptations as gymnosperms plus flowers, which promote animal pollination, double fertilization, and seed dispersal via fruits.

7. The kingdom Fungi consists of multicellular organisms, with eukaryotic cells (enclosed within cell walls), that acquire food by releasing digestive enzymes onto organic material. The body of a fungus is a mycelium formed of hyphae. Fungi include the yeasts, molds, and mushrooms.

8. The kingdom Animalia consists of multicellular organisms, with eukaryotic cells, that acquire food by ingesting organic material. All but one of the many phyla consist of invertebrates, which have no backbone. Sexual reproduction involves a swimming sperm and in most cases a larval stage.

9. The simple invertebrates have radially symmetrical bodies with specialized cells and tissues. Sponges are filter feeders with intracellular digestion. Animals in the phyla Cnidaria and Ctenophora have a body wall, tentacles, and a digestive cavity with one opening. Platyhelminthes, which are intermediate in complexity, have bilateral symmetry and organ systems.

10. The complex invertebrates have a coelom. They form two groups, based on patterns of reproduction: the phyla Mollusca, Annelida, and Arthropoda in one group and the Echinodermata in another group. Most of these animals have organ systems and a digestive tract with two openings. The annelid worms and arthropods have segmented bodies. The echinoderms are advanced animals without body segments, kidneys, or bilateral symmetry in the adult form.

11. The phylum Chordata consists of three subphyla: Vertebrata, Urochordata, and Cephalochordata. All chordates have a notochord, dorsal hollow nerve cord, and pharynx with gill slits. According to the tunicate larva hypothesis, the first fish evolved from a urochordate larva that colonized a river.

12. According to fossil evidence, amphibians evolved from fish with lungs and fleshy fins supported by bones. Adaptations of adult amphibians include lungs, a tongue, and four legs. External fertilization and a fish-like larva, however, tie reproduction to life near water.

13. Reptiles evolved from amphibians that adapted to a dry climate. The adaptations include a waterproof exterior, ribs that ventilate the lungs, a penis, and an amniotic egg.

14. Birds resemble reptiles except they have adaptations for flight: feathers, hollow bones, modified skeleton, acute distance vision, efficient lungs, and endothermy. The first birds probably evolved from some form of dinosaur.

15. Mammals are vertebrates with hair and mammary glands. Other features include endothermy, a diaphragm, and a large cerebral cortex. The first mammals evolved from a therapsid reptile. The three major kinds of mammals are monotremes, marsupials, and placentals.

KEY TERMS

Amniotic (AEM-nee-AH-tik) **egg** (page 611)
Bilateral symmetry (page 603)
Coelom (SEE-lohm) (page 604)
Dehydration synthesis (page 593)
Dorsal hollow nerve cord (page 609)
Eukaryotic (YOO-kaer-ee-AH-tik) **cell** (page 595)
Gill slits (page 609)
Hyphae (HEYE-fay) (page 602)
Larva (LAHR-vah) (page 602)
Lichen (LEYE-kin) (page 602)
Monomer (MAH-noh-muhr) (page 592)
Mycelium (meye-SEE-lee-uhm) (page 602)
Mycorrhiza (MEYE-koh-REYE-zah) (page 602)
Nonvascular plant (page 598)
Notochord (NOH-toh-kord) (page 608)
Pharnynx (FAER-enks) (page 609)
Polymer (PAH-lih-muhr) (page 592)
Prokaryotic (PROH-kaer-ee-AH-tik) **cell** (page 594)
Radial symmetry (page 603)
Segmentation (page 606)
Vascular plant (page 598)

STUDY QUESTIONS

1. Why was the synthesis of polymers so difficult prior to the first cells? Why does it occur so readily within cells?

2. Describe the major difference between a bacterium and a protist.

3. Compare unicellular algae, slime and water molds, and protozoa with regard to how they acquire energy and building materials.

4. Why is sexual reproduction in a fern confined to moist places whereas reproduction in a conifer is not?

5. Describe three ways in which flowers differ from cones.

6. Why are fungi, which look like plants, not placed in the kingdom Plantae?

7. Compare the ways in which food is processed in a sponge, sea anemone, flatworm, and annelid worm.

8. Describe the origin of fishes. Include in your answer (a) a description of their likely ancestor (a urochordate), (b) the kind of environment in which the first fish evolved, (c) the characteristics of a fish that make it a chordate and that make it a vertebrate, and (d) how early fishes differed from modern fishes.

9. Compare reproduction in a frog with reproduction in a lizard.

10. Explain the geographic distribution of monotremes, marsupials, and placentals.

CRITICAL THINKING PROBLEMS

1. If cells formed from inorganic molecules several billion years ago, why don't we see this process occurring today?

2. Why is it likely that cells arose millions of years earlier than 3.5 billion years ago, the age of the oldest fossil cell?

3. Why is it more likely that molecules of RNA, rather than DNA, formed the first genes?

4. If you found a fossil of just a fragment of a plant, how could you tell whether it was a moss or a fern?

5. Why did gymnosperms thrive after formation of Pangaea?

6. If you found a fossil of a multicellular organism, how could you tell whether it was a plant, fungus, or animal?

7. What is the advantage of bilateral symmetry? Why have some animals remained radially symmetrical?

8. What is the evidence that vertebrate animals evolved from an ancestor of modern tunicates?

9. Suppose you discovered a new form of animal, not yet described by biologists. Explain how you would decide the phylum to which it should be assigned.

10. If you found a fossil animal, how could you distinguish between a bird and a reptile?

SUGGESTED READINGS

Introductory Level:

Brusca, R. C., and G. J. Brusca. *Invertebrates*. Sunderland, MA: Sinauer Associates, 1990. A very readable account, phylum by phylum, of the evolution, anatomy, physiology, embryology, and ecology of invertebrate animals.

Laybourn-Parry, J. *A Functional Biology of Free-Living Protozoa*. Berkeley: University of California Press, 1985. Descriptions of all kinds of protozoa, including their modes of locomotion, cell structures, and ways of life.

Margulis, L., and D. Sagan. *Four Billion Years of Evolution from Our Microbial Ancestors*. New York: Simon and Schuster, 1991. Written for the layperson, this thought-provoking book explores the roles of bacteria in the evolution of biochemical pathways, eukaryotic cells, and multicellular forms of life.

Schopf, W. (ed.). *Major Events in the History of Life*. Boston: Jones and Bartlett, 1991. Results of a symposium written for college students. Many papers dealing with chemical evolution and early forms of cells.

Stewart, W. N. *Paleobotany and the Evolution of Plants*. New York: Cambridge University Press, 1983. A review of plant fossils and what they tell us about plant evolution.

Wilson, E. O. *The Diversity of Life*. Boston: Belknap Press of Harvard University, 1992. A discussion of the processes that generate biodiversity, the major periods of mass extinctions, and why biodiversity should be preserved.

Advanced Level:

Deamer, D. W., and G. R. Fleischacker (eds.). *Origins of Life: The Central Concepts*. Boston: Jones and Bartlett, 1994. A carefully chosen set of papers, arranged by research themes, along with commentary, that reviews what we have learned about formation of the first cells.

Knoll, A. H. "The Early Evolution of Eukaryotes: A Geological Perspective," *Science* 256 (May 1, 1992): 622–27. Comparisons of phylogenies of eukaryotes based on molecular similarities and on fossils.

Morris, S. C., J. D. George, R. Gibson, and H. M. Platt (eds.). *The Origins and Relationships of Lower Invertebrates*. New York: Clarendon (Oxford University Press), 1985. Papers on the evolutionary relations among invertebrates, using information about reproduction, development, cell structure, biochemistry, and fossils.

Scagel, R. F., R. J. Bandoni, J. R. Maze, G. E. Rouse, W. B. Schofield, and J. R. Stein. *Nonvascular Plants: An Evolutionary Survey*. Belmont, CA: Wadsworth Publishing, 1982. A sourcebook of information on slime molds, fungi, and algae, written by experts in each area.

Schultze, H. P., and L. Trueb (eds.). *Origins of the Higher Groups of Tetrapods: Controversy and Consensus*. Ithaca, NY: Comstock (Cornell University Press), 1992. A description of what is known about the origin of each class of vertebrate and the relationships among classes, comparing information from fossils and comparative biochemistry.

Taylor, P. D., and G. P. Larwood (eds.). *Major Evolutionary Radiations*. New York: Clarendon Press, 1990. A collection of papers that deals with major episodes of diversification of plants and animals.

✸ EXPLORATION

Interactive Software: Is Anybody Out There?

This interactive exercise explores the probability that we are not alone in the universe–that life exists on planets within other galaxies. The student is provided with a map of the universe and can vary such factors as the age of a galaxy, the mass of a star, the probability that a star has planets, and the probability that a planet has an atmosphere. Each variation changes the number of systems on which life can be expected, and therefore the probability of life elsewhere within the universe.

Population Ecology

Objectives

In this chapter, you will learn
–about niches;
–the factors that influence population
  growth;
–the effects of competition;
–about plant defenses against the
  animals that eat them;
–about predator adaptations for
  obtaining prey;
–about prey adaptations for avoiding
  predators;
–about species that interact in
  mutually beneficial ways.

A population of scarlet ibis in the
Caroni Swamp of Trinidad.

surprising feature of nature is that it appears to be in balance. Populations of plants and animals remain pretty much the same from one year to the next. This constancy is surprising when we consider that every organism is capable of producing more than one offspring, and so every population is capable of rapid growth. What keeps all the populations in check?

Our best clues about how nature stays in balance come from episodes of imbalance. Jackrabbits, for example, have occasionally become extremely abundant in the western states of Nevada, Utah, and Idaho. When this happens, it doesn't take a professional ecologist to notice the change. Anyone driving through these lands at night hears frequent thumps as the car hits jackrabbits.

Humans cause this abundance of jackrabbits by killing their predators. In the West, a rancher's life is dominated by a struggle with wild predators that prey on domestic cattle and sheep. Most of the wolves and mountain lions have been eliminated from ranchlands, but the wiley coyote persists. To rid the ranchlands of coyotes, meat laced with strychnine is set out near the herds. As the coyotes die, the calves and lambs survive better but another problem soon develops—jackrabbits become so abundant and eat so much grass that the cattle and sheep have a hard time finding enough to eat. (Eight jackrabbits eat as much as one sheep, and forty-one eat as much as one cow.) The balance of nature becomes upset.

Under natural conditions, jackrabbits do not become so abundant that they overgraze the land. Jackrabbits produce many young, but their populations remain small because most of the young are eaten by coyotes. The elimination of coyotes from ranches enables most young jackrabbits to survive and reproduce. The population then becomes unnaturally large.

Cases of population outbreaks, such as this one, are dramatic in contrast with the year-to-year stability of other populations. Apparently, populations are normally kept in check by interactions with their physical environment as well as their herbivores predators, parasites, and competitors. Nature gets out of balance when the natural relations among populations are altered, which most often happens as a consequence of human activities.

Populations

A **population** is a group of individuals of the same species that live together in the same geographic region. It is a step above individuals in the hierarchy of life (see inside front cover of this book). **Ecology** is the study of interactions between organisms and their environment, including both living and nonliving components. Population ecology is the study of a population's resources, its interactions with the populations of other species, and how these resources and interactions influence its abundance. Each population has a unique set of ecological traits: a niche, a rate of growth, and a carrying capacity.

The Niche

The set of resources used by a population is its **niche.** A complete list of resources for any population would be very long and would include many items, such as sunlight and oxygen, that are always abundant. Ecologists rarely describe the entire niche. They concentrate, instead, on the resources that are likely to be in short supply and, consequently, to influence population density and competition.

Animal Niches   The most important resources in the niches of animal populations are food and shelter. Acquiring food involves three variables: what the animals eat, when they look for food, and where they search. Acquiring shelter, such as a burrow, crevice, or nest, involves characteristics of the environment, such as softness of the soil or height of the vegetation. Building a shelter requires finding a suitable location as well as materials, such as twigs and grasses.

Not all individuals of an animal population use the same resources in the same way. Differences in bill size within a population of finches, for example, may produce differences in the size of seeds that are eaten. Learned behaviors also contribute to individual differences. An animal that finds food in a particular place, for example, is likely to search in that same place again. Individual differences in resource use tend to enlarge the niche of the population.

Different species that live in the same region have different niches. Sometimes it is chiefly one aspect of the niche that differs. A great horned owl and a red-tailed hawk, for example, both eat small mammals and forage in similar places, but the owl hunts during the night and the hawk hunts during the day. By foraging at different times, they end up taking different species of small mammals. Swifts, swallows, and bats live in the same region but hunt insects in different places and at different times. Swifts and swallows both forage during the day, but swifts fly higher than swallows. Swallows and bats forage at about the same height but bats hunt during the night.

The anatomy of an animal reflects the kinds of resources it uses. Pine martens, for example, hunt for squirrels in trees. Their bodies are short and thick (fig. 32.1a), with sharp claws for climbing. Long-tailed weasels, which are close relatives of pine martens, hunt for mice and pikas in the crevices of rocks. Their bodies are long and narrow (fig. 32.1b), with soft footpads for stealth. Perhaps the most obvious relations between anatomy and niche involve the size and shape of a bird's bill, as shown in figure 32.2.

(a)

(b)

Figure 32.1 The anatomy of an animal reflects its use of resources. (a) A pine marten's short, thick body and long, sharp claws help it catch squirrels in trees. (b) A long-tailed weasel's longer, more narrow body and soft footpads help it sneak up on small mammals in crevices between rocks.

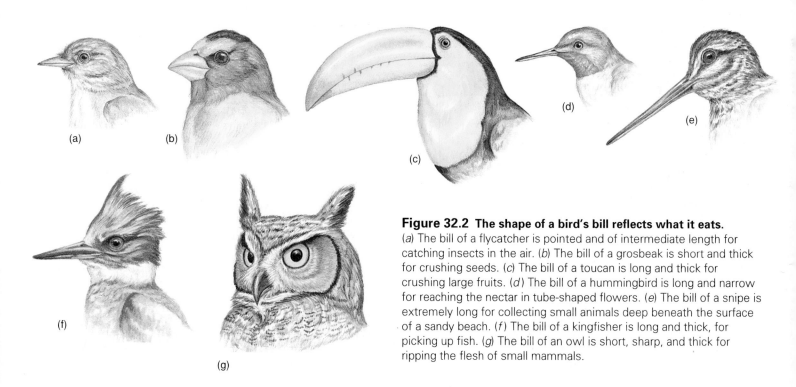

(a)

(b)

(c)

(d)

(e)

(f)

(g)

Figure 32.2 The shape of a bird's bill reflects what it eats. (a) The bill of a flycatcher is pointed and of intermediate length for catching insects in the air. (b) The bill of a grosbeak is short and thick for crushing seeds. (c) The bill of a toucan is long and thick for crushing large fruits. (d) The bill of a hummingbird is long and narrow for reaching the nectar in tube-shaped flowers. (e) The bill of a snipe is extremely long for collecting small animals deep beneath the surface of a sandy beach. (f) The bill of a kingfisher is long and thick, for picking up fish. (g) The bill of an owl is short, sharp, and thick for ripping the flesh of small mammals.

Plant Niches The most important resources in the niches of plant populations are typically moisture, sunlight, a source of nitrogen (ammonium or nitrate), and a source of phosphorus (phosphate). All plants need these materials, but the quantities required depend on specific adaptations. A cactus, for example, requires less soil moisture than a violet.

The anatomy of a plant reflects its use of resources. A barrel cactus lives in open environments, where there is plenty of sunlight but little moisture. It has no leaves (they are modified as spines), an adaptation that conserves water, and has transferred photosynthesis to cells of its broad stem, which also stores water. A fern, by contrast, lives on the forest floor, where there is little sunlight but a lot of moisture. It has large leaves, which lose a lot of water but absorb what little sunlight penetrates to the forest floor. The narrow stem of a fern, largely underground, supports the leaves and transports fluids between the roots and the leaves. Other adaptations of plants to their niches are shown in figure 32.3.

Population Growth

Whether a population grows, remains stable, or declines in number reflects its relations with the environment. The change in number of individuals within a population is a function of the number of births, deaths, immigrants, and emigrants.

The population growth rate, abbreviated r, is the change in population size per individual per unit time:

$$r = \frac{\text{Change in number of individuals per unit time}}{\text{Total number of individuals at beginning of time interval}}$$

It adjusts for different-sized populations (an addition of 10 individuals to a population of 10 is greater than an addition of 10 individuals to a population of 100). The growth rate of the human population, for example, is .017 per year, which means that for every person alive now, there will be .017 more people next year. It is sometimes expressed as a 1.7% growth rate, which means that for every 100 people alive now, there will be 1.7 more people next year.

The birth rate and death rate of a population are governed by life history traits, such as the age at which an organism begins to reproduce and the survival of offspring (described following). The maximum growth rate occurs only when conditions are optimal. At other times, the maximum rate is curbed by environmental factors. A poor food supply, for example, may delay maturity and predators may consume most of the offspring.

Carrying Capacity The **carrying capacity** (abbreviated as **K**) of a population is the maximum density of individuals that the environment can support. This upper limit is set by the particular resource that is in short supply—the one that the growing population runs out of first. It is sel-

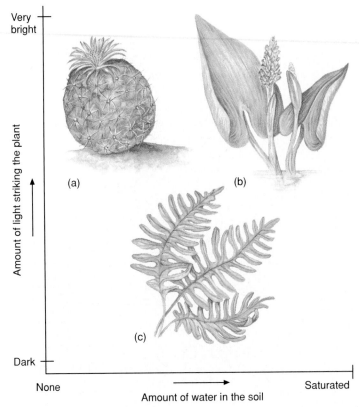

Figure 32.3 The anatomy of a plant reflects its resource use. All plants require water and light, but the quantities required vary from plant to plant, depending on its anatomy and other adaptations. (a) A pincushion cactus requires full exposure to sunlight, because of its small surface (the stem) for absorbing light, but does not require a lot of water, because it has no leaves (a major site of water loss). (b) A pickerel plant requires large quantities of water, because it loses so much out of its leaves, and full exposure to sunlight, because it has just two leaves. (c) A fern requires large quantities of water, because so much water evaporates out of its leaves, but requires little light, because its leaves provide a large surface area for absorption. (No environment is both dry and dark: dry environments do not support plants that are large enough to shade others.)

dom a constant but varies as a function of the weather and the activities of other species. The carrying capacity of algae in a pond may be set by the concentration of phosphates. The carrying capacity of wolves may be set by the availability of elk.

A population remains at carrying capacity when it uses resources at the same rate that they form. Prairie dogs, for example, are at carrying capacity when they eat grasses at the same rate as the grasses grow. A population that exceeds its carrying capacity, with more individuals than the environment can support, will return to its carrying capacity as it experiences lower birth rates and/or greater death rates. A population below its carrying capacity will reach its carrying capacity as it experiences higher birth rates and/or lower death rates.

In territorial birds, birth rates decline as the population nears carrying capacity. Only individuals with territories, which contain food and nesting sites, can reproduce. Birds without territories may survive but they do not reproduce. In forest trees, by contrast, death rates decline as the population nears carrying capacity. A forest at carrying capacity has no spaces left for new trees, so all the young trees die.

Exponential Growth   **Exponential growth** of a population is an important concept in biology. (It is called exponential because r, the population growth rate, is an exponent in the mathematical equations that describe growth.) Every population is capable of this kind of very rapid growth; it occurs whenever the growth rate r is greater than zero for a prolonged period of time. The longer the exponential growth continues, the more rapidly the population grows:

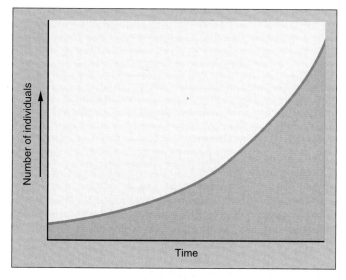

Such rapid growth cannot continue indefinitely, however, for eventually the population runs out of some critical resource.

A population grows at an exponential rate whenever conditions allow. Such a situation may develop when an environmental change raises the carrying capacity. Or it may happen when the population drops well below carrying capacity. A catastrophe may kill so many individuals that the population drops well below its carrying capacity (K):

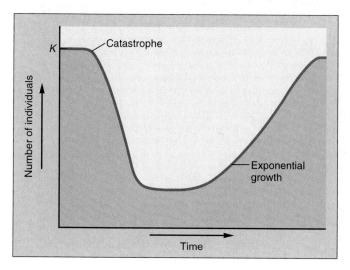

An extremely cold winter, for example, may kill so many warblers that the following spring the few survivors find more caterpillars than they can eat. The well-fed birds lay many eggs and the hatchlings survive well.

Populations that are normally held below their carrying capacity by predators or parasites will grow exponentially when this source of death is removed:

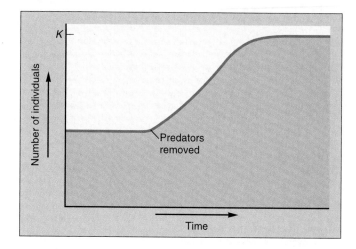

When mountain lions are eliminated, for example, deer populations grow rapidly.

A similar situation develops when species are introduced into new areas where they have no natural enemies. Examples of species that experienced exponential growth after being introduced into North America include starlings, English sparrows, cattle egrets, pheasants, Mediterranian fruit flies, gypsy moths, fire ants, dandelions, and Russian thistles. Rapid growth of introduced species is very disruptive to native species, as described in the accompanying Environment Report.

Populations in northern latitudes, such as Alaska, Canada, and Scandinavia, undergo regular cycles of abundance, with population peaks every three years or ten years, depending on the species. The reason these populations cycle and more southern populations do not has perplexed biologists for many years. Research on one of these cycles is described in the accompanying Research Report.

Selection for Life History Traits   Characteristics of an organism's life that influence population growth rates are called life history traits. They include the age of first reproduction, number of offspring per season, and number of reproductive seasons.

Life history traits are influenced by both genes and environment. Each female inherits the potential for producing a certain number of offspring. This reproductive potential, however, is not realized when environmental conditions are less than optimal. A pair of kit foxes, for example, has life history traits that could generate twenty or thirty pups in a lifetime, but insufficient food prevents them from having so many pups and predators kill many that are born.

An Invasion by Zebra Mussels

The balance of nature is upset by invasions of new species, which happen most often when humans transport organisms from one continent to another or to an island. In the new environment, the transplanted species often find few enemies. Zebra mussels are an example of such an invasion.

The zebra mussel, scientifically named *Dreissena polymorpha*, is a tiny mollusk, about the size of your thumbnail, that has invaded the Great Lakes of North America. Its original home is the Caspian Sea in eastern Europe. It spread, by attaching to boats, into canals, rivers, and lakes of western Europe more than 150 years ago. Then, sometime in 1986, it hitched a ride in water taken on board a cargo ship as ballast (the water provides stability in the open seas). When the ship entered the Great Lakes, and no longer needed the extra weight, the water was released and the zebra mussel found a new home. Within two years the tiny animal had invaded all five of the Great Lakes. It has now spread to the Hudson, Susquehanna, and Mississippi Rivers, as well as many smaller rivers and lakes of the United States and Canada.

Populations of zebra mussels grow rapidly in North America, for they have few enemies and can reproduce rapidly. An adult female lays about 40,000 eggs a year, which, when fertilized, develop into microscopic larvae. The larvae are carried by currents for several weeks until they find a firm surface, such as a rock, boat, or shell of another animal (fig. 32.A). They then attach to the surface and develop into adults. The adults begin to reproduce when they are one year old and continue to reproduce until they are about five years old. They are extremely hardy; they can go without food for almost a year and can live out of water

for as much as two weeks under humid conditions. The ability of these populations to grow is remarkable: in February of 1989, there were between 500 and 1,000 mussels per square meter on the walls of an intake canal at a power plant on Lake Erie. By July, there were more than 700,000—a 700-fold increase in six months.

The zebra mussel is a threat to other species of freshwater animals. Its main food is unicellular algae, which is also the food of other small invertebrates, which, in turn, are the main food of yellow perch, large mouth bass, and small mouth bass. As populations of the zebra mussel grow, populations of algae, small invertebrates, and fishes decline. Moreover, the mussels cover the spawning grounds of walleyes and other fishes, so these animals have no place to lay their eggs. They kill crayfish by attaching to their eyes and pincers. They kill clams by attaching to their shells in such large numbers that the shells become too

heavy to open for feeding. One count revealed as many as 10,000 zebra mussels on a single native mussel.

This tiny animal is a nuisance to industries along the shores of the Great Lakes: huge piles of mussels clog pipes to power plants and water-treatment facilities, as well as the cooling systems of boat engines. They cover the hulls of ships and anchor chains, and they settle so thickly on navigational markers that their weight sinks the markers.

While some evidence indicates that the original zebra mussel has reached its maximum abundance in the Great Lakes, a new species of zebra mussel has now invaded. This species can live in a broader range of temperatures and salt concentrations and may invade warmer lakes and estuaries, where rivers join the ocean. The challenge to biologists is to find some weakness in these zebra mussels—some way of eliminating them without endangering native animals.

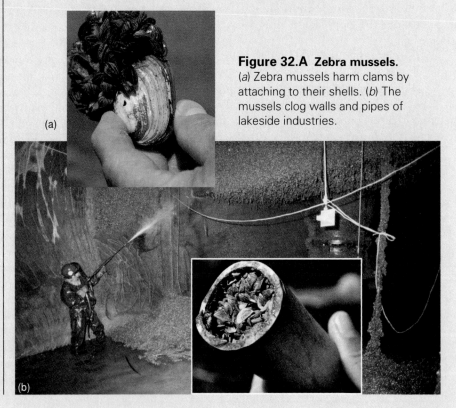

Figure 32.A Zebra mussels.
(*a*) Zebra mussels harm clams by attaching to their shells. (*b*) The mussels clog walls and pipes of lakeside industries.

Populations that oscillate between rapid growth followed by rapid decline are subject to **r selection,** in which *r* stands for rate of population growth. In this type of natural selection, individuals with traits that promote reproductive output (as opposed to survival) leave more genes in the next generation than other individuals in the population. During times of rapid growth, almost every individual survives to old age. The traits that differ among individuals, and form the basis of natural selection, are the traits that influence reproduction. An individual that reproduces more successfully is one that reproduces fast: begins to reproduce at an earlier age and has more offspring each season. Life history traits that are *r* selected are listed in table 32.1.

Table 32.1

Selection for Life History Traits

Trait	*r* selection (population below *K*)	*K* selection (population at *K*)
Body size	Small	Large
Age of first reproduction	Early	Late
Number of offspring per season	Many	Few
Offspring size	Small	Large
Number of reproductive seasons	Few	Many
Life span	Short	Long

The white-footed mouse (fig. 32.4*a*) is an example of an *r*-selected animal. These rodents experience periods of rapid population growth every spring, when food becomes abundant, and as they colonize disturbed habitats. A female has her first litter when she is only six weeks old. She typically has six pups in each litter and can produce three litters a year for up to five years. Plants that colonize disturbed sites, such as roadsides, are *r*-selected species. Lamb's-quarter (fig. 32.4*b*), for example, grows from a seed in the spring and produces between 100,000 and 200,000 tiny seeds before it dies in the autumn.

Populations that remain at their carrying capacity most of the time experience **K selection,** in which *K* stands for carrying capacity. At this maximum density, there are not enough resources to go around and the premium is on getting more than one's share. Selection favors traits that improve competitive ability: an individual that can monopolize resources will be able to live for a long time and reproduce many seasons. Good competitors typically have larger bodies and, consequently, a longer period of growth prior to reproduction. They produce only a few offspring each season, thereby conserving their bodies for additional reproductive seasons. Traits that are *K* selected are listed in table 32.1.

A black bear (fig. 32.5*a*) is an example of a *K*-selected animal. It does not begin to reproduce until at least three or four years old. A female then has a litter of just two pups every other year for its life span of roughly thirty years. Similarly, an oak tree (fig. 32.5*b*) is a *K*-selected plant. It delays reproduction for about a decade, putting all its energy and nutrients into growth in order to

(a)

(b)

Figure 32.4 Organisms that are *r* selected. (*a*) A white-footed mouse weighs about 30 grams (1 oz) and is just 10 centimeters (4 in) long. It can have a hundred offspring within its life span of a few years. (*b*) Lamb's-quarter (also called pigweed) grows to about 60 centimeters (2 ft) and produces more than 100,000 tiny seeds, each less than 0.3 centimeters (0.1 in) long, in its short lifetime of less than a year.

Figure 32.5 Organisms that are *K* selected. (*a*) A black bear weighs approximately 180 kilograms (400 lb) and is 1.8 meters (6 ft) long. It can have at most about twenty-six cubs within its life span of thirty years. (*b*) A burr oak does not begin to reproduce until it is at least ten years old and 24 meters (80 ft) tall, when it has a monopoly over light, moisture, and soil nutrients. An average oak then produces 5,000 acorns, each about 5 centimeters (2 in) long, every year until it is about 450 years old.

(a)

Research Report

Lemmings in the Tundra

The Arctic tundra, with its long, dark winters and short summers, is a harsh place to live. It is home to a variety of grasses and wildflowers, as well as knee-high birch and willow trees. The kinds of animals are few. An abundant herbivore is the lemming, a small mammal that is preyed

Figure 32.B The arctic lemming and its predator, the arctic fox.

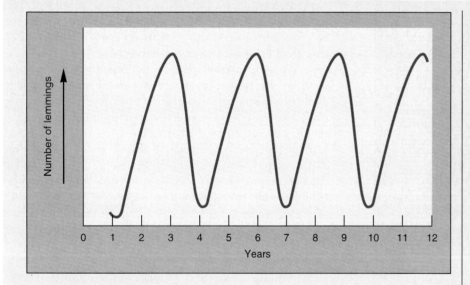

upon by the arctic fox (fig. 32.B). Lemmings are known for their dramatic cycles of abundance. Every three years or so, they become extremely numerous:

What causes these regular rises and declines in population density? Decades of research on lemmings have generated many hypotheses (fig. 32.C) but no definite answer as yet. A

current hypothesis, called the nutrient recovery hypothesis, combines overgrazing with slow rates of decomposition. During a good year, the soil contains sufficient inorganic nutrients, such as calcium and phosphorus, for luxuriant growth of grasses. The lemmings find plenty of nutritious food, they reproduce with great success,

and the population grows larger and larger. They consume tundra plants faster than the plants can grow.

As the lemmings eat most of the vegetation, nutrients are transferred from the plants to the lemmings. In the following summer, most of the inorganic nutrients of the tundra are within the lemmings rather than within the soil. Without these soil nutrients, plants cannot grow.

During that summer of meager plant growth, the lemmings do not get enough to eat. The starving animals wander over the tundra and die of a variety of causes—starvation, disease, and predators. (In Scandinavia, with its long coastline, many lemmings fall from cliffs into the ocean, for they do not see well.) The population dwindles to fewer than 1% percent of the animals alive during the peak periods.

As the lemmings die and their bodies decompose, the inorganic nutrients hoarded within their bodies are released to the soil. There is a time lag in the recovery of nutrients, however. Decomposition is slow in the cold arctic soils; it takes a couple of

(b)

monopolize moisture, nutrients, and sunlight. Once established as a large tree, the oak reproduces every year for several centuries.

Intraspecific Competition

Members of a population compete whenever resources are in short supply, which occurs when the population is at carrying capacity. This type of competition, between members of the same species, is known as **intraspecific competition** (*intra*, within).

Good year

Overgrazed year

Plant-recovery year

years for all the nutrients within dead lemmings to return to the soil.

As inorganic nutrients move out of the dead lemmings and into the soil, the tundra plants again enjoy favorable conditions for new growth. The abundant plant growth enables the lemmings to once again reproduce with great success. Another bumper crop of hungry young are born and another cycle begins. Each cycle takes three years:

The plant-lemming cycles of abundance apparently trigger other cycles in the tundra. The carrying capacity of the arctic fox, which is determined by the supply of lemmings, rises and falls in three-year cycles. Geese and shorebirds, too, have three-year cycles because the foxes feed more on them when lemmings are scarce.

These events emphasize the interdependence of all components of the environment. The relations among soil, plants, and animals are more clearly seen in the tundra, because there are fewer species and the changes are so profound. In warmer environments, where more species live together, dramatic changes are rare because of so many checks and balances among the different species. It is unlikely that a single species, like the lemmings, could become so abundant—other predators, in addition to the main one, would switch their diets to eat the abundant species before it overate its food supply.

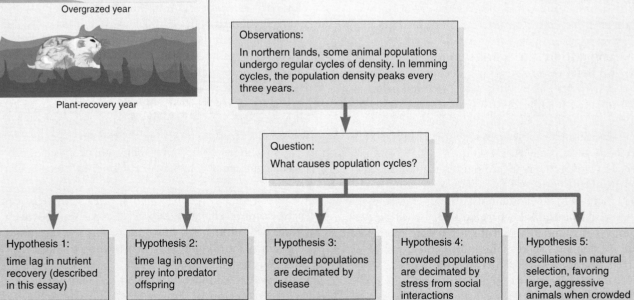

Observations:

In northern lands, some animal populations undergo regular cycles of density. In lemming cycles, the population density peaks every three years.

Question:

What causes population cycles?

Hypothesis 1:
time lag in nutrient recovery (described in this essay)

Hypothesis 2:
time lag in converting prey into predator offspring

Hypothesis 3:
crowded populations are decimated by disease

Hypothesis 4:
crowded populations are decimated by stress from social interactions

Hypothesis 5:
oscillations in natural selection, favoring large, aggressive animals when crowded and small, timid animals when sparse

Figure 32.C Scientific logic regarding population cycles.

Intraspecific competition tends to enlarge the niche of a population. Any individual that can make use of resources that other individuals cannot will experience less intraspecific competition and obtain more resources; it will survive and reproduce more successfully than others.

Niche expansion is seen most often when animals colonize islands. An island has fewer species than a mainland and is likely to have some kinds of food, shelter, and other resources that remain unused. The scrub jay, for example, colonized islands off the coast of southern California. On the mainland it lives in the same areas as the thrasher, where the jay feeds on insects in trees and the thrasher feeds on insects in the ground. Thrashers are poor fliers and have not colonized the islands. In the absence of this ground-feeder, the scrub jays on the islands feed both in trees and on the ground. Their niche expanded as some jays took up feeding on the ground, like thrashers, and discovered a rich supply of food.

Competition among members of the same species comes in two forms: scramble competition and contest competition.

Scramble Competition In **scramble competition** (also known as exploitative competition), there are no rules governing the division of resources; all individuals "scramble" for their share. This type of competition, with no evolved way of resolving it, is common in populations that are seldom near carrying capacity and so have not experienced selection for behaviors that minimize squabbling.

When scramble competition is intense, most members of the population get some of the resources but few get enough to reproduce. This situation is seen most often in populations of herbivores that are normally kept below their carrying capacity by predators, but for some reason the predator population fails to eat enough of its prey. The abnormally large herbivore population may overgraze its food plants. Examples of populations with scramble competition are deer, lemmings, jackrabbits, and grasshoppers.

Contest Competition In **contest competition** (also known as interference competition), certain individuals get first access to resources. This type of competition is seen in *K*-selected populations. Where resources are in short supply, any individual with traits that enable it to take more than its share of the resources will survive and reproduce more successfully than others.

Contest competition in animals is resolved, typically, by ritualized contests. Social behaviors (see page 514), including advertisements, threats, and fights, determine which animals own territories or dominate other animals. An animal that wins a territory monopolizes food, nesting sites, and mates within that space; an animal that dominates others gains first access to resources. The rituals select for greater strength, stamina, and intimidation, as well as for more effective weapons (such as horns and antlers) and larger bodies. Contest

---

Table 32.2

Interactions Between Species

Name of Interaction	Effects on the Two Species
Interspecific competition	Both are harmed
Herbivory	Plant is harmed; herbivore benefits
Predation	Prey is harmed; predator benefits
Parasitism	Host is harmed; parasite benefits
Mutualism	Both benefit

---

competition in plants involves monopoly of space. It selects for individuals that grow more rapidly and form larger roots, taller stems, or broader canopies.

Interactions Between Species

No population lives isolated from populations of other species. Populations of different species compete with one another, eat one another, and interact in ways that are mutually beneficial. The major kinds of interactions are listed in table 32.2.

Interacting species influence one another's evolution. This type of evolution, in which evolutionary changes in one species cause evolutionary changes in the other and vice versa, is known as **coevolution** (see page 571).

Interspecific Competition

Competition between individuals of different species is **interspecific competition** (*inter*, between). Most often, the competition is of the scramble type and selects for increased efficiency of resource use rather than monopoly. The organism that uses its limited amount of resources more efficiently leaves more offspring in the next generation. Competition between species is usually less intense than competition within species, for two individuals of different species are less similar than two individuals of the same species.

Prolonged competition between two species has two possible outcomes: competitive exclusion or niche evolution.

Competitive Exclusion Two species with very similar niches cannot coexist for long. Competition between the species will ultimately cause the population of one species to become more abundant, while the population of the other species becomes less and less abundant until it goes extinct. (Extinction of a population does not necessarily mean extinction of the species.) This phenomenon is known as **competitive exclusion.**

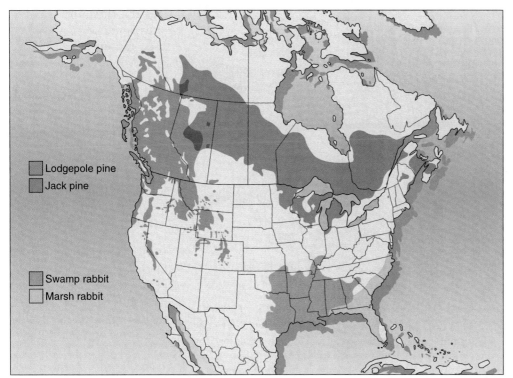

Lodgepole pine
Jack pine

Swamp rabbit
Marsh rabbit

Figure 32.6 Competitive exclusion leads to geographic replacements. Swamp rabbits and marsh rabbits are ecologically too similar to coexist. Their geographic distributions are adjacent but do not overlap. Lodgepole pine and jack pine are ecologically very similar. Their geographic distributions overlap in only two small regions of Canada.

Except for introduced species, such as garden plants and starlings, we seldom see competitive exclusion in action. We see instead the outcome of competitive exclusion: geographic replacements, in which one species occurs in one geographic location and a similar species in another geographic location (fig. 32.6).

Niche Evolution The second possible outcome of interspecific competition is niche evolution, in which the niches of competing species become smaller and less similar. Such changes would reduce the competition and enable similar species to coexist. Niche evolution in response to interspecific competition remains a hypothesis rather than a principle of ecology. While we observe that coexisting species differ to some extent in their use of resources, it is difficult to demonstrate that the differences evolved in response to competition in the past.

Similar species cannot have identical niches; they have different gene pools and so are ecologically different in some way. One species is always more efficient, however slight, at converting resources into offspring. Every generation, the more efficient species will leave more offspring and the less efficient species will leave fewer offspring. In time, only the more efficient species will remain. The less efficient species will be excluded from the region.

Gardeners are familiar with competitive exclusion: unless continually tended, the flowers and vegetables are outcompeted by weeds. Some plants secrete poisons from their roots, a form of chemical warfare called allelopathy. The poison spreads through the soil around the plant and kills seedlings of other species. Walnut trees, for example, kill apple trees and roses, asters kill yellow poplar, ragweed kills sugar maples, and creosote bushes kill burro weed by allelopathy.

Starlings are replacing bluebirds by competitive exclusion. Bluebirds are native to North America, whereas starlings were introduced from Europe. Starlings were brought to New York City in 1890 as part of a romantic plan to have in America all the birds mentioned by Shakespeare. Starlings and bluebirds nest in the same places: tree holes of a certain size. Starlings are much better than bluebirds at obtaining these holes and using them to multiply. Wherever both species occur, the starlings become more abundant and the bluebirds become less abundant. By outcompeting bluebirds and other native birds, starlings have spread throughout North America.

According to the niche evolution hypothesis, interspecific competition selects against mutual use, by individuals of two species, of the same resource. Consider, for example, two finch species that feed on seeds of various sizes. One species has a larger bill and feeds more efficiently on larger seeds; the other species has a smaller bill and feeds more efficiently on smaller seeds. When food is scarce, birds with larger bills that feed (less efficiently) on smaller seeds will not obtain enough food to reproduce, and this type of individual will disappear from the population. Also, birds with smaller bills that feed (less efficiently) on larger seeds will fail to leave genes in the next generation. In time, the niches of the two species will diverge.

A classic example of similar species that coexist, perhaps because of niche evolution, is seen in five species of warblers that live together in the spruce forests of Maine. These species, called MacArthur's warblers after the ecologist who studied them, have similar niches: they all feed in spruce trees at the same time of day and on the same kinds of food (mainly caterpillars). Each species is restricted, however, to foraging in only a small portion of the tree, as shown in figure 32.7. Their niches differ in this respect. In accordance with the niche evolution hypothesis, the history of these warblers may be as follows. All are descended from a single species that foraged in all parts of the tree. Different populations of the species became geographically isolated, probably by glaciers, and formed five species with almost identical niches. The geographical barrier disappeared and the five species came to live

627

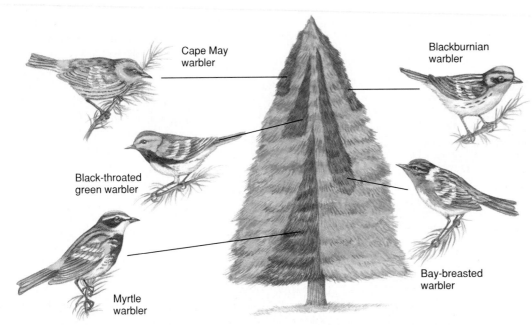

Figure 32.7 Coexisting species. Five species of warblers live in the same spruce trees but forage mainly in different parts of the trees. These different kinds of warblers had a common ancestor, from which they speciated, and once had almost identical niches. According to the niche evolution hypothesis, competition among the species then selected for smaller and more different niches, which enables them to coexist today.

in the same spruce forests. Interspecific competition then led to niche evolution: the niches became smaller and more different, with each species restricted to a particular region of the spruce tree.

We have seen that intraspecific competition tends to enlarge the niche, whereas interspecific competition tends to compress it. A population that is at carrying capacity experiences both kinds of competition, and the size of its niche is a dynamic equilibrium between two opposing forces: intraspecific competition and interspecific competition.

Herbivores and Plants

An herbivore is an animal that eats living plants. Horses, antelope, deer, and prairie dogs are familiar kinds of herbivores. Large animals like these, however, do not affect the vegetation nearly as much as insects. Most vegetation that is eaten is consumed by insect larvae, such as caterpillars (fig. 32.8), because they are so numerous and they eat so much—some caterpillars eat five times their weight each day.

Surprisingly, most vegetation is not eaten by any herbivore. Roughly 90% of all plant leaves die without being eaten. Why don't herbivores eat more food, leave more offspring, have larger populations, and consume more of the plants? Why is the world so green?

Part of the answer to these questions is that herbivore populations do not reach high densities because they are attacked by predators and parasites. The leaves of sweet corn, for example, might be completely consumed by caterpillars were it not for parasitic wasps. These wasps lay their eggs in-

side the caterpillars, and as the young wasps develop they consume the caterpillars. A sweet corn plant actually recruits these wasps: when its leaves are chewed, the plant releases molecules that attract wasps.

Another part of the answer to why the world is green is that plants are not defenseless. Plants have evolved a variety of defenses that protect against herbivores.

Plant Defenses Three kinds of defenses are seen in plants: (1) mechanical defenses on the plant surface, including spines, irritating chemicals, and sticky fluids (fig. 32.9); (2) structural molecules that interfere with digestion, such as tannin and silica; and (3) molecules that are toxic to herbivores. In this section, we look more closely at the effect of plant toxins on insect herbivores.

Plants synthesize a great variety of chemicals that kill insects. Some of these "insect poisons" also affect the human body, and we use them as stimulants, depressants, hallucinogens, spices, and medicines. Examples include nicotine, caffeine, cannabinoids (in marijuana), opium, cocaine, salicin (resembles aspirin), and herbal teas.

Some plants synthesize molecules that mimic insect hormones, particularly the hormones that govern insect development. Fir trees synthesize juvenile hormone, which prevents an insect larva from becoming an adult. The larva grows very large but cannot reproduce, so there are few

Figure 32.8 Insect herbivores have the greatest impact on vegetation. Here, the caterpillar of a gypsy moth (an introduced species) is consuming a leaf. (Note the small holes, where other caterpillars have tasted the leaves.)

Figure 32.9 Mechanical defense of a plant against herbivores. This aphid is caught in a sticky glue secreted by spikes on a potato leaf.

Figure 32.10 Camouflage in an insect herbivore. This katydid is hard to see because it resembles the leaves on which it feeds. If it began to feed on another type of plant, say a sunflower, this particular camouflage would no longer protect it against bird predators.

herbivores of its sort the following year. Ferns and many flowering plants synthesize ecdysone, which causes an insect to become an adult too soon; the abnormally small adult produces few offspring.

Insect Herbivores as Food Specialists A plant and the insects that eat it are coevolved species: the evolution of toxins in plants leads to the evolution in herbivores of ways to counter these compounds, and vice versa.

Every kind of insect herbivore alive has evolved the capacity to deal with at least one kind of plant toxin, either by breaking it down into nontoxic materials or storing it someplace in the body where it cannot damage body cells. An insect eats only the plants that produce the chemical defenses that it can handle. For this reason, insect herbivores are food specialists; each species of insect feeds on just a few species of plants.

A food specialist must spend time looking for its particular kind of food. As the insect wanders in search of its food plants, it is likely to be seen and eaten by a predator. Few individuals find enough food to eat or survive long enough to reproduce; the population rarely becomes so large that it decimates the food plant. (Exceptions do, of course, occur when an herbivore outbreak has a noticeable effect on the abundance of its food plant. Most often these outbreaks are of introduced species with few enemies in their new environments.)

Caterpillars that feed on sugar maple trees demonstrate the difficulties of being an herbivore. The leaves of these trees contain tannin, a molecule that tastes bad and blocks the activity of digestive enzymes. An insect that eats a tannin-loaded leaf not only has a bad experience but cannot digest the leaf. The concentration of tannin varies dramatically from leaf to leaf, since it costs energy to make and the tree conserves energy by not loading each leaf. A caterpillar crawls from one leaf to another, tasting each (leaving a visible hole) in its search for an acceptable meal. More than two-thirds of the leaves are rejected because they contain too much tannin. Caterpillars are typically camouflaged while feeding, but they become conspicuous to birds when moving about in search of edible leaves. Most caterpillars are eaten before they become adult moths and butterflies. A caterpillar's foraging strategy is a compromise between the need to feed and the need to remain concealed from its enemies.

Insect herbivores remain food specialists because it is so difficult to add a new plant species, with a different kind of toxin, to the diet. To do so is likely to require the simultaneous evolution of (1) the ability to detoxify or store the new kind of poison; (2) the ability to recognize the food plant (feeding is typically a genetically programmed behavior triggered by the smell of the plant toxin itself); and (3) a new camouflage, to match the new plant species, for protection against predators (fig. 32.10).

Predators and Their Prey

A predator is an animal that kills and eats other living animals. A prey is an animal eaten by a predator. Predators and their prey are coevolved: the predator population evolves hunting adaptations in response to characteristics of its prey, and the prey evolves antipredator adaptations in response to characteristics of its predator. Because of this never-ending "arms race," the predator population cannot excel at catching its prey and the prey population cannot excel at evading its predator. A wolf and a horse, for example, have evolved in response to each other and can both run at about the same speed—45 kilometers (27 mi) an hour at full gallop and 18 kilometers (11 mi) an hour when trotting.

Adaptations of Predators A predator must be proficient at finding its prey, catching it, and killing it. Each task requires a particular set of adaptations (table 32.3). Some of these adaptations are shown in figure 32.11.

Each predator is restricted, because of its adaptations, to feeding on only certain kinds of prey. A predator with the sprinting ability to catch rabbits, for example, cannot also have the endurance to catch antelope. Predators are

Table 32.3

Adaptations of Predators to Their Prey

Task	Examples
Finding prey	Appropriate sense organs for detecting the prey
	Instinctive knowledge of where to search
	Knowing when to stop searching for one kind of prey, which has become scarce, and begin searching for another (i.e., the ability to make a decision)
Catching prey	Speed, which can be either for short distances (sprinting) or long distances (endurance)
	Stealth
	Movements of appendages that distract and lure the prey
	Social behavior that facilitates group hunting
	Ability to predict the prey's escape movements
Killing prey	Weapons (teeth, claws, and poisons)
	Striking where the prey's body is vulnerable
	Jaws that enable swallowing of the entire, live prey

(a)

(b)

(c)

Figure 32.11 Predator adaptations for capturing prey.
(a) An eagle's beak and talons are adapted for capturing mammals. (b) A snake's jaws can unhinge so it can swallow large items of food (an egg shown here). (c) An anglerfish lures smaller fish near its mouth with an appendage that looks like food to a smaller fish.

not as restricted as herbivores in the kind of food they eat, for the meat of different animals is generally edible. They are restricted, instead, in the kind of animals they can find, catch, and kill.

A familiar example of predator specialization is the domestic dog. Humans have selected varieties of dogs for hunting different game: beagles for hunting rabbits, labradors for ducks, and terriers for gophers. No one variety of dog excels at hunting all these different kinds of prey.

Adaptations of Prey Antipredatory traits can be grouped into three kinds of tasks: to avoid being found, caught, and eaten. Adaptations for these tasks are listed in table 32.4.

Prey populations experience strong selection for traits that help them avoid being eaten, since a large portion of each generation falls victim to predators. Many of the more obvious traits of animals are antipredator adaptations (fig. 32.12).

Mimicry is a special kind of antipredator adaptation. Two common forms of mimicry are Müllerian mimicry and Batesian mimicry, named after the two biologists who first described them.

Müllerian mimicry is an antipredator adaptation in which two or more undesirable prey species appear the same. When a predator has a bad experience with an individual of one species it will afterward avoid anything that appears the same, including individuals of the other species. An undesirable prey is one that tastes bad, induces vomiting, or stings the predator. Undesirable insects have the same bright colors, called warning colors, that make it easier for a bird to recognize prey that give it a bad experience. The yellow and black colors of hornets and wasps, which warn a bird of their sting, are an example of Müllerian mimicry.

Batesian mimicry is an antipredator adaptation in which a good-tasting species (the mimic) looks like a bad-tasting species (the model). A predator that samples one of the models learns not to eat anything that resembles it and will avoid eating the mimics as well as the models. Batesian mimics are especially common among insects (fig. 32.13).

For Batesian mimicry to work, the predator must sample the bad-tasting model more often than it samples the good-tasting mimic. Insect mimics achieve this sampling ratio in four ways: (1) by having smaller populations than their model; (2) by having two appearances, in which some individuals in the population mimic one model and others mimic another; (3) by emerging from their eggs or cocoons later than the models, when the predators have already learned from sampling only the models; and (4) by being smaller than the models, since predators prefer larger prey and, therefore, sample the models more often.

Mutualisms

A **mutualism** is a relationship between two species in which both benefit. The relationship between humans and dogs is a familiar example of a mutualism.

Table 32.4

Adaptations of Prey Animals to Their Predators

Task	Examples
Avoid being found	Avoid being near others in the same population
	Camouflage and freezing posture
	Different color or shape from other prey species
Avoid being caught	Appropriate sense organs for detecting predators
	Group living, which provides more sensory organs for detecting predators
	Speed, either sprinting or endurance
	Unpredictable escape movements
	Flash behaviors: presentation of a color or sound while moving but not when stopped
	Startle behaviors: sudden release of an image, sound, or smell that frightens the predator
Avoid being eaten	Weapons (teeth, claws, armor, quills, poisons)
	Taste bad (combined with conspicuous appearance, slow movements, and aggregated distributions)
	Müllerian mimicry: an undesirable prey species resembles another undesirable prey species
	Batesian mimicry: a good-tasting species resembles a bad-tasting species

---

The best-known mutualisms are between flowering plants and the animals that disperse their pollen and seeds. The plant benefits from the relationship by having its pollen (sperm) transferred to the female parts of another plant and its seeds transferred to sites where they will not compete with the parent. The animal benefits by feeding on the flower's nectar, pollen, or fruits (e.g., cherries and strawberries) around the seeds. These two kinds of mutualisms are described in chapter 15.

A classic example of mutualism occurs between certain species of acacia trees and tropical ants. The acacia provides food and shelter for the ants. It offers proteins,

(a)

(c)

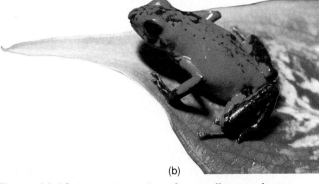

(b)

(d)

Figure 32.12 Prey adaptations for evading predators.
(*a*) A hare can attain speeds of 75 kilometers (47 mi) per hour.
(*b*) The red color of a poisonous frog warns off potential
predators. (*c*) A bombardier beetle squirts toxic fluid at its
enemies. (*d*) A porcupine is protected from predators by its
sharp quills.

Ladybug   Roach

Figure 32.13 Batesian mimicry. A good-tasting roach
(identified by its longer antennae) mimics a bad-tasting ladybug
(identified by its shorter antennae).

fats, and sugars, which accumulate within special organs
on its leaves and leaf petioles. It offers hollow thorns
where the colonies can live, sheltered from weather and
enemies. In return, the ants swarm over and sting any herbivore that attempts to feed on the acacia.

Some butterfly larvae, or caterpillars, have mutualistic relations with ants in which the caterpillars provide ants
with food in return for protection against parasitic wasps.
The caterpillars have three organs that promote this relation: glands that secrete a sugary solution as food for the
ants, glands that secrete airborne molecules that trigger defense behavior in the ants, and organs that vibrate the leaf
as a way of calling ants to the caterpillar (fig. 32.14).

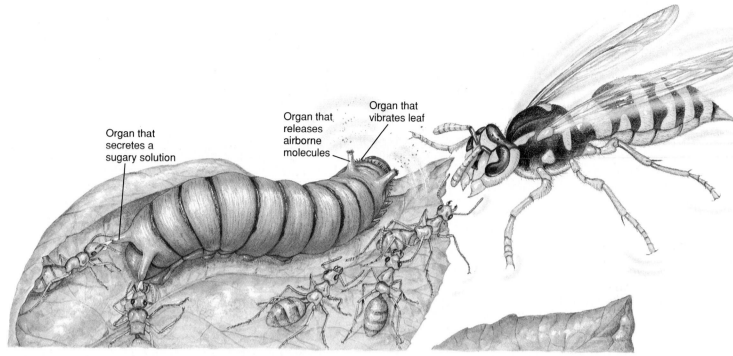

Organ that secretes a sugary solution

Organ that releases airborne molecules

Organ that vibrates leaf

Figure 32.14 Mutualism between caterpillars and ants.
The caterpillar gives ants food in return for protection against wasps. The caterpillar has three kinds of organs for attracting ants: organs that secrete a sugary solution, which the ants feed on; organs that vibrate the leaf, calling ants to the caterpillar; and organs that release airborne molecules, which stimulate the ants to attack the wasp.

SUMMARY

1. Population ecology is the study of interactions between populations and their environment. A population has a niche, which is the set of resources it uses; a carrying capacity (*K*), which is maximum density the environment can support; and a rate of growth (*r*). When the population is below carrying capacity, it may grow exponentially.

2. The population growth rate depends on life history traits, which include the age at first reproduction, number of offspring per season, and number of reproductive seasons. Populations that seldom reach carrying capacity are *r*-selected; populations that are usually at carrying capacity are *K*-selected.

3. When a population is at carrying capacity, its members experience intraspecific competition. Some populations have scramble competition, in which there is no organized way of dividing the resource, and others have contest competition, in which individuals who hold spaces or have higher social rank have first access to the resource.

4. When populations of different species use the same resource, they experience interspecific competition. This type of competition may lead to competitive exclusion or niche evolution.

5. Herbivores and their food plants are coevolved. Plants evolve defenses, mainly poisons, against their herbivores. Herbivores evolve the ability to deal with at least one of these defenses and so to eat at least one kind of plant. Their populations are controlled by the difficulty in finding food and by predators.

6. Predators and their prey are coevolved. Predator adaptations help them find, catch, and kill their prey. Prey adaptations help them avoid being found, caught, and killed by their predators. Mimicry is an antipredator adaptation. Müllerian mimicry is when undesirable prey species look alike, and Batesian mimicry is when good-tasting species resemble bad-tasting ones.

7. Mutualisms are coevolved relations in which both species benefit.

KEY TERMS

Batesian (BAYT-see-un) **mimicry** (page 631)
Carrying capacity (**K**) (page 620)
Coevolution (page 626)
Competitive exclusion (page 626)
Contest competition (page 626)
Ecology (page 618)
Exponential growth (page 621)
Interspecific competition (page 626)
Intraspecific competition (page 625)
K selection (page 623)
Müllerian (muh-LEHR-ee-uhn) **mimicry** (page 631)
Mutualism (MYOO-chew-al-iz-um) (page 631)
Niche (nitch) (page 618)
Population (page 618)
r selection (page 623)
Scramble competition (page 626)

STUDY QUESTIONS

1. Describe, using a real or imaginary animal, how anatomy reflects the niche. Give specific details.
2. Describe a situation that would lead to exponential growth of a population.
3. How would an *r*-selected population differ from a closely related *K*-selected population in terms of body size, number of offspring per season, and life span? Explain why you expect these differences.
4. Many leaves turn color and die in the autumn. Why weren't these leaves consumed by herbivores during the summer?
5. Explain why a predator species does not excel at getting prey and a prey species does not excel at avoiding predators. Use an example to clarify your explanation.
6. What is the difference between Müllerian mimicry and Batesian mimicry? What is the advantage of being a Müllerian mimic? of being a Batesian mimic?

CRITICAL THINKING PROBLEMS

1. Describe an organism, real or imaginary, in a situation that allows exponential growth. Give specific details about the environment.
2. Choose an organism and describe how you would quantify (describe in numbers rather than in words) its niche.
3. Do human populations have carrying capacities? If so, what are the limiting resources and what happens to maintain zero population growth when carrying capacity is reached?
4. The American population of humans has been growing exponentially for several hundred years. Explain the ecological conditions that enable this continuous increase in numbers.
5. You observe that two species coexist and have similar niches. Design an experiment to test whether they are experiencing interspecific competition.
6. Islands have fewer species than areas of similar size on a continent. Explain why these species also have larger niches.

SUGGESTED READINGS

Introductory Level:

Edmunds, M. *Defence in Animals*. New York: Longmont, 1974. A well-illustrated, easy-to-read description of antipredatory adaptations in animals.

Howe, H. F., and L. C. Westley. *Ecological Relationships of Plants and Animals*. New York: Oxford University Press, 1988. An introductory text for college students, with particular emphasis on plant-herbivore interactions and mutualisms.

Owen, D. *Camouflage and Mimicry*. Chicago: University of Chicago Press, 1980. A well-illustrated introduction to two main categories of antipredatory adaptations.

Wickler, W. *Mimicry*. New York: World University Library (McGraw-Hill), 1968. A classic, well-illustrated book on the various kinds of mimicry.

Advanced Level:

Crowley, M. J. (ed.). *Natural Enemies: The Population Biology of Predators, Parasites, and Diseases*. Cambridge, MA: Blackwell Scientific Publications, 1992. A summary of what is known about the coevolution and population dynamics of predator-prey and host-parasite systems, as well as the regulation of populations by their natural enemies.

Pimm, S. L. *The Balance of Nature?* Chicago: University of Chicago Press, 1991. An exploration of factors that affect the balance of nature, including aspects of food-web structure, community structure, variability in space and time, introductions, and extinctions.

Roth, D. A. *The Evolution of Life Histories*. New York: Chapman and Hall, 1992. An analysis of each life history trait, including its selective advantages, constraints, and relations to reproductive effort.

Roughgarden, J., R. M. May, and S. A. Levin (eds.). *Perspectives in Ecological Theory*. Princeton, NJ: Princeton University Press, 1989. Discussions of the theoretical basis of population ecology, including population growth, competition, predator-prey relations, and parasite-host relations.

33

Ecosystems

Objectives

In this chapter, you will learn
–how ecosystems are functional units
  of nature;
–how chemical energy flows through
  the organisms of an ecosystem;
–how nutrients cycle between
  organisms and the physical
  environment of an ecosystem;
–the way in which an ecosystem
  develops from bare rock or soil;
–how similar ecosystems form biomes.

A pond ecosystem. This unit of
nature consists of a set of organisms
and aspects of the physical
environment with which they interact.

W e humans tend to think of ourselves as visitors to nature—places where plants and animals live, but not us. This view of ourselves is an illusion, for we are closely connected to other kinds of organisms in a complex web of life.

The interconnections among all organisms on earth become evident when something goes wrong. An example is the explosion that blew up a nuclear reactor in Chernobyl, Ukraine, in 1986. Nine tons of radioactive atoms, chiefly iodine and cesium, were spewed into the air. This amount is roughly a hundred times more than the amount of radioactivity released by atomic bombs dropped on Japan during World War II.

While the Chernobyl explosion was a major human disaster, it provided an unusual opportunity for researchers to trace the movements of atoms through nature. Migrations of the radioactive atoms were traced by using machines called Geiger counters. The atoms fell to earth in rain and snow, mostly in northern parts of Scandinavia, which are downwind of Chernobyl. In these regions of tundra, lichens were the first to accumulate the radioactive atoms, and they did so in large quantities. Lichens, which are formed of fungi and unicellular algae, turn out to be "radiation sponges," since they acquire nutrients directly from the air. Later, algae and plants also accumulated radioactive atoms, from lake water and soil.

High concentrations of radioactivity then turned up in caribou, which feed mainly on lichens in the tundra. Radioactive atoms also turned up in geese, which feed on algae, and in cows, which feed on plants. Later the atoms appeared in the predators of these animals: wolves, foxes, and humans. In some of the people, the radioactivity caused birth defects, immune deficiencies that resemble AIDS, thyroid cancers, and leukemia.

Atoms move continually from the soil, water, and air to various kinds of organisms, including humans, and then back again to the nonliving environment. One year they are part of an elephant, a few years later part of a shark, and then a palm tree, beetle, chicken, and human. Organisms and their physical environment are interconnected by the atoms they share.

Ecosystem Components

Biologists subdivide nature into ecosystems. Each **ecosystem** consists of a set of interacting populations together with aspects of the physical environment. It is a functional unit of nature, through which energy and atoms move as plants and algae build organic molecules and other organisms eat them. A particular meadow, lake, or woods, for example, is an ecosystem (fig. 33.1).

No ecosystem is isolated from others, since materials travel from place to place within the air, water, and organisms. A molecule of carbon dioxide used by a woodland violet may have been released in the breath of a desert bob-

Figure 33.1 Ecosystems. This photograph shows three ecosystems: spruce-fir forest, meadow, and river. Each ecosystem consists of a unique set of organisms and physical factors but is not isolated from neighboring ecosystems.

Table 33.1

The Components of an Ecosystem

The Organisms	Examples
Producers	Grass, pine tree, green alga
Consumers	
Herbivores	Rabbit, grasshopper, zooplankton (tiny aquatic animals)
Primary carnivores	Coyote, meadowlark, perch
Secondary carnivores	Mountain lion, hawk, bass
Decomposers	Fungus, bacterium, vulture, shark

The Physical Environment	What It Provides
Solar radiation	Light and heat
Gases in the air	Carbon, oxygen, and nitrogen
Water	Hydrogen, oxygen, a liquid for chemical reactions, a way of cooling organisms
The earth's crust	Ions, shelter

cat. Moreover, there are no clear boundaries between adjacent ecosystems. A fox, for example, may be a component of two ecosystems, capturing rabbits in a meadow and chipmunks in a nearby pine forest. The components of an ecosystem are listed in table 33.1 and described following.

The Organisms

The millions of organisms within an ecosystem are grouped according to how they acquire organic monomers. Organic monomers (📖 *see page 40*), are small molecules from which larger molecules, called polymers (📖 *see page 40*), are constructed. Important monomers are glucose, amino acids, fatty acids, and nucleotides. They are used by cells to build proteins, polysaccharides, lipids, and nucleic acids.

An ecosystem has three groups of organisms: producers, consumers, and decomposers.

Producers   A **producer** is an organism that builds its own monomers from inorganic materials. Most producers build sugars by photosynthesis (📖 *see page 106*); they are plants, unicellular algae, and photosynthetic bacteria. (A few build sugars by chemosynthesis, using the energy of inorganic molecules.) Producers use some of the sugar for energy and convert the rest into amino acids, fatty acids, and other kinds of monomers. All the organic monomers within an ecosystem originate with the producers, as only they can build organic molecules from inorganic materials.

In an oak forest ecosystem, for example, mosses, wildflowers, shrubs, and trees synthesize sugar and convert some of it into other kinds of monomers. They join these monomers to form cellulose, starch, enzymes, cell membranes, and other components of cells. The cells, in turn, form leaves, stems, roots, and seeds.

Consumers   A **consumer** acquires its monomers from other live organisms. It consumes other forms of life and digests their polymers into monomers:

The monomers move into the consumer's cells, where they are used as sources of energy and building blocks for constructing the consumer's own kinds of polymers. All animals are consumers, as are some fungi, protista, and bacteria.

An oak forest ecosystem, for example, has many consumers. There are rabbits and caterpillars, which eat leaves; foxes, which eat rabbits; warblers, which eat caterpillars; and hawks, which eat warblers. In the forest soil, insects, spiders, mites, fungi, and protozoa consume tiny soil organisms.

Consumers are grouped according to the general type of organisms they consume. Herbivores eat producers; primary carnivores eat herbivores; secondary carnivores eat primary carnivores; tertiary carnivores eat secondary carnivores; and so on. A carnivore that is not eaten by any other consumer is a top carnivore in the ecosystem. Mountain lions, wolves, eagles, large mouth bass, and tuna are examples of top carnivores.

Decomposers   A **decomposer** acquires monomers from dead organisms, the parts of organisms (e. g., fallen leaves and feathers), and feces. It uses these monomers as sources of energy and for building its own kinds of polymers, excreting what it cannot use into the surrounding environment. Earthworms, many fungi, and many bacteria are important decomposers.

Scavengers are large decomposers, such as hyenas, pigs, vultures, ravens, and sharks. Most scavengers eat living animals as well as dead ones. They are both carnivores and decomposers.

Decomposers transfer organic molecules from dead materials into their own bodies and convert them ultimately to inorganic materials, which they release into the environment. This process, called **decomposition,** makes inorganic molecules and ions available again to the producers. Without decomposers, the atoms of life would become tied up in dead organisms, and the producers would have no raw materials for building new organic monomers.

In the oak forest ecosystem, a dead leaf falls to the ground and is decomposed by soil organisms. Many kinds of decomposers feed on the leaf (fig. 33.2), each specialized for digesting certain kinds of molecules. Fungi arrive early and digest the softer parts, leaving the leaf skeleton. Then come bacteria, which digest the tougher molecules of the skeleton. As the bacteria feed, they are eaten, along with the leaf skeleton, by earthworms. Inside the earthworms, some of the skeleton is digested and some passes out of the digestive tract, along with the bacteria, as earthworm feces. The feces are eaten again and again by earthworms and other soil animals until all the molecules of the leaf are digested.

The Physical Environment

The physical environment provides organisms with energy, atoms, and shelter. All aspects of the physical environment that are required by organisms are components of the ecosystem. They include radiation from the sun, gases in the air, water, and ions in the earth's crust.

Solar Radiation, Gases, and Water   Radiation from the sun supplies an ecosystem with heat and light. Heat warms the organisms and light furnishes the energy for photosynthesis.

Some of the atoms used by organisms come from gases within the air. Carbon dioxide provides carbon and oxygen atoms for building sugars. Nitrogen gas provides nitrogen for building amino acids and other organic molecules built in part from nitrogen. Oxygen is used in cellular respiration, the chemical reaction that releases energy from organic monomers (especially sugars and fatty acids).

Water is a major component of organisms. Its hydrogen atoms are used to build sugar and it is the fluid within which all biochemical reactions take place. Water also cools an

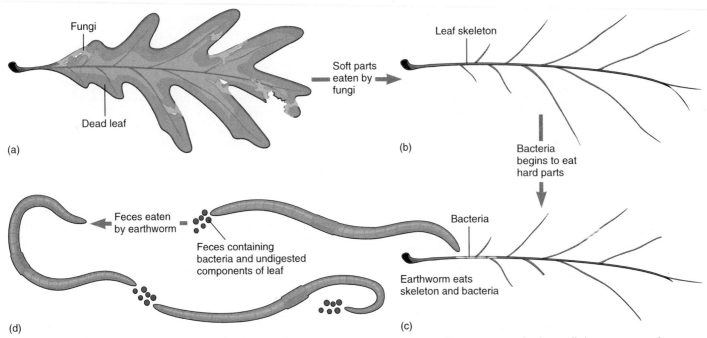

Figure 33.2 Decomposition of a leaf by decomposers.
(*a*) Fungi digest the soft parts of a fallen leaf. (*b*) Bacteria then begin to digest the skeleton of the leaf. (*c*) Earthworms eat the disintegrating leaf skeleton (and the bacteria), digesting some of it and defecating the undigested portions as well as the bacteria. (*d*) The worms and other soil decomposers then eat these feces, again and again, each time digesting more of the skeleton until all its organic molecules have become organic molecules within the decomposers and inorganic molecules within the soil and air.

organism; an overheated plant or animal releases fluids onto its body, and when the water evaporates it removes heat.

Crust of the Earth   Many of life's atoms come from rocks in the crust of the earth. Atoms and small inorganic molecules in the rocks become dissolved in water and form ions in oceans, lakes, rivers, and soil water. These ions, such as sulfate ($SO_4^=$) and phosphate ($PO_4^\equiv$), are absorbed by producers and then used to build organic molecules.

On land, soil forms the upper layer of the earth's crust (fig. 33.3*a*). It is a complex mixture of mineral particles (tiny pieces of rock), ions, water, **humus** (partially decomposed organisms; □ *see page 265*), decomposers, and other soil organisms. The crucial traits of a particular soil are its capacities to hold water and to provide dissolved ions.

Water that is not held in the soil moves downward and beyond the reach of plant roots. The capacity of soil to hold water depends largely on the size of the mineral particles. The two extreme kinds of soil, with regard to water retention, are sandy soils and clay soils. Sandy soils consist of large particles with large gaps between them, through which water drains rapidly downward. Clay soils consist of tiny particles that fit so closely together that most of the water is held within the soil.

Availability of ions depends on the ability of soil components to form bonds with the ions. Clay particles and humus, which carry electric charges, are especially good at holding ions until they are absorbed by plant roots.

In water, the upper layer of the earth's crust consists mainly of inorganic materials rather than the complex mixture of a soil. This layer, composed of bedrock, pebbles, and sand is the bottom substrate of an aquatic ecosystem (fig. 33.3*b*). It contains less organic material than soil, since dead organisms are decomposed as they drift downward through the water. The bottom substrate releases ions to the water, providing producers with raw materials for building monomers. Its crevices, pebbles, and sand provide animals with shelter from predators and strong currents.

Food Chains and Food Webs

An important feature of an ecosystem is who eats whom. We know there are herbivores, primary carnivores, secondary carnivores, and decomposers, but within any one of these categories each kind of organism has a specialized diet. A consumer or decomposer does not eat everything it encounters. Instead, it is restricted to a specific diet determined by its adaptations, which are the products of millions of years of evolution.

A **food chain** is a list of the species through which organic molecules pass. The general sequences are as follows:

Producer $\longrightarrow$ Herbivore $\longrightarrow$ Primary carnivore $\longrightarrow$ Secondary carnivore $\longrightarrow$ Tertiary carnivore

Dead organic material $\longrightarrow$ Decomposer $\longrightarrow$ Primary carnivore $\longrightarrow$ Secondary carnivore $\longrightarrow$ Tertiary carnivore

In a specific food chain, all the species are named. For example:

Dandelion ⟶ Cottontail rabbit ⟶ Gray fox ⟶ Wolf

The many food chains of an ecosystem are interconnected, since each herbivore may eat more than one species of plant and each carnivore eats more than one species of animal:

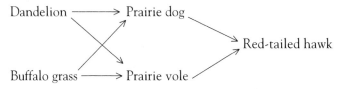

Moreover, an animal may be both a primary carnivore and a secondary carnivore. A hawk, for example, may eat mice (herbivores) and snakes (primary carnivores). A **food web** is a diagram of all the food chains, with their interconnections, within an ecosystem. It is the next step above population in the hierarchy of life (see inside front cover of this book). The food web of a grassland ecosystem is shown in figure 33.4.

Energy Flow

We have seen how food passes through the food web of an ecosystem. Food consists of organic molecules, which in turn have two components: energy (in the bonds) and atoms (in the structure). The movement of energy through a food web, called **energy flow,** is the topic of this section of the chapter. The movement of atoms, called nutrient cycling, is described later.

Energy Flow Through a Food Chain

Energy within the chemical bonds of organic molecules flows through a food chain. It travels along a one-way path, from producer to herbivore to primary carnivore to higher levels, as well as from dead organic material to decomposers to higher levels.

Producers Energy enters a food chain when the producer builds organic monomers. This typically occurs by photosynthesis, in which the energy of sunlight is converted to chemical energy within the bonds of sugar molecules. The producer (a plant, alga, or bacterium) then uses the chemical energy within the sugar to do work and to build new materials.

It takes work to maintain cells. Energy is used to rebuild molecules, to transport materials, and to do many other kinds of cellular work. Energy for cellular work comes from ATP, which is built during cellular respiration (see page 88) from the chemical energy of organic monomers. Cellular respiration is only 40% efficient: 60% of the chemical energy within a monomer becomes

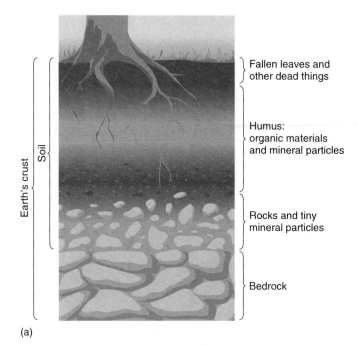

(a)

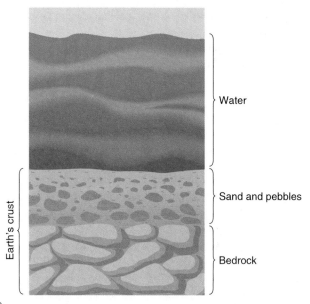

(b)

Figure 33.3 The earth's crust. (*a*) In a terrestrial ecosystem, the upper layers of the earth's crust form soil, which consists of both organic and inorganic materials. Decomposition takes place within the soil. (*b*) In an aquatic ecosystem, the upper layer of the earth's crust is the bottom substrate, which consists of bedrock, pebbles, and sand. It contains little organic material because decomposition takes place within the water above.

heat rather than chemical energy in ATP. And when the ATP is used to do work, some of its chemical energy also becomes heat. Heat energy cannot be passed on (as food) to another organism; it is lost from the food chain and moves into outer space.

Chemical energy retained within organic molecules, rather than used to do work, becomes organic molecules and structures that form as the organism grows and

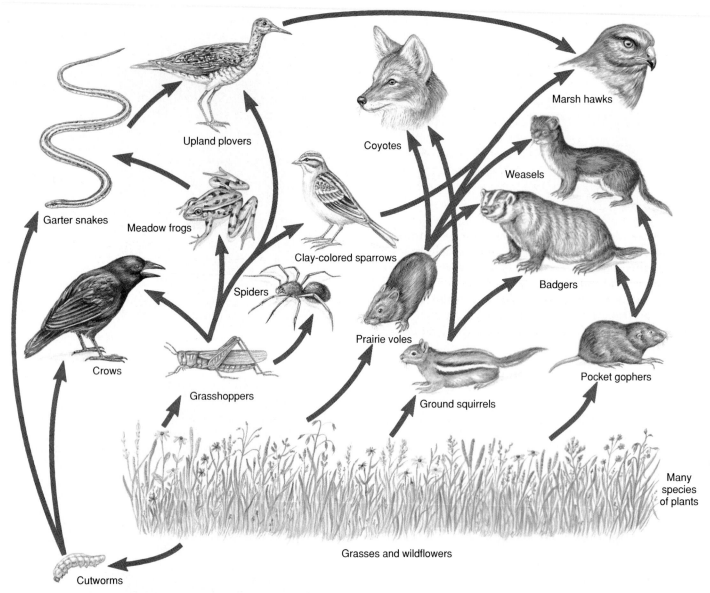

Figure 33.4 The food web of a grassland ecosystem. The food web consists of all the food chains and their interconnections. This sketch does not include all the organisms: a real food web has many more organisms and many more interconnections, as well as food chains involving decomposers.

reproduces. Only this energy, retained as chemical energy in the producer, is available to herbivores:

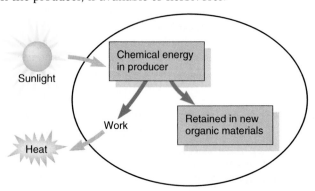

uses this energy to do work and to build new materials. Much of the chemical energy used to do work becomes heat and is lost from the food chain:

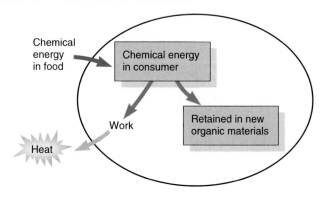

Consumers and Decomposers A consumer or decomposer acquires energy from organic molecules in its food. It

Energy Flow Through an Ecosystem

Ecologists simplify their study of energy flow by grouping all organisms that have the same position in the food chains. Each group forms a **trophic level.** Thus, the trophic levels of an ecosystem include the producers, the herbivores, the decomposers, the primary carnivores, the secondary carnivores, and higher levels when present.

Energy Pyramids   The amount of chemical energy that flows through an ecosystem decreases with each trophic level. This is because each kind of organism converts some of its chemical energy to heat as it builds ATP and then uses the ATP to do work. More chemical energy flows through the producers than the herbivores, more through the herbivores than the primary carnivores, and more through the primary carnivores than the secondary carnivores.

An **energy pyramid** is a diagram of energy flow through a food web (fig. 33.5). The amount of chemical energy flowing out of each trophic level, indicated by the size of the compartment, depends on the productivity of the trophic level.

Productivity   The quantity of new materials formed within a trophic level each year is its **productivity.** In a grassland, for example, productivity is the amount of new leaves, stems, roots, and seeds formed each year. In an unchanging ecosystem, the new materials formed in a trophic level each year do not accumulate because they are consumed by organisms in the next higher trophic level and by decomposers.

The productivity of a trophic level depends on (1) the amount of chemical energy that enters it (through photosynthesis or food) and (2) the percentage of this energy used in forming new material (growth and reproduction) rather than used in doing work:

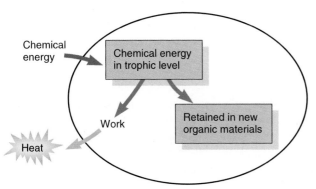

The productivity of producers varies from ecosystem to ecosystem. It depends largely on the rate of photosynthesis, which in turn depends on the amount of light, length of growing season, and availability of inorganic molecules and ions. The productivities of producers in various kinds of ecosystems are listed in table 33.2.

Of the new materials formed by terrestrial plants, about 10% is digested by herbivores and 90% by decomposers. This percentage is quite different in aquatic ecosystems, where about 90% of the new materials formed by producers is digested by herbivores and only 10% goes to decomposers. Why such a difference between ecosystems? Not all plant materials can be eaten by herbivores, because they contain poisons, and much of the plant materials that are eaten are not digested, because they contain cellulose and other indigestible materials that help support the plant in air. By contrast, the photosynthetic bacteria and unicellular algae of aquatic ecosystems seldom contain poisons and have very little structural materials, since these microscopic organisms are suspended in water.

The productivities of higher trophic levels depend chiefly on whether the animal is an ectotherm (cold-blooded; 📖 *see page 398*) or an endotherm (warm-blooded; 📖 *see page 398*). All animals except birds and mammals are ectotherms. An ectotherm uses 80% of the chemical energy within its food to do work, with 20% left for building new materials. An endotherm is more active and converts a large proportion of its chemical energy into heat in order to maintain its higher body temperature; 99% of an endotherm's food energy is used to do work and only 1% is left to build new materials:

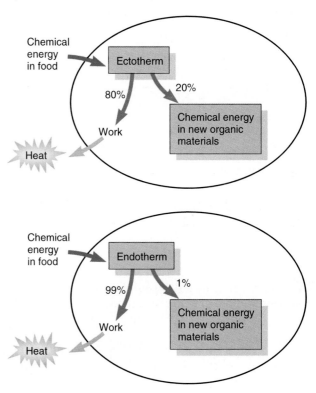

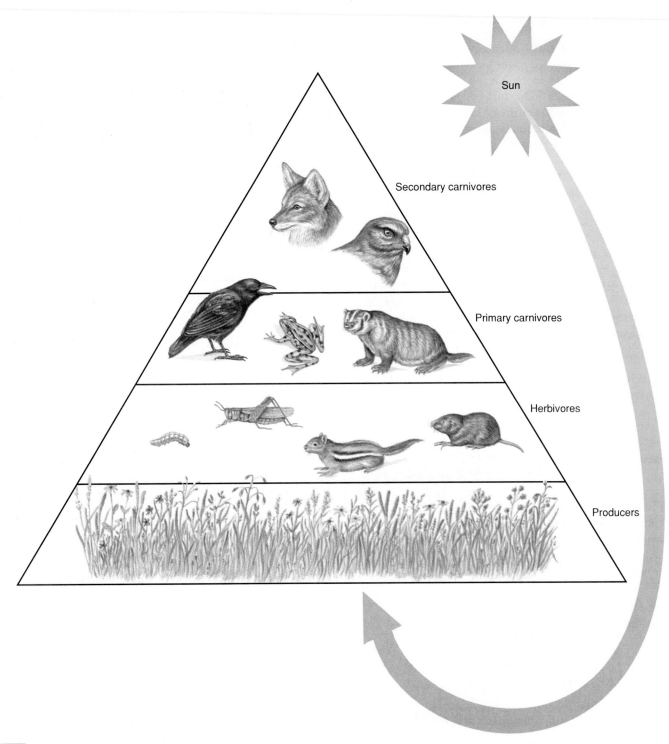

Sun

Secondary carnivores

Primary carnivores

Herbivores

Producers

⬛ Figure 33.5 The energy pyramid of a grassland ecosystem. The size of each compartment is proportional to the amount of energy flowing out of the trophic level. It is a pyramid because more energy flows out of lower levels than higher levels.

Whether an animal is an ectotherm or endotherm has a large effect on the amount of chemical energy that reaches higher trophic levels. In a grassland ecosystem, for example, birds that feed on insects have available to them a larger portion of the chemical energy fixed by plants than birds feeding on mice. Most of the chemical energy acquired by mice is used by the mice to maintain themselves.

Productivity and Food for Humans The human population is growing rapidly, and we already have trouble feeding all the people of the world. Fewer people would starve if everyone ate producers—especially algae, which provide more nutrients because they are more digestible. (Algae contain very little "fiber," as compared with plants.) If no one ate meat, the earth could provide sufficient food for 6

Table 33.2

The Productivities of Ecosystems

Ecosystem Type	Productivity (grams per square meter per year)	Factor That Limits Photosynthesis
Creosote-bush desert	90	Not enough water
Open ocean	125	Not enough light
Arctic tundra	140	Too cold most of the year
Great Plains grassland	600	Not enough water
Spruce-fir forest	800	Too cold most of the year
Maple forest	1,200	Shading by taller plants; too cold most of the year
Tropical rain forest	2,200	Shading by taller plants; atoms tied up in plants
Coral reef	2,500	Not enough space for attachment to bottom of ocean

billion people. (While there are already about 6 billion people on earth, many do not get enough to eat.) If everyone had a typical American diet of 35% meat and 65% plants, the earth could support only 2.5 billion people—less than half the number of people living today.

Biological Magnification   An unfortunate consequence of energy flow is that some toxic substances become increasingly more concentrated as organic material passes through a food chain. This effect is known as **biological magnification.**

Biological magnification occurs with environmental pollutants that organisms do not degrade or excrete. They are typically synthetic molecules that cells do not have enzymes for detoxifying. And they are soluble in fat, rather than water, which means they accumulate in fatty deposits rather than leave the body as urine. These toxic materials stay within the tissues and are passed, through food, from one trophic level to the next. As the biomass within a food chain becomes smaller and smaller, because organic molecules are used in cellular respiration, the toxic materials become more and more concentrated. In a study of DDT, for example, this pesticide (now banned in the United States) was found to be 600 times more concentrated in predatory birds than in the producers of the ecosystem.

Biological magnification has become a problem with industrial wastes, especially polychlorinated biphenyls (PCBs) and dioxins, that are released into aquatic ecosystems. These toxins have been linked with tumors and infectious diseases in fishes, frogs, turtles, and marine mammals. Seals and dolphins are especially vulnerable, for they feed high in the food chain and accumulate large fat deposits. Whenever a seal or dolphin draws on its fat reserves, during times of reproduction or famine, the toxic materials are released into the blood. There they apparently suppress the immune system, making the animal more vulnerable to tumors and infectious diseases. High concentrations of PCBs have been found in the fatty tissues of seals dying from distemper and of bottlenose dolphins suffering from pneumonia, skin lesions, and fungal growths.

Nutrient Cycles

Nutrients are atoms required by organisms; they are components of molecules or biochemical processes. Nutrients travel, along with chemical energy, from organism to organism within a food web. They also travel back and forth between the physical environment and the organisms of the food web. A **nutrient cycle** is the route taken by an atom as it travels from place to place within an ecosystem.

Unlike chemical energy, which is gradually lost as it flows through a food web, atoms cycle repeatedly between the physical environment and the food web. They are not altered by use. An atom in your body was once part of something else—a butterfly, a daisy, a rock, or a dinosaur. After you, it will become part of something else.

Roughly thirty kinds of atoms are found within the molecules and ions of living things. Each kind has its own unique cycle. In this chapter, we describe the cycles of carbon and nitrogen.

The Carbon Cycle

A food web consists mainly of carbon atoms, for these atoms form the framework of all organic molecules. Carbon atoms enter and leave a food web as carbon dioxide. They enter a food web during photosynthesis, when carbon dioxide and water are used to build sugars. The sugars then become frameworks for building other organic monomers. Carbon atoms leave a food web during cellular respiration, when monomers are broken down into carbon dioxide and water:

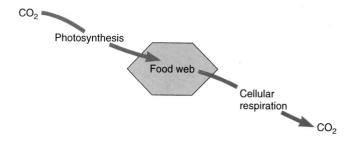

Environment Report

Fossil Fuels, the Carbon Cycle, and Global Warming

Any change in the carbon cycle changes the earth's climate, because the amount of carbon dioxide in the air influences the earth's temperature. Temperature, in turn, influences patterns of rainfall.

Carbon dioxide in the air traps heat near the surface of the earth. Light that strikes the earth is converted into heat and radiated back into the atmosphere. Some of this heat strikes molecules of carbon dioxide (and certain other gases, including methane) in the atmosphere, which radiate the heat back toward the earth's surface (fig. 33.A). The more carbon dioxide in the air, the more heat retained near the earth and the warmer the climate. This phenomenon is known as the greenhouse effect because the effect of carbon dioxide on the temperature of the earth resembles the effect of glass on the temperature of a greenhouse—carbon

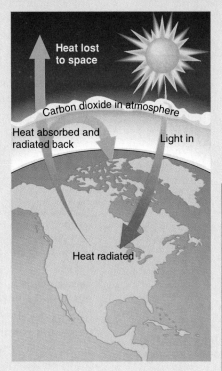

Figure 33.A The greenhouse effect. Light from the sun is converted to heat as it is absorbed by living and nonliving components of the earth. The heat is radiated into space, where some is absorbed by carbon dioxide (as well as water vapor and other gases) and radiated back to earth. The more carbon dioxide in the atmosphere, the warmer the earth.

dioxide and glass allow light to enter but do not allow heat to leave.

Massive quantities of carbon dioxide were taken out of the carbon cycle about 350 million years ago, when the climate was warm and wet. Large tree ferns and other land plants fell into stagnant water, where there was not enough oxygen for decomposers to live. Photosynthetic bacteria and algae accumulated in mud, another place of insufficient oxygen. Over time, these dead but undecomposed producers became fossil fuels—coal, oil, and gas. The carbon in their bodies was taken out of circulation. Consequently, the atmosphere contained less carbon dioxide and the climate became cooler.

Most of the carbon in these undecomposed organisms is now being returned, as carbon dioxide, to the air as

When not part of a food web, carbon exists chiefly as carbon dioxide in air and water. The carbon cycle is diagrammed in figure 33.6.

The amount of carbon dioxide in the air depends on the relative rates of photosynthesis and cellular respiration. Twice during the past 500 million years these rates have been very unequal. These periods and their effects on the earth are described in the accompanying Environment Report.

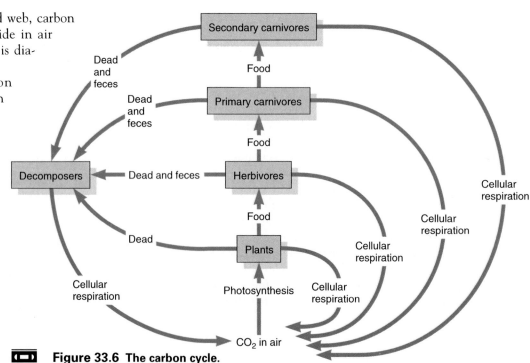

Figure 33.6 The carbon cycle.

we burn fossil fuels in our cars, factories, and power plants. Additional carbon dioxide is added to the air as we remove forests to plant agriculture lands, graze livestock, and build houses. A tree is mostly carbon, and when it burns or decomposes the carbon is released as carbon dioxide.

Since the early 1880s, when we began to burn massive amounts of fossil fuels and to eliminate our forests, the concentration of carbon dioxide in the atmosphere has increased by about 25% and the earth's temperature has risen almost 0.5°C (1°F). The warmer temperatures occur more during the night than the day and more during the winter than the summer.

Experts use computer models to predict what will happen if this trend continues. Information about the earth's environment, acquired largely from satellites, is entered into the model to obtain a prediction. Predictions vary because the models vary, based on how each expert describes the behavior of clouds, effects of air pollutants, and other variables. From these models, some experts predict a doubling of atmospheric carbon dioxide

by the year 2050, which would warm the midlatitudes by almost 2.8°C (5°F) and the poles by about 5.6°C (10°F), with hardly any change at the equator.

What's so bad about a warmer climate? Two serious problems are foreseen: (1) organisms are not adapted to the new climate; and (2) the glaciers will melt and raise the water level of the oceans.

The greenhouse effect will change the climate of most regions. Organisms are adapted to the specific conditions of rainfall and temperature in their region, and they will disappear from their native regions when the climate changes. Tundra will become forests, grasslands will become deserts, and deserts will become forests. The species will move, if they can, following the climate to which they are suited. (They will move not as individuals, but as generations of individuals that disperse into new regions.)

Species that cannot shift with their climate will become extinct. Ecosystems have moved before. As the glaciers retreated, for example, the ecosystems of North America moved northward at a rate of 42 kilometers

(25 mi) a century. With global warming, they will have to move ten times faster and their paths may be blocked by cities, farmlands, large lakes, and other barriers. Species in isolated ecosystems, such as lakes and coral reefs, will not be able to migrate with the climate. Many species will not make it.

Agricultural species will also be affected. As the climate changes, lands will become unsuitable for the crops now grown. In North America, the shift in climate is likely to cause a northeasterly movement. The "breadbasket" of America will shift from the central plains of the United States to Ontario and Quebec in Canada.

The water from melting glaciers will flow into rivers, lakes, and oceans and flood the coastal lands. New York City, New Orleans, Los Angeles, and Chicago will be underwater, as well as most of Florida and entire islands. The oceans may rise a meter or so during the next century.

Global warming will affect you in a personal way. It will alter the landscape, the value of land, and the cost of food. And evidence indicates it is happening right now.

The Nitrogen Cycle

Nitrogen is a component of many organic molecules, including proteins, DNA, RNA, ATP, and chlorophyll. A food web acquires nitrogen from the air. The nitrogen cycle is diagrammed in figure 33.7 and described following.

Nitrogen is abundant as a gas (N_2); 78% of the air is nitrogen gas. Few producers, however, can use this form of nitrogen because the triple covalent bond between the two nitrogen atoms ($N \equiv N$) is extremely difficult to break. Instead, producers acquire nitrogen in the form of ammonium (NH_4^+) and nitrate (NO_3^-). These ions are built from N_2 by nitrogen fixation in bacteria.

Nitrogen Fixation   Nitrogen becomes available to producers when specialized bacteria convert nitrogen gas into ammonia, a process known as **nitrogen fixation** (*fixation*, to put into a useful form). These bacteria have a special enzyme, called nitrogenase, that breaks the triple bond be-

tween the two nitrogen atoms. Each atom then bonds with three hydrogen atoms to form ammonia:

$$N_2 + H_2O + \text{Energy} \longrightarrow \underset{\text{Ammonia}}{NH_3} + O_2$$

In the watery solution of a bacterium, the ammonia picks up a hydrogen ion and becomes ammonium:

$$NH_3 + H^+ \longrightarrow \underset{\text{Ammonium}}{NH_4^+}$$

The ammonium is then used to build the bacterium's proteins, DNA, RNA, and ATP. (Chemists can fix nitrogen, without bacteria, in their laboratories and use this reaction to make commercial fertilizers.)

Some nitrogen-fixing bacteria live within the soil and water. The nitrogen they fix remains inside their bodies, as components of molecules, until they are eaten. When a consumer or decomposer ingests a nitrogen-fixing

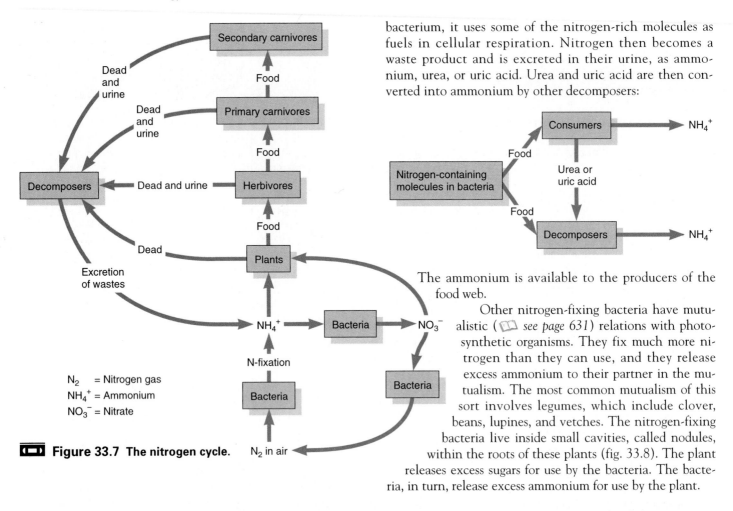

bacterium, it uses some of the nitrogen-rich molecules as fuels in cellular respiration. Nitrogen then becomes a waste product and is excreted in their urine, as ammonium, urea, or uric acid. Urea and uric acid are then converted into ammonium by other decomposers:

Figure 33.7 The nitrogen cycle.

N_2 = Nitrogen gas
NH_4^+ = Ammonium
NO_3^- = Nitrate

The ammonium is available to the producers of the food web.

Other nitrogen-fixing bacteria have mutualistic (☐ *see page 631*) relations with photosynthetic organisms. They fix much more nitrogen than they can use, and they release excess ammonium to their partner in the mutualism. The most common mutualism of this sort involves legumes, which include clover, beans, lupines, and vetches. The nitrogen-fixing bacteria live inside small cavities, called nodules, within the roots of these plants (fig. 33.8). The plant releases excess sugars for use by the bacteria. The bacteria, in turn, release excess ammonium for use by the plant.

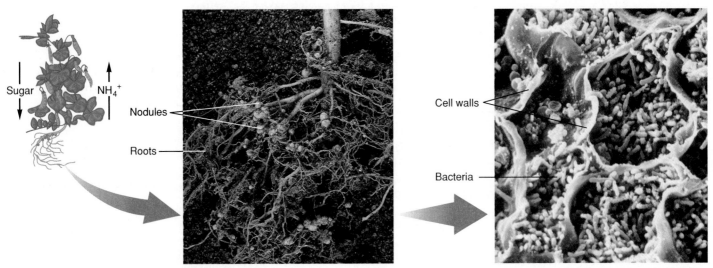

Figure 33.8 Nodules on the root of a legume. A legume, such as the pea shown here, develops cavities, called nodules, on its roots. Nitrogen-fixing bacteria live within cells of the nodules. There they synthesize excess ammonium for the plant and, in return, receive excess sugars synthesized by the plant.

The mutualistic relation between legumes and nitrogen-fixing bacteria enables these plants to thrive in nitrogen-poor soils, such as the acidic soils of mountain meadows and the overworked soils of farmland. An added benefit to the ecosystem is that when a legume dies and is decomposed, its ammonium is available for the growth of other kinds of plants. Legumes enrich the soil with nitrogen.

Routes Into and Out of the Food Web Some of the ammonium formed by bacteria is absorbed directly by producers, and some is converted into nitrate by other kinds of bacteria, called nitrifying bacteria, and then absorbed by producers:

$$\underset{\text{Ammonium}}{NH_4^+} + O_2 \xrightarrow[\text{bacteria}]{\text{Nitrifying}} \underset{\text{Nitrate}}{NO_3^-} + H_2O + \text{Energy}$$

Producers can use either ammonium or nitrate to build molecules, which then pass from organism to organism within the food web.

Nitrogen leaves a food web and returns to the soil or water when consumers and decomposers excrete excess nitrogen in urine. The nitrogen in urea and uric acid is converted into ammonium, by decomposers, and then reabsorbed by producers. Additional steps in the cycle develop when some of the ammonium is converted into nitrate by the nitrifying bacteria. Some of this nitrate is used by producers and some is converted into nitrogen gas by yet another group of bacteria, the denitrifying bacteria:

$$\underset{\text{Nitrate}}{NO_3^-} + \underset{\text{Sugar}}{C_6H_{12}O_6} \xrightarrow[\text{bacteria}]{\text{Denitrifying}} N_2 + CO_2 + H_2O + \text{Energy}$$

The nitrogen gas returns to the air, thus completing the cycle.

Human activities interfere with the nitrogen cycle in several ways. One such activity is the use of commercial fertilizers that add ammonium or nitrate to the soil. Farmers apply these fertilizers periodically in large doses, since adding them continually in tiny doses, which would be more natural, is impractical. Excess ammonium and nitrate, not absorbed by plant roots, move downward into the water table and eventually pollute our drinking water, irrigation water, and lakes. Excess nitrate in drinking water and on foods grown in irrigated fields may be converted, in the digestive tract, into substances that are immediately toxic to infants and that may contribute to cancers in adults. Excess nitrogen in a lake disrupts the food web by causing excessive growth of certain kinds of producers.

Another way in which we interfere with the nitrogen cycle is with automobile engines, which produce nitrates that generate acid rain. The effects of acid rain on ecosystems is discussed in the accompanying Research Report.

Figure 33.9 Lichens on rocks. Primary succession on land begins when lichens colonize bare rock. These organisms, together with mosses, begin the slow process of soil formation.

Succession

The development of an ecosystem is known as **succession.** It begins either with bare rock or soil and involves a regular turnover of organisms in which earlier colonizers are replaced by later ones. Succession culminates with a stable food web called the **climax community.** The composition of this community (whether it contains the organisms of a Ponderosa pine forest, for example, or a tall-grass prairie) depends largely on the local climate. There are two main kinds of succession: primary and secondary.

Primary Succession

The development of an ecosystem for the first time is known as primary succession. After a glacier has scoured the earth, for example, primary succession occurs on the bare rocks and within the newly formed lakes. It may take several centuries for primary succession to form a climax community.

The initial stages of primary succession can be observed on any rock. Lichens and mosses are the first to colonize, for they have no roots and can survive without soil (fig. 33.9). As these first colonizers grow and reproduce, they build organic molecules from nutrients in the air and on the rock surface. When they die, their molecules and ions form a shallow layer of soil. If you could observe such a rock for a few decades, you would notice a gradual increase in soil depth and replacement of the lichens and mosses by grasses and wildflowers. More and more of the rock would become soil, as a consequence of plant activities, until there is enough to support bushes. Centuries later, provided the climate is suitable, trees would grow in a deep soil laid down over the rock.

Research Report

Is Acid Rain Killing Our Forests?

Fifty years ago the evergreen forests of New England were healthy and luxuriant. Today, many of the spruce and fir trees are dead or dying (fig. 33.B). The first symptom of ill health is the appearance of yellow needles, which then turn brown and drop off. The trees weaken and tend to blow down in the strong mountain winds. As more and more trees fall, the survivors become more exposed to the wind, and even healthy trees topple over. The forest gradually dies.

Analyses of dead and dying trees have led to five hypotheses about what is causing the forest damage in New England. The damage may be due to fungal infections, insect infestations, unusual periods of drought, nutrient deficiencies, or acid rain.

How could acid rain kill a tree? Acid rain (more appropriately called acid precipitation, since snow, sleet, hail, dew, and fog also carry acids) is water with a high concentration of hydrogen ions (H^+). The hydrogen ions come from nitric acid (HNO_3) and sulfuric acid (H_2SO_4) that form in the air.

Figure 33.B Dead red spruce in a New England forest.

These molecules release hydrogen ions as they dissolve in water.

Nitric acid forms when nitrogen gas in air near the engine of a car or truck becomes so hot that it interacts with oxygen gas to form NO, which combines with ozone (O_3) in the air to form nitrogen dioxide (NO_2), which is a brown, toxic gas:

$$N_2 + O_2 \xrightarrow{\text{Heat of engine}} NO$$

$$NO + O_3 \longrightarrow NO_2$$

NO_2 then combines with water vapor in the air to form nitric acid:

$$NO_2 + H_2O \longrightarrow HNO_3$$
$$\text{Nitric acid}$$

Sulfuric acid forms when coal is burned. Coal consists of ancient plant materials, mainly cellulose but also some proteins. The proteins contain sulfur. When coal burns, the sulfur is released into the air as SO_2, which combines with oxygen gas to form sulfite:

$$SO_2 + O_2 \longrightarrow SO_3^{\equiv}$$
$$\text{Sulfite}$$

The sulfite combines with water vapor in the air to form sulfuric acid:

$$SO_3^{\equiv} + H_2O \longrightarrow H_2SO_4$$
$$\text{Sulfuric acid}$$

Rain that falls downwind of a large city, where there are many cars and trucks, is likely to contain hydrogen ions mainly from nitric acid. Rain that falls downwind of a coal-burning power

---

The kinds of animals in the food web depend more on the kinds of plants than the kind of soil. A succession of animal species matches the succession of plant species.

Glacial moraines are good places to study primary succession, for the pathway of a retreating glacier exhibits a sequence of successional stages. Far from the glacier, where the ice melted a longer time ago, the successional sequence is advanced. Next to the glacier, where the ice melted more recently, the successional sequence is early:

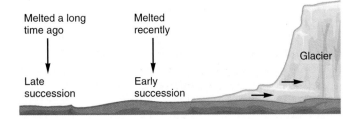

Primary succession on bare rock after retreat of a glacier is outlined in table 33.3.

Another form of primary succession occurs regularly in freshwater lakes and ponds as they are transformed into terrestrial ecosystems (fig. 33.10). Flowing water from rain and melting snow carries materials from the surrounding land into the lake or pond. Soil accumulates beneath the water, and plants colonize. The water becomes more and more shallow with time until eventually the lake or pond is completely filled with soil and plants.

Secondary Succession

Secondary succession is development of an ecosystem after a disturbance, such as a fire (fig. 33.11), hurricane, avalanche, or the plowing of a field. It takes less time than primary succession because a soil is already present, providing nutrients as well as seeds and other dormant stages of organisms.

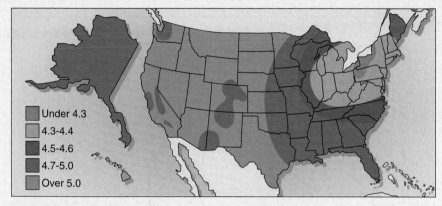

Under 4.3
4.3-4.4
4.5-4.6
4.7-5.0
Over 5.0

Figure 33.C The pH of rain in the United States. The pH scale is a log scale, which means that a pH of 4 is 100 times more acidic than a pH of 6. Normal rainwater has a pH of 5.6.

plant is likely to contain a lot of hydrogen ions mainly from sulfuric acid.

New England is downwind of coal-burning power plants in the Midwest, and rain that falls there contains high concentrations of hydrogen ions from sulfuric acid. The rain is often 100 times, and occasionally 1,000 times, more acidic than normal (fig. 33.C).

The hypothesis that acid rain is destroying our forests is difficult to test. Indirect evidence comes from chemical analyses of the wood of red spruce trees from New England. This wood contains aluminum, which is not usually absorbed by plants because it occurs as an atom within soil particles rather than an ion dissolved in water. Very acidic water, however, dissolves aluminum out of the particles, and the ions are taken up by plant roots. The chemical analyses revealed that tree rings laid down after 1950 contain aluminum, whereas rings laid down before 1950 do not. The year 1950 is about the time we first noticed acid rain. Moreover, soils beneath some of the damaged forests now contain up to 100 times more aluminum ions than usual.

Aluminum ions can kill a plant in two ways. First, they can damage the tiny roots so that the plant cannot absorb enough water to survive. Second, aluminum ions alter the soil chemistry in ways that prevent plants from absorbing calcium, magnesium, and other ions they need to survive.

Biologists studying the effects of acid rain now believe that all five of the hypotheses may be correct. Tree death may begin when acid precipitation impedes the uptake of water and nutrients. Without sufficient water, the tree may die during periods of drought or unusually warm weather. Without sufficient nutrients, the tree may not be able to defend itself against invasions by fungi and insects.

Forest damage similar to the kind seen in the spruce and fir forests of New England is now appearing in the pine forests of the southeastern states and California, the maple forests of the northeastern states, and the oak and hickory forests of the central states. European forests are also severely damaged. We cannot say for certain that acid precipitation is the culprit. We can only hope that by the time we understand what is happening, it will not be too late to save our forests.

The succession of plants after abandonment of an actual farm in North Carolina is outlined in table 33.4. The animals also changed on the abandoned farm. Meadowlarks and sparrows arrived early, for they eat the tiny seeds of early successional herbs (plants without woody stems). Later birds include cardinals and towhees, which eat the larger seeds of shrubs and trees, and woodpeckers, which feed on insects beneath tree bark.

Humans interfere with secondary succession by suppressing forest fires. Regular burning of the forest is natural and beneficial. When fires are not suppressed, they occur frequently (several times a century in most forests) and serve to eliminate dead wood and thick understory. They rarely kill healthy trees. When fires are suppressed, flammable materials accumulate within the forest so that when a fire does occur it is devastating to the entire ecosystem. Natural fires also eliminate forest pests, such as bark beetles, and in this way help maintain healthy trees.

Traits of Successional Species

There is a regular and fairly predictable turnover of species during succession. The kinds of species found during early stages of succession are quite different from the kinds seen later. The contrasting traits of early- and late-successional plant species are described in general here and listed in table 33.5.

Organisms that thrive during the early stages of succession are good colonizers but poor competitors. They arrive quickly, reproduce rapidly, and their young disperse to other disturbed sites. They are known as *r*-selected species (*r* stands for the rate of population growth; *see page 623*), with adaptations for rapid reproduction: small bodies, short life spans, many offspring at a time, and long-distance dispersal. Early successional species are not efficient at using resources, competing, or evading enemies. They are adapted for a short stay in the area.

Table 33.3

Primary Succession on a Glacial Moraine in Alaska

Time Since Glacier	The Plants That Are Present
0 (glacier just melted)	Bare rocks are colonized by lichens and mosses (no roots). As they die, they decompose and form a soil on the rock.
3 years after melt	Some soil has formed and plants with roots begin to colonize. These are grasses, wildflowers, and bushy willow.
10 years	Alder trees colonize. These trees contain nitrogen-fixing bacteria, which add nitrogen to the soil. Alders grow taller than the grasses, wildflowers, and bushy willows, which die because they cannot grow in the shade of alders.
50 years	Spruce trees (a form of conifer) colonize the nitrogen-rich soil. They grow taller than alder. Young alder cannot grow in the shade of spruce trees, but young spruce can.
170 years	Hemlock trees (a form of pine, which is a conifer) colonize and coexist with the spruce trees. The two conifers have slightly different niches: hemlock grows in drier places (higher ground) and spruce grows in wetter places (lower ground).
250 years	A climax forest of spruce and hemlock exists and remains until disturbed by fire or logging. Spruce-hemlock is the climax forest where the climate is cool and dry.

As succession proceeds, there is a gradual transition from species that are more *r*-selected to species that are more *K*-selected (*K* stands for carrying capacity; see page 623). Organisms that thrive during late stages of succession are good competitors but poor colonizers. They arrive late, monopolize resources, and reproduce slowly. The adults tend to have large bodies, long life spans, few offspring at a time, and traits that help them use resources efficiently, compete for resources, and evade enemies. Their young are adapted to remain in the region rather than disperse to new sites. Late successional species are adapted for a long stay in the area.

Each species alters the environment in ways that enable other species to colonize. Environmental changes and interspecific competition then favor the new species over previous ones. A turnover of species occurs until arrival of the climax species, which persist until the ecosystem is again disturbed.

(a)

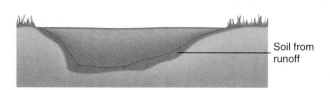

Soil from
runoff

(b)

Soil from
runoff and
decomposition
of plants

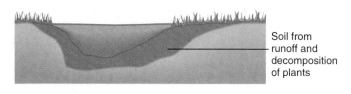

(c)

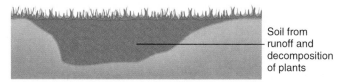

Soil from
runoff and
decomposition
of plants

Figure 33.10 Primary succession in a pond. (*a*) A pond
receives soil in runoff from the surrounding land. Plants then
colonize the soil where the water is shallow. (*b*) The pond
gradually fills with soil and plants. (*c*) Eventually the pond
becomes land.

Figure 33.11 The first stage of secondary succession.
Succession after a fire begins with small, short-lived plants.
This photograph was taken in Yellowstone National Park just
seven months after the enormous fire of 1988.

Table 33.4

Secondary Succession on a Farm in North Carolina

	Years Since Plowing	Plants	Traits That Help Them Colonize	Traits that Cause Their Disappearance
	0 (fall)	Crabgrass (annual)*	Already present; can live in dry soil	Die in the shade of daisies
	1–2	Daisies (annuals and biennials)*	Tiny fruits; can live in dry soil	Young killed by decaying roots of adults; roots too small to compete with grasses and shrubs for water
	3–4	Grasses and shrubs (perennials)*	Tiny fruits; large roots to monopolize water	Young killed by adult shade and poisons from pine needles
	5–15	Transition from grasses and shrubs to pine trees		
	15–50	Pine trees	Monopolize light (tall)	Young killed by shade and poisons from adult needles; roots too shallow to compete with oaks and hickories for water
	50–150	Transition from pines to oaks and hickories		
	150	Oak and hickory trees	Young can grow in ground with adult leaves and in shade; adults have deep roots	None—this forest remains; oaks in drier places than hickories

*An annual grows during one season and then reproduces and dies.

A biennial grows for two years and then reproduces and dies.

A perennial grows for more than two years and then reproduces every year for many years.

Note: The original forest, before it was cut down, consisted of oak and hickory trees.

Table 33.5

Traits of Successional Plants

Early Successional	Late Successional
r-selected traits	*K*-selected traits
Many tiny fruits	Few large fruits
Reproduce once	Reproduce many times
Small body	Large body
Rapid development	Slow development
Cannot grow in shade	Can grow in shade
Large niches	Small niches
Poor competitors	Good competitors
No herbivore poisons	Herbivore poisons

Biomes

A **biome** is a major form of landscape; it consists of many kinds of similar ecosystems. A coniferous forest biome, for example, covers most of Canada and contains all forest ecosystems dominated by conifers—pines, spruces, firs, and Douglas firs. A biome is a step higher than an ecosystem in the hierarchy of life (see inside cover of this book). Biomes, in turn, are components of the biosphere, which includes all portions of the earth where life is found.

On land, the type of biome that develops in a region is controlled mainly by climate, as shown in figure 33.12. Terrestrial biomes are classified according to the physical characteristics of the dominant vegetation: plant form (trees, shrubs, or herbs), vegetation height, layers of leafy vegetation, and continuity of the canopy (the upper layer of leaves). The particular species of organisms are not important. Some terrestrial biomes are shown in figures 33.13 and 33.14. A map of the major terrestrial biomes of the world is provided in figure 33.15.

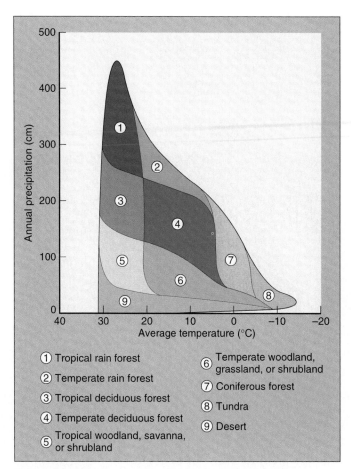

1. Tropical rain forest
2. Temperate rain forest
3. Tropical deciduous forest
4. Temperate deciduous forest
5. Tropical woodland, savanna, or shrubland
6. Temperate woodland, grassland, or shrubland
7. Coniferous forest
8. Tundra
9. Desert

Figure 33.12 The type of biome depends largely on climate.

In water, a biome is classified according to the characteristics of the physical environment—concentration of salts, rate of water flow, and texture of the bottom substrate. Examples of aquatic biomes include lakes, rivers, marshes, coral reefs (fig. 33.16), marine benthic (ocean floor in deep water), and marine pelagic (water above ocean floor).

(a)

Figure 33.13 Two biomes of wet climates. (a) Tropical rain forest in South America. (b) Temperate deciduous forest in northeastern United States. (Tropical refers to regions near the equator; temperate refers to regions between 30° and 60° latitude.)

(b)

Figure 33.14 Two biomes of dry climates. (*a*) Grassland in Nebraska. (*b*) Desert in Arizona.

(b)

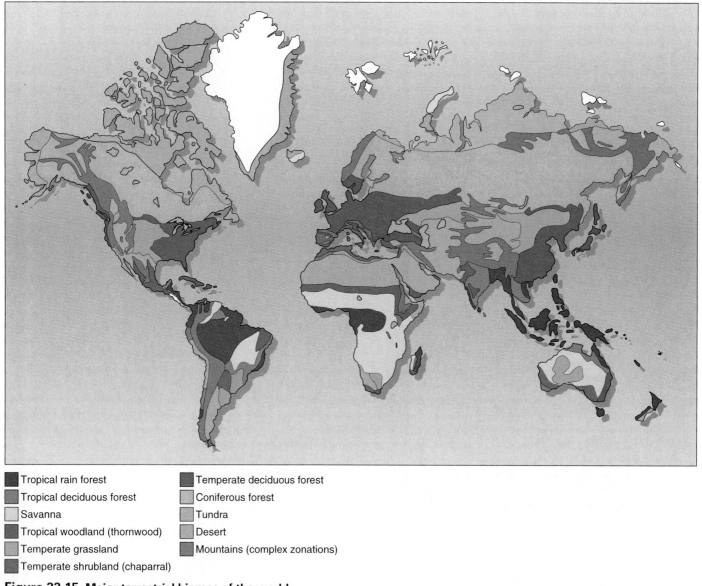

Tropical rain forest

Tropical deciduous forest

Savanna

Tropical woodland (thornwood)

Temperate grassland

Temperate shrubland (chaparral)

Temperate deciduous forest

Coniferous forest

Tundra

Desert

Mountains (complex zonations)

Figure 33.15 Major terrestrial biomes of the world.

SUMMARY

1. An ecosystem is a unit of nature, consisting of a community of interacting populations as well as aspects of the physical environment. The organisms are grouped, according to how they acquire organic monomers, into producers, consumers (herbivores and carnivores), and decomposers. The physical environment provides heat, light, and inorganic materials.

2. A food chain is a list of species through which organic molecules pass. A food web consists of interconnected food chains.

3. Chemical energy flows, within organic molecules, along a one-way path through a food web. Chemical energy enters a food web when a producer builds monomers; it leaves a food web as heat when each organism in the web uses monomers in cellular respiration.

4. An energy pyramid depicts the amount of chemical energy that flows out of the trophic levels. This chemical energy constitutes the productivity of the trophic level. Among ecosystems, the productivities of (a) producers are a function of the physical environment; (b) herbivores are a function of whether they are terrestrial or aquatic; and (c) animals in general are a function of whether they are ectotherms or endotherms. Humans would have more food if they ate more producers. Biological magnification is the increasing concentration of harmful substances in organic materials as they move through a food chain.

5. A nutrient cycle is the route taken by a type of atom as it travels through an ecosystem. Carbon enters a food web as carbon dioxide during photosynthesis, and leaves it as carbon dioxide during cellular respiration. Nitrogen enters a food web when nitrogen-fixing bacteria convert

Figure 33.16 A coral reef in the South Pacific. This type of biome develops in warm tropical oceans where the water is shallow, often on the edges of islands as shown here. The bottom substrate develops from the skeletons of dead coral.

nitrogen gas into ammonium. Some of the ammonium is converted into nitrate by nitrifying bacteria. Plants and algae absorb either ammonium or nitrate. Nitrogen leaves a food web as ammonium in the excretions of consumers and decomposers. It returns to the air when denitrifying bacteria convert ammonium to nitrogen gas.

6. Succession is characterized by a turnover of species until a stable food web, the climax community, is established. Primary succession is development of an ecosystem where one did not previously exist. Secondary succession is development of an ecosystem where a previous one has been disturbed. During succession, *r*-selected species are gradually replaced by *K*-selected species.

7. A biome is a major form of landscape formed of many similar ecosystems. Terrestrial biomes, which develop largely as a function of climate, are classified according to characteristics of the vegetation. Aquatic biomes are classified according to characteristics of the non-living environment.

KEY TERMS

Biological magnification (page 643)
Biome (BEYE-ohm) (page 653)
Climax community (page 647)
Consumer (page 637)
Decomposer (DEE-kum-POH-zur) (page 637)

STUDY QUESTIONS

1. Describe the differences, with regard to food, among herbivores, decomposers, primary carnivores, and secondary carnivores.
2. Describe a particular food chain with which you are familiar, beginning with a producer and ending with a top carnivore.
3. Explain why the amount of chemical energy decreases with each step in a food chain.
4. Explain why the earth could support more people if we all "fed lower in the food chain."
5. Describe how carbon enters and leaves a food web.
6. Describe how nitrogen enters and leaves a food web.
7. Compare the traits of early successional species with those of climax species. Explain the advantage of each trait.

CRITICAL THINKING PROBLEMS

1. Describe the sequence of events that would occur within an ecosystem after extinctions wiped out all its (a) decomposers, (b) producers, (c) consumers, (d) nitrogen-fixing bacteria.
2. What would happen to an ecosystem if the rate of photosynthesis exceeded the rate of cellular respiration? What would happen if the rate of cellular respiration exceeded the rate of photosynthesis?
3. Explain why a food web consists primarily of carbon atoms.

4. Explain why large predators, such as mountain lions, occur in lower densities than large herbivores, such as deer.
5. Why is most agriculture based on early successional plants?

SUGGESTED READINGS

Introductory Level:

Ausubel, J. H. "A Second Look at the Impacts of Climate Change," *American Scientist* (May–June 1991): 210–20. An evaluation of our assumptions and hypotheses relating to the greenhouse effect.

Gates, D. M. *Energy and Ecology*. Sunderland, MA: Sinauer Associates, 1985. An introduction to energy and the ecological principles involving energy flow. A good discussion as well of the effects of fossil fuels, nuclear fuels, and hydroelectric power on the environment.

Hagen, J. B. *An Entangled Bank: The Origins of Ecosystem Ecology*. New Brunswick, NJ: Rutgers University Press, 1992. A history of ecosystems ecology, including the people who contributed to the knowledge, their social interactions, and the present state of this field of research.

Leopold, A. *A Sand County Almanac*. New York: Oxford University Press, 1949. A classic book in which a foremost American conservationist records the ecological changes that take place on an abandoned farm.

Advanced Level:

DeAngelis, D. L. *Dynamics of Nutrient Cycling and Food Webs*. New York: Chapman and Hall, 1992. A review of what is known about nutrient cycling and ecosystem stability.

Pimm, S. L. *Food Webs*. London: Chapman and Hall, 1982. A discussion of food-web patterns, processes that generate them, and their effects on ecosystems. A blend of theory, observations, and experiments.

Post, W. M., T-H. Peng, W. R. Emanuel, A. W. King, V. H. Dale, and D. L. DeAngelis. "The Global Carbon Cycle," *American Scientist* 78 (July–August 1990): 310–26. A report on our current understanding of the carbon cycle, carbon reservoirs, and the impact of land use.

Sprent, J. I. *The Ecology of the Nitrogen Cycle*. New York: Cambridge University Press, 1987. A good reference for details about this nutrient cycle.

☼ EXPLORATION

Interactive Software: Pollution of a Freshwater Lake

This interactive exercise allows students to explore how the addition of certain "harmless" chemicals can pollute a lake. The exercise presents a map of Lake Washington, indicating the location of sewage treatment plants and the chemical composition of their effluent. Students can in-vestigate the nature of pollution by altering the chemicals in the effluent. They will discover that phosphates in the effluent lead to algal growth, which is followed by bacterial growth (they feed on the dead algae) and a precipitous drop in levels of dissolved oxygen. Students can then attempt to "clean up" the lake by altering the nature and amount of effluent.

Human Ecology

Objectives

In this chapter, you will learn
–about the original niche of our
  species;
–about the carrying capacity and
  population growth rate of our species;
–how we may no longer be adapted to
  our environment;
–how human activities are endangering
  life on earth;
–about the status of our species.

Do we have an inborn need for
nature? Our enjoyment of nature may
be a behavioral relict that we inherited
from our distant ancestors, who were
more a part of nature than we are.

What sort of animal are we? What was our original niche? What are our adaptations to this niche? Do our adaptations require a natural environment or can we be healthy and fulfilled in a world of pesticides, food additives, asphalt, and computers? What is the future of our species?

Our distant ancestors had a niche similar to the niches of other animals. They thrived as regular components of food webs for several million years, evolving slowly as they were molded by natural selection to utilize certain foods, evade predators, find shelters, and compete with other species for these resources. Within the past 10,000 years or so, our ancestors escaped the confines of their niche and became more a manager of nature than a participant. We are now molding our environment, rather than being molded by it, to suit our short-term needs. We are taking over the habitats and resources of other species.

Our enormous success in monopolizing nature is due to our remarkable minds. The human intellect is new to nature. Most remarkable is our capacity to change the environment. Will other species be able to adapt to these changes? Will our own bodies be able to cope with the changes generated by our minds?

In this chapter, we examine the ecology of our ancestors. We then compare their way of life, for which we are biologically adapted, to our own way of life, for which we may not be biologically adapted.

The Ecology of Our Ancestors

Human ecology consists of all relations between humans and their environment. As with other species, human ecology includes a niche, a carrying capacity, and a population growth rate. Unlike other species, our ecology depends more on our minds than our physical traits.

Humans have inherited biological adaptations that contribute to survival and reproduction. These anatomical, physiological, and behavioral traits evolved a long time ago and reflect the niche of our distant ancestors, who interacted more directly with nature than we do now. The shapes of our teeth, for example, reflect the variety of plants and animals our ancestors ate, and our extraordinary capacities for aerobic exercise reflect the long distances they ran in pursuit of game animals.

This biological relation between humans and nature has been altered by our culture. Culture consists of learned behaviors that are passed from population to population and from generation to generation by means of spoken or written words. One specific aspect of culture, the applied

sciences, now dominates our relations with the environment. Chemistry, nuclear physics, engineering, agriculture, and medicine have had profound effects on our relations with nature.

Our Ancestral Niche

The ancestral niche (see page 618) of humans, for which we are physically adapted, is the niche of Cro-Magnon people (fig. 34.1), who appeared 70,000 years ago. Our knowledge about these people comes mainly from analyses of their bones, weapons, tools, shelters, garbage, and paintings on the walls of caves. We also make inferences about our ancestors from the life-styles of modern people who have remained unaffected by modern culture because they live deep within rain forests or barren deserts, out of touch with the rest of the world.

The Cro-Magnon niche was large compared with the niches of other animals. Analyses of their garbage indicate they had a varied diet. They hunted wooly mammoths, deer, squirrels, chipmunks, birds, lizards, snakes, and fish; they gathered insects, worms, honey, leaves, roots, and seeds. All evidence indicates that the unusually broad diet of these people was due largely to the fact that they were exceptional learners. They explored their environment for new sources of food, and they learned to use spears to kill large animals, snares to trap small animals, tools to cut and grind food, and fire to cook it.

The Cro-Magnon niche included a wide range of foraging places. They searched the forests, meadows, grasslands, deserts, lakes, marshes, and streams for food to eat. They learned to wear the hides of other animals, so they could extend their hunting into cold environments.

The niche included many kinds of shelter from the cold, wind, rain, snow, and predators. Remains of charcoal, from their campfires, have been found in caves and the remains of shelters they built. The Cro-Magnons apparently learned to construct shelters from whatever was on hand: sod, bushes, trees, stones, and animal skins.

The unusual versatility of our ancestors enabled them to live almost anywhere on land. The first humanlike animals arose in Africa approximately 4.4 million years ago. Within 2.6 million years, their descendants had spread throughout Africa and into Europe and Asia. Then, about 1 million years ago, our own species, *Homo sapiens*, evolved and one of its subspecies, the Cro-Magnons, colonized the rest of the world (fig. 34.2). No other species ever developed such a vast geographic distribution. The history of our species is described in the accompanying Research Report.

Evidence from ancient campsites indicates that our ancestors were a highly social species. They apparently

Figure 34.1 A Cro-Magnon village. These people, who appeared 70,000 years ago, lived in groups of between 30 and 150. Each person contributed in some way to the welfare of others in their group.

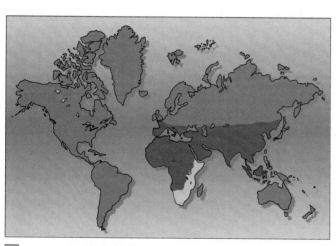

4 to 1.8 million years ago

1.8 million to 40,000 years ago

40,000 years ago to present

Figure 34.2 Our ancestors colonized the world. This history of human colonization is based largely on fossils of bones and tools.

lived in groups of between 30 and 150 people and probably, like modern hunters and gatherers, had a division of labor. Most likely the men formed hunting parties that roamed the land in search of large game and the women remained near the campsite, gathering plants and trapping small animals as well as caring for the children and the sick.

On the average, these people were physically stronger than we are, as indicated by the larger scars on their bones where muscles attached. And, from what we see in contemporary hunter-gatherer societies, they were probably more social than we are, spending much of their time discussing experiences and telling stories.

Primitive societies are not without strife. Each Cro-Magnon group probably defended a territory against other human groups, with fights erupting whenever neighboring groups encroached on one another's hunting grounds. Each society probably had a dominance hierarchy with social rank based on age, wisdom, and particular skills. And there must have been conflicts as younger members of the society struggled to take over roles held by their elders.

The Evolution of Humans

All humans alive today are members of the species *Homo sapiens* (*homo*, man; *sapiens*, wise). We are the only surviving species in the genus *Homo* and the family Hominidae. We are a form of primate. The classification of humans is outlined in table 34.A.

The oldest fossil of a primate is 50 million years old. It is of a mammal that is already quite advanced as a primate, which leads us to believe that the first primate evolved perhaps as early as 60 million years ago. Selection for this unique kind of animal developed when a small mammal began to live in trees, where it found shelter from predators as well as abundant food in the form of fruits and insects.

Life in the trees selected for the traits that now distinguish primates from other mammals: (1) hands and feet adapted for grasping, with longer fingers and toes covered with ridges that prevent slipping, fingernails instead of claws, and opposable thumbs that can move opposite the other fingers; (2) upright posture that frees the hands for manipulating objects; (3) eyes that face forward and provide three-dimensional vision for leaping from branch to branch; and (4) a larger cerebral cortex, the region of the brain associated with learned behaviors. The major primate groups are shown in figure 34.A.

All evidence indicates that the first members of our family, Hominidae, evolved in Africa. Analyses of DNA from modern apes and humans indicate that humans and chimpanzees had a common ancestor sometime between 6 and 8 million

Table 34.A

Classification of the Modern Human

Taxonomic Category	Category That Includes Humans	Other Organisms That Share the Category with Humans
Kingdom	Animalia	Sponge, sea anemone, worm, bee, clam, starfish
Phylum	Chordata	Sea squirt, lancelet, fish
Subphylum	Vertebrata	Fish, frog, snake, bird, horse
Class	Mammalia	Platypus, kangaroo, mouse, dog, cat
Order	Primate	Lemur, tarsier, monkey
Suborder	Anthropoidea	Monkey, chimpanzee, gorilla
Superfamily	Hominoidea	Chimpanzee, gorilla, gibbon
Family	Hominidae	*Australopithecus afarensis* *
Genus	*Homo*	*Homo habilis,* * *Homo erectus* *
Species	*Homo sapiens*	Neanderthals*
Subspecies	*Homo sapiens sapiens*	None

*Extinct

Note: Humans are members of each taxonomic category listed.

years ago. Surprisingly, the DNA of chimpanzees is much more similar to our DNA than to the DNA of other apes. Later evolution of hominids (members of the family Hominidae), based on fossils, is diagrammed in figure 34.B and described here.

The oldest fossil hominids, discovered in 1992 in northeast Africa, are 4.4 million years old. Named *Australopithecus ramidus,* these early humans were intermediate between chimpanzees and later hominids. They were small, standing slightly over a meter tall and weighing about 30 kilograms (66 lb), and appear to have walked upright on two legs. While

their skulls resembled the skulls of chimpanzees, these hominids possessed many traits that make them decidedly human.

The next oldest fossil hominids, also from northeast Africa, are 3.9 million years old. Their scientific name is *Australopithecus afarensis*. The best-known specimen is called Lucy, after the Beatles' song "Lucy in the Sky With Diamonds," which was popular at the time the specimen was discovered. More than 300 fossils of this species have been found. There are two size classes: some, like Lucy, are about a meter tall and others are about 23% taller. If the two forms represent

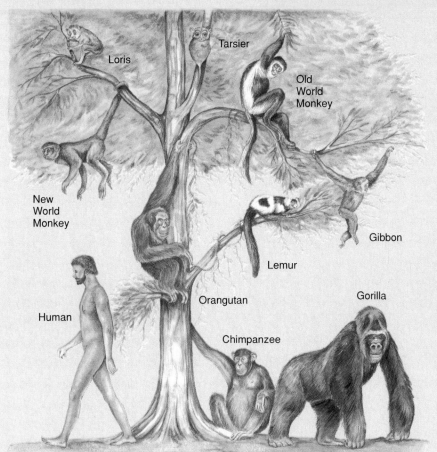

Figure 34.A The various forms of primates. Primates are adapted for life in trees. Apes spend more time on the ground than most primates. Humans are adapted for life on the ground but have retained many features of their ancestral primates.

males and females, which most researchers believe, then the differences are about twice as great as the size differences between modern men and women. In both classes, the arms were longer and the legs were shorter than ours. Analyses of the skeletons and inner ears indicate that these people divided their time between the trees and the ground. They were not fully bipedal. The larynx was positioned too high in the throat for mak-

ing the variety of sounds required for speaking a complex language. They were apparently the only species of our family between 3.9 million years ago and 3 million years ago, when they disappear from the fossil record.

A later type of human, who lived in Africa about 2.4 million years ago, was more similar to us and is placed in the same genus. Not enough fossils have been found of this species, called *Homo rudolfensis*, to infer

much about the way it lived. We have many fossils from a later type, which lived in Africa about 2 million years ago. Its scientific name is *Homo habilis*. The size of the skull, imprints of the brain on the skull, and lower position of the larynx indicate that these people were more intelligent than Lucy and had a more complex form of spoken language. Some of their stone tools have been found.

Homo erectus appears in the fossil record 1.8 million years ago, in both Africa and Asia. These people had larger brains and were taller and slimmer than *habilis*. They were fully bipedal, made stone tools, and probably spoke a language equal in complexity to their *habilis* ancestors. *Homo erectus* was an enormously successful species, spreading into all except the coldest regions of Africa, Europe, and Asia. The species disappears from the fossil record about 300,000 years ago.

We do not know the evolutionary history of our own species, *Homo sapiens,* but it probably arose about 1 million years ago. There are two main hypotheses about where the species arose. The out-of-Africa hypothesis states that *Homo sapiens* originated in Africa, perhaps from *Homo erectus,* and then spread to Europe and Asia. DNA analyses of modern peoples support this hypothesis. The multiregional hypothesis states that *Homo sapiens* evolved in many regions of Africa, Europe, and Asia, from various local populations of *Homo erectus*. There is some fossil evidence in support of this hypothesis.

The oldest fossils of *Homo sapiens* are of the subspecies *neanderthalensis,* named after a valley (the Neanderthal) in Germany where the first fossils were found. The oldest fossils, in Europe, are 400,000 years old. The lower position of the larynx

Continued

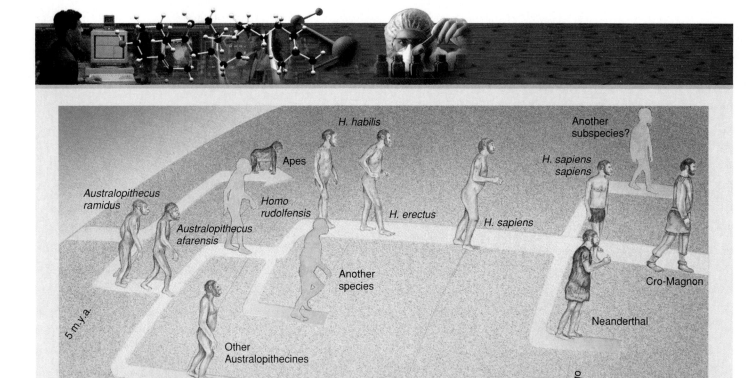

Figure 34.B Ancestral humans. An artist's interpretation, based on fossil bones, of some previous forms of humans. Each species is placed within an evolutionary sequence based on age of the fossils and similarity to other kinds of humans.

indicates these people may have been able to produce all the sounds of a modern language. Physically, the Neanderthals were similar to us. The female pelvis, however, indicates the gestation period was longer than ours (twelve months rather than nine) and the skull indicates their brain was larger than ours. These people disappear from the fossil record about 40,000 years ago.

The first *Homo sapiens sapiens*, our own subspecies, arose sometime between 200,000 and 800,000 years ago. Evidence from fossils and from analyses of the mitochondrial DNA of modern humans indicates this happened in Africa approximately 200,000 years ago. An African origin to the subspecies does not support or contradict either hypothesis regarding the origin of the species: *Homo sapiens sapiens* could have evolved from the original stock of humans, which were still in Africa, or from a local population of humans that remained in Africa while others colonized Europe and Asia. Fossils of this subspecies, about 100,000 years old, have been found in the Middle East and Europe. There they coexisted with the Neanderthals until 40,000 years ago, when the Neanderthals went extinct.

The later populations of *Homo sapiens* are known as Cro-Magnons, after the cave in France where the first fossils were found. Along with the fossils were a variety of tools, cave paintings, and carved objects. The Cro-Magnons of 35,000 years ago were biologically the same as us. If they reappeared today and were taught our culture and language, we would not recognize them as different.

The Carrying Capacity

The carrying capacity (📖 *see page 620*) of the world for humans has increased throughout our history. The total number of people, which reflects the maximum number the world's ecosystems can support, is shown for the past 100,000 years in figure 34.3. At first, the upper limit to human density was set by the amount of food that could be acquired during the cold or dry season. This resource "bottleneck" killed many people, especially children, each year.

Our ancestors then learned to anticipate shortages of food and stockpiled dried meats and seeds in advance, as

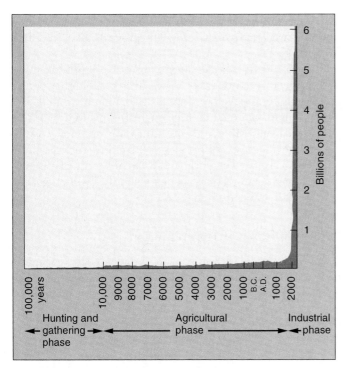

Figure 34.3 Growth of the human population. The human population has increased for the past 100,000 years. It experienced a minor surge associated with the agricultural revolution, 10,000 years ago, and is now experiencing a major explosion, associated with the products of applied science.

indicated by analyses of their refuse heaps. Food storage, a cultural adjustment, permitted more people to survive the winter or dry season. The carrying capacity increased and the population grew.

Approximately 10,000 years ago, humans devised a more reliable way of acquiring food: they learned to grow their own plants and animals. This profound change in human ecology is known as the agricultural revolution. Wildlands covered with largely inedible plants were converted to fields and sown with the seeds of grasses and legumes. These r-selected (see page 623) plants grew quickly and produced large quantities of seeds—cereals and beans, which were stored for winter. Some of the seeds were fed to cattle, sheep, and goats that were captured and kept in pens. The domestication of animals provided fresh milk and meat all year round. Agriculture, a cultural innovation, increased by several fold the carrying capacity of the earth for people (fig. 34.3).

Another time of rapid population growth began during the eighteenth century, largely in response to the industrial revolution and the use of coal as fuel for mass-producing materials. At the present time, we are experiencing a population explosion, which began around 1950, when improved sanitation, vaccines, and antibiotics greatly reduced our death rate.

There are now almost 6 billion people on earth, and the population is growing at a rate (see page 620) of 1.7% a year. While this percentage may seem small, keep

in mind that it is 1.7% of a very large number. Each year there are 94 million or so more people than the previous year. If present trends in birth rates and death rates continue, experts predict the world population will grow to approximately 12 billion people by the year 2050.

The Human Predicament: Limits to Adaptability

The history of humans differs strikingly from the histories of other forms of life, owing to the impact of cultural changes on life-style and environment. Cultural changes take place with amazing speed. New ideas that alter life-styles are adopted right away by enormous numbers of people. Biological changes, by contrast, evolve more slowly. The difference between rate of cultural change and rate of biological change presents a predicament: we may not be biologically adapted to our cultural innovations.

Our biological traits have changed very little since our hunter-gatherer days. These traits evolved as adaptations for acquiring food and shelter and for evading predators. Natural selection acted at all times to modify the biological traits of our hunter-gatherer ancestors; individuals who coped less well with their environment left fewer offspring. As a consequence, the Cro-Magnons were well-adapted to their environment.

In modern societies biological evolution is offset by aspects of culture. Our inventions shelter us from aspects of natural selection: we acquire food from grocery stores, we are sheltered from the cold by central heating, we are protected from parasites by medicines, and we have developed weapons that make us less vulnerable to predators. Most of our inventions promote survival and reproduction regardless of the genes we carry. Natural selection no longer has the impact it had when we lived as hunter-gatherers.

We have old-fashioned bodies in a culturally modified environment. Our biological traits were molded by, and remain best suited to, the environments that prevailed during Cro-Magnon times. Although this mismatch does not mean we should return to the hunter-gatherer lifestyle, it does suggest we ought to consider our biological limitations when planning new ways of life.

The Pace of Life

One aspect of modern life that has changed dramatically in the past 10,000 years is its pace. Consider speed of travel. The maximum speed of Cro-Magnon travel was set by how fast they could run. In today's world of the automobile and the airplane, people travel at incredible speeds simply by pressing down on a gas pedal or boarding a plane, and astronauts hurtle through space at thousands of kilometers an hour.

These new speeds make unusual demands on the human nervous system. When we walk or run, information enters the mind at a rate that gives us time to compare it with memories and make decisions based upon these comparisons. But when we guide a car through fast-moving traffic, information enters the mind at an unnaturally high rate. Decisions that could prove fatal are made instantaneously, based more on reflexes than on comparisons with previous experiences.

Another unnatural aspect of our lives is the flood of information; we are inundated with more facts than we can possibly absorb. An enormous amount of information must be processed in order to participate meaningfully in society. Most of us keep track of events that happen not only locally but all over the world. Students today are asked to learn much more than their parents, for our knowledge grows and our society becomes more complex with each generation. As a consequence, most of our learning comes from lectures, television, and books rather than from direct experience.

Human relationships have also changed. The social interactions of our ancestors probably resembled those of contemporary hunter-gatherer societies. These people maintain lifelong relationships with other members of their village. There is prolonged contact between children and their parents. Babies, for example, sleep with their mothers. (There is evidence that departure from this custom, in which babies sleep alone, may be a cause of sudden infant death.) Modern humans belong to much larger social groups than hunter-gatherer societies. To maintain so many relationships, we often communicate by computer, fax, and answering machines rather than face-to-face. To prevent social overload, our relationships tend to be fleeting or superficial. This form of social life is new to humans.

Mismatches between our life-styles and our biological adaptations generate stress. About 90% of all Americans experience unhealthy levels of stress at least once a week and some of us nearly every day. Of course, periods of stress have always been part of human life. Our hunter-gatherer ancestors endured dangerous hunts, bloody wars, and family squabbles. But these stressful times were offset by tranquil times, in which life was dominated by social traditions, close relationships, and ties to the land and its seasons. What is new to our culture is the unrelenting nature of the stressful state.

Biologically Foreign Materials

Our old-fashioned bodies are unable to cope with pollutants and synthetic materials in our air, water, and food. These materials were not part of our ancestral environment and we have no adaptations for dealing with them.

Our lungs are not adapted to function in smog, for example, nor are our livers adapted to process pesticides and food additives.

Our old-fashioned brains are not adapted to deal with concentrated doses of drugs. Irreversible changes in mental function develop when drugs such as alcohol, LSD, and marijuana are used in large quantities over a prolonged period of time. Mind-altering drugs are not new to humans: our ancestors drank alcohol, smoked tobacco, and chewed a variety of narcotic and hallucinogenic leaves and seeds. What is new is our capacity to purify these drugs and take them in concentrated form.

Primitive Rates of Reproduction

The human population is growing rapidly; each year more people are born than die. Our life history traits, which are generating this growth rate, were established thousands of years ago when humans were hunters and gatherers. A modern woman, like her distant ancestors, has the biological capacity to have a baby every year or so during her reproductive span of approximately thirty years.

Cro-Magnon populations rarely experienced rapid growth. To some extent reproduction was slowed by breast-feeding. In contemporary hunter-gatherer societies, a woman nurses her child for several years, an activity that inhibits ovulation and serves as a means of birth control. But the major factor preventing rapid growth of our ancestors must have been a very high mortality of infants. Young children are particularly vulnerable to starvation, dehydration, and infectious diseases. Most likely the majority of Cro-Magnon infants did not survive to become reproducing adults. In modern populations, cultural innovations (especially sanitation, vaccines, and antibiotics) enable survival of almost all children. Consequently, our inherited capacities for reproduction generate larger families now than they did in ancestral populations. As our culture improved childhood survival, our reproductive capacity became a liability.

An encouraging trend is that as a society develops and enjoys more material goods, its people have fewer children. Members of developed societies tend to control their inherent capacities for reproduction by marrying late and using effective methods of birth control. The current population explosion is due largely to reproduction in developing countries (fig. 34.4), where there are fewer material goods. How soon, if ever, will these countries become developed and control their reproduction? If development does not occur and the population explosion continues, the density of humans will be controlled by starvation, disease, and warfare rather than by voluntary reductions of birth rates.

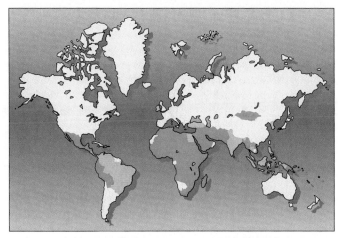

Average number of children per woman

☐ Fewer than 3   ▨ 3 to 5   ▧ More than 5

(a)

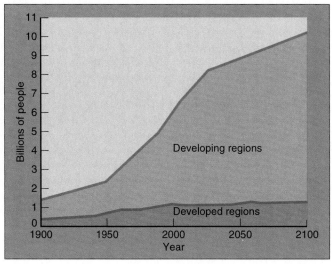

(b)

Figure 34.4 The high rate of growth is due more to reproduction in developing than developed regions.
(a) Global pattern of human birth rate, defined here as the number of children per woman. (b) Contributions by developing and developed regions to total population size, as documented in the past and projected for the future.

The Human Predicament: Environmental Deterioration

Another predicament faced by modern humans is a deteriorating environment. We are fouling our environment with our own wastes and encroaching on wild places that are homes to other species. We may so alter nature that no organisms, including ourselves, can survive on earth.

Table 34.1

Some of the Material Goods Consumed in a Lifetime by a Typical American

12,000 paper grocery bags
28,000 gallons of gasoline
250 shirts
115 pairs of shoes
27,500 newspapers
3,900 magazines
225 pounds of telephone directories
12 automobiles
28,627 aluminum beverage cans
16,610 pounds of coal (as electricity)
750 lightbulbs
69,250 pounds of steel
47,000 pounds of cement

Pollution

Human wastes are polluting the environment. Our culture generates an enormous amount of newspapers, cans, boxes, plastic bottles, diapers, and other forms of solid waste. We are running out of places to dump our trash. Developed countries, with their greater use of material goods, generate much more waste than developing countries. Americans in particular are big consumers, as shown in table 34.1. A baby born in the United States represents twice the destructive impact on the environment as one born in Sweden, 13 times one born in Brazil, and 280 times one born in Nepal!

The most destructive pollutants are gases, heavy metals, pesticides, and other materials that disperse within the soil, water, and air. Coal-burning power plants and gasoline-burning engines form gases that develop into acid precipitation (▭ *see page 648*), which is damaging our lakes and probably our forests. Carbon dioxide also pollutes the air, when released in large quantities from the burning of fossil fuels. Excess carbon dioxide in the atmosphere may produce global warming (▭ *see page 644*), which would shift centers of agriculture, flood coastlines, and eliminate species.

Pollutants that release chlorine or bromine into the upper atmosphere destroy the ozone layer, allowing more ultraviolet radiation to reach the surface of the earth. Some sources of these pollutants are refrigerator fluids, industrial solvents, and the production of styrofoam. Ultraviolet radiation is a highly concentrated form of energy that disrupts biological molecules. Its effects on ecosystems are largely unknown. In humans, an increase in ultraviolet radiation will cause more skin cancers and cataracts.

A wide variety of household and industrial chemicals, which accumulate in soil and water, may affect the reproductive organs of humans and other vertebrate animals. These materials are known as environmental estrogens because they interfere mainly with the activities of estrogen. They may be responsible for the decline in reproductive success that has been observed in lake fish, alligators, and aquatic birds. They may also have caused the decline in sperm production by men as well as the increased incidence of both testicular and breast cancers.

Species Extinctions

While it is true that extinction is a natural process, human activities have greatly increased its rate. Every year, approximately 10,000 species are lost forever. As many as one-fourth of the earth's species may be gone within fifty years. We assault the natural world in many ways: acid precipitation, oil spills, sewage dumps, and air pollution. Our most damaging activities, with regard to extinctions, are habitat destruction and species introductions.

Habitat Destruction   A habitat (see page 512) is a place where wild organisms live. We are eliminating these places by logging and by converting them into farms, pastures, mines, resorts, and cities (fig. 34.5). Most of these changes are irreversible: the soil becomes so altered by erosion, pollutants, and concrete that the ecosystem can no longer return to its original state.

Natural grasslands, for example, have disappeared from central regions of the United States. The native grasses and wildflowers have been plowed under and replaced by wheat and corn. And native animals (fig. 34.6) have no place to live, for they are not allowed in these valuable fields. Every time the land is plowed, some of the soil is washed away by rainwater and blown away by wind. The nutrient cycles have been altered by fertilizers, herbicides, and insecticides, which disrupt chemical balances and poison soil organisms. The land is so modified that even if it were no longer cultivated, it would not provide the foundation for rebuilding the same type of grassland.

Another biome that may soon disappear is tropical rain forest. These regions of dense vegetation, which are home to approximately half the world's species, occur mainly in poor countries with high rates of population growth. Consequently, the people who live there are con-

Figure 34.5 Habitat destruction.

verting wilderness to farms and ranches in order to raise more food. Almost 40% of the world's tropical rain forests were destroyed between 1950 and 1990. Every year, rain forests covering an area half the size of Nebraska are cut or burned. Wherever these forests are removed, the soil becomes infertile, hard, and no longer capable of supporting a forest.

The Everglades of southern Florida (fig. 34.7) is a unique ecosystem that is rapidly disappearing. This broad sheet of slowly flowing water, 80 kilometers (48 mi) broad and 15 centimeters (6 in) deep, supports a thick growth of saw grass bordered by mangrove trees. It is being drained for housing and polluted with commercial fertilizers from nearby farms. The Everglades are home to many organisms unique to this type of wetland, including herons, ibises, spoonbills, wood storks, American alligators, loggerhead turtles, manatees, and Florida panthers. Most of these animals are endangered and may soon go extinct.

Figure 34.6 Native species of the central grasslands of North America. (*a*) Bison no longer live in the wild; they survive only on wildlife preserves. (*b*) The black-footed ferret is almost extinct. (*c*) The burrowing owl is on the decline. (*d*) The swift fox is near extinction. (*e*) The running buffalo clover, once believed extinct, has been found but is extremely rare. (*f*) Mead's milkweed has nearly vanished.

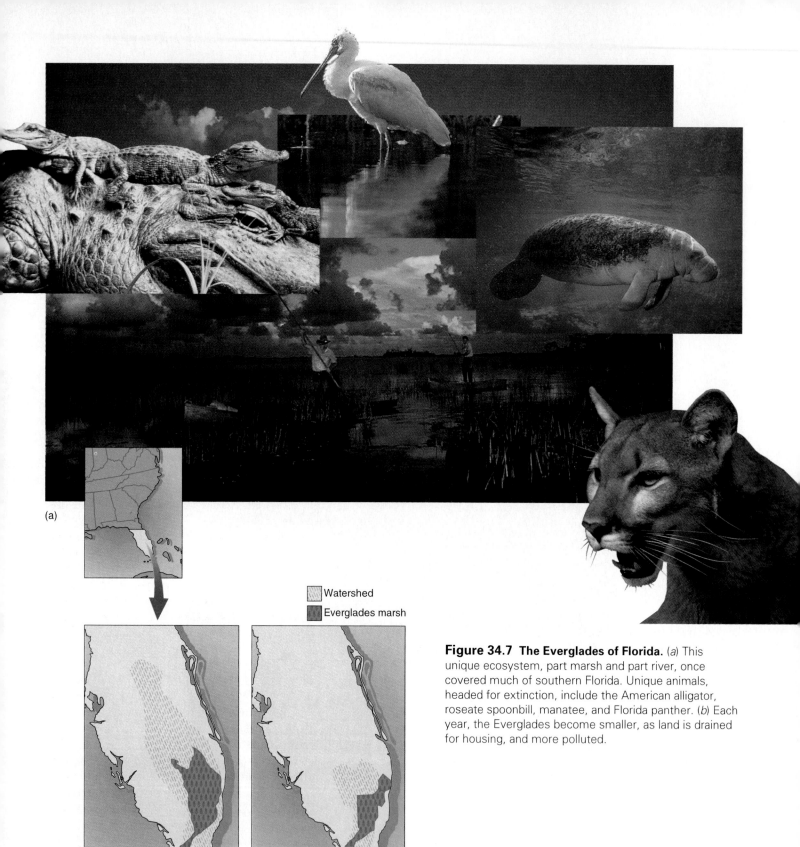

Figure 34.7 The Everglades of Florida. (*a*) This unique ecosystem, part marsh and part river, once covered much of southern Florida. Unique animals, headed for extinction, include the American alligator, roseate spoonbill, manatee, and Florida panther. (*b*) Each year, the Everglades become smaller, as land is drained for housing, and more polluted.

Watershed

Everglades marsh

(a)

Original Everglades
(b)

Everglades today

Species Introductions   A growing problem is the transport of organisms from one part of the world to another. Alien species are introduced deliberately by humans, as in the case of the starling, carp, and water hyacinth in North America, but more often they arrive as stowaways, as in the case of zebra mussels, cockroaches, gypsy moths, and crabgrass.

Most introduced species do not survive long in their new environments. A small percentage, however, have no enemies in their new homes and reproduce without check. These successful invaders are typically r-selected species, with rapid rates of reproduction and superb dispersal, that colonize habitats already altered by human activities. When we interfere in this way with a natural food web, we often alter population densities in ways that enable colonization by alien species. In time the alien species replace the native species.

The Hawaiian Islands are particularly vulnerable to introduced species, for they are the earth's most isolated islands. The nearest continent is roughly 4,000 kilometers (2,500 mi) away. For 70 million years, Hawaiian species evolved in splendid isolation. A new species arrived approximately once every 100,000 years. The Hawaiian Islands have no native snakes or ants and only one mammal (a bat). Most of its native species exist nowhere else on earth. Today the islands are experiencing a full-scale invasion by aliens, chiefly as stowaways on ships and planes. The native species, unused to these foreign competitors, predators, and parasites, are going extinct.

The Future

We are the only species that, as a group, can plan for future generations. We see how our activities are damaging the earth, and we ought to be able to alter our behaviors so that life can persist on this planet. Only humans can annihilate nature, and only humans have the foresight to prevent it from happening.

Status of the Species

How successful are we as a species? According to the criterion of abundance, we are an enormous success. Humans monopolize about 30% of all the energy captured by photosynthesis and our total weight exceeds that of any other species of animal.

Another criterion of species success is length of time until extinction. According to this criterion, we may turn out to be remarkably unsuccessful. The average life span of a species is 4 million years. We are still a young species, approximately 1 million years old, but it is hard to see how we can survive long enough to reach the average life span. It seems more likely that we will soon destroy ourselves with the products of our cultural activities.

We are certainly smart enough to alter our course and prevent our own extinction. The question is, are we wise enough to make sacrifices now to avoid future disasters? Are we altruistic enough to give up certain privileges and luxuries to preserve the quality of life, or even life itself, for future generations?

Possibilities for Change

What will the future bring? Several thousand years ago, the Greeks consulted oracles to find out about the future. At shrines like the Oracle of Apollo at Delphi (fig. 34.8a), the gods would answer the pilgrims' questions through the medium of resident priests and priestesses. These holy people would fall into trances and spew out bizarre phrases that interpreters would then translate.

Today we use computers to look into the future. Masses of information are fed into a computer (fig. 34.8b), which falls into a deep electronic trance and then spews out an almost endless stream of numbers, statistics, and graphs. The resident priests, called futurologists, then translate the computer output into predictions about the future (fig. 34.9).

The ancient oracles were often wrong, and computer versions of the future may be just as inaccurate. Predictions from computer models are based on assumptions about human behavior. In some of the potential problem areas of modern life—overpopulation, pollution, nuclear warfare, and wilderness destruction—predictions of future disasters will come true if we persist with the same behavior patterns of the past.

Our biological inheritance seems to compel us to acquire as much land, food, and shelter as possible and to have as many offspring as we can raise. Our cultural inheritance seems to compel us to exploit nature and accumulate as many material goods as possible. Can we resist these forces and alter our behavior? It seems possible. After all, we are unusually intelligent animals, with the capacity to foresee the consequences of our behavior. Moreover, we are highly social animals, with behaviors that encourage cooperation and mutual effort in the attainment of goals that are beneficial to the group.

(a)

(b)

Figure 34.8 Ancient and modern methods of predicting the future. (*a*) The ancient Greeks consulted oracles to learn about the future. Shown here is the sacred temple of Apollo at Delphi, where the gods occasionally revealed their plans. (*b*) We now predict the future by computer models, which use information from past events.

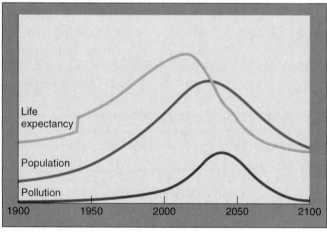

(a)

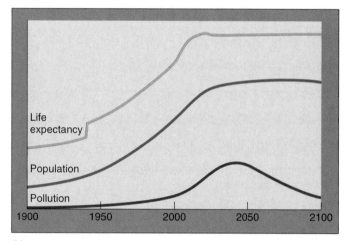

(b)

Figure 34.9 Predictions from computer programs. Computer models are based on certain assumptions. (*a*) Predictions of a model that assumes uncontrolled exploitation of natural resources. This unpleasant future would develop as capital becomes concentrated on extracting increasingly rare resources. (*b*) Predictions of a model that assumes regulation of population growth as well as much lower industrial productivity and consumption. In particular, each person in the world buys no more than $350 worth of consumer goods per year.

SUMMARY

1. Our ancestors, the Cro-Magnons, learned to use a great variety of resources; they had an unusually large niche. These people, who appeared 70,000 years ago, probably lived in small groups, defended group territories, and had a social life governed by division of labor and a dominance hierarchy.

2. The carrying capacity of the earth for humans has increased throughout history as more resources were made available by cultural innovations. The population continues to grow exponentially.

3. A current predicament is that cultural activities may bring about changes that are too rapid for our biological adaptations. The pace of life may be too much for our nervous systems, giving rise to unrelenting stress.

676

The speed of travel may be too fast, the amount of information too much, and our social groups too large. Our old-fashioned bodies cannot deal with synthetic materials or with high doses of mind-altering drugs. Our culturally reduced death rate is linked with a biologically high birth rate, resulting in very rapid population growth.

4. Another predicament is environmental deterioration. Serious environmental problems include acid precipitation, global warming, ozone holes, and environmental estrogens. The ever-expanding human population is eliminating other species, especially through habitat destruction and species introductions.

5. As a species, we are highly successful in terms of abundance but may be one of the least successful in terms of how long we exist prior to extinction. Predictions of future disasters are based on how humans behaved in the past. Our remarkable minds and highly developed societies may, however, alter our behaviors so that the predictions will not come true.

STUDY QUESTIONS

1. Describe the niche of Cro-Magnon peoples and explain why it was so much larger than the niches of other animals.
2. Show on a graph how the number of people on earth has increased during the past 35,000 years. Explain the major increases that occurred about 10,000 years ago, about 250 years ago, and in the present.
3. Explain why changes in our biological traits are lagging far behind changes in our culture.
4. Explain why habitat loss and species introductions are so damaging to wildlife.
5. Discuss the status of the human species.

CRITICAL THINKING PROBLEMS

1. Describe one health problem that may be caused by a mismatch between our biological adaptations and our cultural innovations.
2. Design an optimal life-style, in which our culture is more in line with our biology.
3. Why do you think there is only one species of human alive today?
4. How can you evaluate a prediction developed from a computer model?
5. Discuss the idea that since we are just another species of animal, we should continue to do what comes naturally—monopolize resources, reproduce without social constraints, and outcompete other species wherever we can.

SUGGESTED READINGS

Eaton, S. B., M. Shostak, and M. Konner. *The Paleolithic Prescription*. New York: Harper & Row, 1988. A description of the idea, from a medical point of view but written for the layperson, that our bodies are more suited to Cro-Magnon life-styles than modern life-styles.

Ehrlich, P. R., and A. Ehrlich. *The Population Explosion*. New York: Simon and Schuster, 1990. A discussion of human population growth, including rates of growth, carrying capacity, consequences of exceeding carrying capacity, and ways of preventing overpopulation.

Kellert, S., and E. O. Wilson (eds.). *The Biophilia Hypothesis*. Washington, D.C.: Island Press, 1993. A compilation of ideas and evidence related to the hypothesis that humans have an inborn need for nature, so that contact with the natural world is essential for spiritual and intellectual growth.

Levi-Bacci, M. *A Concise History of World Population*. Cambridge, MA: Blackwell Scientific Publications, 1992. A history of the growth of human populations and its causes, with particular emphasis on the social and economic factors that contribute to population growth in developed regions as compared with developing regions of the world.

McGibbon, B. *The End of Nature*. New York: Random House, 1989. A description of how everything that happens in the world, including the weather, is now influenced by humans. An easy-to-read, informative, and important book.

McKenna, T. *Food of the Gods: The Search for the Original Tree of Knowledge*. New York: Bantam Books, 1992. A look at the historical role of drugs, comparing the use of small doses by our ancestors with the use of mind-destroying doses by modern humans.

Tattersall, I. *The Human Odyssey: Four Million Years of Human Evolution*. New York: Prentice Hall, 1993. A well-illustrated description, aimed at the general public, of human evolution based on the fossil record.

Weiner, J. *The Next One Hundred Years: Shaping the Fate of Our Living Earth*. New York: Bantam Books, 1990. An exceptionally well-written account of research dealing with global warming, acid rain, ozone holes, habitat destruction, and other aspects of environmental change.

Willis, C. *The Runaway Brain: The Evolution of Human Uniqueness*. New York: HarperCollins Publishers, 1993. A description of how coevolution has driven animals to develop more intelligent behavior.

Glossary

A

abscission (aeb-SIH-shun) The controlled severance of a leaf or fruit from the stem of a plant.

absorption The transfer of materials from the digestive tract to the bloodstream.

acid A molecule that adds hydrogen ions to water.

actin (AEK-tin) A globular protein that forms part of a muscle fibril.

action potential A sudden change in the electric charge of a cell, from −55 to +40 to −70 millivolts.

active transport The movement of materials through proteins within a membrane, from regions of low concentration to regions of high concentration.

adaptation An anatomical, physiological, or behavioral trait that is controlled by genes and that improves an individual's chances of surviving and of leaving offspring in a particular environment.

adenosine diphosphate (ah-DEH-noh-seen DEYE-FOHSS-fayte) **(ADP)** An organic molecule that, when combined with energy and a phosphate ion, forms adenosine triphosphate.

adenosine triphosphate (ah-DEH-noh-seen TREYE-FOHSS-fayte) **(ATP)** An organic molecule used by cells to transport energy.

adhesion The tendency of molecules to adhere to a surface.

aerobic (AER-oh-bik) **respiration** Cellular respiration in which oxygen is used to convert glucose into carbon dioxide, water, and energy.

aldosterone (ahl-DAH-stur-ohn) A hormone, secreted by the adrenal glands, that stimulates the transfer of sodium ions out of the nephron and into the bloodstream.

allantois (AL-an-TOH-iss) An extraembryonic membrane that forms blood vessels within the umbilical cord.

allele (ah-LEEL) One version of a gene.

allergy An inappropriate response of the immune system to a harmless material.

allopatric (AEH-loh-PAEH-trik) **speciation** Formation of two or more species from two or more populations that live in different geographic regions.

allopolyploid (AEH-loh-POL-uh-ployd) An organism with more than two sets of chromosomes, inherited from more than one species.

altruistic (AL-troo-ISS-tik) **behavior** An act that increases the chances of another animal's survival or reproduction but decreases the chances of survival or reproduction of the animal that exhibits the behavior.

alveoli (al-VEE-oh-leye) Tiny indentations within the lung tissue.

amino (ah-MEE-no) **acid** A small organic molecule formed of an amino group (NH₂), a carboxyl group (COOH), and an R group.

amniocentesis (AEM-nee-oh-cen-TEE-sus) A procedure for obtaining fetal cells from the amniotic fluid for the diagnosis of genetic abnormalities.

amnion (AEM-nee-on) A extraembryonic membrane that surrounds the embryo.

amniotic (AEM-nee-AH-tik) **egg** The shelled "egg" of a reptile or chicken, within which the embryo develops outside its parent.

anemia (ah-NEE-mee-ah) Any condition in which there are not enough red blood cells or molecules of hemoglobin for adequate transport of oxygen.

angiosperm (AEN-jee-oh-spurm) A flowering plant.

annual An organism with a life span of less than a year.

antagonistic pair Two muscles that work in opposition to each other.

antenna pigment An organic molecule that absorbs light energy and transfers it to the reaction center of a photosystem.

antibody (ANN-tih-bah-dee) A protein that recognizes and binds with a particular molecule that is foreign to the body.

anticodon A sequence of three unpaired bases at one end of a transfer RNA molecule that binds with a sequence of three unpaired bases (a codon) within a messenger RNA molecule.

antidiuretic (AN-teye deye-yur-EH-tik) **hormone** (also known as vasopressin) A hormone, secreted by the pituitary gland, that stimulates movement of water out of the collecting duct and into the bloodstream.

apical dominance Suppression of lateral buds by the apical bud of a plant.

apomixis (AEH-poh-MIX-us) A form of asexual reproduction in which an embryo develops from a single cell in the parent rather than from a fertilized egg.

arteriosclerosis (ar-TEER-ee-oh-sklur-OH-sis) A condition in which the arteries are too rigid and narrow to ease the pressure of blood pumped out of the heart.

asexual (AY-SEK-shoo-uhl) **reproduction** The formation of offspring that receive all their genes from just one parent.

atherosclerosis (AEH-thur-oh-sklur-OH-sis) A form of arteriosclerosis that develops when materials accumulate on the inner walls of the arteries.

atom (AEH-tum) The basis of all physical things, formed of protons, neutrons, electrons, and other particles.

autoimmunity A condition in which the immune system attacks body cells.

autonomic (aw-toh-NAH-mik) **nervous system** The motor neurons that control glands and involuntary muscles.

autosome (AW-tuh-sohm) A chromosome that is not involved in determining whether an offspring is male or female.

axon (AK-sohn) Slender extension of a neuron that carries electrical signals from the cell body to the terminal buttons.

axon hillock The first part of an axon, adjacent to the cell body of a neuron.

B

base A molecule that removes hydrogen ions from water.

Batesian (BAYT-see-un) **mimicry** An antipredator adaptation in which a good-tasting prey species looks like a bad-tasting species.

B cell A lymphocyte that secretes antibodies.

behavioral response An observable activity of an animal that results from a stimulus.

biennial (BEYE-eh-nee-ul) An organism with a life span of two years.

bilateral symmetry A body form with left and right sides that are mirror images of each other.

bile (BEYEL) A complex fluid that is synthesized by the liver and released into the small intestine.

binary fission The form of cell division that occurs in prokaryotic cells.

biochemical pathway A sequence of chemical reactions controlled by a sequence of enzymes.

biological magnification The increase in concentration of a toxic material as it passes through a food chain.

biome (BEYE-ohm) A major form of landscape, consisting of many kinds of similar ecosystems.

blastocyst (BLAEH-stoh-sist) The blastula stage of a developing mammal.

blastopore (BLAEH-stoh-pohr) The indented region of an early embryo through which cells stream inward during gastrulation.

blastula (BLAEH-stoo-lah) An early stage of an animal embryo, consisting of a hollow ball of between 64 and 1,000 cells.

blood-brain barrier A modification of capillaries in the brain that prevents materials from moving freely between the blood and the interstitial fluid.

blood pressure The force exerted by blood on the wall of a blood vessel.

bone A type of connective tissue formed of widely spaced bone cells immersed within a hard material.

bronchiole (BRONK-ee-ohl) A tube that connects a bronchus with an alveolus in the lung.

bronchus (BRONK-us) A tube that connects the trachea with the bronchioles.

buffer A substance that counteracts changes in pH; it removes hydrogen ions from acidic solutions and removes hydroxyl ions from alkaline solutions.

bulbourethral (BUL-boh-yur-EE-thral) **gland** A male gland that secretes a sticky fluid into the urethra just prior to an ejaculation.

bulk transport The movement of large molecules and fluids across a cellular membrane.

bundle sheath Fiber cells that surround the xylem and phloem of a bundle sheath.

C

Calvin-Benson cycle A biochemical pathway that builds a three-carbon sugar from carbon dioxide, hydrogen atoms, and chemical energy.

cardiac output The amount of blood pumped by each heart ventricle each minute.

cardiovascular (KAHR-dee-oh-VASS-kyoo-lahr) **center** A cluster of neurons within the medulla oblongata that adjusts heart rate in response to emotions, level of physical activity, and blood pressure.

cardiovascular (KAHR-dee-oh-VASS-kyoo-lahr) **system** A closed network of tubes within which blood moves through an animal's body.

carnivore (KAR-nih-vor) An animal that eats other animals.

carpel (CAHR-pul) A female reproductive organ within a flower.

carrying capacity (K) The maximum density of individuals of a population that the environment can support.

cartilage (KAHR-tah-lahj) A connective tissue made of cartilage cells and collagen.

Casparian (kah-SPAH-ree-ahn) **strip** A waxy barrier that forces soil water to pass through cells of the endodermis in a plant root.

catalyst (KAEH-tah-list) A material that latches onto reactants and holds them in positions that make it easier for chemical bonds to break or to form.

cell body Rounded portion of a neuron that holds most of the organelles characteristic of an animal cell.

cell cycle A controlled sequence of events in which a small cell grows larger and then divides to form two smaller cells.

cellular membrane A sheet of macromolecules that encloses all cells and most organelles.

cellular respiration A series of biochemical reactions that releases chemical energy from an organic molecule.

cell wall A rigid structure that surrounds a cell, providing it with support and protection.

central nervous system The brain and spinal cord of a vertebrate animal.

centromere (SEN-troh-meer) A constricted region of a chromosome that holds together the sister chromatids during early phases of cell division.

cerebral (sur-EE-bruhl) **cortex** A region of the brain that stores and evaluates information, initiates appropriate behaviors, and suppresses intense emotions.

cervix (SUR-viks) A channel between the uterus and the vagina.

chemical energy Potential energy stored in the positions of electrons within molecules and ions.

chemical messenger A molecule that is secreted by one cell and that alters the activities of another cell.

chemiosmotic phosphorylation (KEE-mee-oz-MAH-tik FOS-for-uh-LAY-shun) Formation of ATP from a phosphate ion and the energy of a proton gradient.

chemoreceptor (KEE-moh-ree-SEP-tuhr) A cell that transduces the presence of molecules into electric signals.

chloroplast (KLOR-oh-plast) A specialized organelle within a cell that contains chlorophyll and carries out photosynthesis.

chorion (KOR-ee-on) An extraembryonic membrane that encloses all structures that develop from the fertilized egg.

chromosomal mutation A change in a large segment of a chromosome.

chromosome (KROH-moh-sohm) A molecule of DNA associated with proteins and other molecules.

cilium (SILL-ee-um) A short projection from a cell that creates movement.

circadian (sur-KAY-dee-un) **rhythm** Daily cycles of biological activity regulated by internal clocks within cells.

class A set of similar orders of organisms.

cleavage (KLEE-vahj) Cell division without cell growth.

climax community A stable food web that marks the culmination of succession.

clitoris (KLIH-tor-is) A female organ in front of the urethra that enlarges during sexual arousal.

cochlea (KOH-klee-ah) An organ of the inner ear that receives vibrations from the middle ear and transduces them into electric signals.

codominant alleles An interaction between two versions of a gene in which the products of both versions are seen in the phenotype.

codon (KOH-don) A sequence of three bases in a messenger RNA molecule that specifies a particular amino acid or a stop signal.

coelom (SEE-lohm) A body cavity within which organs are suspended.

coevolution Evolution of interacting species, in which evolutionary changes in one species cause evolutionary changes in the other species and vice versa.

cohesion The tendency of molecules to cling to each other.

collagen (KAHL-ah-jen) A strong, fibrous protein.

collecting duct A segment of the nephron that returns water from the filtrate to the bloodstream.

collenchyma (koh-LEN-kih-mah) **cell** A plant cell with a thick primary wall that is uneven in thickness.

companion cell A parenchyma cell that synthesizes materials for a sieve tube member.

competition The situation in which use of a resource by one individual reduces the availability of that resource to another individual.

competitive exclusion Extinction of a population of one species due to competition with individuals of another species.

complement A bundle of proteins that kills microbes, guides macrophages to microbes, and assists macrophages in other ways.

complete protein A protein that contains all the essential amino acids in approximately the right proportion for building human proteins.

cone A photoreceptor that detects color and provides fine details in bright light.

connective tissue A tissue formed of loosely packed cells.

consumer An organism that acquires its monomers from the organic molecules within other organisms.

contest competition Competition in which certain individuals have first access to a resource that is in short supply.

convergent evolution Similar environments produce similar adaptations in organisms of very different ancestry.

cork cambium (CAM-bee-um) A band of meristem between the cork cells and the dead phloem in the bark of a tree or shrub.

cornea (KOHR-nee-ah) A transparent, convex tissue that forms the front of the eyeball.

coronary (KOR-oh-nah-ree) **artery** A blood vessel that carries blood from the aorta to capillaries within the heart muscles.

corpus luteum (KOR-pus LOO-tee-um) A gland formed from a follicle within an ovary after ovulation of its egg.

cortex Ground tissue that lies to the outside of vascular tissue in the stem and root of a plant.

cotyledon The first leaf of an embryonic plant.

covalent (koh-VAY-lent) **bond** A mutual attraction between two atoms that share a pair of electrons.

creatine (KREE-ah-tin) **phosphate** An energy-storing molecule found in muscle cells.

cross bridge The region of a myosin molecule that attaches to an actin molecule during a muscle contraction.

crossing-over The exchange of genetic material between similar segments of homologous chromosomes.

cuticle (KYOO-tih-kul) A waxy coat on the leaves and stem of a plant.

cytokinesis (SEYE-toh-kih-NEE-sis) Division of the cytoplasm after mitosis or meiosis.

cytoplasm (SEYE-toh-PLAEH-zum) The region of the cell outside the nucleus.

cytoskeleton (SEYE-toh-SKEL-eh-ton) A web of fibrous proteins that extends throughout the cell.

cytosol (SEYE-toh-sol) The jellylike solution within a cell in which a variety of molecules and ions are suspended or dissolved.

D

deciduous (deh-SIH-dyoo-us) The shedding of leaves at a particular time of the year.

decomposer (DEE-kum-POH-zur) An organism that acquires monomers from dead organisms, the parts of organisms, or feces.

decomposition The process in which organic molecules are converted into inorganic molecules and ions.

deductive logic The thought process that takes a scientist from a hypothesis to a prediction.

dehydration synthesis The chemical reaction by which monomers are joined by covalent bonds; water is released from the monomers as they join together.

dendrite (DEN-dreyet) Short projections of a neuron that receive information.

deoxyribonucleic (DEE-oxy-REYE-boh-noo-KLAY-ik) **acid (DNA)** A nucleic acid formed of two sugar-phosphate strands and pairs of bases.

descriptive hypothesis (heye-PAH-the-siss) An overall statement, or generalization, about a set of observations.

diaphragm (DEYE-ah-frahm) A sheet of muscles that separates the chest cavity from the abdominal cavity. (Also the name of a device that can be placed over the cervix to prevent pregnancy.)

dicotyledon (DEYE-kah-tih-LEE-duhn) A category of flowering plant that forms two embryonic leaves.

differentiation (DIH-fur-en-shee-AY-shun) The embryonic process of cell specialization.

diffusion (dih-FYOO-shun) The tendency of materials within a fluid or gas to move down a concentration gradient.

digestion The breakdown of polymers into monomers.

digestive cavity A food-processing chamber with just one opening.

digestive tract A muscular tube that processes food and has two openings: a mouth and an anus.

dioecious (deye-EE-shus) A species of plant in which some individuals have flowers with only male parts and other individuals have flowers with only female parts.

diploid (DIH-ployd) **cell** A cell with two sets of chromosomes.

directional selection Individuals with an extreme form of a trait survive and reproduce better than other individuals in the population.

disruptive selection Individuals with two or more extreme forms of a trait survive and reproduce better than individuals with intermediate forms of the trait.

distal tubule A segment of the nephron that transfers inorganic ions between the filtrate and the blood.

divergent evolution Different environments produce different adaptations in closely related organisms.

DNA fingerprinting The use of restriction fragment-length polymorphisms to identify people and their relatedness.

DNA probe A colored or radioactively labeled copy of one strand of a small segment of a gene.

DNA sequencing A technique that describes the sequence of bases within a DNA molecule.

dominance hierarchy A set of dominant-submissive relations among members of a group.

dominant allele A version of a gene that masks the effects of another version of the gene.

donor gene A gene transferred to another cell for mass-production of the gene or its protein.

dormancy (DOHR-mahn-see) A genetically programmed reduction in cellular activity.

dorsal hollow nerve cord Part of the central nervous system that develops just above the notochord.

double fertilization Fertilization in which one sperm nucleus unites with an egg and another unites with a cell that develops into endosperm.

dynamic mutation The enlargement, through continued repetitions, of one region of a DNA molecule.

E

ecology The study of interactions between organisms and their environment.

ecosystem (EE-koh-sih-stum) A unit of nature, consisting of a set of interacting populations together with aspects of the physical environment.

ectoderm (EK-toh-durm) An embryonic tissue that will form the nervous system, epidermis of the skin, adrenal medulla, and various other adult tissues.

ectotherm (EK-toh-thurm) An animal that acquires heat from its external environment.

edema (eh-DEE-mah) Accumulation of fluids within the interstitial fluid of an animal.

electron (ee-LEK-trahn) A subatomic particle with a negative electric charge.

electron carrier An organic molecule that can reversibly gain and lose electrons.

electron transport chain A series of molecules along which an electron travels, releasing energy with each movement from one molecule to the next.

embryo (EM-bree-oh) The early stages of a developing animal; in a human, the first two months after conception.

embryonic induction (in-DUK-shun) The process in which one group of embryonic cells governs differentiation of an adjacent group.

endocrine (EN-doh-krin) **gland** An organ that contains hormone-secreting cells.

endocytosis (EN-doh-seye-TOH-sis) The transport, by membrane indentation and vesicle formation, of large molecules or fluids into a cell.

endoderm (EN-doh-durm) An embryonic tissue that will form the linings of internal channels, the liver, certain glands, and various other adult tissues.

endodermis (EN-doh-DUR-miss) A layer of cells in the root that controls what enters the xylem.

endometrium (EN-doh-MEE-tree-um) The lining of the uterus.

endoplasmic reticulum (EN-doh-PLAZ-mik reh-TIK-yoo-lum) A system of membranes that extends throughout the cytoplasm of a cell.

endosperm (EN-doh-spurm) A tissue that provides food for the growing plant embryo.

endosymbiotic (EN-doh-SIM-bee-ah-tik) **theory** An explanation of the origin of eukaryotic cells, in which complex cells arose from the union of simple ones.

endotherm (EN-doh-thurm) An animal that can warm itself from within, by heat generated from cellular respiration.

energy The force that makes it possible to do work—to bring about movement against an opposing force.

energy flow The movement of energy through a food web.

energy of activation The quantity of energy required to boost a chemical reaction so that it proceeds rapidly.

energy pyramid A diagram of energy flow through a food web.

enzyme (EN-zeyem) An organic molecule that increases the rate of a biochemical reaction.

epidermis (EH-pih-DUHR-miss) A layer of flattened cells that forms a surface.

epididymis (EH-pih-DIH-dih-mus) A tube, on top of the testis, in which sperm are stored.

epiglottis (EH-pih-GLAH-tus) A flap of cartilage that prevents food from entering the larynx and trachea.

epistasis (EH-peh-STAY-sis) An inheritance pattern in which the effects of one gene pair mask the effects of other gene pairs.

epithelial (EH-peh-THEE-lee-ul) **tissue** A sheet of tightly packed cells that covers a surface.

essential amino acid An amino acid that an animal requires but cannot make.

essential fatty acid A fatty acid that an animal requires but cannot make.

eukaryotic (YOO-kaehr-ee-AH-tik) **cell** A complex cell, with a nucleus and other membrane-enclosed organelles.

evolution A change in the genetic composition of a population in response to its local environment.

exocytosis (EX-oh-seye-TOH-sis) The transport, by fusion of a transport vesicle with the plasma membrane, of materials out of a cell.

exon (EX-ahn) A segment of precursor RNA that is retained within the messenger RNA.

explanatory hypothesis (heye-PAH-the-siss) A guess about the cause of an observed pattern.

exponential growth A continually increasing enlargement of a population, which occurs whenever the growth rate r is greater than zero for a prolonged period of time.

extracellular fluid The watery solution that lies between the cells of an animal.

extraembryonic membrane A membrane outside the embryo: the amnion, yolk sac, allantois, or chorion.

F

facilitated diffusion The diffusion of ions and large molecules through channels within proteins of a cellular membrane.

family A set of similar genera of organisms.

fast fiber A muscle cell that is specialized for anaerobic work.

fatty acid A long chain of carbon atoms, with attached hydrogen atoms, that terminates with a carboxyl group (COOH).

fermentation Cellular respiration in which glycolysis is followed by the conversion of pyruvic acid and NADH into waste products plus NAD⁺.

fertilization The merging of a sperm and an egg.

fertilized egg A cell formed by union of a sperm and an egg.

fetus (FEE-tus) A later stage of animal development; in a human, the period between two and nine months after conception.

fiber Polysaccharides that an animal eats but does not digest.

fight-or-flight response A rapid adjustment of the body to higher levels of physical activity.

fixed-action pattern A sequence of behaviors that occurs in response to a single stimulus.

flagellum (flah-JEL-um) A long, whiplike structure that propels a cell through its fluid environment.

follicle (FAH-leh-kul) A layer of cells that encloses an egg within an animal ovary.

food chain A sequence of species through which organic molecules pass.

food web A diagram of all the food chains, with their interconnections, within an ecosystem.

G

gamete (GAEH-meet) A cell (sperm or egg) that is specialized for fusing with another cell.

gametophyte (gah-MEE-toh-feyet) An individual plant (or structure within a plant) formed of haploid cells.

gap junction An extension from an animal cell through which materials move into an adjacent cell.

gas exchange The movement of oxygen and carbon dioxide through a body surface.

gastrula (GAEH-stroo-lah) An animal embryo consisting of three layers of cells.

gel electrophoresis (ee-LEK-troh-for-EE-sus) A laboratory technique that arranges molecules according to size.

gene One or more segments of DNA that together produce a functional RNA molecule.

gene cloning The formation of many copies of a donor gene from one copy of the gene.

gene expression The synthesis of a protein from a gene.

gene flow The movement of alleles, within immigrants, from one population to another.

gene locus The physical location of a gene within a chromosome.

gene pair Two copies of a gene at the same gene locus within a homologous pair of chromosomes.

gene pool All the genes, and their alleles, within a population.

general response The widespread defense of the body after a parasite has entered the bloodstream and lymph.

gene therapy A procedure in which a normal gene is transferred into a cell in order to correct for a defective gene in that cell.

genetic code The specific sequence of three bases in a molecule of DNA or RNA that specifies a particular amino acid.

genetic engineering The use of laboratory techniques to manipulate genes.

genome (jeh-NOHM) All the DNA molecules within a cell of an organism.

genomic (jeh-NOH-mik) **imprinting** A pattern of inheritance in which the expression of a gene depends on whether it is inherited from the mother or the father.

genotype (JEE-noh-teyep) The genes that contribute to an individual's phenotype.

genus (JEE-nuss) A set of similar species of organisms.

germination Activation of a dormant plant embryo.

gill slit A narrow opening between the pharynx and the outside environment.

glomerular (gloh-MEHR-yoo-lar) **capsule** A cuplike depression at one end of a nephron through which blood is filtered.

glomerulus (gloh-MEHR-yoo-lus) A bed of capillaries that lies inside the glomerular capsule of a nephron.

glucose (GLOO-kohz) A small organic molecule, formed of six carbons, that as a monomer is a main source of energy and as a polymer forms starch, glycogen, cellulose, and other important polysaccharides.

glycocalyx (GLEYE-koh-KAY-liks) A sugary coat that covers a cell, providing it with adhesion and recognition.

glycolysis (gleye-KAH-lih-sis) The part of cellular respiration in which glucose is split into two smaller molecules.

Golgi (GOHL-jee) **apparatus** A stack of flattened vesicles inside a cell that modifies proteins and ships them to their final destinations.

grana (GRAEH-nuh) Stacks of thylakoids within a chloroplast.

ground tissue Loosely packed parenchyma cells that fill the spaces between dermal tissue and vascular tissue in a plant.

guard cell Epidermal cells that control stomatal size in the leaf of a plant.

H

habitat The kind of environment in which an individual lives.

hair cell A cell that converts sound into electric signals.

haploid (HAEH-ployd) **cell** A cell with one set of chromosomes.

heart rate The number of times the heart contracts each minute.

heat energy Kinetic energy in the form of random movements of molecules and ions.

helper T cell A lymphocyte that coordinates the general response.

hemoglobin (HEE-moh-GLOH-bin) A large protein in red blood cells that transports oxygen gas.

herbivore (HUR-bih-vor) An animal that eats plants.

heterotroph (HEH-tur-oh-TROHF) An organism that acquires energy and nutrients from the organic molecules of other organisms.

heterozygote (HEH-tur-oh-ZEYE-goht) A gene pair consisting of two different alleles.

histamine (HISS-tah-meen) A chemical messenger that triggers an inflammatory response.

homeobox (HOH-mee-oh-box) A set of proteins that controls which part of an embryo develops into the front, middle, and rear.

homeobox (HOH-mee-oh-box) **genes** A set of genes that controls synthesis of the homeobox.

homeostasis (HOH-mee-oh-STAY-sis) The condition in which an organism's internal environment is maintained in an optimal state.

home range A space used predominantly by an individual, a pair, or a group of animals but not actively defended.

homologous (hoh-MAHL-uh-gus) **pair** Two versions of a chromosome within a diploid cell.

homologous (hoh-MAHL-uh-gus) **structures** Structures in different species that are similar because they have been inherited from a common ancestor.

homozygote (HOH-moh-ZEYE-goht) A gene pair consisting of two copies of the same allele.

hormone (HOR-mohn) A chemical messenger that travels within the body fluid of an organism.

host An organism that is harmed by a parasite.

human chorionic gonadotropin (KOR-ee-ON-ik go-NAEH-doh-TROH-pin) A pregnancy hormone that stimulates the corpus luteum to continue secreting estrogen and progesterone.

humus (HYOO-mus) Partially decomposed organisms, mainly plants, within the soil.

hybrid (HEYE-bred) The offspring resulting from a mating between individuals of two species.

hydrogen bond A mutual attraction between the electric charges of two polar molecules.

hydrolysis (heye-DRAH-lih-sis) The chemical reaction in which a polymer is digested into its component monomers; water is added to the monomers as they separate.

hypertension (HEYE-pur-TEN-shun) High blood pressure that persists even when the body is at rest.

hypertonic (HEYE-pur-TAH-nik) The condition in which a solution contains a higher concentration of dissolved materials than another solution.

hypha (HEYE-fah) A chain of cells within a fungus.

hypotension (HEYE-poh-TEN-shun) A gradual drop of blood pressure below normal.

hypothalamus (HEYE-poh-THAL-ah-mus) A major center of homeostasis, located within the brain of a vertebrate animal.

hypotonic (HEYE-poh-TAH-nik) The condition in which a solution contains a lower concentration of dissolved materials than another solution.

I

immunity (ih-MYOO-nih-tee) Long-term protection against a particular disease, generated by an accumulation of T and B cells from a previous encounter with the disease or a vaccination.

implantation The process in which the blastocyst embeds itself into the endometrium of the uterus.

imprinting A genetically programmed behavior that occurs at a specific age and under a particular set of conditions.

incomplete dominance An interaction between two versions of a gene that produces a phenotype intermediate between the two phenotypes produced by genotypes with two copies of each version.

independent assortment The way in which the chromosomes of one pair move into daughter cells has no influence on the way in which the chromosomes of other pairs move.

inductive logic A thought process that takes a scientist from specific observations to a generalized statement (descriptive hypothesis) or explanation (explanatory hypothesis).

inflammatory (in-FLAH-mah-toh-ree) **response** An altered state of tissues in response to an accumulation of fluids and immune materials from the blood.

innate (ih-NAYT) **behavior** A behavior controlled mainly by neural pathways laid down according to instructions from genes.

inner cell mass The embryo of a blastocyst.

inner ear A region of the ear that transduces sound and provides the sense of equilibrium; it consists of the cochlea, semicircular canals, utricle, and saccule.

inorganic (IN-or-GAEH-nik) **molecule** A molecule that was not synthesized by an organism.

insight learning A modification of behavior that occurs when an animal makes new associations between previously learned tasks in order to solve a new problem.

interferon (IN-tur-FEER-ahn) A chemical messenger that signals other cells to begin synthesizing enzymes that prevent viruses from reproducing.

interneuron A neuron that receives information from one neuron and passes it to another neuron.

interphase (IN-tur-fayze) The stage of the cell cycle between divisions, when the chromosomes are too long and thin to be visible.

interspecific competition Competition between individuals of different species.

interstitial (IN-tur-STIH-shul) **fluid** The watery solution located between the cells of an animal.

intraspecific competition Competition between individuals within the same population of a species.

intron (IN-trahn) A segment of precursor RNA that is discarded during formation of a messenger RNA.

ion (EYE-ahn) An atom or molecule with an electric charge, because it has either more electrons or fewer electrons than protons.

ionic (eye-AH-nik) **bond** A mutual attraction between atoms that develops between a positively charged ion and a negatively charged ion.

iris (EYE-ris) A structure in the eye that controls the size of the pupil.

isotonic (EYE-soh-TAH-nik) A condition in which two solutions have the same concentration of dissolved materials.

J

joint A connection between two or more bones.

jumping gene A gene that moves from one chromosome to another within a cell.

K

karyotype (KAER-ee-oh-teyep) An organism's set of chromosomes as seen during cell division.

kidney An organ that forms urine from components of the blood.

killer T cell A lymphocyte that destroys abnormal cells.

kinesis (kih-NEE-sis) A change in locomotion in response to a stimulus but without regard to location of the stimulus.

kinesthesia (KIH-ness-THEE-zee-ah) The sense of knowing the position and movement of body parts.

kinetic (kih-NEH-tik) **energy** Energy within the movement of objects.

kingdom A set of similar phyla of organisms.

kin selection Natural selection for traits that increase the survival and reproduction of an individual's relatives.

Krebs cycle The phase of aerobic respiration in which energy, hydrogen atoms, and electrons are released from an acetyl group.

K-selection Selection for individuals with traits that promote survival and reproduction when resources are in short supply.

L

larva (LAHR-vah) A developmental stage of an animal, which lives on its own and has a body form and life-style that is different from its parent.

larynx (LAEHR-enks) A tube (the voice box) that connects the pharynx with the trachea and produces sounds.

law A generalization, hypothesis, or theory that has been tested repeatedly and not proven false.

lens A transparent disk, located just behind the iris, that focuses light on the retina.

lichen (LEYE-kin) A mutualism between a fungus and a unicellular alga or cyanobacterium.

life cycle The alternation between diploid and haploid cells, brought about by meiosis and fertilization, within an individual.

ligament (LIG-ah-munt) A rope of collagen that holds together the bones of a joint.

limbic (LIM-bik) **system** A region of the brain involved with emotions, sexual responses, memory, sleep, and aspects of homeostasis.

lipid (LIH-pid) A polymer built by dehydration synthesis from several kinds of smaller molecules that consist mainly of carbon and hydrogen atoms.

local response The internal defense of an organism at the site of an invasion by a parasite.

locomotion (loh-koh-MOH-shun) Movement of the entire body, in a controlled fashion, from one place to another.

loop of Henle (HEN-lee) A segment of the nephron that returns water from the filtrate to the blood.

lymph node (limf nohd) An organ that filters microbes out of the lymph and develops an inflammatory response in response to an infection.

lymphocyte (LIM-foh-seyet) A white blood cell (a T or a B cell) that is specific to a particular kind of microbe.

lysosome (LEYE-zoh-sohm) A vesicle within a cell that contains digestive enzymes.

M

macrophage (MAEH-kroh-fahje) A white blood cell that engulfs and digests anything it recognizes as foreign to the body.

malnourishment Condition in which the diet fails to provide enough building materials for the body cells to function optimally.

marrow (MAEH-roh) The soft interior of a bone.

mechanoreceptor A cell that converts mechanical forces into electric signals.

medulla oblongata (meh-DOO-lah AH-blong-GAH-tah) A major center of homeostasis, located within the brain of a vertebrate animal.

meiosis (meye-OH-sis) A form of cell division that converts a single diploid cell into four haploid cells.

membrane potential The difference in electric charge between the inside and the outside of a cell.

Mendelian (men-DEE-lee-ahn) **inheritance** A form of inheritance in which a single gene pair governs the development of a single trait.

meristem (MEHR-ih-stem) Immature tissue located in special regions of a plant.

mesoderm (MEH-zoh-durm) An embryonic tissue that will form muscles, bones, dermis of the skin, kidneys, certain glands, and various other adult tissues.

mesophyll (MEH-zoh-fil) Ground tissue within the leaf of a plant.

messenger RNA (mRNA) An RNA molecule that carries genetic instructions to the site of protein synthesis.

metabolism (meh-TAEH-boh-lizm) All the biochemical reactions that take place within an organism.

middle ear A series of small bones that amplify the vibrations of the tympanic membrane.

migration Regular journeys between two specific places of residence.

mineral An atom that an animal needs and can use in an inorganic form.

mitochondrion (MEYE-toh-KOHN-dree-un) A membrane-enclosed organelle that holds the biochemical pathways of cellular respiration.

mitosis (meye-TOH-sis) Separation of chromosomes during cell division in which two cells are produced, each genetically identical to the other and to the parental cell.

mitotic spindle A set of protein fibers that controls the movements of chromosomes during cell division.

molecule (MAH-leh-kyool) Two or more atoms joined by chemical bonds.

monocotyledon (MAH-noh-kah-tih-LEE-duhn) A category of flowering plant that forms just one embryonic leaf.

monoecious (moh-NEE-shus) An individual plant with two kinds of flowers: one kind with only male reproductive organs and another kind with only female reproductive organs.

monogamous (moh-NAH-gah-muss) The condition in which an animal spends most of its time with just one mate.

monomer (MAH-noh-muhr) A small molecule that is a building block of polymers.

monosaccharide (MAH-noh-SAH-kur-eyed) A small molecule, usually a ring, formed of carbon, oxygen, and hydrogen atoms.

motor neuron A neuron that receives information from another neuron and passes it to a muscle or a gland.

mucus (MYOO-kus) A sticky fluid that lines many of the cavities and tubes within an animal.

Müllerian (muh-LEHR-ee-uhn) **mimicry** An antipredator adaptation in which two or more undesirable prey species appear the same to their predator.

muscle bundle Hundreds of muscle cells arranged in parallel and wrapped within a muscle sheath.

muscle fiber (FEYE-bur) A muscle cell.

muscle fibril (feye-BRIL) Proteins within a muscle cell that shorten in a contraction.

muscle sheath A tough membrane, formed of collagen, that covers a muscle bundle.

muscle tissue A tissue that is specialized for changing its length.

mutation (myoo-TAY-shun) A change in the sequence of bases within a gene.

mutualism (MYOO-chew-al-iz-um) A relationship between two species in which both species benefit.

mycelium (meye-SEE-lee-uhm) The body of a fungus, formed of interwoven hyphae.

mycorrhiza (meye-koh-REYE-zah) A mutualism between a fungus and a plant that enhances uptake of water and soil nutrients through the plant root.

myelin (MEYE-uh-lin) Fatty material that covers the axons of some neurons.

myosin (MEYE-oh-sin) A protein that forms part of a muscle fibril.

N

nastic (NAEH-stik) **response** Movement of an organism in response to an external stimulus but without regard to position of the stimulus.

natural selection Differential reproduction, in which certain individuals in a population leave more of their genes in future generations than other individuals.

negative feedback A system in which a disturbance in one direction triggers a corrective response in the opposite direction.

nephron (NEH-fron) The basic structure of a kidney.

nerve A bundle of neurons carrying information to and from the central nervous system.

nervous tissue A network of cells that carries signals from one part of an animal to another.

neuron (NUR-ahn) A nerve cell; a component of nervous tissue.

neurosecretory (NUR-oh-SEE-krih-tohr-ee) **cell** A neuron that secretes a hormone.

neurotransmitter (NUR-oh-trans-MIH-tuhr) A chemical messenger secreted into a synaptic cleft by a neuron.

neurula (NUR-yoo-lah) An animal embryo with a rudimentary nervous system.

neutron (NOO-trahn) A subatomic particle with no electric charge.

niche (nitch) The set of resources used by a population.

nitrogen fixation The biochemical process that converts nitrogen gas into ammonia.

nondisjunction (NON-dis-JUNK-shun) Failure of the chromosomes within a homologous pair to separate during cell division.

nonvascular plant A plant that has no xylem or phloem for transporting fluids.

notochord (NOH-toh-kord) A flexible rod that runs the length of a vertebrate embryo.

nuclear (NOO-klee-ahr) **envelope** A double membrane that encloses the nucleus of a cell.

nucleic (noo-KLAY-ik) **acid** A polymer formed by dehydration synthesis from nucleotides.

nucleolus (noo-KLEE-oh-lus) A spherical structure within the nucleus that builds ribosomes.

nucleotide (NOO-klee-oh-teyed) A small organic molecule formed of a sugar, a phosphate ion, and a nitrogenous base.

nucleus (NOO-klee-us) A large, spherical region within a cell that contains the chromosomes.

nutrient cycle The route taken by an atom as it travels from place to place within an ecosystem.

O

omnivore (AHM-nih-vore) An animal that feeds on organisms in two or more trophic levels.

order A set of similar families of organisms.

organ (OR-gun) A structure formed of two or more tissues that contributes in different ways to a particular task.

organelle A specialized compartment within the cell that contains the biochemical pathways for performing a particular task.

organic (or-GAEH-nik) **molecule** A molecule synthesized by an organism.

organ system A group of organs, located in different regions of the body, that contributes in different ways to a particular bodily function.

osmosis (ahz-MOH-sis) The diffusion of water through a cellular membrane.

outer ear A funnel that channels sound waves onto the tympanic membrane.

oval window A membrane within the wall of the cochlea that receives vibrations from the stapes of the middle ear.

ovary (OH-vur-ee) In an animal, an organ that forms eggs and female hormones. In a plant, an organ that develops into a fruit.

oviduct (OH-veh-dukt) A tube that carries mature eggs from an ovary to the uterus.

ovulation (AH-vyoo-LAY-shun) Movement of an egg out of an ovary and into an oviduct, where it can be fertilized.

ovule (OH-vyool) A structure that forms the female gametophyte and develops into a seed.

oxidation-reduction (OX-ih-DAY-shun) **reaction** The transfer of an electron from one atom to another.

oxygen debt The amount of oxygen needed to eliminate lactic acid that has formed from anaerobic exercise.

P

pacemaker A patch of specialized cells within the wall of the right atrium that regulates the heart rate.

parasite An organism that lives with an organism of another species and benefits from the relationship at the expense of the other organism.

parenchyma (pahr-EN-kih-mah) **cell** The least specialized of plant cells, with a thin, flexible primary wall and no secondary wall.

pedigree (PEH-deh-gree) A diagram of the phenotypes of family members, over several generations, with regard to a particular trait.

perception (pur-SEP-shun) A mental process in which electric signals in neurons are transformed into meaningful information.

perennial (PUR-eh-nee-ul) An organism with a life span of more than two years.

pericycle (PEHR-ih-seye-kul) Parenchyma cells that give rise to lateral branches of a root.

periderm (PEHR-ih-durm) The outer bark of a tree or shrub.

peripheral nervous system All portions of the nervous system that lie outside the central nervous system of a vertebrate animal.

peristalsis (PAEHR-ih-STAL-sis) Muscular contractions within the digestive tract that move food in a single direction.

petal A structure that forms a whorl between the sepals and the reproductive organs of a flower.

pH (pee-AYTCH) The ratio of hydrogen ions to hydroxyl ions within a watery solution.

phagocytosis (FAY-goh-seye-TOH-sis) The transport, by membrane indentation and vesicle formation, of a solid into a cell.

pharynx (FAER-enks) The first part of the digestive tract after the mouth.

phenotype (FEE-noh-teyep) The physical expression of an individual's genes, resulting from interactions between the genes and the environment.

phloem (FLOH-uhm) A system of tubes that transports organic molecules, especially sugars, from one part of a plant to another.

phospholipid A molecule formed from a fat by substituting a phosphate group for one of the fatty acids.

photoperiod (FOH-toh-peer-yud) The length of day in relation to the length of night.

photoreceptor A cell that transduces light energy into electric signals.

photosynthesis (FOH-toh-SIN-thuh-sis) A sequence of chemical reactions that converts sunlight into chemical energy and then uses the chemical energy to build an organic molecule (a sugar) from inorganic molecules (carbon dioxide and water).

photosystem A group of molecules that carries out the light reactions of photosynthesis.

phylum (FEYE-lum) A set of similar classes of organisms.

phytochrome (FEYE-toh-krohm) A plant protein that changes form when it absorbs certain wavelengths of light.

pigment (PIG-muhnt) An organic molecule that absorbs light.

pinocytosis (PEE-noh-seye-TOH-sis) The transport, by membrane indentation and vesicle formation, of fluids into a cell.

pith Ground tissue that lies to the inside of vascular tissue in the stem and root of a plant.

pituitary (pih-TOO-ih-tahr-ee) An organ, which extends from the hypothalamus of the brain, that secretes hormones, including many that control the activities of other glands.

placenta (plah-SEN-tah) An organ that develops within the wall of the uterus during pregnancy, formed of both embryonic and maternal blood vessels.

plasma (PLAZ-mah) The watery solution that forms the noncellular part of the blood.

plasma (PLAZ-mah) **membrane** A sheet of lipids and proteins that forms the outer boundary of a cell.

plasmid (PLAZ-mid) A tiny loop of DNA that occurs naturally within bacteria and moves from one bacterium to another.

plasmodesmata (PLAEH-zmoh-dez-MAH-tuh) An extension from a plant cell through which materials move into an adjacent cell.

plastid A membrane-enclosed organelle that either holds the biochemical pathways of photosynthesis or stores macromolecules.

pleiotropy (PLEYE-oh-troh-pee) A pattern of inheritance in which a single gene pair controls more than one trait.

pleural (PLUR-ahl) **membrane** A moist membrane that adheres to the lungs and to the inner wall of the chest cavity.

point mutation A substitution, addition, or deletion of a single base within a gene.

pollen grain A male gametophyte.

pollination (PAH-leh-NAY-shun) The transfer of pollen (sperm) from the male part of one flower to the female part of another flower.

pollinator (PAH-leh-NAY-tur) An animal that transports pollen (sperm) from the male part of one flower to the female part of another flower.

polygenic (PAH-lee-JEE-nik) **inheritance** A form of inheritance in which a trait is determined by more than one gene pair.

polymer (PAH-lih-muhr) A large molecule built of one or a few kinds of smaller molecules called monomers.

polymerase (poh-LIH-muhr-ayz) **chain reaction** The continuous replication of a gene within a test tube.

polypeptide (POH-lee-PEHP-teyed) A polymer built by dehydration synthesis from many amino acids.

polysaccharide (POH-lee-SAH-kur-eyed) A polymer built by dehydration synthesis from monosaccharide monomers.

population A group of individuals of the same species that live together in the same geographic region.

potential energy Stored energy that is capable of moving an object.

pressure-flow theory An explanation of how fluids are moved through the phloem of a plant.

primary electron acceptor An organic molecule that captures an electron that has been lifted out of a pigment during photosynthesis.

producer An organism that builds its own monomers from inorganic materials.

productivity The quantity of new materials formed within a trophic level each year.

prokaryotic (PROH-kahr-ee-AH-tik) **cell** A simple cell, without a nucleus or other membrane-enclosed organelles.

prostate (PRAH-stayt) A gland that secretes a fluid into the semen.

protein A molecule formed of one or more polypeptides.

proton (PRO-tahn) A subatomic particle with a positive electric charge.

proto-oncogene (PROH-toh-AHN-koh-jeen) Gene for a protein that stimulates cell division.

proximal tubule A segment of the nephron that returns most of the filtrate to the bloodstream.

Punnett (PUH-net) **square** A chart that illustrates how the different kinds of sperm from a male can fertilize the different kinds of eggs from a female.

pupil (PYOO-pil) The opening into the eye through which light enters.

R

radial symmetry A body form with no distinct left or right sides.

reaction center A chlorophyll molecule that loses electrons when it receives energy from light or an antenna pigment.

receptor (ree-SEP-tuhr) A protein molecule to which a chemical messenger or nutrient binds.

receptor (ree-SEP-tuhr) **cell** A cell that transduces environmental energy into electric signals.

recessive allele A version of a gene that has its effects masked by another version of the gene.

reciprocal altruism (AL-troo-ism) The condition in which an animal that exhibits altruism is likely to benefit from another animal's altruism at another time.

recombinant (ree-KAHM-bih-nant) **DNA technology** The mass production of proteins and genes inside a bacterium.

recombination The formation of new combinations of alleles within an offspring.

reproductive isolating mechanism A structural, behavioral, or biochemical trait that prevents individuals of a species from successfully breeding with individuals of another species.

respiration The process by which oxygen is moved into an animal and carbon dioxide is moved out.

restriction enzyme An enzyme that cuts DNA molecules at a specific base sequence.

restriction fragment-length polymorphism (PAH-lee-MOHR-fiz-um) Variation in the length of a particular DNA fragment, among different individuals, after the DNA has been cut by a particular restriction enzyme.

retina (REH-tih-nuh) A tissue in the eye that contains photoreceptors and sensory neurons.

retrovirus An RNA virus that synthesizes a DNA copy of its genes and then inserts this copy into a DNA molecule of its host cell.

reverse transcription The synthesis of a DNA copy of a gene from an RNA copy of the gene.

R group A small organic molecule bonded to a larger one; it forms different molecules from a common structure.

ribonucleic (REYE-boh-noo-KLAY-ik) **acid (RNA)** A nucleic acid formed of a single sugar-phosphate strand and unpaired bases.

ribosome (REYE-boh-sohm) A spherical structure, formed of RNA and proteins, that serves as a site for protein synthesis.

RNA processing Modification of a precursor RNA molecule to form a messenger RNA molecule.

rod A photoreceptor that detects small differences in light intensity, seen as shades of gray.

r-selection Selection for individuals with traits that promote reproductive success as opposed to traits that promote survival.

S

saccule (SAK-yool) An organ within the inner ear that informs the brain about tilts of the head.

scientific process A systematic way of seeking the truth, involving the processes of observing, hypothesizing, predicting, and testing.

sclerenchyma (skler-EN-kih-mah) **cell** A plant cell with both primary and secondary cell walls.

scramble competition Competition in which there are no rules governing the division of a resource that is in short supply.

scrotum (SKROH-tum) A pouch of skin that holds the testes outside the abdominal cavity.

secretion (seh-KREE-shun) The synthesis and controlled release of special materials for use outside a cell.

segmentation Division of a body or organ into a series of similar parts.

semicircular canals Organs within the inner ear that inform the brain about sudden movements and rotations of the head.

seminal vesicle (SEH-mih-nul VEH-sih-kul) A gland that secretes a fluid into the semen.

seminiferous (SEH-min-IH-fur-us) **tubule** A tube, in the testis, within which sperm develop.

sensory neuron A neuron that carries information from the environment to interneurons within the central nervous system.

sepal (SEE-puhl) A structure that forms the outermost whorl of a flower.

sessile (SEH-sil) Attached to a substrate.

sex chromosome A chromosome involved in determining whether an offspring is male or female.

sexual reproduction The formation of offspring that receive genes from two parents.

sexual selection Individuals with certain traits attract more mates and so leave more offspring than individuals without these traits.

shell A space, within which electrons travel, that is a particular distance from the nucleus of an atom.

sieve tube A tube, formed of living cells, that transports fluid within a plant.

sister chromatids Two copies of a chromosome held together by a chromatid.

slow fiber A muscle cell that is specialized for aerobic work.

sociobiology The study of the biological basis of social behaviors.

soft palate (PAEL-uht) A flap of soft tissue that prevents food from entering the nose.

soil nutrient An inorganic ion that a plant requires for normal growth and reproduction and that it acquires from the soil.

solute (SOL-yoot) A substance that has been dissolved in a solvent.

solvent (SOL-vent) A liquid that dissolves solids.

somatic (soh-MAH-tik) **nervous system** The sensory neurons and motor neurons that govern responses by skeletal muscles.

somite (SOH-meyet) A block of mesoderm in a vertebrate embryo that will later form the bones and muscles of the back as well as other internal structures.

speciation (SPEE-see-AY-shun) The formation of a new species.

species (SPEE-seez) A group of similar organisms capable of mating with one another and producing healthy offspring; a group of organisms with an isolated collection of genes.

sphincter (SFINK-tur) A ring of muscles that contracts to close off a channel.

sporophyte (SPOR-oh-feyet) An individual plant formed of diploid cells.

stabilizing selection Individuals with extreme versions of a trait survive and reproduce less well than individuals with an intermediate version of the trait.

stamen (STAY-men) A male reproductive organ within a flower.

starvation The condition in which fewer calories are consumed than used for a prolonged period of time.

statistics (stah-TIH-stiks) A branch of mathematics, based on probabilities, that is devoted to the analysis and interpretation of measurement data.

steroid A lipid formed of four carbon rings with an attached R group.

stimulus Information that enters the nervous system through the sense organs.

stomata (stoh-MAH-tah) Tiny holes in the epidermis of a leaf, through which gases diffuse.

stroke volume The amount of blood ejected from each ventricle of the heart during each contraction.

substrate A specific ion or molecule that fits into an active site of an enzyme.

succession (suhk-SEH-shun) The development of an ecosystem from bare rock or bare soil.

suppressor gene Gene for a protein that inhibits cell division.

surfactant (sur-FAEK-tunt) A chemical that prevents wet surfaces from sticking together.

sweat gland A structure containing cells that secrete a dilute solution of water and ions.

sympatric (sim-PAEH-trik) **speciation** Formation of two species from a single population located within the same geographic region.

synapse (SIN-aps) The region between a neuron and the next cell in the pathway; it includes the plasma membrane of the neuron, the space between the two cells, and the plasma membrane of the next cell.

synovial (sih-NOH-vee-ahl) **fluid** A thick fluid within the cavity of a joint.

T

target cell A cell that is influenced by a particular chemical messenger.

taste bud Sensory organ that picks up and relays information about dissolved molecules.

taxis (TAK-sis) A change in locomotion that moves an animal directly toward or away from a stimulus.

tendon A rope of collagen that attaches a muscle to a bone.

terminal button Small knob at the end of an axon that secretes neurotransmitters in response to electric signals within the axon.

territory A space that is defended by an individual, a pair, or a group of animals.

testis (TES-tis) An animal organ that produces sperm and testosterone.

theory A set of related hypotheses supported by many observations and experiments.

thermoreceptor A cell that transduces changes in heat energy into electric signals.

thylakoid (THEYE-luh-koyd) Flattened sacs within a chloroplast on which the light-dependent reactions of photosynthesis take place.

tissue (TIH-shoo) A group of cells that lie next to one another and contribute in a similar way to the same task.

trachea (TRAY-kee-ah) A tube (the windpipe) that connects the larynx with the bronchi.

tracheid (TRAY-kee-id) A tube, formed of the walls of dead cells, that transports fluid within a plant.

transcription The synthesis of an RNA copy of a gene from a DNA copy of the gene.

transduction (trans-DUK-shun) The conversion of energy changes in the environment into electric signals within neurons.

transfer RNA (tRNA) A molecule that transports amino acids to a ribosome and arranges them according to instructions within the messenger RNA molecule.

transgenic (trans-JEE-nik) **organism** An organism that has received genes, via genetic engineering, from another species.

translation The process that links together amino acids according to instructions within a messenger RNA molecule.

transpiration Evaporative water loss from a plant, chiefly through the stomata.

transpiration-cohesion theory An explanation of how water moves upward through the xylem of a plant.

trial-and-error learning A modification of behavior that occurs when an animal makes a connection between a particular behavior and its consequences.

trophic (TROH-fik) **level** All organisms of a food web that occupy the same position in the food chains.

trophoblast (TROH-foh-blast) The part of a blastocyst that will form extraembryonic membranes.

tropism (TROH-pism) Movements of an organism toward or away from an external stimulus.

turgor (TUR-gur) **pressure** The outward pressure of a swollen cell against its cell wall.

tympanic (tim-PAEH-nik) **membrane** A membrane (the eardrum) that vibrates as it is displaced by alternating waves of compressed and rarefied air.

U

umbilical (um-BILL-IH-kul) **cord** An organ that connects the bloodstreams of the embryo (or fetus) and its mother and suspends the embryo (or fetus) within the uterus.

ureter (YUR-eh-tur) A tube that carries urine from a kidney to the urinary bladder.

urethra (yur-EE-thrah) A tube that carries urine from the bladder to the outside of the body.

urinary (YUR-in-ahr-ee) **bladder** A hollow, muscular organ that stores urine.

urinary (YUR-in-ahr-ee) **system** A group of organs that work together to form and eliminate urine.

uterus (YOO-tur-us) A muscular chamber that shelters and nourishes the developing baby.

utricle (YOO-trih-kuhl) An organ within the inner ear that informs the brain about tilts of the head.

V

vaccine A harmless form of a disease microbe.

vacuole (VAEH-kyoo-ohl) A relatively large membrane-enclosed bag within a cell.

vagina (vah-JEYE-nah) A muscular tube that extends from the cervix to the outside of the body.

vascular bundle A cluster of xylem and phloem within a plant.

vascular cambium (CAM-bee-um) A band of meristem between the xylem and the phloem.

vascular plant A plant that has xylem and phloem for transporting fluids.

vas deferens (vaz DEH-fur-ens) A tube that carries sperm from the epididymis to the urethra.

vector A DNA molecule used by genetic engineers to transport a gene from one cell to another.

vegetative reproduction A form of asexual reproduction in which a severed part of a plant develops into a new individual.

ventilation The process of moving air or water over a gas-exchange surface.

vertebrate An animal with a backbone (a fish, amphibian, reptile, bird, or mammal).

vesicle (VEE-zih-kahl) A tiny membrane-enclosed bag within a cell.

vessel A tube, formed of the walls of dead cells, that transports fluid within a plant.

virus A cluster of organic molecules that uses the cell of an organism to reproduce itself.

vitamin (VEYE-tah-min) An organic molecule that the cells of an animal cannot synthesize but need in tiny amounts.

W

work To bring about movement against an opposing force.

X

X-linked trait A characteristic determined by a gene within the X chromosome.

xylem (XEYE-lum) A system of tubes, extending from the roots to the leaves, that transports water in a plant.

Y

yolk sac An extraembryonic membrane that, in a mammal, forms blood cells as well as the cells that eventually develop into eggs or sperm.

Z

zygote (ZEYE-goht) A fertilized egg, formed by union of a sperm and an egg.

Credits

LINE ART

Carlyn Iverson
10.3C, 13.9, 13.16, 13.19, 14.2, 14.8, 14.9, 15.2A-B, 15.8, 15.12

Carlyn Iverson/Diphrent Strokes
14.1, 14.3

Diphrent Strokes
I.2, I.3, I.4, I.5, I.10, I.12C, I.14, 1.1, 1.2, 1.3, 1.4, 1.5, 1.6, 1.7, 1.8, 1.9, 1.A, 1.10, 1.11, 2.1, 2.2, 2.3, 2.4, 2.5, 2.6, 2.7, 2.8, 2.9, 2.10, 2.11, 2.12, 3.9, 3.B(a), 3.C, 4.4, 4.5, 4.11, 5.1, 5.2, 5.4, 5.5, 5.7, 5.8, 5.9, 5.11, 5.12, 6.1, 6.3, 6.5, 6.6, 6.7, 6.8, 6.B, 6.10, 6.11, 7.1, 7.3, 7.4, 7.5 (art), 7.6 (art), 7.8A, 7.A, 7.C, 8.4A-B, 8.5(art), 8.6, 8.7, 9.4, 9.7, 9.8, 9.9B, 9.10, 9.11, 9.12, 9.13C, 10.2, 10.4 (art), 10.5, 10.6, 10.7, 10.8 (art), 10.10, 10.11B, 11.1A, 11.2, 11.3, 11.4, 11.5, 11.7, 11.8, 11.9A, 11.B (art), 11.C (art), 11.11, 11.13, 12.2, 12.4, 12.6, 12.7, 12.8, 12.9, 12.B, 12.10, 12.12A, 13.6, 13.7A, 13.8, 13.C, 13.13 (art), 13.15 (art) 13.18A, 14.4A-B, 14.5, 14.6, 14.10, 15.5, 15.9A, 15.10, 15.11, 16.1, 16.2, 16.3B, 16.4A, 16.7, 16.8, 16.9A, 16.A, 16.10, 16.11, 16.12B-C, 16.14, 17.2, 17.4, 17.A, 17.10, 17.12, 18.1, 18.2, 18.3, 18.7, 18.C, 18.12, 19.1, 19.2B, 19.5, 19.6, 19.7, 19.9, 19.B, 19.C, 19.10, 19.11, 20.1, 20.2, 20.5, 20.6, 20.7, 20.10, 21.1, 21.5, 21.6, 21.A, 21.12A-B, 22.3A, 22.4B, 22.5B, 22.6A, 22.7B, 22.8B, 22.9B, 22.B, 22.C, 22.10, 23.1, 23.2, 23.3, 23.4, 23.8, 23.A, 23.B, 23.C, 23.12, 23.15, 24.1, 24.2, 24.3, 24.4, 24.5, 24.6, 24.7 (art), 24.A, 24.B, 25.1, 25.2, 25.6B, 25.10, 26.2 (art), 26.10, 26.11, 26.12, 26.13, 26.14, 27.9, 27.C, 28.6, 28.9, 28.B, 28.10A, 28.13, 28.14, 28.15, 28.16, 29.2, 29.A, 29.12, 30.1, 30.2, 30.6, 30.7, 30.11, 30.12, 30.13, 31.1, 31.2B, 31.3, 31.7, 31.9B-C, 31.A, 31.B (art), 31.10B, 31.11B-E, 31.12B, 31.13B, 31.14B-C, 31.16B, 31.19A-B, 32.6A-B, 32.C, 33.2, 33.3, 33.6, 33.7, 33.8 (art), 33.A, 33.C, 33.10 (art), 33.12, 33.13A (art), 33.13B (art), 33.14A (art), 33.14B (art), 33.15, 33.16A, 34.2, 34.3, 34.4, 34.7B, 34.9 and text art on pages: 25, 26, 27, 30, 32, 34, 35, 41, 42, 43, 44, 45, 48, 49, 50, 60, 65, 75, 76, 86, 90, 93, 94, 106, 107, 108, 112, 113, 114, 115, 124, 126, 128, 132, 133, 137, 145, 146, 147, 148, 149, 156, 157, 158, 160, 161, 162, 163, 164, 168, 172, 176, 178, 183, 186, 187, 188, 192, 193, 194, 198, 203, 204, 209, 213, 215, 219, 223, 233, 238, 240, 241, 247, 248, 250, 251, 252, 256, 257, 260, 261, 263, 266, 273, 274, 276, 278, 279, 286, 299, 300, 302, 303, 304, 308, 314, 315, 316, 317, 321, 326, 328, 334, 335, 339, 355, 357, 358, 359, 360, 366, 374, 375, 385, 390, 391, 392, 395, 396, 398, 446, 447, 451, 459, 474, 477, 480, 490, 497, 498, 528, 548, 583, 548, 549, 551, 570, 583, 593, 594, 601, 606, 609, 621, 640, 641, 643, 648

Ethan Geehr
3.3, 3.4B, 3.5, 3.6, 3.7, 3.8B, 3.10A, 3.11, 4.3, 4.6, 4.7, 5.6 (art), 6.2, 7.7, 8.2A, 11.10, 12.3A (art), 13.2, 13.4, 13.5

Ethan Geehr/Carlyn Iverson
6.4

Ethan Geehr/Diphrent Strokes
4.1, 4.2, 4.A, 4.8, 4.9, 4.10, 4.12, 5.3, 5.10, 5.1, 5.A, 6.2, 6.9, 11.6, 12.1, 12.11, 13.10, 13.17

Laurie O'Keefe
8.1, I.6, I.13, I.15, I.16, 22.12, 24.8, 26.1, 26.3, 26.4, 27.1, 27.3, 27.10, 28.3, 28.11, Chapter Opener 30, 30.3, 30.8 (art), 30.10, 31.8, 31.C, 31.15, 31.17, 32.2, 32.3, 32.7, 32.13, 32.14, 33.4, 33.5, 34.A, 34.B

Laurie O'Keefe/Diphrent Strokes
9.2, 30.14, 30.B

Rictor Lew & Patty Chen
17.1A-B, 17.3 (art), 17.5 (art), 17.6A-B, 17.8, 17.9, 17.11, 18.5, 18.6, 18.8, 18.9, 18.10, 18.11 (art) 19.3, 19.4, 19.8, 19.A, 19.12, 19.13, 20.3, 20.8, 20.9, 20.11, 21.2, 21.3, 21.4, 22.2, 23.5, 23.9, 23.11, 23.13, 23.14, 24.9, 24.10, 24.12, 24.14, 24.15, 24.16, 24.17, 24.18, 25.3, 25.5, 25.7, 25.8, 25.9, 25.11, 25.13, 25.15, 26.5, 26.7, 26.8, 26.9, 28.4, 28.5, 28.7, 28.12, 28.17, 29.3, 29.5, 29.6, 29.7, 29.8, 29.10, 29.13, 29.14

Rictor Lew & Patty Chen/Diphrent Strokes
17.7, 20.4, 21.7, 21.8, 21.10, 21.11, 23.6

PHOTOGRAPHS

Part Openers
I: © Jeanne Drake/Tony Stone Images;
1: © Manfred Kage/Peter Arnold, Inc.;
2: © M. Giles/PIX Elation/Fran Heyl Associates;
3: © Fred Hirschmann; 4: © Daniel J. Cox;
5: © Gregory G. Dimijan, M.D./Photo Researchers, Inc.

Chapter I
I.1a: © Ed Ely/Biological Photo Service;
I.1b: James L. Amos, © National Geographic Society; I.1c: © John C. Sanford, Cornell University; I.7a1: Courtesy of the Armed Forces Institute of Pathology; I.7a2: The Bettmann Archive; I.7b1: © K. Tabro/Visuals Unlimited; I.7b2: © Dwight R. Kuhn; I.7c1: © George Musil/Visuals Unlimited; I.7c2: © Dwight R. Kuhn; I.8a: © Grant Heilman/Grant Heilman Photography; I.8b: © Thomas Kitchin/Tom Stack & Associates; I.9: The Bettmann Archive; I.11a: © J. J. Cardamone, Jr and B. K. Pugashetti/Biological Photo Service; I.11b: © Science VU/Visuals Unlimited; I.12a: Painted by George Richmond, 1840. © Archive/Photo Researchers, Inc.; I.12b: © Christopher Ralling; I.13c: © Comstock; I.17: © Russell A. Mittermeier, Conservation Int'l.

Chapter 1
Opener: © Ken Eward/Science Source/Photo Researchers, Inc.; 1.7c: © Larry Lefever/Grant Heilman Photography

Chapter 2
Opener: Image courtesy of David S. Goodsell, The Scripps Research Institute; Box 2.b: © AAAS, 1993; Box 2.a: Cold Spring Harbor Lab. Archives. From "The Double Helix" by James D. Watson, Athenum Press, NY 1968.

Chapter 3
Opener: © Eric Grave/Science Source/Photo Researchers, Inc.; 3.1a, 3.1b: © Carolina Biological Supply Co./Phototake; 3.1c: © Ed Reschke; 3.1d: © Robert and Linda Mitchell; Box 3.a: © NIBSC/SPL/Photo Researchers, Inc.; 3.2: Jane Hurd © National Geographic Society; 3.4a: © Heather Davies/Chapman & Hall/SPL/Photo Researchers, Inc.; 3.8a: © M. Schliwa/Visuals Unlimited; 3.10b: © Thomas E. Schroeder/Biological Photo Service; Box 3.bb: © Kazuhiko Fujita, Jutendo Univers: School of Medicine, Tokyo, Japan

Chapter 4
Opener: © J. E. Hoffman, Yale University

Chapter 5
Opener: © Jo McBride/Tony Stone Images; 5.3a2: © Lennart Nilsson The Body Victorious; 5.6b: © Daniel Friend, H.M.S.

Chapter 6
Opener: © Larry Ulrich; 6.2a2: © Norma J. Lang/Biological Photo Service; 6.2b4: © E. H. Newcomb and W. P. Wergin/Biological Photo Service; Box 6.a: © Walter H. Hodge/Peter Arnold, Inc.

Chapter 7

Opener: © David M. Phillips/Visuals Unlimited; **7.2a:** © Scott Camazine/Photo Researchers, Inc.; **7.2b:** © Don Kelly/Grant Heilman Photography; **7.5a2, 7.5b2, 7.5c2, 7.5d2, 7.5e2:** © Andrew S. Bajer/Univ. Oregon; **7.6a2:** © David M. Phillips/Visuals Unlimited; **7.6b2:** © B. A. Palevitz and E. H. Newcomb/BPS/Tom Stack & Associates; **Box 7.b:** © Moredun Animal Health/SPL/Photo Researchers, Inc.; **7.8b:** © Lee D. Simon/Photo Researchers, Inc.

Chapter 8

Opener: © David M. Phillips/Visuals Unlimited; **8.2b:** © M. Abbey/Photo Researchers, Inc.; **8.3a, 8.3b:** Hans Pfletschinger/Peter Arnold, Inc.; **8.5b1:** © Cabisco/Visuals Unlimited

Chapter 9

Opener: © Biophoto Associates/Science Source/Photo Researchers, Inc.; **9.1:** © Bob Sacha; **9.3:** © Don Kelly/Grant Heilman Photography; **Box 9.a:** The Bettmann Archive; **9.5:** © Erika Stone/Photo Researchers, Inc.; **9.6:** © Timothy O'Keefe/Tom Stack & Associates; **9.9a:** © PIX Elation/Fran Heyl Associates; **9.13a:** © Erika Stone/Photo Researchers, Inc.; **9.13b:** © Biophoto Associates/Photo Researchers, Inc.

Chapter 10

Opener: © Bruce Berg/Visuals Unlimited; **10.1a:** © Mark Burnett/Photo Researchers, Inc.; **10.1b:** © Richard H. Gross; **10.3a1, 10.3a2:** © Bob Coyle; **10.3b:** © Joseph Nettis/Photo Researchers, Inc.; **10.4a2, 10.4a3:** © Biophoto Associates/Science Source/Photo Researchers, Inc.; **10.4b2:** Bernstein; **Box 10.aa:** The Bettmann Archive; **Box 10.ab:** Toulouse-Lautrec, Henride. *Divan Japonais* 18 Lithograph, printed in color, 317/XX 24 1/2". The Museum of Modern Art, New York. Abby Aldrich Rockefeller Fund. Photos © 1995 The Museum of Modern Art, New York; **10.8a:** Northwind Picture Archives; **10.9:** © Leonard Lessin/Peter Arnold, Inc.; **10.11a:** © Richard Dranitzka/Science Source/Photo Researchers, Inc.

Chapter 11

Opener: From the cover of *Science*, 26 February, 1993, by J. B. Lawrence, K. Carter, Y. Xing, J. M. McNeil, and C. Johnson © AAAS 1993; **11.1b:** © Network Prod./The Image Works, Inc.; **Box 11.a:** The Bettmann Archive; **11.9b, 11.9c:** © Bill Longcore/Photo Researchers, Inc.; **Box 11.bb:** Lee D. Simon/Science Source/Photo Researchers, Inc.; **Box 11.ca:** Cold Spring Harbor Laboratory Archives; **11.12:** © Boehringer Ingelheim International GmbH. Photo by Lennart Nilsson

Chapter 12

Opener: From David W. Ow, Keith V. Wood, Marlene DeLuca, Jeffrey R. DeWet, Donald R. Helinski, Stephen H. Howell, "Transient and Stable Expression of the Firefly Luciferase Gene In Plant Cells and Transgenic Plants", *Science* 234:856-859, 14 Nov. 1986 © AAAS 1986; **12.3a2:** © SPL/Photo Researchers, Inc.; **12.3b:** © Dr. Dennis Kunkel/Phototake; **Box 12.a:** © Grant Heilman/Grant Heilman Photography; **12.5:** Courtesy of Perkin-Elmer Corporation; **12.12b:** © Ralph L. Brinster, University of Pennsylvania; **12.13:** Research conducted at Oregon State University-Hermiston AS Research & Extension Center. Photo by Photography Plus, Inc., Umatilla, Oregon

Chapter 13

Opener: © Dwight Kuhn; **13.3a:** © Ed Reschke; **13.3b:** © Biophoto Associates/Science Source/Photo Researchers, Inc.; **13.3c:** © Larry Mellichamp/Visuals Unlimited; **13.3d:** © BioPhoto Associates/Photo Researchers, Inc.; **Box 13.a:** © Michael J. Balick; **13.6a1:** © Barry L. Runk/Grant Heilman Photography; **13.6b2:** © E. J. Cable/Tom Stack & Associates; **13.7b:** © BioPhoto Associates/Photo Researchers, Inc.; **Box 13.Ba, 13.bb, 13.bc:** © Robert and Linda Mitchell; **13.11a, 13.11b:** © Dr. Jeremy Burgess/SPL/Photo Researchers, Inc.; **13.12a:** © Barbara J. Miller/Biological Photo Service; **13.12b:** © E. S. Ross; **13.12c:** © Grant Heilman/Grant Heilman Photography; **13.13a2:** © Robert Bornemann/Photo Researchers, Inc.; **13.13b2:** © Francois Gohier/Photo Researchers, Inc.; **13.13c2:** © A. W. Ambler/Nat'l Audubon Society/Photo Researchers, Inc.; **13.14:** © Mark Moffett/Minden Pictures; **13.15a2:** © Bruce Iverson; **13.15b2:** © Ed Reschke/Peter Arnold, Inc.; **13.18b, 13.18c, 13.18d:** © Omikron/Photo Researchers, Inc.

Chapter 14

Opener: © Grant Heilman/Grant Heilman Photography; **Box 14.aa:** © E. S. Ross; **Box 14.ab:** © Dwight R. Kuhn; **14.7:** © BioPhoto Associates/Science Source/Photo Researchers, Inc.; **Box 14.b:** © Johnny Autery/Dixon Mills, Alabama

Chapter 15

Opener: © David Muench; **15.1:** Ruth Bernstein; **15.2c:** © Heather Angel; **Box 15.aa:** © Lefever/Grushow/Grant Heilman Photography; **Box 15.ab:** © M. Timothy O'Keefe/Bruce Coleman, Inc.; **15.3a:** © Tim Daniel/Photo/Nats; **15.3b:** © David Stone/Photo/Nats; **15.3c:** © Kjell B. Sandved/Visuals Unlimited; **15.3d:** © Merlin D. Tuttle, Bat Conservation Int'l/Photo Researchers, Inc.; **15.4a, 15.4b:** © Thomas Eisner/Cornell University; **Box 15.ba:** © Bill Beatty/Visuals Unlimited; **Box 15.bb:** © David Scharf/Peter Arnold, Inc.; **15.6:** © John Colwell/Grant Heilman Photography; **15.7a:** © Susan Waaland/Biological Photo Service;

15.7b: © William E. Ferguson; **15.7c:** © Harry Engles/Animals Animals/Earth Scenes; **15.7d:** © Dwight R. Kuhn; **15.9b:** © Ed Reschke/Peter Arnold, Inc.; **15.13a:** © Ardea, London, Ltd.; **15.13b:** © John D. Cunningham/Visuals Unlimited; **15.14a:** © John Serrao/Photo Researchers, Inc.; **15.14b:** © Grant Heilman/Grant Heilman Photography; **15.14c:** © Renee Purse/Photo Researchers, Inc.

Chapter 16

Opener: © 1994 ZEFA-Spoenlein; **16.3a:** © Runk/Schoenberger/Grant Heilman Photography; **16.4b:** Sylvan H. Wittwer, Agricultural Experimental Station Michigan State University East Lansing; **Box 16.b:** © Philip Jones Griffiths/Magnum Photos; **16.5:** © Frans Lanting/Minden Pictures; **16.6:** From the cover of *Science*, Oct. 18, 1991 by Stephen Gladfeller, Stanford University Medical Center © AAAS; **16.9b:** © Ed Reschke; **16.12a:** © E. R. Degginger; **16.13:** © Randy Moore/Biophot; **16.15a:** © John Kaprellian/Photo Researchers, Inc.; **16.15b:** © John Kaprellian/Photo Reserachers, Inc.

Chapter 17

Opener: © Y. Arhtus Bertrand/Agence Vandy-stadt/Photo Researchers, Inc.; **17.3a:** © Ed Reschke/Peter Arnold, Inc.; **17.3b:** © G. W. Willis/Biological Photo Service; **17.3c, 17.3e:** © Ed Reschke; **17.3f:** © M. I. Walker/Photo Researchers, Inc.; **17.5a:** © Chuck Brown/Photo Researchers, Inc.; **17.5b:** © John D. Cunningham/Visuals Unlimited; **17.5c:** © Dwight R. Kuhn; **17.5d:** © Fred Hossler/Visuals Unlimited; **17.5e1:** © Dwight R. Kuhn; **17.5f:** © Ed Reschke; **17.6c1:** © Runk/Schoenberger/Grant Heilman Photography; **17.6c2, 17.6c3, 17.7b2:** © Ed Reschke

Chapter 18

Opener: © Larry Lefever/Grant Heilman Photography; **18.a:** © Louise Williams/SPL/Photo Researchers, Inc.; **18.4:** Courtesy of the Food and Agricultural Organization of the United Nations; **Box 18.b:** © Professor J. James/SPL/Photo Researchers, Inc.; **18.11a2:** © G. Shih & R. Kessel/Visuals Unlimited

Chapter 19

Opener: © Dennis Kunkel/CNRI/Phototake, NYC; **19.2a:** © Lennart Nilsson *Behold Man*, 1974, Albert Bonniers Forlag and Little, Brown and Co.

Chapter 20

Opener: © David Madison/Bruce Coleman, Inc.; **Box 20.aa, 20.ab:** © Martin M. Rotker; **20.12:** © SPL/Photo Researchers, Inc.

Chapter 21

Opener: © Jim Volkert/The Stock Market;
21.9: © Fred Bruemmer

Chapter 22

Opener: © Lowell Georgia/Science Source/Photo Researchers, Inc.; **22.1a:** © Lennart Nilsson; **22.1b:** © BioPhoto Associates/Photo Researchers, Inc.; **22.1c:** © CDC/Science Source/Photo Researchers, Inc.; **22.1d:** © G. Musil/Visuals Unlimited; **22.3b:** Photo, "Orbital Cellulitis" from MAYO *Clinic Family Health Book*, David E. Larson, M.D., Editor in Chief. By permission of William Moorow & Company, Inc.; **22.4a:** © Boehringer Ingelheim International GmbH. Photo by Lennart Nilsson; **22.5a:** © David S. Goodsell; **22.6b:** © CNRI/SPL/Photo Researchers, Inc.; **22.7a, 22.8a, 22.9a:** © Boehringer Ingelheim International GmbH. Photo By Lennart Nilsson; **22.a:** © Carl Purcell/Photo Researchers, Inc.; **22.11:** © Boehringer Ingelheim International GmbH. Photo by Lennart Nilsson; **22.13:** © David Weintraub/Science Source/Photo Researchers, Inc.

Chapter 23

Opener: © Jean Francois Causse/Tony Stone Images; **23.7a:** AP Laserphoto/Wide World Photos, Inc.; **23.7b:** © Irven DeVore/Anthro Photos; **23.10a:** © John Paul Kay/Peter Arnold, Inc.; **23.10b:** Courtesy F. A. Davis Company, Philadelphia, and Dr. R. H. Kampmeier

Chapter 24

Opener: © 1993 Robert Cerri/The Stock Market; **24.7b:** Courtesy John E. Heuser/Washington Univ. School of Medicine, St. Louis, MO; **24.11:** © Dr. Marcus E. Raichle-Washington University's McDonnell Center for High Brain Function/Peter Arnold, Inc.; **24.13:** © Dr. Richard M. Restak, Washington D.C.

Chapter 25

Opener: © Adolfo Previdere/Bruce Coleman, Inc.; **25.4:** © Peter J. Bryant/Biological Photo Service; **25a 1-3:** Derek Boyes, *Life Magazine*; **25.6a:** © Dr. Scott Mittman, Maria Maglio, and Dr. David Copenhagen; **25.9c:** © Dr. Richard M. Costanzo; **25.12:** © Agence Nature/NHPA; **25.b:** © 1992 Group III/Bruce Coleman, Inc.; **25.c:** © E. R. Lewis, Y. Y. Zeevi, & T. E. Everhart/Biological Photo Service; **25.14:** © Richard Hamilton Smith

Chapter 26

Opener: © Gerard Lacz/Peter Arnold, Inc.; **26.2a:** © Peter J. Bryant/Biological Photo Service; **26.2c:** Lawrence Migdale/Photo Researchers, Inc.; **26.6:** © Michael Klein/Peter Arnold, Inc.; **26.aa, 26.ab, 26.ac, 26.ad:** © Charles Gurche

Chapter 27

Opener: © Jim Stamates/Tony Stone Images; **27.2:** © Dana Hyde/Photo Researchers, Inc.; **27.4:** Thomas McAvoy, *Life Magazine* © Time Warner, Inc.; **27.5:** © Mauritius GmbH/Phototake, NYC; **27.6:** © Charlie Palek/Animals Animals/Earth Scenes; **27.7:** © S. Maslowski/Visuals Unlimited; **27.8:** © John Hendrickson; **27.a:** © Lishman/Duff Photo; **27.11:** © Art Wolfe/Allstock; **27.b:** Courtesy of Lisa Davis & Lowell Getz; **27.12a:** © Roy Morsch/Bruce Coleman, Inc.; **27.12b:** © Jonathan Scott/Planet Earth Pictures; **27.13:** © Ned H. Kalin, U. of Wisconsin Medical School; **27.14:** © Jim Brandenburg/Minden Pictures; **27.15:** © Patricia D. Moehlman 1980; **27.16:** © D. W. MacDonald/OSF/Animals Animals/Earth Scenes

Chapter 28

Opener: © Philip Matson/Tony Stone Images; **28.1:** © Hans Pfletschinger/Peter Arnold, Inc.; **28.2:** © John Cancalosi/Peter Arnold, Inc.; **28.8:** © Lennart Nilsson *A Child Is Born*; **28.10b:** © Francis Leroy Biocosmes/SPL/Photo Researchers Inc.; **28.14c:** © Lennart Nilsson *A Child is Born*; **28.a:** © BioPhoto Assoc/Photo Researchers Inc.; **28.c:** © Susan E. Lanzendorf, Eastern Virginia Medical School

Chapter 29

Opener: © Jean-Claude Lejeune; **29.1a, 29.1b, 29.1c, 29.1d, 29.1e:** © Cabisco/Visuals Unlimited; **29.4:** © Ken Wagner/Phototake, NY; **29.9:** © Photo Researchers, Inc.; **29.11:** © Lennart Nilsson *A Child is Born*; **29.15:** © SIU/Visuals Unlimited

Chapter 30

30.a: © Erwin and Peggy Bauer/Bruce Coleman, Inc.; **30.4a:** © Wendy Shaltil/Bob Rozinski/Tom Stack & Associates; **30.4b:** © Martha Cooper/Peter Arnold, Inc.; **30.5a, 30.5b:** © Breck P. Kent/Animals Animals/Earth Scenes; **30.8a1:** © Senckenberg Museum of Natural History, Frankfurt am Main; **30.8b:** © Yale Peabody Museum of Natural History; **30.8c:** © Kjell Sandved/Butterfly Alphabet; **Box 30.ba2:** © Galen Rowell/Peter Arnold, Inc.; **30.16:** © David L. Pearson/Visuals Unlimited

Chapter 31

Opener: © Ed Reschke; **31.2a:** © Karl Aufderheide/Visuals Unlimited; **Box 31.bb1:** © Richardo Geurrero and Isabel Esteve; **31.4a:** © John D. Cunningham/Visuals Unlimited; **31.4b:** © Dwight R. Kuhn; **31.4c:** © Dr. E. R. Degginger/Color-Pic, Inc.;

31.5: © Runk/Schoenberger/Grant Heilman Photography; **31.6:** © Ray Coleman/Photo Researchers, Inc.; **31.9a:** © Jan Took/Biofotos; **31.10a:** © Steve Earley/Animals Animals/Earth Scenes; **31.11a:** © Dwight Kuhn; **31.12a:** © Alvin E. Staffan/National Audubon Society/Photo Researchers, Inc.; **31.13a, 31.14a:** © Marty Snyderman/Visuals Unlimited; **31.16a:** © Zig Leszczynski/Animals Animals/Earth Scenes; **31.18:** © Joe McDonald/Bruce Coleman, Inc.

Chapter 32

Opener: © Art Wolfe; **32.1a:** © Manfred Danegger/Peter Arnold, Inc.; **32.1b:** © V. J. Anderson/Animals Animals/Earth Scenes; **Box 32.aa:** © Runk/Schoenberger/Grant Heilman Photography; **Box 32.ab1:** Courtesy of Detroit Edison; **Box 32.ab2:** © Pat Beck/Detroit Free Press; **Box 32.ba:** © Bob & Clara Calhoun/Bruce Coleman, Inc.; **Box 32.bb:** © Joe McDonald/Animals Animals/Earth Scenes; **32.4a:** © Jeff Lepore/Photo Researchers, Inc.; **32.4b:** © John A. Lynch/Photo/Nats; **32.5a:** © Wayne Larkinen/Bruce Coleman, Inc.; **32.5b:** © Jim Strawser/Grant Heilman Photography; **32.8:** © Ray Pfortner/Peter Arnold, Inc.; **32.9:** © Ward M. Tingey/Cornell University Agricultural Experiment Station, Ithaca, NY; **32.10:** © Michael Fogden/Bruce Coleman, Inc.; **32.11a:** © Arthur C. Smith III/Grant Heilman Photography; **32.11b:** © Michael & Patricia Fogden; **32.11c:** © Fred Bavendam; **32.12a:** © Manfred Danegger/Peter Arnold, Inc.; **32.12b:** © Heather Angel/Biofotos; **32.12c:** © Thomas Eisner and Daniel Aneshansley, Cornell University; **32.12d:** © Robert Winslow/Tom Stack & Associates

Chapter 33

Opener: © David Molchos; **33.1:** © Charlie Ott/Photo Researchers, Inc.; **33.8a:** © Runk/Schoenberger/Grant Heilman Photography; **33.8b:** © C. P. Vance/Visuals Unlimited; **Box 33.b:** © Michael P. Gadomski/Bruce Coleman, Inc.; **33.9:** © Grant Heilman/Grant Heilman Photography; **33.10a1, 33.10b1, 33.10c1:** © Dwight R. Kuhn; **33.11:** © Grant Heilman/Grant Heilman Photography; **33.13ab:** © IFA/Peter Arnold, Inc.; **33.13ac:** © Partridge Productions Ltd./OSF/Animals Animals/Earth Scenes; **33.13ad:** © Brian Rogers/Biofotos; **33.13ae:** © John Angell Grant/Bruce Coleman, Inc.; **33.13af:** © John Shaw/Tom Stack & Associates; **33.13bb:** © Tom Brakefield/The Stock Market; **33.13bc:** © John Shaw/Bruce Coleman, Inc.; **33.13bd:** © Edward R. Degginger/Bruce Coleman, Inc.; **33.13bf:** © Ralph Reinhold/Animals

Index